D0918749

Chemistry of Tin

Chemistry of Tin

Edited by

P.G. HARRISON, PhD DSc*Lond.*
Reader in Chemistry
University of Nottingham

Blackie

Glasgow and London

Published in the USA by
Chapman and Hall
New York

Blackie & Son Limited.
Bishopbriggs, Glasgow G64 2NZ
and
7 Leicester Place, London WC2H 7BP

Published in the USA by
Chapman and Hall
a division of Routledge, Chapman and Hall, Inc.
29 West 35th Street, New York, NY 10001-2291

First published 1989

British Library Cataloguing in Publication Data

Chemistry of tin.
1. Tin compounds
I. Harrison, P.G. (Phillip Geoffrey), *1943–*
546′.6862

ISBN 0-216-92496-0

For the USA, International Standard Book Number is
0-412-01751-2

Phototypesetting by Thomson Press (India) Ltd., New Delhi
Printed in Great Britain by Bell & Bain (Glasgow) Ltd.

Preface

Since the dawn of recorded history, the element tin has played a major role in the development of human civilisation. Indeed, the discovery that smelting tin with copper produces bronze was a major technological landmark. It has been over the last quarter of a century, however, that the tremendous potential offered by the metallurgy and chemistry of tin has begun to be realised. Metallurgical uses account for 90% of the volume of tin produced, but numerous and diverse chemical applications have also been discovered, for both inorganic and organometallic tin compounds. These include uses as stabilising additives for PVC plastics, biocidal agents, catalysts, and gas-sensing devices. Clearly there is a wealth of chemistry intrinsic to tin.

Knowledge of tin chemistry is, in general, scattered in the primary and review literature. The organometallic chemistry of tin was covered extensively in two early monographs[1,2], but sadly these now need to be updated.

In this book, I together with my co-authors—all eminent in their own particular areas—have reviewed the present state of knowledge of the chemistry of tin. Early chapters are devoted to more general aspects of the chemistry—such as structure, bonding and spectroscopy for example—whilst later chapters cover inorganic and organometallic chemistry, reaction mechanisms, radical chemistry, the use of organotin compounds as reagents in organic synthesis, tin in biological systems, and the industrial uses of tin chemicals.

In an exercise such as this, where the magnitude of the literature has to be accommodated in a relatively short space, it has not been possible to be totally comprehensive. This volume does, however, give the reader an excellent panoramic overview of the subject, whilst at the same time providing a tremendous amount of pertinent detail.

My personal involvement with tin chemistry began in 1965 when I had the great privilege of studying for a PhD under the supervision of Professor Alwyn Davies at University College, London. My own interests impinge on most aspects of tin chemistry, and this has led to my meeting a great many other workers in the area. A major landmark occurred in 1973 with the first *Conference on the Organometallic and Coordination Chemistry of Germanium, Tin and Lead* at Marseille, organised by Professor Jean-Claude Maire. Subsequent conferences in the triennial series have been held at Nottingham, Dortmund, Montreal, and Padua. These conferences have fostered a sense of friendship within the Group 4 community which has greatly assisted in the development of the subject.

References

1. W.P. Neumann, *Ferdinand Enke Verlage*, Stuttgart, 1967
2. R.C. Pouer, *The Chemistry of Organotin Compounds*, Logos Press, London, 1970.

PGH

Contributors

Professor A.G. Davies	Department of Chemistry, University College London, 20 Gordon Street, London WC1H 0AJ
Professor J.D. Donaldson	Department of Chemistry, City University, Northampton Square, London EC1V 0HB
Dr C.J. Evans	International Tin Research Institute, Kingston Lane, Uxbridge, Middlesex UB8 3PJ
Professor F. Glockling	Inorganic Chemistry Laboratory, University of Oxford, South Parks Road, Oxford OX1 3QR
Dr S.M. Grimes	Department of Chemistry, City University, Northampton Square, London EC1V 0HB
Dr P.G. Harrison	Department of Chemistry, University of Nottingham, University Park, Nottingham NG7 2RD
Dr P.D. Lickiss	School of Chemistry and Molecular Sciences, University of Sussex, Falmer, Brighton BN1 9QJ
Dr K.C. Molloy	School of Chemistry, University of Bath, Claverton Down, Bath BA2 7AY
Dr M.J. Selwyn	School of Biological Sciences, University of East Anglia, Norwich NR4 7TJ
Dr J.L. Wardell	Department of Chemistry, University of Aberdeen, Meston Walk, Old Aberdeen AB9 1FX

Contents

1 Tin – the element

P.G. HARRISON

1.1 From earliest times

The discovery, around 3500 BC, that copper, easily smelted but rather soft, could be made harder and stronger by alloying with tin, can be considered to be one of the great milestones in man's technological development, and heralded the advent of the Bronze Age. Because of its relative scarcity, tin has always been a strategic metal and remains so to the present day. Isolation of the pure metal probably only dates from about 800 BC. The earliest recorded reference to tin occurs in the Old Testament in the book of Numbers (31:22): 'Only the gold, and the silver, the brass, the tin, and the lead.' Substantially later (*c.* 600 BC) Ezekiel (27:12) relates: 'Tarshish was thy merchant by reason of the multitude of all kind of riches; with silver, iron, tin, and lead, they traded in their fairs.' Both suggest that tin was regarded as valuable commodity, and indeed a strip of tin free from silver and lead has been found in the wrappings of an Egyptian mummy dating from not later than 600 BC.

The Bronze Age of the ancient Western world not only represented the first production of a truly useful metal but also, because of its relative scarcity compared to copper, resulted in the development of substantial trading contact. Around 2000 BC, tin for the eastern Mediterranean could be obtained from Central Europe, and by 1800 BC trading links allowed bronze manufacture over much of the then known world. Knowledge of bronze is reputed to have travelled to other ancient civilizations such as China from the west, and the oldest Chinese bronze artefacts date from 2200 BC.

The early sources of tin are shrouded in uncertainty. The Greek chronicler Herodotus referred, around 440 BC, to 'the islands called the Cassiterides, from which we are said to have our tin'. The actual location of these islands has been the subject of much debate, and has been identified with various places including Cornwall, the Scilly Isles, or supposed islands off the French coast, but gradually Spain, Britain and Brittany became known as the sources of tin. Not until 8 BC can Cornwall can be definitely identified in Diodorus Siculus' *Bibliothecca historica*, although the Phoenicians may have worked tin mines in Cornwall as early as 1000 BC.

Some confusion between tin and lead appears to have occurred in the ancient world. Pliny in his *Historia naturalis* writes:

> There are two kinds of plumbum – nigrum and candidum or album [i.e. black lead and shining or white lead]. The plumbum candidum is the most valuable and it was called cassiteros by the Greeks. There is a fabulous story told of their going in quest of it to the islands in the Atlantic Ocean, and of its being brought in boats made of osiers covered with hides. The plumbum candidum occurs as a black sand found on the surface of the earth, and is to be detected only by its weight, it is mingled with small pebbles, particularly in the dried beds of rivers. The miners wash the sand, which is then melted in the furnace and becomes converted into plumbum album.

Without doubt Pliny's *plumbum candidum* was tin and his *plumbum nigrum* lead. He further described Roman vessels made of copper and coated with *stannum*, and also states that *plumbum candidum* was esteemed even in the days of the Trojan War. Homer writes of it in the *Iliad* under the name of *cassiteros*.

The early Greek alchemists called tin 'Hermes', but by about AD 500 alchemists called it 'Zeus' or 'Jupiter', and represented it by the symbol ♃ (derived from a corruption of the first and last letters of Ζεύς), which was understood to mean the thunderbolts of the king of the gods. It was also referred to as *diabilus metallorum*, the devil of metals, because of the embrittling effect it had on alloying with other metals. Geber, in his *Summa Perfectionis Magisterii*, probably written about the twelfth century, writes: 'Of Jupiter or Tin. We ſignifie to the Sons of Learning, the Tin is a Metallick Body, white, not pure [white], livid, and ſounding little, partaking of little Earthineſs; Poſſeſſing in its Root Harſhness, Softneſs, and ſwiftness of Liquefaction, without ignition, and not abiding the Cupel, or Cement, but Extensſible under the Hammer.... Its vice is, that it breaks every [metallic] Body, but Saturn [lead], and moſt pure Sol [gold]' [i.e. when alloyed with them it makes them brittle].

Compounds of tin appear to date back to the Copts of Egypt, who reportedly used basic tin citrate in dye preparation. Modern chemistry of tin appears to start with the classic experiments of Libavius in 1605, who described a fuming liquid which he called *liquor argenti vivi sublimati* or *spiritus argenti vivi sublimati* prepared by distillation from a mixture of tin or tin amalgam and mercuric chloride. The liquid, obviously tin(IV) chloride, was thereafter referred to as *spiritus fumans Libavii*. It is possible, however, that the same experiment was carried some three hundred years earlier! Although there are other reports originating in the seventeenth century, it was principally in the eighteenth century that inorganic tin compounds began to be studied in earnest. The first organotin compound, diethyltin diiodide, was prepared by Frankland in 1849, when tin chemistry came of age.[1]

1.2 Occurrence

Important tin-producing countries are Malaysia, Bolivia, Indonesia, Nigeria, Thailand, Zaire, and China, with smaller quantities from the United Kingdom, Burma, Japan, Canada, Portugal, Spain and Australia. By far the most important ore, and the only one of major commercial importance, is cassiterite, a naturally occurring form of tin(IV) oxide. Other ores are sulphidic in nature and include stannite, $SnS_2 \cdot Cu_2S \cdot FeS$; herzenbergite, SnS; teallite, $SnS \cdot PbS$; franckeite, $2SnS_2 \cdot Sb_2S_3 \cdot 5PbS$; cylindrite, $Sn_6Pb_6Sb_2S_{11}$; plumbostannite, $2SnS_2 \cdot 2PbS \cdot 2(Fe,Zn)S \cdot Sb_2S_3$; and canfieldite, $4Ag_2S \cdot SnS_2$. These ores are of commercial significance only in Bolivia.

1.3 Physical properties of metallic tin

Tin has the largest number (10) of stable isotopes of any element, and in addition many unstable isotopes with half-lives varying from 2.2 minutes to $\sim 10^5$ years (Table 1.1). The normal form of the element is β- or white tin, but α- or grey tin is the thermodynamically stable modification below 13.2°C. The existence of a third form referred to as γ-tin and said to be stable at temperatures in excess of 161°C has not been substantiated. Physical, crystallographic, thermal and spectroscopic data for elemental tin are collated in Tables 1.2–1.5, respectively. The β-form of tin is a silvery-white metal which is

Table 1.1 Stable and unstable isotopes of tin.

Stable isotopes

	Mass	Abundance (%)
^{112}Sn	111.90494	0.95
^{114}Sn	113.90296	0.65
^{115}Sn	114.90353	0.34
^{116}Sn	115.90211	14.24
^{117}Sn	116.90306	7.57
^{118}Sn	117.90179	24.01
^{119}Sn	118.90339	8.58
^{120}Sn	119.90213	32.97
^{122}Sn	121.90341	4.71
^{124}Sn	123.90524	5.98

Unstable isotopes

	Half-life	Mode of decay	Decay energy (MeV)
^{108}Sn	9m	EC	
^{109}Sn	18.1m	β^+, EC	
^{110}Sn	4.0h	EC	
^{111}Sn	35m	β^+, EC	2.52
^{113m}Sn	20m	IT, EC	0.079, 1.1
^{113}Sn	115d	EC	1.02
^{117m}Sn	14d	IT	0.317
^{119m}Sn	250d	IT	0.089
^{121m}Sn	76y	β^-	0.45
^{121}Sn	27h	β^-	0.383
^{123}Sn	125d	β^-	1.42
^{123}Sn	42m	β^-	1.46
^{125m}Sn	9.7m	β^-	2.39
^{125}Sn	9.4d	β^-	2.34
^{126}Sn	$\sim 10^5$y	β^-	$\sim$0.3
^{127}Sn	2.1h	β^-	
^{127}Sn	4m	β^-	$\sim$ 3.1
^{128}Sn	59m	β^-	1.3
^{130}Sn	2.6m		
^{131}Sn	3.4m		
^{132}Sn	2.2m		

Table 1.2 Physical properties of tin.

Density		
α-Tin	(measured at 288 K)	7.29 g cm^{-3}
	(from X-ray at 299 K)	7.2867 ± 0.0024 g cm^{-3}
β-Tin	(measured at 288 K)	5.77 g cm^{-3}
	(from X-ray at room temperature)	5.765 g cm^{-3}
Liquid	(measured at m.p.)	6.968 ± 0.005 g cm^{-3}
		6.978 ± 0.022 g cm^{-3}
	(measured at 600 K)	6.70 g cm^{-3}
	(measured at 1200 K)	6.29 g cm^{-3}
Hardness	(Moh scale)	1.5–1.8
	at 293 K	3.9 HB
	at 373 K	2.3 HB
	at 473 K	0.9 HB

Table 1.2 (*Contd.*)

Resistivity	
α-Tin (at 293 K)	12.6 μΩ cm
β-Tin (at 273 K)	300 μΩ cm
Young's modulus (at 293 K)	49.9 kN mm^{-2}
Bulk modulus (at 293 K)	58.2 kN mm^{-2}
Shear strength (room temperature)	12.3 N mm^{-2}

Table 1.3 Crystallographic data.

α-Tin	Crystal system	Cubic
	Space group	Fd3m (Type A4)
	Z	2
	a	64.912 ± 0.005 pm
	Nearest neighbour distances	4 at 280 pm
		12 at 459 pm
	Lattice '*d*' spacings (relative intensities in parentheses)	37.3(100) pm
		22.8(80) pm
		19.5(70) pm
		16.1(30) pm
		14.8(50) pm
		13.2(60) pm
		12.4(40) pm
		11.4(20) pm
		10.9(40) pm
β-Tin	Crystal system	Tetragonal
	Space group	14_1/amd(141)
	Z	4
	a (299 K)	58.315 ± 0.008 pm
	c (299 K)	31.813 ± 0.006 pm
	Cell volume (299 K)	108.18 ± 0.04 × 10^{24} cm^{-3}
	Nearest-neighbour distances	4 at 302 pm
		2 at 318 pm
		4 at 377 pm
		8 at 441 pm
	Twinning plane	(301)
	Lattice '*d*' spacings (relative intensities in parentheses)	29.153(100) pm
		27.93(90) pm
		20.62(34) pm
		20.17(74) pm
		16.59(17) pm
		14.84(23) pm
		14.58(13) pm
		14.42(20) pm
		13.04(15) pm
		12.92(15) pm
		12.05(20) pm
		10.95(13) pm

Table 1.4 Thermal data.

Fusion point	231.9681°C
Enthalpy of fusion	7.06 kJ g atom^{-1}
Entropy of fusion	3.35 e.u.
Boiling point	2270°C
Enthalpy of vaporization	296.4 kJ g atom^{-1}
Vapour pressure	
at 1096 K	10^{-5} mm Hg
at 1196 K	10^{-4} mm Hg
at 1315 K	10^{-3} mm Hg
at 1462 K	10^{-2} mm Hg
at 1646 K	10^{-1} mm Hg
at 1882 K	1 mm Hg
Entropy at 295 K	12.3 e.u.
Specific heat (C_v) at 298 K	
α-tin	215.5 J kg^{-1} K^{-1}
β-tin	223.3 J kg^{-1} K^{-1}
Thermal conductivity at 273.2 K	
Crystal $\parallel$ to c axis	0.527 W cm^{-1} K^{-1}
Crystal $\perp$ to c axis	0.759 W cm^{-1} K^{-1}
Polycrystalline	0.682 W cm^{-1} K^{-1}
Coefficient of expansion at 273 K	
Linear	19.9×10^6
Cubical	59.8×10^6
Single crystal ($\parallel$ to c axis)	28.4×10^6
Single crystal ($\perp$ to c axis)	15.8×10^6
Expansion on melting	2.3%
Surface tension at melting point	544 mN m^{-1}
Viscosity at melting point	1.85 mNs m^{-2}
Gas solubility in liquid tin	
(oxygen at 809 K)	0.00018%
(oxygen at 1023 K)	0.0049%
(hydrogen at 1273 K)	0.04%
(hydrogen at 1573 K)	0.36%

Table 1.5 Spectroscopic data.

X-ray atomic energy levels		
	K	29200.0 ± 0.4 eV
	L_I	4464.7 ± 0.3 eV
	L_{II}	4156.1 ± 0.3 eV
	L_{III}	3928.8 ± 0.3 eV
	M_I	883.8 ± 0.3 eV
	M_{II}	756.4 ± 0.3 eV
	M_{III}	714.4 ± 0.4 eV
	M_{IV}	493.3 ± 0.3 eV
	M_V	484.8 ± 0.3 eV
	N_I	136.5 ± 0.4 eV
	N_{II}, N_{III}	88.6 ± 0.4 eV
	N_{IV}, N_V	23.9 ± 0.3 eV
	O_I	0.9 ± 0.5 eV
	O_{II}, O_{III}	1.1 ± 0.5 eV

Table 1.5 (*Contd.*)

X-ray emission wavelengths

Designation	λ(Å)	Energy (keV)
$\alpha_1 KL_{II}$	0.495053	25.0440
$\alpha_1 KL_{III}$	0.490599	25.2713
$\beta_2 KM_{II}$	0.435877	28.4440
$\beta_2 KM_{III}$	0.435236	28.4860
$\beta_2 KM_{II,III}$	0.425915	29.1093
$KO_{II,III}$	0.42467	29.195
$\beta_1^{II} KM_{IV}$	0.43184	28.710
$\beta_5^{I} KM_{V}$	0.43175	28.716
$\beta_4 KN_{IV,V}$	0.42495	29.175
$\beta_4 L_I M_{II}$	3.34335	3.7083
$\beta_3 L_I M_{III}$	3.30585	3.7500
$\gamma_{2,3} L_I N_{II,III}$	2.8327	4.3768
$\gamma_4 L_I O_{II,III}$	2.7775	4.4638
$\eta L_{II} M_I$	3.78876	3.27234
$\beta_1 L_{II} M_{IV}$	3.38487	3.66280
$\gamma_5 L_{II} N_I$	3.08475	4.0192
$\gamma_1 L_{II} N_{IV}$	3.00115	4.13112
$\zeta L_{III} M_I$	4.07165	3.04499
$\alpha_2 L_{III} M_{IV}$	3.60891	3.43542
$\alpha_1 L_{III} M_V$	3.59994	3.44398
$\beta_1 L_{III} N_I$	3.26901	3.7926
$\beta_{2,15} L_{III} N_{IV,V}$	3.17505	3.90486
$\beta_7 L_{III} O_I$	3.1546	3.9279
$\beta_{10} L_I M_{IV}$	3.12170	3.9716
$\beta_9 L_I M_V$	3.11513	3.9800

relatively soft and ductile, and can emit a characteristic cracking noise when subjected to stress deformation (the so-called 'cry' of tin) due to plastic deformation along the [301] crystal twinning plane. Transformation from the β- to the α-form is quite slow and does not proceed at an appreciable rate except at temperatures of *c.* -40°C. The transformation of tin into the grey form under extreme of cold was first observed by Aristotle, and came to be known as 'tin pest' or 'tin disease'. When tin or a tin alloy is affected by tin pest, grey-coloured spots appear and the metal becomes brittle. Because of the large density change, the metal expands, producing pustule-like or nodular excrescences at the affected points. Transformation extends radially outward from the spots until the whole mass is infected. The metal then breaks down and disintegrates to a brittle powder. The change can be retarded or even inhibited by the presence of trace amounts of impurity metals such as aluminium, zinc, antimony or bismuth. The two modifications have quite different structures, which accounts for many of the differences in their properties. Thus β-tin is a typical metallic conductor (though much poorer than copper), whereas, in contrast, α-tin exhibits high resistivity and is semiconducting.

The α-form has the diamond structure, with each tin atom having four nearest neighbours at 280 pm. The structure of β-tin is much less symmetrical and is related to that of α-tin in an interesting way. In Figure 1.1*a* the α-tin structure is shown referred to a tetragonal unit cell (*a* axes at 45° to the axes of the conventional cubic unit cell). In the diamond structure, the six interbond angles are equal (109.47°), but in β-tin the coordination tetrahedron is severely flattened so that two of these angles are enlarged

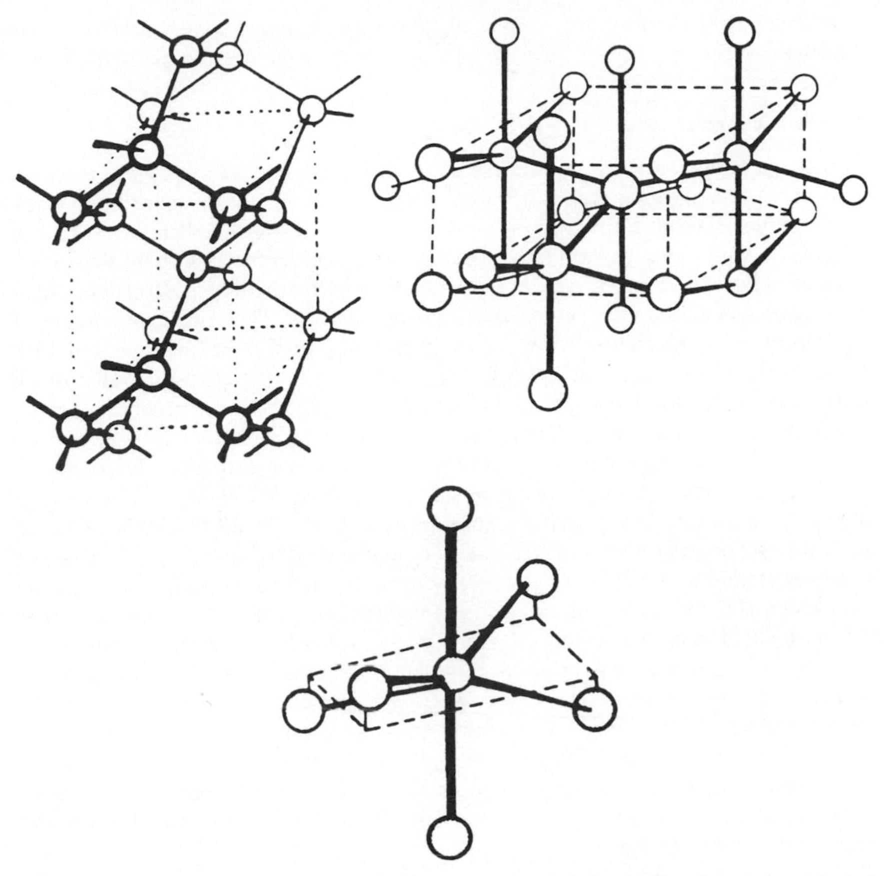

Figure 1.1 (*a*) The structure of α-tin. (*b*) The structure of β-tin. (*c*) The distorted octahedral coordination geometry in β-tin.

to 149.5° while the other four are reduced to 94° (Figure 1.1*b*). In this process, the nearest neighbours change from four at 280 pm in α-tin to four at 302 pm and a further two at 318 pm, and coordination at tin in the β-modification is best regarded as distorted octahedral (Figure 1.1*c*). This structural change is remarkable, not only because of the 26% increase in density (causing the phenomenon known as 'tin pest'), but also because the more dense β-tin modification is the high-temperature form. The relationship between the two forms has been discussed at length[2]. The transformation between the two modifications is accompanied by a redistribution of the valence electrons, as shown by the Mössbauer isomer shifts (α-tin, $\delta = 2.10\,\mathrm{mm\,s^{-1}}$, β-tin, $\delta = 2.65\,\mathrm{mm\,s^{-1}}$), the higher value corresponding to a higher electron density in the tin 5*s* orbital for β-tin. Significantly, the redistribution of valence electron density is reflected in the reaction of the two modifications with hydrochloric acid, which with β-tin gives tin(IV) chloride pentahydrate, whereas α-tin produces tin(II) chloride dihydrate. Fusion occurs at a relatively low temperature for a metal (232°C), with a very large liquid range before a boiling point of 2270°C. Mechanically, tin is relatively weak at ambient temperatures, but tin-plating and tin-rich alloys have found extensive application[3].

1.4 Chemical reactions

At ambient temperatures, tin is relatively unreactive and is inert to oxygen, nitrogen, hydrogen, water and ammonia. Surface oxidation does occur at elevated temperatures, resulting in significant tarnishing of the surface at 200°C. At white heat, tin is said to burn with a white flame to form tin (IV) oxide. Water vapour reacts with the tin surface at temperatures in excess of 700°C to give oxide films and hydrogen. Reagents which are oxidizing in nature show significantly greater reactivity. Thus, tin metal is attacked by chlorine and bromine to form the corresponding tin (IV) halide. Reaction with fluorine and iodine is slow below 100°C. Elemental sulphur and selenium react only on heating, when the reaction is reported to be vigorous, giving the corresponding tin (II) or tin (IV) chalcogenide, depending upon the mole ratio of reactants employed. Tin (II) telluride is formed with tellurium. Little reaction is observed with dilute hydrochloric or sulphuric acids, but dilute nitric is said to produce $Sn(NO_3)_2$ and NH_4NO_3, although the constitution of the tin-containing product as a bivalent tin nitrate must be in doubt. Hot concentrated acids react rapidly to give tin (II) chloride, tin (II) sulphate, and 'metastannic acid', a hydrated tin (IV) oxide (see Chapter 12), respectively. The rate of reaction is dramatically enhanced in the presence of oxygen. Tin dissolves in hot aqueous alkali solutions to afford alkali metal hydroxystannates, $M_2[Sn(OH)_6]$ ($M = Na, K$). Under near-neutral conditions, tin metal is essentially inert to corrosion, owing to the formation of a surface oxide film. However, some attack at the metal surface can occur when in contact with solutions containing chloride, sulphate or nitrate anions. Attack is much more rapid in the presence of oxidizing reagents such as persulphate. Organic acids react much more slowly, and attack by acetic or citric acids is not significant in the absence of oxygen. Bimetallic systems involving tin can also react much more rapidly.

References

1. For a recent authoritative account, see R.D. Penhallurick, *Tin in Antiquity*, The Institute of Metals, London 1986.
2. R.D.M.J.P. Musgrave, *Proc. Roy. Soc.* 1963, **272A**, 503.
3. B.T.K. Barry and C.J. Thwaites, *Tin and its Alloys and Compounds*, Ellis Horwood, Chichester 1983.

2 Compounds of tin: general trends

P.G. HARRISON

Tin lies at the centre of the Periodic Table, and has the ability to form stable, relatively strong bonds with the majority of elements. Most of the bonds are largely σ-covalent in character, but essentially ionic bonds are formed with the heavier alkali metals and alkaline earths. Perhaps most surprising for a typical Main Group element is the rapidly developing chemistry of 'π-bonded' polyhapto compounds exhibited by bivalent tin, an area traditionally considered to be confined to the '*d*'-block metals. Coupled with this tremendous versatility in bonding is the availability of a wide range of coordination geometries and lattice structural types, leading to a wealth of solid-state chemistry. In this chapter is presented some of the fundamental background upon which the rich tapestry of tin chemistry is based.

2.1 Electronegativity values, ionization potentials and elemental radii

Electronegativities values, ionization potential data, and estimates of the radii of the Group 14 elements in various chemical situations are collected in Table 2.1[1]. The more recently devised electronegativity scales of Sanderson and Hargattai-Bliefelt, which are based on analysis of infrared band intensities for $MHCl_3$ and MX_4 molecules and bond distance data in bivalent and tetravalent derivatives, respectively, concur with the alternating scale of Allred-Rochow, rather than the original Pauling scale. While it must be borne in mind that the absolute electronegativity varies significantly upon the particular chemical environment, it can be seen that the electronegativity varies little down the Group. The 'inert pair' effect has a much greater effect upon the electronegativity, and the values for M^{II} are significantly lower than those for M^{IV}, and increase in the reverse order. Values of the various radii increase with increasing atomic number, so that the covalent radius of tin is almost twice that of carbon. 'Ionic' radii are substantially smaller than the corresponding covalent radii, with the M^{II} 'ionic' radii being significantly larger than the corresponding M^{IV} values. The sum of the relevant van der Waals' radii is often used to assess the presence or otherwise of significant bonding interactions. These for tin with a variety of elements are collected in Table 2.2 together with typical covalent bond distances. It must be stressed that the values quoted are to only considered as typical, and that observed bond distances can vary substantially from compound to compound.

Whereas all the elements of the Group can form compounds with four covalent bonds, the first and second ionization energies for germanium, tin and lead are similar to those of elements which readily form bivalent cations, for example (1st and 2nd ionization energies (eV) in parentheses): magnesium (7.464, 15.035); calcium (6.113, 11.871); manganese (7.435, 15.640); iron (7.870, 16.18) and cobalt (7.86, 17.06). Hence it is not surprising to find the emergence of a stable +II oxidation state with these elements. In the case of tin, the +IV state is more stable than the +II state, but the

Table 2.1 Properties of the Group 14 elements.

Element	Electronegativity scales: Pauling	Allred–Rochow	Sanderson	Hargittai–Bliefelt	Sanderson M^{IV}	Sanderson M^{II}
C	2.5	2.5	2.5	2.6		
Si	1.8	1.7	1.7	1.9		
Ge	1.8	2.0	2.3	2.5	2.62	0.56
Sn	1.8	1.7	2.0	2.3	2.30	1.49
Pb	1.9				2.29	1.92

Element	Elemental Radii (pm): Non-bonded radius	M^{IV} covalent radius	M^{IV} 'ionic' radius	M^{II} 'ionic' radius	van der Waals' radius
C	125	77.2			153
Si	155	117.6	40		193
Ge	158	122.3	53	73	198
Sn	182	140.5	69	118	217
Pb		146	78	119	202

Element	Ionization potentials (eV): 1st	2nd	3rd	4th
C	11.260	24.383	47.887	64.492
Si	8.151	16.345	33.492	45.141
Ge	7.899	15.934	34.22	45.71
Sn	7.344	14.632	30.502	40.734
Pb	7.416	15.032	31.937	42.32

Data taken from *Inorganic Chemistry of the Main Group Elements*, The Chemical Society, 1977, vol. 4, p. 145; I. Hargittai and C. Bliefelt, *Z. Naturforsch., Teil B*, 1983, **38**, 1304; N.N. Greenwood and A. Earnshaw, *Chemistry of the Elements*, Pergamon, Oxford, 1984, p. 431; C. Glidewell, *Inorg. Chim. Acta*, 1979, **36**, 135; A. Bondi, *J. Phys. Chem.*, 1964, **68**, 441; R.T. Sanderson, *Inorg. Chem.*, 1986, **25**, 1856; and *Handbook of Chemistry and Physics*, 61st edn., CRC Press, Boca Raton, 1981.

Table 2.2 Typical bond distance data together with corresponding sums of van der Waals' radii.

Bond	Bond distance r(Sn–X)(pm)	Sum of van der Waals' $r_w(Sn) + r_w(X)$(pm)
Sn–H	215	337
Sn–F	197	352
Sn–Cl	239	397
Sn–Br	255	412
Sn–O	215	357
Sn–S	240	402
Sn–N	215	367
Sn–P	250	407
Sn–C(σ)	215	417
Sn–C(π)	274	403
Sn–Sn	275	434

Van der Waals' data from Table 2.1 and *Handbook of Chemistry and Physics*, 61st edn., 1980–81, CRC Press, Boca Raton, Florida.

energy difference between the two oxidation states is quite small (electrochemical potential for the reaction $Sn^{2+} + 2e^- \rightarrow Sn^{4+}$ 0.1364 V), whereas for germanium and lead the + IV and + II states, respectively, are dominant. Hence for tin, extensive and varied chemistries are found for both oxidation states, although the majority of bivalent tin compounds are susceptible to facile aerobic oxidation. The + III oxidation state is generally considered unstable with respect to disproportionation. However, kinetically stable + III species such as $\cdot Sn[E(SiMe_3)_2]_3$ (E = N, CH) have been obtained and have extremely long and perhaps indefinite lifetimes in solution.

2.2 Enthalpies of formation and bond energy data

Observed enthalpies of formation for a wide variety of compounds are collected in Table 2.3, together with calculated values for some compounds and transient species. Both bivalent and tetravalent compounds of tin are thermodynamically stable with

Table 2.3 Enthalpy of formation data[a,b].

Compound/species	$\Delta H_f(c \text{ or } l)$	$\Delta H_f(g)$	$\Delta H_f(g)$
Sn (white, tetragonal)	0.00[c]		
Sn (grey, cubic)	2.51[c]		
Sn		301.8	
SnO	– 285.9[c]		
SnS	– 77.7[c]		
SnF_2		– 484.9	– 479.0
$SnCl_2$	– 349.4[c]	– 235.8	– 343.6
$SnBr_2$	– 265.8[c]	– 121.2	– 159.3
SnI_2	– 143.8[c]	8.4	23.8
$SnCl_2 \cdot H_2O$	– 944.3[c]		
$Sn(C_5H_5)_2$			525.4
SnO_2	– 578.1[c]		
SnH_4		162.6	133.3
$SnCl_4$	– 544.7[d]	– 471.1	– 421.3
$SnBr_4$	– 405.9[c]	– 313.5	– 127.5
SnI_4	– 214.3[c]		
$SnCl_4 \cdot 2py$	– 221.1[c]		
$SnCl_4 \cdot 2isoqu$	– 156.3[c]		
$SnBr_4 \cdot 8H_2O$	– 2744.6[c]		
$Sn(SO_4)_2$	– 1644.2[c]		
Me_4Sn		– 19.2; – 14.0	– 68.6
Et_4Sn		– 44.7	– 132.9
$Me_3SnCHCH_2$		90.7	38.5
Me_3SnCH_2Ph		88.2	55.6
Me_3SnPh		104.5	84.4
Me_3Sn^iPr		– 46.8	– 85.7
Me_3Sn^tBu		– 66.9; – 104.1	– 66.0
Me_3SnH		21.7	– 18.8
Me_3SnOH	– 379.6[c]	– 316.8	– 201.5
Me_3SnCl	– 244.1[c]	– 194.0	– 194.8
Me_3SnBr	– 204.0[c]	– 140.4	– 99.5
Me_3SnI	– 130.4[d]	– 82.3	– 35.9
Me_3SnOEt	– 305.6[d]	– 263.8	– 203.1
Me_3SnS^nBu	– 196.9[d]	– 155.1	
Me_3SnNMe_2	– 55.6[d]	– 18.0	14.2

Table 2.3 (*Contd.*)

Compound/species	ΔH_f(*c* or *l*)	ΔH_f(g)	ΔH_f(g)
$(Me_3Sn)_2NMe$	− 131.7[d]	− 81.5	
$(Me_3Sn)_3N$	− 122.1[c]	− 59.4	
$Me_3SnSiMe_3$		− 207.7	
$Me_3SnGeMe_3$		− 309.7	
$Me_3SnSnMe_3$		− 26.8; − 29.8	
Me_2SnH_2		87.8	
Et_2SnH_2		42.6	− 10.9
Me_2SnCl_2		− 296.8	− 286.7
$Me_2ClSnSnClMe_2$			− 306.4
Ph_3SnEt	380		
Ph_3SnMe	380		
Ph_3SnI	380		
Ph_3SnSPh	543		
$Ph_3SnSnMe_3$	359		
Sn^+			1008.7
$Me_3Sn^·$			113.8; 33.6; 127.5
Me_3Sn^+			770.2; 781.7; 756.6
$Me_2Sn{-}CH_2^·$			317.7
$Me_2Sn{=}CH_2$			129.6
Me_2Sn			1.05
$Me_2ClSn^·$			− 100.7

(*a*) Data from M. Cartwright and A.A. Wolf, *J. Chem. Soc., Dalton Trans.*, 1976, 829; M.F. Lappert, J.B. Pedley, J. Simpson and T.R. Spalding, *J. Organometal. Chem.*, **29**, 1971, 195; *Selected Values of Chemical Thermodynamic Properties*, Nat. Bur. Stand. Tech. Note 270–3, US Government Printing Office, Washington DC, 1968; J.D. Cox and G. Pilcher, *Thermochemistry of Organic and Organometallic Compounds*, Academic Press, New York, 1970; J.C. Baldwin, M.F. Lappert, J.B. Pedley and J.S. Poland, *J. Chem. Soc., Dalton Trans.*, 1972, 1943; J.M. Miller and M. Onyszchuk, *J. Chem. Soc. (A)*, 1967, 1132; *Handbook of Chemistry and Physics*, 61st. edn., CRC Press, Boca Raton, Florida, 1981; M.J.S. Dewar, J.E. Friedheim and G.L. Grady, *Organometallics*, 1985, **4**, 1784; M.J.S. Dewar, G.L. Grady and J.J.P. Stewart, *J. Amer. Chem. Soc.*, 1984, **106**, 6771; M.J.S. Dewar, G.L. Grady, D.R. Kuhn and K.M. Merz, *J. Amer. Chem. Soc.*, 1984, **106**, 6773; M.J.S. Dewar, G.L. Grady and D.R. Kuhn, *Organometallics*, 1985, **4**, 1041; D.B. Chambers and F. Glockling, *Inorg. Chim. Acta*, 1970, **4**, 150; C. Glidewell, *J. Organometal. Chem.*, 1985, **294**, 173; W.J. Pietro and W.J. Henre, *J. Amer. Chem. Soc.*, **104**, 4329.
(*b*) $kJ\,mol^{-1}$.
(*c*) $\Delta H_f(c)$.
(*d*) $\Delta H_f(l)$.

fairly large negative enthalpies of formation. In all the three series of compounds, SnX_2, SnX_4 and Me_3SnX (X = halogen), the enthalpy of formation becomes more negative with increasing electronegativity/decreasing size. Formation of high-coordination-number complexes also has a dramatic stabilizing effect. Increasing the number of tin-carbon bonds in tetravalent tin compounds causes an increase in ΔH_f, and some tetraorganostannanes have positive enthalpies of formation, as do all tin hydrides. Not surprisingly, the enthalpies of formation of the transient species are usually very positive, although that of dimethylstannylene is calculated to be only $+1.05\,kJ\,mol^{-1}$ and that of the tin-centred radical $Me_2ClSn^·$ is calculated to be $-100.7\,kJ\,mol^{-1}$ (cf. $Me_3Sn^·$).

Single-bond energy data are listed in Table 2.4. Bond dissociation values vary somewhat between individual molecules, and hence only general trends in the data should be noted. The strongest bonds are formed with the more electronegative elements, oxygen and chlorine (no data appear to be available for the Sn–F bond). Strengths of bonds involving tin are, however, only moderately strong and, therefore, under appropriate conditions, easily cleaved.

2.3 Multiple bond formation[2]

The reluctance of the heavier elements of Group IV to form stable compounds possessing multiple bonds is in sharp contrast to the behaviour of carbon, where $p\pi$–$p\pi$ multiple bond formation is a major feature of its chemistry. Until recently, no stable silicon-containing analogues were known, and examples of $p\pi$–$p\pi$ bonding were

Table 2.4 Bond energy data[a,b].

Compound	Bond	$D(R_3Sn–X)$	$E(Sn–X)$
Me_4Sn	Sn–C	272; 273	201
Me_3Sn^tBu	Sn–C	246	222
Ph_3SnMe	Sn–C	259	
Ph_3SnEt	Sn–C	251	
Ph_4Sn	Sn–C	347	
Me_3SnH	Sn–H	309	
Me_3SnOH	Sn–O	460	322
Me_3SnOEt	Sn–O	351	276
Me_3SnCl	Sn–Cl	422	314
$SnCl_4$	Sn–Cl		315
Me_3SnBr	Sn–Br	355	255
Me_3SnI	Sn–I		188
Me_3SnS^nBu	Sn–S		217
Ph_3SnSPh	Sn–S	288	
Me_3SnNMe_2	Sn–N		171
$(Me_3Sn)_2NMe$	Sn–N		201
$(Me_3Sn)_3N$	Sn–N		176
$Me_3SnSiMe_3$	Sn–Si	285	235
$Ph_3SnSiMe_3$	Sn–Si	280	
$Me_3SnGeMe_3$	Sn–Ge	289	225
$Ph_3SnGeMe_3$	Sn–Ge	297	
$Me_3SnSnMe_3$	Sn–Sn	234; 258	160
$Ph_3SnSnPh_3$	Sn–Sn	259	

(*a*) Data from M.F. Lappert, J.B. Pedley, J. Simpson and T.R. Spalding, *J. Organometal. Chem.*, 1971, **29**, 195; J.C. Baldwin, M.F. Lappert, J.B. Pedley and J.S. Poland, *J. Chem. Soc., Dalton Trans.*, 1972, 1943; D.B. Chambers and F. Glockling, *Inorg. Chim. Acta*, 1907, **4**, 150; R.A. Jackson, *J. Organometal. Chem.*, 1979, **166**, 17; A.C. Baldwin, K.E. Lewis and D.M. Golden, *Int. J. Chem. Kin.*, 1979, **11**, 529; T.J. Burkey, M. Majewski and D. Griller, *J. Am. Chem. Soc.*, 1986, **108**, 2218; and *Selected Values of Chemical Thermodynamic Properties*, Nat. Bur. Stand. Tech. Note 270–3, US Government Printing Office, Washington DC, 1968.
(*b*) $kJ\ mol^{-1}$.

restricted to transient species which could be generated either at high temperatures or in low-temperature matrices. However, routes to stable compounds have been devised, and the chemistry of compounds containing Si=Si, Si=C, Si=N, and Si=P, as well as some germanium analogues, is quite extensive.

A handful of compounds exhibiting multiple bonds involving tin has been described. However, in many cases a problem arises: when is a double bond a formal double bond? Molecular geometry and shortening of bond distances compared to typical single-bond distances may be invoked as corroboration of multiple bonding. Further, albeit indirect, evidence, such as stretching force constant and nmr chemical shift and one-bond coupling constant values, may be obtained from spectroscopic data. In the simplest case for which data is available, molecular SnO_2, which is formed along with Sn_2O_2 and SnO (minor product) when tin vapour is co-condensed with oxygen at 20 K in krypton or nitrogen matrices, has the same linear $D_{\infty h}$ structure as CO_2. Tin–oxygen stretching vibration force constants and Sn–O bond distances are similar for both SnO_2 and SnO (SnO_2: force constant 5.36×10^2 Nm^{-1}, Sn–O distance 181 pm; SnO: force constant 5.62×10^2 Nm^{-1}, Sn–O distance 183 pm, cf. Sn_2O_2 (containing single bonds): force constant 2.59×10^2 Nm^{-1}, Sn–O distance 205 pm), from which a bond order of 1.9 was estimated for the tin-oxygen double bond[3–6].

1, 1-Dimethylstannaethene, $Me_2Sn{=}CH_2$, has been generated in the gas phase by proton abstraction from Me_3Sn^+, and the Sn=C π-bond energy estimated to be 188 kJ mol^{-1}[7]. MNDO calculations, however, cast some doubt on this figure, and predict singlet (Me_2Sn and CH_2 residues parallel) and triplet (Me_2Sn and CH_2 residues orthogonal) forms to be very similar in energy, with the triplet slightly (by 4.6 kJ mol^{-1}) more stable! In both, the tin is predicted to be pyramidal, with Sn–C bond distances of 192 pm and 199 pm for the singlet and triplet forms, respectively[8] (cf. 198.2 pm calculated[9] for $H_2Sn{=}CH_2$).

As with silicon and germanium, stable compounds containing multiple bonds involving tin (or their equivalent) have been prepared by steric crowding at the reactive double bond. Examples of three types of double bond, Sn=C, Sn=P, and Sn=Sn, have been obtained. The tin atom in the stable stannaethene (**1**) is only slightly pyr-

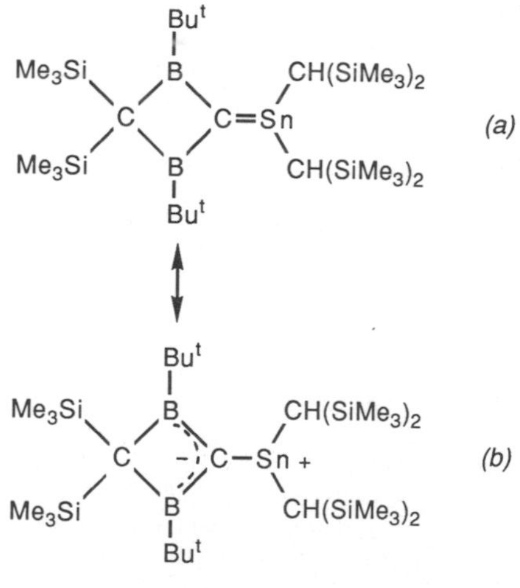

(a)

(b)

(**1**)

amidalized, whereas the C(=Sn) carbon atom is significantly so, and the average twist angle between the SnC_2 and CB_2 residues is 61° (cf. calculated geometries for singlet and triplet $Me_2Sn{=}CH_2$). The tin–carbon double bond distance is somewhat longer (202.5 pm) than those predicted for $Me_2Sn{=}CH_2$ and $H_2Sn{=}CH_2$, but is significantly shorter than the single tin–carbon distances within the molecule (215.2 and 217.2 pm). However, although the ^{13}C (142 ppm) and ^{119}Sn (427.3 ppm) nmr chemical shifts are typical of tricoordinated carbon and tin atoms in the bond, the ^{11}B shift (64 ppm) is indicative of negative π charge on the boron atoms and therefore a significant contribution of the ylidic form (**1b**) to the bonding[10].

No structural data is as yet available for the stannaphosphene (**2**). Nevertheless, nmr data are again consistent with the description as a formal Sn=P double bond ($\delta(^{31}P)$ + 204.7 ppm, $\delta(^{119}Sn)$ 658.3 ppm). Significantly, the magnitude of the one-bond $^{1}J(^{31}P{-}^{117,119}Sn)$ coupling constant (2191, 2295 Hz) is much larger than that observed for singly-bonded compounds, and expected for a high degree of π-character in the bond. The presence of a Sn=P double bond is also reflected in the reactions, which demonstrate high reactivity of the double bond towards reagents such as methanol and hydrogen chloride[11].

Two distannenes, $R_2Sn{=}SnR_2$ (R = $CH(SiMe_3)_2$ and $C_6H_2{}^iPr_3$-2, 4, 6), have been prepared, although bis[bis(trimethylsilyl)methyl]tin, $Sn[CH(SiMe_3)_2]_2$, has been much more widely studied. Whilst having the monomeric angular structure (**3**) in the gas phase[12], in the solid dimerization to the *trans*-bent structure (**4**) occurs, and in solution an equilibrium between the two forms is established[13–17]. The nature of the tin–tin bond is somewhat controversial, and the initial interpretation of the bonding in terms of a double 'banana' bond as in (**5**) has been challenged[18], albeit supported by theoretical calculations on the prototype distannene, $H_2Sn{=}SnH_2$[15,19]. A more recent theoretical description of distannene, however, favours the singly-bonded biradical structure (**6**), in which the unpaired electrons can interact hyperconjugatively across

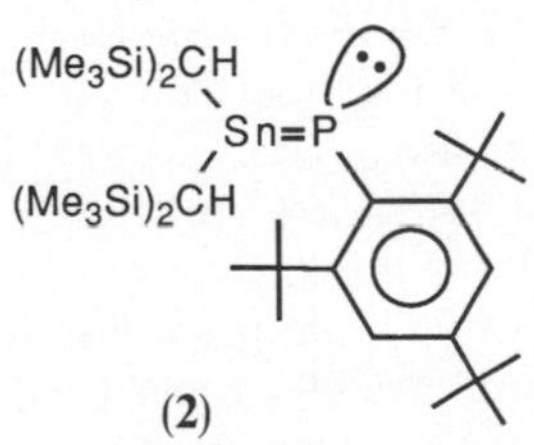

(Me₃Si)₂CH Sn=P (Me₃Si)₂CH

(**2**)

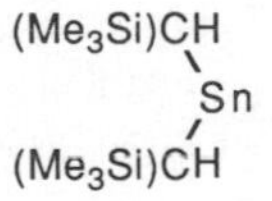

(Me₃Si)CH Sn (Me₃Si)CH

(**3**)

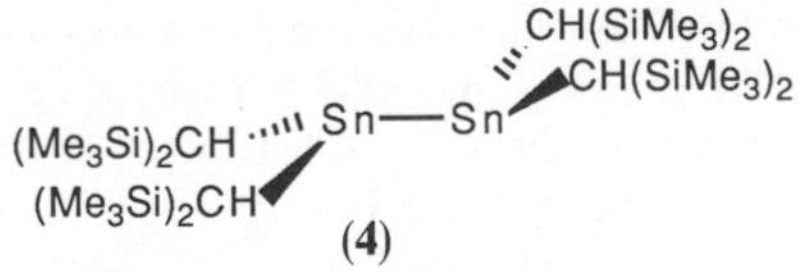

(Me₃Si)₂CH Sn—Sn CH(SiMe₃)₂ CH(SiMe₃)₂ (Me₃Si)₂CH

(**4**)

(Me₃Si)₂CH Sn Sn CH(SiMe₃)₂ CH(SiMe₃)₂ (Me₃Si)₂CH

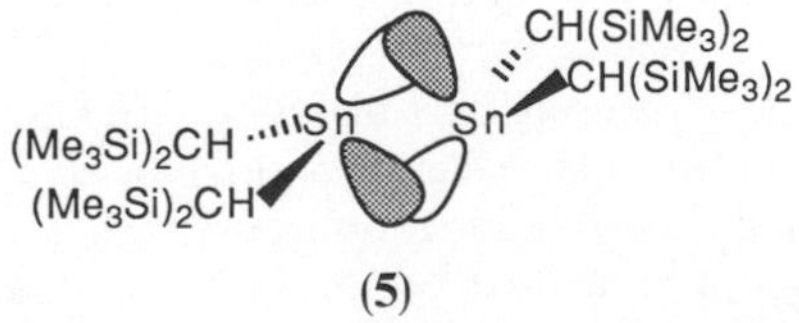

(**5**)

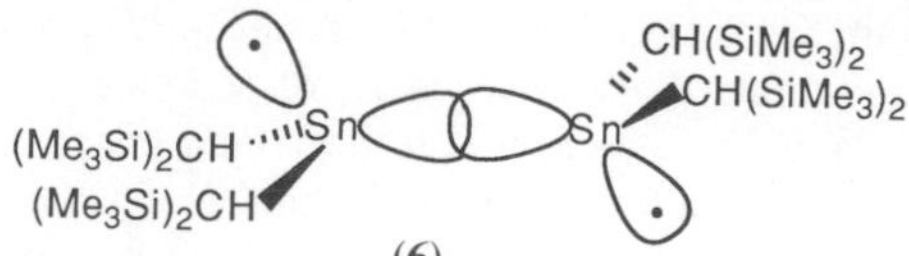

(6)

space giving rise to a π or π-type bond, or σ-conjugatively via the Sn–Sn σ-bond[19]. This description is more consistent with the tin–tin bond distance (276.8(1) pm), which is essentially the same as than that in typical tin–tin bonded tetravalent tin compounds such as $Ph_3SnSnPh_3$ (278.0(4) and 275.9(4) pm)[21] and not consistent with multiple bonding. Both dimers exhibit the very low-field ^{119}Sn chemical shifts ($R = CH(SiMe_3)_2$[22]: solid 692 ppm, solution 725, 740 ppm; $R = C_6H_2{}^iPr_3$-2, 4, 6[23]: solution 427 ppm) usually associated with multiply-bonding tin atoms. However, the dissimilarity of the two chemical shifts, together with the vastly different one-bond $^1J(^{117}Sn-^{119}Sn)$ coupling constants ($R = CH(SiMe_3)_2$: $J = 1340$ Hz; $R = C_6H_2{}^iPr_3$-2, 4, 6: $J = 2930$ Hz; cf. $Me_3SnSnMe_3$: $J = 4211$ Hz[24]) and the differing solution behaviour ([tetrakis(triisopropyl)phenyl]distannene retains its structural integrity in solution), suggests that the tin–tin bonding in the two distannenes is different. That in bis{bis[bis(trimethylsilyl)-methyl]tin} is very weak with a very small enthalpy of dissociation (53.5 kJ mol^{-1}), and the nmr coupling constant data would be consistent with either **(5)** or **(6)**, in which tin 5*s* contributions to the bonding would be small. The bonding in [tetrakis(triisopropyl)phenyl]distannene is obviously much stronger and may involve a more formal double bond.

2.4 Structure and bonding in bivalent compounds[24,26]

2.4.1 *General considerations*

Application of simple VSEPR theory predicts an angular geometry for bivalent tin compounds, and this is indeed observed for the tin (II) halides in the gas phase[27] or isolated in a low-temperature matrix[28] and for other derivatives where the steric requirements of the ligands precludes increase in coordination number at tin. The chemistry of simple monomeric bivalent tin species, SnX_2 (X = H, halogen, Me), usually referred to as stannylenes (IUPAC nomenclature stannanediyls), are of considerable interest as carbene analogues[29,30]. Electronically, stannylenes behave quite differently to carbenes, which invariably exhibit a triplet ($^3B_1$) ground state with the singlet ($^1A_1$) state to higher energy. Like the analogous silylenes and germylenes, a singlet ground state has been calculated for the prototype stannylene[31], SnH_2, and dimethylstannylene[32], $SnMe_2$, and stannylenes behave as singlet species[33,34]. Some structural data are collected in Table 2.5. Particularly diagnostic of the singlet electronic state is low XSnX bond angle of *c.* 90–100°. This is calculated to be significantly higher in the triplet first excited state for SnH_2. In addition, the Sn–H bond distance is reduced in the excited triplet state, which would appear to indicate a greater *p* character for the bond in the ground state.

Bivalent tin compounds wherever possible adopt structures in which the metal achieves coordination numbers higher than two, either by complexation, chelation, or by bridging. Examples of the coordination geometries found are illustrated in Table 2.6. The basic unit is usually recognizable as a trigonal pyramid **(7)**, but

Table 2.5 Structural data for stannylene molecules.

Molecule	M–X (pm)	XMX (°)	Ref.
SnH_2 ($^1A_1$)	(175.6)	(92.7)	*a, b, c*
($^3B_1$)	(170.7)	(118.2)	
	(177)	(93)	
SnF_2		94(5)	*d*
$SnCl_2$	234.7(7)	99(1)	*e*
$SnBr_2$	255(2)	95^f	*g*
$SnMe_2$ ($^1A_1$)	(203)	(99.1)	*a, h, i*
	(270)	(96)	
$Sn[CH(SiMe_3)_2]_2$	222(2)	97(2)	*c*
$Sn[N(SiMe_3)_2]_2$	209.6(1), 208.8(6)	104.7(2)	*j*
$Sn(OC_6H_2Me^tBu_2)_2$	199.5(4), 202.2(4)	88.8(2)	*k*
$Sn(SC_6H_2{}^tBu_3\text{-}2,4,6)_2$	243.5(1)	85.4(1)	*l*

(*a*) Calculated values.
(*b*) G. Olbrich, *Chem. Phys. Lett.*, 1980, **73**, 110.
(*c*) T. Fjeldberg, A. Haaland, B.E.R. Schilling, M.F. Lappert and A.J. Thorne, *J. Chem. Soc., Dalton Trans.*, 1976, 1551.
(*d*) R.F. Hauge, J.W. Hastie and J.L. Margrave, *J. Mol. Spectrosc.*, 1973, **45**, 420.
(*e*) A.A. Ishchenko, L.S. Ivashkevich, E.Z. Zasorin, V.P. Spiridonov and A.A. Ivanov, *Sixth Austin Symp. on Gas Phase Molecular Structure*, Austin, Texas, 1976.
(*f*) Assumed value.
(*g*) M. Lister and L.E. Sutton, *Trans. Faraday Soc.*, 1941, **37**, 393, 406.
(*h*) M.J.S. Dewar, J.E. Friedheim and G.L. Grady, *Organometallics*, 1985, **4**, 1784.
(*i*) P. Bleckmann, H. Maly, R. Minkwitz and G. Olbrich, *Tetrahedron Lett.*, 1982, **23**, 4655.
(*j*) T. Feldberg, H. Hope, M.F. Lappert, P.P. Power and A.J. Thorne, *J. Chem. Soc., Chem. Commun.*, 1983, 639.
(*k*) B. Cetinka, I. Gümrükgu, M.F. Lappert, J.L. Atwood, R.D. Rogers and M.J. Zawototko, *J. Amer. Chem. Soc.*, 1980, **102**, 2088.
(*l*) P.B. Hitchcock, M.F. Lappert, B.J. Samways and E.L. Weinberg, *J. Chem. Soc., Chem. Commun.*, 1983, 1492.

additional bonds or contacts are often present, leading to distorted pseudo-trigonal bipyramidal (**8**), square-based pyramidal (**9**), octahedral (**10**), distorted octahedral (**11**), or facially-capped trigonal prismatic (**12**) coordination geometries. The most common geometries are the three-coordinate trigonal pyramid and the four-coordinate pseudo-trigonal bipyramid and square-based pyramid. The bond angles at tin in the trigonal pyramidal geometry are usually *c.* 90°, indicative of essentially p^3 hybridization for tin

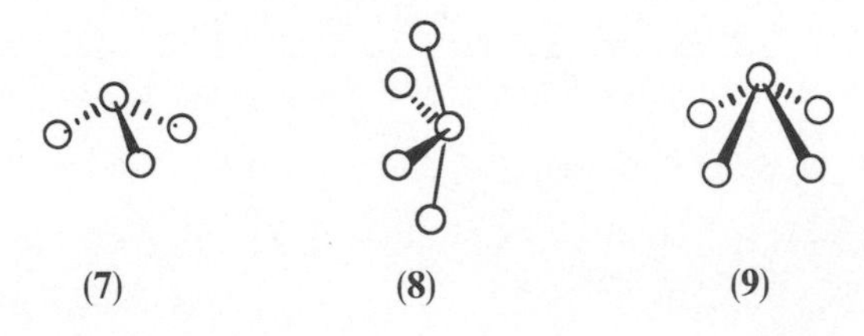

(**7**) (**8**) (**9**)

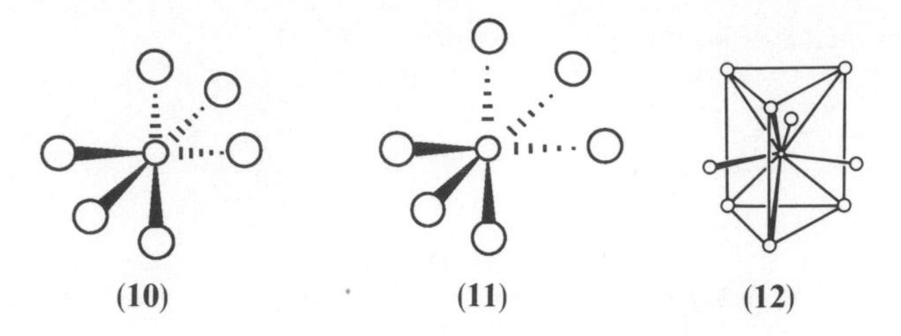
(10) (11) (12)

Table 2.6 Examples of coordination geometries exhibited by bivalent tin compounds.

Coordination number	Coordination geometry	Examples	Ref.
2	Angular	$Sn[CH(SiMe_3)_2]_2$	*a*
		$Sn[N(SiMe_3)_2]_2$	*b*
		$Sn(OC_6H_2Me^tBu_2)_2$	*c*
		$Sn(OC^tBu_3)_2$	*d*
		$Sn(SC_6H_2{}^tBu_3\text{-}2,4,6)_2$	*e*
3	Trigonal planar	$(OC)_5Cr\cdot Sn[CH(SiMe_3)_2]_2$	*f*
3	Pyramidal	SnS	*g*
		Orthorhombic SnSe	*h*
		$SnSO_4$	*i*
		$SnHPO_4$	*j*
		$[Sn(EPh)]_3^-$ (E = S, Se)	*k*
		$[Sn(O_2CH)]_3^-$	*l*
4	Square pyramidal	SnO	*m*
		$[Sn_3F_{10}]^{4-}$	*n*
		(Phthalocyaninato)Sn	*o*
4	ψ-Trigonal bipyramidal	SnFCl	*p*
		$[SnCl_4]^-$	*q*
		$Sn(O_2CH)_2$	*r*
		$Sn(OCMeCHCPhO)_2$	*s*
		$Sn_6O_4(OMe)_4$	*t*
		$Sn(S_2COMe)_2$	*u*
		$Sn[CPMe_2)_3]_2$	*v*
5	Square pyramidal	[Tris(pyrazolyl)borato]Sn	*w*
		$[Sn_2O(O_2CCF_3)_4\cdot O(OCCF_3)_2]_2$ (Sn^{II} sites)	*x*
6	Octahedral	Cubic SnSe	*y*
		SnTe	*y, z*
6	Pentagonal bipyramidal	$[Sn_2O(O_2CC_6H_4NO_2\text{-}o)_4\cdot(thf)_2]_2$ (Sn^{II} sites)	*aa*
6	Irregular	$\{(C_5H_5)Co[P(OEt)_2O]_3\}_2Sn$	*bb*
		$Sn_2(edta)\cdot 2H_2O$ (Sn1 site)	*cc*
7	Irregular	$Sn_2(edta)\cdot 2H_2O$ (Sn2 site)	*cc*
9	Trifacially capped trigonal prismatic	SnX_2 (X = Cl, Br)	*dd, ee*
		$Sn(NCS)_2$	*ff*
		$Sn(SbF_6)_2(AsF_3)_2$	*gg*

(*a*) T. Fjeldberg, A. Haaland, B.E.R. Schilling, M.F. Lappert and A.J. Thorne, *J. Chem. Soc., Dalton Trans.*, 1976, 1551.

(*b*) T. Feldberg, H. Hope, M.F. Lappert, P.P. Power and A.J. Thorne, *J. Chem. Soc., Chem. Commun.*, 1983, 639.

(*c*) B. Cetinka, I. Gümrükgu, M.F. Lappert, J.L. Atwood, R.D. Rogers and M.J. Zawototko, *J. Amer. Chem. Soc.*, 1980, **102**, 2088.

(*d*) T. Fjeldberg, P.B. Hitchcock, M.F. Lappert, S.J. Smith and A.J. Thorne, *J. Chem. Soc., Chem. Comm.*, 1985, 939.
(*e*) P.B. Hitchcock, M.F. Lappert, B.J. Samways and E.L. Weinberg, *J. Chem. Soc., Chem. Commun.*, 1983, 1492.
(*f*) D.E. Goldberg, D.H. Harris, M.F. Lappert and K.M. Thomas, *J. Chem. Soc., Chem. Commun.*, 1976, 261.
(*g*) W. Hoffmann, *Z. Krist.*, 1935, **92**, 161.
(*h*) A. Okazaki and I. Ueda, *J. Phys. Soc. Jpn.*, 1956, **11**, 470.
(*i*) J.D. Donaldson and D.C. Puxley, *Acta Crystallogr.*, 1972, **B28**, 864.
(*j*) R.C. McDonald and K. Eriks, *Inorg. Chem.*, 1980, **19**, 1237.
(*k*) P.A.W. Dean, J.J. Vittal and N.C. Payne, *Can. J. Chem.*, 1985, **63**, 394.
(*l*) A. Jelen and O. Lundquist, *Acta Chem. Scand.*, 1969, **23**, 3071.
(*m*) W.J. Moore and L. Pauling, *J. Amer. Chem. Soc.*, 1941, **63**, 1392.
(*n*) G. Bergerhoff and L. Goost, *Acta Crystallogr.*, 1970, **B26**, 19.
(*o*) M.K. Friedel, B.F. Hoskins, R.L. Martin and S.A. Mason, *Chem. Commun.*, 1970, 400.
(*p*) C. Geneys, S. Vilminot and L. Cot, *Acta Crystallogr.*, 1976, **B32**, 3199.
(*q*) H.J. Haupt, F. Huber and H. Preut, *Z. anorg. allg. Chem.*, 1976, **422**, 97, 255.
(*r*) P.G. Harrison and E.W. Thornton, *J. Chem. Soc., Dalton Trans.*, 1978, 1274.
(*s*) P.F.R. Ewings, P.G. Harrison and T.J. King, *J. Chem. Soc., Dalton Trans.*, 1975, 1455.
(*t*) P.G. Harrison, B.J. Haylett and T.J. King, *Chem. Commun.*, 1978, 112.
(*u*) P.F.R. Ewings, P.G. Harrison and T.J. King, *J. Chem. Soc., Dalton Trans.*, 1976, 1399.
(*v*) H.H. Karsch, A. Appelt and G. Müller, *Angew Chem., Int. Edn. Engl.*, 1985, **118**, 2020.
(*w*) A.H. Cowley, R.L. Geerts, C.M. Nunn and C.J. Carrano, *J. Organomet. Chem.*, 1988, **341**, C27.
(*x*) T. Birchall and J.P. Johnson, *Inorg. Chem.*, 1982, **21**, 1679.
(*y*) R. Rundle and D. Olsen, *Inorg. Chem.*, 1964, **3**, 596.
(*z*) H. Krebs, K. Grun and D. Kallen, *Z. anorg. Chem.*, 1961, **312**, 307.
(*aa*) P.F.R. Ewings, P.G. Harrison, A. Morris and T.J. King, *J. Chem. Soc., Dalton Trans.*, 1976, 1602.
(*bb*) E.M. Holt, W. Kläui and J.J. Zuckerman, *J. Organomet. Chem.*, 1987, **335**, 29.
(*cc*) F.P. van Remoortere, J.J. Flynn, F.P. Boer and P.P. North, *Inorg. Chem.*, 1971, **10**, 1511.
(*dd*) J. Anderson, *Acta Chem. Scand.*, 1975, **A29**, 956.
(*ee*) J.M. van den Berg. *Acta Crystallogr.*, 1961, **14**, 1002.
(*ff*) A.G. Filby, R. Howie and W. Moser, *J. Chem. Soc., Dalton Trans.*, 1978, 1997.
(*gg*) A.J. Edwards and K.I. Khallow, *J. Chem. Soc., Chem. Commun.*, 1984, 50.

(thus placing the non-bonding electron pair in the 5*s* orbital). Distortions from ideal geometry are frequently encountered, and a distinction between the two four-coordinate geometries is sometimes semantic. Coordination numbers higher than four are relatively rare, and in some cases, such as the $PbCl_2$-type structure exhibited by tin (II) chloride and bromide, the apparent coordination number does not reflect the number of actual bonding contacts. Indeed, in that case the stereochemistry is probably better regarded as distorted octahedral, with the basic trigonal pyramidal $[SnX_3]$ unit being supplemented by three additional longer secondary bonds in the three *trans* octahedral positions.

For the majority of compounds, and probably all with bonds to light atoms such as fluorine, oxygen and nitrogen, the tin lone pair is stereochemically active. This is usually diagnosed by the presence of an apparent vacancy in the primary coordination sphere, together with the consequent structural distortions, and also by the ability to function as a two-electron donor ligand. The only crystallographically corroborated example[35] of a complex involving coordination of the tin (II) lone pair to a Main Group Lewis acid is the complex $(SnN^tBu)_4 \cdot 2AlCl_3$, although the complexes $(C_5H_5)_2Sn{-}AlX_3$ are also proposed[36] to contain a formal Sn–Al bond. The analogous adduct with boron trifluoride is not a true stannylene complex, but rather has a complex structure with no tin–boron bonding[37]. In contrast, many complexes are known in which a tin(II) species functions as a donor to transition metal Lewis acids, and several examples are illustrated in Figure 2.1[38–48]. The bonding between the transition metal and tin in these complexes is $\sigma + \pi$ in character, with synergistic $p\pi$–$d\pi$ retrodative bonding supplementing the σ bond. Complex formation appears to

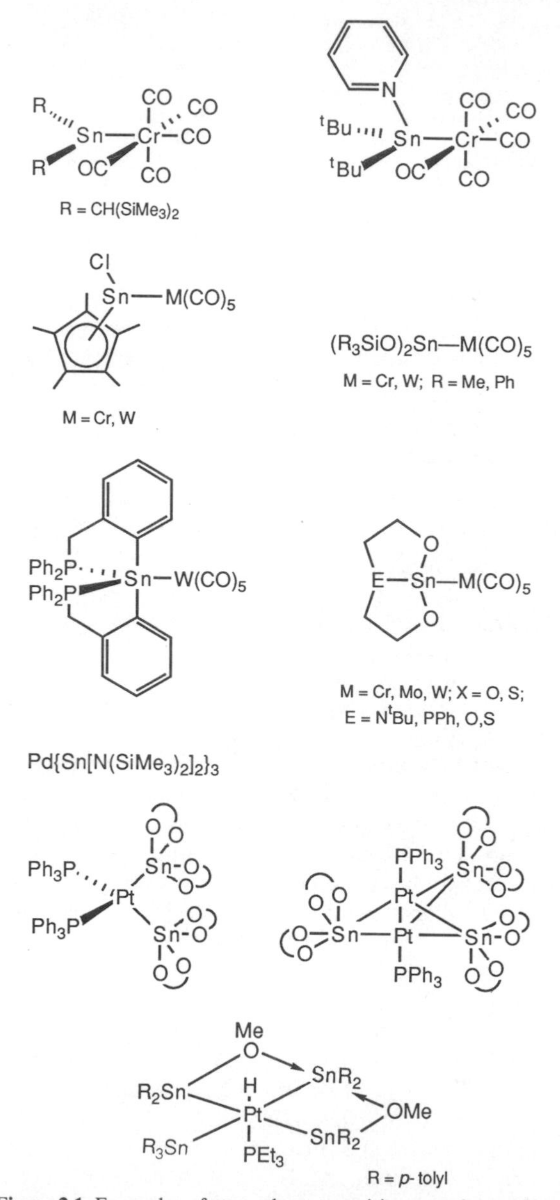

Figure 2.1 Examples of stannylene–transition metal complexes.

have little effect on the geometry at tin in stannylene complexes. The $[C_2SnCr]$ fragment in the $(OC)_5Cr \leftarrow Sn[CH(SiMe_3)_2]_2$ is coplanar[44], but that in the related $(OC)_5Cr \leftarrow Sn^tBu_2(py)$ complex is not so, due to the coordination of the pyridine nitrogen into the vacant *p* orbital on tin[43]. In lattices in which the tin(II) enjoys near or exact octahedral coordination, the non-bonding electron pair is no longer stereochemically active, but rather is delocalized into conduction band levels.

2.4.2 *Tin(II) halides and halide complexes*

In the solid state, the tin(II) halides form variants of close-packed lattices with distortions arising because of the metal non-bonding pair of electrons. The $[SnF_3]$ trigonal pyramid is the basic unit of both the orthorhombic and monoclinic modifications of tin (II) fluoride. However, strongly fluorine-bridging results in pseudo trigonal bipyramidal geometry at tin for the former and highly distorted octahedral geometry for the latter[49,50]. Anhydrous tin(II) chloride and bromide exhibit the prototype $PbCl_2$-type lattice (**12**)[51–53] in which the tin atom is surrounded by nine halogen atoms from three different layers, six at the apices of a trigonal prism and three in facially-capping positions. The coordination is, however, far from regular, and the metal forms three short bonds (the typical $[SnX_3]$ structural unit). The remaining halogen atoms are at much longer distances, some of which cannot represent bonding interactions. Crystals of tin(II) iodide possess a unique layer lattice, with the metal atoms occupying two different sites. Two-thirds occupy sites based on the $PbCl_2$-type lattice, while the remainder are in $PdCl_2$-type chains, which interlock with the $PbCl_2$-type part of the structure to give almost perfect octahedral coordination[54].

A similar range of coordination geometries is also observed for halide complexes and complex anions. Although many complexes of stoichiometries $MX_2 \cdot L$ and $MX_2 \cdot 2L$ have been synthesized, the structures of only a few are known. The basic unit of the $SnCl_2 \cdot 2H_2O$ and the thiourea complex of tin(II) chloride is again the pyramidal $[SnCl_2 \cdot L]$ ($L = H_2O$, $S{=}C(NH_2)_2$). However, in both these units are tighly bound into three-dimensional structures by hydrogen bonding and/or chlorine and sulphur bridging[55–57]. In contrast, the structure of the 1,4-dioxan complex consists of an angular $[SnCl_2]$ unit linked by dioxan ligands to form polymeric chains with pseudo-trigonal bipyramidal geometry for tin[58]. In both the 1,10-phenanthroline and 2,2′-bipyridine complexes of tin(II) chloride, the nitrogen donor ligand chelates the tin atom with two longer and weaker contacts to chlorine resulting in the formation of chain structures[59]. The structures of the tin(II) bromide hydrates $2SnBr_2 \cdot H_2O$, $3SnBr_2 \cdot H_2O$, and $6SnBr_2 \cdot 5H_2O$ are all based on the trigonal prismatic geometry (**12**), in which the tin is surrounded by a trigonal prism of six bromine atoms, and further bromines, water molecules or vacancies at the facially-capping sites[60].

The particular structure adopted by a halogenostannate(II) salt is very dependent on the cation and halogen. Several examples of pyramidal $[SnX_3]^-$ anions have been characterized, although most involve additional longer contacts. Crystals of $[NH_4][SnF_3]$ contain essentially discrete $[SnF_3]^-$ anions[61,62]. Discrete isolated $[SnCl_3]^-$ anions occur[63,64] in $[Co(Ph_2CH_2CH_2PPh_2O)][SnCl_3]$ and $[Co(en)_3][SnCl_3]Cl_2$. No interaction occurs between the two anions in the latter complex, whereas in the closely related compound, $[Co(NH_3)_6][SnCl_4]Cl$, contains $[SnCl_4]^{2-}$ anions of the pseudo-trigonal bipyramidal geometry[64]. An intermediate situation is present in the structure of $[NH_4]_2[SnCl_3]Cl \cdot H_2O$, where significant interaction between the chloride and trichlorostannate ions occurs. However, neighbouring $[SnCl_3 \cdots Cl]$ units are also connected by two longer $Sn \cdots Cl$ bridges forming a chain structure completing a severely distorted octahedral environment at tin[65]. A similar distorted octahedral coordination with three short and three much longer metal-halogen distances is found[61,62] in $[NH_4][SnF_3]$, and the low-temperature form[66] of $CsSnCl_3$. The caesium trihalogenostannates undergo phase transitions to modifications with the cubic perovskite lattice in which the tin has ideal cubic coordination[67–71]. Ideal octahedral coordination is also found in the perovskite methylammonium salts, $[MeNH_3][SnBr_xI_{3-x}]$ ($x = 0$ to 3) and $[MeNH_3][Sn_{1-n}Pb_nX_3]$ (X = Cl, Br, I)[72], and $CsSnI_3$[73]. In these, the tin (II) lone pair is no longer stereochemically active, and populates the lattice conduction bands, giving rise to intense colour and electrical conductivity.

The strong tendency to form Sn–F–Sn bridges observed in SnF_2 is also a prominent and dominating factor in the structural chemistry of other fluorotin(II) compounds. Only the three-coordinate pyramidal and four-coordinated pseudo-trigonal bipyramidal and square-pyramidal geometries are found. Amongst the known fluorostannate salts, the $[NH_4][SnF_3]$ salt (see above) appears to be exceptional in not exhibiting fluorine bridging. The $[Sn_2F_5]^-$ anion comprises two corner-sharing pyramids as in (**13**), although a fourth fluorine is located at a distance of only 253 pm from tin, and hence the structure may be regarded as containing infinite $[Sn_2F_5]_\infty^{\infty-}$ anions[74]. The $[Sn_3F_{10}]^{4-}$ anion consists of three edge-sharing square pyramidal $[SnF_4]$ units as in (**14**)[75]. Illustrative of the profound effect on structure that the cation can have are those[76] of $[NH_4][SnF_3]$ and $K[SnF_3]\cdot\frac{1}{2}H_2O$. Whereas the former comprises isolated $[SnF_3]^-$ anions, the latter forms infinite chains of edge-sharing $[SnF_4]$ square pyramids (**15**), i.e. a $[SnF_3]_\infty^{\infty-}$ polyanion. Cationic fluorotin (II) networks such as those found in Sn_2F_3Cl[77], Sn_3F_5Br[78], Sn_2F_3I[79] and $[Sn_5F_9][BF_4]$[80] have typical pyramidal $[SnF_3]$ or distorted square pyramidal $[SnF_4]$ coordination at tin with halide or tetrafluoroborate anions occupying large holes in the networks. The simple mixed tin (II) fluoride chloride[81], SnFCl, and isothiocyanate[82], SnF(NCS), both have ribbon-like structure (**16**) propagated by four-membered $[Sn_2F_2]$ rings, with pseudo-trigonal bipyramidal tin.

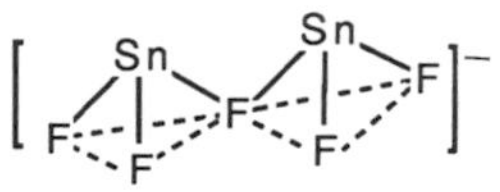

(**13**)

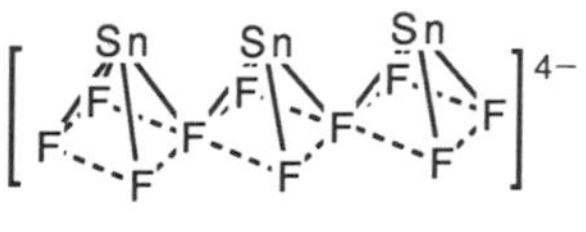

(**14**)

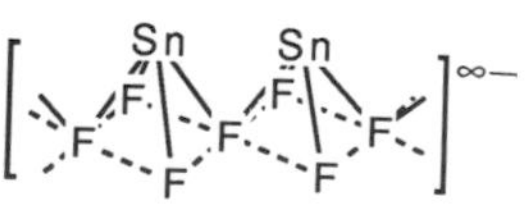

(**15**)

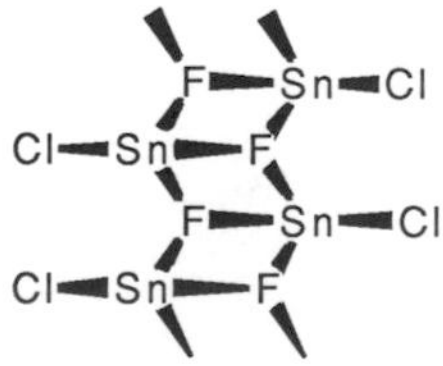

(**16**)

2.4.3 *Chalcogenide derivatives*

Similar trends are observed with the chalcogenide derivatives. Tetragonal tin (II) oxide forms the layer lattice (**17**) with square pyramidal coordination and equal tin-oxygen distances[83]. Tin (II) sulphide has a structure with parallel zigzag –Sn–S–Sn–S–Sn–S chains, which are connected by short interchain Sn···S contacts, resulting in a basic pyramidal [SnS_3] structural unit. Tin (II) selenide exhibits one phase which is isomorphous with SnS and a second cubic phase with the NaCl lattice, which is also adopted by tin (II) telluride[84–89].

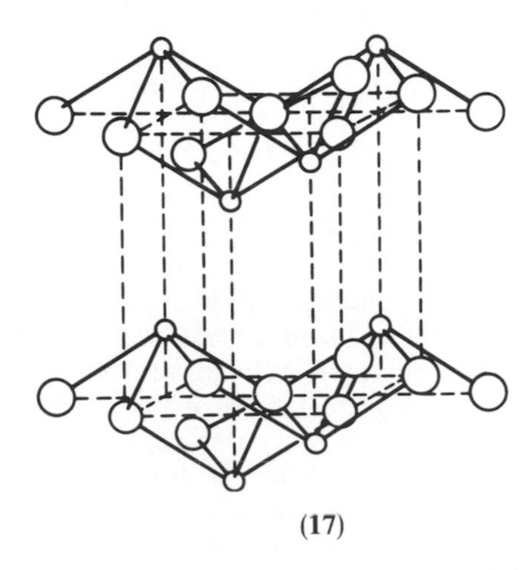

(**17**)

Hydrolysis of tin (II) salts yields oxotin ions of several different types depending upon the conditions, particularly pH, some of which have been isolated and characterized. Structurally, they exhibit the same basic features. Thus, the sodium salts, $Na_2[Sn_2O(OH)_4]$ and $Na_4[Sn_4O(OH)_{10}]$, contain the $[Sn(OH)_3]^-$ and oxo-bridged $[(HO)_2SnOSn(OH)_2]^{2-}$ anions, respectively[90], while the barium salt, $Ba[SnO(OH)]_2$, contains the one-dimensional $[SnO(OH)]_\infty^{\infty-}$ polyanion[91]. All three anions exhibit typical pyramidal coordination at tin. The discrete $[Sn_3O(OH)_2]^{2+}$ cation, found in basic tin (II) sulphate, $[Sn_3O(OH)_2]SO_4$, contains one tin in pyramidal coordination with two in distorted square-pyramidal four-coordination[92,93]. Unlike polymeric anhydrous tin (II) oxide, the monohydrate $3SnO{\cdot}H_2O$, which can be isolated from neutral solutions, contains adamantane-type [Sn_6O_8] clusters[94], very similar to that (**18**) found in $[Sn_6O_4(OMe)_4]$ (from the controlled hydrolysis of tin (II) methoxide)[95], but again the local geometry at tin is distorted pseudo-trigonal bipyramidal.

2.4.4 *Molecular compounds containing bonds to elements of Groups 14, 15 and 16, and related derivatives*

Simple σ-bonded bivalent dialkyl- or diaryltin compounds of the type R_2Sn are not stable. Rather, attempts to generate such compounds always results in the formation of isomeric metal–metal bonded cyclic oligomers, $(R_2Sn)_n$, containing tetravalent tin atoms[96–100]. Predictably, this facile polymerization process may be precluded by increasing the steric bulk of the organic groups attached to tin, as in $Sn[CH(SiMe_3)_2]_2$

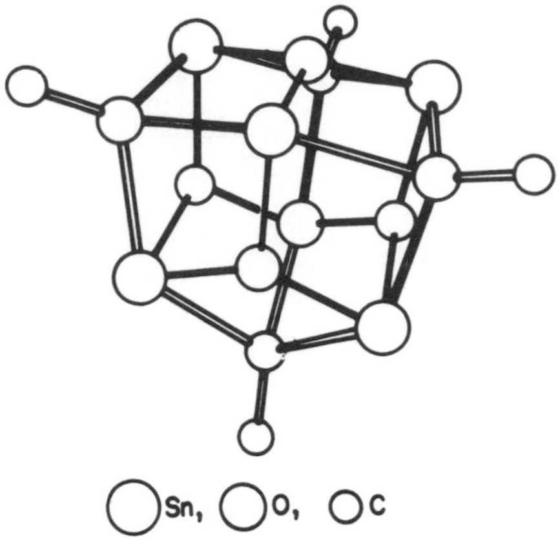

Sn, O, C

(**18**)

which is monomeric in the gas phase and in solution but dimeric in the solid[12–17] (see section 2.3). The steric bulk of the ligand is, however, no guarantee of two-coordination, and the tin atom in the very sterically crowded bis(silyl)tin (II) derivative, $[(Me_3Si)_3Si]_2Sn(\mu\text{-}Cl)Li(thf)_3$, is rendered three-coordinate (**19**) by the bridging chlorine atom. The Si–Sn–Si bond angle is nevertheless opened[101] to 114.2(4)°.

Monomeric amide, alkoxide, and thiolate derivatives of bivalent tin are also only obtained with very sterically demanding ligands which preclude the possibility of bridging. Thus, the bis(trimethylsilyl)amide[102], $Sn[N(SiMe_3)_2]_2$, the bis(alkoxide)[103], $Sn[O(CMe_3)_3]_2$; bis(aryloxide)[104], $Sn(OC_6H_2Me\text{-}4\text{-}{}^tBu_2\text{-}2,6)_2$; and bis(arenethiolate)[105], $Sn(SC_6H_2Me\text{-}4\text{-}{}^tBu_2\text{-}2,6)_2$ are all two-coordinated with a angular V-shaped geometry. Otherwise, for smaller ligands, association takes place via nitrogen, oxygen, or sulphur bridging to give dimers, trimers, or polymers depending upon the steric requirements of the particular ligand, with either pyramidal or distorted pseudo-trigonal bipyramidal geometry at tin. Thus, tin (II) bis(dimethylamido)tin (II) exists as the dimer (**20**)[106], whereas a polymeric structure such as (**21**) has been proposed for bis(aziridinyl)tin (II)[107]. Crystals of the cyclic tin (II) amide, $[SnN({}^tBu)SiMe_2N({}^tBu)]$, comprise both monomeric and dimeric units[108]. Similarly, bis(tert-butoxy)tin (II)[109]. bis(triphenylsiloxy)tin (II)[110], and 5-tert-butyl-5-aza-2, 8-dithia-1-stanna(II)bicyclo[3.3.0^{1,5}]octane[110], $[Sn(SC_2H_4)_2N{}^tBu]_2$,

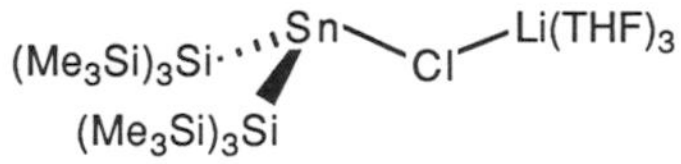

$(Me_3Si)_3Si$ – Sn – Cl – $Li(THF)_3$; $(Me_3Si)_3Si$

(**19**)

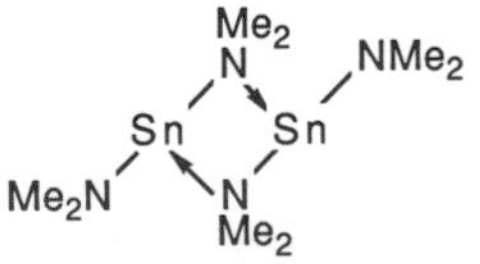

Me_2N, Sn, NMe_2, Sn, $N Me_2$, $N Me_2$

(**20**)

(21)

are dimeric, while tin(II) bis(2,6-diisopropylbenzenethiolates) are trimeric[111], and small alkoxides, $Sn(OR)_2$ (R = Me, Et) are polymeric[112,113]. A relatively large number of tin–nitrogen cage structures have been characterized[114], and typical members are illustrated in Figure 2.2. Both bis(di-*tert*-butylphosphino)tin(II), $Sn(P^tBu_2)_2$, and bis(*tert*-butylthio)tin(II), $Sn(S^tBu)_2$, are dimeric in solution[115], but

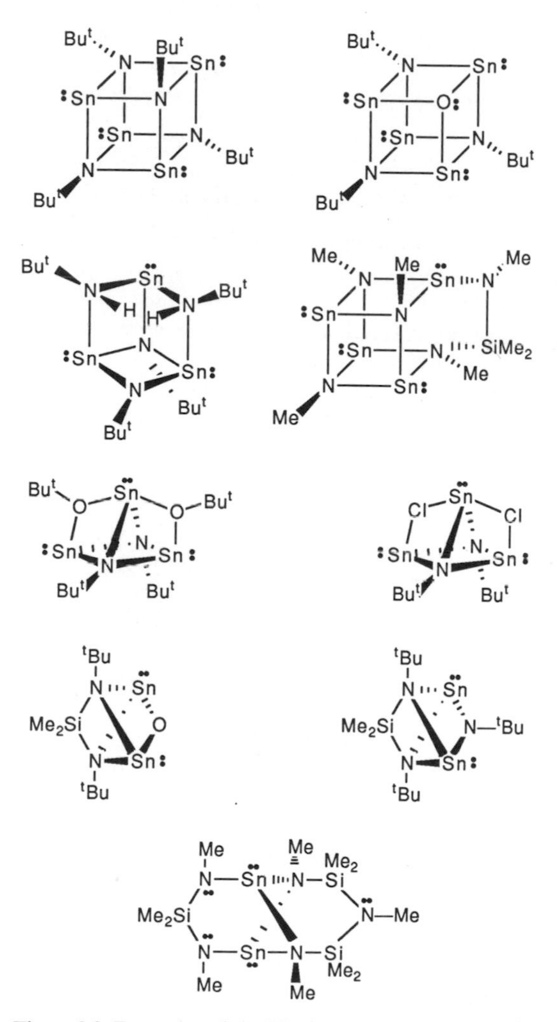

Figure 2.2 Examples of tin(II)-nitrogen cage compounds.

their solid-state structures are not yet known. The complex thilato- and selenato-stannate(II) anions $[Sn(EPh)_3]^-$ (E = S, Se)[116] have pyramidal three-coordination. The analogous lithium phosphinostannate(II) $(^tBu_2P)Sn(\mu\text{-}^tBu_2P)_2Li\cdot thf$ is similar, but two of the three phosphorus atoms chelate the lithium[117].

When the ligand incorporates additional donor sites, coordination saturation at tin can be achieved either intramolecularly or intermolecularly, resulting in the formation, respectively, of discrete monomers or polymers. Thus, $Sn[N(SiMe_3)_2][C(SiMe_3)_2C_5H_4N\text{-}2]$[118], $SnCl[C(SiMe_3)_2C_5H_4N\text{-}2]$[118], and $Sn[CH(PPh_2)_2][(PPh_2)_2CH]$[119–121] are monomeric and pyramidal, whilst $Sn[C(SiMe_3)_2C_5H_4N\text{-}2]_2$[118], $Sn(S_2COMe)_2$[122], N, N′-ethylenebis(acetylideneimino)-tin(II)[123], the tin(II) derivatives of 1, 3-diones and related compounds[124–126], and the tetraphosphane complex, $Sn[C(PMe_2)_3]_2$[121,127] are all monomeric with distorted pseudo-trigonal bipyramidal geometry. Tin (II) carboxylates[128–134], sulphates[135–137], and phosphates[138–143] on the other hand, are usually polymeric in nature. Nevertheless, in spite of the large changes in gross structure, typical three- and four-coordinate geometries about the tin atom are found, and the anionic groups function as bridging ligands.

2.5 Structural behaviour of tetravalent tin compounds[25,26]

Tetrahedral four-coordination is only observed for tin in R_4Sn (R = organic group) compounds and R_nSnX_{4-n} derivatives which are either sterically crowded or in which the ligand is weakly electronegative and a poor donor, and tin strives to achieve coordination numbers greater than four wherever possible. The most common higher coordination numbers are five and six, although examples of seven and eight are also known. Representative examples are shown in Table 2.7. Of the two possible five-coordinate geometries, trigonal bipyramidal and square-based pyramidal, the former predominates.

2.5.1 *Hydrides and tetraorganostannanes*

Stannane, SnH_4, organotin(IV) hydrides, (SnR_nH_{4-n} where $n = 1$ to 3; R = alkyl), and most compounds with four tin-carbon bonds are tetrahedral at tin, although minor distortions can arise to crystal packing effects[25]. Bonds to electronegative organic ligands are generally longer and weaker, and hence more susceptible to cleavage. For example, the Sn–C bond distances in $Sn(CF_3)_4$ are lengthened by 6 pm compared with those in $SnMe_4$[144]. However, two tin atoms in compounds containing four tin–carbon bonds appear to be reluctant to form higher coordination numbers. Although the complex $Me_3SnCF_3\cdot P(NMe_2)_3$ has been reported[145], tetraorganostannanes, R_4Sn, do not in general form donor-acceptor complexes with Lewis bases. Two five-coordinated examples, **(22)**[146] and **(23)**[147], and one six-coordinated example **(24)**[148] with four tin–carbon bonds, all involving intramolecular N→Sn coordination, have been characterized.

2.5.2 *Halides and halide complexes*

Tin (IV) halides, SnX_4, and organotin (IV) halides (SnR_nX_{4-n} where $n = 1$ to 3; R = alkyl, alkenyl, aryl) are tetrahedral in the vapour and liquid[25], but in the solid they

Table 2.7 Examples of coordination geometries exhibited by tetravalent tin compounds.

Coordination number	Coordination geometry	Examples	Ref.
3	Trigonal planar	$[SnE_3]^{2-}$ (E = Se, Te)	*a*
4	Tetrahedral	Me_4Sn	*b*
		$(CF_3)_4Sn$	*b*
		Ph_4Sn	*c*
		Ph_3SnCH_2I	*d*
		$[(Me_3Si)_3C]Me_2SnF$	*e*
		$(PhMe_2CCH_2)_3SnCl$	*f*
		$[(Me_3Si)_2N]_3SnBr$	*g*
		$Ph_3SnSnPh_3$	*h*
		$Ph_3MnSn(CO)_5$	*i*
		K_4SnO_4	*j*
5	Trigonal bipyramidal	Me_3SnF	*k*
		$Me_3SnCl{\cdot}py$	*l*
		Me_3SnO_2CMe	*m*
		$Me_3SnONPh{\cdot}CO{\cdot}Ph$	*n*
		[3-(2-py)-2-thienyl](*p*-tol)$_3$Sn	*o*
		$Ph_3SnNO_3{\cdot}OEPh_3$ (E = P, As)	*p*
		$[(C_6H_4OS)_2SnCl]^-$	*q*
		$[SnCl_5]^-$	*r*
5	Square pyramidal	$[(MeC_6H_3S_2)_2SnX]^-$ (X = Cl, Br)	*q, s*
		$(PhCH_2)_3Sn(2\text{-}SC_5H_4NO)$	*t*
6	Octahedral	SnO_2	*u*
		$MgSnO_4$	*v*
		$CaSnO_3$	*w*
		Me_2SnF_2	*x*
		$Me_2Sn(acac)_2$	*y*
		[3-(2-py)-2-thienyl]$_2$Ph$_2$Sn	*z*
		$[SnX_6]^{2-}$	*r, aa, bb*
		(X = F, Cl, Br, I)	*cc*
6	Skew trapezoidal	$Me_2Sn(NO_3)_2$	*dd*
		$Me_2Sn(ONMe{\cdot}CO{\cdot}Me)_2$	*ee*
		$Me_2Sn(trop)_2$	*ff*
		$Me_2Sn(S_2CNEt_2)_2$	*gg*
7	Pentagonal bipyramidal	$Me_2Sn(NCS)_2(ter)$	*hh*
		$[Me_2Sn(O_2CMe)_3]^-$	*ii*
		$MeSn(NO_3)_3$	*jj*
		$MeSn(S_2CNEt_2)_3$	*kk*
8	Dodecahedral	$Sn(NO_3)_4$	*ll*
		$Sn(O_2CMe)_4$	*mm*
8	Square antiprismatic	Bis(phthalocyaninato)tin	*nn*

(*a*) R.C. Burns, L.A. Devereux, P. Granger and G.J. Schrobilgen, *Inorg. Chem.*, 1985, **24**, 2615.
(*b*) R. Eujen, H. Bürger and H. Oberhammer, *J. Mol. Struct.*, 1981, **71**, 109.
(*c*) P.C. Chieh and J. Trotter, *J. Chem. Soc. (A)*, 1970, 911.
(*d*) P.G. Harrison and K.C. Molloy, *J. Organomet. Chem.*, 1978, **152**, 53.
(*e*) S.S. Al-Juaid, S.M. Dhaher, C. Eaborn, P.B. Hitchcock and J.D. Smith, *J. Organomet. Chem.*, 1987, **325**, 117.
(*f*) D. Schomburg, M. Link, H. Linoh and R. Tacke, *J. Organomet. Chem.*, 1988, **339**, 69.
(*g*) M.F. Lappert, M.C. Misra, M. Onyszchuk, R.S. Rowe, P.P. Power and M.J. Slade, *J. Organomet. Chem.*, 1987, **330**, 31.
(*h*) H. Preut, H.J. Haupt and F. Huber, *Z. anorg. allg. Chem.*, 1973, **396**, 81.
(*i*) H.P. Weber and R.F. Bryan, *Acta Crystallogr.*, 1967, **22**, 822.

Table 2.7 (*Contd.*)

(*j*) R. Marchand, Y. Piffard and M. Tournoux, *Acta Crystallogr.*, 1975, **B31**, 511.
(*k*) H.C. Clark, R.J. O'Brien and J. Trotter, *J. Chem. Soc.*, 1964, 2332.
(*l*) R. Hulme, *J. Chem. Soc.*, 1963, 1524.
(*m*) H. Chih and B.R. Penfold, *J. Cryst. Mol. Struct.*, 1973, **3**, 285.
(*n*) P.G. Harrison, T.J. King and K.C. Molloy, *J. Organomet. Chem.*, 1980, **185**, 199.
(*o*) V.G. Kumar Das, L.K. Mun, C. Wei, S.J. Blunden and T.C.K. Mak, *J. Organomet. Chem.*, 1987, **322**, 163.
(*p*) M. Nardelli, C. Pelizzi and G. Pelizzi, *J. Organomet. Chem.*, 1977, **125**, 161.
(*q*) R.R. Holmes, S. Shafieezad, V. Chandrasekhar, A.C. Sau, J.M. Holmes and R.O. Day, *J. Amer. Chem. Soc.*, 1988, **110**, 1168.
(*r*) J. Shamir, S. Luslki, A. Bino, S. Cohen and D. Gibson, *Inorg. Chem.*, 1985, **24**, 2301.
(*s*) A.C. Sau, R.O. Day and R.R. Holmes, *J. Amer. Chem. Soc.*, 1981, **103**, 1264; *Inorg. Chem.*, 1981, **20**, 3076.
(*t*) S.W. Ng, C. Wei, V.G. Kumar Das and T.C.K. Mak, *J. Organomet. Chem.*, 1987, **334**, 283.
(*u*) J. Pannetier and G. Denes, *Acta Crystallogr.*, 1980, **B36**, 2763.
(*v*) P. Poix, *Ann. Chim.*, 1965, **10**, 49.
(*w*) B. Durand and H. Loiseleur, *J. Appl. Chem.*, 1978, **11**, 289.
(*x*) E.O. Schlemper and W.C. Hamilton, *Inorg. Chem.*, 1966, **5**, 995.
(*y*) G.A. Miller and E.O. Schlemper, *Inorg. Chem.*, 1973, **12**, 677.
(*z*) V.G. Kumar Das, L.K. Mun, C. Wei and T.C.W. Mak, *Organometallics*, 1987, **6**, 10.
(*aa*) M.J. Durand, J.L. Galigne and A. Lari-Lavassani, *J. Solid State Chem.*, 1976, **16**, 157.
(*bb*) K.B. Dillon, J. Halfpenny and A. Marshall, *J. Chem. Soc., Dalton Trans.*, 1985, 1399.
(*cc*) W. Werker, *Rec. Trav. Chim.*, 1939, **58**, 257.
(*dd*) J.J. Hilton, E.K. Nunn and S.C. Wallwork, *J. Chem. Soc., Dalton Trans.*, 1973, 173.
(*ee*) P.G. Harrison, T.J. King and J.A. Richards, *J. Chem. Soc., Dalton Trans.*, 1975, 826.
(*ff*) T.P. Lockhart and F. Davidson, *Organometallics*, 1987, **6**, 2471.
(*gg*) J.S. Morris and E.O. Schlemper, *J. Cryst. Mol. Struct.*, 1980, **9**, 13.
(*hh*) D.V. Naik and W. Scheidt, *Inorg. Chem.*, 1973, **12**, 272.
(*ii*) T.P. Lockhart, J.C. Calabrese and F. Davidson, *Organometallics*, 1987, **6**, 2479.
(*jj*) G.S. Brownlee, A. Walker, S.C. Nyburg and J.T. Szymanski, *Chem. Commun.*, 1971, 1073.
(*kk*) J.S. Morris and E.O. Schlemper, *J. Cryst. Mol. Struct.*, 1978, **8**, 295.
(*ll*) C.D. Garner, P. Sutton and S.C. Wallwork, *J. Chem. Soc. (A)*, 1967, 1949.
(*mm*) N.W. Alcock and V.L. Tracey, *Acta Crystallogr.*, 1979, **B35**, 80.
(*nn*) W.E. Bennett, D.E. Broberg and N.C. Baenziger, *Inorg. Chem.*, 1973, **12**, 930.

exhibit a preponderance for the formation of halogen-bridged lattices. The vapour-phase data for the tetrahedral halides warrant little comment, although it is notable that the metal–halogen distance increases while the metal–carbon bond decreases in the series $SnMe_nX_{4-n}$ as the halogen content of the molecule increases, consistent with a redistribution of *p* character to the more electronegative ligands.

As noted for bivalent tin, fluorotin(IV) compounds exhibit a strong tendency for the formation of strong Sn–F–Sn bridges in the solid. Thus, in contrast to the heavier tin (IV) halides which are tetrahedral, tin (IV) fluoride, SnF_4, forms a very strongly bridged, two-dimensional sheet polymer with octahedrally coordinated tin (**25**)[149]. Dimethyltin difluoride[150] (and almost certainly methyltin trifluoride[151]) has a similar sheet structure in which the non-bridging fluorine atoms are replaced by methyl groups. Fluorine bridging in trimethyltin fluoride results in a one-dimensional chain polymer with a planar [Me_3Sn] moiety and trigonal bipyramidal coordination at tin (**26**)[152]. Consistent with their structures, these materials are infusible and insoluble. Other organotin fluorides are similar, and only when the organic groups are very large, as in $(Me_3Si)_3CSnMe_2F$, $(Me_3Si)_3CSnPh_2F$, and $(PhMe_2Si)_3CSnMe_2F$[153],

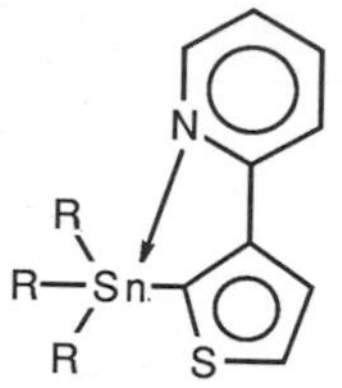

R = *p*-tolyl
(22)

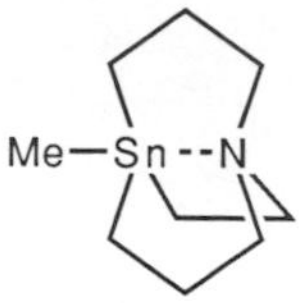

(23)

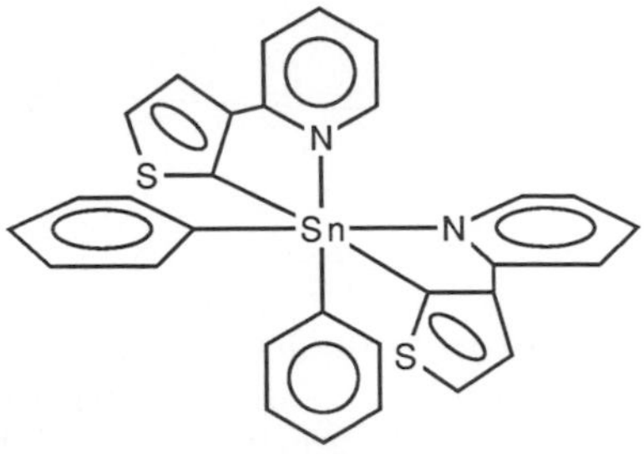

(24)

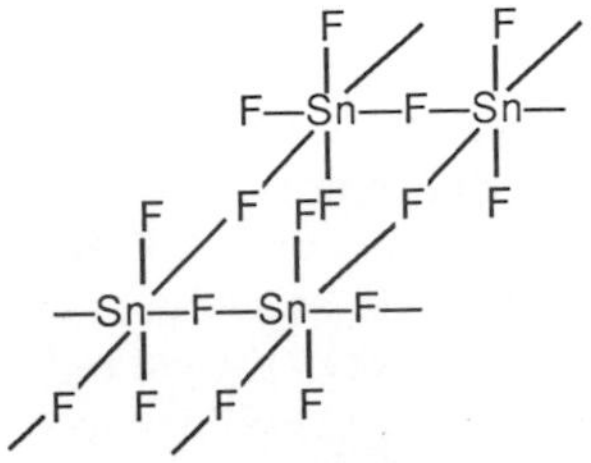

(25)

does steric hindrance preclude association. Bridging by other halogens is much weaker. A loosely associated one-dimensional chain structure similar to that of Me_3SnF is formed by trimethyltin chloride at 135 K, whereas crystals of Ph_3SnCl[155], Ph_3SnBr[156], $[(Me_3Si)_2CH_2]_3SnCl$[157], $[\eta\text{-}C_5H_5Fe(CO)_2]_2(p\text{-tolyl})SnBr$[158], and $(Me_3SiCH_2)_3SnI$[159] contain discrete, isolated molecules. The distortion away from tetrahedral geometry in tricyclohexyltin chloride suggests some intermediate state[160]. Association into one-dimensional chains occurs for dialkyltin dihalides when the alkyl group is small, and two types of structure have been characterized. In dimethyltin dichloride[161] and diethyltin dichloride and dibromide[162], the chains are formed by the two covalently bonded halogen atoms on each tin bridged to two different tin atoms in the chain as in (**27**), whereas in diethyltin diiodide[162] and bis(chloromethyl)tin dichloride[163] the two halogen atoms essentially chelate the next tin atom in the chain as in (**28**). In both, the geometry at tin is highly distorted, and is best regarded as

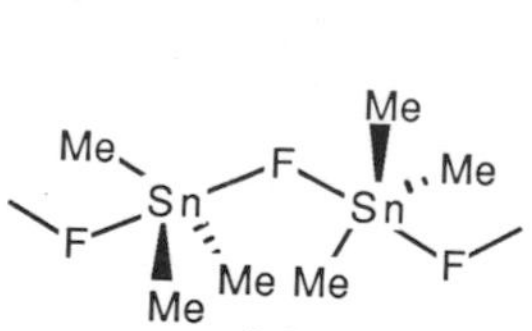

(26)

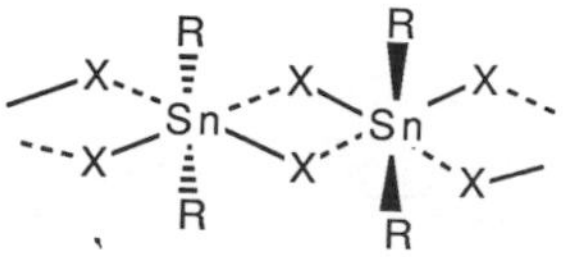

(27)

(28)

intermediate between tetrahedral and octahedral. No intermolecular interaction appears to be present in bis(biphenylyl-2)tin dichloride[164] or bis(ferrocenyl)tin dichloride[165]. Diphenyltin dichloride[166,167] and methylphenyltin dichloride[168] form loose tetrameric fragments which can be regarded as intermediate between the associated and unassociated extremes.

Higher coordination numbers at tin in organotin halides can also arise by intramolecular coordination of a donor atom remote in the organic ligand. Examples are shown in Figure 2.3[169–175]. When two nitrogen atoms are present on the organic group, ionization of halide from tin can occur, as in **(29)**[176]. The methylphenyltin homologues (Figure 2.3) contain chiral tin atoms and exhibit high optical stability, since the intramolecular coordination blocks the pathways for stereoisomerization. However, dynamic processes do occur which have been examined by variable-temperature nmr, and involve Sn–N bond dissociation and inversion at the then uncoordinated nitrogen atom[175,177]. Analogous process have been observed in the cationic species **(29)**[176].

The tin(IV) halides, MX_4, and organotin halides, $SnR_{4-n}X_n$, are good Lewis acids and acceptors of halide and neutral donor molecules. Lewis acidity in these halides increases in the order $I < Br < Cl < F$ and $1 < 2 < 3 < 4$. The pyridine adduct of trimethyltin chloride, $Me_3SnCl{\cdot}C_5H_5N$, was the first authenticated example of a tin

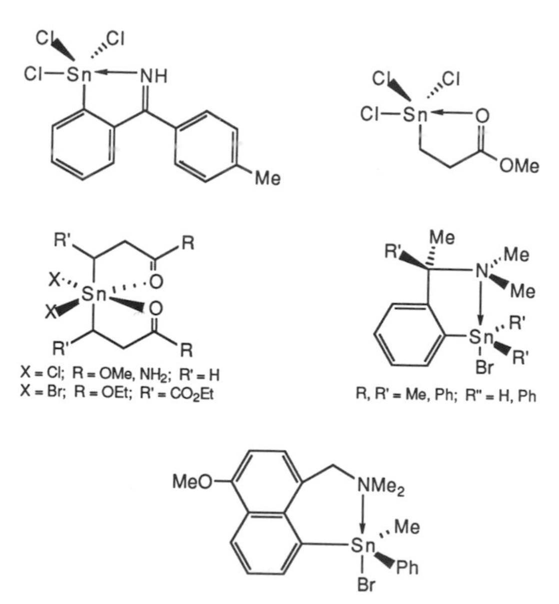

Figure 2.3 Examples of intramolecularly-coordinated organotin halides.

compound with coordination number greater than four[178]. Coordination saturation appears to be reached in the trigonal bipyramidal arrangement (**30**) with a planar or near-planar [SnR_3] moiety for the monohalides, but the dihalides, trihalides, and tetrahalides can form 1:2 adducts with an octahedral geometry as well as 1:1 adducts with the trigonal bipyramidal geometry. In the latter, the organic groups always occupy equatorial sites.

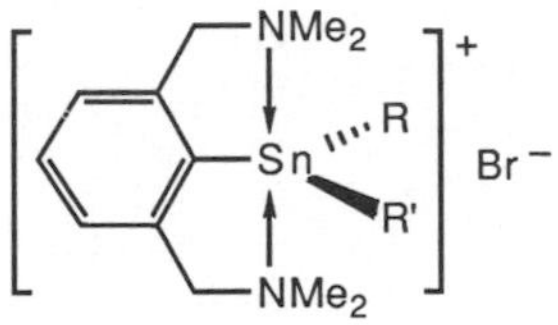

(**29**)

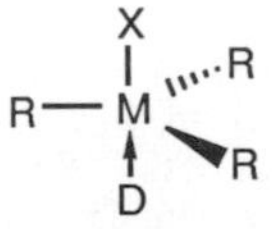

(**30**)

Typical examples of five- and six-coordinate complexes and complex anions are listed in Table 2.8. The 1:1 adduct of dimethyltin dichloride and diphenylcyclopropenone is unusual, and comprises a chlorine-bridged dimer with distorted octahedral geometry for tin (**31**)[179], rather than the usual trigonal bipyramidal geometry found in other 1:1 complexes such as Ph_2SnCl_2·benzthiazole[180], i.e. the small steric requirements of the

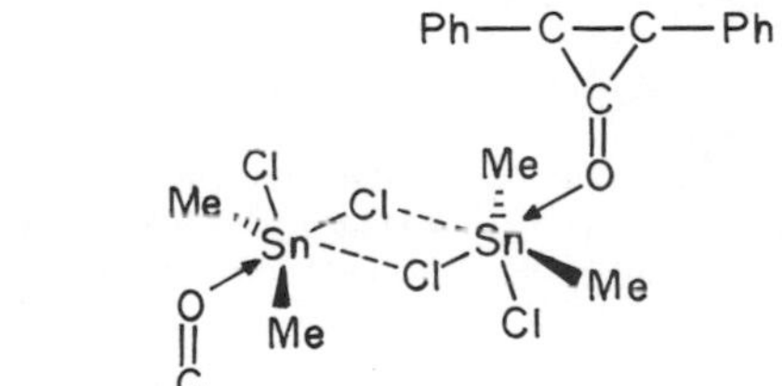

(**31**)

donor permit close approach of bridging chloride. Similar bridged anions such as (**32**) have also been isolated[181,182]. Two types of behaviour are observed with terdentate ligands. In some isolated cases, seven-coordinated adducts with a distorted pentagonal geometry such as (**33**) are formed[183,184], but terpyridine promotes the displacement of halide from tin and the formation of ionic complexes such as $[Me_2Sn(ter)Cl]^+[Me_2SnCl_3]^-$ containing six-coordinate cations and five coordinate anions[185]. Ionization of halide takes place more readily with bromine and iodine. Thus, whereas hexamethylphosphoric triamide forms a neutral, trigonal bipyramidal adduct with trimethyltin chloride, the ionic complex $[Me_3Sn(hmpt)_2]^+[Me_3SnBr_2]^-$ is formed with dimethyltin dibromide[186].

Table 2.8 Examples of five- and six-coordinate tin(IV) halide complexes and complex anions.

Complex	Ref.	Complex	Ref.
$[SnF_6]^{2-}$	*a, b*	$[MeSnCl_4]^-$	*p*
$SnF_4 \cdot bipy$	*c*	$[Me_2SnCl_3]^-$	*q*
$[SnCl_5]^-$	*d, e, f*	$[Me_2SnCl_3]_2^{2-}$	*r*
$[SnCl_6]^{2-}$	*f, g, h, i*	$[Me_6Sn_3Cl_8]^{2-}$	*s*
$[SnCl_5(OPCl_3)]^-$	*j*	$[Me_2SnCl_4]^{2-}$	*p*
$SnCl_4 \cdot 2CH_3CN$	*k*	$[Me_3SnCl_2]^-$	*q*
$SnCl_4 \cdot 2dmso$	*l*	$Me_3SnCl \cdot Ph_3PCHCOMe$	*t*
$[Sn_2Cl_{10}]^{2-}$	*f*	$Me_2SnCl_2 \cdot 2dmso$	*u, v*
$[SnBr_6]^{2-}$	*m*	$Ph_2SnCl_2 \cdot bipy$	*w*
$SnI_4 \cdot 2dpso$	*n*	$MeSnCl_3 \cdot 2py$	*x*
$[SnCl(S_2C_6H_3Me)]^-$	*o*	$MeSnI_3 \cdot 2dpso$	*y*

(*a*) I.A. Baidina, V.V. Bakakin, S.V. Borisov and N.V. Podberezskaya, *J. Struct. Chem.*, 1976, **17**, 434.
(*b*) K.O. Christie, V.J. Schack and R.D. Wilson, *Inorg. Chem.*, 1977, **16**, 849.
(*c*) A.D. Adley, P.H. Bird, A.R. Fraser and M. Onyszchuk, *Inorg. Chem.*, 1972, **11**, 1402.
(*d*) R.F. Bryan, *J. Amer. Chem. Soc.*, 1964, **86**, 744.
(*e*) A.G. Ginsburg, N.G. Bokii, A.I. Yanovskii, Yu. T. Struchkov, V.N. Setkina and D.N. Kursanov, *J. Organomet. Chem.*, 1977, **136**, 45.
(*f*) J. Shamir, S. Luski, A Bino, S. Cohen and D. Gibson, *Inorg. Chem.*, 1985, **24**, 2301.
(*g*) J.A. Lerbscher and J. Trotter, *Acta Crystallogr., Sect. B*, 1976, **32**, 2671.
(*h*) K. Nielsen and R.W. Berg, *Acta Chem. Scand., Sect. A*, 1980, **34**, 153.
(*i*) M.H. Ben Ghozlen, A. Daoud and J.W. Bats, *Acta Crystallogr., Sect. B*, 1981, **37**, 1415.
(*j*) A.J. Bannister, J.A. Durrant, I. Raymond and H.M.M. Shearer, *J. Chem. Soc., Dalton Trans.*, 1976, 928.
(*k*) M. Webster and H.E. Blayden, *J. Chem. Soc. (A)*, 1969, 2443.
(*l*) J.M. Kisenyi, G.R. Willey and M.G.B. Drew, *Acta Crystallogr.*, 1985, **C41**, 700.
(*m*) K.B. Dillon, J. Halfpenny and A. Marshall, *J. Chem. Soc., Dalton Trans.*, 1983, 1091.
(*n*) A.V. Jatsenko, S.V. Medvedev, K.A. Paseshnitchenko and L.A. Aslanov, *J. Organomet. Chem.*, 1985, **284**, 181.
(*o*) A.C. Sau, R.O. Day and R.R. Holmes, *Inorg. Chem.*, 1981, **20**, 3076.
(*p*) L.E. Smart and M. Webster, *J. Chem. Soc., Dalton Trans.*, 1976, 1924.
(*q*) A.J. Buttenshaw, M. Duchene and M. Webster, *J. Chem. Soc., Dalton Trans.*, 1975, 2231.
(*r*) M. Matsubayashi, K. Ueyama and T. Tanaka, *J. Chem. Soc., Dalton Trans.*, 1985, 465.
(*s*) K. Shimiizu, G.E. Matsubayashi and T. Tanaka, *Inorg. Chim. Acta*, 1986, **122**, 37.
(*t*) J. Buckle, P.G. Harrison, T.J. King and J.A. Richards, *J. Chem. Soc., Dalton Trans.*, 1975, 1552.
(*u*) N.W. Isaacs and C.H.L. Kennard, *J. Chem. Soc. (A)*, 1970, 1257.
(*v*) L.A. Aslanov, V.M. Ionov, W.M. Attiya, A.B. Permin and V.S. Petrosyan, *J. Organomet. Chem.*, 1978, **144**, 39.
(*w*) P.G. Harrison, T.J. King and J.A. Richards, *J. Chem. Soc., Dalton Trans.*, 1974, 1723.
(*x*) L.A. Aslanov, V.M. Ionov, W.M. Attiya and A.B. Permin, *J. Struct. Chem.*, 1978, **119**, 166.
(*y*) H. Lindemann and F. Huber, *Z. anorg. allg. Chem.*, 1972, **394**, 101.

Various stereochemical permutations are observed for octahedral tin complexes. Adducts of tin (IV) chloride and bromide exhibit the *cis* geometry (**34**)[187–195], although some examples of the *trans* geometry (**35**) are known[196–200]. The preference for the *cis* configuration has been rationalized by simple ligand field theory whereby the splitting between the d_{z^2} and $d_{x^2-y^2}$ would be minimized[201]. When the steric bulk of the two donor atoms is too great, the *trans* geometry is adopted. The structures adopted by the complexes of organotin trihalides and diorganotin dihalides also vary with the

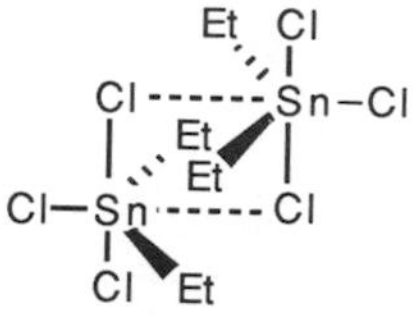

(32) 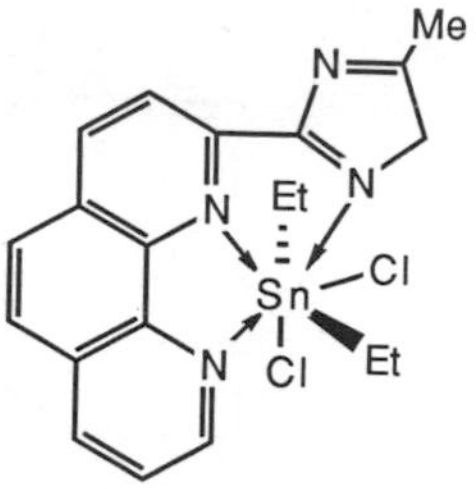 **(33)**

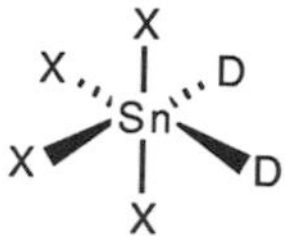

(34) 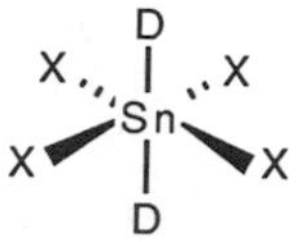 **(35)** 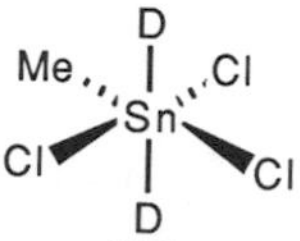 **(36)**

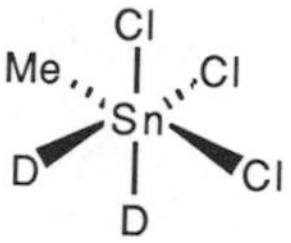

(37) 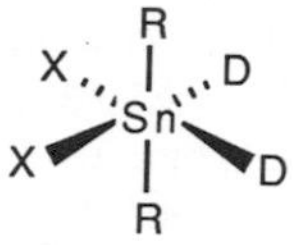 **(38)** 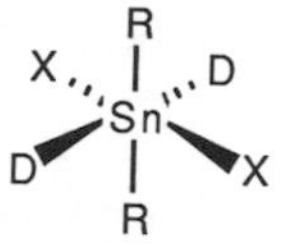 **(39)**

particular donor ligands. The two donors are *trans* in the 2:1 pyridine and hmpt complexes of methyltin trichloride as in (**36**), but the analogous dmf complex adopts the *cis* structure (**37**)[202–203]. The two organic groups in $R_2SnX_2 \cdot 2L$ complexes are invariably *trans*, although the isolation of both *cis* and *trans* isomers for (4-$ClC_6H_4)_2SnCl_2 \cdot$(4,4′-dimethyl-2,2′-bipyridyl) demonstrates the small differences in energy between the two forms[204]. The two halide and two donor molecules can adopt either configuration (**38**) or (**39**), depending upon the nature of the substituents[205–214].

2.5.3 *Chalcogenides and chalcogenide complex ions*

Tin (IV) oxide crystallizes in the rutile lattice (Figure 2.4) in which the tin atoms enjoy almost perfect octahedral coordination with only a small tetragonal distortion[215]. Tin (IV) sulphide and selenide both crystallize with the hexagonal CdI_2-layer-lattice also with six-coordinated tin[211], and many polytypes have been characterized[217–219].

The structures adopted by stannates are profoundly different to those of silicate and germanates, in spite of the similarity in stoichiometries. Tin is general six-coordinated in a regular or slightly distorted octahedral fashion by oxygen atoms, although four- and five-coordination do occur. Octahedral oxygen coordination about tin is present in all the hydroxystannates(IV), $M[Sn(OH)_6]$ (M = 2Na, 2K, Mg, Ca, Mn, Fe, Co, Ni, Cu, Zn, Cd)[220–228]. The structures of the metal hexahydroxystannates and of the crystalline metal stannates, $MSnO_3$, obtained by calcination of the hexahydroxy-stannates above 500–600°, are dependent upon the radius of the M^{2+} cation. The former are cubic with a NaCl-type lattice below 126 pm and hexagonal above this size, whilst the latter have an ilmenite-type structure when the M^{2+} radius is less than

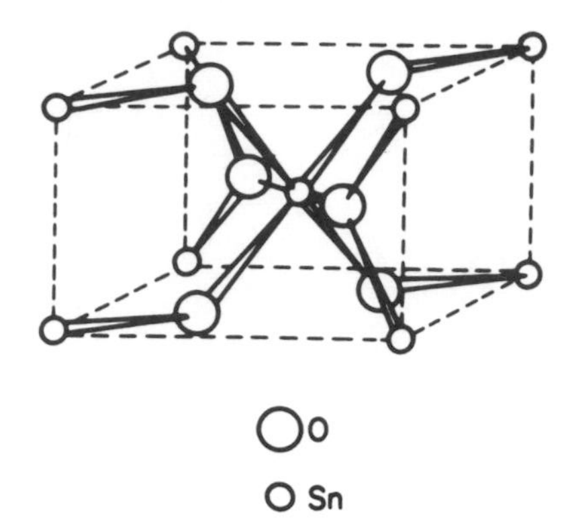

Figure 2.4 The unit cell of tin(IV) oxide.

109 pm and a perovskite-type above 126 pm[229]. Potassium orthostannate, K_4SnO_4, contains isolated $[SnO_4]$ tetrahedra[230–232], but other materials of the stoichiometry M_2SnO_4 (M = Mg, Ca, Sr, Ba, Mn, Co, Cd, Zn) are not true orthostannates, and contain $[SnO_6]$ octahedra which share common edges to give chain or layer structures[233–238]. Octahedrally-coordinated tin is also present in the 'distannates', $M_2Sn_2O_7$(M = lanthanide, Bi)[239–241], and most 'metastannates', $MnSnO_3$ (M = 2Li, 2Tl, Mg, Ca, Sr, Ba, Mn, Cd, Pb)[242–248]. The potassium and rubidium metastannates, however, have unusual square-pyramidal five-coordination for tin, and the solid contains infinite $[SnO_3]^{\infty -}_{\infty}$ polyanions as shown in Figure 2.5[249,250].

The structures of the heavier chalocogen analogues of the stannates are largely based upon the tetrahedral $[SnX_4]$ unit. All three isolated orthostannate anions $[SnX_4]^{4-}$ (X = S, Se, Te) have been characterized[251–260]. The trigonal planar $[SnX_3]^{2-}$ anions are also known[259]. Three types of binuclear anion are known. The $[Sn_2S_7]^{6-}$ [252,261,262] and $[Sn_2S_6]^{4-}$ [263,264] are formed by vertex- (**40**) and edge-sharing (**41**), respectively, of two $[SnS_4]$ tetrahedra, but the tellurium derivative, $[Sn_2Te_6]^{6-}$ contains the Sn–Sn bonded arrangement (**42**) with a staggered conformation[265–267]. The metathiostannates, $K_2SnS_3 \cdot 2H_2O$[268] form infinite $[SnS_3]^{2-}$ chains. Coordination numbers

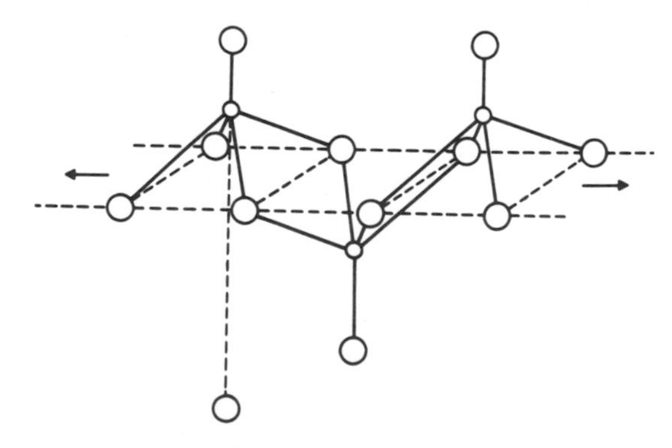

Figure 2.5 The structure of the infinite $[SnO_3]^{\infty -}_{\infty}$ polyanions in M_2SnO_3 (M = K, Rb).

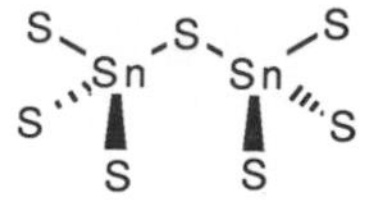

(40) 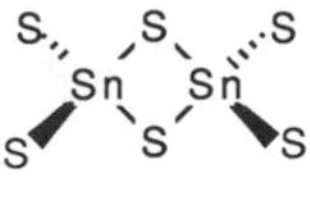 **(41)**

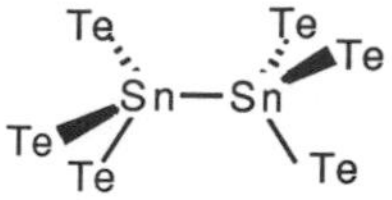

(42) 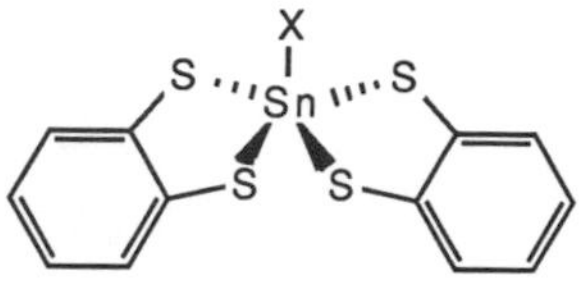 **(43)**

greater than four are less frequent than for the stannates. The tin atom has a distorted octahedral geometry in the lanthanide thiostannates, Ln_2SnS_5[269,270], Fe_2SnS_4[271], and Mn_2SnS_4[272].

Complex anions derived from chelating oxo- and thio-ligands exhibit five- or six-coordination depending upon the nature of the chelating ligand. The two bis(toluene-3, 4-dithiolato)halogenostannate anions, $[(C_7H_6S_2)_2XSn]^-$ (X = Cl, Br) (**43**), are two of the very few authentic examples of square pyramidal five-coordination at tin, being very close to ideality along the Berry pseudorotation coordinate (X = Cl: 76.9%; X = Br: 94.2%). Replacement of one of the sulphur atoms in the chelate ring, or of the halogen by hydroxyl, causes a change in stereochemistry to trigonal bipyramidal five-coordination as in the $[(C_6H_4OS)_2SnCl]^-$ anion or octahedral as in the $[((CN)_2C_2S_2)_2SnOH]^-$ anion[273,274].

2.5.4 *Organotin derivatives with bonds to elements of Groups 15 and 16*

As a result of the greater effective nuclear charge at tin coupled with the high donor capacity of oxygen, few organotin compounds containing bonds to oxygen remain tetrahedral at tin. Rather, association via intermolecular oxygen bridging occurs in the solid wherever possible. Only when steric hindrance precludes association does four-coordination occur in such compounds as hexaphenyl- and hexabenzyldistannoxanes, $R_3SnOSnR_3$ (R = Ph, $PhCH_2$)[275,276]. The former compound is expectedly bent at oxygen, but the SnOSn skeleton in the latter is surprising linear, a phenomenon which has been attributed to the low electronegativity of the metal coupled with the electron-donating character of the organic group[276].

Other triorganotin derivatives of oxygen ligands such as hydroxide[277,278], alkoxides[279–281] and oximes[282] show a marked tendency to associate into one-dimensional chain polymers in which planar or near planar $[R_3Sn]$ units are bridged by oxygen atoms as in (**44**). Steric crowding at tin reduces this tendency towards association, but increasing the size of the substituent at oxygen has less effect. Thus, whereas $Me_3SnON{=}C_6H_{11}$ has the structure (**44**), very similar to those of Me_3SnOH and Me_3SnOMe, as the organic group attached to tin increases in size, the association along the series $R_3SnON{=}C_6H_{11}$ (R = Me, Et, ^{n}Pr, Ph) becomes weaker and the $[R_3Sn]$ moiety becomes more pyramidal (**45**) until, for $Ph_3SnON{=}C_6H_{11}$, the solid comprises isolated tetrahedral (**46**) molecules[283,284]. Similarly, homologous methoxides such as tri-*n*-butyltin methoxide exist as oils where association is negligible.

(44) (45)

(46)

However, higher triorganotin hydroxides such as tri-*n*-butyltin and triphenyltin hydroxides are still associated[285,286].

Similar polymeric chain structures involving planar bridged [R_3Sn] moieties are adopted by several other types of triorganotin derivative including carboxylates, oxy-acid derivatives such as sulphates, sulphonates, phosphates and nitrates, and pseudo-halides, and examples are listed in Table 2.9. Some of the chains are linear, others helical, but the most unusual, and at present unique, is the cyclic hexamer former by

Table 2.9 Examples of triorganotin compounds exhibiting one-dimensional associated structures.

Compound	Ref.	Compound	Ref.
Me_3SnOH	*a, b*	$Me_3SnNCO \cdot Me_3SnOH$	*w*
Ph_3SnOH	*c*	Ph_3SnNCO	*x*
Me_3SnOMe	*d*	$Me_3SnN(CH_3)NO_2$	*y*
$Me_3SnON{=}C_6H_{10}$	*e*	$Me_3Sn{-}N{\cdot}S{\cdot}N{:}S{:}NSO_2$	*z*
$Me_3SnO_2CCH_3$	*f*		
$Me_3SnO_2CCF_3$	*f*		
$(Me_3SnO_2C)CH_2$	*g*	Me_3Sn—N (succinimide ring)	*aa*
$Me_3SnO_2CH_2NH_2$	*h*		
$(CH_2{=}CH)_3SnO_2CCl_3$	*i*		
$(PhCH_2)_3SnO_2CCH_3$	*j*		
$Ph_3SnO_2CCH_3$	*k*		
Me_3SnNO_3	*l*		
$Me_3SnO_2SCH_3$	*m, n*	Me_3Sn—N (phthalimide ring)	*bb*
$Me_3SnO_2SeCH_3$	*o*		
$Me_3SnO_2P(OH)Ph$	*p*		
$Me_3SnO_2PCl_2$	*q*		
$Me_3SnO_2P(CH_3)_2$	*q*	Me_3Sn—N (triazole ring)	*cc*
Me_3SnN_3	*r*		
$Me_3SnN(CN)_2$	*s*		
$Me_3SnN{:}C{:}NSnMe_3$	*t*	Me_3Sn (O, S ring)	*dd*
Me_3SnNCS	*u*		
Ph_3SnNCS	*v*		

(*a*) N. Kasai, K. Yasuda and R. Okawara, *J. Organomet. Chem.*, 1965, **3**, 172.
(*b*) A.G. Davies and S.D. Slater, *Main Group Chem.*, 1986, **9**, 87.
(*c*) C. Glidewell and D.C. Liles, *Acta Crystallogr., Sect. B*, 1978, **34**, 129.
(*d*) A.M. Domingos and G.M. Sheldrick, *Acta Crystallogr., Sect. B*, 1974, **30**, 519.
(*e*) P.F.R. Ewings, P.G. Harrison, T.J. King, R.C. Phillips and J.A. Richards, *J. Chem. Soc., Dalton Trans.*, 1975, 1950.

(*f*) H. Chih and B.R. Penfold, *J. Cryst. Mol. Struct.*, 1973, **13**, 285.
(*g*) U. Schubert, *J. Organomet. Chem.*, 1978, **155**, 285.
(*h*) B.Y.K. Ho, K.C. Molloy, J.J. Zuckerman, F. Reidinger and J.A. Zubieta, *J. Organomet. Chem.*, 1980, **187**, 213.
(*i*) S. Calogero, D.A. Clemente, V. Peruzzo and G. Tagliavini, *J. Chem. Soc., Dalton Trans.*, 1979, 1172.
(*j*) N.W. Alcock and R.E. Times, *J. Chem. Soc. (A)*, 1968, 1873.
(*k*) K.C. Molloy, T.G. Purcell, K. Quill and I.W. Nowell, *J. Organomet. Chem.*, 184, **267**, 237.
(*l*) H.C. Clark, private communication quoted in D. Potts, H.D. Sharma, A.J. Carty and A. Walker, *Inorg. Chem.*, 1974, **13**, 1205.
(*m*) R. Hengel, U. Kunzo and J. Strähle, *Z. anorg. allg. Chem.*, 1976, **423**, 35.
(*n*) G.M. Sheldrick and R. Taylor, *Acta Crystallogr.*, 1977, **B33**, 135.
(*o*) U. Ansorge, E. Lindner and J. Strähle, *Chem. Ber.*, 1978, **111**, 3048.
(*p*) K.C. Molloy, M.B. Hossain, D. van der Helm, D. Cunningham and J.J. Zuckerman, *Inorg. Chem.*, 1981, **20**, 2402.
(*q*) F. Weller and A.F. Shiada, *J. Organomet. Chem.*, 1987, **322**, 185.
(*r*) R. Allmann, R. Hohlfeld, A. Waskawska and J. Lorberth, *J. Organomet. Chem.*, 1980, **192**, 353.
(*s*) Y.M. Chow, *Inorg. Chem.*, 1971, **10**, 1938.
(*t*) R.A. Forder and G.M. Sheldrick, *J. Chem. Soc. (A)*, 1971, 1107.
(*u*) R.A. Forder and G.M. Sheldrick, *J. Organomet. Chem.*, 1970, **21**, 115.
(*v*) A.M. Domingos and G.M. Sheldrick, *J. Organomet. Chem.*, 1974, **67**, 257.
(*w*) J.B. Hall and D. Britton, *Acta Crystallogr.*, 1972, **B28**, 2133.
(*x*) T.N. Tarkhova, E.V. Chuprunov, M.A. Simonov and N.V. Belov, *Kristallografiya*, 1977, **22**, 1004.
(*y*) A.M. Domingos and G.M. Sheldrick, *J. Organomet. Chem.*, 1974, **69**, 207.
(*z*) H.W. Roesky, M. Witt, M. Diehl, J.W. Bats and H. Fuess, *Chem. Ber.*, 1979, **112**, 1372.
(*aa*) F.E. Hahn, J.S. Dory, C.L. Barnes, M.B. Hossain, D. van der Helm and J.J. Zuckerman, *Organometallics*, 1983, **2**, 969.
(*bb*) E.V. Chuprunov, T.N. Tarkhova, Yu. T. Korallova and N.V. Belov, *Dokl. Akad. Nauk SSSR*, 1978, **242**, 606.
(*cc*) I. Hammann, K.H. Büchel, K. Bungartz and L. Born, *Pflanzenschutznachrichten Bayer* (Engl. edn.), 1978, **31**, 61.
(*dd*) P.A. Bates, M.B. Hursthouse, A.G. Davies and S.D. Slater, *J. Organomet. Chem.*, 1987, **325**, 129.

triphenyltin diphenylphosphate, $Ph_3SnO_2P(OPh)_2$[287] (Figure 2.6). Observed structures of triorganotin carboxylates fall into three categories. The majority, including trimethyl-, trivinyl-, triphenyl-, and tribenzyltin carboxylates possess the rather tightly-bound one-dimensional polymeric structure (**47**), with the carboxylato groups strongly bridging the triorganotin residues in the usual *syn, anti* fashion. Tricyclohexyltin carboxylates, although having been previously misrepresented as examples of monomeric triorganotin carboxylates with unidenate carboxylato ligands and four-coordinate tin as in (**48**), have a similar structure (**49**), although the *syn, anti* bridging is longer and weaker. Few examples of monomeric triorganotin carboxylates have been characterized, and these involve a very high degree of steric crowding. Except for $(C_5H_5)Re(NO)(PPh_3)(CO_2SnPh_3)$, the chelation of the carboxylato ligand is unsymmetrical. Triphenyltin *o*-(2-hydroxy-5-methylphenylazo)benzoate adopts a distorted trigonal bipyramidal geometry where the carboxylate group chelates via an axial and an equatorial sites[288–292]. Hydrated triorganotin derivatives retain the trigonal bipyramidal geometry at tin with the water molecule occupying an equatorial site, with $[R_3SnX \cdot H_2O]$ units linked by hydrogen bonding[293,294].

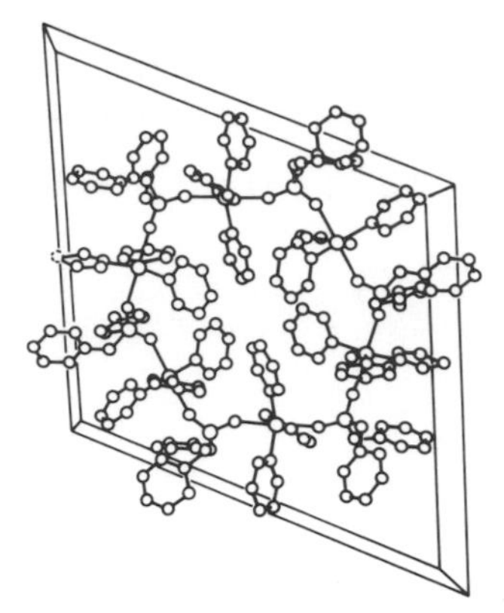

Figure 2.6 The hexameric structure of $Ph_3SnO_2P(OPh)_2$. Phenyl rings on the phosphate group have been omitted for clarity. Redrawn with permission from ref. 287.

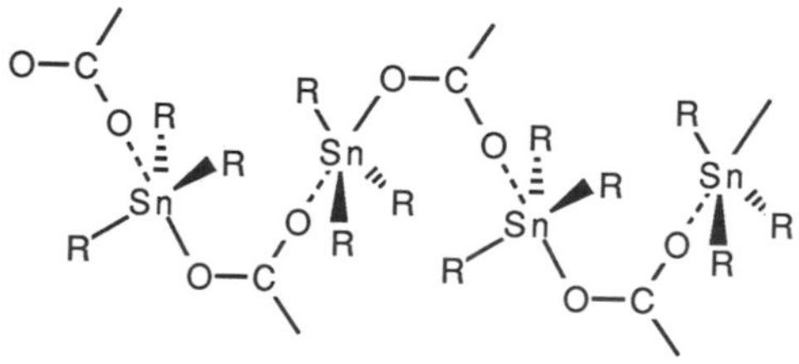

(47)

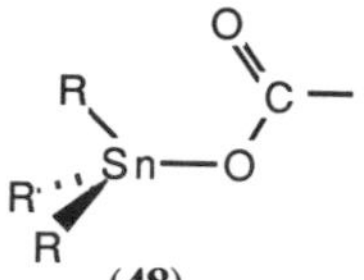

(48)

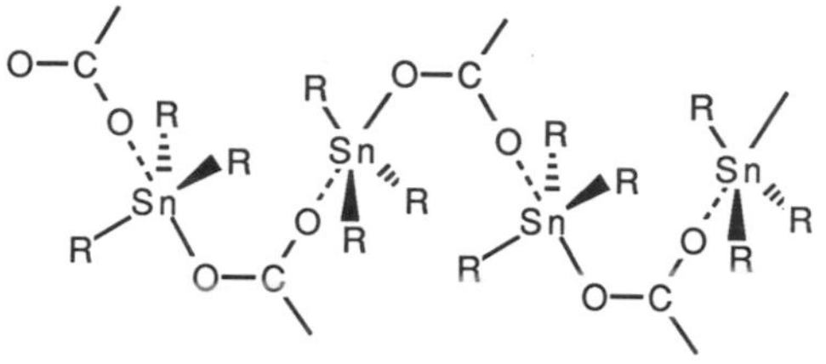

(49)

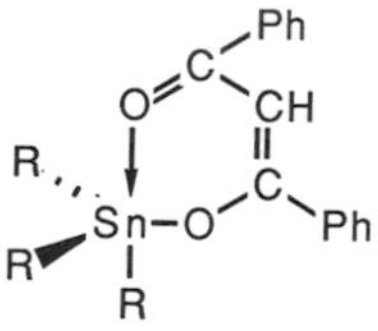

(50)

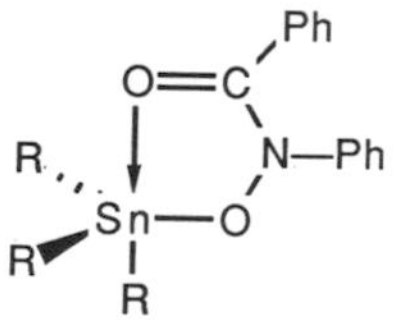

(51)

Triorganotin derivatives of strongly chelating oxygen ligands such as acetylacetonates[295] and N-acylhydroxylaminates[296,297] exhibit contrasting behaviour, and exhibit the *cis* trigonal bipyramidal geometry (**50**) and (**51**). [Tris(pyrazoyl)borato]-trimethyltin, $[HB(pz)_3]SnMe_3$, (**52**) is unique, being the only authenticated example amongst triorganotin compounds of six-coordination[298].

Few examples of four-coordinate organotin compounds containing two or three tin–

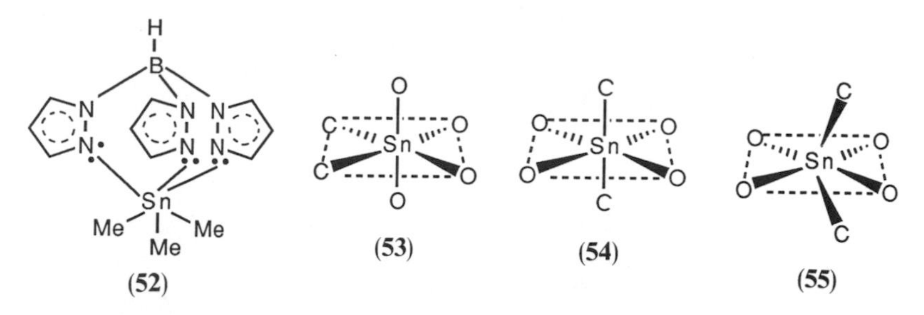

(52) (53) (54) (55)

oxygen bonds have been characterized, although it is highly likely that this geometry occurs in sterically-crowded compounds such as $Ph_nSn(OSiPh_3)_{4-n}$ ($n = 1, 2$). Di- and trialkoxides with less sterically demanding substituents appear to be associated[299]. The favoured coordination geometries for diorganotin derivatives of carboxylic acids, inorganic oxyacids, and related hydroxy-species are *cis* (**53**) and *trans* (**54**) octahedral, and skew-trapezoidal (**55**) (intermediate between tetrahedral four-coordination and octahedral six-coordination), although examples of seven- and eight-coordination geometries have been found. Both bridging and chelation can occur with bifunctional ligands, but *a priori* predictions concerning the solid-state structure can be unwise, as the particular structure can vary tremendously within the same class of compound. For instance, dimethyltin diacetate[300] is monomeric in the crystal, adopting the skew-trapezoidal structure with chelating acetato groups. Dimethyltin dipicolinate, on the other hand, has a polymeric structure in which both picolinate ligands chelate the tin atom via one oxygen and the nitrogen, with seven-coordination being completed by a bridging oxygen[301]. Dimethyltin derivatives of oxyacids appear to adopt bridged structures. For example, dimethyltin bis(fluorosulphonate), $Me_2Sn(OSO_2F)_2$, has a polymeric sheet structure[302], and dimethyltin phosphate, $(Me_2Sn)_3(PO_4)_3 \cdot 8H_2O$, adopts an infinite 'ribbon' structure[303]. The structures of a large number of diorganotin derivatives, R_2SnL_2, of potentially chelating groups (L) are known, all of which exhibit $[SnC_2O_4]$ six-coordination with the CSnC bond angle varying from the *cis*-octahedral geometry (**53**), through varying degrees of the skew-trapezoidal structure (**55**) to the *trans* octahedral geometry (**54**). As can be seen from Tables 3.19 and 3.22, the majority of compounds in this group possess the skew-trapezoidal geometry, intermediate between the two extremes of *cis*-octahedral, exemplified by dimethyltin bis(oxinate)[304] and $Me_2Sn(ONHCOMe)_2$[305], and *trans*-octahedral, exemplified by $Me_2Sn(acac)_2$[306]. Several factors obviously contribute to the particular geometry adopted in these compounds. However, the series of closely related compounds, $Me_2Sn(ONMeCOMe)_2$ (CSnC angle 145.8°)[307], $Me_2Sn(ONHCOMe)_2$ (CSnC angle 109.1°)[305], and $Me_2Sn(ONHCOMe)_2 \cdot H_2O$ (CSnC angle 156.8°)[305], illustrate the rather large effect that hydrogen-bonding and crystal packing effects can have upon the stereochemistry adopted about the tin atoms.

Little is known concerning the structural chemistry of monoorganotin derivatives. Infrared and colligative data support the seven-coordinate structure (**56**) for monoorganotin tris(carboxylates)[308], similar to that determined for methyltin trinitrate (**57**)[309]. Methyltin trimethoxide forms infinite $-(O-Sn-)_\infty$ chains[310]. Intramolecular N–Sn coordination occurs in methylstannaturane, $MeSn(OCH_2CH_2)_3N$, and association into trimeric units (**58**) results in six- and seven coordination at tin[311–314]. In donor solvents, the association is broken down, and mononuclear species

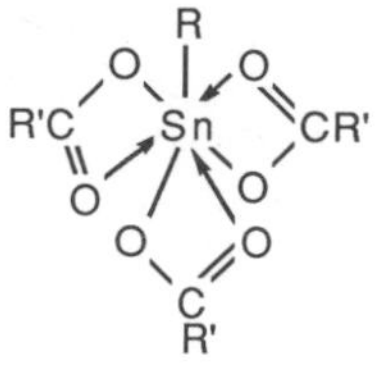

(**56**) 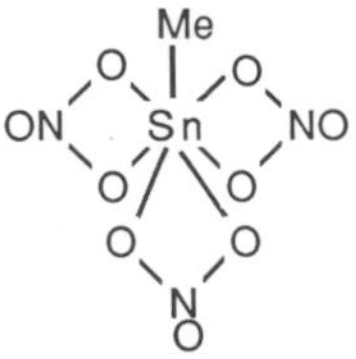 (**57**)

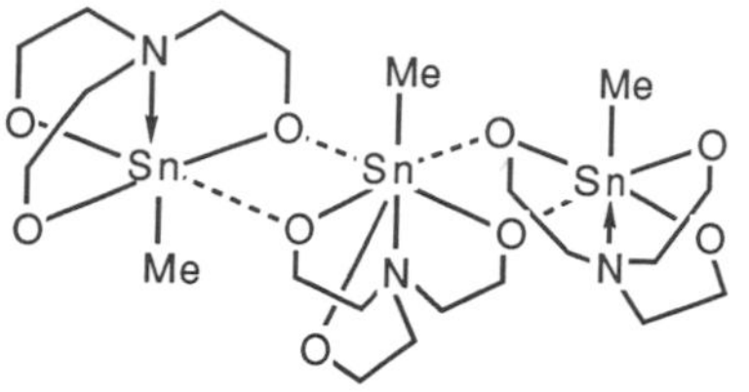

(**58**)

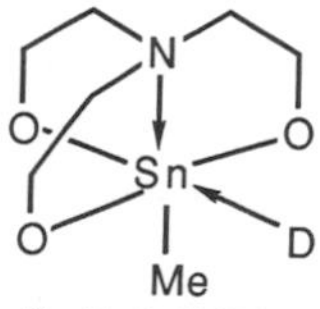

$D = H_2O$, DMSO

(**59**) (**60**)

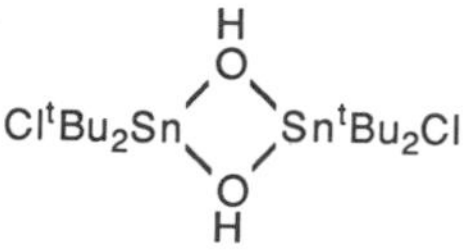

(**61**) 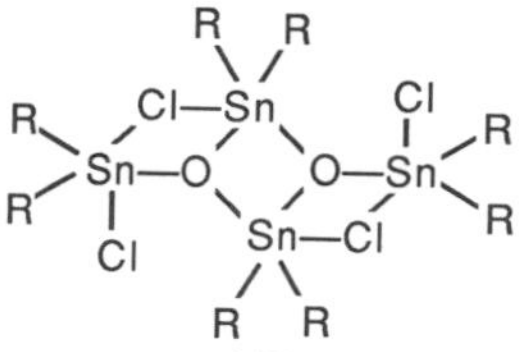 (**62**)

(**59**) exist in water and DMSO. Tin (IV) nitrate[315] and tin (IV) acetate[316] have an eight-coordinated dodecahedral structure (**60**).

Hydrolysis of di- and monoorganotin halides or carboxylates yields a number of structural types which are formal intermediates en route to the products of exhaustive hydrolysis, the diorganotin oxides, R_2SnO, and monoorganotin oxides, $RSnO_{3/2}$, (or 'stannonic acids', RSn(OH)O). Hydrolysis of di-*tert*-butyltin dihalides, $^tBu_2SnX_2$ (X = F, Cl, Br), affords the dimeric halide hydroxides, $[^tBu_2Sn(OH)X]_2$, (**61**) which are characterized by a central four-membered $[Sn_2O_2]$ ring[317]. The structural feature is a dominant feature in the structural chemistry of oxotin compounds, and its occurrence is very widespread (Table 2.10). With less bulky organic groups, the intermediate halide hydroxide is not stable, and condensation to a dimeric 1, 3-dichlorodistannoxane (**62**) occurs spontaneously. Several examples of these and other similar 1, 3-dicarboxylato- (**63**) and 1-chloro-3-hydroxydistannoxanes (**64**), 1, 3-dihydroxydistannoxanes (**65**), and tristannoxanes have been characterized, and exhibit 'ladder' structures composed

Table 2.10 Examples of compounds containing the four-membered $[Sn_2O_2]$ ring.

Compound	Structural type	Ref.
$[ClMe_2SnOSnMe_2Cl]_2$	Ladder	*a*
$[ClPh_2SnOSnPh_2Cl]_2$	Ladder	*b*
$[ClPh_2SnOSnPh_2(OH)]_2$	Ladder	*b*
$[(SCN)Me_2SnOSnMe_2(NCS)]_2$	Ladder	*c*
$[(F_3CCO_2)Me_2SnOSnMe_2(O_2CCF_3)]_2$	Ladder	*d*
$[(Cl_3CCO_2)^nBu_2SnOSn^nBu_2(O_2CCCl_3)]_2$	Ladder	*e*
$[(Me_3SiCH_2)_2(HO)SnOSn(OH)(CH_2SiMe_3)_2]_2$	Ladder	*f*
$[(^nBuSn(O)O_2CPh)_2\text{-}^nBuSn(Cl)(O_2CPh)_2]_2$	Ladder	*g*
$[(^nBuSn(O)O_2CR)_2(^nBuSn(O_2CR)_2]_2$ (R = Me, Ph, C_6H_{11})	Ladder	*g, h*
$[(MeSn(O)O_2CC_6H_{11})_2\text{-}MeSn(O_2CC_6H_{11})_3]_2$	Ladder	*h*
$[PhSn(O)O_2CC_6H_{11}]_6$	Drum	*i*
$[^nBuSn(O)O_2CR]_6$ (R = C_5H_9, C_6H_{11}, $C_6H_4NO_2$-2)	Drum	*g, h*
$[(^nBuSn(OH)O_2PPh_2)_3O][Ph_2PO_2]$	Cluster	*j*
$[^nBuSn(O)O_2P(C_6H_{11})_2]_4$	Cubane drum	*k*
$[SnCl_3(OH)\cdot H_2O)]_2\cdot 4H_2O$	Ladder	*l*
$[SnCl_3(OH)\cdot H_2O)]_2\cdot$3diox	Ladder	*l, m*
$[SnCl_3(OMe)\cdot MeOH]_2$ (R = Me, Et)	Ladder	*n, o*
$[SnCl_3(PO_2Cl_2)\cdot POCl_3]_2$	Ladder	*p*
$[Sn_3O_2Cl_4(ClO_4)_4]_2$	Ladder	*q*
$[Sn_2(O_2CC_6H_4NO_2\text{-}2)_4O\cdot thf]_2$	Cluster	*r*
$[^tBu_2Sn(OH)X]_2$ (X = F, Cl, Br)	Dimer	*s*
$[RSn(OH)(OH_2)Cl_2]_2$ (R = Et, ^{n}Bu)	Dimer	*t, u*
$[(C_2H_4OS)_2SnCl]_2[H][Et_4N]$	Dimer	*u*
$[((NC)_2C_2S_2)_2SnOH]_2(Et_4N)_2$	Dimer	*u*
$[Me_2Sn(O_2C_2H_4)]_n$	Ribbon polymer	*v*
SnO	Sheet polymer	*w*
SnO_2	3-D polymer	*x*
$[Sn(O^tBu)_2]_2$	Dimer	*y*
$Sn_6O_4(OMe)_4$	Cluster	*z*
$[Sn_8O_4](SO_4)_4$	Cluster	*aa*

(*a*) P.G. Harrison, M.J. Begley and K.C. Molloy, *J. Organomet. Chem.*, 1980, **186**, 213.
(*b*) J.F. Vollano, R.O. Day and R.R. Holmes, *Organometallics*, 1984, **3**, 745.
(*c*) Y.M. Chow, *Inorg. Chem.*, 1971, **10**, 673.
(*d*) R. Faggiani, J.P. Johnson, I.D. Brown and T. Birchall, *Acta Crystallogr.*, 1978, **B34**, 3743.
(*e*) R. Graziani, G. Bambieri, E. Forsellini, P. Furlen, V. Peruzzo and G. Tagliavini, *J. Organomet. Chem.*, 1977, **125**, 43.
(*f*) H. Puff, E. Friedrichs and F. Visel, *Z. anorg. allg. Chem.*, 1981, **477**, 50.
(*g*) V. Chandrasekhar, C.G. Schmid, S.D. Burton, J.M. Holmes, R.O. Day and R.R. Holmes, *Inorg. Chem.*, 1987, **26**, 1050.
(*h*) R.R. Holmes, C.G. Schmid, V. Chandrasekhar, R.O. Day and J.M. Holmes, *J. Amer. Chem. Soc.*, 1987, **109**, 1408.
(*i*) V. Chandrasekhar, R.O. Day and R.R. Holmes, *Inorg. Chem.*, 1987, **24**, 1970.
(*j*) R.O. Day, J.M. Holmes, V. Chandrasekhar and R.R. Holmes, *J. Amer. Chem. Soc.*, 1987, **109**, 940.
(*k*) K.C.K. Swamy, R.O. Day and R.R. Holmes, *J. Amer. Chem. Soc.*, 1987, **109**, 5546.
(*l*) J.C. Barnes, H.A. Simpson and T.J.R. Weakley, *J. Chem. Soc., Dalton Trans.*, 1980, 949.
(*m*) N.G. Bokii and Yu. T. Struchkov, *J. Struct. Chem.*, 1971, **12**, 253.
(*n*) G. Sterr and R. Mattes, *Z. anorg. Chem.*, 1963, **322**, 319.

Table 2.10 (*Contd.*)

(*o*) M. Webster and P.H. Collins, *Inorg. Chim. Acta*, 1974, **9**, 157.
(*p*) D. Moras, A. Mitschler and R. Weiss, *Acta Crystallogr.*, 1969, **B25**, 1720.
(*q*) C. Belin, M. Chaabouni, J.L. Pascal, J. Potier and J. Roziere, *Chem. Commun.*, 1980, 105.
(*r*) P.F.R. Ewings, P.G. Harrison, A. Morris and T.J. King, *J. Chem. Soc., Dalton Trans.*, 1976, 1602.
(*s*) H. Puff, H. Hevendehl, K. Höfer, H. Reuter and W. Schuh, *J. Organomet. Chem.*, 1985, **287**, 163.
(*t*) C. Lecompte, J. Protas and M. Devaud, *Acta Crystallogr.*, 1976, **B32**, 923.
(*u*) R.R. Holmes, S. Shafieezad, V. Chandrasekhar, J.M. Holmes and R.O. Day, *J. Amer. Chem. Soc.*, 1988, **110**, 1174.
(*v*) A.G. Davies and S.D. Slater, *Main Group Chem.*, 1986, **9**, 87.
(*w*) W.J. Moore and L. Pauling, *J. Amer. Chem. Soc.*, 1941, **63**, 1392.
(*x*) J. Pannetier and G. Denes, *Acta Crystallogr.*, 1980, **B36**, 2763.
(*y*) T. Fjeldberg, P.B. Hitchcock, M.F. Lappert, S.J. Smith and A.J. Thorne, *J. Chem. Soc., Chem. Commun.*, 1985, 939.
(*z*) P.G. Harrison, B.J. Haylett and T.J. King, *Chem. Commun.*, 1978, 112.
(*aa*) C. Lundgren, G. Wernfors and T. Yamaguchi, *Acta Crystallogr.*, 1982, **38B**, 2357.

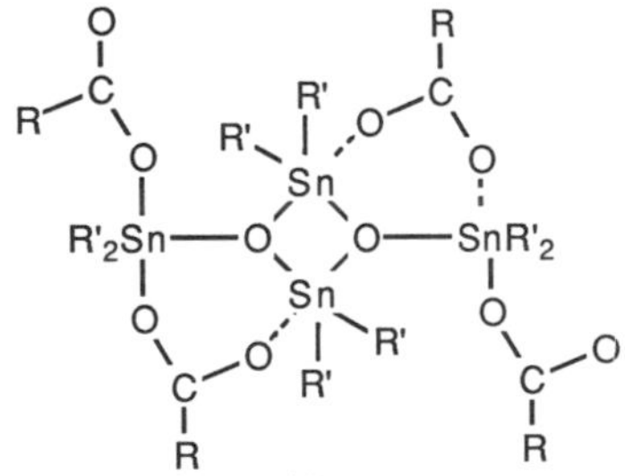

(63) 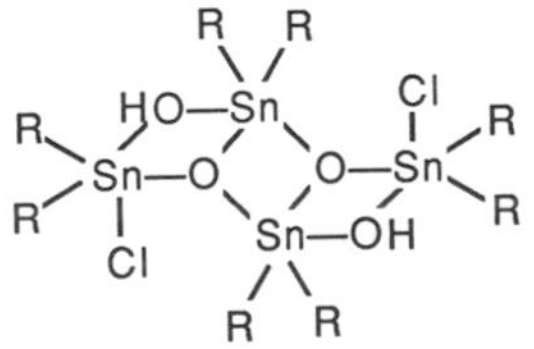 **(64)**

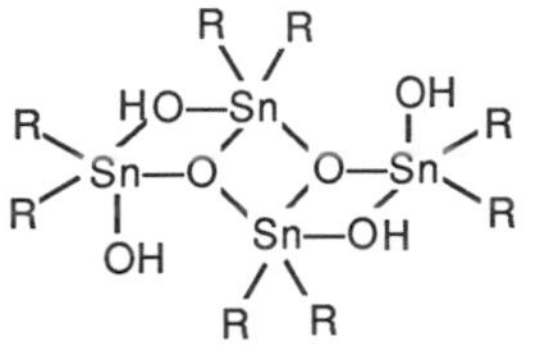

(65) 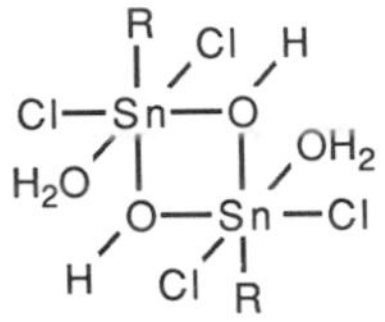 **(66)**

of fused four-membered $[Sn_2O_2]$ rings[318–326]. The initial product in the hydrolysis of monoalkyltin trichlorides is the dimeric species $[RSn(OH)(OH_2)Cl_2]_2$ (**66**)[327,328]. Two further structural types, 'ladder' (**67**) and 'drum' (**68**) structures, have also been discerned in the monoorganotin system (Table 2.10), and are interconvertible in solution. Related to the 'drum' structure are two other types of cage exhibited by monobutyltin oxide phosphinates. That of $[^nBuSn(O)O_2P(C_6H_{11})_2]_4$ is characterized by a central $[Sn_4O_4]$ cube (Figure 2.7)[329], while in $[(^nBuSn(OH)O_2PPh_2)_3O][Ph_2PO_2]$ a tricoordinating oxygen atom caps a tristannoxane ring in a 'chair' conformation (Figure 2.8)[330]. Few unequivocal structural data are available for diorganotin oxides, R_2SnO, and monoorganotin oxides, $RSnO_{3/2}$, which usually exist as in-

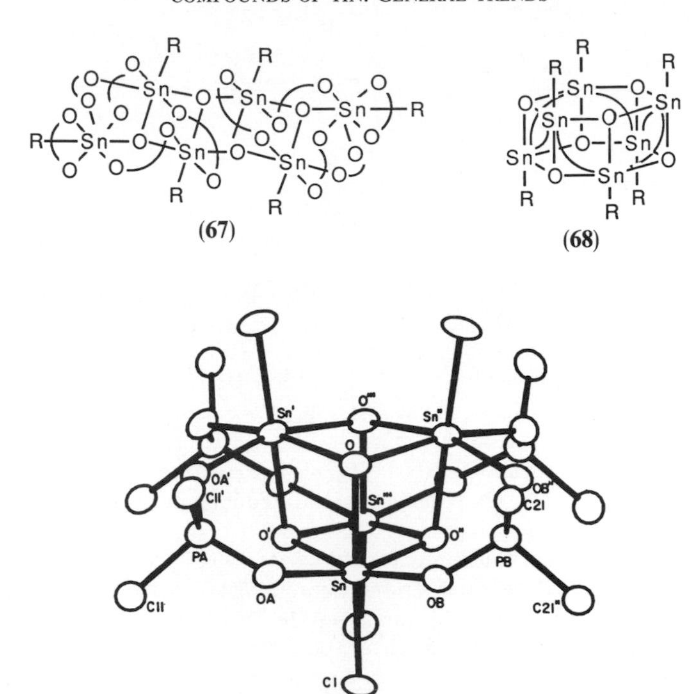

Figure 2.7 The structure of $[(^nBuSn(O))O_2P(C_6H_{11})_2]_4$ illustrating the $[Sn_4O_4]$ cubane core. Reproduced with permission from ref. 329.

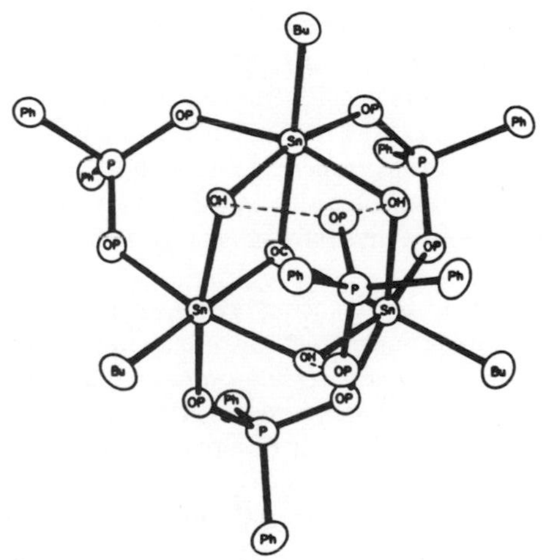

Figure 2.8 The structure of $[(^nBuSn(OH)O_2PPh_2)_3O][Ph_2PO_2]$. Reproduced with permission from ref. 330.

soluble and infusible amorphous solids obtained by the exhaustive hydrolysis of the corresponding organotin halides or carboxylates. From the foregoing data, it would seem highly probable that the principal structural element in these compounds is the four-membered $[Sn_2O_2]$ ring. Crystallographic data are available for only two

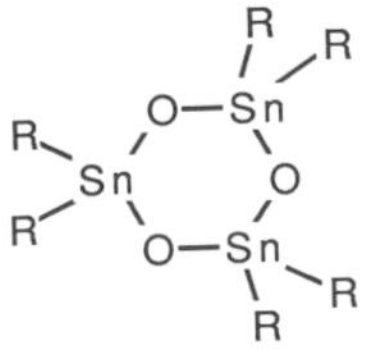

(69)

rather esoteric cases, di-*tert*-butyl- and bis(2,6-diethylphenyl)tin oxides, which crystallize as trimeric molecules (**69**) with a planar six-membered $[Sn_3O_3]$ ring[331,332], the bulky organic groups presumably precluding further association. Spectroscopic data, however, indicate a cross-linked, three-dimensional network for less sterically-hindered oxides, in which the stereochemistry at tin is distorted trigonal bipyramidal with the organic groups occupying equatorial sites. Two models which have been proposed involving networks of fused [4 + 6]- or [4 + 8]-membered stannoxane rings are illustrated in Figure 2.9. The structures of the monoorganotin oxides are expected to be similar.

Associated and chelated structures are far less prevalent in compounds containing bonds to the heavier chalcogenides. Triorganotin thiolates, R_3SnSR', are monomeric and tetrahedral[333–337], as are the tin atoms in $(Me_3SnS)_3P{=}S$[338]. Weak intermolecular Sn–N interactions do, however, occur in $Ph_3SnSC_5H_4N$-4[339], causing severe distortions of the geometry at tin towards a trigonal bipyramid. Likewise, bis(triphenyltin) sulphide[340] and selenide[341] are monomeric molecules which are bent at sulphur or selenium and contain tetrahedral tin. In contrast to the diorganotin oxides, the diorganotin chalcogenides form ring structures, most commonly a cyclic trimer with a twist-boat conformation[342]. However, with bulky substituents, a four-membered ring may be stabilized, and di-*tert*-butyltin sulphide, selenide, and telluride exist as the dimers (**70**) with a planar central $[Sn_2X_2]$ (X = S, Se, Te) ring[343]. Some diorganotin sulphides may also exist in a polymeric modification of lower solubility and higher melting point than the trimeric form, and probably have the same linear polymeric structure (**71**) as that characterized for diisopropyltin sulphide[344]. The monoorganotin sesquisulphides generally crystallize with the adamantane skeleton (**72**)[345–349].

In contrast to carboxylato and other oxyanion ligands, dithio ligands such as dithiocarbamate and dithiphosphate function as unidentate ligands in their triorganotin derivatives[350–352]. When two or fewer tin–carbon bonds are present, chelation can occur give skew-trapezoidal (e.g. dimethyltin bis(dithiocarbamates), $Me_2Sn(S_2CNR_2)_2$ (R = Me, Et)[353,354] and diphenyltin bis(dithiophosphate), $Ph_2Sn[S_2P(OEt)_2]_2$[355], or *trans* octahedral structure (e.g. $Ph_2Sn[S_2P(O^iPr)_2]_2$[356]. Methyltin tris(diethyldithiocarbamate) has a distorted pentagonal bipyramidal geometry, in which two dithiocarbamate groups chelate via equatorial sites while the third spans an equatorial and axial site[357]. However, the coordination in dimethyltin bis(diethylphosphinate), $Me_2Sn(S_2PEt_2)_2$, is best regarded as a distorted tetrahedral since the secondary Sn···S contacts are quite long[358].

Tetrahedral four-coordination is generally the rule for organotin derivatives of amines and phosphines, although trimethyltin aziridine is associated in the solid via N–Sn bridging and dimeric in solution (**73**)[359]. Unusually, but like the silicon and

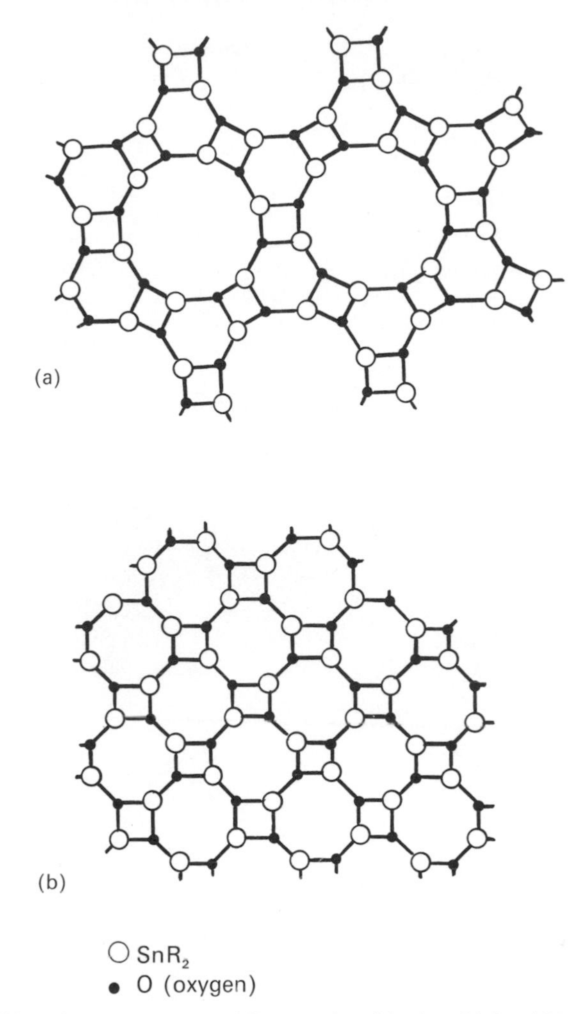

Figure 2.9 Possible solid-state structures of diorganotin oxides: (*a*) with fused [4 + 6] stannoxane rings or (*b*) with fused [4 + 8] stannoxane rings.

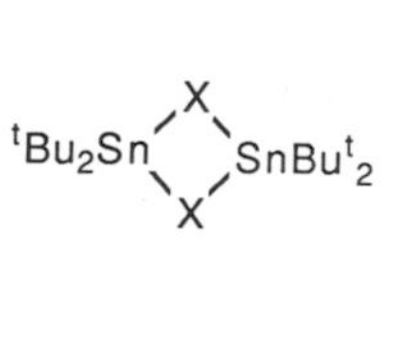

(70) 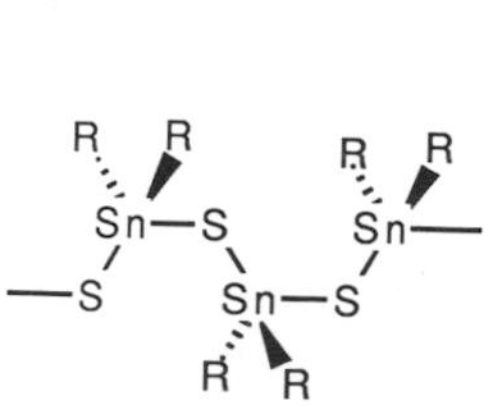 **(71)** 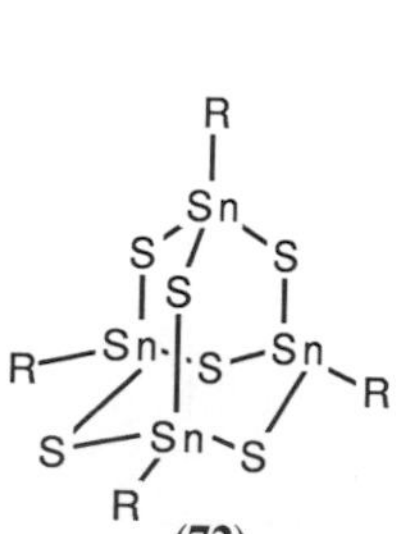 **(72)**

N

Me_3Sn $SnMe_3$

N

(73)

germanium analogues, $(Me_3Sn)_3N$ is planar at nitrogen[360], which has been ascribed principally to electrostatic interactions[361,362]. With phosphorus, a diverse range of ring and cage frameworks are formed, with ring sizes varying from three to six, although few structures have been determined. Examples are illustrated in Figure 2.10[363–369].

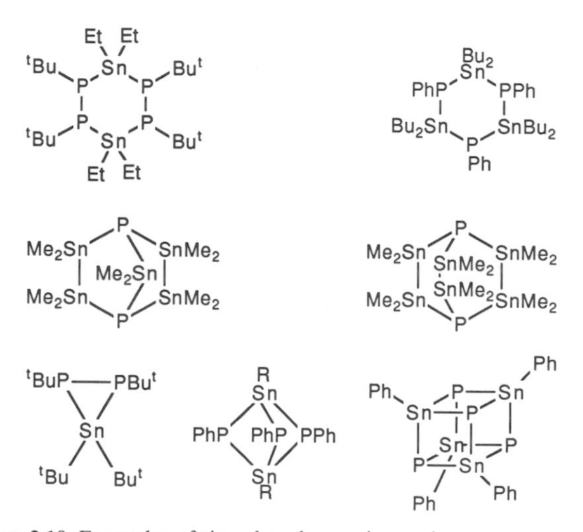

Figure 2.10 Examples of tin–phosphorus ring and cage compounds.

2.6 Mixed-valence compounds

Compounds encorporating tin atoms in both valence states exhibit structures which in general endeavour to accommodate the stereochemical and bonding requirements of each, although in some cases unusual coordination geometries are found for bivalent tin. Thus, simple binary compounds such as $Sn^{IV}Sn^{II}_2F_8$[370] and $Sn^{IV}Sn^{II}S_3$[371,372] have polymeric lattices with trigonal pyramidal tin(II) and octahedral tin(IV) atoms. With more sophisticated compounds such as the mixed-valence carboxylates, $[Sn^{II}Sn^{IV}(O_2CC_6H_4NO_2\text{-}o)_4O\cdot thf]_2$[373], $[Sn^{II}Sn^{IV}(O_2CCF_3)_4O]_2\cdot C_6H_6$[374] and $[Sn^{II}_4Sn^{IV}O_2(O_2CCF_3)_6]$[375], where the tin(IV) atoms exhibit the expected octahedral coordination, the tin(II) atoms exhibit pentagonal pyramidal or square-based pyramidal coordination polyhedra. The former represents the only example of this stereochemistry in Group 14. Similarly, in the triphenyltin derivatives, $Sn^{IV}(C_6H_4PPh_2\text{-}o)_3Sn^{II}Cl$[40], $Ph_3Sn^{IV}\text{–}Sn^{II}(NO_3)$[376] and $\{Ph_3Sn^{IV}\text{–}Sn^{II}(NO_3)\cdot Ph_3As\}_2$[376], the tin(IV) atoms are tetrahedrally coordinated whereas the coordination of the tin(II) varies substantially.

2.7 Compounds with metal–metal bonds

Catenation is not extensive in tin chemistry, and chains or rings can only be obtained if substituted by organic groups. Linear chains containing up to six tin atoms and several six-membered cyclostannanes have been characterised[377–379], but smaller three- or four-membered rings such as (**74**)[380] and (**75**)[381,382] can only be obtained with very bulky groups. Compounds containing bonds to other metals are abundant, as Chapter 8 demonstrates. The tin atoms are generally tetrahedrally coordinated, but examples with coordination numbers of five and six with bonds to both Main Group and Transition metals such as (**76**)[383], (**77**)[384], $(acac)_2Sn[Co_2(CO)_7]$[385] and $Cl_3Sn[Mo(CO)_3(dth)Cl]$[384] have also been obtained.

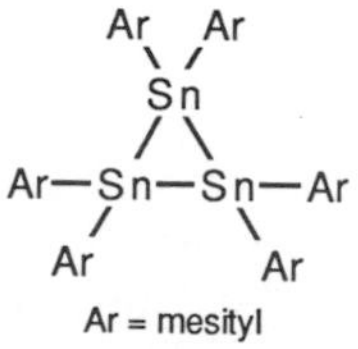

Ar = mesityl

(**74**)

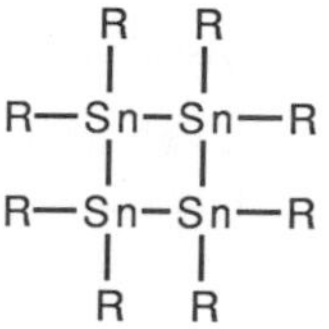

(**75**)

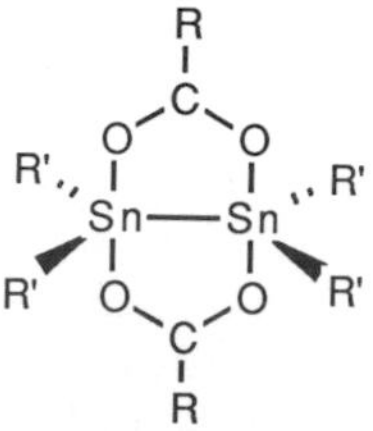

(**76**)

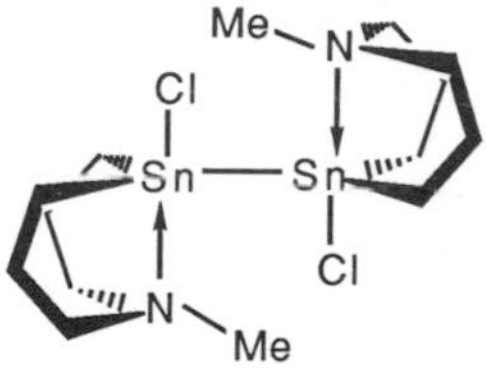

(**77**)

Complexes containing tin–metal bonds formed by the coordination of a tin (II) lone pair to a Main Group or transition metal have been described in Section 2.4.1 and also in Chapter 7. More unusual are the polytin Zintl anions such as $[Sn_4]^{4-}$, $[Sn_5]^{2-}$ and $[Sn_9]^{4-}$, which have tetrahedral, trigonal bipyramidal (**78**) and mono-capped square antiprismatic (**79**) structures, respectively[260,387–391]. These species are generally not very stable and require large cations for their isolation, but appear to have significant future application. For example, thin films of AuSn alloy may be grown by immersing gold foil in a solution containing $[Sn_9]^{4-}$[392]. Mixed Zintl anions have also been characterised. The $[Sn_2Bi_2]^{2-}$ is the first heteroatomic example of the P_4 family of 20 electron tetrahedral clusters. The two anions $[TlSn_9]^{3-}$ and $[TlSn_8]^{3-}$ occur

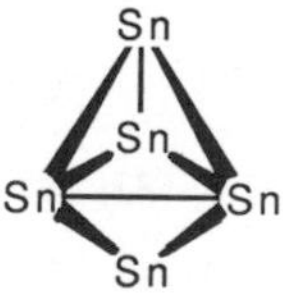

(**78**)

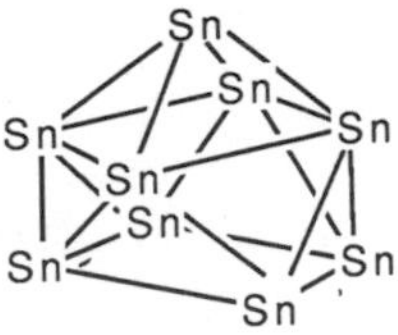

(**79**)

(80) **(81)**

together in the same phase, and have the bicapped square antiprismatic (**80**) and tricapped trigonal prismatic (**81**) geometries, respectively[393,394].

2.8 Theoretical studies of reaction pathways[8,32]

Apart from studies of the cleavage of the tin–carbon bond (Chapter 5), few studies concerning the mechanistic aspects of the reactions of tin compounds have been carried out. The alternative theoretical approach to this problem, employing even fairly simple methods such as MNDO calculations, has been applied with some success, and could probably be used much more widely.

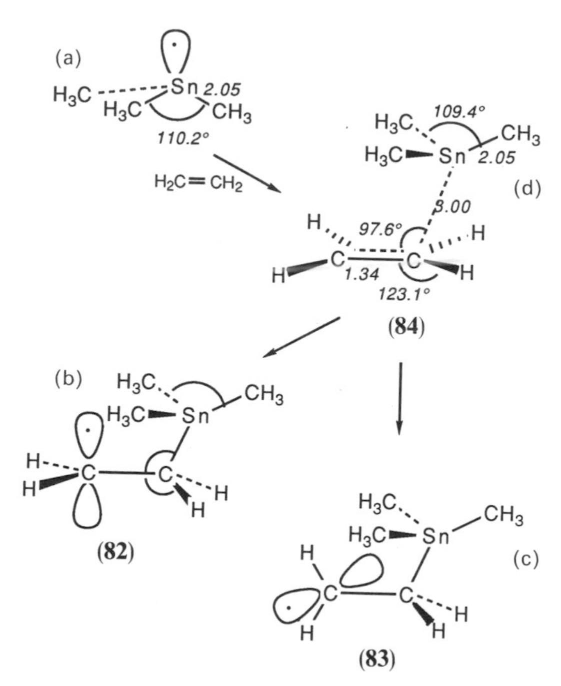

Figure 2.11 Calculated geometries for the $Me_3Sn^{\cdot}$ radical (*a*), the adducts with ethene with the Sn–C bond parallel (*b*) or perpendicular (*c*) to the axis of the singly occupied atomic orbital on carbon, and the transition state (*d*).

2.8.1 *Hydrostannylation*

The chain-propagating step of the free-radical hydrostannation of alkenes involves the reversible addition of a tin-centred radical, $R_3Sn^{\cdot}$, to the alkene, $R'_2C{=}CR'_2$, to form the β-radical, $R_3SnCR'_2CR'^{\cdot}_2$. MNDO calculations of the model reaction between the $Me_3Sn^{\cdot}$ radical and ethene predict it to be exothermic by 67.7 kJ mol^{-1}, which is *c.* 33 kJ mol^{-1} more negative than experimental estimates. Calculated geometries and enthalpies of formation for the species participating in the reaction are shown in Figure 2.11. Of the two possible geometries for the β-radical, that in which the tin–carbon bond is parallel to the axis of the singly occupied carbon *p* orbital (**82**) is calculated to be slightly more stable (by 10.5 kJ mol^{-1}) than the alternative rotamer (**83**). The barrier to reaction is very low (20.5 kJ mol^{-1}) in agreement with the rapid and reversible nature of the reaction, and leads to an equilibrium constant value of 5×10^4. The structure calculated for the transition state (**84**) involves little distortion from the two reactants, with a very long nascent Sn–C bond.

2.8.2 *Insertion of stannylenes into reactive bonds*

Dimethylstannylene is known to undergo facile insertion into the Sn–Cl bond of Me_2SnCl_2 and the C–I bond of methyl iodide, but only polymerizes in the presence of Me_3SnCl. This reluctance to insert into the Sn–Cl bond of Me_3SnCl was attributed to kinetic factors ascribed to the lower positive charge on tin in Me_3SnCl. Two reaction pathways are possible for such insertion reactions, one involving a two-step process involving radicals, or alternatively a direct concerted insertion. MNDO calculations

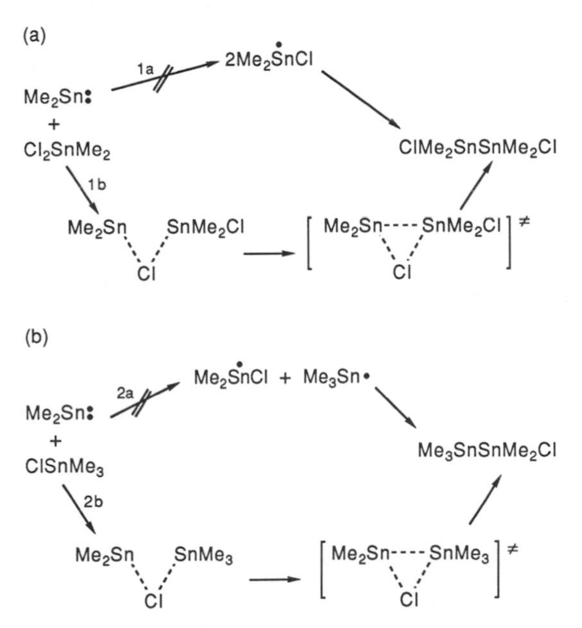

Scheme 1 Calculated reaction pathways for the insertion of Me_2Sn: into the Sn–Cl bonds of (*a*) Me_2SnCl_2 and (*b*) Me_3SnCl.

predict the concerted process to occur for dimethylstannylene insertions into the Sn–Cl bonds of both Me_3SnCl and Me_2SnCl_2 via an initial adduct formed by association of the stannylene with the chlorine of the methyltin chloride, i.e. pathways 1*a* and 2*a*, rather than 1*b* and 2*b* (Scheme 1). In each case, the activation energy for the two-step process was significantly greater (Me_3SnCl: 59.4 kJ mol^{-1}; Me_2SnCl_2: 41.8 kJ mol^{-1}) than for the concerted process, and since the activation energy for reaction with Me_3SnCl is higher, the reaction rate is slower allowing polymerization to occur. In contrast, the two-step process is predicted for insertion into the C–I bond of methyl iodide, i.e. pathway 3*b* rather than 3*a* (Scheme 2).

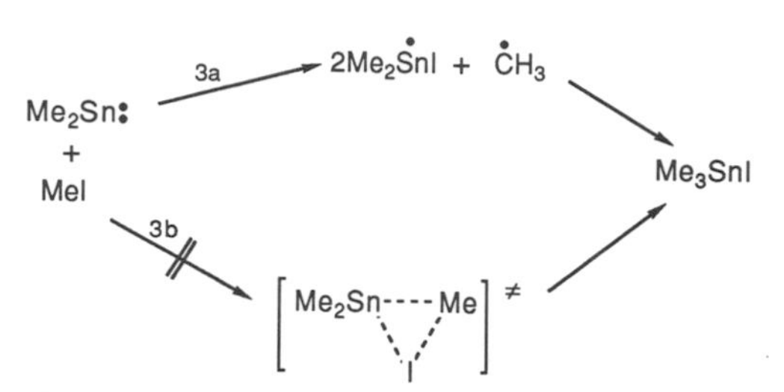

Scheme 2 Calculated reaction pathways for the insertion of Me_2Sn: into the C–I bond of MeI.

2.8.3 *Cycloaddition reactions of stannylenes*

MNDO calculations on the cycloaddition of tin (II) bromide to butadiene, which is reported to proceed by a concerted disrotary process, predict it to take place *syn* and in a synchronous manner via a symmetrical transition state (Scheme 3). The calculated activation energy (81.5 kJ mol^{-1}) is quite low, and is consistent with the observation that the reaction occurs at ambient temperatures.

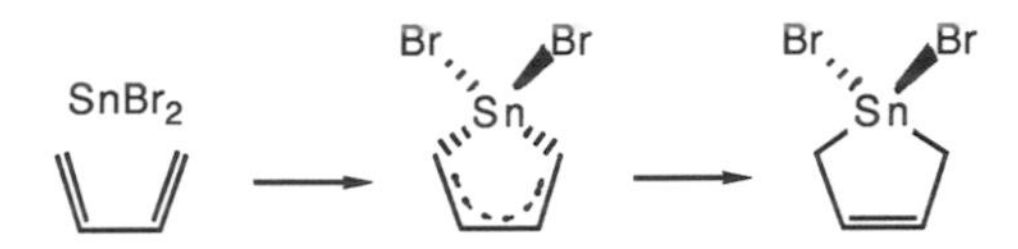

Scheme 3 Calculated reaction pathway for the cycloaddition of $SnBr_2$ to butadiene.

References

1. See also E.A.V. Ebsworth, in *The Organometallic Compounds of Group IV Elements*, ed. A.G. MacDiarmid, Vol. 1, Part 1, Marcel Dekker, New York, 1968.
2. For reviews in this area see A.H. Cowley, *Polyhedron*, 1984, **3**, 389; and A.H. Cowley, *Acc. Chem. Res.*, 1984, **17**, 386.
3. A. Bos and J.S. Ogden, *J. Phys. Chem.*, 1973, **77**, 1513.
4. G. Herzberg, *Molecular Spectra and Molecular Structure*. Vol. 1., *Spectra of Diatomic Molecules*, Van Nostrand, New York, 1959.
5. A. Bos, J.S. Ogden and L. Ogree, *J. Phys. Chem.*, 1974, **78**, 1763.
6. H. Schnöckel, *Angew. Chem., Int. Edn. Engl.*, 1978, **17**, 616.
7. W.J. Pietro and W.J. Henre, *J. Amer. Chem. Soc.*, 1982, **104**, 4329.

8. M.J.S. Dewar, G.L. Grady, D.R. Kuhn and K.M. Merz, *J. Amer. Chem. Soc.*, 1984, **106**, 6773.
9. K.D. Dobbs and W.J. Henre, *Organometallics*, 1986, **5**, 2057.
10. H. Meyer, G. Baun, W. Massa, S. Berger and A. Berndt, *Angew. Chem., Int. Edn. Engl.*, 1987, **26**, 546.
11. C. Couret, J. Escudié, J. Satge, A. Raharinirina and J.D. Andriamizaka, *J. Amer. Chem. Soc.*, 1985, **107**, 8280.
12. T. Fjeldberg, P.B. Hitchcock, M.F. Lappert, K.M. Thomas and A.J. Thorne, *J. Chem. Soc., Dalton Trans.*, 1986, 1551.
13. D.E. Goldberg, P.B. Hitchcock, M.F. Lappert, K.M. Thomas, A.J. Thorne, T. Fjeldberg, A. Haaland and B.E.R. Schilling, *J. Chem. Soc., Dalton Trans.*, 1986, 2387.
14. P.J. Davidson, D.H. Harris and M.F. Lappert, *J. Chem. Soc., Dalton Trans.*, 1976, 2268.
15. T. Fjeldberg, A. Haaland, B.E.R. Schilling, H.V. Volden, M.F. Lappert and A.J. Thorne, *J. Organomet. Chem.*, 1984, **276**, C1.
16. T. Fjeldberg, A. Haaland, M.F. Lappert, B.E.R. Schilling, R. Seip and A.J. Thorne, *J. Chem. Soc., Chem. Commun.*, 1982, 1407.
17. P.B. Hitchcock, M.F. Lappert, S.J. Miles and A.J. Thorne, *J. Chem. Soc., Chem. Commun.*, 1984, 480.
18. L. Pauling, *Proc. Natl. Acad. Sci. USA*, 1983, **80**, 3871.
19. G. Trinquier, J.P. Malneu and P. Rivière, *J. Amer. Chem. Soc.*, **104**, 4529.
20. H. Prut, H.J. Haupt and F. Huber. *Z. anorg. allg. Chem.*, 1973, **396**, 81.
21. K.W. Zilm, G.A. Lawless, R.M. Merrill, J.M. Millar and G.G. Webb, *J. Amer. Chem. Soc.*, 1987, **109**, 7236.
22. S. Masamune and L.R. Sita, *J. Amer. Chem. Soc.*, 1985, **107**, 6390.
23. T.N. Mitchell, *J. Organometal. Chem.*, 1974, **70**, C1.
24. P.G. Harrison, *Coord. Chem. Rev.*, 1976, **20**, 1.
25. J. Zubieta and J.J. Zuckerman, *Progr. Inorg. Chem.*, 1978, **24**, 251.
26. P.A. Cusack, P.J. Smith, J.D. Donaldson and S.M. Grimes, *A Bibliography of X-ray Crystal Structures of Tin Compounds*, I.T.R.I Publication, No. 588.
27. I. Hargittai, J. Tremmel, E. Vadja, A.A. Ischenko, A.A. Ivanov, L.S. Ivashkevich and V.P. Spiridonov, *J. Mol. Struct.*, 1977, **42**, 147.
28. G.A. Ozin and A Vander Voet, *J. Chem. Phys.*, 1972, **56**, 4768.
29. O.M.Nefedov, S.P. Kolesnikov and A.I. Ioffe, *J. Organomet. Chem.*, 1977, **5**, 181.
30. W.P. Neumann, *The Organometallic and Coordination Chemistry of Germanium, Tin and Lead*, eds. M. Gielen and P.G. Harrison, Freund, Tel-Aviv, 1978, 51.
31. G. Olbrich, *Chem. Phys. Lett.*, 1980, **73**, 110.
32. M.J.S. Dewar, J.E. Friedheim and G.L. Grady, *Organometallics*, 1985, **4**, 1784.
33. M. Schriewer and W.P. Neumann, *Angew. Chem., Int. Edn. Engl.*, 1981, **20**, 1019.
34. K.H. Scherping and W.P. Neumann, *Organometallics*, 1982, **1**, 1017.
35. M. Veith and W. Franks, *Angew. Chem. Int. Edn. Engl.*, 1985, **24**, 223.
36. J. Doe, S. Borkett and P.G. Harrison, *J. Organomet. Chem.*, 1973, **52**, 343.
37. T.S. Dory, J.J. Zuckerman and C.L. Barnes, *J. Organomet. Chem.*, 1985, **281**, C1.
38. P. Jutzi. B. Hampel, K. Stroppel, C. Krüger, K. Angermund and P. Hofman, *Chem. Ber.*, 1985, **118**, 2789.
39. P.B. Hitchcock, M.F. Lappert and M.C. Misra, *J. Chem. Soc., Chem. Commun.*, 1985, 863.
40. K. Jurkschat, H.P. Abicht, A. Tzschach and R. Mahieu, *J. Organomet. Chem.*, 1986, **309**, C47.
41. U. Baumeister, H. Hartung, K. Jurkschat and A. Tzschach, *J. Organomet. Chem.*, 1986, **304**, 107.
42. A Zschunke, M. Scheer, M. Völke and A. Tzschach, *J. Organomet. Chem.*, 1986, **308**, 325.
43. M.D. Brice and F.A. Cotton, *J. Amer. Chem. Soc.*, 1973, **95**, 4529.
44. J.D. Cotton, P.J. Davison, D.E. Goldberg, M.F. Lappert and K.M. Thomas, *Chem. Commun.*, 1974, 893.
45. M. Grenz and W.W. Du Mont, *J. Organomet. Chem.*, 1983, **241**, C5.
46. A. Tzschach, K. Jurkschat, M. Scheer, J. Meunier-Piret and M. van Meerssche, *J. Organomet. Chem.*, 1983, **259**, 165.
47. G.W. Bushnell, D.T. Eadie, A. Pidcock, A.R. Sam., R.D. Holmes-Smith, S.R. Stobart, E.T. Brennan and T.S. Cameron, *J. Amer. Chem. Soc.*, 1982, **104**, 583.
48. J.L. Almeida, K.R. Dixon, C. Eaborn, P.B. Hitchcock and A. Pidcock, *J. Chem. Soc., Chem. Commun.*, 1982, 1315.
49. J.D. Donaldson and R. Oteng, *Inorg. Nucl. Chem. Lett.*, 1967, **3**, 163.

50. R.C. McDonald, H. Ho-Kuen Hau and K. Eriks, *Inorg. Chem.*, 1976, **15**, 762.
51. I. Naray-Szabo, *Inorganic Crystal Chemistry*, Akademiai Kaido, Budapest, 1969, 215.
52. J.M. van den Berg, *Acta Crystallogr.*, 1961, **14**, 1002.
53. J. Anderson, *Acta Chem. Scand., Ser. A*, 1975, **29**, 956.
54. R.A. Howie, W. Moser and I.C. Moser, *Acta Crystallogr.*, 1972, **B28**, 2965.
55. N. Kamenar and D. Grdenic, *J. Chem. Soc.*, 1961, 3954.
56. H. Kiriyama, K. Kitahama, O. Nakamura and R. Kiriyama, *Bull. Chem. Soc. Jpn.*, 1973, **46**, 1389.
57. P.G. Harrison, B.J. Haylett and T.J. King, *Inorg. Chim. Acta*, 1983, **75**, 265.
58. E. Hough and D.G. Nicholson, *J. Chem. Soc., Dalton Trans.*, 1976, 1782.
59. S.J. Archer, K.R. Koch and S. Schmidt, *Inorg. Chim. Acta*, 1987, **126**, 209.
60. J. Anderson, *Acta Chem. Scand.*, 1972, **26**, 1730, 2543, 3813; J. Anderson and G. Lundgren, *Acta Chem. Scand.*, 1970, **24**, 2670.
61. G. Bergerhoff and L. Goost, *Acta Crystallogr., Sect. B*, 1973, **29**, 632.
62. G. Bergerhoff and H. Namgung, *Acta Crystallogr., Sect. B*, 1978, **34**, 699.
63. J.K. Stalik, P.W.R. Corfield and D.W. Meek, *Inorg. Chem.*, 1973, **12**, 1668.
64. H.J. Haupt, F. Huber and H. Preut, *Z. Anorg. Allg. Chem.*, 1976, **422**, 97, 255.
65. P.G. Harrison, B.J. Haylett and T.J. King, *Inorg. Chim. Acta*, 1983, **75**, 265.
66. F.R. Poulsen and S.E. Rasmussen, *Acta Chem. Scand.*, 1970, **24**, 150.
67. J. Barrett, S.R.A. Bird, J.D. Donaldson and J. Silver, *J. Chem. Soc. (A)*, 1971, 3105.
68. S.R.A. Bird, J.D. Donaldson and J. Silver, *J. Chem. Soc., Dalton*, 1972, 1950.
69. J.D. Donaldson and J. Silver, *J. Chem. Soc., Dalton*, 1973, 666.
70. J.D. Donaldson, D. Laughlin, S.D. Ross and J. Silver, *J. Chem. Soc., Dalton*, 1973, 1985.
71. J.D. Donaldson, J. Silver, S. Hadjiminolis and S.D. Ross, *J. Chem. Soc., Dalton*, 1975, 1500.
72. D. Weber, *Z. Naturforsch., Teil B*, 1978, **33**, 862, 1443; 1979, **34**, 939.
73. P. Mauersberger and F. Huber, *Acta Crystallogr.*, 1980, **B36**, 782.
74. R.R. McDonald, A.C. Jarson and D.T. Cramer, *Acta Crystallogr.*, 1964, **17**, 1104.
75. G. Bergerhoff and L. Goost, *Acta Crystallogr., Sect, B*, 1970, **26**, 19.
76. G. Bergerhoff, L. Goost and E. Schultze-Rhonhof, *Acta Crystallogr., Sect. B*, 1968, **24**, 803.
77. G. Bergerhoff and L. Goost, *Acta Crystallogr., Sect. B*, 1974, **30**, 1362.
78. J.D. Donaldson and D.C. Puxley, *J. Chem. Soc., Chem. Commun.*, 1972, 289; S. Vilminot, W. Gramer and L. Cot, *Acta Crystallogr., Sect. B*, 1978, **34**, 35.
79. S. Vilminot, W. Gramer, Z. Al Oraibi and L. Cot, *Acta Crystallogr., Sect. B*, 1978, **34**, 3308.
80. J. Brörisch and G. Bergerhoff, *Acta Crystallogr., Sect. C*, 1984, 40, 2005.
81. C. Geneys, S. Vilminot and L. Cot, *Acta Crystallogr., Sect. B*, 1976, **32**, 3199.
82. S. Vilminot, W. Gramer, Z. Al Oraibi and L. Cot, *Acta Crystallogr., Sect. B*, 1978, **34**, 3306.
83. W.J. Moore and L. Pauling, *J. Amer. Chem. Soc.*, 1941, **63**, 1392.
84. W. Tremel and R. Hoffmann, *Inorg. Chem.*, 1987, **26**, 118.
85. W. Hoffman, *Z. Kristallogr.*, 1935, **92**, 161.
86. R.E. Rundle and D.H. Olsen, *Inorg. Chem.*, 1964, **3**, 596.
87. A. Okazaki and I. Ueda, *J. Phys. Soc. Jpn.*, 1956, **11**, 470.
88. E. Zeipel, *Ark. Mat. Astron. Fys.*, 1935, **25A**, 1.
89. J.W. Earley, *Amer. Mineral.*, 1950, **35**, 338.
90. H.G. von Schnering, R. Nesper and H. Pelshenke, *Z. Anorg. Allg. Chem.*, 1983, **499**, 117.
91. R. Nesper and H.G. von Schnering, *Z. Anorg. Allg. Chem.*, 1983, **499**, 109.
92. S. Grimvall, *Acta Chem. Scand.*, 1973, **27**, 1447.
93. C.G. Davies, J.D. Donaldson, D.R. Laughlin, R.A. Howie and R. Beddoes, *J. Chem. Soc., Dalton Trans.*, 1975, 2241.
94. R.A. Howie and W. Moser, *Nature (London)*, 1968, **219**, 372.
95. P.G. Harrison, B.J. Haylett and T.J. King, *J. Chem. Soc., Chem. Commun.*, 1978, 112.
96. L. Ross and M. Dräger, *Z. Naturforsch., Teil B*, 1983, **38**, 665.
97. M. Dräger, B. Mathiasch, L. Ross and M. Ross, *Z. Anorg. Allg. Chem.*, 1983, **506**, 99.
98. D.H. Olsen and R.E. Rundle, *Inorg. Chem.*, 1963, **2**, 1310.
99. W.P. Neumann and J. Fu, *J. Organomet. Chem.*, 1984, **273**, 295.
100. L.C. Willemsens and G.J.M. van der Kerk, *Investigations in the Field of Organolead Chemistry*, International Lead Zinc Research Organization Inc., 1965.
101. A.M. Arif, A.H. Cowley and T.M. Elkins, *J. Organomet. Chem.*, 1987, **325**, C11.

102. T. Fjeldberg, H. Hope, M.F. Lappert, P.P. Power and A.J. Thorne, *J. Chem. Soc., Chem. Commun.*, 1983, 639.
103. T. Fjeldberg, P.B. Hitchcock, M.F. Lappert, S.J. Smith and A.J. Thorne, *J. Chem. Soc., Chem. Commun.*, 1985, 939.
104. B. Cetinkay, I. Gümrükgu, M.F. Lappert, J.L. Atwood, R.D. Rogers and M.J. Zawototko, *J. Amer. Chem. Soc.*, 1980, **102**, 2088.
105. P.B. Hitchcock, M.F. Lappert, B.J. Samways and E.L. Weinberg, *J. Chem. Soc., Chem. Commun.*, 1983, 1492.
106. M.M. Olmstead and P.P. Power, *Inorg. Chem.*, 1984, **23**, 413.
107. K.C. Molloy, M.P. Bigwood, R.H. Herber and J.J. Zuckerman, *Inorg. Chem.*, 1982, **21**, 3709.
108. M. Veith, *Z. Naturforsch., Teil B*, 1978, **33**, 7.
109. W.W. du Mont and M. Grenz, *Z. Naturforsch., Teil B*, 1983, **38**, 113.
110. K. Jurkschat, M. Scheer, A. Tzschach, J. Meunier-Piret and M. van Meersche, *J. Organomet. Chem.*, 1985, **281**, 173.
111. P.B. Hitchcock, M.F. Lappert, B.J. Samways and E.L. Weinberg, *J. Chem. Soc., Chem. Commun.*, 1983, 1492.
112. L.D. Silverman and M. Zeldin, *Inorg. Chem.*, 1980, **19**, 272.
113. B.J. Haylett, Ph.D. Thesis, University of Nottingham, 1981.
114. M. Veith, *Angew. Chem., Int. Edn. Engl.*, 1987, **26**, 1.
115. W.W. du Mont and M. Grenz, *Chem. Ber.*, 1985, **118**, 1045.
116. P.A.W. Dean, J.J. Vittal and N.C. Payne, *Can. J. Chem.*, 1985, **63**, 394.
117. A.M. Arif, A.H. Cowley, R.A. Jones and J.M. Power, *J. Chem. Soc., Chem. Commun.*, 1986, 1446.
118. L.M. Engelhardt, B.S. Jolly, M.F. Lappert, C.L. Raston and A.H. White, *J. Chem. Soc., Chem. Commun.*, 1988, 336.
119. H.H. Karsch, A. Appelt and G. Müller, *Organometallics*, 1986, **5**, 1664.
120. H.H. Karsch, A. Appelt and G. Hanika, *J. Organomet. Chem.*, 1986, **312**, C1.
121. A.L. Balch and D.E. Oram, *Organometallics*, 1986, **5**, 2159.
122. P.F.R. Ewings, P.G. Harrison and T.J. King, *J. Chem. Soc., Dalton Trans.*, 1976, 1399.
123. P.F.R. Ewings, P.G. Harrison and A. Mangia, *J. Organomet. Chem.*, 1976, **114**, 35.
124. P.F.R. Ewings, P.G. Harrison and D.E. Fenton, *J. Chem. Soc., Dalton Trans.*, 1975, 821.
125. P.F.R. Ewings, P.G. Harrison and T.J. King, *J. Chem. Soc., Dalton Trans.*, 1975, 1455.
126. A.B. Cornwell and P.G. Harrison, *J. Chem. Soc., Dalton Trans.*, 1975, 1722.
127. H.H. Karsch, A. Appelt and G. Müller, *Angew. Chem., Int. Edn. Engl.*, 1985, **24**, 402.
128. P.G. Harrison and E.W. Thornton, *J. Chem. Soc., Dalton Trans.*, 1978, 1274.
129. A.D. Christie R.A. Howie and W. Moser, *Inorg. Chim. Acta*, 1979, **36**, L447.
130. N.Kh. Dzhafarov, I.R. Amiraslanov, G.N. Nadzhafov, E.M. Movsumov, F.R. Kerimova and K.S. Mamedov, *J. Struct. Chem.*, 1981, **22**, 245.
131. I.R. Amiraslanov, N.Kh. Dzhafarov, G.N. Nadzhafov, Kh.S. Mamedov, E.M. Movsumov and B.T. Usubaliev, *J. Struct. Chem.*, 1980, **21**, 109.
132. N.Kh. Dzafarov, I.R. Amiraslanov, G.N. Nadzhafov, E.M. Movsumov and Kh.S. Mamedov, *J. Struct. Chem.*, 1981, **22**, 242.
133. I.R. Amiraslanov, N.Kh. Dzhafarov, G.N. Nazhafov, Kh.S. Mamedov, E.M. Movsumov and B.J. Usubakiev, *J. Struct. Chem.*, 1980, **22**, 104.
134. P.G. Harrison and A.T. Steel, *J. Organomet. Chem.*, 1982, **239**, 105.
135. G. Lundgren, C. Wernfors and T. Yamaguchi, *Acta Crystallogr., Sect. B*, 1982, **38**, 2357.
136. J.D. Donaldson and S.M. Grimes, *J. Chem. Soc., Dalton Trans.*, 1984, 1301.
137. J.D. Donaldson, S.M. Grimes, A. Nicolaides and P.J. Smith, *Polyhedron*, 1985, **4**, 391.
138. R. Herak, B. Prelesnik, M. Curic and P. Vasic, *J. Chem. Soc., Dalton Trans.*, 1978, 566.
139. T.R.J. Weakly and W.W.L. Watt, *Acta Crystallogr., Sect. B*, 1979, **35**, 3023.
140. R.C. McDonald and K. Eriks, *Inorg. Chem.*, 1980, **19**, 1237.
141. T.H. Jordan, B. Dickens, L.W. Schroeder and W.E. Brown, *Inorg. Chem.*, 1980, **19**, 2551.
142. M. Hala, F. Maruno, S.I. Iwai and H. Aoki, *Acta Crystallogr., Sect. B*, 1980, **36**, 2128.
143. P. Vasic, B. Prelesnik, R. Herak and M. Giric, *Acta Crystallogr., Sect. B*, 1981, **37**, 660.
144. R. Eujen, H. Bürger and H. Oberhammer, *J. Mol. Struct.*, 1981, **71**, 109.
145. V.S. Petrosyan and O.A. Reutov, *Pure Appl. Chem.*, 1974, **37**, 147.
146. V.G. Kumar Das, L.K. Mun, C. Wei, S.J. Bunden and T.C.W. Mak, *J. Organomet. Chem.*, 1987, **322**, 163.

147. A. Tzschach and K. Jurkschat, *Pure Appl. Chem.*, 1986, **58**, 639.
148. V.G. Kumar Das, L.K. Mun, C. Wei and T.C.W. Mak, *Organometallics*, 1987, **6**, 10.
149. R. Hoppe and W. Dähne, *Naturwissenschaften*, 1962, **49**, 254.
150. E.O. Schlemper and W.C. Hamilton, *Inorg. Chem.*, 1966, **5**, 995.
151. L.E. Levchuk, J.R. Sams and F. Aubke, *Inorg. Chem.*, 1972, **11**, 43.
152. H.C. Clark, R.J. O'Brien and J. Trotter, *J. Chem. Soc.*, 1964, 2332.
153. S.S. Al-Juaid, S.M. Dhaher, C. Eaborn, P.B. Hitchcock and J.D. Smith, *J. Organomet. Chem.*, 1987, **325**, 117.
154. M.B. Hossain, J.L. Lefferts, K.C. Molloy, D. van der Helm and J.J. Zuckerman, *Inorg. Chim. Acta*, 1979, **36**, L409.
155. N.G. Bokii, G.N. Zakharova and Yu.T.T. Struchkov, *J. Struct. Chem.*, 1970, **11**, 828.
156. H. Preut and F. Huber, *Acta Crystallogr., Sect. B*, 1979, **35**, 744.
157. M.J.S. Gynane, M.F. Lappert, S.J. Miles, A.J. Carty and N.J. Taylor, *J. Chem. Soc., Dalton Trans.*, 1977, 2009.
158. Z.T. Wang, H.K. Wang, X.Y. Liu, X.K. Yao and H.G. Wang, *J. Organomet. Chem.*, 1987, **331**, 263.
159. L.N. Zakharov, B.I. Petrov, V.A. Lebedev, E.A. Kuz'min and N.V. Belov, *Kristallografiya*, 1978, **23**, 1049.
160. S. Calogero, P. Ganis, V. Peruzzo and G. Tagliavini, *J. Organomet. Chem.*, 1979, **179**, 145.
161. A.G. Davies, H.J. Milledge, D.C. Puxley and P.J. Smith, *J. Chem. Soc. (A)*, 1970, 2862.
162. N.W. Alcock and J.F. Sawyer, *J. Chem. Soc., Dalton Trans.*, 1977, 1090.
163. N.G. Bokii, Yu.T. Struchkov and A.K. Prokof'ev, *J. Struct. Chem.*, 1972, **13**, 619.
164. J.L. Baxter, E.M. Holt and J.J. Zuckerman, *Organometallics*, 1985, **3**, 255.
165. N.G. Bokii, Yu.T. Struchkov and A.K. Prokof'ev, *Sov. J. Coord. Chem.*, 1975, **1**, 965.
166. P.T. Green and R.F. Bryan, *J. Chem. Soc. (A)*, 1971, 2549.
167. N.G. Bokii, Yu.T. Struchkov and A.K. Prokof'ev, *J. Struct. Chem.*, 1972, **13**, 619.
168. M.M. Amini, E.M. Holt and J.J. Zuckerman, *J. Organomet. Chem.*, 1987, **327**, 147.
169. P.G. Harrison, T.J. King and M.A. Healy, *J. Organomet. Chem.*, 1979, **182**, 17.
170. T. Kimura, T. Ueki, N. Yasuoka, N. Kasai and M. Kakudo, *Bull. Chem. Soc. Jpn.*, 1969, **42**, 2479.
171. M. Yoshida, T. Ueki, N. Yasuoka, N. Kasai, M. Kakudo, I. Omae, Sl Kikkawa and S. Matsuda, *Bull. Chem. Soc. Jpn.*, 1968, **41**, 1113.
172. B.W. Fitzsimmons, D.G. Othen, H.M.M. Shearer, K. Wade and G. Whitehead, *J. Chem. Soc., Chem. Commun.*, 1977, 215.
173. G. van Koten, J.G. Noltes and A.L. Spek, *J. Organomet. Chem.*, 1976, **118**, 183.
174. G. van Koten, J.T.B.H. Jastrzebskii, J.G. Noltes, W.M.G.F. Potenagel, J. Kroon and A.L. Spek, *J. Amer. Chem. Soc.*, 1978, **100**, 5021.
175. G. van Koten, J.T.B.H. Jastrzebskii, J.G. Noltes, G.J. Verhoeck, A.L. Spek and J. Kroon, *J. Chem. Soc., Dalton Trans.*, 1980, 1352.
176. G. van Koten, J.T.B.H. Jastrzebskii, J.G. Noltes, A.L. Spek and J.C. Schoone, *J. Organomet. Chem.*, 1978, **148**, 233.
177. G. van Koten and J.G. Noltes, *J. Amer. Chem. Soc.*, 1976, **98**, 5393.
178. R. Hulme, *J. Chem. Soc.*, 1963, 1524.
179. S.W. Ng, C.L. Barnes, M.B. Hossain, D. van der Helm, J.J. Zuckerman and V.G. Kumer Das, *J. Amer. Chem. Soc.*, 1982, **104**, 5359.
180. P.G. Harrison and K.C. Molloy, *J. Organomet. Chem.*, 1978, **152**, 63.
181. M. Matsubayashi, K. Ueyama and T. Tanaka, *J. Chem. Soc., Dalton Trans.*, 1985, 465.
182. G. Matsubayashi, R. Shimizu and T. Tanaka, *J. Chem. Soc., Dalton Trans.*, 1987, 129.
183. C. Pelizzi, G. Pelizzi and P. Tarasconi, *Polyhedron*, 1983, **2**, 145.
184. R. Graziani, U. Casllato, R. Eltone and G. Plazzogna, *J. Chem. Soc., Dalton Trans.*, 1982, 805.
185. F.W.B. Einstein and B.R. Penfold, *J. Chem. Soc. (A)*, 1968, 3019.
186. L.A. Aslanov, W.M. Attiya, V.M. Ionov, A.B. Permin and V.S. Petrosyan, *J. Struct. Chem.*, 1977, **18**, 884.
187. A.D. Adley, P.H. Bird, A.R. Fraser and M. Onyszchuk, *Inorg. Chem.*, 1972, **11**, 1402.
188. C.I. Brandon, *Acta Chem. Scand.*, 1963, **17**, 759.
189. Y. Hermodsson, *Acta Crystallogr.*, 1960, **13**, 656; *Ark. Kemi*, 1970, **31**, 73.
190. P. Domiano, A. Musatti, M. Nardelli, C. Pelizzi and G. Predieri, *Inorg. Chim. Acta*, 1980, **38**, 9.

191. M. Webster and H.E. Blayden, *J. Chem. Soc. (A)*, 1969, 2443.
192. D.M. Barnhart, C.N. Caughlan and M. Ul-Haque, *Inorg. Chem.*, 1968, **7**, 1135.
193. J.C. Barnes and T.J.R. Weakley, *J. Chem. Soc., Dalton Trans.*, 1976, 1786.
194. H.W. Roesky, M. Kuhn and J.W. Bates, *Chem. Ber.*, 1982, **115**, 3025.
195. S.E. Denmark, B.R. Henke and E. Weber, *J. Amer. Chem. Soc.*, 1987, **109**, 2512.
196. M.M. Olmstead, K.A. Williams and W.K. Kusker, *J. Amer. Chem. Soc.*, 1982, **104**, 5567.
197. L.A. Aslanov, V.M. Ionov, W.M. Attia, A.B. Permin and V.S. Petrosyan, *J. Organomet. Chem.*, 1978, **144**, 39.
198. I.R. Beattie, M. Milne, M. Webster, H.E. Blayden, P.J. Jones, R.C.G. Killean and J.L. Lawrence, *J. Chem. Soc. (A)*, 1969, 482.
199. G.G. Mather, G.M. McLaughlin and A. Pidcock, *J. Chem. Soc., Dalton Trans.*, 1973, 1823.
200. I.R. Beattie, R. Hulme and L. Rule, *J. Chem. Soc.*, 1965, 1581.
201. J.C. Hill, R.S. Drago and R.H. Herber, *J. Amer. Chem. Soc.*, 1969, **91**, 1644.
202. L.A. Aslanov, V.M. Ionov, W.M. Attiya and A.B. Permin, *J. Struct. Chem.*, 1978, **19**, 166.
203. L.A. Aslanov, V.M. Ionov, W.M. Attiya, A.B. Permin and V.S. Petrosyan, *J. Organomet. Chem.*, 1978, **144**, 39.
204. V.G. Kumar Das, Y.C. Keong, C. Wei, P.J. Smith and T.C.W. Mak, *J. Chem. Soc., Dalton Trans.*, 1987, 129.
205. V.G. Kumar Das, C. Wei, C.K. Yap and T.C.W. Mak, *J. Organomet. Chem.*, 1986, **299**, 41.
206. V.G. Kumar Das, C.K. Yap and P.J. Smith, *J. Organomet. Chem.*, 1985, **291**, C17.
207. E.A. Blom, B.R. Penfold and W.T. Robinson, *J. Chem. Soc. (A)*, 1969, 913.
208. L. Randaccio, *J. Organomet. Chem.*, 1973, **55**, C58.
209. N.W. Isaacs and C.H.L. Kennard, *J. Chem. Soc. (A)*, 1970, 1257.
210. S.L. Chadha, P.G. Harrison and K.C. Molloy, *J. Organomet. Chem.*, 1980, **202**, 247.
211. P.G. Harrison, T.J. King and J.A. Richards, *J. Chem. Soc., Dalton Trans.*, 1974, 1723.
212. L. Coghi, C. Pellizzi and G. Pelizzi, *Gazz. Chem., Ital.*, 1974, **104**, 873.
213. P. Ganis, V. Peruzzo and G. Valle, *J. Organomet. Chem.*, 1983, **256**, 245.
214. L. Prasad, Y.Le Page and F.E. Smith, *Inorg. Chim. Acta*, 1983, **68**, 45.
215. J. Pannetier and G. Denes, *Acta Crystallogr.*, 1980, **B36**, 2763.
216. A.F. Wells, *Structural Inorganic Chemistry*, 4th edn., Oxford University Press, Oxford, 1975.
217. B. Palosz, W. Palosz and S. Gierlotka, *Acta Crystallogr.*, 1985, **C41**, 807.
218. B. Palosz, W. Palosz and S. Gierlotka, *Acta Crystallogr.*, 1985, **C41**, 1402.
219. B. Palosz, S. Gierlotka and F. Levy, *Acta Crystallogr.*, 1985, **C41**, 1404.
220. C.O. Björling, *Arkiv. Kemi, Min. Geol.*, 1941, **15B**, 1.
221. H. Strune and B. Contag, *Acta Crystallogr.*, 1960, **13**, 601.
222. I. Morgenstern-Badarau, P. Poix and A. Michel, *Compt. Rend.*, 1964, **258C**, 3036.
223. I. Morgenstern-Badarau, Y. Billiet, P. Poix and A Michel, *Compt. Rend.*, 1965, **260**, 3668.
224. C. Cohen-Addad, *Bull. Soc. Miner. Crist.*, 1967, **90**, 32.
225. A.N. Christensen and R.G. Hazell, *Acta Chem. Scand.*, 1969, **23**, 1219.
226. I. Morgenstern-Badarau, *J. Solid-State Chem.*, 1976, **17**, 399.
227. I. Morgenstern-Badarau, C. Levy-Clement and A. Michel, *Compt. Rend.*, 1969, **268**, 696.
228. E. Dubler, R. Hess and H.R. Oswald, *Z. Anorg. Allg. Chem.*, 1976, **421**, 61.
229. M. Inagaki, T. Kuroishi, Y. Xamashita and M. Urata, *Bull. Chem. Soc. Jpn*, 1985, **58**, 1292.
230. R. Marchand, Y. Piffard and M. Tournoux, *Acta Crystallogr., Sect. B*, 1975, **31**, 511.
231. R. Olacuaga, J.M. Reau, M. Devalette, G. Le Flem and P. Hagenmuller, *J. Solid-State Chem.*, 1975, **13**, 275.
232. B. Nowitzki and R. Hoppe, *Z. Anorg. Allg. Chem.*, 1983, **505**, 105.
233. P. Poix, *Ann. Chim.*, 1965, **10**, 49.
234. M. Trömel, *Z. Anorg. Allg. Chem.*, 1969, **371**, 237.
235. R. Weiss and R. Faivre, *Compt. Rend.*, 1959, **248**, 106.
236. G. Wagner and H. Binder, *Z. Anorg. Allg. Chem.*, 1959, **298**, 12.
237. M. Nogues and P. Poix, *Ann. Chim.*, 1968, **3**, 335.
238. L. Siegel, *J. Appl. Cryst.*, 1978, **11**, 284.
239. F. Brisse and O. Knop, *Can. J. Chem.*, 1968, **46**, 859.
240. C.G. Whinfrey and A. Tauber, *J. Amer. Chem. Soc.*, 1961, **83**, 755.
241. G. Vetter, F. Queyroux and J.C. Gilles, *Mater. Res. Bull.*, 1978, **13**, 211.
242. G. Kreuzburg, F. Stewner and R. Hoppe, *Z, Anorg. Allg. Chem.*, 1970, **379**, 242.
243. A. Verbaere, M. Dion and M. Tournoux, *J. Solid-State Chem.*, 1974, **11**, 184.

244. C. Levy-Clement, I. Morgenstern-Badarau, Y. Billier and A. Michel, *Compt. Rend.*, 1967, **265C**, 585.
245. B. Durand and H. Loiseleur, *J. Appl. Cryst.*, 1978, **11**, 289.
246. R.S. Roth, *J. Res. Nat. Bur. Stand.*, 1957, **58**, 75.
247. G. Wagner and H. Binder, *Z. Anorg. Allg. Chem.*, 1959, **298**, 12.
248. R.D. Shannon, J.L. Gillson and R.J. Bouchard, *J. Phys. Chem. Solids*, 1977, **38**, 877.
249. B.M. Gatehouse and D.J. Lloyd, *J. Solid-State Chem.*, 1970, **2**, 410.
250. R. Hoppe and K. Seeger, *Z. Anorg. Allg. Chem.*, 1970, **375**, 264.
251. H. Vincent, E.F. Bertaut, W.H. Baur and R.D. Shannon, *Acta Crystallogr., Sect. B*, 1976, **32**, 1749.
252. J. Mandt and B. Krebs, *Z. Anorg. Allg. Chem.*, 1976, **420**, 31.
253. B. Krebs and H.J. Jacobson, *Z. Anorg. Allg. Chem.*, 1976, **421**, 97.
254. N. Ryanck, P. Larvelle and A. Katty, *Acta Crystallogr., Sect. B*, 1976, **32**, 672.
255. W. Schiwy, S. Pohl and B. Krebs, *Z. Anorg. Allg. Chem.*, 1973, **402**, 77.
256. R.A. Beskrovnaya, L.D. Dyatlova and V.V. Serebrennikov, *Tr. Tomsk. Univ.*, 1971, 403; *Chem. Abstr.*, 1973, **78**, 131485.
257. B. Krebs and H.U. Hürter, *Z. Anorg. Allg. Chem.*, 1980, **462**, 143.
258. B. Eisenmann, H. Schäfer and H. Schrod, *Z. Naturforsch., Teil B*, 1983, **38**, 921.
259. R.C. Barns, L.A. Devereux, P. Granger and G.J. Schrobilgen, *Inorg. Chem.*, 1985, **24**, 2615.
260. R.G. Teller, L.J. Krause and R.C. Haushalter, *Inorg. Chem.*, 1983, **22**, 1809.
261. B. Krebs and W. Schiwy, *Z. Anorg. Allg. Chem.*, 1973, **398**, 63.
262. J.C. Jumas, J. Olivier-Fourcade, F. Vermont-Gaud-Daniel, M. Ribes, E. Philippot and M. Maurin, *Rev. Chim. Minerale*, 1974, **11**, 13.
263. B. Krebs, S. Pohl and W. Schiwy, *Z. Anorg. Allg. Chem.*, 1972, **393**, 241.
264. B. Krebs and H. Müller, *Z. Anorg. Allg. Chem.*, 1983, **496**, 47.
265. G. Dittmar, *Acta Crystallogr., Sect. B*, 1978, **34**, 2390.
266. G. Dittmar, *Z. Anorg. Allg. Chem.*, 1979, **453**, 68.
267. B. Eisenmann, H. Schwerer and H. Schäfer, *Z. Naturforsch., Teil B*, 1981, **36**, 2538.
268. W. Schiwy, C. Blutau, D. Gathje and B. Krebs, *Z. Anorg. Allg. Chem.*, 1975, **412**, 1.
269. S. Jaulmes, *Acta Crystallogr., Sect. B*, 1974, **30**, 2283.
270. M. Julien-Pouzol and S. Jaulmes, *Acta Crystallogr., Sect. B*, 1979, **35**, 2672.
271. J.C. Jumas, E. Philippot and M. Maurin, *Acta Crystallogr., Sect. B*, 1977, **33**, 3850.
272. J.C. Jumas, M. Ribes, M. Maurin and E. Philippot, *Ann. Chim. (Paris)*, 1978, **3**, 125.
273. R.R Holmes, S. Shafieezad, V. Chandrasekhar, A.C. Sau, J.M. Holmes and R.O. Day, *J. Amer. Chem. Soc.*, 1988, **110**,1168.
274. A.C. Sau, R.O. Day and R.R. Holmes, *J. Amer. Chem. Soc.*, 1981, **103**, 1264; *Inorg. Chem.*, 1981, **20**, 3076.
275. C. Glidewell and D.C. Liles, *Acta Crystallogr., Sect. B*, 1978, **34**, 1693.
276. C. Glidewell and D.C. Liles, *J. Chem. Soc., Chem. Commun.*, 1979, 93; *J. Organomet. Chem.*, 1979, **174**, 275; *Acta Crystallogr., Sect. B*, 1979, **35**, 1689.
277. N. Kasai, K. Yasuda and R. Okawara, *J. Organomet. Chem.*, 1965, **3**, 172.
278. C. Glidewell and D.C. Liles, *Acta Crystallogr., Sect. B*, 1978, **34**, 129.
279. A.M. Domingos and G.M. Sheldrick, *Acta Crystallogr., Sect. B*, 1974, **30**, 579.
280. A.G. Davies and R.J. Puddephatt, *J. Chem. Soc. (C)*, 1967, 2663.
281. E. Amberger and R. Hönigschmidt-Grossich, *Chem. Ber.*, 1965, **98**, 3795; 1969, **102**, 3589.
282. P.F.R. Ewings, P.G. Harrison, T.J. King, R.C. Phillips and J.A. Richards, *J. Chem. Soc., Trans.*, 1975, 1950.
283. P.G. Harrison, *Inorg. Chem.*, 1970, **9**, 175.
284. P.G. Harrison, R.C. Phillips and E.W. Thornton, *J. Chem. Soc., Chem. Commun.*, 1977, 603.
285. C. Glidewell and D.C. Liles, *Acta Crystallogr.*, 1978, **B74**, 129.
286. J.M. Brown, A.C. Chapman, R. Harper, D.J. Mowthorpe, A.G. Davies and P.J. Smith, *J. Chem. Soc., Dalton Trans.*, 1972, 338.
287. K.C. Molloy, F.A.K. Nasser, C.L. Barnes, D. van der Helm and J.J. Zuckerman, *Inorg. Chem.*, 1982, **21**, 960.
288. P.G. Harrison, K. Lambert, T.J. King and B. Majee, *J. Chem. Soc., Dalton Trans.*, 1983, 363; *et loc. cit.*
289. K.C. Molloy, T.G. Purcell, K. Quill and I.W. Nowell, *J. Organomet. Chem.*, 1984, **267**, 237.

290. K.C. Molloy, K. Quill and I.W. Nowell, *J. Chem. Soc., Dalton Trans.*, 1987, 101.
291. J.F. Vollano, R.O. Day, D.N. Rau, V. Chandrasekhar and R.R. Holmes, *Inorg. Chem.*, 1984, **23**, 3153.
292. D.R. Senn, J.A. Gladysz, K. Emmerson and R.D. Larsen, *Inorg. Chem.*, 1987, **26**, 2737.
293. P.G. Harrison and R.C. Phillips, *J. Organomet. Chem.*, 1979, **182**, 37.
294. V.G. Kumar Das, C. Wei, S.W. Ng and T.C.W. Mak, *J. Organomet. Chem.*, 1987, **322**, 33.
295. R. Kniep, D. Mootz, U. Severin and H. Wunderlich, *Acta Crystallogr., Sect. B*, 1982, **38**, 2022.
296. W.J. Moore and L. Pauling, *J. Amer. Chem. Soc.*, 1941, **63**, 1392.
297. A. Byström, *Arkiv. Kemi.*, 1943, **17B**, No. 8.
298. S. Grimvall, *Acta Chem. Scand.*, 1973, **27**, 1447.
299. J.D. Kennedy, *J. Mol. Struct.*, 1976, **31**, 207; *J. Chem. Soc., Perkin Trans.*, 1977, **2**, 242.
300. T.P. Lockhart, J.C. Calabrese and F. Davidson, *Organometallics*, 1987, **6**, 2479.
301. T.P. Lockhart and F. Davidson, *Organometallics*, 1987, **6**, 2471.
302. F.H. Allen, J.A. Lerbscher and J. Trotter, *J. Chem. Soc. (A)*, 1971, 2507.
303. J.P. Ashmore, T. Chivers, K.A. Kerr and J.H.G. van Roode, *Inorg. Chem.*, 1977, **16**, 191.
304. E.O. Schlemper, *Inorg. Chem.*, 1967, **6**, 2012.
305. P.G. Harrison, T.J. King and R.C. Phillips, *J. Chem. Soc., Dalton Trans.*, 1976, 2317.
306. G.A. Miller and E.O. Schlemper, *Inorg. Chem.*, 1973, **12**, 677.
307. P.G. Harrison, T.J. King and J.A. Richards, *J. Chem. Soc., Dalton Trans.*, 1975, 826.
308. H.H. Anderson, *Inorg. Chem.*, 1964, **3**, 912.
309. G.S. Brownlee, A. Walker, S.C. Nyburg and J.J. Szymanski, *J. Chem. Soc., Chem. Commun.*, 1971, 1073.
310. A.M. Domingos and G.M. Sheldrick, *Acta Crystallogr.*, 1974, **B30**, 519.
311. M. Zeldin and J. Ochs, *J. Organomet. Chem.*, 1975, **86**, 369.
312. K. Jurkschat, C. Mügge, A. Tzschach, A. Zschunke and G.W. Fischer, *Z. Anorg. Allg. Chem.*, 1980, **463**, 123.
313. A. Tzschach, K. Jurkschat and C. Mügge, *Z. Anorg. Allg. Chem.*, 1982, **492**, 135.
314. R.G. Swisher, R.O. Day and R.R. Holmes, *Inorg. Chem.*, 1983, **22**, 3692.
315. C.D. Garner, P. Sutton and S.C. Wallwork, *J. Chem. Soc. (A)*, 1967, 1949.
316. N.W. Alcock and V.L. Tracy, *Acta Crystallogr., Sect. B*, 1979, 80.
317. H. Puff, H. Hevendehl, K. Höfer, H. Reuter and W. Schuh, *J. Organomet. Chem.*, 1985, **287**, 163.
318. P.G. Harrison, M.J. Begley and K.C. Molloy, *J. Organomet. Chem.*, 1980, **186**, 213.
319. H. Puff, I. Bung, E. Friedrichs and A. Jansen, *J. Organomet. Chem.*, 1983, **254**, 23.
320. C.D. Garner, B. Hughes and T.J. King, *Inorg. Nucl. Chem. Lett.*, 1976, **12**, 859.
321. R. Graziani, G. Bambieri, E. Forsellini, P. Furlan, V. Peruzzo and G. Tagliavini, *J. Organomet. Chem.*, 1977, **125**, 43.
322. R. Faggiani, J.P. Johnson, I.D. Brown and T. Birchall, *Acta Crystallogr., Sect. B*, 1978, **34**, 3743.
323. J.F. Vollano, R.O. Day and R.R. Holmes, *Organometallics*, 1984, **3**, 745.
324. Y.M. Chow, *Inorg. Chem.*, 1971, **10**, 673.
325. H. Matsada, A. Kashina, S. Matsuda, N. Kasai and K. Jitsumori, *J. Organomet. Chem.*, 1972, **34**, 341.
326. R. Hämäläinen and U. Turpeinen, *J. Organomet. Chem.*, 1987, **333**, 323.
327. C. Lecompte, J. Protal, and M. Devaud, *Acta Crystallogr.*, 1976, **B32**, 923.
328. R.R. Holmes, S. Shafieezad, V. Chandrasekhar, J.M. Holmes and R.O. Day, *J. Amer. Chem. Soc.*, 1988, **110**, 1174.
329. K.C.K. Swamy, R.O. Day and R.R. Holmes, *J. Amer. Chem. Soc.*, 1987, **109**, 5546.
330. R.O. Day, J.M. Holmes, V. Chandrasekhar and R.R. Holmes, *J. Amer. Chem. Soc.*, 1987, **109**, 940.
331. H. Puff, W. Schuk, R. Sievers and R. Zimmer, *Angew. Chem., Int. Edn. Engl.*, 1981, **20**, 591.
332. S. Masamune, L.R. Sita and D.J. Williams, *J. Amer. Chem. Soc.*, 1983, **105**, 630.
333. G.D. Andreeth, G. Bocelli, G. Calestani and P. Sgarabotto, *J. Organomet. Chem.*, 1984, **273**, 31.
334. N.G. Bokii, Yu.T. Struchkov, D.N. Kravtsov and E.M. Rokhlina, *J. Struct. Chem.*, 1973, **14**, 258.
335. M.E. Cradwick, R.D. Taylor and J.L. Wardell, *J. Organomet. Chem.*, 1974, **66**, 43.

336. P.L. Clarke, M.E. Cradwick and J.L. Wardell, *J. Organomet. Chem.*, 1973, **63**, 279.
337. N.G. Furmanova, A.S. Batsanov, Yu.T. Struchkov, D.N. Kravtsov and E.M. Rokhlina, *J. Struct. Chem.*, 1979, **20**, 245.
338. A.-F. Shihada and F. Weller, *J. Organomet. Chem.*, 1988, **342**, 177.
339. N.G. Bokii, Yu.T. Struchkov, D.N. Kravtsov and E.M. Rokhlina, *J. Struct. Chem.*, 1973, **14**, 458.
340. O.A. D'yachenko, A.B. Zolotoi, L.O. Atovmyan, R.G. Mirskov and M.G. Voronkov, *Dokl. Phys. Chem.*, 1977, **237**, 1142.
341. B. Krebs and H.J. Jacobsen, *J. Organomet. Chem.*, 1979, **178**, 301.
342. B. Menzbach and P. Blackmann, *J. Organomet. Chem.*, 1975, **91**, 291.
343. H.J. Jacobsen and B. Krebs, *J. Organomet. Chem.*, 1977, **136**, 333.
344. M. Dräger, A. Blecher, H.J. Jacobsen, and B. Krebs, *J. Organomet. Chem.*, 1978, **161**, 319.
345. H. Puff, R. Gattermeyer, R. Hundt and R. Zimmer, *Angew. Chem., Int. Edn. Engl.*, 1977, **16**, 547.
346. H. Puff, A. Bongartz, R. Sievers and R. Zimmer, *Angew. Chem., Int. Edn. Engl.*, 1978, **17**, 939.
347. R.H. Benno and C.J. Fritchie, *J. Chem. Soc., Dalton Trans.*, 1973, 543.
348. A. Haas, H.-J. Kutsch and C. Knüger, *Chem. Ber.*, 1987, **120**, 1045.
349. H. Berwe and A. Haas, *Chem. Ber.*, 1987, **120**, 1175.
350. S. Pohl, *Angew. Chem., Int. Edn. Engl.*, 1976, **15**, 162.
351. D. Kobett, E.F. Paulus and H. Scherer, *Acta Crystallogr., Sect. B*, 1972, **28**, 2323.
352. G.M. Sheldrick and W.S. Sheldrick, *J. Chem. Soc. (A)*, 1970, 490.
353. G.M. Sheldrick, W.S. Sheldrick, R.F. Dalton and K. Jones, *J. Chem. Soc. (A)*, 1970, 493.
354. K.C. Molloy, M.B. Hossain, D. van der Helm, J.J. Zuckerman and I. Haiduc, *Inorg. Chem.*, 1979, **18**, 3507.
355. T. Kimura, N. Yasuoka, N. Kasai and M. Kakudo, *Bull. Chem. Soc. Jpn.*, 1972, **45**, 1649.
356. J.S. Morris and E.O. Schlemper, *J. Cryst. Mol. Struct.*, 1980, **9**, 13.
357. B.W. Liebich and M. Tomassini, *Acta Crystallogr., Sect. B*, 1978, **34**, 944.
358. C. Silvestru, I. Haiduc, S. Klima, U. Thewalt, M. Gielen and J.J. Zuckerman, *J. Organomet. Chem.*, 1987, **327**, 181.
359. P. Livant, M.L. Mckee and S.D. Worley, *Inorg. Chem.*, 1983, **22**, 895.
360. E.A.V. Ebsworth, E.K. Murray, D.W.H. Rankin and H.E. Robertson, *J. Chem. Soc., Dalton Trans.*, 1981, 1501.
361. R. Varma, K.R. Ramaprasad and J.F. Nelson, *J. Chem. Phys.*, 1975, **63**, 915.
362. J.R. Durig, M. Jalilian, Y.S. Li and R.O. Cater, *J. Mol. Struct.*, 1979, **55**, 177.
363. M. Baudler and H. Suchomel, *Z. Anorg. Allg. Chem.*, 1983, **505**, 39.
364. M. Baudler and H. Suchomel, *Z. Anorg. Allg. Chem.*, 1983, **505**, 39.
365. B. Mathiasch and M. Dräger, *Angew. Chem., Int. Edn. Engl.*, 1978, **17**, 767.
366. B. Mathiasch, *J. Organomet. Chem.*, 1979, **165**, 295.
367. M. Dräger and B. Mathiasch, *Angew. Chem., Int. Edn. Engl.*, 1981, **20**, 1029.
368. H. Schumann and H. Banda, *Angew. Chem., Int. Edn. Engl.*, 1969, **8**, 989; 1968, **7**, 813.
369. M. Baudler and H. Suchomel, *Z. anorg. allg. Chem.*, 1983, **505**, 39.
370. M.F.A. Dove, R. King and T.J. King, *J. Chem. Soc., Chem. Commun.*, 1973, 944.
371. D. Mootz and H. Puhl, *Acta Crystallogr.*, 1967, **23**, 471.
372. R. Kniep, D. Mootz, U. Severin and H. Wunderlich, *Acta Crystallogr.*, 1982, **838**, 2022.
373. P.F.R. Ewings, P.G. Harrison, A. Morris and T.J. King, *J. Chem. Soc., Dalton Trans.*, 1976, 1602.
374. T. Birchall and J.J. Johnson, *J. Chem. Soc., Dalton Trans.*, 1981, 69.
375. T. Birchall, R. Faggiani, C.J.L. Lock and V. Manivannan, *J. Chem. Soc., Dalton Trans.*, 1987, 1675.
376. M. Nardelli, C. Pelizzi, G. Pelizzi and P. Tarasconi, *Z. anorg. allg. Chem.*, 1977, **431**, 250.
377. H. Puff, C. Bach, H. Reuter and W. Schauch, *J. Organomet. Chem.*, 1984, **277**, 17.
378. S. Adams and M. Dräger, *Angew. Chem., Int. Edn. Engl.*, 1987, **26**, 1255.
379. P.G. Harrison, in *The Chemistry of Inorganic Homo- and Heterocycles*, Vol. 1, eds. I. Haiduc and D.B. Sowerby, Academic Press, New York, 1987, 377.
380. M. Dräger, B. Mathiasch, L. Ross and M. Ross, *Z. anorg. allg. Chem.*, 1983, **506**, 99.
381. V.K. Belsky, N.N. Zemlyansky, N.D. Kolosora and I.V. Borisova, *J. Organomet. Chem.*, 1981, **215**, 41.
382. M.F. Lappert, W.P. Leung, C.L. Raston, A.J. Thorne, B.W. Skelton and A.H. White, *J. Organomet. Chem.*, 1982, **233**, C25.

383. S. Adams, M. Dräger and B. Mathiasch, *J. Organomet. Chem.*, 1987, **326**, 173.
384. K. Jurkschat, A. Tzschach, C. Mügge, J. Piret-Meunier, M. van Meerssche, C. van Binst, C. Wynants, M. Gielen and R. Willem, *Organometallics*, 1988, **7**, 593.
385. R.D. Ball and D. Hall, *J. Organomet. Chem.*, 1973, **56**, 209.
386. R.A. Anderson and F.W.B. Einstein, *Acta Crystallogr.*, 1976, **B32**, 966.
387. P.A. Edwards and J.D. Corbett, *Inorg. Chem.*, 1977, **16**, 903.
388. S.C. Critchlow and J.D. Corbett, *J. Chem. Soc., Chem. Commun.*, 1981, 236.
389. J.D. Corbelt and P.A. Edwards, *J. Amer. Chem. Soc.*, 1977, **99**, 3313.
390. L. Diehl, K. Khodadadeh, D. Kummer and J. Strachle, *Z. Naturforsch.*, 1976, **31B**, 522.
391. R.C. Burns and J.D. Corbelt, *Inorg. Chem.*, 1985, **24**, 1489.
392. R.C. Haushalter, M.M. J. Treacy and S.B. Rice, *Angew. Chem., Int. Edn. Engl.*, 1987, **26**, 1155.
393. S.C. Critchlow and J.D. Corbett, *Inorg. Chem.*, 1982, **21**, 3286.
394. R.C. Burns and J.D. Corbett, *J. Amer. Chem. Soc.*, 1982, **104**, 2804.

3 Investigating tin compounds using spectroscopy

P.G. HARRISON

The late Professor Jerold Zuckerman, who himself made substantial contributions to the spectroscopy of tin compounds, once remarked that tin compounds could be studied by more techniques than the compounds of any other element. Whilst this statement might be regarded by some as controversial and even provocative, nevertheless there is no doubt that, with distinctive infrared and Raman spectra, two spin-$\frac{1}{2}$ nmr nuclei of moderate abundance, a Mössbauer active nuclide, and ten stable naturally-occurring isotopes giving rise to a very characteristic isotopic distribution for mass spectrometry, tin compounds are relatively easy to investigate spectroscopically. The intention in this chapter is not to dwell upon background theory, only the necessary minimum of which is included, but to describe the utility of these four major spectroscopic techniques in the investigation of structure, bonding, and other processes which can be probed thereby. The use of e.s.r. spectroscopy in tin chemistry is discussed in Chapter 9.

3.1 Infrared spectroscopy

The positions of typical tin-element stretching vibrations are listed in Table 3.1. Band positions are dominated by the mass effect, falling to lower frequencies as the element mass increases. Thus, whereas the Sn–H stretching mode is found in the approximate range 1820–1920 cm^{-1}, stretching vibrations involving first-row elements (C, N, O, F) characteristically occur in the range *c.* 500–600 cm^{-1}, those involving second-row elements (Si, P, S, Cl) in the range *c.* 300–380 cm^{-1}, those involving third-row elements (Ge, As, Se, Br) in the range 180-260 cm^{-1}, and those involving fourth-row elements (Sn, Sb, Te, I) in the range *c.* 150–210 cm^{-1}. It must, however, always be borne in mind that these ranges are only guidelines for isolated molecules, and other effects such as the nature of the ligand, nature of the substituents at tin or the other element, or involvement in further coordination can sometimes modify the band position quite significantly. For example, the position of ν(Sn–H) in the stannyl halides, SnH_3X, decreases in the order Cl (1948 cm^{-1}) > Br (1928 cm^{-1}) > I (1905 cm^{-1})[1], and the ν(Sn–O) vibrations observed in complexes such as $SnX_4 \cdot 2Ph_3EO$ (E = P, As) (ν(Sn–O) 380–390 cm^{-1} (E = P), 310–320 cm^{-1} (E = As)[2]) or organotin derivatives of more electronegative oxygen ligands such as the methyltin nitrates, $Me_nSn(NO_3)_{4-n}$ ($n =$ 2, 3) (ν(Sn–O) 280 cm^{-1} ($n = 2$), 220 cm^{-1} ($n = 3$))[3], occur at much lower frequencies than those in organotin alkoxides[4] (ν(Sn–O) 500–525 cm^{-1}). Increasing steric bulk of the organic groups has the effect of weakening the tin–tin bond in hexaorganodistannanes, with concomitant reduction in the ν(Sn–Sn) position from 200 cm^{-1} in $Me_3SnSnMe_3$ to only 92 cm^{-1} in hexakis(2, 4, 6-triethylphenyl)distannane[5]. Intermolecular association can have a great effect upon band position. Organotin fluorides in particular are strongly associated via fluorine bridging in the solid state, and the

Table 3.1 Examples of typical ν(Sn–X) stretching frequencies.

Bond	Compound	ν(Sn–X)(cm^{-1})	Ref.
Sn–H	SnH_4	1906	*a*
	SnH_3Cl	1948	*b*
	$MeSnH_3$	1874	*c*
	Me_2SnH_2	1863	*a*
	Me_3SnH	1846	*a, d*
	Me_2SnClH	1877	*e*
Sn–D	$MeSnD_3$	1352	*c*
	Me_3SnD	1325	*d*
Sn–C	$MeSnH_3$	527	*c*
	Me_3SnH	521, 514	*a, d*
	Me_4Sn	526, 506	*f*
	$Me_3SnC{\equiv}CH$	537	*g*
	$Sn(C_3H_5)_4$	493, 455	*h*
	$Sn(CH{:}CH_2)_4$	527, 514	*i*
	$Sn(CH_2CH{:}CH_2)_4$	487, 464	*j*
Sn–Si	$Sn(SiMe_3)_4$	328, 311	*k*
	$Me_3SnSiMe_3$	322	*l*
Sn–Ge	$Me_3SnGePh_3$	225	*l*
	$Et_3SnGePh_3$	230	*l*
Sn–Sn	$Me_3SnSnMe_3$	192	*m*
	$Ph_3SnSnPh_3$	138	*n*
	$Sn(SnPh_3)_4$	103	*n*
	$Me_4Sn_2(O_2CR)_2$ ($R = H, CH_3, CH_2Cl, CHCl_2, CCl_3$)	211–216	*o*
Sn–N	Me_3SnNMe_2	618	*p*
	Bu_3SnNRR'	584–603	*q*
	$Sn(NMe_2)_2$	440	*r*
	$^tBu_3SnNH_2$	527	*s*
	$Sn(NMe_2)_4$	535	*t*
Sn–P	$(Me_3Sn)_3P$	284, 351	*u*
	$(Ph_3Sn)_3P$	296, 347	*u*
	$Me_3SnP^tBu_2$	293, 370	*v*
Sn–As	$R^1_3SnAsR^2_2$ ($R^1, R^2 = Me, Ph$)	180–203	*w*
	$(R_3Sn)_3As$ ($R = Me$ or Ph)	240, 211	*w*
Sn–Sb	$Sb(SnMe_3)_3$	183	*x*
Sn–O	Bu_3SnOR	500–525	*y*
	$Sn(OMe)_2$	570	*z*
	$Sn(OEt)_2$	578	*z*
Sn–S	$Me_3SnSSnMe_3$	367, 322	*aa*
	$Me_2Sn(SMe)_2$	340	*aa*
	$(Me_2SnS)_3$	319, 342, 359, 363	*aa*
	Me_3SnSMe	338	*aa*
	$Ph_4Sn_4S_6$	319, 333, 340	*aa*
	Ph_3SnSPh	348	*bb*
	$Ph_3SnSSnPh_3$	376, 330	*bb*
	$(Ph_2SnS)_3$	371, 321	*bb*
	$Sn(SMe)_2$	361	*cc*
Sn–Se	$Me_nSn(SeMe)_{4-n}$	263–226	*dd*
	$Me_3SnSeSnMe_3$	240, 225	*bb, ee*
	$Me_3SnSeSnMe_3 \cdot M(CO)_5$ ($M = Cr, Mo, W$)	232–210	*ee*

Table 3.1 (*Contd.*)

Bond	Compound	ν(Sn–X)(cm^{-1})	Ref.
Sn–Se	$(Me_2SnSe)_3$	267, 255	*bb*
	$Ph_3SnSePh$	241	*bb*
Sn–Te	$Me_3SnTeSnMe_3$	190.182	*ff*
	$Me_3SnTeSnMe_3 \cdot M(CO)_5$ (M = Cr, Mo, W)	187–180	*ff*
	$(Me_2SnTe)_3$	173–203	*gg*
Sn–F	R_3SnF (R = Me, Et, ^{n}Pr, ^{n}Bu, Ph)	340–377 (solid) 555–588 (gas phase)	*hh*
	Me_2SnF_2	365	*ii*
Sn–Cl	R_nSnCl_{4-n} (R = Me, Et, ^{n}Bu)	328–382	*jj*
Sn–Br	R_nSnBr_{4-n} (R = Me, Et, ^{n}Bu)	222–264	*jj*
Sn–I	R_nSnI_{4-n} (R = Me, Et, ^{n}Bu)	174–207	*jj*
Sn–TM	$Me_3SnMn(CO)_5$	182	*l*
	$Cl_3SnMn(CO)_5$	201	*l*
	$Me_3SnFe(CO)(C_5H_5)$	185	*l*
	$Me_3SnCo(CO)_4$	176	*l*

(*a*) D.C. McKean, A.R. Morrison and P.W. Clark, *Spectrochim. Acta*, 1985, **41A**, 1467.
(*b*) J.M. Bellama and R.A. Gsell, *Inorg. Nucl. Chem. Lett.*, 1971, **7**, 365.
(*c*) H. Kimel and C.R. Dillard, *Spectrochim. Acta*, 1968, **24A**, 909.
(*d*) Y. Imai and K. Aida, *Bull. Chem. Soc. Jpn*, 1982, **55**, 999.
(*e*) A.K. Sawyer, J.E. Brown and G.S. May, *J. Organomet. Chem.*, 1968, **11**, 192.
(*f*) F. Watari, *Spectrochim. Acta*, 1978, **34A**, 1239.
(*g*) A.V. Belykov, E.T. Bogoradovskii, V.S. Zavgorodnii, G.M. Apal'kova, V.S. Nikitin and L.S. Khaikin, *J. Mol. Struct.*, 1983, **98**, 27.
(*h*) B. Busch and K. Dehnicke, *J. Organomet. Chem.*, 1974, **67**, 237.
(*i*) G. Masetti and G. Zerbi, *Spectrochim Acta*, 1970, **16A**, 1891.
(*j*) G. Davidson, P.G. Harrison and E.M. Reilly, *Spectrochim. Acta*, 1972, **29A**, 1265.
(*k*) H. Bürger, U. Goetze and W. Sawodny, *Spectrochim Acta.*, 1970, **26A**, 685.
(*l*) N.A.D. Carey and H.C. Clark, *Chem. Commun.*, 1967, 292.
(*m*) B. Fontal and T.G. Spiro, *Inorg. Chem.*, 1971, **10**, 9.
(*n*) P.A. Bulliner, C.O. Quicksall and T.G. Spiro, *Inorg. Chem.*, 1971, **10**, 13.
(*o*) B. Mathiasch and T.N. Mitchell, *J. Organomet. Chem.*, 1980, **185**, 351.
(*p*) A. Marchand, M.T. Forel and M. Riviere-Baudet, *J. Organomet. Chem.*, 1978, **156**, 341.
(*q*) A. Marchand, C. Lemerle, M.T. Forel and M.H. Soulard, *J. Organomet. Chem.*, 1972, **42**, 353.
(*r*) P. Foley and M. Zeldin, *Inorg. Chem.*, 1975, **14**, 2264.
(*s*) H.J. Götze, *Angew. Chem., Int. Edn. Engl.*, 1974, **13**, 88.
(*t*) H. Bürger and W. Sawodny, *Spectrochim. Acta*, 1967, **23A**, 2841.
(*u*) G. Englehardt, P. Reich and H. Schumann, *Z. Naturforsch.*, 1967, **22B**, 352.
(*v*) H. Schumann and L. Rösch, *Chem. Ber.*, 1974, **107**, 854.
(*w*) H. Schumann and A. Roth, *Chem. Ber.*, 1969, **102**, 3713.
(*x*) H. Schumann, H.J. Breunig and V. Frank, *J. Organomet. Chem.*, 1973, **60**, 279.
(*y*) A. Marchand, J. Mendelsohn and J. Valade, *Compt. Rend.*, 1964, **259**, 1737.
(*z*) J.S. Morrison and H.M. Haendler, *J. Inorg. Nucl. Chem.*, 1967, **29**, 393.
(*aa*) P.G. Harrison and S.R. Stobart, *J. Organomet. Chem.*, 1973, **47**, 89.
(*bb*) H. Schumann and P. Reich, *Z. Anorg. allg. Chem.*, 1970, **375**, 73; **377**, 63.
(*cc*) P.G. Harrison and S.R. Stobart, *Inorg. Chim. Acta*, 1973, **7**, 306;
(*dd*) J.W. Anderson, G.K. Barker, J.E. Drake and M. Rodger, *J. Chem. Soc., Dalton Trans.*, 1973, 1716.
(*ee*) H. Schumann, R. Mohtachemi and V. Frank, *Chem. Ber.*, 1973, **106**, 1555.

(*ff*) H. Schumann, R. Mohtachemi and V. Frank, *Chem. Ber.*, 1973, **106**, 2049.
(*gg*) A. Blecher and B. Mathiasch, *Z. Naturforsch.*, 1978, **33**, 246.
(*hh*) K. Licht, H. Geissler, P. Koehler, K. Hottmann, H. Schnorr and H. Kriegsmann, *Z. anorg. allg. Chem.*, 1971, **385**, 271.
(*ii*) M. Goldstein and W.D. Unsworth, *J. Chem. Soc. (A)*, 1971, 2121.
(*jj*) R.J.H. Clark, A.G. Davis and R.J. Puddephatt, *J. Chem. Soc. (A)*, 1968, 1828.

position of the ν(Sn–F) band for isolated molecules in the gas phase occurs at 555–588 cm^{-1}, but in the solid ν(Sn–F) is shifted[6] to 340–377 cm^{-1}. Association is less pronounced for organotin chlorides and the heavier halides, but nevertheless present to some extent in the solid, and hence spectra of isolated molecules can only usually be obtained from solution[7], in the gas phase[6], or by matrix isolation[8,9].

In cases where assignment is difficult, ambiguities may be resolved by ^{116}Sn–^{124}Sn isotopic substitution. A particular example is for phenyltin compounds, where isotopic substitution produces shifts of the order of 5 cm^{-1}, permitting unequivocal assignments of vibrations involving tin atoms (Table 3.2)[10].

The effect of temperature is dependent on the nature of the compound. For solids comprising isolated molecules (e.g. Ph_4Sn) or weakly bound small aggregates (e.g. Ph_3SnCl), the spectra are essentially temperature-independent. However, appreciable frequency shifts have been observed for compounds with strongly associated lattices (e.g. Me_3SnF, Me_2SnF_2, and Me_2SnCl_2). In particular, for these associated com-

Table 3.2 Far infrared data for isotopically pure [^{116}Sn] and [^{124}Sn] phenyltin compounds (cm^{-1})[a,b].

$^{116}SnPh_4$	$^{124}SnPh_4$	Assignment
396w	396w	w_1
389w	389w	} w
270s	**265**s	} ν_{asym}(SnPh)
265 (sh)	**259** (sh)	
221w	221w	ν_{sym}(SnPh)
209m	209m	} δ(SnPh)
192m	192m	} (or phenyl *u*)
151w	151w	
$^{116}SnPh_3I$	$^{124}SnPh_3I$	
271s	**266**s	ν_{asym}(SnPh)
262 (sh)	262 (sh)	
242m	**238**m	ν_{sym}(SnPh)
235m	235m	phenyl *u*
159w	**154**w	ν(SnI)
$^{116}SnPh_3(O_2CMe)$	$^{124}SnPh_3(O_2CMe)$	
322w	322w	δ(OCO)
284 (sh)	**280** (sh)	}
274s	**269**s	} ν_{asym}(SnPh)
265 (sh)	**260** (sh)	}
211m	211m	phenyl *u*
205w	205w	
450m	450m	phenyl *y*
612w	**608**w	ν(SnO)

(*a*) Data from N.G. Dance, W.R. McWhinnie and R.C. Poller, *J. Chem. Soc. (A)*, 1968, 1828.
(*b*) Bands shifting on isotopic substitution are in bold.

pounds, both symmetric and antisymmetric ν(Sn–C) modes shift to higher frequency as the temperature is lowered, whereas the ν(C–H) modes move in the opposite direction[11].

The quality and extent of vibrational data reported in the literature varies enormously from brief selected data (which are sometimes assigned by guesswork!) to full assignments with normal coordinate analyses. The most closely studied groups of compounds are tetraorganostannanes, tin hydrides, and tin halides and halide anions; compounds which have been subjected to detailed assignment or normal coordinate analysis are listed in Table 3.3, together with force constant data. Detailed assignments have also been made for several other compounds including several tetrahedral molecules (Table 3.4), trigonal bipyramidal $[SnX_5]^-$ (X = Cl, Br) anions (Table 3.5),

Table 3.3 Bond stretching force constant data (Nm^{-1})[a].

Compound/anion	*f*(Sn–C)	*f*(Sn–X)	Ref.
SnH_4		2.27	*a*
$MeSnH_3$	2.124	2.241, 2.217	*b*
$EtSnH_3$	2.10	2.06	*c*
Me_3SnH	2.18	1.99	*a*
Me_4Sn	2.19		*d*
$(CF_3)_4Sn$	1.86		*e*
$Me_3SnC{\equiv}CH$	2.27 (Sn–Me) 2.51 (Sn–C≡)		*f*
Me_3SnNMe_2	2.27	2.850	*g*
$Sn(NMe_2)_4$		3.11	*h*
SnO_2 (N_2 matrix)		5.57	*i*
(Kr matrix)		5.36	*i*
SnO (gas phase)		5.62	*i*
Sn_2O_2 (matrix)		2.59	*i*
Me_3SnCl	2.12		*j*
$^nPrSnCl_3$	2.0	2.13	*k*
$(CH_2{=}CH)_3SnCl$	1.96	1.78	*l*
$(CH_2{=}CH)_3SnBr$	1.95	1.38	*l*
$(CH_2{=}CH)_3SnI$	1.93	1.031	*l*
$Sn(SiMe_3)_4$		1.66	*m*
$Me_3SnSnMe_3$	2.08	1.39	*n*
$Ph_3SnSnPh_3$	3.03	1.17	*o, n*
$[Me_2SnF_4]^{2-}$	2.54	1.24	*p*
$[Me_2SnCl_4]^{2-}$	2.44	0.81	*p*
$[Me_2SnBr_4]^{2-}$	2.37	1.07	*p*
$[SnF_6]^{2-}$		2.30	*p*
$[SnCl_6]^{2-}$		1.09	*p*
$[SnBr_6]^{2-}$		0.84	*p*
$SnCl_4{\cdot}2L$ (L = Me_2O, Et_2O, thf, Me_2S, tht, Et_2S, Me_2Se, tmpa, hmpa)		1.34–1.71	*q*
$SnBr_4{\cdot}2L$ (L = Me_2S, tht, Et_2S, Me_2Se, tmpa, hmpa)		1.16–1.39	*q*
$[NH_4][SnCl_3]$		0.98	*r*
$K[SnCl_3]$		1.16, 1.11	*r*
$M[Sn_2Cl_5]$		1.10 (terminal) 1.11 (bridging)	*r*

(*a*) Y. Imai and K. Aida, *Bull. Chem. Soc. Jpn*, 1982, **55**, 999.
(*b*) H. Kimmel and C.R. Dillard, *Spectrochim. Acta*, 1968, **24A**, 909.
(*c*) J.R. Durig, Y.S. Li, J.F. Sullivan, J.S. Church and C.B. Bradley, *J. Chem. Phys.*, 1983, **78**, 1046.
(*d*) F. Wateri, *Spectrochim. Acta*, 1978, **34A**, 1239.
(*e*) R. Eujen, H. Bürger and H. Oberhammer, *J. Mol. Struct.*, 1981, **71**, 109.
(*f*) A.V. Belyakov, E.T. Bogoradovskii, V.S. Zavgorodnii, G.M. Apal'kova, V.S. Nikitin and L.S. Khaikin, *J. Mol. Struct.*, 1983, **98**, 27.
(*g*) A. Marchand, M.T. Forel and M. Riviere-Baudet, *J. Organomet. Chem.*, 1978, **156**, 341.
(*h*) H. Bürger and W. Sawodny, *Spectrochim. Acta*, 1967, **23A**, 2841, 2827.
(*i*) A. Bos and J.S. Ogden, *J. Phys. Chem.*, 1973, **77**, 1513.
(*j*) H. Kriegsman and S. Pischtschan, *Z. anorg. allg. Chem.*, 1961, **308**, 212.
(*k*) H. Geissler, C. Peuter, R. Heess and H. Kriegsmann, *Z. anorg. allg. Chem.*, 1972, **393**, 230.
(*l*) E. Vincent, L. Verdonck, L. Naessens and G.P. van der Kelen, *J. Organomet. Chem.*, 1984, **277**, 235.
(*m*) H. Bürger, U. Goetze and W. Sawodny, *Spectrochim. Acta*, 1970, **26A**, 685.
(*n*) B. Fontal and T.G. Spiro, *Inorg. Chem.*, 1971, **10**, 9.
(*o*) P.A. Bulliner, C.O. Quicksall and T.G. Spiro, *Inorg. Chem.*, 1971, **10**, 13.
(*p*) C.W. Hobbs and R.S. Tobias, *Inorg. Chem.*, 1970, **9**, 1037.
(*q*) S.J. Ruzicka and A.E. Merbach, *Inorg. Chim. Acta*, 1976, **20**, 221.
(*r*) S.R.A. Bird, J.D. Donaldson, S.D. Ross and J. Silver, *J. Inorg. Nucl. Chem.*, 1974, **36**, 934.

Table 3.4 Vibrational assignments for tetrahedral SnX_4 molecules (cm^{-1}).

Compound	ν_1	ν_2	ν_3	ν_4	Ref.
SnH_4	–	758	1902	678	*a*
SnD_4	–	539	1368	487	*a*
$SnCl_4$(liq)	367	104	403	129	*b – h*
$SnCl_4$ (solid)	359.2	105.6	398	124.6	
	361.8	109.4	405.6	128.7	
	363.8	115.3	410.4	130.7	
	366.5			135.9	
	369.6				
$SnBr_4$ (liq)	220	54	279	63	
(solid)	218.1	67.0	274.5	84.5	
	218.9	71.0	278.0	86.5	
	219.6		282.5	88.5	
	220.4		285.5		
			287.5		
SnI_4 (CCl_4 soln)	149	47	216	63	
(solid)	–	–	208	64.5, 60.5	
$Sn(NMe_2)_4$	516	–	535		*i*
$Sn(SiMe_3)_4$	311	58	330	78	*j*

(*a*) D.R. Lide and D.E. Mann, *J. Chem. Phys.*, 1956, **25**, 1128.
(*b*) M.L. Delwaulle, F. Francois, M.B. Delhaye-Buisset and M. Delhaye, *J. Phys. Radium*, 1954, **15**, 206.
(*c*) R.J.H. Clark and B.J. Hunter, *J. Chem. Soc.* (*A*), 1971, 2999.
(*d*) I.W. Levin, *Spectrochim. Acta*, 1969, **25A**, 1157.
(*e*) H.F. Shurvell, Canad. *J. Spectrosc.*, 1972, **17**, 109.
(*f*) J.A. Creighton and T.J. Sinclair, *Spectrochim. Acta*, 1979, **35A**, 137.
(*g*) H. Stammreich, R. Forneris and Y. Tavares, *J. Chem. Phys.*, 1956, **25**, 1278.
(*h*) H. Stammreich, Y. Tavares and D. Bassi, *Spectrochim. Acta*, 1961, **17**, 661.
(*i*) H. Bürger and W. Sawodny, *Spectrochim. Acta*, 1967, **23A**, 2841.
(*j*) H. Bürger and U. Goetze, *Angew. Chem., Int. Edn. Engl.*, 1968, **7**, 212.

Table 3.5 Vibrational assignments for $[SnX_5]^-$ (X = Cl and Br) anions (cm^{-1})[a].

Anion	ν_1	ν_2	ν_3	ν_4	ν_5	ν_6	ν_7	ν_8
$[SnCl_5]^-$	338	256	330	148	351	160	59	–
$[SnBr_5]^-$	202	154	208	106	257	111	–	103

(*a*) Data from J.A. Creighton and J.H.S. Green, *J. Chem. Soc. (A)*, 1968, 808.

Table 3.6 Vibrational assignments for $[SnX_6]^{2-}$ (X = F, Cl, Br, I and OH) anions (cm^{-1}).

Anion	ν_1	ν_2	ν_3	ν_4	ν_5	ν_6	Ref.
$Na_2[SnF_6]$	592	477	559	300	252		*a*
$Rb_2[SnCl_6]$	316	241	312	173	171	107	*b*
$[Et_4N]_2[SnCl_6]$	309	232	306, 291	163	159		*c*
$[PCl_4][SnCl_6]$	292	220	358, 344 328	162, 148 138	165, 161		*d*
$Mg[SnBr_6]$	183	144			69		*e*
$[Et_4N]_2[SnBr_6]$	182	137	203	111	101		*c*
$[SnBr_6]$(soln)	185	137			95		*e*
$[Et_4N]_2[SnI_6]$			156	90, 79			*c*
$Na_2[Sn(OH)_6]$			530	260	150		*f*

(*a*) G. Begun and A.C. Rutenberg, *Inorg. Chem.*, 1967, **6**, 2212.
(*b*) D.M. Adams and D.M. Morris, *J. Chem. Soc. (A)*, 1967, 1666.
(*c*) R.J.H. Clark, L. Maresca and R.J. Puddephatt, *Inorg. Chem.*, 1968, **7**, 1603.
(*d*) J. Shamir, S. Luski, A. Bino, A. Cohen and D. Gibson, *Inorg. Chem.*, 1985, **24**, 2301.
(*e*) J. Hiraishi, I. Nakagawa and T. Shimanouchi, *Spectrochim. Acta*, 1964, **20**, 819.
(*f*) V. Lorenzelli, T. Dupuis and J. Lecompte, *Compt. Rend.*, 1964, **259**, 1057.

Table 3.7 Tin–carbon and tin–halogen stretching frequencies for alkyltin halide anions (cm^{-1})[a].

Anion	ν(Sn–C)	ν(Sn–Cl)	ν(Sn–Br)	ν(Sn–I)
$Me_3SnCl_2^-$	552vs	227s		
$Me_3SnBr_2^-$	555vs		140m	
$Me_3SnI_2^-$	553vs			134m
$Me_2SnCl_3^-$	573m 518m	313s, 256s sh, 242vs br		
$Me_2SnBr_3^-$	567m 512m		228s sh, 218s	
$Ph_2SnCl_3^-$	229vs 218s	332vs, 281s, 242vs		
$Me_2SnCl_4^{2-}$	580m	227s		
$Me_2SnBr_4^{2-}$	572m		220s	
$Me_2SnI_4^{2-}$	559m			
$MeSnCl_5^{2-}$	534m	318s, 258vs, 215m		186m
$BuSnCl_5^{2-}$	560m 539m	305s, 250vs, 235vs		
$MeSnCl_3Br_2^{2-}$	530m	312s, 294s, 267s	210m, 160s	
$MeSnCl_3I_2^{2-}$	530w	328s, 297vs, 267m sh		187m sh, 163s
$MeSnBr_5^{2-}$	524m		212m, 191m, 160s br	
$MeSnBr_3Cl_2^{2-}$	528m	306s, 248s	209m, 188s sh, 160s br	
$MeSnCl_4I^{2-}$	534m	307s, 261w		163s

(*a*) Data from M.K. Das, J. Buckle and P.G. Harrison, *Inorg. Chim. Acta*, 1972, **6**, 17.

Table 3.8 Tin–carbon and tin–halogen stretching frequencies for alkyltin halides (cm^{-1})[a].

Compound	X = Cl			X = Br			X = I		
ν(Sn–C)									
Me_4Sn		528							
Me_3SnX	542		513	539		511	536		508
Me_2SnX_2	560		542	554		518	542		511
$MeSnX_3$		551			539			527	
Et_4Sn		508							
Et_3SnX	518		489	510		484	506		482
Et_2SnX_2	531		497	528		493	520		490
$EtSnX_3$		520			511				
$Bu^n{}_4Sn$	592		503						
$Bu^n{}_3SnX$	601		513	599s		503	598		501
$Bu^n{}_2SnX_2$	602		515	600m		511	592		508
Bu^nSnX_3	602		522	596w		513			
ν(Sn–X)									
Me_3SnX		331			234			189	
Me_2SnX_2	361		356	250		240	204		186
$MeSnX_3$	382		368	264		235	207		174
Et_4Sn									
Et_3SnX		337			222			182	
Et_2SnX_2	359		352	260		240	198		176
$EtSnX_3$	377		366	260		~253			
$Bu^n{}_4Sn$									
$Bu^n{}_3SnX$		328			228			184	
$Bu^n{}_2SnX_2$	356		340	248		240	196		180
Bu^nSnX_3	378		367	261		~247			
SnX_4		407			280			219	

(*a*) Data from R.J.H. Clark, A.G. Davies and R.J. Puddephatt, *J. Chem. Soc. (A)*, 1968, 1828.

octahedral $[SnX_6]^{2-}$ (X = F, Cl, Br, I, OH) anions (Table 3.6), $[Me_3SnX_2]^-$ (X = Cl, Br, I), $[Me_2SnCl_3]^{-1}$, $[Me_2SnX_4]^{2-}$ and $[RSnX_5]^{2-}$ anions (X = Cl, Br, I) (Table 3.7), alkyltin halides, R_nSnX_{4-n} (R = Me, Et, ^{n}Bu; X = Cl, Br, I) (Table 3.8), tetravinyltin[12], tetraallyltin[13], tetracyclopropyltin[14], tetrabenzyltin, tribenzyltin chloride, dibenzyltin dichloride and dibromide[15], *n*-butyltin halides, ${}^nBu_nSnX_{4-n}$ (X = F, Cl, Br, I; $n = 1$–3)[16], triorganotin fluorides, R_3SnF (R = Me, Et, ^{n}Pr, ^{n}Bu, Ph)[6], Me_2SnF_2[17], Me_3SnSMe[18], $Sn(SR)_4$ (R = Me, Et, ^{i}Pr)[19], tin(II) thiolates, $Sn(SR)_2$ (R = Me, Ph, CH_2Ph) and $SnSCH_2CH_2S$[20], $Me_nSn(SMe)_{4-n}$ ($n = 0$–3)[21], $({}^nBu_3Sn)_2NEt$[22], and ${}^nBu_3SnNRR'$ (R, R′ = alkyl or phenyl)[22].

Matrix isolation can be an extremely useful technique for investigating the spectra of otherwise inaccessible molecules such as monomeric tin (II) halides, $SnCl_2$, $SnBr_2$ and SnClBr, (Table 3.9) and monomeric SnO, SnO_2 and Sn_2O_2 molecules[23,24] which cannot be obtained in any other way. The spectra show that monomeric SnO_2 is linear, and normal coordinate analysis shows that the Sn–O stretching force constants for SnO and SnO_2 correspond to multiple bonds. Hence the bonding in these isolated molecules is not too dissimilar to that in CO and CO_2. Molecules of SnO_2 are formed

Table 3.9 Vibrational assignments for isolated tin (II) halide molecules (cm^{-1})[a].

Molecule	Argon matrix	N_2 matrix	Gas phase
$SnCl_2$ ν_1	353	341	352(p)
ν_2	124		120(p)
ν_3	332	320	
$SnBr_2$ ν_1	244	237	
ν_2	82	84	
ν_3	231	223	
SnClBr ν(Sn–Cl)		328	
ν(Sn–Br)		228	240(p)
δ(ClSnBr)			100(p)

(*a*) Data from G.A. Ozin and A. Vander Voet, *J. Chem. Phys.*, 1972, **56**, 4769.

by the direct insertion of a tin atom into an O_2 molecule, and that the major route to Sn_2O_2 is by the subsequent reaction of SnO_2 with a second tin atom[23], although the aggregation of SnO molecules to give higher Sn_nO_n has also been proposed[24]. Matrix isolation can also be employed in the study of reactions, and in this medium weak interactions between $SnCl_2$ and $SnBr_2$ with CO, NO, and N_2 can be detected[25]. Spectra for matrix-isolated Me_nSnCl_{4-n} ($n = 0$–4) have also been obtained[8,26].

Application of group theoretical principles to tin-ligand vibrations can be employed with some effect to assign stereochemistry. In most cases, the 'local' symmetry approximation applies. Group theoretical predictions for both *cis* and *trans* isomers of five-coordinate trigonal bipyramidal $[MX_3Y_2]$ and six-coordinate octahedral $[MX_4Y_2]$, both common geometries in tin chemistry, are shown in Table 3.10. Thus it is apparent that the five-coordinate anions $[Me_3SnX_2]^-$ (X = Cl, Br, I), with only one tin–carbon and one tin–halogen stretching frequency active in the infrared, have *trans* halogen atoms with a planar $[Me_3Sn]$ moiety. Deviation from planarity results in infrared activity for two tin–carbon stretching modes, and two bands are observed in the infrared. Similarly, linear or bent configurations for a $[Me_2Sn]$ unit will give rise to either one or two infrared-active tin–carbon stretching modes, respectively. Application of these simple guidelines has been used widely to assign the geometries of dimethyl- and trimethyltin derivatives. Typical data are illustrated in Table 3.7 together with the corresponding tin–halogen stretching mode values. *Cis* and *trans* structures for six-coordinated SnX_4L_2 complexes have been distinguished similarly,

Table 3.10 Normal stretching modes of *cis* and *trans* $[MX_3Y_2]$ and $[MX_4Y_2]$ geometries[a].

	Symmetry point group	$\nu_{(M-X)}$	$\nu_{(M-Y)}$
cis-MX_3Y_2	C_{2v}	$2\mathit{A}_1 + \mathit{B}_2$	$\mathit{A}_1 + \mathit{B}_1$
trans-MX_3Y_2	D_{3h}	$A_1' + \mathit{E}'$	$A_1' + \mathit{A}_2'$
cis-MX_4Y_2	C_{2v}	$2\mathit{A}_1 + \mathit{B}_1 + \mathit{B}_2$	$\mathit{A}_1 + \mathit{B}_1$
trans-MX_4Y_2	D_{4h}	$A_{1g} + B_{1g} + \mathit{E}_u$	$A_{1g} + \mathit{A}_{2u}$

(*a*) Infrared-active species are italicized.

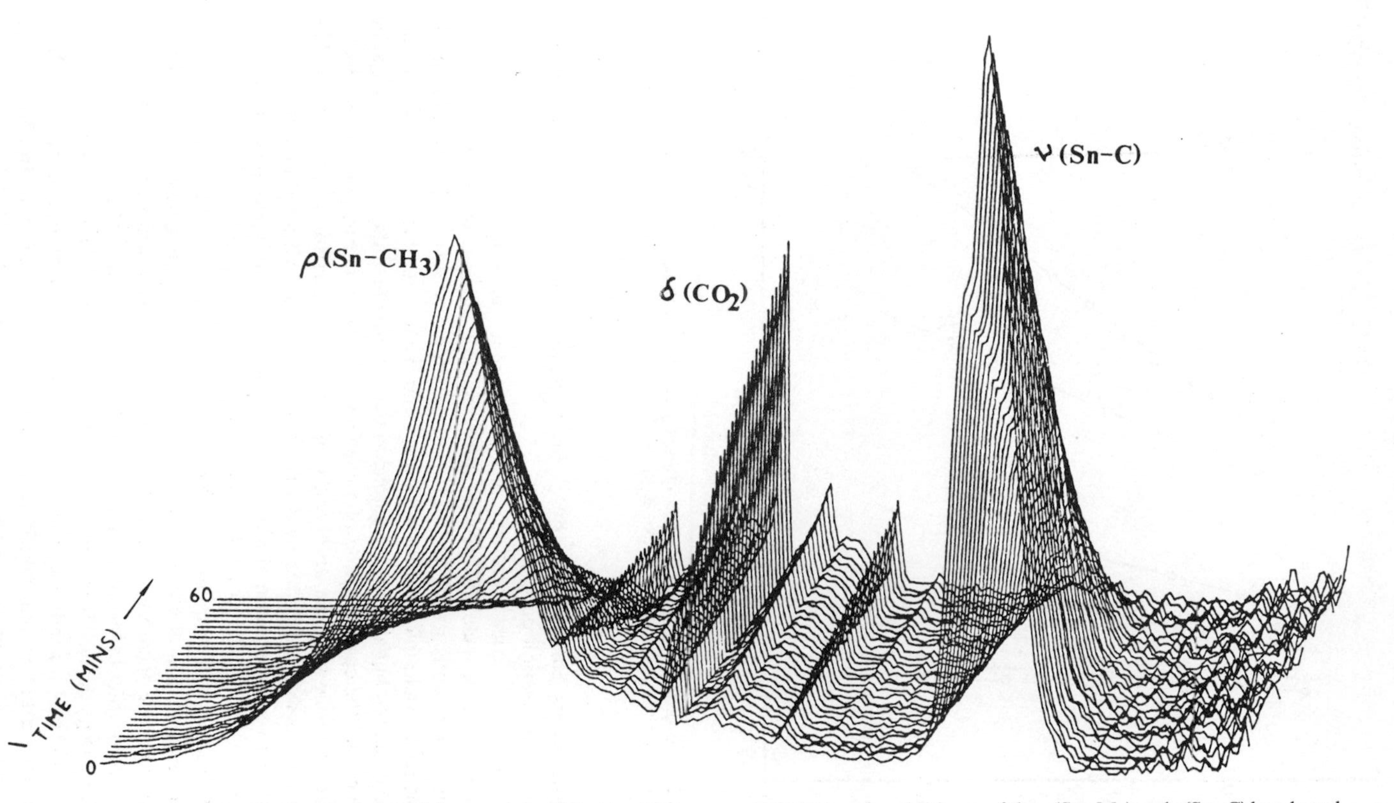

Figure 3.1 Time-resolved infrared spectra of the reaction of Me_4Sn with oxygen at 573 K showing the decay of the ρ(Sn–Me) and ν(Sn–C) bands and the growth of the $\delta(CO_2)$ band.

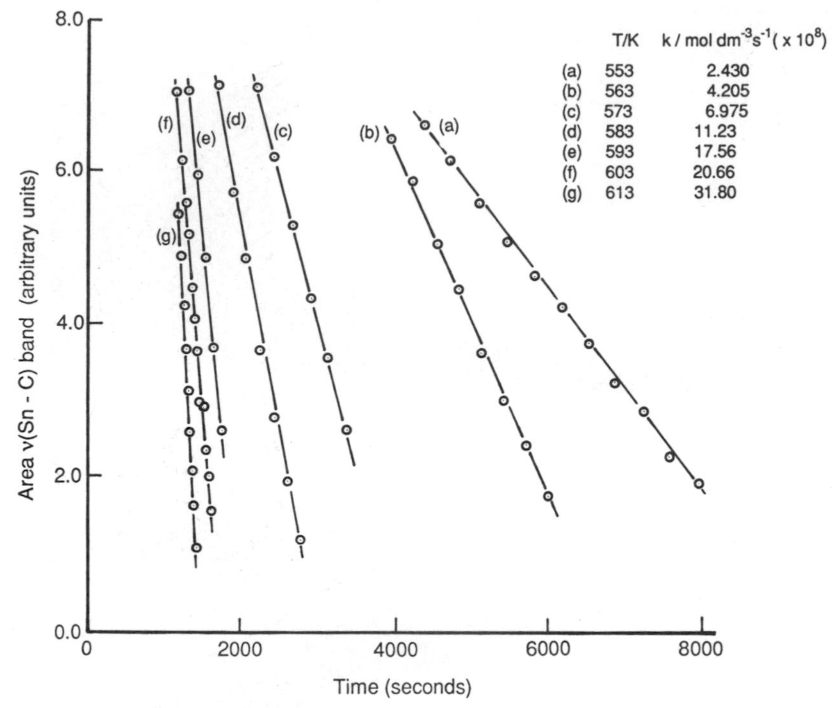

Figure 3.2 Zero-order rate plots of the δ(Sn–C) band absorbance versus time at various temperatures for the reaction of Me_4Sn with oxygen.

with assignments corroborated by ^{116}Sn–^{124}Sn isotopic shifts[27,28]. In solution, *cis–trans* isomerization occurs[29].

The advent of Fourier transform infrared has made possible the facile and rapid collection of spectra at elevated temperatures enabling the thermal decompositions of organotin compounds to be studied[30]. The technique is very useful, since a single experiment can yield information not only about the nature of the reaction products and intermediates, but also reaction stoichiometry, rate constants and thermodynamic data. Typically the concentration of a particular reactant, intermediate or product is monitored by following the absorbance of bands characteristic of the species. The methodology is well illustrated by the reaction of tetramethyltin with oxygen, which proceeds to give as gas-phase products not CO_2 and H_2O, the expected products from exhaustive oxidation, but rather CO_2 and CH_4 in a 1:3 mole ratio. Figure 3.1 shows time-resolved spectra obtained at 583 K showing the decay of the ρ(Sn–CH_3) and ν(Sn–C) fundamentals of tetramethyltin with the concomitant growth of the $\delta(CO_2)$ deformation mode of carbon dioxide. Reaction kinetics in the temperature range 553–613 K are zero-order in loss of tetramethyltin (Figure 3.2), which, together with the low activation energy derived from the linear Arrhenius plot (119 kJ mol^{-1}, cf. $\bar{D}$(Sn–C) for Me_4Sn 273 kJ mol^{-1}) indicates that the reaction is not a homogeneous gas-phase process, but rather is one which is surface-mediated proceeding by dissociative chemisorption of tetramethyltin on the walls of the infrared cell.

3.2 Nuclear magnetic resonance

Tin has three naturally occurring isotopes with spin $\frac{1}{2}$ (all others have spin zero), and details of these are listed in Table 3.11. In practice, only the two more abundant isotopes are of any consequence. Compounds containing tin are usually amenable to study by several nuclei, and spectra can afford a substantial amount of information concerning bonding, stereochemistry and dynamic processes. Because of the differing abundances of the different nmr nuclei present, spectra in which tin is involved in coupling to other nuclei are readily recognizable. This is illustrated by the ^{1}H, ^{13}C, and ^{119}Sn spectra of tetramethyltin shown in Figure 3.3. Both the ^{1}H and ^{13}C spectra comprise a central intense peak which is flanked on either side by two pairs of satellites due to the two-bond coupling to the ^{117}Sn and ^{119}Sn nuclei. In addition, in the ^{1}H spectrum two much weaker pairs of satellites can also be seen, due to two-bond coupling to ^{115}Sn and one-bond coupling to ^{13}C nuclei. The ^{119}Sn spectrum (Figure 3.3*c*) comprises a thirteen-line spectrum due to coupling to the twelve equivalent hydrogen nuclei. The weak features are due to ^{13}C coupling. Since the spectra of organotin compounds containing more complex organic groups will exhibit complicated spectra due to the many couplings present, ^{119}Sn chemical shifts are normally obtained from broad band decoupled spectra (Figure 3.3*d*).

Although both ^{117}Sn and ^{119}Sn are amenable to study, the latter is usually chosen because of its marginally superior abundance and receptivity (note that both are vastly easier than ^{13}C!). The first observations of ^{119}Sn resonances were achieved by CW techniques, but the difficulties allowed the measurement of few spectra. INDOR (Indirect Double Resonance) proved a useful technique for a recording spectra of organotin compounds where coupling between ^{1}H and ^{119}Sn could be identified (or fluorotin compounds with identifiable ^{19}F–^{119}Sn coupling). This method afforded substantial advantages over CW methods, and spectra of over 700 compounds were obtained in this way. Unfortunately, the negative value for the magnetogyric ratio of ^{119}Sn leads to negative nuclear Overhauser factors, and hence the effect cannot be beneficial for the enhancement of signal intensities in double resonance experiments, although it can be useful for assignments. Nowadays, because of the far superior sensitivity and ease of accumulation, all ^{119}Sn spectra are obtained by Fourier transform methods, and ^{119}Sn signals can be significantly enhanced by polarization transfer[32]. However, the INDOR technique is still the most effective method for the determination of the signs of coupling constants, and the signs and magnitudes of reduced coupling constants between tin and other directly bound nuclei are collected in Table 3.12[33].

Table 3.11 NMR properties of tin nuclei with spin $\frac{1}{2}$.

Nucleus	Sn^{115}	Sn^{117}	Sn^{119}
Natural abundance (C/%)	0.35	7.61	8.58
Magnetic moment (μ/μ_N)	−1.590	−1.732	−1.8119
Magnetogyric ratio ($\gamma/10^7$ rad T^{-1} s^{-1})	−8.792	−9.578	−10.021
NMR frequency (Ξ/MHz)	(32.86)	35.632295	37.290662
Relative receptivity			
D^p	1.24×10^{-4}	3.49×10^{-3}	4.51×10^{-3}
D^c	0.705	19.8	25.6
Standard		Me_4Sn	

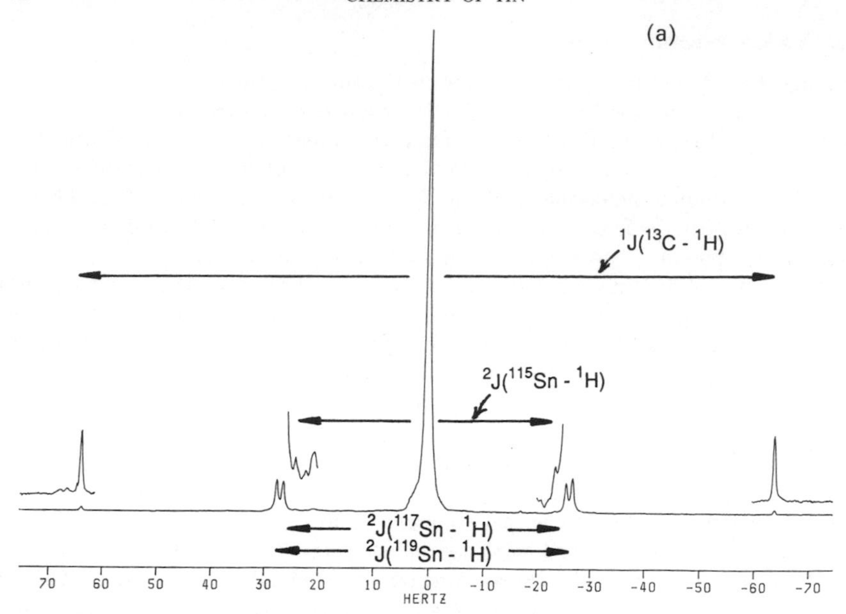

(a)
$^1J(^{13}C - ^1H)$
$^2J(^{115}Sn - ^1H)$
$^2J(^{117}Sn - ^1H)$
$^2J(^{119}Sn - ^1H)$
70 60 50 40 30 20 10 0 -10 -20 -30 -40 -50 -60 -70
HERTZ

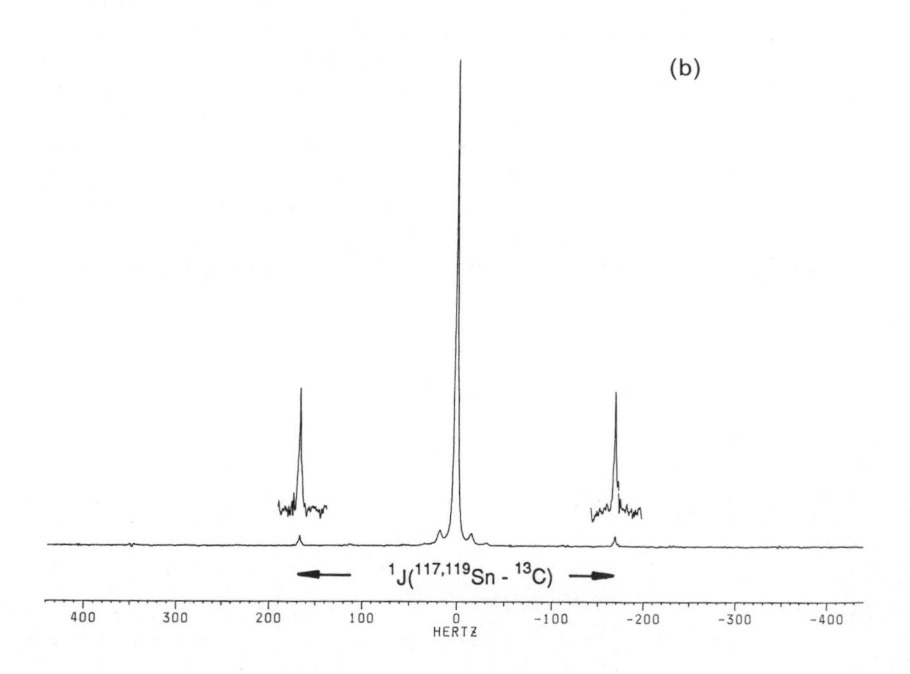

(b)
$^1J(^{117,119}Sn - ^{13}C)$
400 300 200 100 0 -100 -200 -300 -400
HERTZ

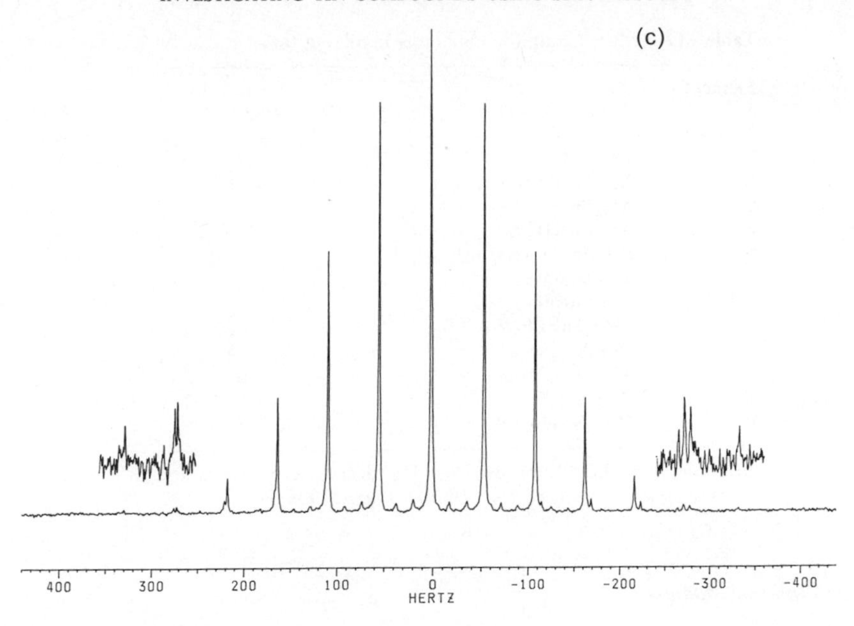

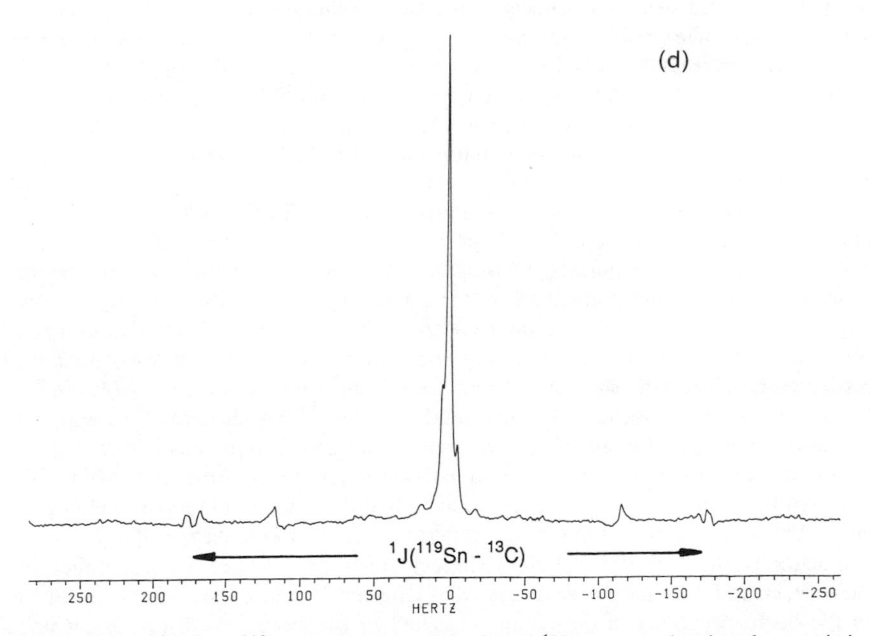

Figure 3.3 ^{1}H-, ^{13}C-, and ^{119}Sn-nmr spectra of Me_4Sn. (*a*) ^{1}H-spectrum showing characteristic satellites from the two-bond couplings to ^{117}Sn and ^{119}Sn as well as the one-bond coupling to ^{13}C. The two-bond coupling to ^{115}Sn can also be seen. (*b*) ^{13}C-spectrum showing one-bond coupling to ^{119}Sn. (*c*) ^{119}Sn-spectrum showing all thirteen lines due to coupling to the twelve methyl hydrogen nuclei. (*d*) As for (*c*), but with broad band decoupling.

Table 3.12 Reduced coupling constants involving tin[a].

Element X	Tin compound	$^1K(Sn{-}X)$ (nm^{-3})
H	Me_3SnH	+38.9
B	$Me_3SnB(NMe_2)_2$	+66.3
C	Me_4Sn	+29.3
N	$Me_3SnNHPh$	−5.85
F	$(PhMe_2CCH_2)_3SnF$	−54.6
Si	$Me_3SnSiMe_3$	+73.7
P^{III}	Me_3SnPPh_2	−32.8
P^{V}	$[Me_3SnPPh_2W(CO)_5]$	−2.8
Se	$(Me_3Sn)_2Se$	−124
Sn	$Me_3SnSnMe_3$	+268
Te	$(Me_3Sn)_2Te$	−98.1
W	$[Me_3SnW(CO)_3(\pi\text{-}C_5H_5)]$	+81

(*a*) Data from J.D. Kennedy, W. McFarlane, G.S. Pyne and B. Wrackmeyer, *J. Chem. Soc., Dalton Trans.*, 1975, 386.

3.2.1 *Solution studies*

3.2.1.1 *Chemical shifts and coupling constants.* Relatively little information can be extracted from ^1H spectra. Chemical shift and coupling constant data for organotin hydrides and corresponding anions[34] are shown in Table 3.13, and illustrate the general positions that the hydrogen directly bound to tin, or in methyl or phenyl groups bound to tin, may be found. The ^{13}C chemical shift data[35–46] are sensitive to various factors, including position of the carbon atom in the alkyl or aryl group, the other substituents attached to tin, the coordination number of the tin, and the donor ability of the solvent.

Tetramethyltin with an absolute resonance frequency Ξ of 37290665 ± 3 Hz is the universally adopted standard for ^{119}Sn chemical shifts. Conventionally, negative chemical shift values are upfield from $Me_4Sn = 0$. Data are to be found in a number of sources[40–57] and, being dominated by the paramagnetic contribution, span a very large range (> 4000 ppm). Some primary isotope effects have been observed, the largest being with the $[SnH_3]^-$ anion, where the ^{119}Sn shift moves by 9.842 ppm on perdeuteration[58] (cf. the shift of 2.86 ppm on going from Me_4Sn to d_{12}-Me_4Sn[59]). Close correlations have been demonstrated between ^{119}Sn chemical shift and the corresponding ^{29}Si, ^{73}Ge, and ^{207}Pb values for analogous compounds[60,61]. T_1 values are rather short, whereas T_2 values are quite varied, and the approximate correlation observed between T_1 and the paramagnetic contribution to the observed shielding of the ^{119}Sn nucleus indicates that the dominant ^{119}Sn T_1 mechanism is spin rotation interaction[62]. As with ^{13}C, ^{119}Sn chemical shifts are affected by a number of parameters, and these have been discussed at length[53,54]. The major factors appear to be the electronegativity of the groups attached to tin, geometric distortions which modify the interbond angles at tin, and the coordination number at tin, although other factors such as the effect of ring currents and local electric fields as well as ligand polarizability and changes in the excitation energy may also be significant in certain cases. The effect of ligand electronegativity is illustrated by the change of chemical shift

Table 3.13 Nmr parameters for organotin hydrides and anions derived therefrom[a].

Compound	δ/ppm			J/Hz		
	C_6H_5	Sn–H	SnC–H	^{117,119}Sn–H	^{117,119}Sn–C–H	H–Sn–C–H
SnH_4		−3.85		1846, 1931		
$[SnH_3]^-$		−1.68		104.9, 109.4		
$[MeSnH_3]$		−4.14	−0.27	1770, 1852	62	2.7
$[MeSnH_2]^-$		−2.78	+0.44	117.8, 122.1	27.7	4.6
$[Me_2SnH_2]$		−4.76	−0.17	1682, 1758	55.5, 58.0	2.55
$[Me_2SnH]^-$		−3.59	+0.24		18.2	4.0
$[Me_3SnH]$		−4.73	−0.18	1664, 1744	54.5, 56.5	2.37
$[Me_3Sn]^-$			+0.57		14.2	
$[(PhCH_2)_2SnH_2]$	−6.93	−5.21	−2.11	1733, 1814	63.1, 66.2	1.7
$[(PhCH_2)_2SnH]$	−6.89, −6.25	−4.22	−2.01			3.8
$[(PhCH_2)_3SnH]$	−6.86	−5.70	−2.19	1697, 1772	59.7, 62.7	1.5
$[PhSnH_3]$		−4.93		1836.7, 1921.5		
$[PhSnH_2]^-$	−7.53, −6.78	−4.28				
$[Ph_2SnH_2]$		−6.09		1842.0, 1927.8		
$[Ph_2SnH]^-$	−7.59, −6.88	−6.25		142.0, 148.5		
$[Ph_3SnH]$	−7.83, −7.49	−6.83		1850.7, 1935.8		
$[Ph_3Sn]^-$	−7.43, −6.8					
$[PhMeSnH_2]$	−7.23, −6.99	−5.10	−0.61	1771, 1835	57.5, 60.2	2.6
$[Ph_2MeSnH]$	−7.50, −7.13	−6.17	−0.34	1788, 1872	56.0, 58.5	2.2

(*a*) Data from T. Birchall and A.R. Pereira, *J. Chem. Soc., Dalton Trans.*, 1975, 1087.

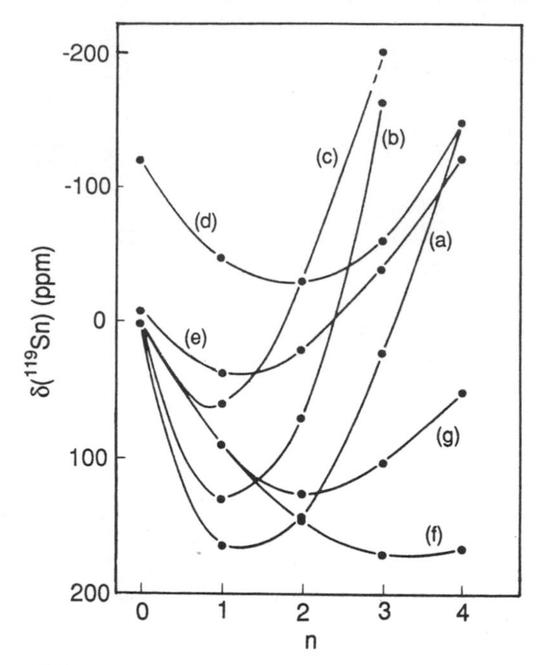

Figure 3.4 Plots of $\delta(^{119}Sn)$ versus *n* for various $R_{4-n}SnX_n$ series. R, X = Me, Cl (*a*); Me, Br (*b*); Bu, O^{t}Bu (*c*); Ph, Cl (*d*); Bu, NEt_2 (*e*); Me, SMe (*f*); and Me, S^{t}Bu (*g*).

along the series $Me_3SnCH_nCl_{3-n}$ as the value of *n* changes from 3 to 0 when $\delta(^{119}Sn)$ decreases by 85 ppm[63], and by the correlation between $\delta(^{119}Sn)$ and the Hammett σ-constant of the aromatic substituent in aryltin compounds[64]. Although in these series the effect of electron-withdrawing ligands is to cause a decrease in shielding at tin, plots of series of organotin compounds, $R_{4-n}SnX_n$ ($n = 0$–4), invariably exhibit a characteristic 'U' shape in which the minimum shielding occurs when $n = 2$ (Figure 3.4). This phenomenon is usually attributed to a change in the paramagnetic term due to *p*-electron imbalance, which is at a maximum when $n = 2$. The effect of changes in interbond angles at tin is exemplified by a comparison of tin-containing heterocycles with their open-chain analogues. Thus, the chemical shift of the heterocycle $Me_2\overline{Sn[CH_2]_3CH_2}$ (+ 54 ppm) is 50 ppm less shielded than is Me_2Et_2Sn, and increase in the ring size results in greater shielding ($Me_2\overline{Sn[CH_2]_4CH_2}$ $\delta(^{119}Sn) = -42$ ppm; $Me_2\overline{Sn[CH_2]_5CH_2}$ $\delta(^{119}Sn) = -2$ ppm)[63,62]. Change in coordination number at tin can have a dramatic effect upon the shielding and cause changes of several hundred ppm. Approximate ranges for four-, five- and six-coordinate tin compounds are + 200 to − 60 ppm, − 90 to − 330 ppm, and − 125 to − 515 ppm, respectively[41,44,54]. Changes in coordination can arise not only from complexation of donor molecules or anions to organotin Lewis acids, but also from self-association as in the case of organotin alkoxides[64,66,67]. The underlying causes of the observed very large increases in shielding are not fully understood, but the major contribution must arise from the paramagnetic contribution due to a reduction in effective nuclear charge and perhaps the use of tin 5*d* orbitals in bonding. Some selected ^{119}Sn chemical shift data, including some more recent values, are collected in Table 3.14.

Table 3.14 Representative solution ^{119}Sn chemical shift data.

Compound	$\delta(^{119}Sn)$	Solvent	Ref.
$(C_5H_5)_2Sn$	−2199	C_6H_{12}	*a*
$(MeC_5H_4)_2Sn$	−2171.1	$CDCl_3$	*b*
$(Me_5C_5)_2Sn$	−2129	C_6D_6	*c*
$(Me_5C_5)SnBr$	−1630	$CDCl_3$	*c*
$[(Me_5C_5)Sn][CF_3SO_3]$	−2180	CH_2Cl_2	*c*
$[(Me_5C_5)Sn][BF_4]$	−2247	CD_2Cl_2	*c*
$Sn[(Me_3Si)_2]C_2B_4H_4$	−205.9		*d*
$Sn(Me_3Si)(Me)C_2B_4H_4$	−170.9		*d*
$[(Me_3Si)_2CH]_2Sn$	2315	$CD_3C_6D_5$	*e*
$Sn[C(PMe_2)_3]_2$	−247.7	$CD_3C_6D_5$	*f*
$Sn[CH(PPh_2)_2]_2$	175.6	$CD_3C_6D_5$	*f*
$[Ph_4As][Sn(SPh)_3]$	146	CH_2Cl_2	*g*
$[Ph_4As][Sn(SePh)_3]$	208	CH_2Cl_2	*g*
Me_3SnOH	118		*h*
$Me_2Sn(bzac)_2$	−353.5	$CDCl_3$	*i*
$Me_2Sn(dbm)_2$	−348.0	$CDCl_3$	*i*
$Me_2Sn(ox)_2$	−235.1	$CDCl_3$	*i*
$Me_2Sn(O_2Cpy\text{-}2)_2$	−451	DMSO	*j*
$^{n}Bu_2Sn$ (O, O ring)	−189		*k*
$^{n}Bu_2Sn$ (O, S ring)	−35		*k*
$^{n}Bu_2Sn$ (S, S ring)	194		*k*
$^{t}Bu_2Sn$ (O, O ring)	−225		*k*
$^{t}Bu_2Sn$ (O, S ring)	−52		*k*
$^{t}Bu_2Sn$ (S, S ring)	171		*k*
$^{n}Bu_3SnOSn^{n}Bu_3$	82.7	Toluene	*l*
$^{n}Bu_3SnSSn^{n}Bu_3$	81.6	Toluene	*l*
$Ph_3SnOSnPh_3$	−82.7	Toluene	*l*
$Ph_3SnSSnPh_3$	−51.0	Toluene	*l*
$(C_6H_{11})_3SnOH$	1.5	Toluene	*l*
$(C_6H_{11})_3SnOSn(C_6H_{11})_3$	−7.6	Toluene	*l*
$(C_6H_{11})_3SnSSn(C_6H_{11})_3$	21.1	Toluene	*l*
[3(2-py)-2-(thienyl)]$SnPh_3$	−181.6	$CDCl_3$	*m*
$[3(2\text{-py})\text{-}2\text{-(thienyl)}]_2SnPh_2$	−245.5	$CDCl_3$	*m*
(2-thienyl)$SnPh_3$	−135.5	$CDCl_3$	*m*
$(2\text{-thienyl})_4Sn$	−147.0	$CDCl_3$	*m*
$Na_2[Sn(OH)_6]$	−595.3	$2MH_2O$	*n*
$K_2[Sn(OH)_6]$	−595.5	$2MH_2O$	*n*
$[NR_4][SnCl_6]$	−731.9	CH_2H_2	*o*
$[NR_4][SnBr_6]$	−2066.5	CH_2Cl_2	*o*
$[NR_4][Sn(CN)_6]$	−916.2	CH_2Cl_2	*o*
$[NR_4][Sn(NCS)_6]$	−842.5	CH_2Cl_2	*o*

Table 3.14 (*Contd.*)

(*a*) J.M. Basset, B.W. Fitzsimmons, F.C. Fowler, D. Harris, S. Keppie, M.F. Lappert, W. McFarlane, J. Poland, G.S. Pyne and D.S. Rycroft, unpublished data, quoted in *NMR and the Periodic Table*, eds. R.K. Harris and B.E. Mann, Academic Press, London, 1978, 351.
(*b*) A. Bonny, A.D. McMaster and S.R. Stobart, *Inorg. Chem.*, 1978, **17**, 935.
(*c*) P. Jutzi and B. Hielscher, *Organometallics*, 1986, **5**, 1201.
(*d*) N.S. Hosmane, N.N. Sirmokadam and R.H. Herber, *Organometallics*, 1984, **3**, 1665.
(*e*) K.W. Zilm, G.A. Lawless, R.M. Merrill, J.M. Miller and G.W. Webb, *J. Amer. Chem. Soc.*, 1987, **109**, 7236.
(*f*) H.H. Karsch, A. Appelt and G. Müller, *Organometallics*, 1986, **5**, 1664.
(*g*) P.A.W. Dean, J.J. Vittal and N.C. Payne, *Can. J. Chem.*, 1985, **63**, 394.
(*h*) J.D. Kennedy and W. McFarlane, *J. Organomet. Chem.*, 1975, **94**, C7.
(*i*) W.F. Howard, R.W. Crecely and W.H. Nelson, *Inorg. Chem.*, 1985, **24**, 2204.
(*j*) T.P. Lockhart and F. Davidson, *Organometallics*, 1987, **6**, 2471.
(*l*) S.J. Blunden and R. Hill, *J. Organomet. Chem.*, 1987, **333**, 317, 107.
(*k*) S.D. Slater, PhD thesis, University of London, 1988.
(*m*) V.G. Kumar Das, L.K. Mun, C. Wei, S.J. Blunden and T.C.W. Mak, *J. Organomet. Chem.*, 1987, **322**, 163.
(*n*) P.G. Harrison, unpublished data.
(*o*) K.B. Dillon and A. Marshall, *J. Chem. Soc., Dalton Trans.*, 1987, 315.

Coupling constants are related via the Fermi contact term to the *s* electron density in the bond. The magnitudes of coupling constants can, therefore, be used to infer valence electron distribution and hence bonding and structure. Both the one-bond $^1J(^{13}C-^{119}Sn)$ and two-bond $^2J(^1H-^{119}Sn)$ coupling constants have been employed to measure changes in electronic distribution in tin–carbon bonds, and indeed have been shown to be directly related[35,39], indicating that minimal changes occur in the C–H bonds. The increase of the two-bond $^2J(^1H-^{119}Sn)$ coupling constant along the four-coordinate series Me_4Sn (54.7 Hz), Me_3SnCl (58.2 Hz), Me_2SnCl_2 (68.9 Hz), $MeSnCl_3$ (96.9 Hz) has been interpreted in terms of a redistribution of *s* electron density in the bonds to carbon, with a concomitant increase in *p* character in the bonds to chlorine in accordance with Bent's isovalent hybridization principle[68]. Similar changes are observed on increase in coordination number both with $^1J(^{13}C-^{119}Sn)$ and $^2J(^1H-^{119}Sn)$[39,41,43,68–70]. The magnitudes of $^1J(^{13}C-^{119}Sn)$ and $^2J(^1H-^{119}Sn)$ have been related in a quantitative manner to the CSnC bond angles in methyl- and butyltin (IV) compounds,[44,69–71] and several sets of data are collected in Table 3.15. For butyltin (IV) compounds, the linear relationship

$$|^1J(^{13}C-^{119}Sn)| = (9.99 \pm 0.73)\theta - (746 \pm 100) \quad (3.1)$$

applies[70], but for methyltin (IV) compounds, the relationship between $^2J(^1H-^{119}Sn)$ and θ is of the form[69]

$$\theta = 0.0161|^2J(^1H-^{119}Sn)|^2 - 0.799|^2J(^1H-^{119}Sn)| + 133.4 \quad (3.2)$$

Couplings to other directly bound nuclei can be used similarly. For example, analysis of the one-bond $^1J(^{119}Sn-^{31}P)$ coupling constants for six-coordinated complexes of methyltin trihalides and tin (IV) halides with uni- and bidentate tertiary phosphines

Table 3.15 Magnitudes of the $^1J(^{119}Sn-^{13}C)$ and $^2J(^{119}Sn-^1H)$ coupling constants and estimated C–Sn–C bond angles for methyl and butyltin compounds.[a]

Compound	Solvent[b]	$\vert ^1J(^{119}Sn, ^{13}C)\vert$, Hz	$\vert ^2J(^{119}Sn, ^1H)\vert$, Hz	Estd C–Sn–C angle, deg.
Me_4Sn	[c]	336.3	54.7	109.5
$(Me_3Sn)_2Se$	CH_2Cl_2	340	56.1	106.6
$(Me_3Sn)_2S$	CH_2Cl_2	356	57.1	108.0
Me_3SnOAc	$CDCl_3$	401	58.5	111.9
Me_3SnCl	CCl_4	379.7	58.1	110.1
Me_3SnCl	acetone-d_6	433.4	64.4	114.8
Me_3SnCl	pyridine	472	67.0	118.2
Me_3SnCl	DMF-d_7	513.4	70.0	121.8
Me_3SnBr	CCl_4	368.9	57.8	109.1
Me_3SnBr	DMF-d_7	490.8	69.8	119.3
$Me_3Sn(o\text{-}OC_6H_4NMe_2)$	nr	422	56.6	113.8
Me_3Sn(oxinate)	CD_2Cl_2, $CDCl_3$	427	57.0	114.2
$Me_2SnBr(O_2CC_6H_5)$	$CDCl_3$	500	70.5	120.6
$Me_2SnCl(O_2CC_6H_5)$	$CDCl_3$	568	76.1	126.6
Me_2SnCl(pen)	D_2O	614.4	79.1	130.6
$Me_2Sn(oxinate)_2$	$CDCl_3$	632	71.2	132.2
$Me_2Sn(trop)_2$	$CDCl_3$	643	72.2	133.2
$Me_2Sn(S_2PMe_2)_2$	$CDCl_3$	553	78.8	125.3
$Me_2Sn[S_2CN(CH_2)_4]_2$	$CDCl_3$	655	85.9	134.2
$Me_2Sn(O_2CC_6H_5)_2$	$CDCl_3$	660	84.0	134.6
$Me_2Sn(S_2CNMe_2)_2$	$CDCl_3$	664	84.0	135.0
$Me_2Sn(S_2CNEt_2)_2$	$CDCl_3$	664	84.0	135.0
$Me_2Sn(OAc)_2$	C_6D_6, CCl_4	665	82.5	135.1
$Me_2Sn(koj)_2$	Me_2SO	748	83.3	142.4
$Me_2Sn(dbm)_2$	$CDCl_3$	913	97.0	156.8
$Me_2Sn(bac)_2$	$CDCl_3$	931	98.4	158.4
$Me_2Sn(acac)_2$	$CDCl_3$	977	99.3	162.5
Me_2SnCl_2	C_6H_6	469.4	69.0	117.9
Me_2SnCl_2	CD_3CN	584.9	81.2	128.1
Me_2SnCl_2	acetone-d_6	601.8	85.0	129.5
Me_2SnCl_2	DMF-d_7	886.9	104.5	154.6

Table 3.15 (*Contd.*)

Compound	Solvent[b]	$\lvert {}^{1}J({}^{119}Sn, {}^{13}C) \rvert$, Hz	$\lvert {}^{2}J({}^{119}Sn, {}^{1}H) \rvert$, Hz	Estd C–Sn–C angle. deg.
Me_2SnCl_2	Me_2SO	1009	113	165.3
Me_2SnBr_2	C_6H_6	442.6	67.0	115.6
Me_2SnBr_2	CD_3CN	523.9	77.3	122.7
Me_2SnBr_2	acetone-d_6	533.4	79.1	123.5
Me_2SnBr_2	DMF-d_7	820.8	101.5	148.8
Bu_4Sn	nr	307		109.5
$Bu_2Sn(SC_5H_3\text{-}5\text{-}NO_2)_2$	nr	522.5		127.0
$[Bu_3SnCl_2]^-$	nr	494.2		124.1
$Bu_2SnCl_2 \cdot$dppoe	nr	598.2		134.6
$(Bu_2SnOC(O)CCl_3)_2O$	nr	696.0		144.3
		646.3		139.4
$Bu_2Sn(OCH_2CH_2O)$	nr	653		140.0
Bu_2Sn(morf·dtc)$_2$	nr	600.2		134.8
$Bu_2SnCl_2 \cdot$phen	nr	1016		176.4
Bu_2Sn(dbzm)$_2$	nr	880		162.8

(*a*) Data from T.P. Lockhart and W.F. Manders, *Inorg. Chem.*, 1986, **25**, 892; J. Holecek and A. Lycka, *Inorg. Chim. Acta*, 1986, **118**, L15; and W.F. Howard, R.W. Crecely and W.H. Nelson, *Inorg. Chem.*, 1985, **24**, 2204.
(*b*) nr = not recorded.
(*c*) *J* Values averages of values recorded in 9 solvents.

shows that the Sn–P bond situated *trans* to the more electron-donating ligand is strengthened, whereas the *cis* Sn–P bond is weakened[57].

3.2.1.2 *Dynamic processes.* Dynamic processes which occur at rates of the same order as the nmr time-scale are amenable to study by nmr. The first such to be investigated were the metallotropic fluxional molecules such as $Me_3SnC_5H_5$ and stannyl-indenes[72], but several other molecules including (σ-7-cycloheptatrienyl)triphenyltin[73], (σ-5-cyclohepta-1, 3-dienyl)triphenyltin[73], (cyclononatetraenyl)trimethyltin[74], and the bis(trimethylstannyl)dihydropentalenes[75] undergo similar rearrangements. Activation parameters for observed processes are collected in Table 3.16. The metallotropic

Table 3.16 Activation parameters for fluxional metallotropic rearrangements.

Compound	Log *A*	E_a	$\Delta S^{\neq}_{300}$	$\Delta G^{\neq}_{300}$	Ref.
$Me_3SnC_5H_5$	57.7 ± 4	32.6 ± 4	8.4 ± 16	27.6 ± 4	*a*
		28.5 ± 3	12.6 ± 13	29.7 ± 3	*b*
		26.8 ± 4		30.1 ± 4	*c*
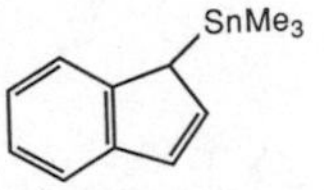 $SnMe_3$		50.6 ± 3	−55.6 ± 13	69.4 ± 3	*b*
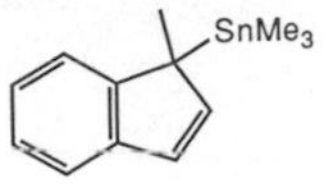 $SnMe_3$		48.9 ± 3	57.7 ± 3		*d*
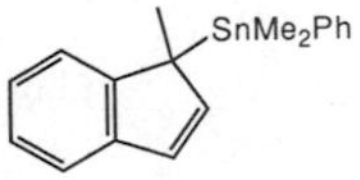 $SnMe_2Ph$		49.7 ± 1	58.9 ± 1		*d*
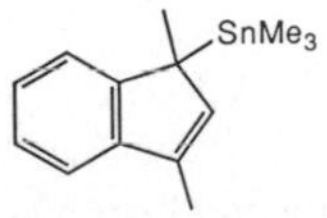 $SnMe_3$		52.3 ± 2	59.4 ± 2		*d*
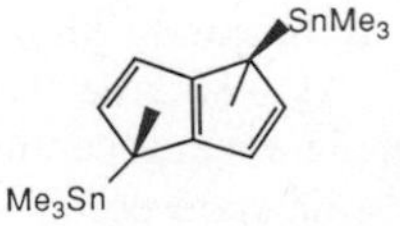 $SnMe_3$, Me_3Sn		42.2 ± 1	39.8 ± 1	−6.2 ± 4	e
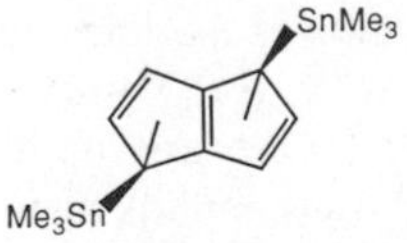 $SnMe_3$, Me_3Sn		31.2 ± 1	28.7 ± 1	−87.0 ± 3	e

(*a*) A.V. Kisin, V.A. Korenevski, N.M. Sergeyev and Yu.A. Ustynyuk, *J. Organomet. Chem.*, 1972, **34**, 93.
(*b*) Yu.N. Luzikov and Yu.A. Ustynyuk, *J. Organomet. Chem.*, 1974, **65**, 303.
(*c*) Yu.K. Grishin, N.M. Sergeyev and Yu.A. Ustynyuk, *Org. Mag. Res.*, 1972, **4**, 377.
(*d*) N.M. Sergeyev, Yu.K. Grishin, Yu.N. Luzikov and Yu.A. Ustynyuk, *J. Organomet. Chem.*, 1972, **38**, C1.
(*e*) Yu.A. Ustynyuk, A.K. Shestikova, V.A. Chertkov, N.N. Zemlyansky, I.V. Borisova, A.I. Gusev, E.B. Tchuklanova and E.A. Chernyshev, *J. Organomet. Chem.*, 1987, **335**, 43.

rearrangements in the case of the cyclopentadienyl derivatives occur via a series of (5, 1) shifts of the tin atom around the ring (eqn 3.3) rather than the alternative (1, 3) shift.

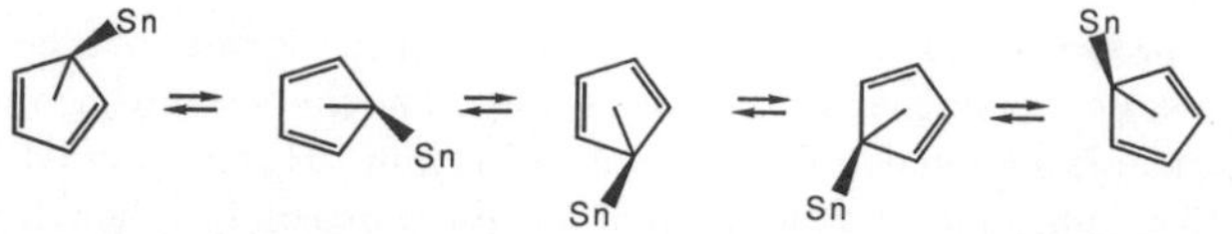

(3.3)

Ring migration in (σ-7-cycloheptatrienyl)triphenyltin and (cyclononatetraenyl)trimethyltin occur via either (1, 4) or (1, 5) shifts in the case of the former, and (1, 9) shifts in the latter. In the case of the stannylindenes, however, a (1, 2) migratory pathway via undetectable quantities of the isoindene (eqn 3.4) is experimentally indistinguishable from a (1, 3) migration. Similar processes occur with the *E* and *Z* isomers of bis(trimethylstannyl)dihydropentalene (eqns 3.5, 3.6).

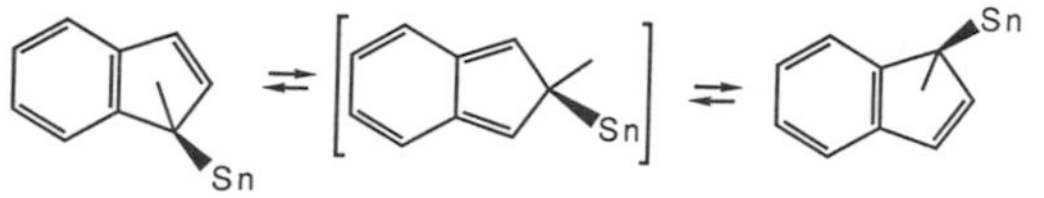

(3.4)

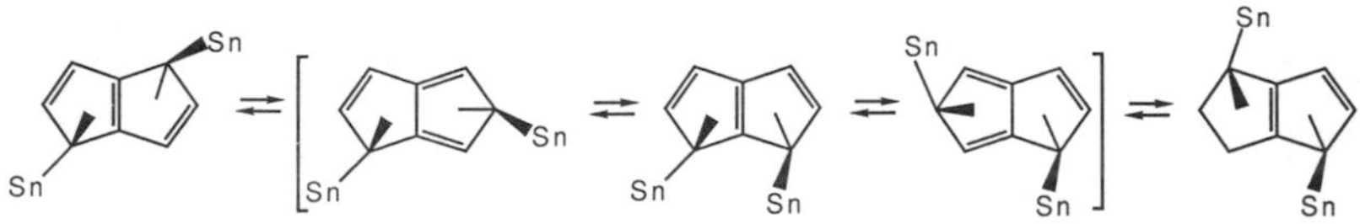

(3.5)

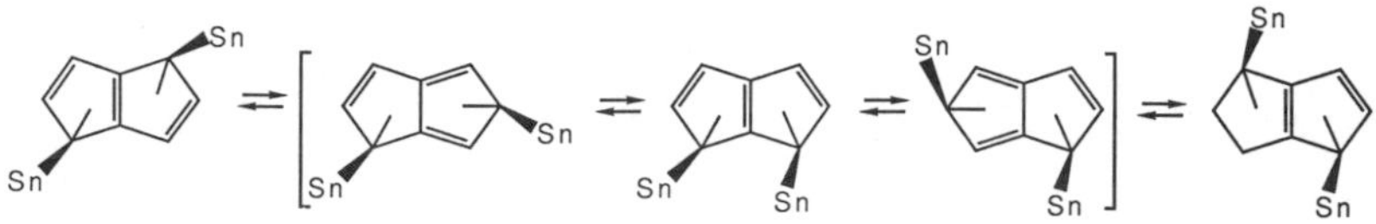

(3.6)

$Sn \equiv SnMe_3$

A second type of fluxional process occurs with stereochemically non-rigid geometries, and is exemplified by the spectra of tin(II) complexes of ambidentate phosphanido ligands, such as $Sn[C(PMe_2)_3]_2$ and $Sn[CH(PPh_2)_2]_2$[76]. At $-80°$, the ^{31}P spectrum of $Sn[C(PMe_2)_3]_2$ exhibits a triplet and a quintet expected for a degenerate $A_2A'_2XX'$ (apparent A_4X_2) system, in accord with the observed solid-state pseudo-trigonal bipyramidal structure (**1**), in which a rapid pseudorotation is assumed to equilibrate the axial and equatorial P_A and P_B at tin ($\delta(P_A)$ -17.2; $\delta(P_B)$ -41.5; J 13.7 Hz). The P_A signal exhibits $^{117,119}Sn$ satellites (J 742, 776 Hz). On warming to $+20°$, however, the spectrum collapses to a single line, which also shows coupling to tin with J 522, 552 Hz as all six phosphorus atoms become equivalent on the nmr time-scale by a rapid P_A/P_X exchange process ($\Delta G_{+20°}$ 53.1 kJ mol^{-1}). This phenomenon is rationalized in terms of the three-coordinate intermediate (**2**), where both $P_{A'}$ and P_X may occupy the fourth coordination site by rotation around the Sn–P and/or P–C axis. In the ^{119}Sn spectrum, at $-78°$ a quintet of triplets is observed (δ -258.0) due to one-bond coupling to the four equivalent phosphorus donors and a three-bond coupling to the two equivalent P_X atoms (Figure 3.5*a*). On heating to $+70°$, all the phosphorus

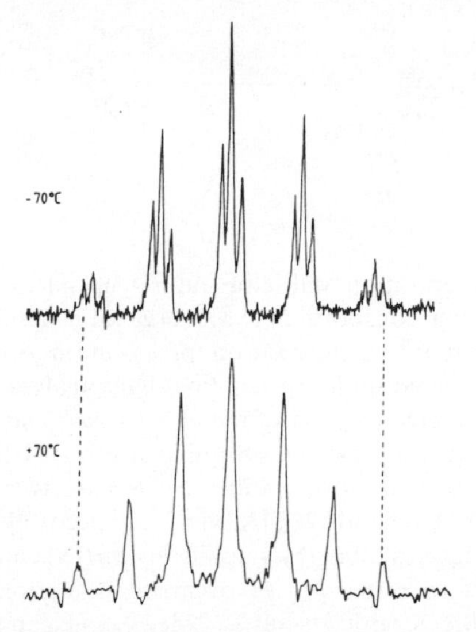

Figure 3.5 ^{119}Sn-nmr spectra of $Sn[C(PMe_2)_3]_2$ at (*a*) $-70°$ and (*b*) $+70°$ (adapted from ref. 76).

atoms become equivalent and a septet is observed ($\delta -247.7$, J 551 Hz) (Figure 3.5*b*). The behaviour of $Sn[CH(PPh_2)_2]_2$ is quite different, and the nmr spectra indicate the three-coordinated $[P_2SnC]$ structure (**3**). Thus, at $-15°$ the ^{31}P spectrum comprises two triplets consistent with an A_2B_2 system ($\delta(P_A) -12.5$, $\delta(P_B) -17.9$, J 24 Hz), and the ^{119}Sn spectrum is a triplet of triplets ($\delta(^{119}Sn) +175.6$, J 1172, 163 Hz). The inequivalence of the two methine carbon atoms is evident from the ^{13}C spectrum

(**1**) (**2**) $P = PMe_2$ (**3**)

($\delta(^{13}C)$ 17.8, 24.2, assignment arbitrary). Spectral changes occur both on heating and on cooling. At $-90°$, further splitting is seen in both the ^{31}P and ^{119}Sn spectra due to hindered rotation around the P_A–C bonds ($\Delta G_{-70°}$ 38.5 kJ mol^{-1}) which gives rise to inequivalent P_A and $P_{A'}$ atoms (**4**), ($\delta(P_A) -9.5$, $\delta(P_{A'}) -16.1$). On warming to $+50°$, the A_2B_2 pattern in the ^{31}P spectrum disappears, and at $+70°$ a new single broad signal ($\delta(^{31}P) -14.5$) arises as all four phosphorus atoms become equivalent ($\Delta G_{+50°}$ 43.9 kJ mol^{-1}), and it would appear that the C-bonded two-coordinated structure (**5**) is present at this temperature.

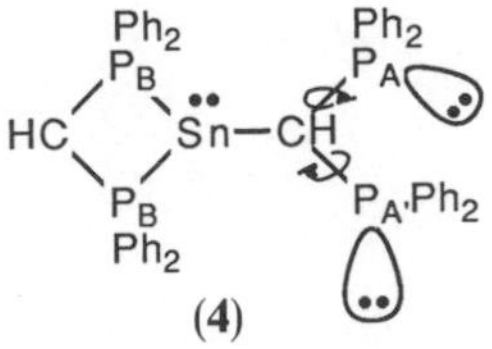

(4) 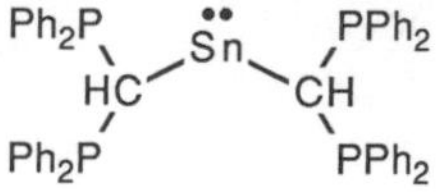 (5)

Interpretation of concentration and temperature-dependent nmr data can lead to the evaluation of binding constants between organotin halides and donor molecules[77,78] or chloride ion[79]. Isomerization processes in octahedral tin chloride complexes can be investigated similarly, and line-shape analysis of the ^{1}H spectra has demonstrated that the interconversion of the *cis, cis, trans* and *cis, trans, cis* isomers (with respect to Cl, N and O atoms) of bis(8-quinolato)tin(IV) dichloride and bis(5, 7-dichloro-8-quinolato)tin(IV) dichloride undergo *intramolecular* interconversion with activation energies E_a of 54 ± 1 and 52 ± 1 kJ mol^{-1}, respectively[80]. In an elegant study of the complexation of (excess) dimethylsulphide by tin(IV) chloride using magnetization transfer techniques, the fastest process occurring in solution has been shown to be the exchange of free Me_2S with *cis*-$SnCl_4 \cdot 2Me_2S$, and combined ^{1}H and ^{119}Sn experiments show that the isomerization reaction between *trans*- and *cis*-$SnCl_4 \cdot 2Me_2S$ is 10^4–10^5 times slower. The rate of exchange of free Me_2S with *trans*-$SnCl_4 \cdot 2Me_2S$ is too slow to be measured, and is at least one order of magnitude smaller than the rate of isomerization[81,82]. Two-dimensional nmr quite literally adds a second dimension to the study of dynamic systems. 2-D ^{31}P nmr has been applied to exchange processes in trichlorostannate complexes of platinum(II)[83], whilst 2-D ^{119}Sn nmr has demonstrated that the ditin compounds, $CH_2[PhSn(SCH_2CH_2)_2NMe]_2$ and $[MeSn(CH_2CH_2CH_2)_2NMe]_2$ isomerize at the tin centre in an uncorrelated way[84,85].

Information concerning radical reactions obtained from chemically induced nuclear polarization (CIDNP) spectroscopy[86]. For organotin compounds the effect has been observed in both ^{1}H and ^{119}Sn spectra, and the technique is well illustrated by the photolysis of Me_3SnNEt_2 in C_6D_6[87], where the overall reaction is

$$2Me_3SnNEt_2 \rightarrow Me_3SnSnMe_3 + Et_2NH + Me{-}CH{=}NEt \qquad (3.7)$$

Figure 3.6 shows the ^{1}H nmr spectrum during uv irradiation, where the observed CIDNP effects are enhanced absorption of the Me–*CH*=N–CH_2–CH_3 (δ 7.4 quartet), CH_3–CH=N–CH_2–CH_3 (δ 3.3 quartet), and Me_3SnH (δ 4.8 decet) resonances, and an emission (i.e. negative) appearance for the $(CH_3{-}CH_2)_2NH$ (δ 2.5 quintet) resonance. No effects are observed in the other parts of the spectrum. These observed nuclear polarizations indicate that the reaction proceeds via $\overline{Me_3Sn \cdot\cdot NEt_2}^{s}$ radical pairs formed by homolytic Sn–N bond cleavage from singlet states of Me_3SnNEt_2. Other reactions which have been examined include exchange between Me_3SnH and $Me_3Sn^{\cdot}$ radicals[88], photolysis of hexaorganodistannanes[89], tristannanes[90], and benzylstannanes[91], and the reaction of allyltriethyltin with $CBrCl_3$[92].

3.2.2 *Solid-state studies*

The theory of solid-state nmr has been elegantly presented elsewhere[93], and only the briefest background is presented here. Three principal problems have to be overcome

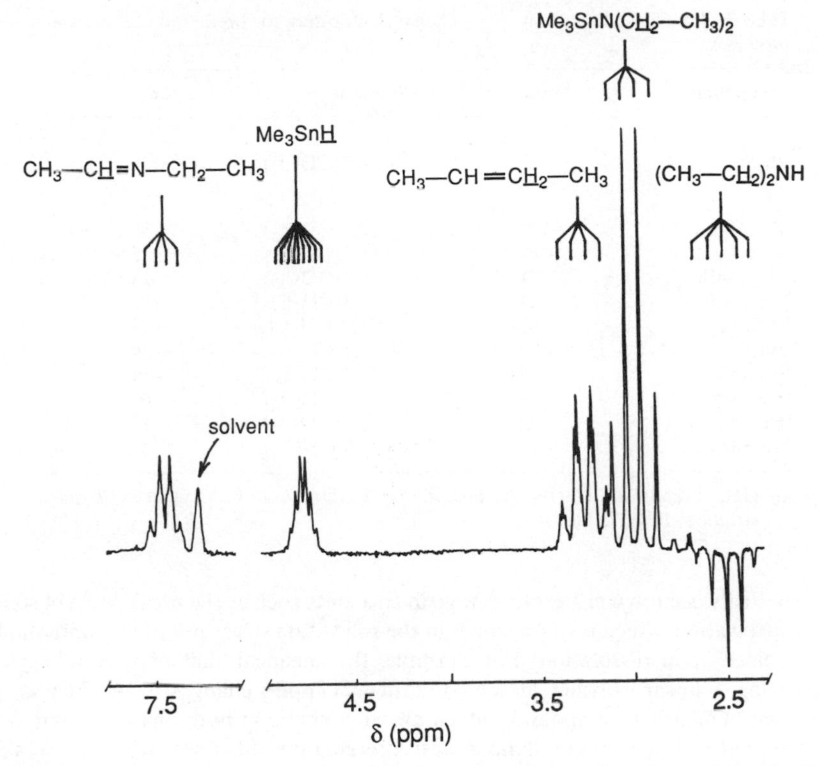

Figure 3.6 ^{1}H-nmr spectrum of Me_3SnNEt_2 upon irradiation illustrating CIDNP effects (adapted from ref. 87).

in order to obtain high resolution nmr spectra from solid samples: (i) line broadening due to anisotropic dipole–dipole and quadrupole–field gradient interactions; (ii) chemical shift anisotropy; and (iii) long spin-lattice relaxation times. Anisotropic interactions are also present in liquid samples, but in that case are averaged to zero by rapid Brownian motion. For the study of dilute nuclei such as ^{13}C or ^{119}Sn, the elimination of dipolar broadening is relatively simple, and can be accomplished by applying a second decoupling rf field. Chemical shift anisotropic effects are removed by so-called *magic-angle spinning (MAS)*, which entails spinning the sample at an angle of 54.73° to the magnetic field, but at a very high speed (of the order of kHz). If rates lower than the shielding anisotropy (expressed in Hz) are used, satellite spinning sidebands are observed in the spectra. The problems of long spin-lattice relaxation times can be overcome by cross-polarization (CP) from a second nucleus in the system, usually ^{1}H, which is more abundant and has shorter relaxation times, and this technique is most commonly used for organotin compounds.

Like ^{119}Sn Mössbauer spectroscopy, data from solid-state nmr can be correlated directly with structural data obtained by X-ray crystallography, and hence can provide an interesting and sometimes illuminating comparison between solid and solution phase structure. The greatest changes in shielding between solid and solution not

Table 3.17 Solid and solution ^{119}Sn chemical shift data for organotin chlorides and bromides.[a]

Compound	Solid	Solution	Δδ
Ph_3SnCl	{−33, −35}	−44.7 ($CDCl_3$)	~10
$(PhCH_2)_3SnCl$	+105	+52.5 ($CDCl_3$)	52.5
$^{c}Hex_3SnCl$	+82	+66.2 (C_6D_6)	15.8
$^{c}Hex_3SnBr$	+94	+79.2 ($CDCl_3$)	14.8
Me_2SnCl_2	+74.5	+137.0 (CH_2Cl_2)	−62.5
Et_2SnCl_2	+82.0	+121.0 (CH_2Cl_2)	−39
$^{n}Bu_2SnCl_2$	+84	+123.4 (CH_2Cl_2)	−39
Bz_2SnCl_2	+55.0	+36.4 (CH_2Cl_2)	+19
$^{c}Hex_2SnCl_2$	+7	+73.8 ($CDCl_3$)	−67
$^{c}Hex_2SnBr_2$	+82	+70.7 ($CDCl_3$)	11
Me_2SnBr_2	+90	+70.0 ($CHCl_3$)	20

(*a*) Data from R.K. Harris, A. Sebald, D. Furlani and G. Tagliarini, *Organometallics*, 1988, **7**, 388.

unexpectedly occur when a gross change in structure, such as the breakdown of solid-state association, which is so prevalent in the solid-state structures of tin compounds, takes place upon dissolution. For example, the chemical shift of tricyclohexyltin hydroxide, a linear polymer in the solid, ($\delta - 217$ ppm) changes to $+11.6$ ppm in solution. In contrast, compounds which are monomeric in both phases exhibit very similar shifts. Triorganotin halides and diorganotin dihalides are only weakly associated in the crystal, and the association is readily broken down in solution, and so only relatively small chemical shift changes are observed (Table 3.17)[94]. Nevertheless, the solid↔solution shifts observed even in these cases do demonstrate the sensitivity towards even quite small changes in geometry at tin. Other ^{119}Sn chemical shift data are collected in Table 3.18.

^{13}C CPMAS nmr has been employed to probe the solid-state structure or organotin compounds, and data for a number of methyltin compounds are shown in Table 3.19.

Table 3.18 Solid state ^{119}Sn chemical shift data.

Compound	$\delta(^{119}Sn)$	Ref.
Me_3SnOH	−99, −152	*a*
Me_3SnCl	174	*b*
$(Me_3Sn)_2CO_3$	32	*b*
Me_3SnCN	−142	*b*
Me_3SnO_2CMe	−52	*b*
$(PhCH_2)_3SnO_2CMe$	−106	*b*
Me_3SnOMe	−48.5	*b*
Me_3SnF	35	*b*
Et_3SnO_2CMe	−43, −51	*b*
$^{n}Bu_3SnO_2CMe$	−48, −54.2	*b*
$[^{n}Bu_3Sn(H_2O)_2]^+[C_5(CO_2Me)_5]^-$	48	*b*
Me_2SnO	−152	*c*
$^{n}Bu_2SnO$	−177	*c*

Table 3.18 (*Contd.*)

Compound	$\delta(^{119}Sn)$	Ref.
$(Me_3Sn)_4C$	48.2	*d*
Ph_4Sn	−120.7	*e*
$^{n}Bu_2Sn$ O, O	−230	*b*
$^{n}Bu_2Sn$ O, S	−42	*b*
$^{n}Bu_2Sn$ S, S	173	*b*
$^{t}Bu_2Sn$ O, O	−225	*b*
$^{t}Bu_2Sn$ O, S	−96, −101	*b*
$^{t}Bu_2Sn$ S, S	171	*b*
$\{[(Me_3Si)_2CH_2Sn\}_2$	692	*f*
$Na_2[Sn(OH)_6]$	−563.4, 564.4	*g*
$K_2[Sn(OH)_6]$	−569.7	*g*
SnO_2	−603.4	*g*

(*a*) R.K. Harris, K.J. Packer and P. Reams, *J. Mag. Res.*, 1985, **61**, 564.
(*b*) S.D. Slater, PhD thesis, University of London, 1988.
(*c*) R.K. Harris and A. Sebald, *J. Organomet. Chem.*, 1987, **331**, C9.
(*d*) R.K. Harris, T.N. Mitchell and G.J. Nesbitt, *Mag. Res. Chem.*, 1985, **23**, 1080.
(*e*) A.G. Davies, A.J. Price, H.M. Dawes and M.B. Hursthouse, *J. Chem. Soc., Dalton Trans.*, 1986, 297.
(*f*) K.W. Zilm, G.A. Lawless, R.M. Merrill, J.M. Millar and G.W. Webb, *J. Amer. Chem. Soc.*, 1987, **109**, 7236.
(*g*) P.G. Harrison and A. Sebald, unpublished data.

The ^{13}C chemical shifts of methyl groups bound to tin are sensitive to slight variations in bond angles and bond distances, and polymorphism has been demonstrated to occur in some compounds[95]. Shifts generally increase, becoming more deshielded as the coordination number at tin increases and as the number of methyl groups bound to tin decreases. However, as with solution data it is the one-bond $^1J(^{119}Sn{-}^{13}C)$ coupling constant which has attracted most attention, and is related quantitatively and linearly to the CSnC bond angle θ by the relationship[96–103]:

$$|^1J(^{119}Sn{-}^{13}C)| = 10.7\theta - 778 \quad (3.8)$$

Exceptions invariably arise to such semi-empirical relationships, and in particular $Me_2Sn(oxinate)_2$ and $Me_2Sn(ONHCONMe)_2$ lie well of the calculated line. Using the

Table 3.19 ^{13}C chemical shift, $^{1}J(^{119}Sn-^{13}C)$ coupling constant, and Me–Sn–Me bond angle data.[a]

Compound	Sn–methyl chemical shift (ppm)	$\lvert ^{1}J(^{119}Sn, ^{13}C)\rvert$ (Hz)	Me–Sn–Me angle (deg)
Tetracoordinated			
Me_4Sn	0	336[e]	109.5
$[Me_2SnS]_3$	12.2, 8.3, 7.5	430 (av)	118 (av)
$Me_2SnPh_2 \cdot 2Cr(CO)_3$	−6.7	380	115.5
$MeSnPh_3$	−8.8, −6.9	510[g]	
Pentacoordinated			
$Me_3Sn[ON(Ph)COPh]$	5.6	410	109.7 (av)
Me_3SnCl	4.1	470	117.2
$Me_3SnNO_3 \cdot H_2O$	1.0	490	120 (av)
$Me_3Sn[O_2C(1\text{-}Np)]$	3.9, 1.8	490	119.6 (av)
Me_3SnNO_3	0.9	500	
$Me_3Sn[O_2C(c\text{-}C_6H_{11})]$	2.2, 0.1	510 (av)	119.6 (av)
$Me_3Sn[O_2C(p\text{-}C_6H_4NH_2)]$	1.5	530	119.5 (av)
Me_3SnOAc	1.7, −0.1	540	120
Me_3SnF	2.3	550	
$[(Me_3Sn)_2CO_3]_n$	0.4	590	
Me_3SnOH	6.2, 3.5	600	
$Me_3Sn(S_2CNMe_2)$	3.5		
$Me_2PhSnOAc$	2.1	610	128.1
$Me_2Sn(Cl)(S_2CNMe_2)$	13.3	580	128
$Me_2Sn(Cl)$(cysteine ethyl ester)	10.1	600	119.5
Me_2Sn(glycylmethioninate)	−0.1	640	123.8
$Me_2Sn(NO_3)(OH)$	9.6	730	139.9
$Me_2Sn(Cl)(o\text{-}C_6H_4NMe_2)$	1		
$[Me_2SnO]_n$	6.4	660	
Hexacoordinated			
$Me_2Sn(S_2PMe_2)_2$	16.9	470	122.6
$Me_2Sn(S_2COEt)_2$	9.9	570	130.1
$Me_2Sn[ON(H)COCH_3]_2$	7.3	600	109.1
$Me_2Sn(\text{oxinate})_2$	10.7, 7.5	630	110.7
$Me_2Sn(OAc)_2$	4.0	660	
$Me_2Sn(NCS)_2$	14.1	670	147.4
$Me_2Sn(S_2CNMe_2)_2$	17.9	670	136
$Me_2Sn(S_2CNEt_2)_2$	15.2, 16.5	680	135.6
$[Me_2SnCl_2 \cdot \text{salicylaldehyde}]_2$	14.6	680	131.4
$Me_2Sn[S_2CN(CH_2)_4]_2$	18.0	705	137.4
$Me_2Sn(\text{laurate})_2$	5.2	720	
Me_2SnPO_4H	13.0, 11.5, 10.6	780	
$[Me_2SnCl_2 \cdot \text{lutN–O}]_2$	19.0	810	145.3
$[Me_2Sn(O_2CCH_2Cl)]_2O$	11.7	820	152 (av)
$Me_2SnCl_2 \cdot 2\text{dmf}$	24.2	990	165.0
$Me_2SnCl_2 \cdot 2\text{dmso}$	28.5, 23.5	1060	170
$Me_2SnCl_2 \cdot 2\text{pyrN–O}$	19.6	1120	180.0
$Me_2Sn(\text{acac})_2$	11.4	1175	180.0
$MeSn(Cl)(S_2CNMe_2)_2$	31.6	950 ± 50	
$[MeSn(O)OH]_n$	16.2, 10.0	1160, 1030	
Heptacoordinated			
$MeSn(S_2CNEt_2)_3$	37.1	1015	

(*a*) Data from T.P. Lockhart and W.F. Manders, *J. Amer. Chem. Soc.*, 1987, **109**, 7015.

relationship a value of 135° has been predicted for the CSnC bond angle in the intractable polymer dimethyltin oxide.

3.3 Mössbauer spectroscopy

The study of recoilless nuclear resonant fluorescence, better known as Mössbauer spectroscopy, has become an important technique for probing the structures of tin materials. The importance of the technique lies not only in the considerable quantity of information concerning, *inter alia*, structure and bonding, but also in the very wide spectrum of application, which includes in addition to pure compounds, mixtures, ores and minerals[104], catalysts[105–107], coal hydrogenation[108,109], pigments[110], glasses[111], thin films[112–116], adsorbed species and surfaces[117–119], intercalation compounds[120–125], PVC plastics[126], matrix isolated molecules[127–130], frozen solutions[131–134], sols[135], and biological samples[136,137]. Early tin data has been reviewed by Parish[138], Zuckerman[139] and Bancroft and Platt[140].

The basic principles of Mössbauer spectroscopy have been described comprehensively by other authors[141–143], and only a brief summary of the theoretical background is presented here. The criteria for the observation of a Mössbauer resonance are fulfilled by the 23.875 KeV ($\pm\frac{1}{2}\leftrightarrow\pm\frac{3}{2}$) transition for ^{119}Sn, and the nuclear decay scheme for this nuclide is shown in Figure 3.7. The energy level diagram illustrated in Figure 3.8 shows the transitions for tin atoms in various electronic and magnetic environments. The half-life $t_{1/2}$ of the $\pm\frac{3}{2}$ excited state has been measured as 17.75 ± 0.12 ns, which puts the natural (or lower limiting) resonance linewidth, Γ_{nat}, as 0.325 $\pm$ 0.002 mm s^{-1}.[144] Source materials, therefore, necessarily require to be of the lowest linewidth possible, and sources are usually fabricated from barium or calcium stannate which have a linewidth of 0.33 mm s^{-1} with a half-life of *c.* 250 days. Several parameters are available from the Mössbauer spectra of tin materials which can be translated into chemically useful information: the chemical, isomer or centre shift, the quadrupole

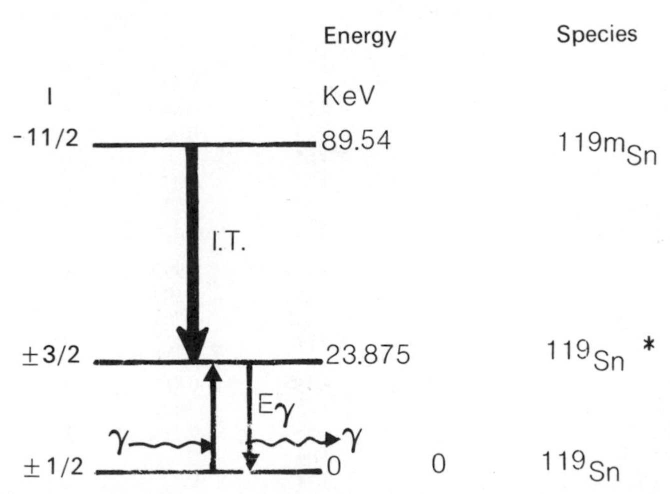

Figure 3.7 Part of the nuclear decay scheme for ^{119}Sn showing the Mössbauer relevant transitions.

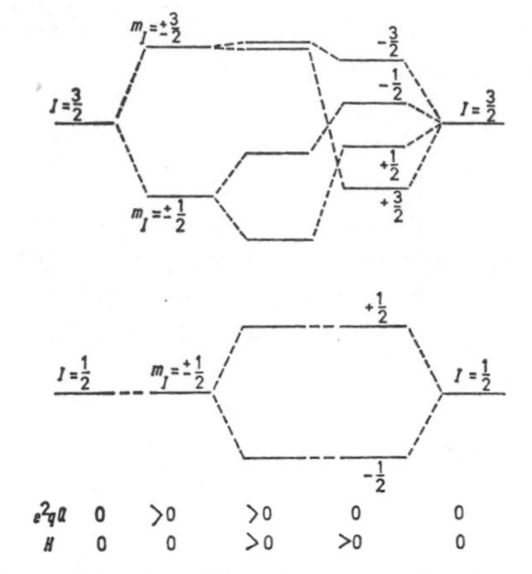

Figure 3.8 Energy level diagram showing effect of non-zero electric and magnetic fields on the ^{119m}Sn $\pm\frac{1}{2}\leftrightarrow\pm\frac{3}{2}$ transitions.

splitting, the resonance area and its temperature coefficient, and the ratio of the peak areas of quadrupole-split spectra. Several compilations of isomer shift and quadrupole splitting data have been produced[138–143,145], and only a selection of some more recent data is presented in Table 3.20.

Although by far the majority of spectra have been obtained using conventional transmission optical geometry, both emission[146–150] and back-scattering (conversion electron Mössbauer spectroscopy) have also been used. The latter is an extremely useful technique for the investigation of thin films and surfaces[113–116].

Table 3.20 Examples of typical Mössbauer isomer shift (δ) and quadrupole splitting (Δ) data (mm s^{-1}).

Compound	δ	Δ	Ref.
Bivalent compounds			
α-SnF_2	3.462	1.561	*a*
$Sn(O_3SF)_2$	4.18	0.68	*b*
$Sn(O_3SCF_3)_2$	4.15	0.84	*b*
$Sn(SbF_6)_2$	4.44	0	*b*
$Sn(SbF_6)_2\cdot 2AsF_3$	4.66	0	*b*
$Sn(NMe_2)_2$	2.72	2.07	*c*
$Sn(NCH_2CH_2)_2$	2.72	2.03	*c*
$Sn[N(SiMe_3)_2]_2$	2.88	3.52	*c*
(18-crown-6)Sn_2Cl_4	{3.300	0.992	*d*
	3.889	2.109	
(18-crown-6)$Sn_2(SCN)_4$	{3.262	1.704	*d*
	4.14	1.423	
(18-crown-6)$Sn(ClO_4)_2$	4.445	0.88	*d*
$(C_5H_5)_2Sn$	3.73	0.65	*e*
$(MeC_5H_4)_2Sn$	3.83	0.78	*f*
$(^tBuC_5H_4)_2Sn$	3.64	0	*g*
$(Me_3SiC_5H_4)_2Sn$	3.58	0	*g*
$(Me_5C_5)_2Sn$	3.53	0.99	*h*
$(Ph_5C_5)_2Sn$	3.74	0.58	*i*
C_5H_5SnCl	3.71	1.04	*j*

Table 3.20 (*Contd.*)

Compound		δ	Δ	Ref.
C_5H_5SnBr		3.40	0.99	*k*
C_5H_5SnI		3.90	0	*k*
$C_5H_5SnAlCl_4$		3.73	0	*j*
$[(Me_5C_5)Sn]^+[CF_3SO_3]^-$		3.81	–	*l*
$[(^tBuC_5H_4)Sn]^+[BF_4]^-$		3.67	–	*m*
$[(Me_3Si)_2]C_2B_4H_4$		3.25	2.74	*n*
$(Me_3Si)(Me)C_2B_4H_4$		3.28	2.74	*n*
$(C_6H_6)Sn(AlCl_4)_2\cdot C_6H_6$		3.93	0	*o*
$[(Me_3Si)_2CH]_2Sn$		2.16	2.31	*p*
$[(Me_3Si)_2CH]_2Sn\cdot Cr(CO)_5$		2.21	4.43	*q*
$[(Me_3Si)_2CH]_2Sn\cdot Mo(CO)_5$		2.15	4.57	*q*
$(C_5H_5)_2Sn\cdot Cr(CO)_5$		1.86	2.60	*r*
$(C_5H_5)_2Sn\cdot Mo(CO)_5$		1.96	2.71	*r*
$[(C_5H_5)_2SnFe(CO)_4]_2$		1.86	0	*r*
$^tBu_2SnFe(CO)_4\cdot$dmso		1.87	3.45	*s*
$^tBu_2SnFe(CO)_4\cdot$py		1.82	3.06	*t*
$^tBu_2SnCr(CO)_5\cdot$dmso		1.98	3.50	*t*
$^tBu_2SnCr(CO)_5\cdot$py		2.01	3.44	*t*
Tetravalent compounds				
SnF_4		−0.359	1.82	*a*
$Sn(O_2CCF_3)_4$		−0.04	1.56	*u*
cis-$SnCl_4\cdot$2dmso		0.40	0.41	*v*
trans-$SnCl_4\cdot$2dmso		0.40	0.57	*v*
cis-$SnCl_4\cdot$2dmf		0.39	0.53	*v*
trans-$SnCl_4\cdot$2dmf		0.38	0.73	*v*
cis-$SnBr_4\cdot$2dmf		0.66	0.44	*v*
trans-$SnBr_4\cdot$2dmf		0.66	0.83	*v*
Me_2SnF_2		1.23	4.52	*w*
$Me_2Sn(NbF_6)_2$		1.77	5.51	*w*
$Me_2Sn(O_3SF)_2$		1.80	5.55	*w*
$Me_2Sn(SbF_6)_2$		2.04	6.04	*w*
cis-(*p*-$ClC_6H_4)_2SnCl_2\cdot Me_2$bipy		0.84	1.99	*x*
trans-(*p*-$ClC_6H_4)_2SnCl_2\cdot Me_2$bipy		1.14	3.49	*x*
$Cs_2[Me_2Sn(O_3SF)_4]$		1.83	5.50	*y*
$Ba[Me_2Sn(O_3SF)_4]$		1.76	5.37	*y*
$Me_3SnO_2CC_6H_4OMe$-2		1.36	3.71	*z*
$Me_3SnO_2CC_6H_4OH$-2		1.40	3.47	*z*
$Me_3SnO_2CC_6H_4Me$		1.30	3.68	*z*
$Ph_3SnO_2CC_6H_5$		1.24	2.55	*z*
$Ph_3SnO_2CC_6H_4Cl$-4		1.24	2.36	*z*
$Ph_3SnO_2CC_6H_4OMe$-2		1.25	2.30	*z*
[3-(2-py)-2-thienyl]$_2Ph_2Sn$		1.03	0.73	*aa*
[3-(2-py)-2-thienyl]Ph_3Sn		1.11	0.63	*bb*
[3-(2-py)-2-thienyl]$(C_6H_{11})_3Sn$		1.35	0.96	*bb*
(2-thienyl)Ph_3Sn		1.16	0	*bb*
Mixed valence compounds				
α-Sn_2F_6	(Sn^{IV})	−0.392	0.680	*a*
	(Sn^{II})	4.103	0.493	
Sn_3F_8	(Sn^{IV})	−0.315	0.510	*a*
	(Sn^{II})	3.822	1.265	
Sn_7F_{16}	(Sn^{IV})	−0.325	0	*a*
	(Sn^{II})	3.696	1.312	

Table 3.20 (*Contd.*)

Compound		δ	Δ	Ref.
$Sn_{10}F_{34}$	(Sn^{IV})	−0.337	0.995	*a*
	(Sn^{II})	4.510	0	
$[Sn_2(O_2CMe)_6]_2$	(Sn^{IV})	0.14	0.47	*u*
	(Sn^{II})	3.48	1.83	
$[Sn_2(O_2CCF_3)_6]_2$	(Sn^{IV})	−0.03	0.53	*u*
	(Sn^{II})	3.94	1.11	
$Sn[Me_2Sn(O_3SF)_4]$	(Sn^{IV})	1.87	5.38	*y*
	(Sn^{II})	3.84	0	
$(Me_3SnC_5H_4)_2Sn$	(Sn^{IV})	1.30	0	*cc*
	(Sn^{II})	3.58	0.89	

(*a*) J. Fournes, J. Grannes, Y. Potin and P. Hagenmuller, *Solid State Chem.*, 1986, **59**, 833.
(*b*) S.P. Mallela, S.T. Tomic, K. Lee, J.R. Sams and F. Aubke, *Inorg. Chem.*, 1986, **25**, 2939.
(*c*) K.C. Molloy, M.P. Bigwood, R.H. Herber and J.J. Zuckerman, *Inorg. Chem.*, 1982, **21**, 3709.
(*d*) R.H. Herber and A.E. Smelkinson, *Inorg. Chem.*, 1978, **17**, 1023.
(*e*) P.G. Harrison and J.J. Zuckerman, *J. Amer. Chem. Soc.*, 1969, **91**, 6885.
(*f*) S.G. Baxter, A.H. Cowley, J.G. Lasch, M. Lattman, W.P. Sharum and C.A. Stewart, *J. Amer. Chem. Soc.*, 1982, **104**, 4064.
(*g*) R. Hani and R.A. Geanangel, *J. Organomet. Chem.*, 1985, **293**, 197.
(*h*) P. Jutzi and R. Dickbreder, *Chem. Ber.*, 1986, **119**, 1750.
(*i*) R.L. Williamson and M.B. Hall, *Organometallics*, 1986, **5**, 2142.
(*j*) P.G. Harrison and J.A. Richards, *J. Organomet. Chem.*, 1976, **108**, 35.
(*k*) J.W. Connelly and C. Hoff, *Adv. Organomet. Chem.*, 1981, **19**, 123.
(*l*) T.S. Dory and J.J. Zuckerman, *J. Organomet. Chem.*, 1984, **264**, 295.
(*m*) R. Hani and R.A. Geanangel, *J. Organomet. Chem.*, 1985, **293**, 197.
(*n*) N.S. Hosmane, N.N. Sirmokadam and R.H. Herber, *Organometallics*, 1984, **3**, 1665.
(*o*) P.F. Rodesiler, Th. Auel and E.L. Amma, *J. Amer. Chem. Soc.*, 1975, **97**, 7405.
(*p*) J.D. Cotton, P.J. Davidson, M.F. Lappert, J.D. Donaldson and J. Silver, *J. Chem. Soc., Dalton Trans.*, 1976, 2286.
(*q*) P.J. Davidson, D.H. Harris and M.F. Lappert, *J.* Chem. Soc., *Dalton Trans.*, 1976, 2268.
(*r*) A.B. Cornwell, P.G. Harrison and J.A. Richards, *J. Organomet. Chem.*, 1976, **108**, 47.
(*s*) T.J. Marks and A.R. Newman, *J. Amer. Chem. Soc.*, 1973, **95**, 769.
(*t*) G.W. Grynkewich, B.Y.K. Ho, T.J. Marks, D.L. Tomaja and J.J. Zuckerman, *Inorg. Chem.*, 1973, **12**, 2522.
(*u*) T. Birchall and J.P. Jackson, *Inorg. Chem.*, 1982, **21**, 3724.
(*v*) D. Tudela, V. Fernandez and J.D. Tornero, *J. Chem. Soc., Dalton Trans.*, 1985, 1281.
(*w*) S.P. Mallela, S. Yap, J.R. Sams and F. Aubke, *Inorg. Chem.*, 1986, **25**, 4329.
(*x*) V.G. Kumar Das, Y. Chee-Keong and P.J. Smith, *J. Organomet. Chem.*, 1987, **327**, 311.
(*y*) S.P. Mallela, S. Yap, J.R. Sams and F. Aubke, *Inorg. Chem.*, 1986, **25**, 4074.
(*z*) P.J. Smith, R.O. Day, V. Chandrasekhar, J.M. Holmes and R.R. Holmes, *Inorg. Chem.*, 1986, **25**, 2495.
(*aa*) V.G. Kumar Das, L.K. Mun, C. Wei and T.C.K. Mak, *Organometallics*, 1987, **6**, 10.
(*bb*) V.G. Kumar Das, L.K. Mun, C. Wei, S.J. Blunden and T.C.K. Mak, *J. Organomet. Chem.*, 1987, **322**, 163.
(*cc*) E.J. Bulten and H.A. Budding, *J. Organomet. Chem.*, 1978, **157**, C3.

3.3.1 *The isomer shift*

The precise value of the emitted γ-ray energy, E_γ, depends upon the electronic environment of the emitting nucleus and, similarly, the energies of photons absorbed by a ground state nucleus also depend upon its electronic environment. The value of the isomer shift (δ) of the absorber relative to the source is given by the expression

$$\delta = \frac{Ze^2R^2}{5\varepsilon_0} \cdot \frac{\delta R}{R} \cdot \{|\psi(0)|^2_{abs} - |\psi(0)|^2_{source}\} \tag{3.9}$$

and depends upon two parameters, $\delta R/R$, the change in nuclear radius upon excitation or de-excitation, and the difference in total electron probability at the nucleus ($r = 0$) upon going from the source nucleus to the absorber nucleus. Various estimates of the value of $\delta R/R$ for tin have been made. A value of 1.61×10^{-4} has been obtained from MO calculations[151,152], compared with values of 1.34×10^{-4} using the scalar-relativistic linear muffin-tin-orbital method[153] and $0.87 \pm 0.25 \times 10^{-4}$ using the internal conversion method[154]. A positive sign indicates that the nucleus shrinks upon de-excitation.

The greatest contribution to the isomer shift will be due to the 5*s* valence electrons, since the lower *s* electrons will remain essentially isoenergetic from compound to compound. Contributions to the change in electron density will also be largely independent of *p*, *d*, and *f* electrons, since their orbitals have nodes at the nucleus and thus this effect is confined to interpenetration shielding of the *s* electrons. Consistent with this, simple SCF–MO calculations for simple tetrahedral and octahedral tin species using tin 5*s*, 5*p* and 5*d* orbitals show that the isomer shifts bear an approximately linear relationship with both the 5*s* occupation number and the 5*s* electron density at the nucleus[155,156]. Values of δ are conventionally positive for closing velocities, and hence higher isomer shifts for correspond to higher *s* electron densities. Because of this sensitivity of the isomer shift towards *s* electron density, δ can be employed to distinguish between the two different oxidation states of tin. The generally accepted dividing point between the two states is that of α-tin (2.10 mm s^{-1}), rather than β-tin (2.65 mm s^{-1}), and the range of δ observed for tin (IV) compounds is *c.* -0.5 to $+2.1$ mm s^{-1} and that for tin (II) compounds *c.* $+2.5$ to $+5.0$ mm s^{-1}. The large isomer shift differences between the two oxidation states has been shown by MO calculations to be due to the lone pair orbital, which is present in bivalent compounds but absent in the higher oxidation state[151,152]. For tin (IV) compounds the isomer shift is reduced as the total electronegativity of the groups attached to tin (or the charge on tin) increases. This effect is illustrated in Figure 3.9 for the two series SnX_4 and $[SnX_4Y_2^{2-}]$ (X and Y = F, Cl, Br, I). It is noteworthy that the values for both α- and β-tin fall on the respective tetrahedral and octahedral lines[157]. The converse is the case for bivalent tin compounds, where the isomer shift increases with increasing electronegativity of the ligands and the highest shifts are found for compounds such as $Sn(SbF_6)_2(AsF_3)_2$ ($\delta = 4.66$ mm s^{-1})[158], $Sn[Sn(SO_3CF_3)_6]$ (Sn^{II} sites)[159] ($\delta = 4.69$ mm s^{-1}), and $Sn(ClO_4)_2$(15-crown-5)$_2$ ($\delta = 4.53$ mm s^{-1})[160].

3.3.2 *The quadrupole splitting*

Interaction of the nuclear quadrupole moment Q with an electric field gradient lifts the degeneracy of the $\pm\frac{3}{2}$ state, giving rise to a doublet spectrum whose centroid is the isomer shift δ, and whose separation is the quadrupole splitting Δ. The expression for Δ is

$$\Delta = \tfrac{1}{2}e^2qQ(1 + \eta^2/3)^{1/2} \tag{3.10}$$

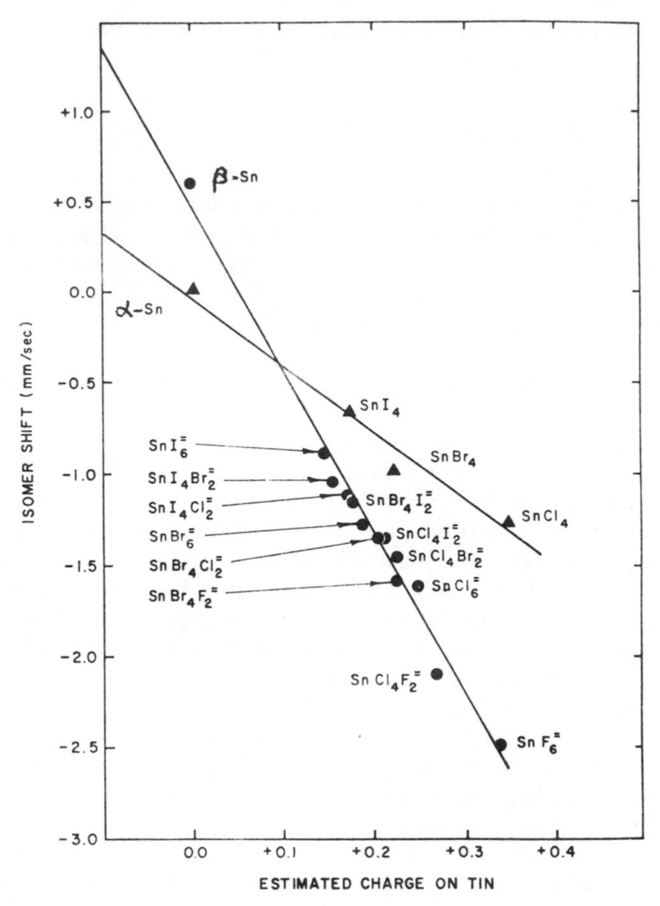

Figure 3.9 Plots of isomer shift versus estimated charge on tin for tetrahedral and octahedral halogenotin species (adapted from ref. 51).

where Q is the quadrupole moment of the nucleus, $eq = V_{zz}$ ($= -z$ component of the electric field gradient), and η is the asymmetry parameter defined as

$$\eta = (V_{xx} - V_{yy})/V_{zz} \tag{3.11}$$

Electric field gradients arise from three sources: (i) non-cubic coordination geometries; (ii) imbalances in the σ-framework of cubic (tetrahedral and octahedral) geometries; and (iii) distortions from regular cubic geometries. In tin(II) compounds, the dominant contribution to the electric field gradient arises from the p_z electron density in the tin non-bonding orbital. Lattice contributions are expected to be negligibly small, as has been shown for bivalent tin compounds[161]. Thus, tin atoms in cubic environments, e.g. $SnMe_4$, $SnCl_4$, and $[SnCl_6^{2-}]$, have no quadrupole splitting, but $[SnCl_5^-]$, $[SnCl_3^-]$, and SnO exhibit quite large splittings. Distortions from otherwise regular cubic coordination geometries, as in SnO_2, result in small splittings. Where the electronegativity difference between different ligands bonded to tetrahedrally or octahedrally coordinated tin is large, e.g. alkyl or aryl versus oxygen or halogen, the imbalance in the σ-framework is such as to give rise to a large quadrupole splitting, as in Ph_3SnCl and $[Me_2SnCl_4^{2-}]$. Exceedingly high values have been observed[162] for organotin derivatives of very strong inorganic acids such as $Me_2Sn(SbF_6)_2$ ($\Delta = 6.04$ mm s^{-1}) and $Me_2Sn(SO_3F)_2$ ($\Delta = 5.55$ mm s^{-1}). Smaller differences lead to smaller splitting values as in $Ph_3SnC_6F_5$[163], $SnCl_4 \cdot 2PPh_3$[164] and $SnCl_4 \cdot 2$dmso[165]. Different ligands of similar electronegativity produce electric field gradients too small to give rise to observable splittings, and organotin hydrides, Me_nSnH_{4-n} ($n = 0$–3), compounds containing bonds to Group 14 metals and carbon such as $Ph_3SnSnPh_3$ and the mixed anions $[SnX_4Y_2^{2-}]$ (X and Y = halogen) all exhibit single lines. The spectrum of tetrakis(adamantyl)tin is rather unusual. Rather than the expected single line, the spectrum comprises two tin sites with two different quadrupole splittings of 0.76 and 2.70 mm s^{-1} resulting from elongation of the Sn–C bonds due to steric crowding[166].

The sign of the electric field gradient has been determined for a relatively large number of compounds (Table 3.21). This is usually accomplished by the application of

Table 3.21 Signs of the quadrupole coupling in various compounds.[a]

Compounds with a positive sign	
cis-Sn(edt)$_2 \cdot$bipy	$\{Fe(SnCl_3)[P(OMe)_3]_5\}[BPh_4]$
$RSnCl(ox)_2$ (R = Bu, Ph)	$[Fe(C_5H_5)(CO)_2]_2SnX_2$ (X = Cl, NCS)
$BuSn(ox)_3$	$Fe(C_5H_5)(CO)_2SnCl_3$
$Ph_2SnCl(ox)$	$Mn(CO)_5SnCl_3$
$R_2Sn(ox)_2$ (R = Me, Ph)	$Mn(CO)_5SnCl_2Me$
Me_2Sn(sal-N-2-OC_6N_4)	SnO
$Bu_2Sn(O_2CCH{:}CHCO_2)_2$	SnS
Me_2SnO	SnF_2
Me_2SnMoO_4	$NaSnF_3$
Me_2SnCl_2	$NaSn_2F_5$
$[Et_4N][Me_2SnCl_3]$	$Sn(O_2CH)_2$
$Cs_2[Me_2SnCl_4]$	$Sn(O_2CMe)_2$
$[Me_4N][EtSnCl_5]$	SnC_2O_4
$K_2[Me_2SnF_4]$	$K_2Sn(C_2O_4)_2 \cdot H_2O$
$[Et_4N][Me_2SnBr_3]$	$SnSO_4$
$Ph_2Sn(S_2CNEt_2)_2$	$Sn_3(PO_4)_2$
$Ph_2Sn(NCS)_2 \cdot$phen	
Compounds with a negative sign	
$R_3SnC_6F_5$ (R = Me, Ph)	$[Me_4N][Me_3SnCl_2]$
$[Me_3Sn(bipyO)][BPh_4]$	$[Ph_3CH_2PPh_3][Et_3SnCl_2]$
Ph_3Sn(bzbz)	$[Me_4N][Ph_3SnCl_2]$
Ph_3Sn(ONPhCOPh)	Et_3SnCN
Ph_3Sn(ox)	$Bu_3SnOSnBu_3$
Ph_3Sn(sal-N-2-HOC_6H_4)	*trans*-Sn(edt)$_2 \cdot 2Et_2SO$
$Me_2SnCl(ox)$	$Fe(C_5H_5)(CO)_2SnBu_3$
Me_3SnO_2CMe	

(*a*) Data taken from R.V. Parish and C.E. Johnson, *J. Chem. Soc. (A)*, 1971, 1907; N.E. Erickson, *Chem. Commun.*, 1970, 1349; J.D. Donaldson, E.J. Filmore and M.J. Tricker, *J. Chem. Soc. (A)*, 1971, 1109; B.A. Goodman, R. Greatrex and N.N. Greenwood, *J. Chem. Soc. (A)*, 1971, 1868; S.R.A. Bird, J.D. Donaldson, A.F. LeC. Holding, B.J. Senior and M.J. Tricker, *J. Chem. Soc. (A)*, 1971, 1616; G.M. Bancroft, V.G. Kumar Das, T.K. Sham and M.G. Clark, *J. Chem. Soc., Dalton Trans.*, 1976, 643; R.C. Poller and J.N.R. Ruddick, *J. Chem. Soc., Dalton Trans.*, 1972, 555; E.T. Libbey and G.M. Bancroft, *J. Chem. Soc., Chem. Commun.*, 1973, 503; T.C. Gibb, B.A. Goodman and N.N. Greenwood, *Chem. Commun.*, 1970, 774; J.N.R. Ruddick and J.R. Sams, *J. Chem. Soc., Dalton Trans.*, 1974, 470; and P.G. Harrison and T.J. King, *J. Chem. Soc.*, 1974, 2298.

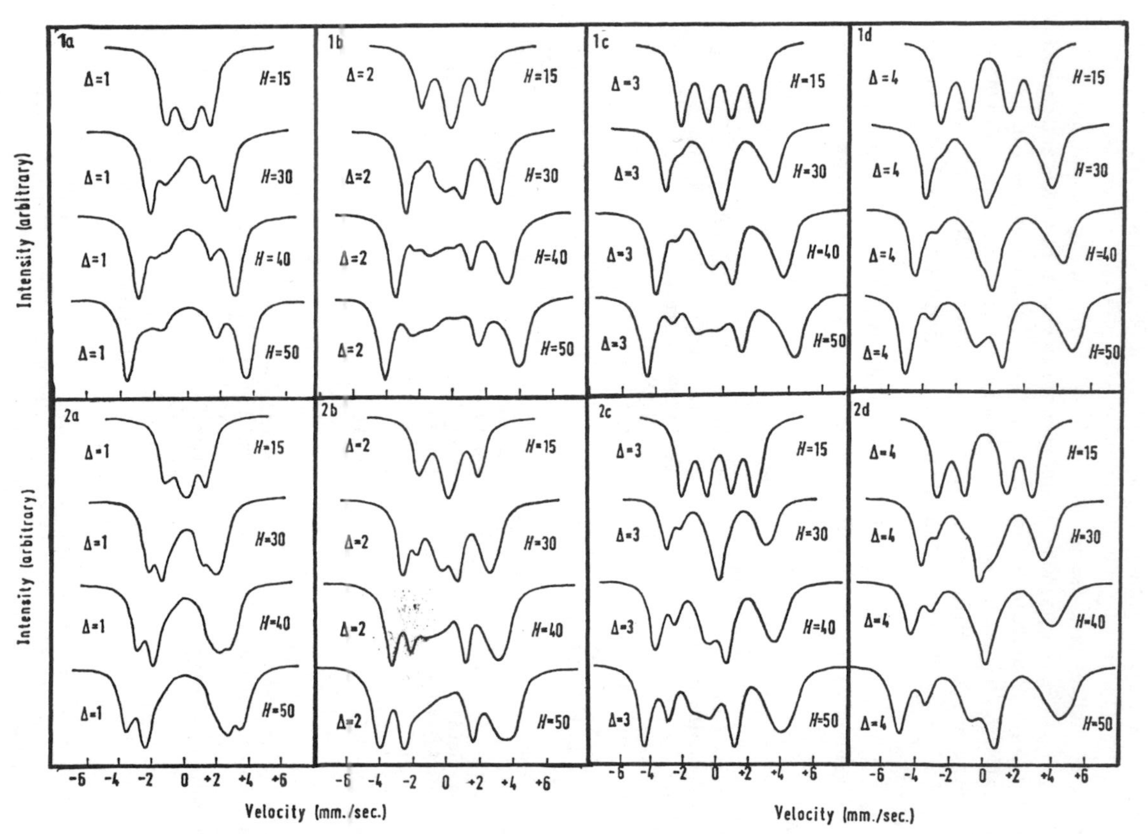

Figure 3.10 Calculated Mössbauer spectra for various magnitudes of magnetic field strength and (positive) quadrupole splittings. Adapted from ref. 167.

large external magnetic field which produces characteristic magnetically perturbed spectra. It should be noted that V_{zz} and the quadrupole splitting Δ are of opposite signs due to the negative quadrupole moment of ^{119}Sn. Gibb[167] has published calculated spectra for various magnitudes of quadrupole splitting and field strength (Figure 3.10), and several actual spectra have also been published[163,168–177]. In organotin(IV) compounds, positive quadrupole coupling constants are found when organic groups are situated along the principal (z) axis. Conversely, when the organic groups lie in the xy-plane, the coupling is negative. The positive values of Δ found for bivalent tin compounds are consistent with a dominating contribution from the tin non-bonding pair of electrons. It should be noted that the principal *electric* axis will not always coincide with the principal axis of 'local' symmetry, and therefore some care should be always exercised in interpretation.

Several attempts have been made to calculate quadrupole splitting values for various ligand configurations based upon a simple additive model in which the electric field gradient at the tin nucleus is made up from contributions from each ligand[140,178,180]. Each ligand is considered to contribute a 'partial quadrupole splitting', $[X]_{pqs}$, and the quadrupole splitting comprises the sum of these individual contributions. In spite of the inherent simplicity of the approach, the model gives quite good agreement with experimental data. Although the method is effective for the rationalization of quadrupole splitting values for compounds of known geometry, application of the technique as a predictive tool should be approached with some caution, since it can lead to erroneous conclusions. For example, the observed quadrupole splitting of tricyclohexyltin chloride (3.49 mm s^{-1}) would suggest a polymeric structure and trigonal bipyramidal coordination geometry at tin with the organic groups in the equatorial sites and bridging chlorines. In fact, crystals comprise discrete tetrahedral molecules, and the high value of the quadrupole splitting arises gross distortions from ideal tetrahedral geometry[181]. Nevertheless, in some cases the technique can be applied very usefully, and can be used to estimate the CSnC bond angle in six-coordinated dimethyltin compounds. The structures of these vary in geometry from *cis* (CSnC = *c.* 90°), through skew-trapezoidal (90° < CSnC < 180°), to *trans* (CSnC = *c.* 180°), for which predicted values of the quadrupole splitting increase smoothly from *c.* 2 mm s^{-1} to *c.* 4 mm s^{-1}. The calculated variation of the quadrupole splitting with CSnC bond angle against quadrupole splitting is shown in Figure 3.11, together with data for known compounds (Table 3.22)[182,183], from which the CSnC angle for dimethyltin compounds of unknown structure may be estimated with reasonable accuracy. The quite different values predicted for *cis*- and *trans*-octahedral $[SnX_4Y_2]$ configurations ($\Delta_{trans} = -2\Delta_{cis}$) has been used widely to assign geometries[164,165,170,184]. The predicted change in sign has been confirmed in the case of *cis*-$[Sn(edt)_2 \cdot bipy]$ ($\Delta = +1.17$ mm s^{-1}) and *trans*-$[Sn(edt)_2 \cdot 2Et_2SO)]$ ($\Delta = -1.94$ mm s^{-1})[170].

3.3.3 *The recoil-free fraction and its temperature coefficient*

The recoil-free fraction of the absorber, $f_a(T)$, can be expressed in terms of the vibrational properties of the crystal lattice according to

$$f_a(T) = \exp(-E_\gamma^2 \langle x_{iso}^2(T) \rangle / (\hbar c)^2) \tag{3.12}$$

where $\langle x_{iso}^2(T) \rangle$ is the mean-square amplitude of vibration of the nucleus in the

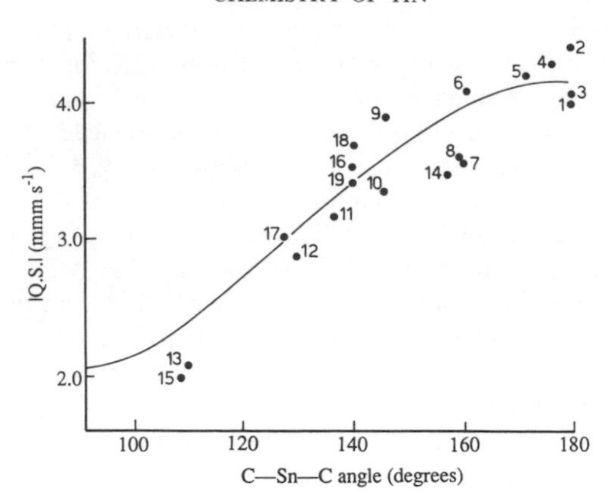

Figure 3.11 Plot of quadrupole splitting versus C–Sn–C bond angle for dimethyltin compounds (adapted from ref. 183). Compound numbers refer to Table 3.22.

direction of the γ-ray. Thus, $f_a(T)$ will be large if E_γ is small and if $\langle x_{iso}^2(T)\rangle$ is also small, i.e. if the tin atom is tightly bound in the lattice.

The precise form of $\langle x_{iso}^2(T)\rangle$ is complex, but is usually adequately described using the Debye theory of solids, which yields a more elaborate expression for $f_a(T)$:

$$f_a(T) = \exp\left\{\frac{-3E_\gamma^2}{Mc^2k\theta_D}\cdot\left[\frac{1}{4}+\left(\frac{T}{\theta_D}\right)^2\cdot\int_0^{\theta_D/T}\frac{x\,dx}{(e^x-1)}\right]\right\} \tag{3.13}$$

Table 3.22 CSnC angle and Mössbauer quadrupole splitting data for dimethyltin compounds.[a]

No.	Compound	C–Sn–C angle (deg)	Quadrupole splitting (mm s^{-1})
1	$Me_2SnCl_2\cdot 2pyO$	180	3.96
2	Me_2SnF_2	180	4.38
3	$Me_2Sn(acac)$	180	4.02
4	$Me_2SnCl_2\cdot(salenH_2)$	175–177	4.27
5	$Me_2SnCl_2\cdot 2dmso$	172	4.16
6	$Me_2SnCl_2\cdot Ni(salen)$	161.0	4.06
7	$Me_2Sn(salen)$	160.0	3.46
8	$Me_2SnCl(terpy)$	159	3.55
9	$Me_2Sn(NCS)_2$	145.9	3.87
10	$Me_2Sn(NMeCOMe)_2$	145.8	3.31
11	$Me_2SnSn(dtc)_2$	136	3.14
12	$Me_2Sn[S_2NC(CH_2)_4]_2$	130.0	2.85
13	$Me_2Sn(ox)_2$	110.7	2.02
14	$Me_2Sn(NHCOMe)_2\cdot H_2O$	156.8	3.42
15	$Me_2Sn(NHCOMe)_2$	109.1	1.99
16	$[Me_2SnCl_3]^-$	140	3.51
17	$Me_2SnCl(S_2NCR_2)$	128	2.98
18	$[Me_2Sn(OH)(NO_3)]_2$	139.9	3.67
19	$[Me_2Sn(NCS)_2O]_2$	140.5	3.38

(*a*) Data from T.K. Sham and G.M. Bancroft, *Inorg. Chem.*, 1975, **14**, 2281, *et loc. cit*; P.G. Harrison, T.J. King and J.A. Richards, *J. Chem. Soc., Dalton Trans.*, 1975, 826; and P.G. Harrison, T.J. King and R.C. Phillips, *J. Chem. Soc., Dalton Trans.*, 1976, 2317.

where ω_D is the maximum vibrational frequency of the lattice and θ_D is the characteristic Debye temperature of the lattice defined as $\hbar\omega_D/k$. In this model, $f_a(T)$ will be large for small E_γ, large M (the mass of the vibrating body), high θ_D (indicative of a strongly bound lattice), and at low temperatures. Accurate measurement of absolute values of $f_a(T)$ is far from easy. That for β-tin has been determined[185] using the technique of $X-\gamma$ delayed coincidences over the temperature range 2–360 K. Two other methods can be employed for the determination of absolute values of $f_a(T)$ for other compounds. The first utilizes the more accessible parameter, $A(T)$, the resonance area. For a thin absorber, the resonance area and $f_a(T)$ are related by the expression

$$A = \tfrac{1}{2}\cdot f_a \cdot \Gamma_a \cdot \pi \cdot L(T_a) \cdot t_a \tag{3.14}$$

where $L(T_a)$ is a function which shows how the area saturates as T_a, the 'effective thickness' of the absorber, is increased, and has been evaluated elsewhere[186], Γ_a is the half-height width of the absorber, and t_a is the number of resonant nuclei per unit area of cross-section in the absorber. Hence a comparison of the resonance area of the unknown sample with that of a standard, usually β-tin, under identical conditions will yield $f_a(T)$ for the unknown. The other involves calibration of the expression involving $\langle x^2_{iso}(T)\rangle$ (eqn 3.12) by an independent method, such as X-ray crystallography. Some values of $f_a(T)$ for inorganic compounds are given in Table 3.23. High values are observed for 'ionic' type lattices such as SnO_2, $BaSnO_3$ and $CaSnO_3$. Values for typical monomeric organotin (IV) compounds will be much smaller.

In the temperature range *c.* 77–150 K, tin compounds usually behave as Debye solids in the high-temperature limit ($T \geqslant \theta_D$). In this limit, the integral in eqn (3.13) reduces to θ_D/T, and $\ln f_a(T)$ is given by the expression

$$\ln(f_a(T)) = \frac{-3E_\gamma^2}{Mc^2k\theta_M^2}\cdot T - \frac{3E_\gamma^2}{4Mc^2k\theta_M} \tag{3.15}$$

which is linear in T, and where θ_M is a characteristic temperature, which for an ideal monatomic isotropic cubic solid is equivalent to the Debye temperature θ_D. Since for a 'thin' absorber $N\sigma_0 < 1$, where N = number of absorber atoms per cm^2 and σ_0 = resonant cross-section for ^{119}Sn, $A(T)$ is directly related to $f_a(T)$ (eqn 3.14), the temperature coefficient of the recoil-free fraction a is given by

$$a = \frac{d}{dT}\ln A(T) = \frac{d}{dT}\ln f_a(T) = \frac{-3E_\gamma^2}{Mc^2k\theta_M^2} \tag{3.16}$$

Since the determination of a only involves monitoring *relative* values of the resonance area $A(T)$ over a range of temperatures, usually 78 – ~150 K, on the same sample, it can usually be measured with a greater degree of confidence, and has therefore been employed much more extensively than has $f_a(T)$ in the investigation of lattice dynamics and lattice structure[187–197]. Data is usually normalized to the value at 78 K for convenience of intercomparison of samples, and a selection is presented in Table 3.24.

Several attempts have been made to correlate $-a$ with lattice structure. Observed values fall in the approximate range $0.5–3.0 \times 10^{-2}\,K^{-1}$, and as expected the lowest values are exhibited by the strongly bound 'ionic' type lattices exemplified by SnO_2, while solids composed of isolated, non-interacting molecules exhibit values > $\sim 2 \times 10^{-2}\,K^{-1}$. There is a general trend towards lower values of $-a$ as the degree of intermolecular association increases, and the lattice becomes more tightly bound.

Table 3.23 Absolute recoil-free fraction values.[a]

Compound	f	T(K)
β-Sn	0.057	290
SnO	0.35	295
$Sn(OMe)_2$	0.12	77
$Sn(OEt)_2$	0.15	77
$Sn(O^iPr)_2$	0.076	77
$Sn(C_6H_4Me)_2$	0.078	77
$Sn_6O_4(OMe)_4$	0.015	77
SnF_2	0.195	77
$SnCl_2$	0.116	77
$SnCl_2 \cdot 2H_2O$	0.16	77
$SnCl_2 \cdot SC(NH_2)_2$	0.145	77
$CsSnCl_3$	0.114	77
$(NH_4)_2SnCl_3 \cdot Cl \cdot H_2O$	0.13	77
SnO_2	0.55	290
$CaSnO_3$	0.57	290
$BaSnO_3$	0.59	290
	0.55	295
$Sn(TPP)(OH)_2$	0.100	77
	0.040	153
$Bu_2SnCl_2 \cdot$dppoe	0.44	77
	0.003	290
$(Bu_3Sn)_2SO_4$	0.240	77
$(Bu_3Sn)_2SeO_4$	0.246	77
$(Bu_3Sn)_2CrO_4$	0.302	77

(a) Data from R.H. Herber, *Phys. Rev. (B)*, 1983, **27**, 4013; P.G. Harrison, N.W. Sharpe, C. Pelizzi and P. Tarasconi, *J. Chem. Soc., Dalton Trans.*, 1983, 921, 1687; K.P. Mitrofanov, V.P. Gor'kev, M.V. Plotnikova and S.I. Reiman, *Nucl. Instrum. Methods*, 1978, **155**, 539; H. Sano and R.H. Herber, *J. Inorg. Nucl. Chem.*, 1968, **30**, 409; P.G. Harrison and B.J. Haylett, unpublished data; P.G. Harrison, K.C. Molloy and E.W. Thornton, *Inorg. Chim. Acta*, 1979, **33**, 137; and H. Sano and Y. Mekata, *Chem. Lett.*, 1975, 155.

However, as the data in Table 3.24 show, great care needs to be exercised, because values for lattices comprising efficiently packed monomeric, non-associated molecules can often exhibit values similar to those of related associated compounds. Indeed, for phenyltin compounds the range of $-a$ values observed for monomeric compounds encompasses the ranges for all other types of associated lattices[197].

3.3.4 *The effective vibrating mass model*

Substituting $\hbar\omega_L/k$ for θ_M in the expression for a (eqn 3.16) gives the relationship

$$a = \frac{-3E_\gamma^2 k}{M_{eff} c^2 \hbar^2 \omega_L^2} \tag{3.17}$$

where M_{eff} is the effective mass of the recoiling species, i.e. the $[SnL_n]$ species which vibrates as an integral unit within the lattice, and ω_L is the acoustic phonon vibration of

Table 3.24 Recoil-free fraction temperature coefficient (*a*) Values (K^{-1})[a].

Compound	$-a(\times 10^2)$	Structure
Alkyltin compounds		
Me_3SnO_2CH	1.90	1D-polymer
Me_3SnO_2CMe	1.62	1D-polymer
$Me_3SnO_2CC_5H_4N \cdot 2H_2O$	1.27	H-bonded 3D-polymer
$Me_3Sn(gly)$	1.15	1D-polymer
$Me_3SnO_3SPh \cdot H_2O$	1.71	H-bonded 1D-polymer
$Me_3SnONPhCOPh$	1.74	Monomer
$Me_3SnONC_6H_{10}$	0.97	1D-polymer
$Et_3SnONC_6H_{10}$	1.16	1D-polymer?
$^nPr_3SnONC_6H_{10}$	1.43	1D-polymer?
Me_2SnO	0.87	3D-polymer?
$Me_2Sn(ONMeCOMe)_2$	1.85	Monomer
$Me_2Sn(ONHCOMe)_2$	0.92	H-bonded 3D-polymer
$[ClMe_2SnOSnMe_2Cl]_2$	1.15	Cl-bridged 2D-polymer
$ClMe_2SnSnMe_2Cl$	1.93	Weakly associated
$^nPr_2SnCl_2 \cdot dppoe$	0.76	Cl-bridged
$^nBu_2SnCl_2 \cdot dppoe$	1.23	Cl-bridged
$^nPr_2SnCl_2 \cdot dppoet$	1.76	Monomer
$^nBu_2SnCl_2 \cdot dppoet$	1.56	Monomer
Phenyltin compounds		
Ph_4Sn	1.37	Monomer
$Ph_3SnC_6F_5$	2.17	Monomer
Ph_3SnH	2.10	Monomer
Ph_3SnF	1.49	1D-polymer
Ph_3SnCl	1.56	Monomer
Ph_3SnCN	1.73	Monomer
Ph_3SnNCS	1.84	1D-polymer
$Ph_3Sn(triaz)$	1.04	1D-polymer
$(Ph_3Sn)_2$	2.09	Monomer
$(Ph_3Sn)_2O$	1.56	Monomer
Ph_3SnOH	1.10	1D-polymer
$Ph_3SnONC_6H_{11}$	1.82	Monomer?
Ph_3SnO_2CCH	1.15	1D-polymer
Ph_3SnO_2CCMe	1.91	1D-polymer
$Ph_3Sn[O_2CCH_2(8\text{-}C_9H_6NO)] \cdot H_2O$	1.99	H-bonded 1D-polymer
$Ph_3SnS_2P(OEt)_2$	1.43	Monomer
$Ph_3SnS_2P(O^iPr)_2$	1.40	Monomer
Ph_2SnH_2	2.19	Monomer
Ph_2SnCl_2	1.54	Monomer
$Ph_2SnCl_2 \cdot H_2salen$	1.27	1D-polymer
$Ph_2SnCl_2 \cdot 0.75pyr$	1.79	Monomer, 1D-polymer
$Ph_2SnCl_2 \cdot dppoe$	1.04	1D-polymer
Ph_2SnO	1.15	3D-polymer?
$Ph_2Sn[S_2P(O^nPr)_2]_2$	1.06	Monomer
$PhSnCl_3$	1.97	Monomer
$PhSnCl_3 \cdot H_2salen$	1.94	1D-polymer
Cyclohexyltin compounds		
$(C_6H_{11})_3SnF$	0.91	1D-polymer?
$(C_6H_{11})_3SnCl$	1.41	Monomer
$(C_6H_{11})_3SnBr$	1.64	Monomer
$(C_6H_{11})_3SnI$	1.60	Monomer
$(C_6H_{11})_3SnOH$	0.66	1D-polymer
$(C_6H_{11})_3SnO_2CH$	0.93	1D-polymer

Table 3.24 (*Contd.*)

Compound	$-a(\times 10^2)$	Structure
$(C_6H_{11})_3SnO_2CMe$	1.59	1D-polymer
$(C_6H_{11})_3SnO_2CC_6H_4(o\text{-}N_2R)$	1.30	Monomer?
$(C_6H_{11})_3SnO_2PPh_2$	1.58	1D-polymer
$(C_6H_{11})_3Sn(triaz)$	1.31	1D-polymer
$(C_6H_{11})_3SnNCS$	1.34	1D-polymer
Inorganic compounds		
$Sn(porph)(OH)_2$	0.12	Monomer
SnF_2	0.65	2D-polymer
$SnCl_2$	1.11	$PbCl_2$-type lattice
$SnCl_2 \cdot 2H_2O$	0.89	2D-polymer
$SnCl_2 \cdot SC(NH_2)_2$	1.28	3D-polymer
$[NH_4]_2[SnCl_3]Cl \cdot H_2O$	1.29	3D-polymer
$Cs[SnCl_3]$	1.16	3D-polymer
$[Me_4N][SnCl_3]$	3.17	Unassociated
$[Me_4N][Sn(NCS)_3]$	2.42	Unassociated
(18-crown-6)$Sn(ClO_4)_2$	2.25	Unassociated
(18-crown-6)Sn_2Cl_4 (2 sites)	1.66, 2.83	Unassociated
$Sn(NCS)_2$	1.35	Associated?
$Sn(NMe_2)_2$	1.55	Dimer
$Sn(NCH_2CH_2)_2$	2.03	Associated?
$Sn[N(SiMe_3)_2]_2$	1.95	Monomer
SnO	0.23	2D-layer lattice
$Sn(OMe)_2$	2.38	Associated?
$Sn(OEt)_2$	2.79	Associated?
$Sn(O^iPr)_2$	1.38	Associated?
$Sn_6O_4(OMe)_4$	1.85	Monomer
$Sn(O_2CH)_2$	0.84	2D layer lattice
Organotin(II) compounds		
$Sn(C_5H_5)_2$	3.13	Monomer
$Sn(C_5H_4Me)_2$	1.02, 1.99[b]	Monomer?

(*a*) Data from R.H. Herber and A.E. Smelkinson, *Inorg. Chem.*, 1978, **17**, 1023; 202; P.G. Harrison, K.C. Molloy and E.W. Thornton, *Inorg. Chim. Acta*, 1979, **33**, 137; R.H. Herber, A.E. Smelkinson, M.J. Sienko and L.F. Schneemeyer, *J. Chem. Phys.*, 1978, **68**, 3705; R.H. Herber, *Phys. Rev.*, 1983, **27B**, 4013; R.H. Herber and R.F. Davis, *J. Inorg. Nucl. Chem.*, 1980, **42**, 1577; R.H. Herber, F.J. DiSalvo and R.B. Frankel, *Inorg. Chem.*, 1980, **19**, 3135; R.H. Herber and M. Katada, *J. Solid State Chem.*, 1979, **27**, 137; P.G. Harrison, M.J. Begley and K.C. Molloy, *J. Organomet. Chem.*, 1980, **186**, 213; P.G. Harrison, N.W. Sharp, C. Pelizzi, G. Pelizzi and P. Tarasconi, *J. Chem. Soc., Dalton Trans.*, 1983, **921**, 1687; P.G. Harrison, K. Lambert, T.J. King and B. Majee, *J. Chem. Soc., Dalton Trans.*, 1983, 363; K.C. Molloy, M.P. Bigwood, R.H. Herber and J.J. Zuckerman, *Inorg. Chem.*, 1982, **21**, 3709; V.G. Kumar Das, C. Wei, S.W. Ng and T.C.W. Mak, *J. Organomet. Chem.*, 1987, **322**, 33; K.C. Molloy, K. Quill, S.J. Blunden and R. Hill, *J. Chem. Soc., Dalton Trans.*, 1986, 875; and K.C. Molloy and K. Quill, *J. Chem. Soc., Dalton Trans.*, 1985, 1417.

(*b*) Exhibits a discontinuity at a transition temperature of 154.2 K.

this species which can be observed at low wavenumber in the Raman spectrum. Therefore, rearranging eqn (3.17), M_{eff} is given by

$$M_{eff} = \frac{-3E_\gamma^2 k}{ac^2\hbar\omega_D^2} \quad \text{kg molecule}^{-1} \tag{3.18}$$

which for ^{119}Sn can be reduced to

$$M = -\frac{1.0292 \times 10^4}{a\omega_D^2} \text{ g mole}^{-1} \tag{3.19}$$

where a is in units of K^{-1} and ω_L is in cm^{-1}. This method of determining the effective molecularity in the solid state has been used widely by Herber[198]. Although it obviously has uses, there remains some uncertainty regarding the correct assignment of the lattice vibrational modes.

3.3.5 *Anisotropic nuclear vibrations and the Goldanskii–Karyagin effect*

The anisotropies in the electric field gradient in the vicinity of the Mössbauer nucleus which give rise to quadrupole splittings also cause the nucleus to vibrate in an anisotropic manner, resulting in an inequality in the populations of the $\pm\frac{1}{2}$ and $\pm\frac{3}{2}$ excited states. This occurrence is manifested in the spectrum as an inequality in the areas of the resonance lines of a quadrupole split doublet, and is known as the Goldanskii–Karyagin effect[199,200]. It reflects the variation in the recoil-free fraction of the nuclide with θ, the angle between the direction of the γ-ray and the principal component of the electric field gradient, V_{zz}. For axial symmetry of the vibrational ellipsoid, the ratio of intensities of the two resonance lines A is given by

$$A = \frac{M_\pi}{M_\sigma} = \frac{\int_0^\pi (\exp(-\varepsilon\cos^2\theta)).(1+\cos^2\theta)\cdot\sin\theta\cdot d\theta}{\int_0^\pi (\exp(-\varepsilon\cos^2\theta))\cdot(\frac{5}{3}-\cos^2\theta)\cdot\sin\theta\cdot d\theta} \tag{3.20}$$

where M_π and M_σ refer to the $(\pm\frac{1}{2}\leftrightarrow\pm\frac{3}{2})$ and the $(\pm\frac{1}{2}\leftrightarrow\pm\frac{1}{2})$ transitions, respectively, and the vibrational asymmetry parameter is defined as

$$\varepsilon = \frac{-E_\gamma^2}{(\hbar c)^2}\cdot(\langle x_\parallel^2\rangle - \langle x_\perp^2\rangle) \tag{3.21}$$

where the subscripts $\parallel$ and $\perp$ refer to vibrations parallel and perpendicular to V_{zz}. Thus, an analysis of the variation of the intensity ratio A can yield a measure of the vibrational anisotropy of the nucleus in the lattice. However, the problem is to relate A to the ratio I_+/I_- ($= R$), the ratio of the intensities of the resonance lines at higher and lower isomer shifts, which requires a knowledge of the sign of V_{zz} (i.e. which of the resonance lines represents the M_π and M_σ transitions). In addition, the function in eqn (3.20) has to be evaluated. The form of this function is shown in Figure 3.12 and numerical values are listed elsewhere[201]. The sign of V_{zz} is determined by the shape of the electric field gradient. Because for ^{119}Sn the nuclear quadrupole moment Q is negative, for an oblate field V_{zz} is positive and the M_π transition occurs at a less positive isomer shift (lower energy) than the M_σ transition. Conversely, for a prolate electric field gradient, V_{zz} is negative and the M_σ transition is at lower energy. The sign of V_{zz} can be determined either directly by the application of a large external magnetic field to the sample (section 3.3.2), or by comparison of the unknown compound with one for the sign has been determined (Table 3.21). However, in some cases, the correct sign may

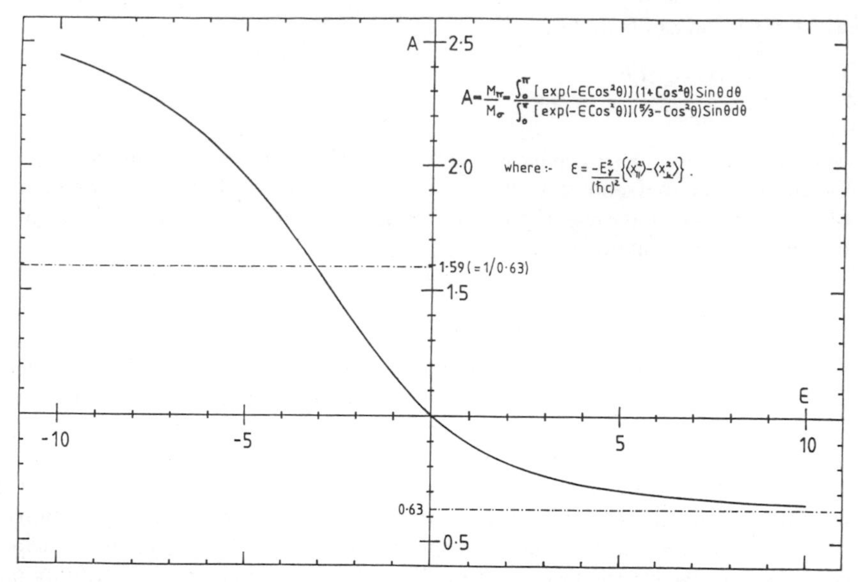

Figure 3.12 Plot relating the Goldanskii-Karyagin ratio, A, to the vibrational asymmetry parameter, ε, of the Mössbauer nuclide.

be deduced by calculating separate values of $[\langle x_\parallel^2 \rangle - \langle x_\perp^2 \rangle]$ for both $A = R$ and $A = 1/R$, whereby one value will be obviously unacceptably large.

The application of the method is well illustrated by $\alpha,\beta,\gamma,\delta$-tetraphenyldihydroxytin(IV) (TPP)$Sn(OH)_2$[202], whose crystal structure determination could not be refined anisotropically, i.e. no information concerning the anisotropic vibration of the tin atom was obtained from the crystallographic data. However, the vibrational anisotropy can be deduced from the Goldanskii–Karyagin effect data. Since V_{zz} which is collinear with the O–Sn–O axis, is negative, the electric field is prolate with $A = R$, which, being greater than unity, implies that ε is negative also. Thus $\langle x_\parallel^2 \rangle$ is less than $\langle x_\perp^2 \rangle$, i.e. the tin atom vibrates with the greater amplitude in the plane of the porphinato ring (an oblate vibration), an observation not wholly expected in view of the constraining nature of the porphinato ring. However, the amplitudes of vibration are consistent with the observed bond lengths (Sn–N (in plane) 209 pm; Sn–O (out of plane) 201.5 pm). Comparison with the analogous dichloride, (TPP)$SnCl_2$, where V_{zz} is negative (indicating a prolate electric field) but ε positive (indicating a prolate vibration), shows that the greater amplitude of vibration is now parallel with V_{zz} and along the Cl–Sn–Cl axis (Sn–Cl 242 pm). Hence, even with very similar molecules, V_{zz} and ε do not necessarily have the same sign, and the tin atom tends to vibrate with greater amplitude in the direction in which the bond distances are longer. In addition, the temperature coefficient of the vibrational amplitude along the O–Sn–O direction is approximately three times that within the porphinato plane, reflecting the rigid constraining nature of the ring compared with the more easily translated hydroxyl groups.

3.4 Mass spectra

In natural abundance tin has ten stable isotopes, the characteristic abundance distribution (see Table 1.1) of which makes the identification of tin-containing ions very facile. Further, ions containing two or more tin atoms as well as tin with other polyisotopic elements can similarly be readily identified, drastically simplifying the assignment of spectra. Examples are illustrated in Figure 3.13, with exact mass and abundance data in Table 3.25. Electron impact (EI) is generally the preferred method of examination, but other techniques such as chemical ionization (CI) and fast atom bombardment (FAB) have also been employed.

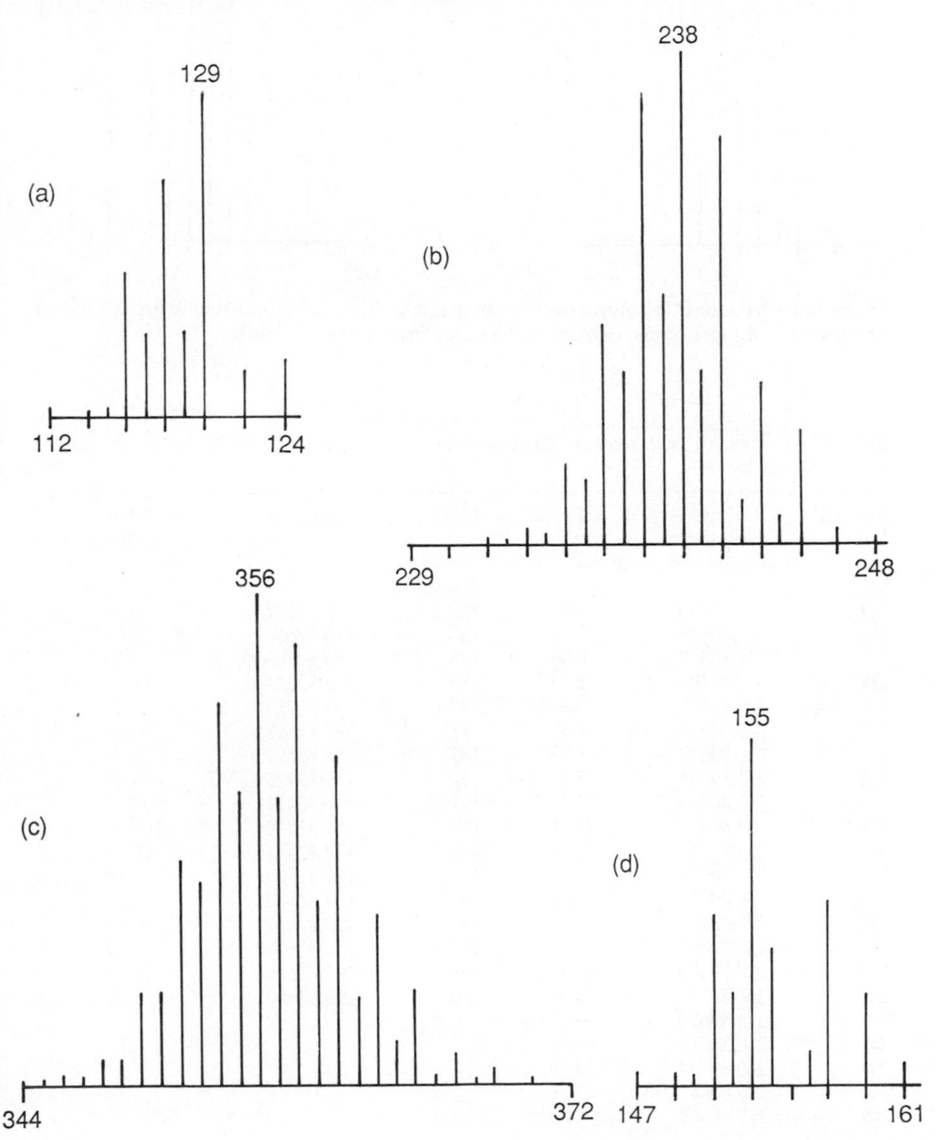

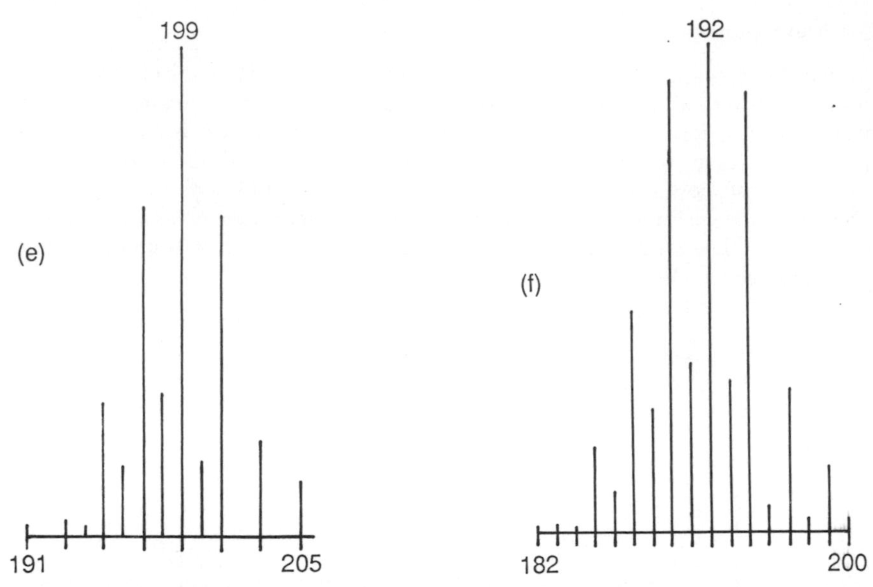

Figure 3.13 Characteristic isotopic distribution patterns for tin in natural abundance (*a*), and for combinations of polyisotopic elements: (*b*) Sn_2; (*c*) Sn_3; (*d*) SnCl; (*e*) SnBr; (*f*) SnGe.

Table 3.25 Calculated peak masses and relative abundances for SnCl, SnBr, SnGe, Sn_2 and Sn_3 combinations.

Nominal mass	Peak mass (wtd. mean)	Relative abundance	Nominal mass	Peak mass (wtd. mean)	Relative abundance
Sn_2:			SnGe:		
224	223.80988	0.05	182	181.82922	0.95
226	225.80790	0.07	184	183.82687	1.92
227	226.80847	0.04	185	184.82806	0.70
228	227.80703	1.53	186	185.82627	16.81
229	228.80796	0.82	187	186.82719	8.28
230	229.80625	3.57	188	187.82508	44.69
231	230.80697	1.99	189	188.82594	24.70
232	231.80491	16.78	190	189.82456	93.46
233	232.80523	13.51	191	190.82477	34.13
234	233.80419	44.41	192	191.82372	100.00
235	234.80513	35.04	193	192.82492	30.60
236	235.80423	92.44	194	193.82388	79.89
237	236.80519	50.83	195	194.82547	5.03
238	237.80417	100.00	196	195.82477	28.80
239	238.80565	35.64	197	196.82860	2.27
240	239.80476	82.47	198	197.82616	12.40
241	240.80759	9.53	200	199.82660	2.26
242	241.80626	33.23			
243	242.80863	5.70			
244	243.80734	23.16			
246	245.80865	3.13			
248	247.81048	1.99			

Table 3.25 (*Contd.*)

Nominal mass	Peak mass (wtd. mean)	Relative abundance	Normal mass	Peak mass (wtd. mean)	Relative abundance
SnCl:			Sn_3:		
147	146.87380	2.33	344	343.70944	0.52
149	148.87150	2.35	345	344.71009	0.50
150	149.87239	0.83	346	345.70878	1.92
151	150.87093	35.46	347	346.70951	1.77
152	151.87188	18.85	348	347.70793	6.75
153	152.87022	70.24	349	348.70822	6.65
154	153.87152	27.07	350	349.70703	18.82
155	154.87036	100.00	351	350.70750	19.15
156	155.86929	6.82	352	351.70669	45.59
157	156.86932	37.77	353	352.70728	41.27
159	158.87312	18.42	354	353.70652	77.95
161	160.87114	4.75	355	354.70735	59.19
			356	355.70663	100.00
SnBr:			357	356.70764	57.39
191	190.82329	1.68	358	357.70690	89.12
193	192.82129	2.80	359	358.70847	37.75
194	193.82188	0.60	360	359.70770	65.05
195	194.82041	26.34	361	360.70993	18.01
196	195.82134	13.99	362	361.70883	35.56
197	196.81952	67.20	363	362.71084	8.71
198	197.82066	28.32	364	363.70976	19.73
199	198.81950	100.00	365	364.71257	1.64
200	199.81973	14.88	366	365.71123	5.97
201	200.81889	67.51	367	366.71387	0.67
203	202.82192	18.75	368	367.71255	2.85
205	204.82158	10.37	370	369.71389	0.37
			372	371.71572	0.15

(*a*) Data from D.B. Chambers, F. Glockling and M. Weston, *J. Chem. Soc. (A)*, 1967, 1759; D.B. Chambers and F. Glockling, *J. Chem. Soc. (A)*, 1968, 735; and P.G. Harrison, *J. Organomet. Chem.*, 1973, **47**, 89.

By far the majority of mass spectral data for tin compounds refers to organotin derivatives, for which the principal fragmentation routes have been established. Most decomposition processes are of the type $SnR^+ \rightarrow Sn^+ + R$ rather than $SnR^+ \rightarrow R^+ + Sn$, so that most of the ion current is carried by tin-containing species.

Tetraethynyltin, $Sn(C{\equiv}CH)_4$, fragments in a very simple manner with stepwise loss of ethynyl groups[203]. However, other tetraorganostannanes exhibit more complex behaviour. Fragmentation patterns for Me_4Sn, Et_4Sn, $(CH_2{=}CH)_4Sn$, Ph_4Sn, Ph_3SnX (X = halogen), and $(C_6F_5)_2SnR_2$ are shown in Figure 3.14, and serve to illustrate most of the decomposition pathways which can occur[204–207]. These are dominated by the influence of the odd- or even-electron nature of the ions involved. The most abundant ions are the tri-coordinated even-electron ions $[SnR_3^+]$, which together with other tri-coordinated ions and $[SnR^+]$ ions carry most of the ion current. Because of the very much weaker Sn–C bonds in the molecular ions, parent molecular ions are invariably of very low abundance, and in some cases (e.g. for tetravinyltin) are not observed at all. Parent ions decompose mainly by the elimination of an odd-electron neutral fragment, whereas $[SnR_3^+]$ ions lose even-electron fragments.

(a)

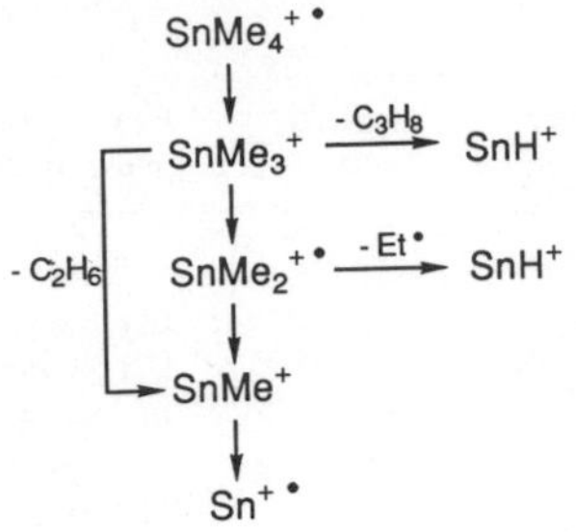

$SnMe_4^{+\bullet}$

$SnMe_3^{+}$ $\xrightarrow{-C_3H_8}$ SnH^{+}

$-C_2H_6$

$SnMe_2^{+\bullet}$ $\xrightarrow{-Et^{\bullet}}$ SnH^{+}

$SnMe^{+}$

$Sn^{+\bullet}$

(b)

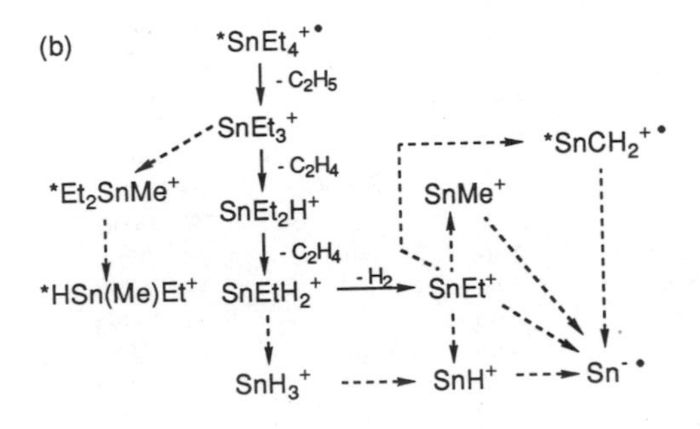

(c)

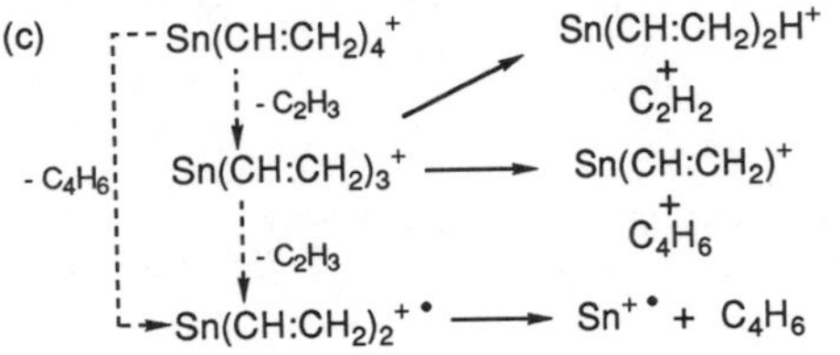

$Sn(CH{:}CH_2)_4^{+}$ $\rightarrow$ $Sn(CH{:}CH_2)_2H^{+}$ + C_2H_2

$-C_2H_3$

$-C_4H_6$

$Sn(CH{:}CH_2)_3^{+}$ $\rightarrow$ $Sn(CH{:}CH_2)^{+}$ + C_4H_6

$-C_2H_3$

$Sn(CH{:}CH_2)_2^{+\bullet}$ $\rightarrow$ $Sn^{+\bullet}$ + C_4H_6

(d)

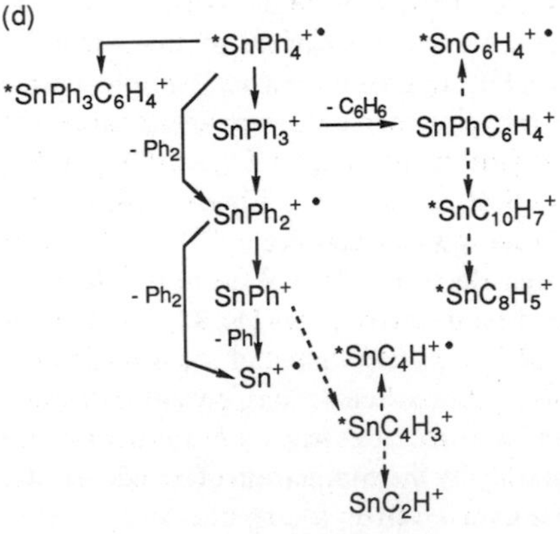

$^{*}SnPh_4^{+\bullet}$ $\qquad$ $^{*}SnC_6H_4^{+\bullet}$

$^{*}SnPh_3C_6H_4^{+}$

$SnPh_3^{+}$ $\xrightarrow{-C_6H_6}$ $SnPhC_6H_4^{+}$

$-Ph_2$

$SnPh_2^{+\bullet}$ $\qquad$ $^{*}SnC_{10}H_7^{+}$

$-Ph_2$

$SnPh^{+}$ $\qquad$ $^{*}SnC_8H_5^{+}$

$-Ph$

$Sn^{+\bullet}$ $\qquad$ $^{*}SnC_4H^{+\bullet}$

$^{*}SnC_4H_3^{+}$

SnC_2H^{+}

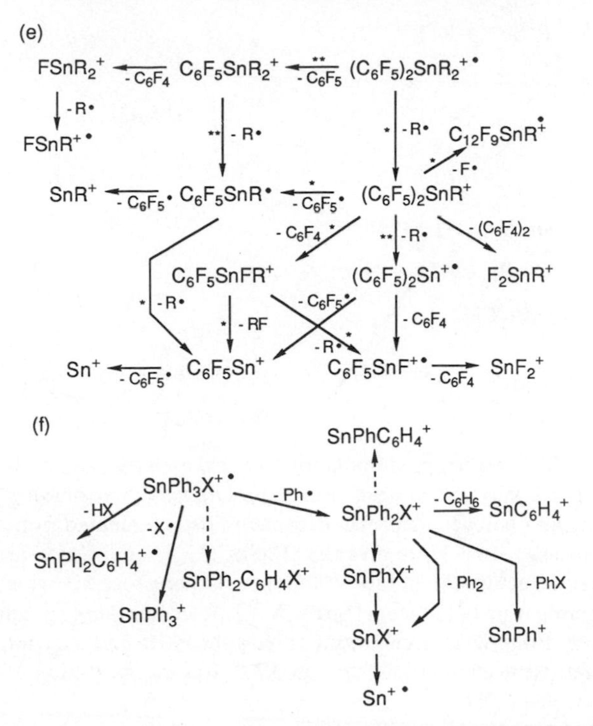

Figure 3.14 Fragmentation patterns for (*a*) Me_4Sn; (*b*) Et_4Sn; (*c*) $(CH_2{:}CH)_4Sn$; (*d*) Ph_4Sn; (*e*) $R_2(C_6F_5)_2Sn$; and (*g*) Ph_3SnX (adapted from refs. 204–207).

Several modes of decomposition have been identified. Elimination of a neutral radical is the major process for odd-electron ions (eqn 3.22), hence the high abundance of $[SnR^+]$ ions from the parent ion. Alkene elimination (eqn 3.23) is probably general for even-electron ions containing the $SnCH_2CHR_2$ grouping, and is a predominant process for ethyltin ions. The proportion of tin hydride ions increases with the number of ethyl groups bound to tin, and the process most probably occurs via a β-hydrogen transfer (eqn 3.24). A similar process eliminating ethyne is observed with from the $[Sn(CH{:}CH_2)_3]^+$ ion (eqn 3.25).

$$\geq Sn{-}R^{+\cdot} \longrightarrow \geq Sn^+ + R^\cdot \qquad (3.22)$$

$$> SnEt^+ \longrightarrow > SnH^+ + C_2H_4 \qquad (3.23)$$

$$\geq Sn(CH_2CH_2H) \longrightarrow > SnH^+ + C_2H_4 \qquad (3.24)$$

$$(CH_2{:}CH)_2\overset{+}{Sn}(CH{=}CH{-}H) \longrightarrow (CH_2{:}CH)_2SnH^+ + C_2H_2 \qquad (3.25)$$

$$R''SnRR'^{+} \rightarrow R''Sn^{+} + R\text{-}R' \quad (3.26)$$

$$SnPh_3^{+} \longrightarrow [PhSnC_6H_4]^{+} + C_6H_6 \quad (3.27)$$

(6)

$$SnPh_3F^{+\cdot} \rightarrow SnPh_2C_6H_4^{+\cdot} + HF \quad (3.28)$$

$$SnXPh_3^{+\cdot} \rightarrow XSnPh_2C_6H_4^{+\cdot} + H^{\cdot} \quad (3.29)$$

$$SnEt_3^{+} \rightarrow SnEt_2CH_3^{+} + CH_2 \quad (3.30)$$

$$SnPh^{+} \xrightarrow{-C_2H_2} SnC_4H_3^{+} \xrightarrow{-C_2H_2} SnC_2H^{+}; \quad SnC_4H_3^{+} \xrightarrow{-H_2} SnC_4H^{+} \quad (3.31)$$

Elimination of R–R′ occurs predominantly from even-electron ions when R and R′ are hydrogen, phenyl, vinyl or halogen (eqn 3.26). Diphenyl(halogeno)tin ions eliminate both biphenyl and halogenobenzene. Benzene is also eliminated from tricoordinated phenyltin ions (e.g. eqn 3.27) to give the $[PhSnC_6H_4]^+$ ion, which probably has the tricoordinated structure (**6**), in low abundance. Other processes which are observed include the elimination of HX from $[SnPh_3X^{+\cdot}]$ molecular ions (e.g. eqn 3.28), loss of a hydrogen atom from the molecular ions of tetraphenyltin and the triphenyltin halides (eqn 3.29), methylene elimination (e.g. eqn 3.30), and fragmentation of phenyl groups attached to tin (eqn 3.31).

The fragmentation behaviour of fluorophenylstannanes is generally similar, although some additional decomposition pathways are also observed. In particular, the neutral species C_6F_4 and RF molecules can be eliminated from tri-coordinated tin ions (e.g. eqn 3.32, 3.33).

$$(C_6F_5)_3Sn^{+} \rightarrow (C_6F_5)_2SnF^{+} + C_6F_4 \quad (3.32)$$

$$C_6F_5Sn(F)C_6H_5^{+} \rightarrow C_6F_5Sn^{+} + C_6H_5F \quad (3.33)$$

Molecular ions are always observed for metal–metal bonded compounds such as $Et_3SnSnEt_3$, $Ph_3SnSnPh_3$, $Ph_3SnSnMe_3$, and $Me_3SnGePh_3$[204,206]. Radical loss from the molecular ion is a major process, and gives rise to four ions (eqn 3.34). Rearrangement occurs in the molecular ions derived from the unsymmetrical molecules, either before elimination of a neutral species, or as a synchronous process yielding rearranged ions (eqn 3.35). Subsequent fragmentation is dominated by the elimination of neutral molecules from the even-electron ions thus produced. Bond cleavage of the molecular ion proceeds largely in the direction expected for the weakest bond in the molecule. The most pronounced effect is the low abundance of ions arising from phenyl–metal cleavage from the parent. For $Me_3SnGePh_3$ and $Ph_3SnSnMe_3$ primary cleavage of an Sn–Me bond gives an ion of high abundance, and bond dissociation energies suggest that in these molecules the Sn–Me bond is weaker than the Sn–Ge or Sn–Sn bonds. Similarly, in $Ph_3SnGeMe_3$ the low abundance of SnGe species is consistent with the Sn–Ge bond being the weakest. For hexaethyldistannane, the loss of ethene is a process of very low activation energy, and almost 70% of the ion current is carried by ditin ions. In contrast, for hexaphenyldistannane less than 10% of the current is carried by ions containing tin–tin bonds, and the most favoured

Table 3.26 Abundances of tin-containing fragments in the mass spectra of triorganotin oximes.[a]

Assignment	$(CH_3)_3SnON{=}C(CH_3)_2$ Mass	Rel. intens.	$(CH_3)_3SnON{=}C_6H_{10}$ Mass	Rel. intens.
$(CH_3)_3SnON{=}C<^+$	236	23.7	276	14.4
$(CH_3)_2SnON{=}C<^+$	221	59.2	261	34.4
$CH_3SnON{=}C<^+$	206	1.32	...	...
$SnON{=}C<^+$	191	17.9	231	6.2
$(CH_3)_3Sn^+$	164	100	164	100
$(CH_3)_2Sn^+$	149	12.1	149	15.2
CH_3Sn^+	134	19.7	134	30.3

(*a*) Data from P.G. Harrison and J.J. Zuckerman, *Inorg. Chem.*, 1970, **9**, 175.

decomposition modes involve tin–tin bond cleavage in the molecular ion.

$$R_3Sn{\cdot}MR_3'^{+\cdot} \rightarrow R_3Sn{\cdot}MR_2'^{+} + R_2Sn{\cdot}SnR_3'^{+} + R_3Sn^+ + R_3'Sn^+ \quad (3.34)$$

$$R_3Sn{\cdot}MR_3'^{+\cdot} \rightarrow R_2R'Sn^+ + RR_2'Sn^+ + R_3'Sn^+ + RR_2'M^+ + R_2R'M^+ + R_3M^+ \quad (3.35)$$

Fragmentation patterns derived from functionally-substituted organotin compounds of the types R_nSnX_{4-n} comprise ions arise from the sequential loss of organic groups R and electronegative groups X from tin as illustrated by spectra of trimethyltin oxime derivatives[209] (Table 3.26). Spectra are invariably dominated by tricoordinated even-electron ions, and loss of R–R and R–X are also sometimes observed. When the electronegative ligands attached to tin are complex, several additional decomposition routes can also be observed as the ligand also undergoes fragmentation. For example, in the spectra of methyltin nitrates loss of $NO_2\cdot$ from $[SnNO_3^+]$ fragments is widespread[210], and ring contraction processes involving the formal loss of [O] and [PhN] from the heterocyclic $\overline{O{\cdot}NPh{\cdot}CO{-}Sn}$ heterocyclic fragments in triorganotin *N*-acylhydroxylamines[211].

Tin(II) compounds also decompose by the sequential loss of both ligands, with the unicoordinated ion $[SnL^+]$ being the most abundant. Again, much decomposition of the ligands bound to tin is observed. Thus, in the spectrum of $(C_5H_5)_2Sn$, besides the parent molecular ion, $[(C_5H_5)_2Sn^+]$, and the $[C_5H_5Sn^+]$ and $[Sn^+]$ ions, the cyclopropyltin, $[SnC_3H_3^+]$, and ethynyltin, $[SnCCH^+]$, ions are seen[212]. Bis(pentane-2, 5-dionato)tin (II) derivatives exhibit acetatotin fragments derived from ring contraction reactions involving loss of an alkyne (eqn 3.36). Fragmentation patterns for the perfluoromethyl homologues are quite complex (e.g. Figure 3.15) due to the presence of fluorine in the molecule[213].

(acac)$_n$Sn(O–C(CH$_3$)=CH–C(CH$_3$)–O)$^+$ → [(acac)$_n$Sn(O$_2$C–CH$_3$)]$^+$ + CH$_3$C≡CH

n = 0,1 (3.36)

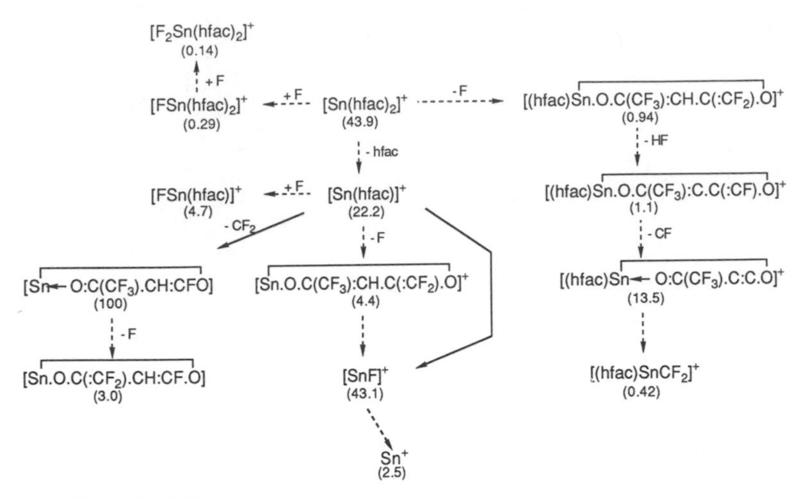

Figure 3.15 Fragmentation pattern for $Sn(hfac)_2$ (adapted from ref. 213).

Softer ionization techniques such as positive ion fast atom bombardment (FAB)[214] and chemical ionization (CI) mass spectrometry have met with mixed success. The former should in principle be an excellent method for studying thermally unstable or labile compounds which are difficult to examine by usual electron impact methods. Several types of matrix liquids, including sulfolane, glycerol, glycerol/hmpa, thioglycerol, diamylphenol, 18-crown-6/tetraglyme, and *p*-nitrophenyl-octylether have been employed, the latter being a particularly effective matrix. Unfortunately, early results from the application of this technique to organotin compounds have been disappointing. Parent molecular ions are no more intense than in usual EI spectra, and spectra from both techniques are quite similar. The major difference between FAB and EI spectra for the phenyltin chlorides, Ph_nSnCl_{4-n} ($n = 1$–3), is the preferential loss of halogen compared to phenyl groups, the reverse to that observed for EI, suggesting that primary ionization occurs in the condensed phase. A similar trend is also observed for the FAB spectra of hmpa adducts of the phenyltin halides, and also in this case a greater proportion of ions containing the hmpa ligand bound to tin is found[215,216].

In chemical ionization (CI) mass spectrometry, sample molecules are ionized via ion–molecule reactions in the gas phase. Thus, reagent gas ions generated by reactions of high-energy electrons with the reagent gas, e.g. CH_5^+ and $C_2H_5^+$ from CH_4 or Me_3C^+ from isobutane, react with the sample, causing the sample molecules to ionize in the ion sources. The method has been applied with some success in the study of electrophilic cleavage reactions of organotin compounds in the gas phase free from solvation effects. Thus, cleavage reactions of simple organotin compounds of the types R_4Sn, R_3SnR', $R_2SnR'_2$, R_3SnX, R_2SnX_2, and $R_2SnR'(X)$ have been examined using methane as the reagent gas, and the results obtained parallel to some extent solution data using the proton as the electrophile[217,218].

References

1. J.M. Bellama and R.A. Gsell, *Inorg. Nucl. Chem. Lett.*, 1971, **7**, 365.
2. J.P. Clark, V.M. Langford and C.J. Wilkins, *J. Chem. Soc. (A)*, 1967, 792.

3. D. Potts, H.D. Sharma, A.J. Carty and A. Walker, *Inorg. Chem.*, 1974, **13**, 1205.
4. A. Marchand, J. Mendelsohn and J. Valade, *Compt. Rend.*, 1964, **259**, 1737.
5. H.U. Ulrich, W.P. Neumann and T. Apoussidis, *Liebigs Ann. Chem.*, 1981, 1190.
6. K. Licht, H. Geissler, P. Koehler, K. Hoffmann, H. Schnorr and H. Kriegsmann, *Z. anorg. allg. Chem.*, 1971, **385**, 271.
7. R.J.H. Clark, A.G. Davies and R.J. Puddephatt, *J. Chem. Soc. (A)*, 1968, 1828.
8. J.D. Brown, D. Tevault and K. Nakamoto, *J. Mol. Struct.*, 1977, **40**, 43.
9. G.A. Ozin and A. Vander Voet, *J. Chem. Phys.*, 1972, **56**, 4769.
10. N.G. Dance, W.R. McWhinnie and R.C. Poller, *J. Chem. Soc., Dalton Trans.*, 1976, 2349.
11. R.H. Herber, *Polyhedron*, 1985, **4**, 1969.
12. G. Masetti and G. Zerbi, *Spectrochim. Acta*, 1970, **26A**, 1891.
13. G. Davidson, P.G. Harrison and E.M. Riley, *Spectrochim. Acta*, 1972, **29A**, 1265.
14. B. Busch and K. Dehnicke, *J. Organomet. Chem.*, 1974, **67**, 237.
15. L. Verdonck and Z. Eeckhaut, *Spectrochim. Acta*, 1972, **28A**, 433.
16. B. Mathiasch, *Z. anorg. allg. Chem.*, 1974, **403**, 225.
17. M. Goldstein and W.D. Unsworth, *J. Chem. Soc. (A)*, 1971, 2121.
18. D.F. van de Vondel, E.V. van den Berghe and G.P. van der Kelen, *J. Organomet. Chem.*, 1970, **23**, 105.
19. J.W. Ypenburg and H. Gerding, *Rec. Trav. Chim. Pays-Bas*, 1972, **91**, 1117.
20. P.G. Harrison and S.R. Stobart, *Inorg. Chim. Acta*, 1973, **7**, 306.
21. J.W. Anderson, G.K. Barker, J.E. Drake and M. Rodger, *J. Chem. Soc., Dalton Trans.*, 1973, 1716.
22. A. Marchand, C. Lemerle, M.T. Forel and M.H. Soulard, *J. Organomet. Chem.*, 1972, **42**, 353.
23. A. Bos and J.S. Ogden, *J. Phys. Chem.*, 1973, **77**, 1513.
24. A. Bos and A.T. Howe, *J. Chem. Soc., Faraday II Trans.*, 1975, **71**, 28.
25. D. Tevault and K. Nakamoto, *Inorg. Chem.*, 1976, **15**, 1282.
26. F. Königer and A. Müller, *J. Mol. Spectrosc.*, 1975, **56**, 2009.
27. N. Ohkaku and K. Nakamoto, *Inorg. Chem.*, 1973, **12**, 2440.
28. N. Ohkaku and K. Nakamoto, *Inorg. Chem.*, 1973, **12**, 2446.
29. S.J. Ruzicka and A.E. Meerbach, *Inorg. Chim. Acta*, 1976, **20**, 221.
30. A. Ashworth, E. Clark and P.G. Harrison, *J. Chem. Soc., Chem. Commun.*, 1987.
31. In some instances the ^{117}Sn isotope has to be used owing to extralaboratory effects. Attempts to record ^{119}Sn spectra at the Royal Holloway College, University of London, were found to be affected by spurious signals. These were attributed to interference from aircraft at Heathrow Airport which use the same frequency (*c.* 109 MHz) as the spectrometer! A.G. Davies and S. Slater, personal communication.
32. D.M. Doddrell, D.T. Pegg, W. Brooks and M.R. Bendall, *J. Amer. Chem. Soc.*, 1981, **103**, 727.
33. J.D. Kennedy, W. McFarlane, G.S. Pyne and B. Wrackmeyer, *J. Chem. Soc., Dalton Trans.*, 1975, 386.
34. T. Birchall and A.R. Pereira, *J. Chem. Soc., Dalton Trans.*, 1975, 1087.
35. H.G. Kuivila, J.L. Considine, R.H. Sarma and R.J. Mynott, *J. Organomet. Chem.*, 1976, **111**, 179.
36. M. Bullpitt, W. Kitching, W. Adcock and D. Doddrell, *J. Organomet. Chem.*, 1976, **116**, 187.
37. T.N. Mitchell and G. Walter, *J. Organomet. Chem.*, 1976, **121**, 177.
38. R.M. Davidson, H.G. Grant and D. O'Smith, *Spectrochim. Acta*, 1985, **41A**, 581.
39. T.A.K. Al-Allaf, *J. Organomet. Chem.*, 1986, **306**, 337.
40. T.N. Mitchell, *Org. Mag. Res.*, 1976, **8**, 34.
41. J. Holecek, M. Nadvornik, K. Handlir and A. Lycka, *J. Organomet. Chem.*, 1983, **241**, 177.
42. J. Holecek, K. Handlir, M. Nadvornik and A. Lycka, *J. Organomet. Chem.*, 1983, **258**, 147.
43. M. Nadvornik, J. Holecek, K. Handlir and A. Lycka, *J. Organomet. Chem.*, 1984, **275**, 43.
44. J. Holecek, M. Nadvornik, K. Handlir and A. Lycka, *J. Organomet. Chem.*, 1986, **315**, 299.
45. A. Lycka, J. Jirman, A. Kolonicny and J. Holecek, *J. Organomet. Chem.*, 1987, **333**, 305.
46. J. Holecek, K. Handlir, M. Nadvornik, S.M. Teleb and A. Lycka, *J. Organomet. Chem.*, 1988, **339**, 61.
47. P.J. Smith and L. Smith, *Inorg. Chim. Acta Rev.*, 1973, **7**, 11.
48. P.J. Smith and A. Tupciauskas, *Ann. Rep. NMR Spectrosc.*, 1978, **8**, 291.
49. R. Hani and R.A. Geanangel, *Coord. Chem. Rev.*, 1982, **44**, 229.
50. V.S. Petrosyan, *Progr. NMR Spectrosc.*, 1977, **11**, 115.

51. B. Wrackmeyer, *Ann. Rept. NMR Spectrosc.*, 1985, **16**, 73.
52. J.D. Kennedy and W. McFarlane, *Revs. Si, Ge, Sn and Pb Compds.*, 1974, **1**, 235.
53. R.K. Harris, J.D. Kennedy and W. McFarlane, *N.m.r. and the Periodic Table*, eds. R.K. Harris and B.E. Mann, Academic Press, London, 1978, 310.
54. J. Otera, *J. Organomet. Chem.*, 1981, **221**, 57.
55. J.D. Kennedy and W. McFarlane, in *Multinuclear NMR*, ed. J. Mason, Plenum, New York, 1987, 305.
56. K.B. Dillon and A. Marshall, *J. Chem. Soc., Dalton Trans.*, 1987, 315.
57. O.A. Reutov, V.S. Petrosyan, N.S. Yahsina and E.I. Gefel, *J. Organomet. Chem.*, 1988, **341**, C31.
58. R.E. Wasylishen and N. Burford, *J. Chem. Soc., Chem. Commun.*, 1987, 1414.
59. C.R. Lassigne and E.J. Wells, *J. Mag. Res.*, 1978, **31**, 195.
60. T.N. Mitchell, *J. Organomet. Chem.*, 1983, **255**, 279.
61. P.J. Watkinson and K.M. Mackay, *J. Organomet. Chem.*, 1984, **275**, 39.
62. C.R. Lassigne and E.J. Wells, *Can. J. Chem.*, 1977, **55**, 927.
63. A.G. Davies, P.G. Harrison, J.D. Kennedy, R.J. Puddephatt, T.N. Mitchell and W. McFarlane, *J. Chem. Soc. (A)*, 1969, 1136.
64. H.J. Kroth, H. Schumann, H.G. Kuivila, C.D. Schaeffer and J.J. Zuckerman, *J. Amer. Chem. Soc.*, 1975, **97**, 1754.
65. J.D. Kennedy, W. McFarlane and G.S. Pyne, *Bull. Soc. Chim. Belg.*, 1975, **84**, 289.
66. J.D. Kennedy, *J. Mol. Struct.*, 1976, **31**, 207.
67. J.D. Kennedy, *J. Chem. Soc., Perkin Trans. II*, 1977, 242.
68. M.K. Das, J. Buckle and P.G. Harrison, *Inorg. Chim. Acta*, 1972, **6**, 17.
69. T.P. Lockhart and W.F. Manders, *Inorg. Chem.*, 1986, **25**, 892.
70. J. Holecek and A. Lycka, *Inorg. Chim. Acta*, **118**, L15.
71. W.F. Howard, R.W. Crecely and W.H. Nelson, *Inorg. Chem.*, 1985, **24**, 2204.
72. For a review of the area see R.B. Larrabee, *J. Organomet. Chem.*, 1974, **74**, 313.
73. B.E. Mann, B.F. Taylor, N.A. Taylor and R. Wood, *J. Organomet. Chem.*, 1978, **162**, 137.
74. A. Bonny and S.R. Stobart, *J. Chem. Soc., Dalton Trans.*, 1979, 786.
75. Yu. A. Ustynyuk, A.K. Shestakova, V.A. Chertkov, N.N. Zemlyansky, I.V. Borisova, A.I. Gusev, E.B. Tchuklanova and E.A. Chernyshev, *J. Organomet. Chem.*, 1987, **335**, 43.
76. H.H. Karsch, A. Appelt and G. Müller, *Organometallics*, 1986, **5**, 1664.
77. C.H. Yoder, D. Mokrynka, S.M. Coley, J.C. Otter, R.E. Haines, A. Grushow, L.J. Ansel, J.W. Hovick, J. Mikus, M.A. Shermak and J.N. Spencer, *Organometallics*, 1987, **6**, 1679.
78. H. Fujiwara, F. Sakai and Y. Sasaki, *J. Chem. Soc., Perkin Trans. II*, 1983, 11.
79. M. Newcomb, A.M. Madonik, M.T. Blanda and J.K. Judice, *Organometallics*, 1987, **6**, 145.
80. J.L. Nieto, F. Galindo and A.M. Gutierrez, *Polyhedron*, 1985, **4**, 1611.
81. C.T.G. Knight and A.E. Merbach, *J. Amer. Chem. Soc.*, 1984, **106**, 804.
82. C.T.G. Knight and A.E. Merbach. *Inorg. Chem.*, 1985, **24**, 576.
83. H. Ruêgger and P.S. Pregosin, *Inorg. Chem.*, 1987, **26**, 2912.
84. C. Wynants, G. Van Binst, C. Múgge, K. Jurkschat, A. Tzschach, H. Pepermans, M. Gielen and R. Willem, *Organometallics*, 1985, **4**, 1906.
85. K. Jurkschat, A. Tzschach, C. Múgge, J. Piret-Meunier, M. Van Meerssche, G. Van Binst, C. Wynants, M. Gielen and R. Willem, *Organometallics*, 1988, **7**, 593.
86. H.T. Muus, P.W. Atkins, K.A. McLaughlin and J.B. Petersen (eds.), *Chemically Induced Magnetic Polarization*, Reidel, Dordrecht, 1977.
87. M. Lehnig, *Tetrahedron Lett.*, 1974, 3323.
88. M. Lehnig, *Tetrahedron. Lett.*, 1977, 3663.
89. M. Lehnig, W.P. Neumann and P. Seifert, *J. Organomet. Chem.*, 1978, **162**, 145.
90. C. Grugel, M. Lehnig, W.P. Neumann and J. Sauer, *Tetrahedron Lett.*, 1980, **21**, 273.
91. T.V. Leshina, R.Z. Sagdeev, N.E. Polyakov, M.B. Taraban, V.I. Valyaev, V.I. Rakhlin, R.G. Mirskov, S. Kh. Khangazheev and M.G. Voronkov, *J. Organomet. Chem.*, 1983, **259**, 295.
92. A. Standt and H. Dreeskamp, *J. Organomet. Chem.*, 1987, **322**, 49.
93. R.K. Harris, *Nuclear Magnetic Resonance Spectroscopy*, Pitman, London, 1983.
94. R.K. Harris, A. Sebald, D. Furlani and G. Tagliavini, *Organometallics*, 1988, **7**, 388.
95. T.P. Lockhart and W.F. Manders, *Inorg. Chem.*, 1986, **25**, 583.
96. W.F. Manders and T.P. Lockhart, *J. Organomet. Chem.*, 1985, **297**, 143.
97. T.P. Lockhart, W.F. Manders and J.J. Zuckerman, *J. Amer. Chem. Soc.*, 1985, **107**, 4546.

98. T.P. Lockhart, W.F. Manders and E.O. Schlemper, *J. Amer. Chem. Soc.*, 1985, **107**, 7451.
99. T.P. Lockhart and W.F. Manders, *J. Amer. Chem. Soc.*, 1985, **107**, 5863.
100. T.P. Lockhart and W.F. Manders, *Inorg. Chem.*, 1986, **25**, 1068.
101. T.P. Lockhart and F. Davidson, *Organometallics*, 1987, **6**, 2471.
102. T.P. Lockhart, J.C. Calabrese and F. Davidson, *Organometallics*, 1987, **6**, 2479.
103. T.P. Lockhart and W.F. Manders, *J. Amer. Chem. Soc.*, 1987, **109**, 7015.
104. J. Du, J. Deng, D. Yao, L. Zen, R. Chen, Z. Yu and Y. Zheng, *Zhongshan Daxue Xuebao, Ziran Kexueban*, 1985, **3**, 104.
105. F.J. Berry and C. Hallett, *J. Chem. Soc., Dalton Trans.*, 1985, 45.
106. H. Berndt, H. Mehner, H. Völter and W. Meisel, *Z. anorg. Chem.*, 1977, **429**, 47.
107. R. Frety, M. Guenin, P. Bussiere and Y.L. Lam, *Master. Sci., Monogr.*, 1985, **28**, 1055.
108. P.S. Cook, J.D. Cashion and P.J. Cassidy, *Fuel*, 1985, **64**, 1121.
109. P.P. Vaishnava, H.J. Shyu and P.A. Montano, *Fuel*, 1981, **60**, 624.
110. J. Rotsche, U. Reichel, R. Schroeder and C. Pietzsch, *Silikattechnik*, 1985, **36**, 316.
111. K. Burger and A. Vertes, *Nature*, 1983, **306**, 353; *Inorg. Chim. Acta*, 1983, **76**, L247.
112. E. Leja, J. Korecki, K. Krop and K. Toll, *Thin Solid Films*, 1979, **59**, 147.
113. M. Grozdana, Ts. Bonchev and G. Georgiev, *Dokl. Bulg. Akad. Nauk*, 1978, **31**, 1281.
114. M. Grozdana, Ts. Bonchev and V. Lilkev, *Nucl. Instrum. Methods*, 1979, **165**, 231.
115. C.L. Lau and G.K. Wertheim, *J. Vac. Sci. Technol.*, 1978, **15**, 622.
116. T.D. Chan, T.S. Bonchev, C.T. Nguyen and S. Peneva, *Bulg. J. Phys.*, 1977, **4**, 399.
117. S. Bukshpan, *Phys. Lett. (A)*, 1977, **62**, 109.
118. S. Bukshpan, T. Sonnino and J.G. Dash, *Surface Sci.*, 1975, **52**, 466.
119. N.A. Stepanova, A.A. Maligin, G.N. Kuznetsova, T.K. Egoiov, A.V. Evdokimov and T.M. Katkova, *Zh. Obshch. Khim.*, 1984, **54**, 111.
120. J.E. Phillips and R.H. Herber, *Inorg. Chem.*, 1986, **25**, 3081.
121. R. Schogl and H.P. Boehm, *Z. Naturforsch., Teil B*, 1984, **39**, 112.
122. V.L. Solozhenko, P.B. Fabrichnyi, M.E. Leonovg and Y.A. Kalashnikov, *Khim. Fiz.*, 1983, 711.
123. V.L. Solozhenko, I.V. Arkhangel'skii, A.M. Gos'kov, Ya. A. Kalashnikov, *Zh. Fiz. Khim.*, 1983, **57**, 2265.
124. R.H. Herber and R.F. Davis, *J. Chem. Phys.*, 1975, **63**, 3668.
125. N. Karnazos, L.B. Welsh and M.W. Shafer, *Phys. Rev. (B)*, 1975, **11**, 1808.
126. J.S. Brooks, D.W. Allen and J. Unwin, *Polym. Degrad. Stab.*, 1985, **10**, 79.
127. A. Bos, A.T. Howe, *J. Chem. Soc., Faraday II Trans.*, 1974, **70**, 440.
128. A. Bos, A.T. Howe, B.W. Dale and L.W. Becker, *J. Chem. Soc., Faraday II Trans.*, 1974, **70**, 451.
129. A. Schichl, F.J. Litterst, H. Micklitz, J.P. Devort and J.M. Friedt, *Chem. Phys.*, 1977, **20**, 371.
130. H. Micklitz and P.H. Barrett, *Appl. Phys. Lett.*, 1972, **20**, 387; *Phys. Rev. (B)*, 1972, **5**, 1704.
131. A. Vertes, I. Nagy-Czako and K. Burger, *J. Phys. Chem.*, 1976, **80**, 1314.
132. A. Vertesand I. Nagy-Czako, *Magyar Kem. Folyoirat*, 1976, **82**, 200.
133. R.H. Herber, *J. Inorg. Nucl. Chem.*, 1973, **35**, 67.
134. V.S. Petrosyan, N.S. Yashina, S.G. Sacharov, O.A. Reutov, V. Ya. Rechev and V.I. Goldanskii, *J. Organomet. Chem.*, 1973, **52**, 333.
135. R.L. Cohen and K.W. West, *Chem. Phys. Lett.*, 1972, **16**, 128; *J. Electrochem. Soc.*, 1972, **119**, 433.
136. J. Akashi, M. Chiba and H. Sano, *Bunseki Kagaku*, 1983, **32**, E123.
137. A.P. Dawson, B.G. Farrow and M.J. Selwyn, *Biochem. J.*, 1982, **202**, 163.
138. R.V. Parish, *Progr. Inorg. Chem.*, 1972, **15**, 101.
139. J.J. Zuckerman, *Adv. Organomet. Chem.*, 1970, **9**, 21.
140. G.M. Bancroft and R.H. Platt, *Adv. Inorg. Chem. Radiochem.*, 1972, **15**, 59.
141. N.N. Greenwood and T.C. Gibb, *Mössbauer Spectroscopy*, Chapman and Hall, London, 1968.
142. T.C. Gibb, *Principles of Mössbauer Spectroscopy*, Chapman and Hall, London, 1976.
143. G.M. Bancroft, *Mössbauer Spectroscopy*, McGraw-Hill, London, 1973.
144. N. Benczer-Keller and T. Fink, *Nuclear Phys. (A)*, 1971, **161**, 123.
145. J.N.R. Ruddick, *Revs. Si, Ge, Sn, and Pb Compds.*, 1976, **2**, 115.

146. S. Ichiba and T. Sakamoto, *Bull. Chem. Soc. Jpn.*, 1985, **58**, 1323.
147. M.M. Kolosova, S.A. Simanova, E.N. Yurchenko, E.S. Boichinova and V.I. Kuznetsov, *Zh. Neorg. Khim.*, 1985, **30**, 1493.
148. H. Muramatsu, T. Miura, N. Nakahara and M. Fujicka, *Radiochem. Radioanal. Lett.*, 1983, **55**, 169.
149. T. Okada, H. Sekizama, F. Amko and S. Amber, *Proc. Ind. Nat. Sci. Acad. Phys. Sci., Special Vol.*, 1982, 450.
150. S. Ichiba, M. Yamada and H. Negita, *Radiochem. Radioanal. Lett.*, 1978, **35**, 31.
151. V. Maenning and M. Grodzicki, *Theor. Chim. Acta*, 1986, **70**, 189.
152. M. Grodzicki, V. Maenning, R. Blaes and A.X. Trautwein, *Hyperfine Interact.*, 1986, **29**, 1547.
153. A. Svane and E. Antoncik, *Phys. Rev. B*, 1986, **34**, 1944.
154. H. Muramatsu, T. Miura, H. Nakahara, M. Fujioka, E. Tanaka and A. Hashizume, *Hyperfine Interact.*, 1984, **20**, 305.
155. N.N. Greenwood, P.G. Perkins and D.H. Wall, *Symp. Faraday Soc.*, **1**, 1968, 51.
156. N.N. Greenwood, P.G. Perkins and D.H. Wall, *Phys. Lett.*, 1968, **28A**, 339.
157. J.E. Huheey and J.C. Watts, *Inorg. Chem.*, 1971, **10**, 1553.
158. A.J. Edwards and K.I. Khallow, *J. Chem. Soc., Chem. Commun.*, 1984, 50.
159. R.J. Batchelor, J.N.R. Ruddick, J.R. Sams and F. Aubke, *Inorg. Chem.*, 1977, **16**, 1414.
160. R.H. Herber and G. Carrasquillo, *Inorg. Chem.*, 1981, **20**, 3693.
161. J.D. Donaldson, D.C. Puxley and M.J. Tricker, *Inorg. Nucl. Chem. Lett.*, 1972, **8**, 845.
162. S.P. Mallela, S. Yap, J.R. Sams and F. Aubke, *Inorg. Chem.*, 1986, **25**, 4327.
163. R.V. Parish and C.E. Johnson, *J. Chem. Soc. (A)*, 1971, 1907.
164. P.G. Harrison, B.C. Lane and J.J. Zuckerman, *Inorg. Chem.*, 1972, **11**, 1537.
165. D. Tudela, V. Fernandez and J.D. Tornero, *J. Chem. Soc.*, 1985, 1281.
166. C.S. Frampton, R.M.G. Roberts, J. Silver, J.F. Warmsley and B. Yavari, *J. Chem. Soc., Dalton Trans.*, 1985, 169.
167. T.C. Gibb, *J. Chem. Soc. (A)*, 1970, 2503.
168. N.E. Erickson, *Chem. Commun.*, 1970, 1349.
169. J.D. Donaldson, E.J. Filmore and M.J. Tricker, *J. Chem. Soc. (A)*, 1971, 1109.
170. B.A. Goodman, R. Greatrex and N.N. Greenwood, *J. Chem. Soc. (A)*, 1971, 1868.
171. S.R.A. Bird, J.D. Donaldson, A.F.LeC. Holding, B.J. Senior and M.J. Tricker, *J. Chem. Soc. (A)*, 1971, 1616.
172. G.M. Bancroft, V.G. Kumar Das, T.K. Sham and M.G. Clark, *J. Chem. Soc., Dalton Trans.*, 1976, 643.
173. R.C. Poller and J.N.R. Ruddick, *J. Chem. Soc., Dalton Trans.*, 1972, 555.
174. E.T. Libbey and G.M. Bancroft, *J. Chem. Soc., Chem. Commun.*, 1973, 503.
175. T.C. Gibb, B.A. Goodman and N.N. Greenwood, *Chem. Commun.*, 1970, 774.
176. J.N.R. Ruddick and J.R. Sams, *J. Chem. Soc., Dalton Trans.*, 1974, 470.
177. P.G. Harrison and T.J. King, *J. Chem. Soc., Dalton Trans.*, 1974, 2298.
178. R.V. Parish and R.H. Platt, *Inorg. Chim. Acta*, 1970, **4**, 65.
179. G.M. Bancroft and K.D. Butler, *J. Chem. Soc., Dalton Trans.*, 1973, 1694.
180. M.G. Clark, A.G. Maddock and R.H. Platt, *J. Chem. Soc., Dalton Trans.*, 1972, 281.
181. S. Calogero, D.A. Clemente, V. Peruzzo and G. Tagliavini, *Dalton* 1979, 1172.
182. P.G. Harrison, *Organotin Compounds: New Chemistry and Applications*, ed. J.J. Zuckermann, Adv. Chem. Ser. **157**, American Chemical Society, Washington DC, 1976, 258.
183. T.K. Sham and G.M. Bancroft, *Inorg. Chem.*, 1975, **14**, 2281.
184. V.G. Kumar Das, Y. Chee-Keong and P.J. Smith, *J. Organomet. Chem.*, 1987, 311.
185. C. Hohenemser, *Phys. Rev.*, 1965, **139**, A185.
186. D.W. Hafemeister and E. Brooks-Shera, *Nucl. Instr. Meth.*, 1966, **41**, 133.
187. R.H. Herber and A.E. Smelkinson, *Inorg. Chem.*, 1978, **17**, 1023.
188. R.H. Herber, A.E. Smelkinson, M.J. Sienko and L.F. Schneemeyer, *J. Chem. Phys.*, 1978, **68**, 3705.
189. R.H. Herber, *Phys. Rev.*, 1983, **27B**, 4013.
190. R.H. Herber and R.F. Davis, *J. Inorg. Nucl. Chem.*, 1980, **42**, 1577.
191. R.H. Herber, F.J. DiSalvo and R.B. Frankel, *Inorg. Chem.*, 1980, **19**, 3135.
192. R.H. Herber and M. Katada, *J. Solid State Chem.*, 1979, **27**, 137.
193. P.G. Harrison, M.J. Begley and K.C. Molloy, *J. Organomet. Chem.*, 1980, **186**, 213.

194. P.G. Harrison, N.W. Sharp, C. Pelizzi, G. Pelizzi and P. Tarasconi, *J. Chem. Soc., Dalton Trans.*, 1983, 921, 1687.
195. P.G. Harrison, K. Lambert, T.J. King and B. Majee, *J. Chem. Soc., Dalton Trans.*, 1983, 363.
196. K.C. Molloy, M.P. Bigwood, R.H. Herber and J.J. Zuckerman, *Inorg. Chem.*, 1982, **21**, 3709.
197. K.C. Molloy and K. Quill, *J. Chem. Soc., Dalton Trans.*, 1985, 1417.
198. R.H. Herber and M.F. Leahy, *Organotin Compounds: New Chemistry and Applications*, ed. J.J. Zuckermann, Adv. Chem. Ser. **157**, American Chemical Society, Washington DC, 1976, 155.
199. V.I. Goldanskii, G.M. Gorodinskii, S.V. Karyagi, L.A. Kovytko, L.M. Krizhauskii, E.F. Makarov, I.P. Suzdalev and V.V. Khrapov, *Dokl. Akad. Nauk SSSR*, 1962, **147**, 127.
200. S.V. Karyagin, *Dokl. Akad. Nauk SSSR*, 1963, **148**, 1102.
201. B.J. Haylett, PhD thesis, University of Nottingham, 1981.
202. P.G. Harrison, K.C. Molloy and E.W. Thornton, *Inorg. Chim. Acta*, 1979, **33**, 137.
203. W. Davidsohn and M.C. Henry, *J. Organomet. Chem.*, 1966, **5**, 29.
204. D.B. Chambers, F. Glockling and M. Weston, *J. Chem. Soc. (A)*, 1967, 1759.
205. F. Glockling, M.A. Lyle and S.R. Stobart, *J. Chem. Soc., Dalton Trans.*, 1974, 2537.
206. M. Fishwick and M.G.H. Wallbridge, *J. Chem. Soc. (A)*, 1971, 57.
207. T. Chivers, G.F. Lanthier and J.M. Miller, *J. Chem. Soc. (A)*, 1971, 2556.
208. D. Chambers and F. Glockling, *J. Chem. Soc. (A)*, 1968, 735.
209. P.G. Harrison and J.J. Zuckerman, *Inorg. Chem.*, 1970, **9**, 175.
210. D. Potts and J.M. Miller, *J. Chem. Soc., Dalton Trans.*, 1975, 393.
211. P.G. Harrison, *Inorg. Chem.*, 1973, **12**, 1545.
212. P.G. Harrison and J.J. Zuckerman, *J. Amer. Chem. Soc.*, 1970, **92**, 2577.
213. P.F.R. Ewings, P.G. Harrison and D.E. Fenton, *J. Chem. Soc., Dalton Trans.*, 1975, 821.
214. For a review of the application of FAB mass spectrometry to organometallic compounds see J.M. Miller, *Adv. Inorg. Chem. Radiochem.*, **28**, Academic Press, New York, 1984, 1.
215. J.M. Miller and A. Fulcher, *Can. J. Chem.*, 1985, **63**, 2308.
216. J.M. Miller, H. Mondal, I. Wharf and M. Onyszchuck, *J. Organomet. Chem.*, 1986, **306**, 193.
217. R.H. Fish, R.L. Holmstead and J.E. Casida, *Tetrahedron Lett.*, 1974, **14**, 1303.
218. R.H. Fish, R.L. Holmstead, M. Gielen and B. De Poorter, *J. Org. Chem.*, 1978, **43**, 4969.

4 The inorganic chemistry of tin

J.D. DONALDSON and S.M. GRIMES

Tin forms inorganic compounds in the IV+ oxidation state by using all of its valence shell electrons in bonding, and in the II+ oxidation state in which, formally, only the *p*-electrons are used leaving a non-bonding electron pair. Many compounds in both oxidation states are known. The IV+ oxidation state is the more stable state and tin (II) compounds are moderately strong reducing agents.[1] The Sn–Sn (II)–Sn (IV) potentials are:

$$\text{in acid solution:}\quad Sn \xrightarrow{0.136V} Sn(II) \xrightarrow{-0.15V} Sn(IV),$$

and

$$\text{in basic solution:}\quad Sn \xrightarrow{0.91V} [Sn(OH)_3] \xrightarrow{0.93V} [Sn(OH)_6]^{2-}$$

These potentials explain why tin (IV) in acid solution can be readily reduced to tin metal and why the disproportionation of $[Sn(OH)_3]^-$ to elemental tin and $[Sn(OH)_6]^{2-}$ occurs to an appreciable extent in alkaline tin (II) solutions. The inorganic chemistry of tin in this chapter is discussed under the headings tin (II) chemistry, tin (IV) chemistry, and mixed-valence tin compounds.

4.1 Tin (II) chemistry

4.1.1 *Solution chemistry of tin (II)*[1]

Tin (II) oxide is amphoteric and dissolves in aqueous solutions of both acids and alkalis. The predominant species present in acid solutions containing complexing anions are the pyramidal triligandstannate(II) ions, $[SnX_3]^-$ e.g. with $X = F^-$, Cl^-, Br^-, I^-, HCO_2^-, $CH_3CO_2^-$ and NCS^-. There is very little evidence for the formation of complexes with coordination greater than three. This is to be expected because the formation of $[SnX_3]^-$ from a normal tin (II) compound (SnX_2) results in the filling of the empty *p*-orbital on the tin. Since the *s*–*d* energy gap (~ 14ev) in tin is much greater than the *s*–*p* gap (~ 7ev), use of *d*-orbitals to increase the coordination above three is less likely. The $[SnX_3]^-$ ions can therefore be regarded as being based on a tetrahedral distribution of electron pairs around the tin with one orbital occupied by a non-bonding electron pair. The $[SnF_3]^-$ anion is the stable and predominant ion in solutions containing an excess of F^- but, when insufficient F^- is present to complex all of the tin as $[SnF_3]^-$, the polynuclear ion $[Sn_2F_5]^-$ is formed in which the preferred three coordination for Sn(II) is preserved by a bridging F atom. The only other Sn(II)-F species for which there is any evidence are $[SnF]^+$, $[(Sn_3F_5)_n]^+$ and SnF_2. Various techniques have been used to establish the complexes present in Cl^--, Br^--

and NCS^--containing solutions, and evidence has been found for $[SnX]^+$ and SnX_2 in addition to the predominant $[SnX_3]^-$ ions. Nitrogen is the donor atom in the NCS complexes. Mixed halide complexes of the type $[SnF_2Cl]^-$ are also known. The main species in tin(II)-monocarboxylic acid solutions are the tricarboxylatostannate(II) ions, $[Sn(RCO_2)_3]^-$, but there is also evidence for the formation of less stable complexes, for example, $[SnHCO_2]^+$, $Sn(HCO_2)_2$, $[Sn_2(HCO_2)_5]^-$, in formate and $[Sn(CH_3CO_2)]^+$, $Sn(CH_3CO_2)_2$, $[Sn_2(CH_3CO_2)_5]^-$ and $[Sn_3(CH_3CO_2)_7]^-$ in acetate solutions. It has been suggested that the ions present in citric acid solution are $[Sn(C_6H_5O_7)]^-$, $[Sn(OH)(C_6H_5O_7)]^{2-}$ and $[Sn(C_6H_5O_7)_2]^{4-}$, and that 1:1:1 tin(II):M:citrate or tartrate chelates are formed in solutions containing Sn(II) with M = Fe(III) and Cu(II). There is some evidence for complex formation in solutions of Sn (II) in mineral oxyacids. Complexes of the type $[Sn(HSO_4)]^+$, $Sn(HSO_4)_2$, $SnSO_4$ and $[Sn(SO_4)_2]^{2-}$ have been reported, and there is evidence for a number of pyrophosphate complexes in solutions containing tin(II) and pyrophosphate ions. The stable species in phosphorous acid solutions is $[Sn(HPO_3)_3]^{4-}$.

The main species present in alkaline Sn (II) solutions is the $[Sn(OH)_3]^-$ anion, but there is also evidence for the polynuclear ion $[Sn_2(OH)_4O^{2-}]$ in which the stable pyramidal Sn (II) environment would be maintained by a bridging oxygen atom. The predominant basic species in Sn–OH solutions at low pH is $[Sn_3(OH)_4]^+$, and the basic tin (II) salts precipitated at about pH 2 are derivatives of this ion or of a closely related, often partially dehydrated species.

A number of tin (II) compounds are water-soluble. The solutions are susceptible to hydrolysis giving hydrous tin (II) oxide and to oxidation to Sn (IV). The stability of the solutions depends upon the nature of the anions present and the pH. Solutions containing anions such as F^- or $CH_3CO_2^-$ which form strong complexes with Sn (II) are relatively stable to hydrolysis and oxidation. Ease of oxidation increases with pH, and in very alkaline solutions spontaneous disproportionation to Sn and Sn (IV +) can occur. Tin (II) fluoride, SnF_2, and chloride, $SnCl_2$, are very soluble in water ($SnCl_2$, 83.9 g in 100 ml H_2O at 0°; SnF_2, 41 g in 100 ml at 25°), and the solubilities (g 100 ml^{-1} of solution) of some other tin (II) compounds are $SnSO_4$ (35.2 at 20°), NH_4SnF_3 (59 at 25°), and $Sn(CH_3CO_2)_2 \cdot 2CH_3CO_2H$ (1.9 at 25°). A number of Sn (II) compounds are soluble in non-aqueous solution. Anhydrous $SnCl_2$, for example, is readily soluble in acetone, glycol, methanol, pyridine and tetrahydrofuran, whilst $Sn(NCS)_2$ is soluble in ethanol, $Sn(HCO_2)_2$ in propylene glycol, and $SnBr_2$ in alcohol, tetrahydofuran and pyridine. The main complex species in these solutions is probably $[SnX_2 \cdot \text{ligand}]$ where the monodentate donor atom completes the pyramidal three-coordination of tin. Many solid derivatives of the type $[SnX_2 \cdot 2\ \text{ligand}]$ are known, but only one of the ligand atoms is normally bonded to the tin, the other ligand molecule being present only for lattice packing purposes. Sn (II) is also stable in acetonitrile solutions in which the complex cations, $[Sn(CH_3CN)_2]^{2+}$, $[Sn(CH_3CN)_3]^{2+}$ and $[Sn(CH_3CN)_6]^{2+}$ are said to exist. It is also likely that the ion $[Sn(NH_2)_3]^-$ is present in solutions of Sn (II) compounds in liquid ammonia.

4.1.2 *Structures of tin (II) compounds*

The most important aspect of tin (II) chemistry that has been studied over the past ten to fifteen years has been concerned with lone-pair effects[2,3]. Particular attention has been paid to the distortion of tin (II) environments by the non-bonding electron pair and to compounds in which the non-bonding electrons are delocalized in solid state bands or clusters.

Surveys of the literature[2,4] on known crystal structures of tin (II) compounds show that the tin atoms are nearly all in low symmetry, lone-pair-distorted environments. There are, however, a number of compounds in which the distorting effects of the non-bonding orbitals are reduced or eliminated by solid-state effects. Although the detailed distribution of the atoms around tin (II) in the solid state can be complicated, most of them can be described in terms of the environments illustrated by structures (7)–(**12**) in Chapter 2. The most common tin (II) environment is a trigonal pyramidal arrangement of three nearest neighbour tin–ligand bonds with three longer essentially non-bonding contacts completing a distorted octahedral coordination. These longer contacts arise because close approach of ligands to the tin is prevented by the lone-pair orbitals.

A large number of tin (II) compounds contain tin atoms with pyramidal three-coordination. Tin (II) sulphate[5], for example, has three short Sn–O bonds (2.25–2.27 Å) in a pyramidal arrangement with three longer Sn–O distances (2.92–2.99 Å) completing a distorted octahedral coordination. Other tin (II) compounds[4] that have tin in this type of site include monoclinic SnF_2, NH_4SnF_3, $NaSn_2F_5$, Sn_2F_3Cl, Sn_3F_5Br, $Sn_3F_3PO_4$, $SnCl_2$, $KCl \cdot SnCl_3 \cdot H_2O$, $CsSnCl_3$, $SnCl(H_2PO_2)$, $K_2Sn_2O_3$, $Sn_3(PO_4)_2$, $SnHPO_4$, $Sn_2(OH)PO_4$, α-$SnWO_3$, $KSn(HCO_2)_3$, $Ca[Sn(CH_3CO_2)_3]_2$, $SnHPO_3$, $SnFPO_3$, SnS, $BaSnS_2$, $Sn(NCS)_2$, $Sn[S_2P(OC_6H_5)_2]_2$, SnSe and $[CH_3B\{NSi(CH_3)_3\}_2Sn]_2$. The environment of tin in $Ca[Sn(CH_3CO_2)_3]_2$ is shown in Figure 4.1 as a typical example of the trigonal pyramidal tin (II) environment[6] and the

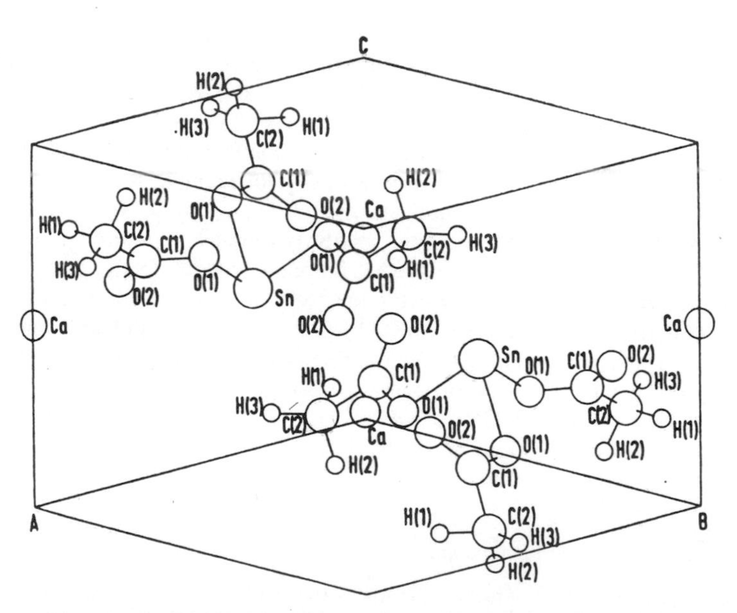

Figure 4.1 The unit cell of $Ca[(Sn(O_2CCH_3)_3]_2$ illustrating the trigonal bipyramidal geometry at tin (reproduced by permission from ref. 6).

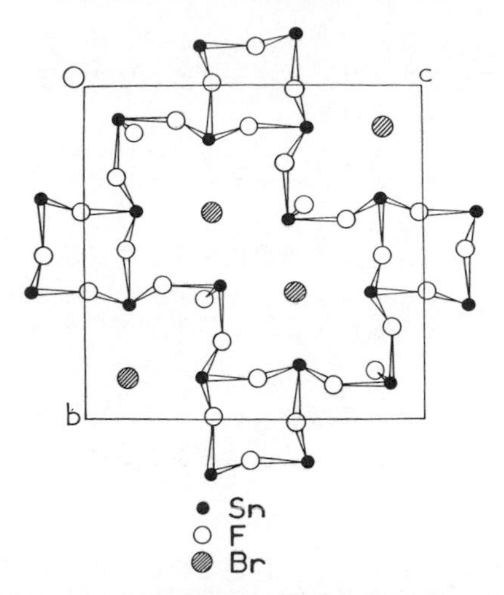

Figure 4.2 (100) Projection of the unit cell of Sn_3BrF_5 (reproduced by permission from ref. 7).

structure of Sn_3F_5Br (Figure 4.2) shows how the pyramidal three-coordination is maintained by bridging atoms in the formation of a tin (II) fluoride cationic network[7]. Evidence[8] for the predominance of trigonal pyramidal environments in tin (II) chemistry, even in the presence of some potentially bidentate ligands, comes from the structure determination of potassium hydrogen bis(maleato)stannate(II) in which the discrete ions $[Sn(CHCO_2{:}CHCO_2)(CHCO_2{:}CHCO_2H)]_2^{2-}$ are found. These ions have tin (II) in a trigonal pyramidal environment with two Sn–O bonds to an anisobidentate maleate ligand and one Sn–O bond to a terminal unidentate monoprotomaleate ligand. Table 4.1 lists the bond length and pyramidal bond angle data for crystal structures of tin (II) compounds, determined since 1980, which contain three-coordinated $[SnX_3]$ environments. Table 4.2 contains the bond length and angle data for compounds with trigonal pyramidal $[SnX_2Y]$ sites. Details of tin (II) structures with pyramidal three-coordinated sites that were published before 1980 are to be found in reference 4.

The second most common type of tin (II) environment has a distorted four-coordinated pyramidal arrangement of atoms around the tin. The main structural feature found in almost all four-coordinated tin environments is the existence of two bonds of considerably greater length than those normally found in tin (II) compounds. The environment of tin in $Na_2Sn(C_2O_4)_2$ shown in Figure 4.3 provides a good example of this type of tin (II) site[24]. Table 4.3 provides bond length and angle data for tin (II) structures determined since 1980, which have distorted four-coordinated tin (II) environments. Again examples of structures determined before 1980 are to be found in reference 4. Table 4.4 contains similar data for compounds with $[SnX_3Y]$ and $[SnX_2Y_2]$ sites. In some tin (II) compounds the distorted tin (II) environments have more complicated geometries, including those for β-SnS and β-SnSe[38], $Sn_2Sb_2S_5$[39], and $Bi_xSb_{2-x}Sn_2S_5$[40]. Although most tin (II) compounds contain the element in a lone-pair distorted site, there are a number of examples where the tin is in a high-symmetry undistorted site. Compounds with high-symmetry tin (II) sites include $CsSnBr_3$[41], the high-temperature modification of $CsSnCl_3$[42], phases in the

Table 4.1 Bond lengths and angles in tin (II) materials with $[SnX_3]$ sites.

Compound	Sn	X	Bond lengths (Å) 1	2	3	Next nearest	Sn–X	Bond angles (deg)			Ref.
$Co(SnF_3)_2 \cdot 6H_2O$	Sn1	F	2.03	2.05	2.06	3.26	Sn–F	84.1	84.1	86.1	9
$Sn_2F_3BF_4$	Sn1	F	2.08	2.10	2.12	2.98	Sn–F	78.9	84.7	84.9	10
	Sn2	F	2.07	2.10	2.10	2.96	Sn–F	81.6	86.4	86.8	–
$[C_5H_{12}N][SnCl_3]$	Sn1	Cl	2.53	2.55	2.56	3.33	Sn–Cl	88.6	88.8	91.0	9
$KCl \cdot KSnCl_3 \cdot H_2O$	Sn1	Cl	2.55	2.55	2.55	3.16	Sn–Cl	88.1	88.1	89.1	11
$NH_4Cl \cdot NH_4SnCl_3 \cdot H_2O$	Sn1	Cl	2.55	2.56	2.56	3.22	Sn–Cl	88.1	88.5	88.8	12
$NH_4Br \cdot NH_4SnBr_3 \cdot H_2O$	Sn1	Br	2.72	2.72	2.73	3.40	Sn–Br	89.0	89.0	90.0	9
$[C_5H_{12}N][SnBr_3]$	Sn1	Br	2.71	2.72	2.72	3.40	Sn–Br	89.3	89.8	91.3	9
KSn(CHCOO:CHOO)(CHCOO:CHCOOH)	Sn1	O	2.20	2.20	2.21	2.64	Sn–O	79.7	79.7	80.9	8
$K_2Sn_2[CH_2(CO_2)_2]_3 \cdot H_2O$	Sn1	O	2.18	2.18	2.20	2.92	Sn–O	83.7	84.0	74.6	15
	Sn2	O	2.16	2.21	2.24	2.72	Sn–O	83.3	78.2	73.2	–
$[Li(OBu)_3Sn]_2$	Sn1	O	2.08	2.10	2.10	3.07	Sn–Li	79.7	80.1	97.6	14
$[Na(OBu)_3Sn]_2$	Sn1	O	2.10	2.11	2.11	3.06	Sn–C	83.4	83.5	93.5	14
$K(OBu)_3Sn$	Sn1	O	2.06	2.07	2.07	3.62	Sn–K	81.5	90.7	91.9	14
$Rb_2Sn_2O_3$	Sn1	O	2.04	2.04	2.04	3.55	Sn–Rb	96.5	96.6	96.6	15
$Cs_2Sn_2O_3$	Sn1	O	2.02	2.04	2.04	3.54	Sn–Sn	96.0	96.0	96.7	16
$Na_4[SnO_3]$	Sn1	O	2.01	2.02	2.04	3.11	Sn–Na	98.5	99.5	99.7	17
$[NH_4][Sn(OCOCH_2Cl)_3]$	Sn1	O	2.15	2.15	2.17	2.86	Sn–C	78.3	81.5	86.9	9
$[NH_4]_2[Sn(HPO_3)_2]$	Sn1	O	2.11	2.12	2.17	2.70	Sn–O	83.6	84.6	89.8	18
α-SnS	Sn1	S	2.62	2.66	2.66	3.29	Sn–S	89.0	89.0	89.0	19
Sn_2S_3	Sn1	S	2.65	2.65	2.77	3.14	Sn–S	83.6	83.6	83.6	20
$Ga_2Sn_2S_5$	Sn1	S	2.63	2.64	2.74	3.15	Sn–S	85.4	86.8	92.0	21
	Sn2	S	2.64	2.69	2.75	3.10	Sn–S	82.9	89.4	92.7	–
$BaSn_2S_3$	Sn1	S	2.56	2.57	2.75	3.15	Sn–S	90.2	91.4	94.6	22
	Sn2	S	2.59	2.64	2.64	3.37	Sn–S	91.0	91.0	95.5	–
	Sn3	S	2.55	2.59	2.68	3.35	Sn–S	90.2	92.8	97.5	–
	Sn4	S	2.44	2.71	2.71	3.27	Sn–S	92.0	95.6	95.6	–
$[Sn(NMe_2)_2]_2$	Sn1	N	2.07	2.67	2.67	2.98	Sn–C	80.0	99.8	100.4	23

Table 4.2 Bond lengths and angles in tin (II) materials with $[SnX_2Y]$ sites.

Compound	Sn	X	Y	Bond lengths (Å) X	X	Y	Next nearest	Sn–X	Bond angles (deg)			Ref.
$Sn(tu)Cl_2$	Sn1	Cl	S	2.49	2.61	2.70	3.19	Sn–Cl	88.0	90.6	93.5	12
$NH_4Br \cdot NH_4SnBr_2Cl \cdot H_2O$	Sn1	Br	Cl	2.70	2.72	2.56	3.28	Sn–Br	87.3	–90.5		11

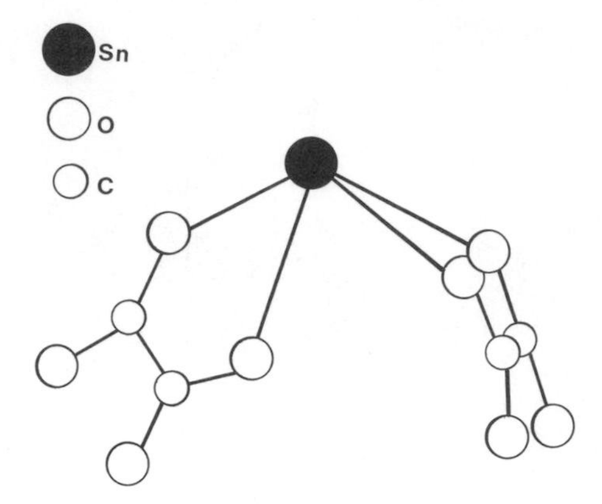

Figure 4.3 The environment of tin in $Na_2(C_2O_4)_2$ (reproduced by permission from ref. 24).

system $CsSnBr_{3-x}Cl_x$[42], Cs_4SnBr_6[43], SnTe[2], and the high-temperature forms of SnS[2] and SnSe[2].

In all of these compounds the high symmetry arises because of delocalization of the non-bonding electrons into solid-state bands. Similar solid-state effects are also seen in blue-black tin oxide in which the tin atom, although in a distorted environment, has regular square pyramidal coordination of oxygen atoms. Studies on the reasons for the formation of high-symmetry tin (II) sites have provided some of the more important advances in our knowledge of the effects of non-bonding electron pairs on the chemistry of Main Group elements in recent years, and are discussed in the next section of this chapter.

4.1.3 *Bonding in tin (II) compounds*[2]

One of the problems associated with the description of bonding in tin (II) compounds is the difficulty of including in any model those derivatives that have tin (II) in undistorted regular octahedral sites and have unusual solid-state properties. Covalent models have to assume that the non-bonding electrons in high-symmetry structures are in stereochemically inactive orbitals, that is, that they behave essentially as a $5s^2$ pair. An alternative suggestion[44] is that, since the *ns* orbital (A_{1g}) of the central atom plays little part in the bonding, the extra pair of electrons can be accommodated in the A_{1g} antibonding molecular orbital, without distorting the O_h symmetry. None of the models based on covalency can, however, explain why non-bonding orbitals can be stereochemically active in one site and inactive in a similar site in another compound. More importantly, none of these approaches accounts for the unusual optical and electrical solid-state properties associated with high-symmetry tin (II) environments. The electrostatic arguments of Orgel[45], on the other hand, do predict that distortion will be greatest for tin (II) bonding to light atoms and least for tin (II) bonding to heavy atoms. This is consistent with the observation that the most distorted tin (II) sites are generally found when tin is bonded to oxygen and fluorine atoms. Orgel's approach,

Table 4.3 Bond lengths and angles in tin (II) materials with $[SnX_4]$ sites.

Compound	Sn	X	Bond lengths (Å) 1	2	3	4	Next nearest	Sn–X	Bond angles (deg)	Ref.
$RbSn_2Br_5$	Sn1	Br	2.76	2.76	3.11	3.11	3.49	Sn–Br	79.0 91.8 148.1	25
$InSn_2Br_5$	Sn1	Br	2.79	2.79	3.06	3.06	3.41	Sn–Br	78.2 91.3 146.0	26
$InSn_2I_5$	Sn1	I	3.05	3.05	3.27	3.27	3.62	Sn–I	77.7 90.9 144.7	26
$Sn(CH_2C_2O_4)$	Sn1	O	2.18	2.23	2.26	2.56	2.95	Sn–O	77.1–78.3	27
$Sn(H_2PO_2)_2$	Sn1	O	2.16	2.16	2.35	2.35	3.36	Sn–O	79.0–151.4	9
$SnC_2H_4O_2$	Sn1	O	2.07	2.11	2.31	2.34	2.95	Sn–C	67.2–141.4	28
$Sn_2(S_2O_4)_2$	Sn1	O	2.24	2.26	2.26	2.32	3.39	Sn–O	72.6–78.9	29
Sn_2OSO_4	Sn1	O	2.15	2.26	2.35	2.56	2.83	Sn–O	71.1–142.9	30
	Sn2	O	2.14	2.23	2.34	2.52	2.91	Sn–O	78.0–158.4	—
$Sn[C_5(COOMe)_5]_2$	Sn1	O	2.24	2.24	2.27	2.27	2.93	Sn–O	69.1–144.2	31
$In_5Sn_{0.5}S_7$	Sn1	S	2.99	2.99	3.02	3.02	3.24	Sn–S	79.8–136.7	32
$Tl_2Sn_2S_3$	Sn1	S	2.68	2.89	2.93	3.11	3.44	Sn–S	79.4–162.0	33
Eu_2SnS_5	Sn1	S	2.37	2.37	2.43	2.46	3.50	Sn–S	95.5–143.6	34
	Sn2	S	2.32	2.35	2.74	2.74	3.01	Sn–S	93.4–167.6	—
$Sn[C(PMe_2)_3]_2$	Sn1	P	2.60	2.60	2.79	2.84	3.27	Sn–C	62.9–142.5	35
$Sn[S_2P(OC_6H_5)_2]_2$	Sn1	S	2.62	2.65	2.83	3.04	3.39	Sn–S	74.2–170.7	36

Table 4.4 Bond lengths and angles in tin(II) materials with $[SnX_3Y]$ and $[SnX_2Y_2]$ sites.

Compound	Sn	X	Y	Bond lengths (Å) X	X	X	Y	Next nearest	Sn–X	Bond angles (deg)	Ref.
Sn_4OF_6	Sn1	F	F	2.05	2.13	2.27	2.32	3.47	Sn–F	77.4–149.3	9
	Sn2	F	O	2.12	2.14	2.55	2.07	3.49	Sn–F	77.7–142.5	
	Sn3	F	O	2.11	2.24	2.39	2.07	2.99	Sn–F	77.7–155.9	
	Sn4	F	O	2.11	2.24	2.39	2.03	2.95	Sn–F	77.9–154.3	
				X	X	Y	Y				
$CsSn_3F_{5.5}Br_{1.5}$	Sn1	F	Br	2.05	2.33	3.25	3.25	2.48	Sn–F	82.7–136.7	11
	Sn2	F	Br	1.98	2.34	3.20	3.20	2.45	Sn–F	86.1–136.5	
	Sn3	F	F	2.09	2.26	2.31	2.50	3.62	Sn–Br	69.2–87.6	
$Sn^{II}Sn^{V}F_4(O_2CCF_3)_8 2CF_3COOH$	Sn1	F	O	2.19	2.30	2.47	2.50	2.57	Sn–O	72.7–100.4	37
	Sn2	F	O	2.16	2.47	2.36	2.44	2.47	Sn–O	75.6–147.4	

however, suffers from the same disadvantage as the covalent approach in so far as localization of the description to crystal field effects on the tin atom alone necessarily means that the model cannot explain any three-dimensional solid-state effects. The high-symmetry tin(II) compounds are normally coloured and have interesting electrical and optical properties. For example, $CsSnBr_3$ crystallizes in an undistorted perovskite lattice. It is a black solid with a metallic lustre, an electrical conductor and a material which shows interesting photoemission effects. The ^{119}Sn Mössbauer spectrum of the material has a very narrow linewidth consistent with the undistorted octahedral symmetry of the tin atom, but its chemical shift ($\delta = 3.95\,mm\,s^{-1}$) is much lower than would be expected from a material with relatively long Sn–Br bonds (2.94 Å). An explanation for the differences observed in the stereochemical activity of the lone pairs in tin(II) compounds can, however, be obtained by considering the energy levels of the bonding of orbitals of both tin and its ligands. Consideration of the overall symmetry of the lattice then enables us to extend the arguments used to account for the solid-state effects in high-symmetry tin(II) compounds.

A simple rationale for differences in the *s*:*p* ratios in tin(II) compounds can be provided by consideration of orbital energy matching[2]. The rationale employs derivatives of the trihalogenostannate(II) ions (SnX_3^-; X = F, Cl, Br, I) as examples and is shown to be entirely consistent with their structures and Mössbauer data[46]. The stereochemical activity of the tin lone-pair will depend on the directional tin $5p_z$ orbital character in the non-bonding lone-pair orbital and the Mössbauer parameters, in the Townes–Dailey approximation, on the fractional 5*s* and 5*p* character of the molecular orbitals. We have suggested[2] that, in order to form a strong a_1 bond, tin 5*s* and $5p_z$ orbitals will premix such that the energy of the hybrid orbital so formed will match that of the halogen *np* orbital. The binding energies of the tin 5*s* and 5*p* and halogen *np* orbitals are shown in Figure 4.4, and it can be seen that in order to obtain a good energy

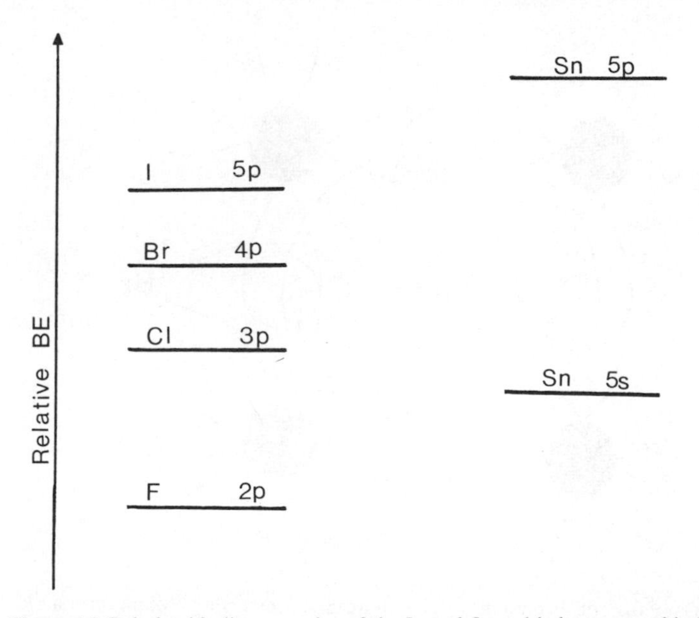

Figure 4.4 Relative binding energies of tin 5*s* and 5*p* and halogen *np* orbitals.

match with the appropriate halogen group orbital, the tin a_1 bonding orbital should contain high $5s$ character when bound to fluorine but high $5p_z$ character when bound to iodine. The non-bonding orbital (because of orthogonality requirements) thus contains high tin $5p_z$ and low $5s$ character in $[SnF_3]^-$. This orbital occupancy would lead to a highly directional lone pair containing high tin $5p_z$ character and hence distorted tin (II) environments, large negative Mössbauer quadrupole coupling constants and small isomer shifts for $[SnX_3]^-$ ions containing electronegative ligands such as fluorine. In contrast, for $[SnX_3]^-$ ions containing less electronegative ligands such as iodine, the lone pair would contain high tin $5s$ character and would therefore be much less stereochemically active, the isomer shifts would be large and the quadrupole coupling constants small. One of the strengths of the orbital matching model is the ease with which it can be extended to describe the bonding in and properties of the tin (II) compounds with high-symmetry structures.

The increase in the tin $5s$ character of the non-bonding pair in going from fluorides to iodides is paralleled by a decrease in distortion. As the distortion decreases, it becomes possible to remove the distorting effects of the non-bonding electron pair by solid-state effects as well as by direct alteration in the *p*-character of the electrons in the orbital. The removal of distortion by solid-state effects can be explained in terms of direct population of empty delocalized bands by the non-bonding electrons. This phenomenon is best illustrated by discussion of the undistorted perovskite, $CsSnBr_3$. If we look at a projection of the perovskite lattice showing the positions of the tin and the bromine atoms, it is easy to see how a delocalized empty band system can arise from the mutual overlap of empty bromine t_2 $4d$-orbitals (Figure 4.5). Calculation of the size of the $5s^2$ orbital on the tin atom leads to the conclusion that there will be an overlap

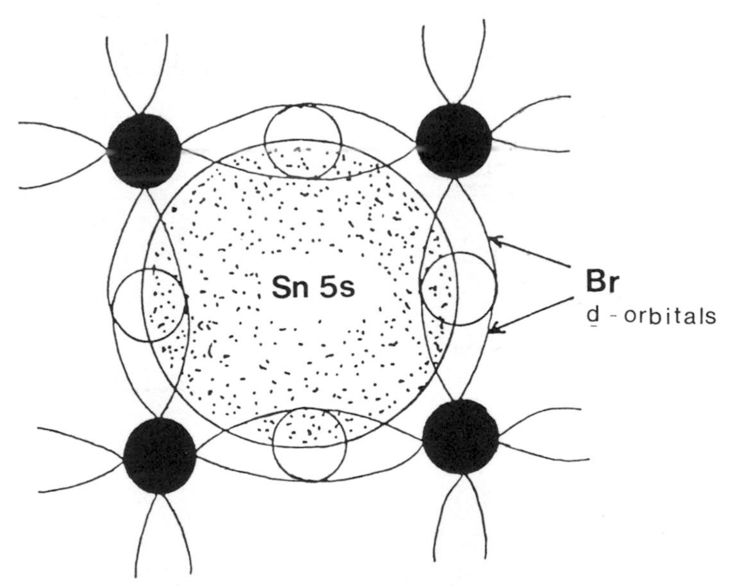

Figure 4.5 Overlap of tin $5s$-orbitals with bands formed from empty bromine orbitals leading to delocalization of tin electrons into solid state bands.

between the $5s^2$ orbital and the empty bromine $4d$-orbitals. It then becomes possible for the distortive effects of the non-bonding orbital to be reduced by the direct population of the solid state band with the non-bonding tin electrons. The vibrational spectra for the compound can only be rationalized if Br–Br interactions such as those that would arise from population of the bromine $4d$-bands are considered. It seems to be particularly significant that $Cs_2Sn^{IV}Br_6$, which has a closely related structure to that of $CsSnBr_3$, is white and shows none of the optical and electrical properties associated with $CsSnBr_3$. The caesium and bromine atoms are in the same relative position in both compounds, but in the tin(IV) complex only one-half of the perovskite tin sites are occupied. The cell volume for Cs_2SnBr_6 ($a = 10.8$ Å) is less than that for the corresponding number of caesium and bromine atoms in $CsSnBr_3$ ($a = 11.6$ Å). The tin(IV) complex must therefore contain a lower energy band like $CsSnBr_3$ but, unlike $CsSnBr_3$, the complex has no high-energy non-bonding electrons with which to populate these empty bands. For this reason Cs_2SnBr_6, in common with similar hexabromo- and hexachlorostannates(IV), is white and insulating.

The concept of direct population of solid-state bands can explain all of the apparently anomalous properties of $CsSnBr_3$. Its undistorted perovskite structure must arise because the mutual overlap of the bromine $4d$ orbitals would be at a maximum in such a structure, and this in turn would permit maximum transfer of electron density from the potentially distorting non-bonding tin orbitals. The electrical properties of $CsSnBr_3$ must arise as a result of the population of the empty bromine t_2 bands with electrons from the tin atoms. The very low ^{119}Sn Mössbauer chemical isomer shift for $CsSnBr_3$ and its variation with temperature arise because tin s-electron density is lost to the solid-state band structure and because the amount of loss is at a maximum at about room temperature (see section 4.1.4). This loss of s-electron density is at the same time reflected in the change in electrical conductivity of the material, in which a decrease in the chemical isomer shift corresponds to an increase in the conductivity (Table 4.5).

The results of direct population also explain the following optical properties of $CsSnBr_3$[47]. At 300 K, $CsSnBr_3$ shows an absorption edge at 1.80 eV. Excitation to high energy of this edge produces an intense red emission with a maximum at 1.78 eV and a symmetrical Lorentzian shape. At higher energies the intensity of this band decreases

Table 4.5 Variable temperature ^{119}Sn Mössbauer parameters and electrical conductivity for $CsSnBr_3$.

Temperature (K)	δ^* ($mm\,s^{-1}$)	Γ ($mm\,s^{-1}$)	Conductivity ($ohm^{-1}\,cm^{-1} \times 10^4$)
4.2	4.14	1.00	
80	3.97	0.84	7.91
295	3.89	0.77	9.34
295 (^{119}Sn-enriched)	3.89	0.77	9.34
323 (^{119}Sn-enriched)	3.85	1.00	8.66
348 (^{119}Sn-enriched)	3.89	1.28	7.85
383 (^{119}Sn-enriched)	4.05	1.30	
418 (^{119}Sn-enriched)	4.23	1.38	

*Relative to $BaSnO_3$

rapidly until at 430 K it reaches 1% of its 300 K intensity. This high-temperature quenching may be associated with the loss of the Sn (II) Mössbauer resonance and its replacement with a Sn (IV) resonance at *c.* 400 K. As the temperature is reduced to below ambient, the luminescence spectrum shows the usual sharpening and shift to lower energy, the band maximum being at 1.72 eV at 80 K. Further cooling results in a decrease in intensity of this band and its replacement by a broader asymmetric feature with a maximum at 1.75 eV. This band and the emission at higher temperatures has a lifetime of less than 10^{-6} s. At 10 K, a second intense emission is observed at 1.16 eV. This second band has not been observed at higher temperatures with optical excitation, but it can be observed at room temperature using electron beam excitation. No photoconductivity was detectable.

The similarity of the position of the absorption edge of $CsSnBr_3$ to the onset of the emission band indicates that the optical properties in the visible region are to be associated with direct band-to-band transitions. The bandwidths and lifetimes are consistent with this. The band gap (1.7 eV) is much larger than that responsible for the change in electrical conductivity with temperature (0.34 eV). Failure to observe photoconductivity supports the existence of at least two bands.

A proposed band model which explains the observed electrical, and optical properties of $CsSnBr_3$ as well as Mössbauer data, is shown in Figure 4.6. In this model the tin 5*s* orbitals form a narrow band which overlaps with an empty band formed by the bromine 4*d* orbitals in the forbidden energy region, allowing direct population of the empty band by the tin 5*s* electrons.

The high-symmetry solid-state environment for tin in its lower oxidation state is only found for those compounds in which the distorting effect of the lone-pair orbitals can be reduced by causing the electrons to populate a low-energy delocalized band in the structure. This model can be extended to explain colour in tetragonal SnO and other materials which have relatively short metal–metal contacts because, in these cases, the ns^2 element is not only supplying the donor electrons but also forming the band by mutual overlap of its empty *d*-orbitals.

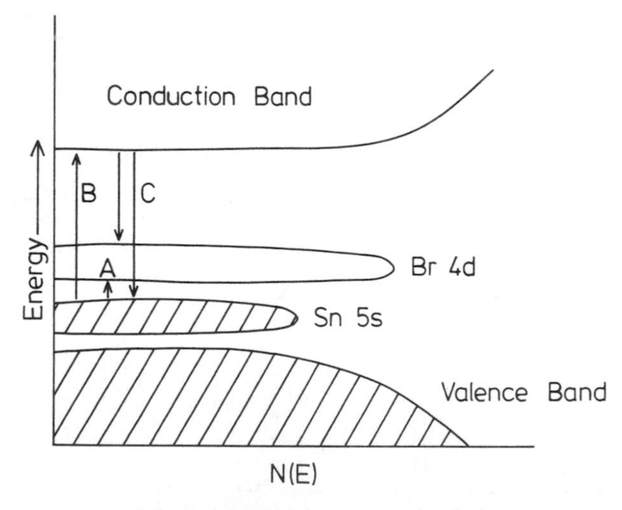

Figure 4.6 Band d structure of $CsSnBr_3$.

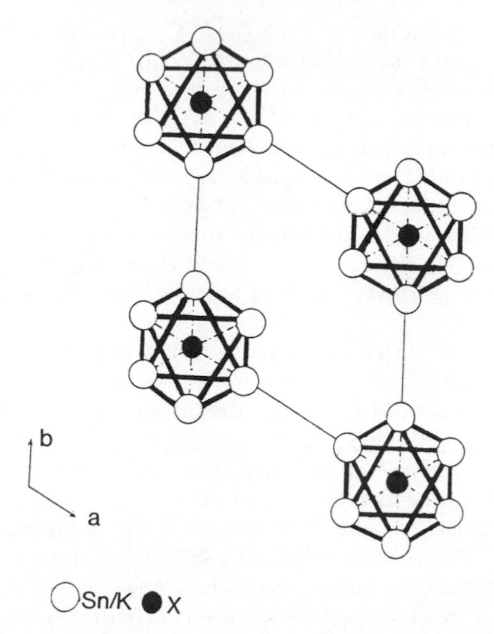

Figure 4.7 Cluster formation in $K_3Sn_2(SO_4)_3X$ species.

Recent work has shown that the non-bonding orbitals can also be delocalized in clusters. For example, $K_3Sn_2(SO_4)_3Br$ consists[48] of a three-dimensional network of tin atoms and bridging sulphate groups. The tin atoms occupy only two-thirds of the symmetry-related positions available for them in the network, and the remaining one-third is taken up by potassium atoms. The remaining potassium and halogen atoms lie in special positions and are discrete ions. The Sn–O and Sn–X bond lengths are unusually long and the Mössbauer chemical shift is unexpectedly low in view of the length of the bonds around the tin (see section 4.1.4). Taken with the crystal structure, these observations are consistent with cluster formation of tin atoms around the bromine atoms (Figure 4.7). Because of the distribution of Sn–O bonds around the tin atoms, the non-bonding electron pairs must be directed towards the centre of the octahedron and the bromine atom. These non-bonding electron pair orbitals must then overlap empty *d* orbitals of adjacent tin atoms giving rise to the weak Sn–Sn cluster interactions and Sn–Sn distances in the range 4.62–4.64 Å. The fact that the non-bonding electron pairs are all directed towards the centre of an octahedron must lead, even if they are being delocalized into cluster orbitals, to a build-up of electron density. In this respect it is interesting that the clusters contain only four tin atoms, the remaining two atoms being potassium. The effect of the potassium atoms and possibly the central halogen atom must be to reduce the electron build-up within the octahedron and to stabilize the cluster. The effects of the cluster electron-delocalization interactions would be threefold: (i) to remove the distorting effects of lone-pair orbitals pointing towards a halide ion; (ii) to delocalize the tin electron density into cluster orbitals giving rise to its anomalous ^{119}Sn Mössbauer chemical shift, and (iii) to weaken the tin–ligand bond outside the cluster. Cluster formation of metal atoms around the halogens therefore provides an explanation for all the apparent anomalous structural and Mössbauer data for the $K_3Sn_2(SO_4)_3X$ compounds. Cluster formation involving tin (II) atoms is a novel example of the Sn–Sn interactions that have important consequences in certain tin (II) compounds. The types of Sn–Sn interactions range from dimer formation in $Sn[CH(SiMe_3)_2]_2$[49] where the stability and structure of the material arise from interactions between the lone pair of one tin atom with the empty orbital on the other tin atom in the dimer, to the interactions that account for the optical and electrical properties of compounds such as $CsSnBr_3$ and SnO where the non-bonding orbitals are delocalized into three-dimensional and adjacent-layer tin bands respectively. The Sn–Sn interactions in the $K_3Sn_2(SO_4)_3X$ clusters must be intermediate in nature between those of $Sn[CH(SiMe_3)_2]_2$ and the full delocalized systems.

In the vapour state, tin (II) compounds of the type SnX_2 should have V-shaped angular structures, and it has long been known[1] that $SnCl_2$ and $SnBr_2$ do in fact have this type of lone-pair distorted structure. Recent work on the chemistry of molecular tin (II) compounds with bulky groups attached to the tin[50] has shown that $Sn[N(SiMe_3)_2]_2$ is also a V-shaped angular structure with Sn–N bond lengths of 2.09 Å and a N–Sn–N bond angle of 96°. In the solid state this compound retains its monomeric molecular structure, although the N–Sn–N bond angle in the solid is widened to 100.47. The molecule $Sn(OC_6H_2MeBu^t_2)_2$[51] is also a V-shaped monomer in the vapour phase. Gaseous tin (II) *t*-butoxide is, however, a *trans*-dimer with a very small endoyalic O–Sn–O angle (76°), while solid tin (II) (tri-*t*-butyl)methoxide contains monomeric angular $[Sn(OCBu^t_3)_2]$ molecules[52]. Tin-119 nmr can be used[52] to distinguish between compounds with a tin coordination of two and those in which the coordination is raised to three or four by intermolecular interactions.

4.1.4 *Mössbauer spectroscopy and bonding in tin (II) compounds*

The ^{119}Sn Mössbauer effect has proved to be a valuable tool in providing information on the ways in which tin uses its bonding electrons in tin (II) compounds. Mössbauer spectroscopy has, for example, been used to compare the covalent character of the bonds in different modifications of the same material, e.g. $CsSnCl_3$[53]. The Mössbauer data for adducts of normal tin (II) compounds and for tin (II) complexes show lower chemical shifts[2] than the corresponding parent tin (II) materials. This is to be expected, because of the greater *s*-electron involvement in the bonding of the complexes[1] and can be explained in terms of the change in bonding as the tin environment alters from SnX_2 to $[SnX_3]^-$ as the ligand X(3) forms a bond to tin creating a discrete group, and no longer acts as a bridge in a polymeric chain. The chemical shifts for MSn_2X_5 compounds such as $NaSn_2F_5$ ($\delta = 3.32\,mm\,s^{-1}$) are intermediate between that of SnX_2 and the corresponding trihalostannate (II) (for example SnF_2, $\delta = 3.62\,mm\,s^{-1}$, and $NaSnF_3$, $\delta = 3.12\,mm\,s^{-1}$). This is consistent with the presence of a bridging F atom in $[Sn_2F_5]^-$. The anomalous nature of the Sn (II)–F bond is clearly shown in the Mössbauer data for tin (II) fluorides and the fluorostannates (II) which are anomalously low, but which can be explained by high use of tin *s*-electron density in the bonding as predicted by the orbital energy matching concept (see section 4.1.3).

It has been found that, as expected, formation of adducts of the type $SnX_2 \cdot L$ results in a lowering of chemical shifts when compared to their parent compounds. The strength of the tin–ligand bond can be inferred from the chemical shift for the complex, and from the difference between this value and the shift of the parent compound. The relatively small lowering in shift in going from the tin(II) halides, $SnCl_2$, $\delta = 4.12\,mm\,s^{-1}$, $SnBr_2$, $\delta = 3.98\,mm\,s^{-1}$, to the 1, 4-dioxan complexes, $SnCl_2$-1, 4-dioxan, $\delta = 3.71\,mm\,s^{-1}$, $SnBr_2$-1, 4-dioxan, $\delta = 3.71\,mm\,s^{-1}$, for example, shows that the tin–ligand bonds are weak in these materials. It is, however, difficult to obtain detailed information on the relative acceptor strengths of all tin(II) compounds towards a given ligand because the chemical shifts, which measure total *s*-electron densities at the tin nucleus will also be very dependent on the atoms, other than the donor ligand atoms, bonded to the metal. The largest change in the chemical shift on complex formation occurs with the formation of the 1:1 adduct, addition of further ligand molecules to form poly-ligand complexes generally has a much smaller effect on the shift. These observations are consistent with the expected predominant monofunctional acceptor properties of tin(II) compounds.

There is no general relationship between the chemical shifts and quadrupole splittings for tin(II) compounds. The presence of the non-bonding electron pair does not in itself appear to lead to an electronic imbalance. $CsSnBr_3$, in which the tin is in a regular octahedral site, has a very narrow line as would be expected, and the $[Sn_2F_5]^-$ anion has a large quadrupole splitting because of the asymmetry of this ion and the nature of the Sn–F bond. The signs of the quadrupole coupling constants are positive for several bivalent tin compounds[53], which is consistent with the main contribution to the field gradient arising from the *p*-electron excess in the lone pair. However, in order to explore the relative sizes of quadrupole splittings in tin(II) compounds, it has been shown to be necessary to consider the *p*-electrons in the bonding orbitals as well as the lone pair, because they too can have a considerable effect. It is significant that the orbital energy matching concept predicts that quadrupole splitting should be greatest for tin(II) bonded to light atoms such as fluorine or oxygen where high use of tin *s*-electrons in Sn–F and Sn–O bonds leaves *p*-electron density in the non-bonding electron pair, thus creating the *p*-electron imbalance responsible for the quadrupole splitting.

A considerable amount of attention has recently been given to the relationship between crystal structure and ^{119}Sn Mössbauer data. Two aspects of the relationship can be considered: (i) the general relationship between the main features of the structure of a tin(II) compound and its Mössbauer data, and (ii) a specific relationship between Mössbauer chemical shifts and bond lengths. The Mössbauer parameters (Table 4.6)

Table 4.6 Mössbauer parameters of ternary tin halides.

	Shift ($mm\,s^{-1}$)	Splitting ($mm\,s^{-1}$)
SnClF	3.73	1.10
Sn_2ClF_3	3.35	1.29
SnBrF	3.59	0.96
Sn_3BrF_5	3.69	1.18
SnIF	3.61	0.76
Sn_2IF_3	3.57	1.34
Sn_2BrCl_3	3.99	0
SnICl	3.71	0.60
SnIBr	3.71	0

for the ternary halides $SnXX'$, $Sn_2XX'_3$ and $Sn_3XX'_5$ serve to illustrate the general relationship. The Mössbauer data for these compounds cannot be consistent with the presence of both Sn–X and Sn–X′ sites and, although it is possible to devise an $SnCl_2$-type structure with only one tin site for $SnXX'$ and $Sn_2XX'_3$, the same cannot be done for $Sn_3XX'_5$. The crystal structure of Sn_3BrF_5 (Figure 4.2) is consistent with the Mössbauer data in that it is built up from a polymeric Sn–F anionic network with Br^- ions, not bonded to the tin, occupying spaces in the lattice to balance the charge. The tin sites in the ternary structure are not identical crystallographically, but are sufficiently alike to give similar Mössbauer resonances which appear as a quadrupole-split doublet. The crystal structure of Sn_2ClF_3 also comprises polymeric fluorotin cations and chloride anions, and is best represented as $[Sn_2F_3]_n^{n+}Cl_n^{n-}$ with no Sn–Cl bonds, in agreement with the Mössbauer data.

The Mössbauer isomer shift for $Sn[CH(SiMe_3)_2]_2$ has a value just above that for α-tin – the borderline between Sn (II) and Sn (IV). The crystal structure and Mössbauer data for this compound distinguish it from the formally similar compounds, $Sn(C_5H_5)_2$, $Sn(C_5H_5)Cl$ and $(Sn^nBu_2)_n$. The cyclopentadienyl compounds have no Sn–Sn bonding and their isomer shift values, which are only just below that for $SnCl_2$ (4.12 mm s^{-1}), are characteristic of tin (II) compounds with a non-bonding electron pair on each tin atom. In contrast, $(Sn^nBu_2)_n$ is polymeric with strong Sn–Sn interactions in which the formally non-bonding pairs on the tin atoms are used in polymerization, and in which the chemical shift and the zero quadrupole splitting are indicative of a compound containing both Sn–C and Sn–Sn bonds. The Mössbauer parameters for $Sn[CH(SiMe_3)_2]_2$ lie between those for $Sn(C_5H_5)_2$ where there is no Sn–Sn interaction and $(Sn^nBu_2)_n$ where there is strong Sn–Sn interaction. Although the shift for $Sn[CH(SiMe_3)_2]_2$ is on the borderline between tin (II) and tin (IV), the value of the quadrupole splitting is not consistent with the formation of Sn–Sn interactions to give a Sn (IV) polymer. The value of the quadrupole splitting is -2.31 mm s^{-1}, the negative sign indicating that there is a deficiency of electrons in the direction of the major component of the field gradient. This in turn means that the *z* direction of the field is in the direction of an empty or nearly empty *p*-type orbital and not in the direction of a lone pair, consistent with a tin (II) formulation for the compound. In the chromium pentacarbonyl adduct, $(OC)_5Cr \cdot Sn[CH(SiMe_3)_2]_2$, however, the empty *p*-orbital gives rise to a quadrupole splitting of -4.43 mm s^{-1}. The reduction in the size of the quadrupole splitting from 4.43 mm s^{-1} to 2.31 mm s^{-1} can only arise because of weak Sn–Sn interactions between neighbouring R_2Sn moieties, and indeed crystallographic studies show that molecules are arranged in pairs in the crystal with a very short Sn–Sn distance of only 2.76 Å[54]. This dimeric structure arises from overlap between the lone pair on each tin atom and the empty *p*-orbital of the neighbouring tin atom. On each tin atom there is a direction in which any electron density on the tin must arise from the weak interaction with the lone pair on the neighbouring tin atom, and this produces the field gradient.

For tin (II) compounds with the tin in trigonal pyramidal $[SnL_3]$ sites, with Sn–L bonds to the same ligand L, a close connection has been found between the Mössbauer chemical isomer shift and the Sn–L bond distances. For tin compounds with single $[SnF_3]$ and $[SnO_3]$ sites, for example, there is an increase in shift with increasing average bond length (Table 4.7), that is, with increasing electrostatic character. Crystals of $Sn_6O_6(OMe)_4$ have been shown to consist of two crystallographically non-interacting $[Sn_6O_4(OMe)_4]$ units

Table 4.7 Mössbauer chemical isomer shift and bond length in tin–oxygen, tin–fluorine and tin–chlorine compounds.

Compound	Average bond length (Å)	Shift ($mm\,s^{-1}$)
NH_4SnF_3	2.08	3.25
$NaSn_2F_5$	2.12	3.32
Sn_3F_8	2.17	3.82
$CaSn(CH_3CO_2)_3$	2.14	2.90
$KSn(CH_2ClCO_2)_3$	2.16	2.96
$KSn(HCO_2)_3$	2.16	3.08
$SnHPO_3$	2.17	3.15
$SnSO_4$	2.26	3.95
$Co(dpe)_2 \cdot Cl \cdot SnCl_3 \cdot nC_6H_5Cl$	2.43	3.08
$Co(dpe)_2 \cdot Cl \cdot SnCl_3$	2.44	3.10
$CsSnCl_3$ (monoclinic)	2.52	3.64
$KCl \cdot KSnCl_3 \cdot H_2O$	2.57	3.70
$SnCl_2$	2.74	4.12

with a central adamantane-like $[Sn_6O_4]$ core. The chemical shift value of 2.78 $mm\,s^{-1}$ is consistent with tin being in the + II oxidation state and with the very short Sn–O (2.05–2.08 Å) in the $[Sn_6O_4]$ skeleton. Those compounds with pyramidal $[SnCl_3]$ groups and containing only Sn–Cl bonds as nearest and next nearest neighbours illustrate the relation between Sn–Cl and length and shift. The similarity in the data between $CsSnCl_3$ and $[(NH_4)_5CoSO_2C_6H_5][Cl_3SnOClO_3]H_2O$ ($\delta = 3.52\,mm\,s^{-1}$, average bond length = 2.52 Å) show that the next nearest bond in the perchlorate complex, the Sn–O bond of 2.91 Å, produces at the most a very weak perturbation of the tin electron density.

The use of ^{119}Sn Mössbauer spectroscopy, particularly in conjunction with other techniques, has had and will continue to have an important part to play in the interpretation of the bonding characteristics of tin in its II + oxidation state.

4.2 Tin (IV) chemistry

Tin (IV) oxide is amphoteric and dissolves in aqueous solutions of both acids and alkalis. The predominant tin species present in acid solutions containing complexing anions are the octahedral $[SnX_6]^{2-}$ ions e.g. with X = F, Cl, Br, I, NCO and NCS. There is also evidence for the formation of mixed-ligand species such as $[SnCl_4Br_2]^{2-}$, $[SnF_4(OH)_2]^{2-}$ and $[SnCl_4NO_3]^-$. Tin-119 nmr studies have identified[55] all of the possible bromochlorostannates (IV) $[SnCl_nBr_{6-n}]^{2-}$ and most of the chlorofluorostannates (IV), $[SnCl_nF_{6-n}]^{2-}$, including $[SnCl_5F]^{2-}$, *cis*- and *trans*-$[SnCl_4F_2]^{2-}$, *fac*-$[SnCl_3F_3]^{2-}$, *cis*- and possibly *trans*-$[SnCl_2F_4]^{2-}$ and $[SnClF_5]^{2-}$. Several thiocyanato- and cyano-derivatives, $[SnX_{6-n}Y_n]^{2-}$, or X = Cl or Br and Y = CN or NCS, have also been identified in aqueous solution[56].

The main species present in alkaline solutions containing an excess of hydroxide ion is $[Sn(OH)_6]^{2-}$, but in less basic solutions dehydration may occur to give ions such as $[SnO_3]^{2-}$. SnS_2 is soluble in aqueous alkaline solutions containing sulphide ions because of the formation of $[SnS_3]^{2-}$. Some tin (IV) compounds (e.g. $SnCl_4$ and

$Sn(SO_4)_2$) are soluble in water but are subject to hydrolysis and, in fact, in the absence of strong complexing anions, all aqueous solutions of Sn (IV) tend to hydrolyse to give a precipitate of hydrous tin (IV) oxide.

A number of tin (IV) materials are soluble in non-aqueous solvents. Tin (IV) chloride and bromide are soluble in glycol, acetone, pyridine, benzene, dimethylsulphoxide, and many other organic solvents, and the complex ions, $[SnCl_5]^-$ and $[SnBr_5]^-$ can be obtained from solutions of the halides in dichloroethane. Tin (IV) iodide is soluble in a wide range of non-aqueous solvents including carbon tetrachloride, chloroform, benzene, heptane, diethylether, toluene and carbon disulphide. A number of complex tin (IV) species have been identified in non-aqueous solutions including $[Sn(NH_2)_6]^{2-}$ in liquid ammonia, $[Sn(HSO_4)_6]^{2-}$ in 100% H_2SO_4 and $SnX_4 \cdot 2$ligand for many solutions of tin (IV) halides in donor solvents which can act as ligands. Tin (IV) chloride in liquid N_2O_4 gives the ionic salt, $NO^+[SnCl_4(NO_3)]^-$. Both tin (IV) chloride and bromide can, because of their Lewis acid properties, act as non-aqueous solvents.

Tin (IV) carboxylates are not well known, although $Sn(CH_3CO_2)_4$ can be prepared from SnI_4 and acetic anhydride or from tetravinyltin in anhydrous acetic acid. The cleavage of tetravinyltin by carboxylic acids is a general reaction and the compounds $Sn(RCO_2)_2$ (R = H, C_2H_5, *iso*-C_4H_9 *n*-C_4H_9 and *n*-$C_{11}H_{23}$) have been prepared. Tin (IV) nitrate, $Sn(NO_3)_4$, is prepared by the action of N_2O_5 with $SnCl_4$ followed by vacuum sublimation at 40°. It reacts readily with aliphatic hydrocarbons and vigorously with diethylether, probably because of the formation of a $NO_3^{\bullet}$ radical. It also forms the adduct, $Sn(NO_3)_4 \cdot 2py$, in which the nitrate groups are unidentate. The action of hydrous tin (IV) oxide on phosphoric acid results in the precipitation of a gel of approximate formula $Sn(P_2O_5)_{0.6}(H_2O_4)_{4.9}$. This is probably a basic salt and is known to act as an ion exchange material. Reaction of $K_4P_2O_7$ with SnO_2 gives the pyrophosphate SnP_2O_7, and the mixed phosphate $KSn_2(PO_4)_3$ has also been reported. Tin (IV) hypophosphite $Sn(H_2PO_2)_4$ is obtained as a white crystalline solid by the oxidation of solutions of tin (II) hypophosphite. The mixed salt $Sn(H_2PO_2)_4 \cdot SnCl_4$ has also been prepared. $Sn(SO_4)_2 2H_2O$ has been obtained as deliquescent white crystals from Sn (IV) solutions. Anhydrous $Sn(SO_4)_2$ is said to exist, and the mixed sulphates $K_2Sn(SO_4)_3$ and $CaSn(SO_4)_3 \cdot 3H_2O$ have also been prepared.

4.2.1 *Structure and bonding in tin (IV) chemistry*

The structural chemistry of inorganic tin (IV) derivatives is relatively simple, being for the most part based on tetrahedral coordination in molecular compounds and on octahedral geometry in compounds with more ionic bonding and in most of its complexes. There are, however, examples of 5-, 7- and 8-coordination.

The halides SnX_4 (X = Cl, Br, I) are tetrahedral molecules in the vapour phase. Solid $SnBr_4$ and SnI_4 have structures consisting of hexagonally close-packed halogen atoms with tin occupying one-eighth of the tetrahedral holes. In both, the tin atoms have a distorted tetrahedral environment, the distortion being greatest for $SnBr_4$. Stannane, SnH_4, is also tetrahedral, and reacts with sodium in liquid ammonia to give $NaSnH_3$ and Na_2SnH_2. Halostannanes such as $ClSnH_3$ are also known but are stable only at lower temperatures. Tetrahedral or near tetrahedral tin sites are also found in $Sn(NCPh_2)_4$, $[(Me_3Si)_2N]_3SnBr$, K_4SnO_4 and in a range of sulphide derivatives including $Sn(SCH_2CH_2S)_2$, K_2SnS_3, Na_4SnS_4, M_2SnS_4, Cu_2MSnS_4 (M = Ca, Sr, Ba), Eu_2SnS_5. $Na_4Sn_2S_6$, $Na_6Sn_2S_7$, $Eu_3Sn_2S_7$, $BaCdSnS_4$ and $M_3Sn_2S_7$ (M = Ca,

Ba). Some of these compounds, for example Na_4SnS_4, contain discrete $[SnS_4]^{4-}$ tetrahedra, while other compounds are said to contain discrete condensed anions such as $[Sn_2S_6]^{4-}$ (in $Na_4Sn_2S_6 \cdot 14H_2O$) and $[Sn_2S_7]^{4-}$ (in $Na_6Sn_2S_7$). Two tetrahedral tin(IV) sites have also been found in the recently determined structure[57] of $Eu_2Sn_2S_5$. Both sites are distorted, one having bond lengths in the range 2.37–3.50 Å and the other in the range 2.32–3.01 Å.

In most of its solid-state compounds tin is in a six coordinated site. It has nearly regular octahedral symmetry in SnO_2 which has the rutile structure, and in SnS_2 which crystallizes in the CdI_2 lattice. Tin forms a wide range of stannates(IV) in which it is generally to be found in six coordination. Typical stannates(IV) include $BaSnO_3$, K_2SnO_3, $Na_2Sn_2O_5$, $Ca_2Sn_2O_7$, Ca_2SnO_4 and Co_2SnO_4. The amount of distortion of tin sites can vary from negligible in SnO_2 to considerable in compounds such as $CaSnSiO_5$, in which the tin is in a tetragonally distorted site with two short axial bonds of 1.947 Å and four longer equatorial bonds of about 2.09 Å. Tin is also in octahedral environments in the sulphides, Sn_2S_3, Cu_2SnS_4 and Cu_2FeSnS_4.

Solid tin(IV) fluoride, SnF_4, has a distorted octahedral environment with two short and four long Sn–F bonds and is isostructural with PbF_4. Most of the solid derivatives of the $[SnX_6]^{2-}$ ions (X = Cl, Br, I) have tin in regular six coordination, and the alkali metal salts have the K_2PtCl_6 structure. The crystal structures of hexahalogenostannates(IV) determined recently include $[ttf]_2[SnCl_6]$, $[Hpy]_2[SnBr_6]$ and $[NH_2Me_2]_2[SnBr_6]$. The reaction of tetrafulvalene (ttf) with $SnCl_4$ gives $[ttf][SnCl_6]$ which contains[58] the $[SnCl_6]^{2-}$ anion in which Sn–Cl bond lengths (2.437–2.438 Å) are longer than those normally found in this ion (2.401–4.243 Å). The $[SnBr_6]^{2-}$ ion in $[NH_2Me_2][SnBr_6]$ has[59] two bonds of 2.609 Å and four of 2.601 Å in contrast to $[Hpy]_2[SnBr_6]$[60] which has three distinct Sn–Br distances, 2.63, 2.58 and 2.56 Å. $BaSnF_6$ is said to have a regular and Na_2SnF_6 a distorted octahedral environment for tin. Six-coordination tin environments are also present in K_3HSnF_8 and in most hexahydroxystannates(IV) (e.g. $MSn(OH)_6$ M = Ca, Mn, Fe, Co and Zn).

Tin(IV) forms a wide range of six-coordinated complexes with chelating ligands such as edta, acetylacetone, salicylaldehyde, dithiols and 8-hydroxyquinoline, and complexes of the following compositions have been prepared: bis(toluene-3,4-dithiol)Sn(IV), tetrakis(8-hydroxyquinolate)Sn(IV), SnY_2X_2 (X = Cl, Br, I; Y = 8-hydroxyquinolate, salicylaldehyde, acetylacetonate).

The tin(IV) halides and many other tin(IV) materials act as acceptors and form complexes with neutral donor ligands[61]. Six coordinated tin(IV) complexes of the type $SnX_4 \cdot 2L$ are known in which ligand atoms can either be in *cis-* or *trans-* positions. The *cis* adducts are usually distorted in such a way that the SnX_4 residue is very similar to the original tetrahedral SnX_4 and most $SnX_4 \cdot$2ligand complexes appear to have this structure. The compounds $SnCl_4 \cdot 2CH_3CN$, $SnCl_4 \cdot 2POCl_3$, $SnCl_4 \cdot 2SeOCl_2$, $SnBr_4 \cdot$2oxH (oxH = 8-hydroxyquinoline) are known to be *cis*. The ligands are, however, thought to be in *trans*-positions in $SnBr_4 \cdot$2salicyladehyde, $SnCl_4 \cdot 2P(C_2H_5)_3$ and $SnCl_4 \cdot 2As(C_2H_5)_3$. Di-ligand complexes of tin(IV) halides with a very large variety of donor molecules are known including those with N-donors (pyridine, NH_3, aniline, nitriles and many amines), P-donors (triethyl- and triphenyl-phosphines and PH_3), O-donors (carboxylic acids, aldehydes, ketones, alcohols, esters, H_2O, phosphine oxides and sulphoxides). S-donors (thiourea, and alkyl sulphides). Complexes with As- and Sb-donors are also known. A number of 1:1 complexes have been reported including $SnBr_4 \cdot N(CH_3)_3$, $SnCl_4 \cdot P(C_6H_5)_3$, $SnF_4 \cdot$thf and $SnCl_4 \cdot$oxH. Among the complexes found with other tin(IV) compounds are $Sn(NO_3)_4 \cdot 2$py, $Sn(NO_3)_4 \cdot 6N(C_2H_5)_3$, and bis(ethane-1-2-dithiolato)Sn(IV)·2L (L = pyridine, dimethylsulphoxide).

In addition to the more common four- and six-coordination found for tin in its IV + oxidation state, there are some compounds in which the tin environments are more easily described in terms of five-, seven-, or eight-coordination. The $[SnCl_5]^-$ ions in its 3-chloro-1, 2, 3, 4-tetraphenylcyclobutenium salt has a five-coordinated trigonal bipyramidal structure with equatorial bond lengths and angles of 2.30–2.40 Å and 114–126°, axial bond lengths of 2.37–2.39 Å and equatorial-axial bond angles of 90°. Five-coordination has also been reported for tin(IV) in $SnTa_2O_7$, K_2SnO_3, and for one site in $Eu_3Sn_3S_{12}$. Seven-coordination has been ascribed to tin in a number of compounds including Sn(edta)·H_2O, $K_6Sn_2[C_2O_4]_7 \cdot 4H_2O$ and $(Me_4N)[Sn(OCOCH_3)_5]$. The eight-coordinated tin structures found for tin in $Sn(CH_3CO_2)_4$ and $Sn(NO_3)_4$ are illustrated in Chapter 2 (structure (**60**)). It is likely that tin is also eight-coordinated in $Sn(oxine)_4$, $Sn(tropolonate)_4$ and $Sn(C_2O_4)_4^{4-}$ derivatives.

4.2.2 *Mössbauer spectroscopy of tin(IV) compounds*

The ^{119}Sn Mössbauer chemical shifts (δ) for most tin(IV) materials follow the trends expected in that they increase as the covalent character of the bonding increases[62]. For example, shifts of the tetrahalides are more positive (Table 4.8) with increasing covalent character of the Sn–X bond, that is with the trend away from the $4d^{10}Sn^{4+}$ ion. Similar relationships have been found for all halogeno-tin compounds, and there is a linear relationship between δ for the $[SnX_4Y_2]^{2-}$ complexes (X, Y = Cl, Br, I) and the sum of the electronegativities of the halogens bonded to tin. The increase in the coordination number from four in the tetrahalides to six in complexes also represents an increase in the electrostatic character of the bonds to tin(IV), and is reflected in a decrease in the shift for the tin(IV) halide complexes with neutral donor ligands, ($SnCl_4$, K_2SnCl_6 and $SnCl_4 \cdot 2NMe_3$ in Table 4.8). The shift becomes greater with increasing donor ligand electronegativity. For example, the shifts for SnCl·2L are 0.22, 0.33, 0.46, 0.08 and 0.71 mm s^{-1}, respectively, for L = $(C_6H_5)_3PO$, $(CH_3)_2SO$, py, $(CH_3)_3N$ and $(C_6H_5)_3P$. Very few inorganic tin(IV) compounds show a quadrupole splitting (Δ) of the resonance line, which indicates little imbalance in the *p*- or *d*-electron distribution around tin. Zero splittings are expected for compounds with high-symmetry environments (e.g. $SnCl_4$ and $BaSnO_3$), but deviations from ideal geometry do not in themselves give rise to a *p*-electron imbalance and to quadrupole splitting (for example $SnCl_4 \cdot 2NMe_3$, $SnCl_2(oxine)_2$ and $SnCl_2Br_2$ all have zero splitting values). Quadrupole splittings only arise when the bond characteristics of one type of tin-element interaction present is significantly different from the others. The presence of two short and four long bonds in SnF_4 does, for example, produce an electronic imbalance and gives a splitting of 1.80 mm s^{-1}. A few tin(IV) halide complexes with neutral donor molecules also have tin-ligand bonds sufficiently different from Sn–X to produce a field gradient (e.g. *trans*-$SnCl_4 \cdot 2(CH_3)_2C_6H_5P$ (0.97 mm s^{-1}); *cis*-$SnCl_4 \cdot 2CH_3CN$ (0.91 mm s^{-1}) and *cis*-$SnCl_4 \cdot 2Cl_3PO$ (1.12 mm s^{-1}) all have resolvable splittings). The splitting in the nitrile adduct is due to the presence of very weak Sn–N bonds. Since the splitting for a *trans* complex should be twice that of a

Table 4.8 Mössbauer parameters for tin (IV) compounds.

Compound	Shift ($mm\,s^{-1}$)	Splitting ($mm\,s^{-1}$)
SnF_4	−0.47	1.66
$SnCl_4$	0.85	0
$SnBr_4$	1.15	0
SnI_4	1.55	0
K_2SnF_6	−0.59	0
K_2SnCl_6	0.45	0
K_2SnBr_6	0.75	0
K_2SnI_6	1.30	0
$SnCl_4 \cdot 2NMe_3$	0.54	0
$SnCl_4 \cdot 2MeCN$	0.43	0.70
$SnBr_4(oxinH)_2$	0.65	0
$SnBr_4(salH)_2$	0.73	1.22
Ph_2SnCl_2	1.34	2.89
Me_3SnCl	1.43	3.32
$SnZn_2O_4$	0.10	0.75
SnO_2	0	0
Cu_2FeSnS_4	1.48	0
trans-$SnCl_4(tht)_2$	0.72	0.35
cis-$SnCl_4(tht)_2$	0.70	0.24
trans-$SnCl_4(dmso)_2$	0.41	0.57
cis-$SnCl_4(dmso)_2$	0.40	0.41
trans-$SnCl_4(dmf)_2$	0.38	0.73
cis-$SnCl_4(dmf)_2$	0.39	0.53
trans-$SnBr_4(dmf)_2$	0.66	0.83
cis-$SnBr_4(dmf)_2$	0.66	0.44
trans-$SnCl_4(dma)_2$	0.38	0.78
cis-$SnCl_4(dma)_2$	0.38	0.45

corresponding *cis*-complex, it has been possible to distinguish between *cis*-$SnBr_4 \cdot 2$(8-hydroxyquinoline) ($\Delta = 0$) and $SnBr_4 \cdot 2$(salicylaldehyde) ($\Delta = 1.11\ mm\,s^{-1}$). The data for *cis*- and *trans*-$SnCl_4$ complexes from a recent study[63] are in Table 4.8. The five-coordinated species $[SnCl_5]$ is of sufficiently low symmetry to show a splitting of $0.63\ mm\,s^{-1}$. Although no resolvable quadrupole splitting is found for mixed halides $[SnX_nY_{6-n}]^{2-}$, the Mössbauer spectrum for the mixed hexahalogenostannates (IV) found in glassy aqueous mixed halide solutions do have broader resonance lines than the single halide $[SnX_6]^{2-}$ species[64].

Recent variable-temperature Mössbauer studies have been used to provide information on the nature of association of tin moieties in solids. The complexes $SnCl_4 \cdot 2L$ (L = detu and dmtu) are, for example, non-interacting discrete molecules[65], while $LiFeSnO_4$ in both its high- and low-temperature forms, has a three-dimensional lattice but with a considerable amount of covalency in the Sn–O bonds[66].

4.2.3 *Solid-state chemistry of tin (IV) compounds*

Perhaps the most important development in tin (IV) chemistry in the past decade has been the increase in studies of the solid-state properties of tin (IV) compounds. Tin (IV) oxide, stannates (IV), and mixed-metal oxide phases containing tin (IV) have interesting physical properties as bulk solids and as thin films[67]. Tin(IV) oxide itself, for example, is an oxygen-deficient lattice and an *n*-type semiconductor with a band gap of 3.7 eV. The mixed-metal oxide phases antimony–tin oxide and indium–tin oxide have good electrical conductivity leading to commercial applications. Other tin (IV) oxide phases which have been studied include the Sn (IV)-doped $BaPb_{1-x}Bi_xO_3$ perovskite superconductors, SnO_2–Eu phosphors, indium–tin oxide piezoelectric ceramics, and $K_x(In_xSn_{1-x})O_2$ ionic conductors.

A number of recent studies have been concerned with the use of tin (IV) compounds as intercalate host lattices. The chlorotin acetate and phosphate phases $[SnCl(OH)XO_4]\cdot 2H_2O$ (X = P, As) for example[68] will intercalate short-chain fatty acids and some amines, and other tin intercalates have also been described[69].

Another area of current interest in solid-state tin (IV) chemistry is concerned with complexes of crown ethers. Compounds of the type $SnX_4\cdot L\cdot 2H_2O$ and $SnCl_4\cdot L\cdot 4H_2O\cdot nCHCl_3$, where L is a crown ether, have been synthesized[70]. The structure of $Sn(OH_2)_2Cl_4\cdot$18-crown-6$\cdot 2H_2O\cdot CHCl_3$ comprises octahedral $[Sn(OH_2)_2Cl_4]$ units with the water molecules lying *cis* to each other, and involved in an extensive hydrogen-bonding scheme. The crown ether complexes of $2SnCl_4$ dibenzo-24-crown-8[71] and $SnCl_4\cdot$18-crown-6$\cdot 4H_2O$ have been reported, and the structure of the latter determined[72]. The crystal structure of the complex of 15-crown-5 (1, 4, 7, 10, 13-pentaoxocyclopentadecane) with diaquatetrachlorotin (IV) at 120 K has also been determined[73], and consists of octahedral $[SnCl_4(H_2O)_2]$ units linked by bifurcated hydrogen bonds from the water molecules to the 15-crown-5 molecules. The formation of the complex does, however, result in an unusual *trans-* distribution of water molecules with Sn–O distances of 2.11 and 2.13 Å and Sn–Cl distances of 2.36–2.39 Å.

4.3 Mixed-valence tin compounds

In recent years, a considerable amount of work has been carried out on mixed-valence compounds which contain both bivalent and tetravalent tin. The best known is Sn_2S_3, which can be prepared as acicular crystals by heating a powdered mixture of the appropriate proportions of tin and sulphur to 993 K in a sealed tube. Sn_2S_3 contains[74] octahedrally coordinated tin (IV) with Sn–S distances of 2.459–2.604 Å and bond angles of 86.4–91.1°. The tin (II) site has the typical trigonal pyramidal arrangement of atoms around the tin, with two Sn–S bond lengths of 2.645 and 2.765 Å. The compound is a semiconductor and its electrical properties have been studied[75–76]. The mixed-valence sulphide $Sn^{II}{}_4Sb^{III}Sn^{IV}S_9$ consists of a cube face distribution of $[Sn^{IV}S_6]$ octahedra linked through $[(Sn^{II}, Sb^{III})S_8]$ dodecahedra which are lone-pair distorted[77].

The oxidation of SnF_2 in HF with O_2, F_2 or SO_2 yields the insoluble mixed oxidation-state compound Sn_3F_8, whose structure shows *trans*-fluorine-bridged $[Sn^{IV}F_6]$ units linked to polymeric $[Sn^{II}–F]_\infty$ chains[78]. The tin (IV) atoms are octahedrally coordinated by fluorines. The marginally longer bonds are to the *trans*-fluorines which also interact with the tin (II) atoms. The tin (II) atoms are not only linked to tin (IV) by bridging fluorines, but also to one another by the other fluorine atoms. This latter interaction produces nearly coplanar zigzag –Sn–F–Sn–F– chains running through the structure. The overall structure may be represented by the formula $(SnF)_2SnF_6$, the tin (II) sites being pyramidal with Sn–F bonds of 2.10, 2.17 and 2.25 Å, and a next-nearest contact of 2.55 Å. The structure of $(NH_4)_3Sn_3F_{11}$ has also been determined[79], and it has been suggested that the material is best formulated as $(NH_4)_3(Sn_2F_5)(SnF_6)$. Two types of $[SnF_6]^{2-}$ octahedra are present in the lattice, one of which is much more distorted than the other. The $[Sn_2F_5]$ moieties have terminal Sn–F bonds of 2.05 Å and a bridging bond of about 2.13 Å. It is significant that this compound is an ionic conductor and that there is no evidence for any electronic conduction arising from the presence of both tin (II) and tin (IV) species in the lattice. Mixed-valence fluorotin carboxylates have been prepared by reacting SnF_2 with $Sn(RCO_2)_4$. The crystalline compound isolated from SnF_2–$Sn(CF_3CO_2)_4$ solutions has been shown[80] to be $Sn^{II}{}_2Sn^{IV}{}_2F_4(CH_3CO_2)_8 \cdot 2CF_3CO_2H$. Its structure contains two only slightly different *cis*-fluorine-bridged tin (IV) environments, consistent with the single line found in the Mössbauer spectrum of the compound ($\delta - 0.16\,mm\,s^{-1}$). The lattice also contains two lone-pair distorted tin (II) sites which exhibit a single Mössbauer resonance line ($\delta = 4.13\,mm\,s^{-1}$, $\Delta = 0.84\,mm\,s^{-1}$). The total structure comprises two independent centrosymmetric molecules, each molecule consisting of an eight-membered ring with a $-Sn^{II}-F-Sn^{IV}-F-$ arrangement. Adjacent tin (II) and tin (IV) atoms in the molecules are also bridged by either one or two trifluoroacetate groups. There are also unidentate trifluoroacetate groups and acid molecules coordinated to the tin atoms. Eight trifluoroacetate groups are associated with each tetramer, and the two trifluoroacetic acid molecules associated with only one of the molecules and coordinated to two tin (II) atoms.

The mixed-metal oxides $Sn^{II}{}_{2-x}(M^{V}{}_{2-y}Sn^{IV}{}_{y})O_{7-x-y}$ (M = Nb, Ta) are examples of tin (II)–tin (IV) compounds in which the tin (II) environment is lone-pair distorted and similar to that found in tin (II) sulphate[81]. It has also been claimed[82] that octahedral tin (IV) and pyramidal tin (II) sites are also present in Sn_2O_3. The crystal structure of

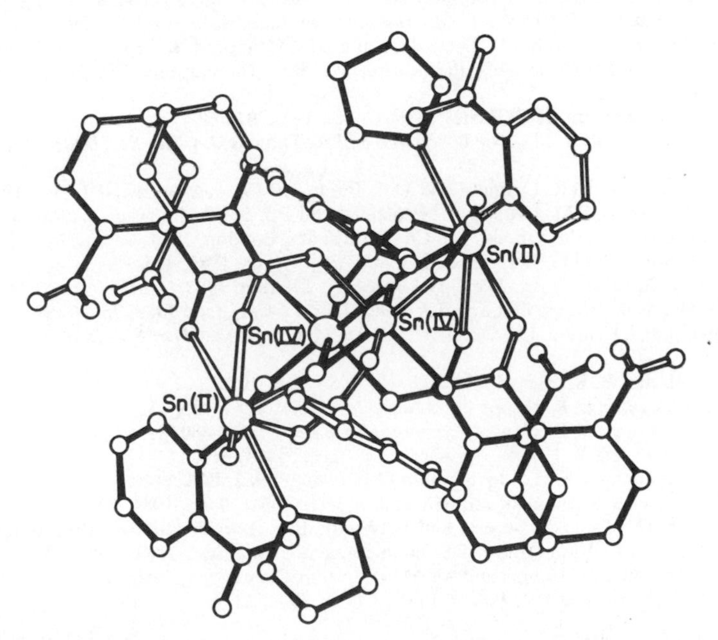

Figure 4.8 Tetranuclear clusters in $[Sn^{II}Sn^{IV}(O_2CC_6H_4NO_2\text{-}2)_4O \cdot thf]_2$ (reproduced by permission from ref. 83).

the mixed valence carboxylate $[Sn^{II}Sn^{IV}(OCOC_6H_4NO_2\text{-}2)_4O\cdot thf]_2$ has been determined.[83] The structure (Figure 4.8) consists of independent tetranuclear clusters containing two tin (II) and two tin (IV) atoms. The main feature of the cluster is a lozenge-shaped four-membered $[Sn^{IV}{}_2O_2]$ ring, with octahedral coordination of the tin being completed by oxygen atoms from four *o*-nitrobenzoate groups which form bridges between the tin atoms in the two oxidation states. The tin (IV)-oxygen bond distances are in the range 2.047–2.067 Å and the geometry at tin (II) is that of a distorted pentagonal pyramid with oxygen atoms from four bridging carboxylate groups and tetrahydofuran occupying equatorial positions at bond distances in the range 2.41–2.66 Å but with a relatively short (2.11 Å) Sn (II)–O distance in an apical position. The variable-temperature Mössbauer data for this compound shows that the tin (IV) atoms are more strongly held in the lattice than the tin (II) atoms.

The most significant feature of the mixed-valence compounds of tin is that, for compounds in which the tin is bonded to oxygen or fluorine atoms, there is no evidence for any inter-valence transfer desorption or of electrical properties arising from hopping mechanisms. Electrical conductivity only arises in Sn_2S_3 where the sulphur atoms have vacant *d*-orbitals, and for which direct population of solid-state bands, as in the case of $CsSnBr_3$ and similar compounds (section 4.1.3), is a more likely explanation for the optical and electrical properties of the compound than inter-valence electron hopping.

References

1. J.D. Donaldson, *Prog. Inorg. Chem.*, 1967, **8**, 287.
2. J.D. Donaldson and S.M. Grimes, *Revs. Si, Ge, Sn, Pb Cmpds.*, 1984, **8**, 1.
3. S.-W. Ng and J.J. Zuckerman, *Adv. Inorg. Chem. Radiochem.*, 1985, **29**, 297.
4. P.A. Cusack, P.J. Smith, J.D. Donaldson and S.M. Grimes, *A Bibliography of X-ray Crystal Structures of Tin Compounds*, Publication No. 588,* International Tin Research Institute, London.
5. J.D. Donaldson and D.C. Puxley, *Acta Cryst.*, 1972, **B28**, 864.
6. J.C. Dewan, J. Silver, J.D. Donaldson and M.J.K. Thomas, *J. Chem. Soc. Dalton Trans.*, 1977, 2319.
7. J.D. Donaldson, D.R. Laughlin and D.C. Puxley, *J. Chem. Soc. Dalton Trans.*, 1977, 865.
8. J.D. Donaldson, S.M. Grimes, A. Nicolaides and P.J. Smith, *Polyhedron*, 1985, **4**, 391.
9. M. Christoforou, PhD thesis, The City University, London, 1988.
10. Von J. Bonish and G. Bergerhoff, *Z. anorg. allg. Chem.*, 1981, **473**, 35.
11. I. Abrahams, PhD thesis, The City University, London, 1987.
12. P.G. Harrison, B.J. Haylett and T.J. King, *Inorg. Chim. Acta*, 1983, **75**, 259.
13. Z. Arifin, E.J. Filmore, J.D. Donaldson and S.M. Grimes, *J. Chem. Soc. Dalton Trans.*, 1984, 1965.
14. M. Veith and R. Roesler, *Z. Naturforsch.*, 1986, **4IB**, 1071.
15. V.R.M. Braun and R. Hoppe, *Z. anorg. allg. Chem.*, 1982, **485**, 15.
16. V.R.M. Braun and R. Hoppe, *Z. anorg. allg. Chem.*, 1981, **480**, 81.
17. B. Nowitzki and R. Hoppe, *Z. anorg. allg. Chem.*, 1984, **515**, 114.
18. T. Yamaguchi and O. Lindqvist, *Acta Crystallogr.*, 1982, **B38**, 1441.
19. S. Del Bucchia, J.C. Jumas and M. Maurin, *Acta Crystallogr.*, 1981, **B37**, 1903.
20. P. Knirp, D. Mootz, U. Severin and H. Wunderlich, *Acta Crystallogr.*, 1982, **B38**, 2022.
21. A. Mazurier, F. Thevet and F.T.S. Jaulmes, *Acta Crystallogr.*, 1983, **C39**, 814.
22. S. Del Bucchia, J.C. Jumas and M. Maurin, *Acta Crystallogr.*, 1980, **B36**, 2935.
23. M.M. Olmstead and P.P. Power, *Inorg. Chem.*, 1984, **23**(4),

**Note added in proof*: Present address of the International Tin Research Institute is Kingston Lane, Uxbridge, Middlesex UB8 3PJ.

24. J.D. Donaldson, M.T. Donoghue and C.H. Smith, *Acta Crystallogr.*, 1976, **B32**, 2098.
25. J. Abrahams, J.D. Donaldson, S.M. Grimes, G. Valle and S. Calogero, *Polyhedron*, 1986, **5**(10), 1593.
26. H.P. Berk, *Z. anorg. allg. Chem.*, 1986, **536**, 45.
27. A. Nicolaides, PhD thesis, The City University, London, 1983.
28. R. Nesper and H.G. Van Schnering, *Z. Naturforsch.*, 1982, **37B**, 1144.
29. A. Magnusson and L.G. Johansson, *Acta Chem. Scand.*, 1982, **A36**, 429.
30. G. Lundgren, G. Wernfors and T. Yamaguchi, *Acta Crystallogr.*, 1982, **B28**, 2357.
31. P. Jutzi, F. Kohl, E. Schluter, M.B. Hursthouse and N.P.C. Walker, *J. Organomet.*, 1984, **271**, 393.
32. A. Likforman, M. Guittard and S. Jaulmes, *Acta Crystallogr.*, 1984, **C40**, 917.
33. S. Del Bucchia, J.C. Jumas, E. Philippot and M. Maurin, *Z. anorg. allg. Chem.*, 1982, **487**, 199.
34. S. Jaulmes, M. Julien-Pouzol, P. Laruelle and M. Guittard, *Acta Crystallogr.*, 1982, **B38**, 79.
35. H.H. Karsch, A. Appelt and G. Muller, *Angew. Chem. Int. Edn. Engl.*, 1985, **24**, 402.
36. J.L. Lefferts, K.C. Molloy, M.B. Hossain, D. Van der Helm and J.J. Zuckerman, *Inorg. Chem.*, 1982, **21**, 1410.
37. T. Birchall and V. Manivannan, *J. Chem. Soc., Chem. Commun.*, 1986, 1441.
38. H.G. Von Schnering and H. Wiedermeir, *Z. Kristallogr.*, 1981, **156**, 143.
39. P.P.K. Smith and B.G.T. Hyde, *Acta Crystallogr.*, 1983, **C39**, 1498.
40. V. Kupcik and M. Wendschuh, *Acta Crystallogr.*, 1982, **B38**, 3070.
41. J.D. Donaldson, J. Silver, S. Hadjiminolis and S.D. Ross, *J. Chem. Soc., Dalton Trans.*, 1975, 1500.
42. J.D. Donaldson, D.R. Laughlin, S.D. Ross and J. Silver, *J. Chem. Soc., Dalton Trans.*, 1973, 1985.
43. R.H. Andrews PhD thesis, University of London, 1977.
44. D.G. Urch, *J. Chem. Soc.*, 1964, 5775.
45. L.F. Orgel, *J. Chem. Soc.*, 1959, 3815.
46. J.D. Donaldson, D.C. Puxley and M.J. Tricker, *J. Inorg. Nucl. Chem.*, 1975, **37**, 655.
47. S.J. Clark, C.D. Flint and J.D. Donaldson, *J. Phys. Chem. Solids*, 1981, **42**, 133.
48. J.D. Donaldson and S.M. Grimes, *J. Chem. Soc., Dalton Trans.*, 1984, 1301.
49. J.D. Cotton, P.J. Davidson, M.F. Lappert, J.D. Donaldson and J. Silver, *J. Chem. Soc., Dalton Trans.*, 1976, 2286.
50. T. Fjeldberg, H. Hope, M.F. Lappert, P.P. Power and A.J. Thorne, *J. Chem. Soc., Chem. Commun.*, 1983, 639.
51. B. Cetinkaya, I. Gumrukcu, M.F. Lappert, J.L. Atwood, R.D. Rogers and M.J. Zaworotko, *J. Amer. Chem. Soc.*, 1980, **102**, 2088.
52. T. Fjeldberg, P.B. Hitchcock, M.F. Lappert, J.J. Smith and A.J. Thorne, *J. Chem. Soc., Chem. Commun.*, 1985, 939.
53. J. Barrett, S.R.A. Bird, J.D. Donaldson and J. Silver, *J. Chem. Soc., Dalton Trans.*, 1971, 3105.
54. T.C. Gibb, B.A. Goodman and N.N. Greenwood, *Chem. Comm.*, 1970, 774.
55. K.B. Dillon and A. Marshall, *J. Chem. Soc., Dalton Trans.*, 1984, 1245.
56. K.B. Dillon and A. Marshall, *J. Chem. Soc., Dalton Trans.*, 1987, 315.
57. R. Colton, D. Dakternicks and C.A. Harvey, *Inorg. Chim. Acta*, 1982, **61**, 1.
58. S. Jaulmes, M. Julien-Pouzol, P. Laruelle and M. Guittard, *Acta Crystallogr.*, 1982, **B38**, 79.
59. K. Kondo, G. Matsubayashi, T. Tanaka, H. Yoshioka and K. Nakatsu, *J. Chem. Soc., Dalton Trans.*, 1984, 379.
60. K.B. Dillon, J. Halfpenny and A. Marshall, *J. Chem. Soc., Dalton Trans.*, 1983, 1091.
61. K.B. Dillon, J. Halfpenny and A. Marshall, *J. Chem. Soc., Dalton Trans.*, 1985, 1399.
62. N.N. Greenwood and T.C. Gibb, *Mössbauer Spectroscopy*, Chapman and Hall, London, 1971.
63. D. Tudela, V. Fernandez and J.D. Tornero, *J. Chem. Soc., Dalton Trans.*, 1985, 1281.
64. M. Katada, H. Kanno and H. Sano, *Polyhedron*, 1983, **2**, 104.
65. V.A. Varnek, O.K. Polesch, L.N. Mazalov and D.M. Kizhner, *Zhur. Strukt. Khim.*, 1982, **23**, 98.
66. R.H. Herber, *Chemical Mössbauer Spectroscopy*, Plenum, New York, 1984, 199.
67. *Proc. Symp. on Properties and Uses of Inorganic Tin Chemicals*, Brussels, 15/16 October 1986.
68. K. Benice and S. Lataly, *Inorg. Chem.*, 1987, **26**, 2537.
69. R.H. Herber, *Acct. Chem. Res.*, 1982, **15**, 216.

70. P.A. Cusack, B.N. Patel, P.J. Smith, D.W. Allen and I.A. Nowell, *J. Chem. Soc., Dalton Trans.*, 1984, 1239.
71. E.N. Gur'yanova, L.A. Ganyushin, I.P. Romm, E.S. Scherbakova and M. Movsum-zade, *J. Gen. Chem. USSR*, 1981, **51**, 356.
72. G. Valle, A. Cassol and U. Russo, *Inorg. Chim. Acta*, 1984, **82**, 81.
73. E. Hough, D.G. Nicholson and A.K. Vasudevan, *J. Chem. Soc., Dalton Trans.*, 1986, 2335.
74. R. Kniep, D. Mootz, U. Severin and A. Wunderlich, *Acta Crystallogr.*, 1982, **38B**, 2022.
75. U. Alpen, J. Van Fenner and E. Gmelin, *Mater Res. Bull.*, 1975, **10**, 175.
76. G. Amthauer, J. Fenner, S. Hafner, W.B. Holzapfel and R. Keller, *J. Chem. Phys.*, 1979, **70**, 4837.
77. J.-C. Jumas, J. Olivier-Fourcade, E. Philippot and N. Maurin, *Rev. Chim. Mineral*, 1979, **16**, 48.
78. M.F.A. Dove, R. King and T.J. King, *Chem. Comm.*, 1973, 944.
79. A. Soufiane and S. Vilminot, *Rev. Chim. Mineral*, 1985, **22**, 799.
80. T. Birchall and V. Manirannan, *J. Chem. Soc., Chem. Commun.*, 1986, 1441.
81. T. Birchall and A.W. Sleight, *J. Solid State Chem.*, 1975, **13**, 118.
82. K. Haiselback, G. Murken and M. Tromel, *Z. anorg. allg. Chem.*, 1973, **397**, 127.
83. P.F.R. Ewings, P.G. Harrison, A. Morris and T.J. King, *J. Chem. Soc., Dalton Trans.*, 1976, 1602.
84. P. Day, *Progr. Inorg. Chem.*, 1967, 8.

5 Formation and cleavage of the tin–carbon bond

J.L. WARDELL

5.1 Preparation of tin (IV)–carbon bonds

5.1.1 *General methods*

5.1.1.1 *Direct synthesis.*[1,6,7] The direct reaction of tin (as foil, powder or alloy) with organic halides has been widely used in the laboratory as a route to organotin species (eqn 5.1). Catalysts and/or promoters are required for all but the most reactive alkyl halides. The directness of the method has clear attractions as an industrial process, but so far only limited use has been made of it. One example of the direct reaction,[8] as a

$$Sn + 2RX \rightarrow R_2SnX_2 \tag{5.1}$$

manufacturing process, is the synthesis of Me_2SnCl_2, from tin foil and MeCl with $Ph_3MeP^+Br^-$ as catalyst and KI as promoter at 180–190°C. The direct synthesis in industry is probably limited because the predominant products are usually diorganotin compounds. The recent report[9] by Holland of the direct and economic formation of triorganotin compounds from tin and alkyl halides in high yields may well change this situation. The Holland process involves addition of the alkyl halide to tin powder dispersed in a molten quaternary halide, such as R_4NX, at 120–140°C, which rapidly produces the trialkyltin halide and R_4NSnX_3. Tin, the quaternary halide and halide can be recovered from R_4NSnX_3 by electrochemical means and can all be reused. The alkylation of tin occurs in a stepwise manner and it is possible to stop the process at the mono- or di-organotin stage.

The initial report of a direct method was by Frankland in 1849 (eqn 5.2). Subsequently, various allyl-, benzyl- and alkyl-tin halides have been obtained by direct synthesis, e.g. functionally substituted alkyl-tin compounds have been obtained[1,6,10] from $YCO(CH_2)_nX$ ($n = 1, 2$; X = halide, $Y = R_2N$, RO or R), $NCCH_2CH_2X$ and $RO(CH_2)_3I$, as well as from halo-succinates and malonates[11]. The reactivity of the alkyl halides is in the sequence RI > RBr > RCl and MeX > EtX > PrX. Particularly reactive halides are benzyl halides; catalysts are not required in the $PhCH_2Cl$-tin reactions in H_2O or in toluene which produce $(PhCH_2)_3SnCl$ or $(PhCH_2)_2SnCl_2$, respectively.[12]

$$2EtI + Sn \xrightarrow[h\nu]{160^\circ C\text{ or}} Et_2SnI_2 \tag{5.2}$$

Various catalysts/promoters have been employed for less reactive halides, including other metals (sodium, magnesium, zinc or copper),[1,6] Lewis bases (amines, alcohols or ethers), halides (iodides, quaternary ammonium halides or cuprous halides), and triorganophosphines or -stibines. While in the main, diorganotin species predominate, the use of certain catalysts can lead to higher alkylations, e.g. tin–sodium alloys provide tri- and even tetra-alkyltins.

Electrochemical methods can be mentioned; oxidation of anionic tin in a solution of RX in benzene provides R_2SnX_2.[13,14] Other electrochemical syntheses have been reported using a tin cathode, e.g. as in the preparation of Me_4Sn (from MeI)[15] and $(NCCH_2CH_2)_4Sn$[16] from $NCCH_2CH_2I$.

The so-called diestertin dichlorides, $Cl_2Sn(CH_2CH_2CO_2R)_2$, have been obtained[17] by reaction of powdered tin, hydrogen chloride and a α, β-unsaturated carbonyl species, e.g. acrylic esters in ether (eqn 5.3).

$$Sn + 2HCl \rightarrow Cl_2Sn(CH_2CH_2CO_2R)_2 \qquad (5.3)$$

Di-β-keto- and -amido- (as well as -cyano)alkyl tin dihalides have also been obtained. It has been suggested that probably intermediates are $H^+SnCl_3^- \cdot Et_2O$ or $(H^+)_2 \cdot SnCl_4^{2-} \cdot 2Et_2O$.

Tin (II) halides have also been used[1,6] in direct synthesis with alkyl halides to give $RSnX_3$. While similar catalysts/promoters have been employed for tin (II) halides as used for tin, particularly effective are R_3Sb,[18] e.g. for use with $C_nH_{2n+1}X$ and $X(CH_2)_nX$, (eqn 5.4). The reactivity sequences established[1,14] for SnX_2 and for RX are $X = I > Br > Cl$, while for C_nH_{2n+1}, the reactivity decreased with increasing n.

$$n\text{-}C_{18}H_{37}Br + SnBr_2 \xrightarrow[150°C]{Et_3Sb} n\text{-}C_{18}H_{37}SnBr_3 \qquad (5.4)$$

Estertin trichlorides, $Cl_3SnCH_2CH_2CO_2R$, are available[17] from tin (II) chloride, hydrogen chloride and acrylic esters, $CH_2{=}CHCO_2R$.

5.1.1.2 *From tin halides and organic derivatives of other metals.*[3,15] Reactions of organotin halides and tin (IV) tetrahalides with organic derivatives of electropositive elements are arguably the most important routes to compounds with new tin–carbon bonds.

$$R_{4-n}SnX_n + nR'M \xrightarrow[\text{hydrocarbons}]{\text{ethers or}} R_{4-n}SnR'_n \qquad (5.5)$$

The metals, M, most frequently employed in the laboratory are magnesium and lithium, with sodium, zinc and aluminium playing more limited roles. Usually tetraorganotin compounds are the target products, particularly as it is difficult to stop the alkylation at a predetermined stage. However, in certain cases this has been achieved,[19] e.g. eqn (5.6). There is a greater probability[20] of partial alkylation occurring with Grignard reagents in Et_2O than with the corresponding lithium species in THF as well as with bulky organic units, e.g. cyclohexyl groups.[20,21]

$$o\text{-}(Me_2NCH_2)C_6H_4CH_2Li + PhMeSnBr_2 \rightarrow o\text{-}(Me_2NCH_2)C_6H_4CH_2SnMePhBr \qquad (5.6)$$

This method is clearly limited by the availability and stability of R′M as well as to functional groups inert to R′M under the chosen reaction conditions. Organolithium compounds are particularly useful. A wide range of substituted alkyl-, alkenyl-, alkynyl- and aryl-lithium species is obtainable from lithium, simple organolithium reagents (e.g. BuLi, Bu^sLi, Bu^tLi, MeLi, etc.) or R'_2NLi, by such reactions as lithiations of hydrocarbons or lithium-halide exchanges with organic halides.[22] Use of low temperatures and masked/protected groups may, however, be neccssary to preserve certain functional groups. Some examples are given[23–25] in eqs (5.7)–(5.9).

$$Me_3SiCH_2SR \xrightarrow[\text{(ii) } Me_3SnCl]{\text{(i) BuLi}} Me_3SiCHSRSnMe_3 \quad (5.7)$$

$$o\text{-}BrC_6H_4CH_2PPh_2 \xrightarrow[\text{(ii) } PhMeBu^tSnCl]{\text{(i) Li}} o\text{-}(PhMeBu^tSn)C_6H_4CH_2PPh_2 \quad (5.8)$$

(5.9)

Transmetallations using organozinc compounds have found specific use in the formation of iodomethyltin compounds[26] (eqn 5.10).

$$CH_2I_2 \xrightarrow{Zn/Cu} ICH_2ZnI \xrightarrow{R_3SnCl} R_3SnCH_2I \quad (5.10)$$

R = Me, Bu, Ph or *cyclo*-C_6H_{11}.

Reformatsky reagents have also been employed[27], e.g. eqn (5.11).

$$Et_2NCOCH_2Br \xrightarrow{Zn} Et_2NCOCH_2ZnBr \xrightarrow{Ph_3SnCl} Et_2NCOCH_2SnPh_3 \quad (5.11)$$

Tin–oxygen and tin–sulphur bonded compounds may be used as alternatives to tin halides.[3] Reactions of tin sulphides with organomercurials have also been used,[28,29] e.g. eqn (5.12).

$$(R_3Sn)_2S + Hg(CHR^1COR^2)_2 \rightarrow 2R_3SnCHR^1COR^2 + HgS \quad (5.12)$$

R = H or COR^2; R^2 = OEt, Me

5.1.1.3 *From triorganostannyl-alkali, -magnesium and -copper.*[3,30,31] Triorganostannyl-metal compounds, stannylanionoids, R_3SnM (M = Li, Na, K, MgX or Cu) are finding extensive use for the preparations of mixed compounds, R_3SnR^1 (R^1 = simple or substituted alkyl, vinyl, allyl, aryl, etc.). Reactions occur with organic halides and tosylates, aldehydes, ketones, β-enones and alkynes as well as with small ring compounds such as epoxides, oxetanes and their sulphur analogues.

R_3SnM (M = alkali metal or MgX) can be obtained[3,30,31] from: (i) R_3SnX (X = halides) and M (Li, Na, Mg) or ArH^-, M^+ (e.g. ArH = naphthalene or phenanthrene; M = Li, Na or K) in ethers (Et_2O, THF, DME, tetraglyme, etc.) or less frequently in liquid ammonia; (ii) R_3SnSnR_3 and M (M = Li, Na or K) in ethers, RLi in HMPA or $R^1{}_2NLi$; or (iii) R_3SnH and MH (M = Na or K), Bu^tMgX (for R = Bu), R_2NLi or $R^1{}_2NM$–$LiOBu^t$ (M = K or Cs).[32] Addition of a copper(I) compound or complex will provide the tin-copper or cuprate species.

Various organic *chlorides*, *bromides*, *iodides* and *tosylates*, R^1X, react with R_3SnM. Particularly clean reactions occur with primary and secondary alkyl halides but eliminations can become troublesome with tertiary alkyl halides. However, it is possible to get tertiary-alkyl tin compounds, as shown by the formation of 1-trimethylstannyladamantane.[33]

$$R_3SnM + R^1X \rightarrow R_3SnR^1 \tag{5.13}$$

The mechanisms of secondary alkyl halide reactions, such as reactions (5.14) and (5.15), have attracted considerable attention.[34]

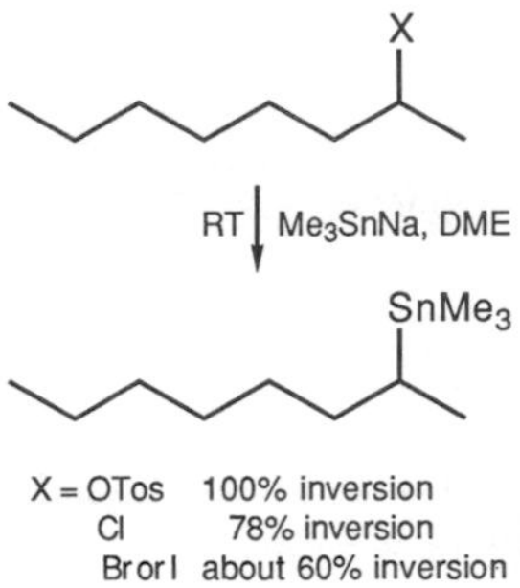

X

RT | Me_3SnNa, DME

$SnMe_3$

X = OTos 100% inversion
Cl 78% inversion
Br or I about 60% inversion

(5.14)

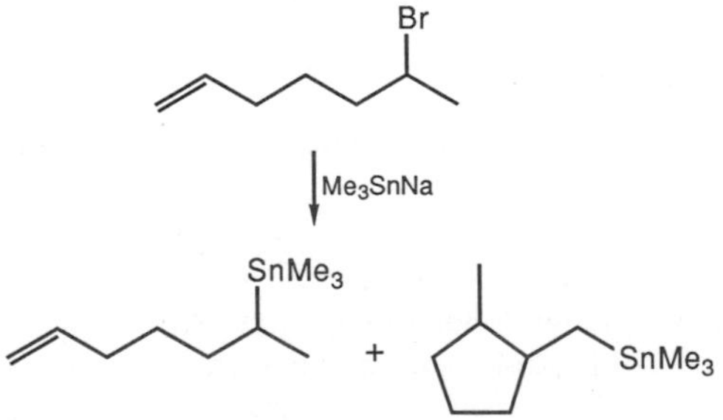

Br

Me_3SnNa

$SnMe_3$ + $SnMe_3$

(5.15)

As shown in eqn (5.15) and in cyclopropylcarbinyl halide/R_3SnM reactions, rearrangements, indicative of radical intermediates, can occur[30] (Scheme 1).

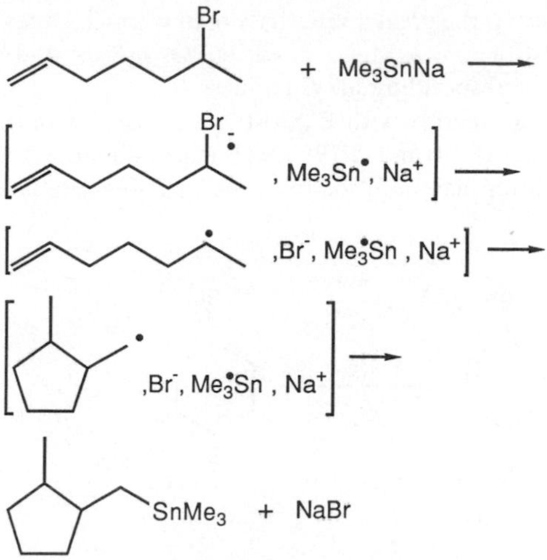

$$\text{CH}_2\text{=CH(CH}_2)_3\text{CHBrCH}_3 + Me_3SnNa \longrightarrow$$

$$[\text{CH}_2\text{=CH(CH}_2)_3\text{CHBr}^{\bullet -}\text{CH}_3, Me_3Sn^{\bullet}, Na^+] \longrightarrow$$

$$[\text{CH}_2\text{=CH(CH}_2)_3\dot{\text{C}}\text{HCH}_3, Br^-, Me_3^{\bullet}Sn, Na^+] \longrightarrow$$

$$[\text{2-methylcyclopentyl-}\dot{\text{C}}\text{H}_2, Br^-, Me_3^{\bullet}Sn, Na^+] \longrightarrow$$

$$\text{2-methylcyclopentyl-CH}_2SnMe_3 + NaBr$$

Scheme 1

Three pathways have been indicated, namely (i) an S_N2 type substitution with inversion of configuration; (ii) an electron transfer (or radical-pair) mechanism for the more readily reduced organic halides, which can lead to rearrangements; and (iii) substitution, with overall retention of configuration, occurring *via* an initial halogen-metal exchange (e.g. as found in reactions of secondary cyclopropyl bromides).[35] Recent work has shown that even electron-transfer processes can proceed with predominant inversion of configuration.[34] Temperature, solvent, halide and method of preparation of R_3SnM (and consequently the presence of additional species, such as RM or M) affect the situation.

Functionally substituted alkyl halides can be used to generate the corresponding tin compounds, e.g. eqns (5.16)–(5.18).

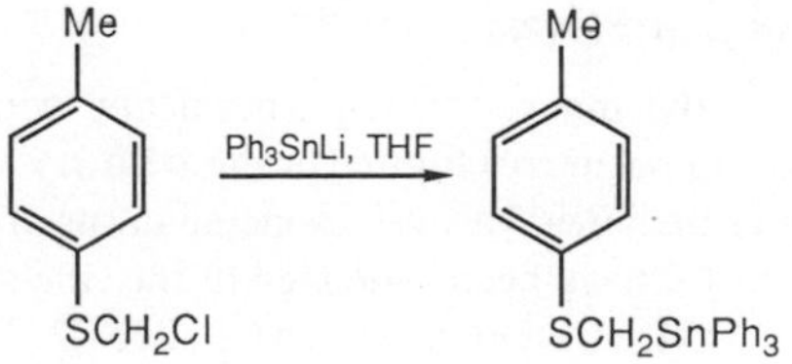

$$p\text{-MeC}_6H_4SCH_2Cl \xrightarrow{Ph_3SnLi,\ THF} p\text{-MeC}_6H_4SCH_2SnPh_3 \qquad (5.16)^{36}$$

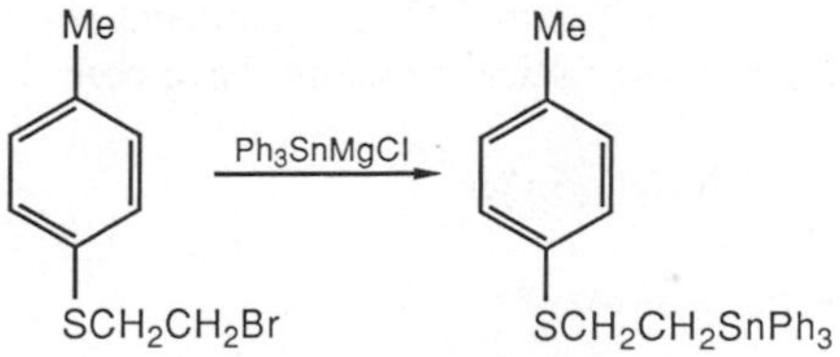

$$p\text{-MeC}_6H_4SCH_2CH_2Br \xrightarrow{Ph_3SnMgCl} p\text{-MeC}_6H_4SCH_2CH_2SnPh_3 \qquad (5.17)^{36}$$

$$Cl(CH_2)_3Br \xrightarrow{Ph_3SnNa} Cl(CH_2)_3SnPh_3 \qquad (5.18)^{37}$$

Reaction (5.18) indicates the greater reactivity of Br over Cl. Other α- and β-substituted alkyl-tins (e.g. substituent = α-MeO, β-$(EtO)_2PO$, β-R_2N and β-RCO) have been obtained from the corresponding alkyl halides.[30]

Alkyl tosylates react readily with R_3SnM and proceed with complete inversion of configuration, e.g. eqns (5.14) and (5.19); *cis*-(**1**) reacts similarly to give *trans*-(**2**). Use of the tosylate substitution has been made in the carbohydrate field[41] (eqn 5.20).

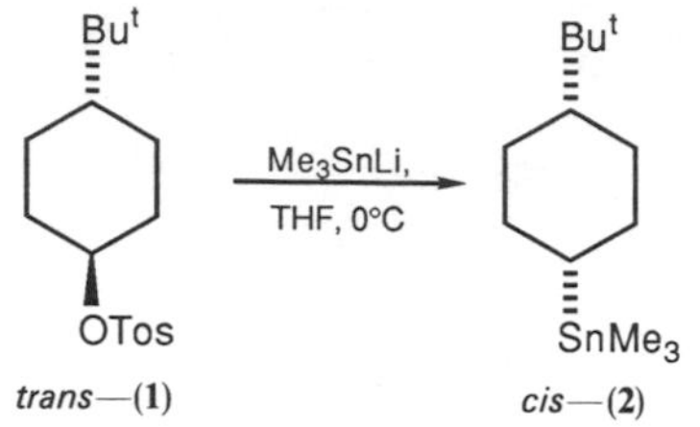

(5.19)

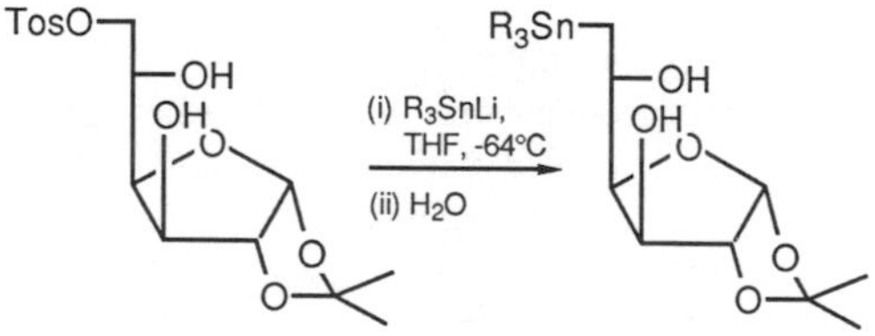

(5.20)

Leaving groups other than halides and tosylates have found limited use,[42] e.g. eqn (5.21):

$$RR^1NCH_2\text{–}X \xrightarrow{Bu_3SnLi, THF} RR^1NCH_2SnBu_3 \quad (5.21)$$

X = SMe, SPh, OMe, $\overset{+}{N}R_3$, or SO_3Na.

Aryl bromides react, in the main,[30,43] regiospecifically with R_3SnLi (eqn 5.22), whereas *cine*-substitution can occur to a limited extent with aryl chlorides or fluorides. Aryne and free radical intermediates[44] as well as metallations of the aryl halide by the RLi, present with the R_3SnLi, have been indicated in the *cine*-substitutions. Various substituents, sensitive to RM, survive these aryl halide–R_3SnM reactions. Some evidence has been found for an initial metal–halogen exchange preceding the formation of the aryl–tin bond[30] (eqn 5.23). For hindered aryl bromides, such as 2,4,6-$Me_3C_6H_2Br$, products of free radical processes have been detected[45].

$$YC_6H_4Br \xrightarrow[\text{tetraglyme}]{Me_3SnNa, 0°C} YC_6H_4SnMe_3 \quad (5.22)$$

Y = *p*-CN, *p*-H_2N, *o*-, *m*- or *p*-MeO

$$ArBr + R_3SnM \rightarrow ArM + R_3SnBr \rightarrow ArSnR_3 + MBr \quad (5.23)$$

Triorganostannyl alkali compounds have been also shown to react with heteroaryl halides, vinyl halides (usually with predominant retention of configuration), propargyl bromides (which provide allenyl-tins) and allyl halides or tosylates, e.g. eqns (5.24) and (5.25). For primary allyl halides, allylic rearrangements do not arise, but these can result with secondary compounds.

TosO H; Me; D; *cis : trans* 94 : 6 —(Bu_3SnLi, THF)→ Bu_3Sn H (D); Me; D (H); *cis : trans* 8 : 92 (5.24)

Cl; *cis : trans* 7 : 93 —(R_3SnLi, THF, 0°C)→ SnR_3 + SnR_3

R = Me; *cis : trans* 32 : 64
R = Ph; *cis : trans* 59 : 29 (5.25)

Acyl chlorides react to give acyltins at low temperature[48] (eqn 5.26).

$$RCOCl + Ph_3SnCl \xrightarrow{THF, -70°C} RCOSnPh_3 \tag{5.26}$$

$$R = Me, Bu^t \text{ or } Ph$$

Aldehydes and ketones react[30,49] with R_3SnM (M = Li or MgCl), to give good yields of α-hydroxyalkytins, especially in Et_2O or THF. The intermediate alkoxide (**3**) can be trapped by electrophiles other than the proton (eqn (5.27)).

RCHO + $Bu_3SnMgCl$ —(THF or Et_2O)→ $RCH(SnBu_3)(OMgCl)$ (**3**)

(**3**) —(H_2O)→ $RCH(OH)SnBu_3$

(**3**) —(E—X)→ $RCH(OE)SnBu_3$

e.g., E—X = MeCO—Cl (5.27)

Use of excess RCHO leads to formation of $RCOSnBu_3$ and RCH_2OH. The bulky Bu_3Sn group goes into the equatorial site on reaction with the cyclo-hexanone[30] (**4**). As simple addition occurs to cyclopropyl methyl ketone, an electron transfer mechanism can be ruled out.[30]

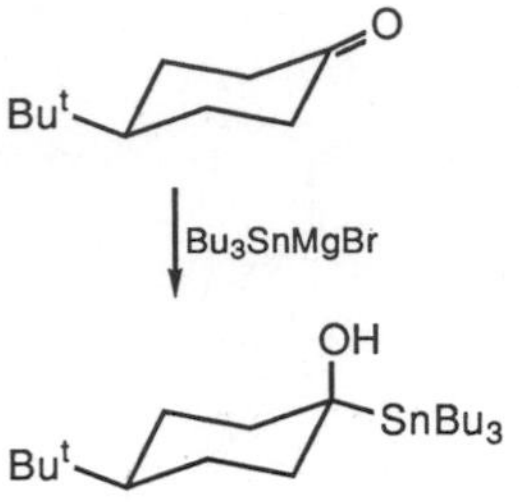

Bu^t … O
$Bu_3SnMgBr$
OH
$SnBu_3$
Bu^t

(**4**) (5.28)

Both 1,2- and 1,4-additions occur with β-enones; to $RCH{=}CHCOR'$; 1,4-additions occur with $Bu_3SnMgCl$ in Et_2O, R_3SnLi (e.g. eqn 29) or $(R_3Sn)_2CuLi$, whereas 1,2-adducts have been isolated from $RR^1C{=}CHCOR^2$. There is some evidence to suggest that the kinetic products are the 1,2-adducts which can rearrange to the thermodynamically more stable 1,4-adducts, especially on raising the reaction temperature and in more polar solvents.[30] For addition of R_3SnNa, the use of a protic solvent is recommended to trap the intermediate enolate as it is formed and before it undergoes condensation reactions.[51]

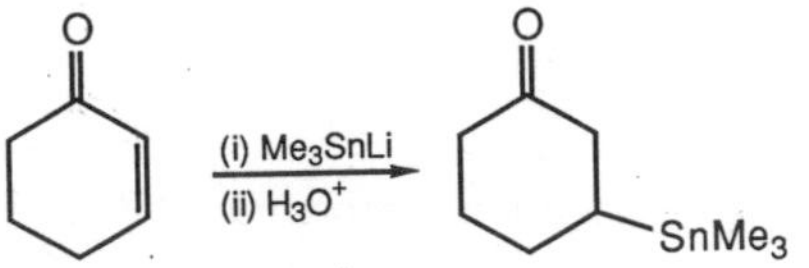

O
(i) Me_3SnLi
(ii) H_3O^+
O
$SnMe_3$

(5.29)[50]

Reasonably good stereoselectivities are observed[53] in the 1, 4-addition reaction as shown in eqns (5.30)–(5.32).

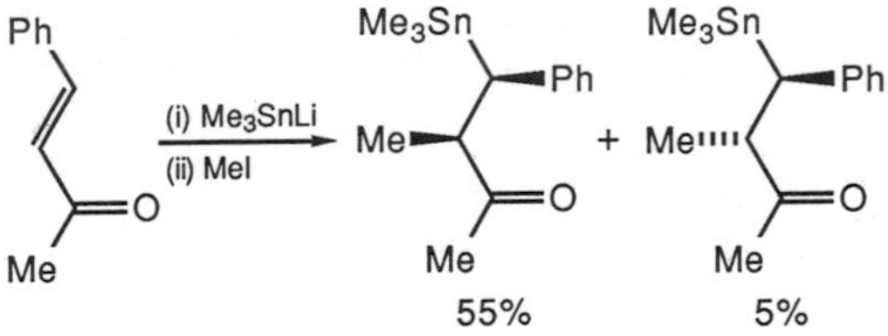

Ph
Me
O
(i) Me_3SnLi
(ii) MeI
Me_3Sn Ph Me O Me
55%
\+ Me_3Sn Ph Me O Me
5%

(5.30)

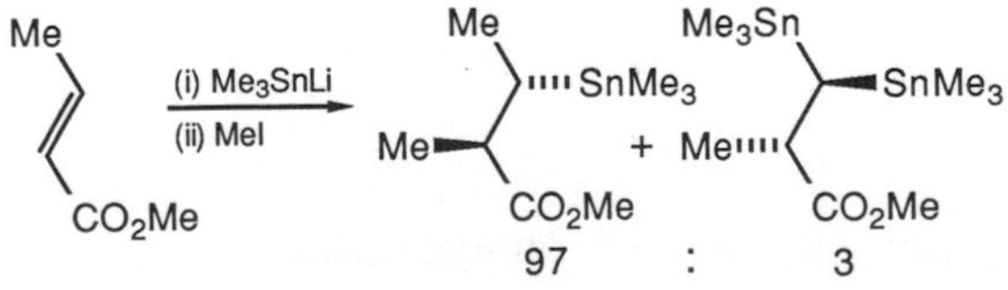

Me
CO_2Me
(i) Me_3SnLi
(ii) MeI
Me $SnMe_3$ Me CO_2Me
97
\+ Me_3Sn $SnMe_3$ Me CO_2Me
: 3

(5.31)

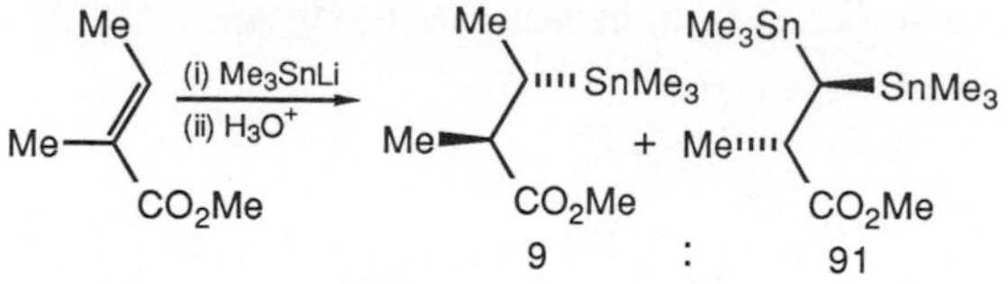

Me, Me, CO_2Me (i) Me_3SnLi (ii) H_3O^+ → Me, Me, $SnMe_3$, CO_2Me 9 : + Me_3Sn, Me, $SnMe_3$, CO_2Me 91

(5.32)

Reactions with epoxides provide β-hydroxyalkyltins, (eqn 5.33).[3,30,31,39,53–57] The regioselectivity is less for trisubstituted epoxides (**5**, $R^3 = R^2 = R^1 =$ alkyl).[55] The *trans*-isomer was obtained from cyclohexene oxide;[39,56] ring opening of the 1, 2-*cis*-dimethyl epoxide (**6**) occurs with inversion of configuration (eqn 5.34).[57]

Use of the epoxide opening has also been made with carbohydrates (eqn 5.35).[53]

Reactions of oxetanes with R_3SnM lead to γ-hydroxyalkyltins[55] (eqn 5.36). For the sulphur ring analogues, desulphurization and attack at sulphur can limit the yields of β- and γ-mercaptoalkyltin products. 1-Methylthiirane, however, provides a good yield of the 2-mercaptopropyltin species[55] (eqn 5.37).

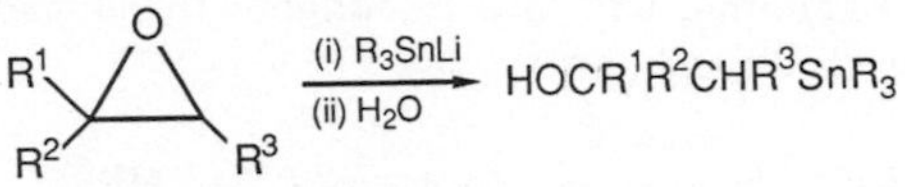

O, R^1, R^2, R^3 (i) R_3SnLi (ii) H_2O → $HOCR^1R^2CHR^3SnR_3$

(**5**)

(R^1 = Me, Ph, $PhCH_2$; $R^2 = R^3$ =H)

(5.33)

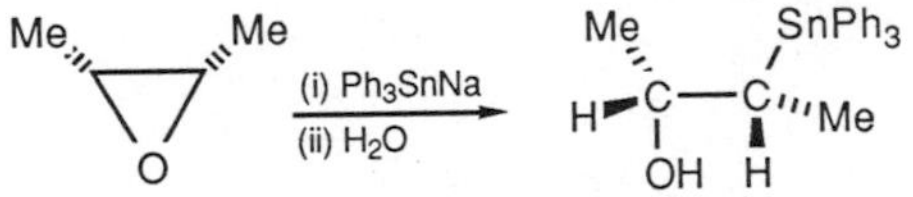

Me, Me, O (i) Ph_3SnNa (ii) H_2O → Me, $SnPh_3$, H–C–C–Me, OH H

(**6**)

(5.34)

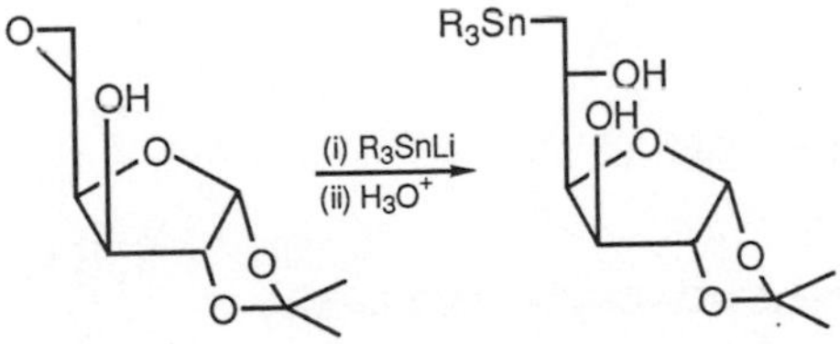

O, OH, O, O, O (i) R_3SnLi (ii) H_3O^+ → R_3Sn, OH, OH, O, O, O

(5.35)

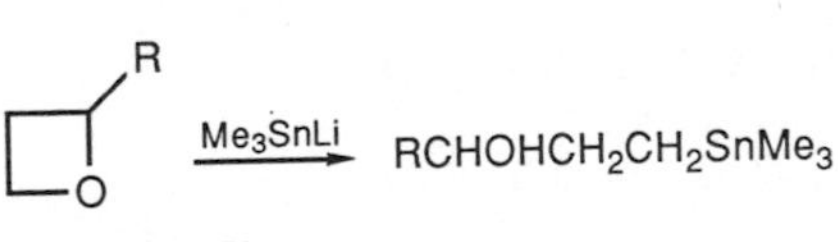

R, O Me_3SnLi → $RCHOHCH_2CH_2SnMe_3$

R = H, Me or Ph

(5.36)

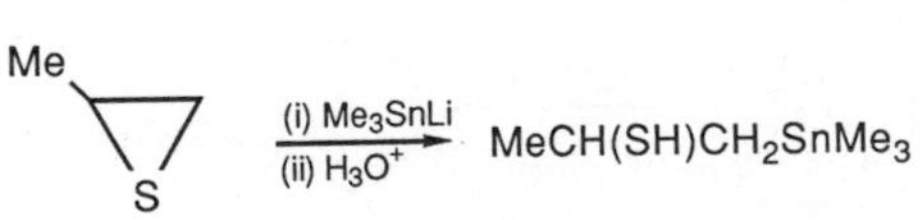

Me, S (i) Me_3SnLi (ii) H_3O^+ → $MeCH(SH)CH_2SnMe_3$

(5.37)

Ring opening[42] also occurs with the heterocycle, (7), eqn (5.38).

$$\text{(7)}\ \xrightarrow[\text{(ii) } H_2O]{\text{(i) } Bu_3SnLi,\ THF}\ Bu_3SnCH_2NMeCH_2CH_2OH \tag{5.38}$$

Triorganostannyl anionic compounds (R_3SnM, M=MgMe, $AlEt_2$, or $Zn(SnR_3)$) add to alkynes in the presence of copper or palladium catalysts.[58] These and related additions of triorganostannyl-copper and cuprates are discussed in section 5.1.2.1.

5.1.1.4 *From other metal–tin compounds.* Both distannanes R_3SnSnR_3,[59,60] and stannylsilanes[61,62] R_3SnSiR_3, add in a *cis*-manner to alkynes, in the presence of $Pd(Ph_3P)_4$, e.g. eqns (5.39) and (5.40).

$$Me_3SnSnMe_3 + RC{\equiv}CH \xrightarrow{Pd(Ph_3P)_4} (Z)\text{–}Me_3SnCR{=}CHSnMe_3 \tag{5.39}$$

$$Me_3SnSnMe_3 + RC{\equiv}CC(O)Y \xrightarrow{Pd(Ph_3P)_4} (Me_3Sn)(R^1)C{=}C(SnMe_3)(C(O)Y) \longrightarrow (Me_3Sn)(R^1)C{=}C(C(O)Y)(SnMe_3) \tag{5.40}$$

Stannylsilanes also add to allenes,[61,63] (e.g. eqn 5.41)[61] and β-enones,[62,64] (eqn 5.42).

$$Me_2C{=}C{=}CH_2 + Me_3SiSnMe_3 \xrightarrow{Pd(Ph_3P)_4} H_2C{=}C(SiMe_3)CMe_2SnMe_3 + Me_2C{=}C(SiMe_3)CH_2SnMe_3 \tag{5.41}$$

$$\text{4,4-R}_2\text{-3-R'-cyclohex-2-enone} \xrightarrow[\text{KCN, 18-crown-6}]{Me_3SiSnBu_3} \text{1-OSiMe}_3\text{-3-SnBu}_3\text{-3-R'-4,4-R}_2\text{-cyclohexene} \tag{5.42}$$

5.1.1.5 *Hydrostannation of alkenes and alkynes.*[1,2,65–67] Organotin hydrides add to alkenes and to alkynes, (eqns 5.43, 5.44).

$$R_3SnH + \gt C{=}C\lt \rightarrow R_3Sn{-}\overset{|}{\underset{|}{C}}{-}\overset{|}{\underset{|}{C}}{-}H \qquad (5.43)$$

$$R_3SnH + {-}C{\equiv}C{-} \rightarrow R_3Sn{-}\underset{|}{C}{=}CH{-} \qquad (5.44)$$

Unless strongly electron withdrawing groups are present in the unsaturated species, the additions are free radical chain processes. The key steps for alkene additions are shown in Scheme 2. Additions to terminal alkenes and alkynes occur readily, usually on UV

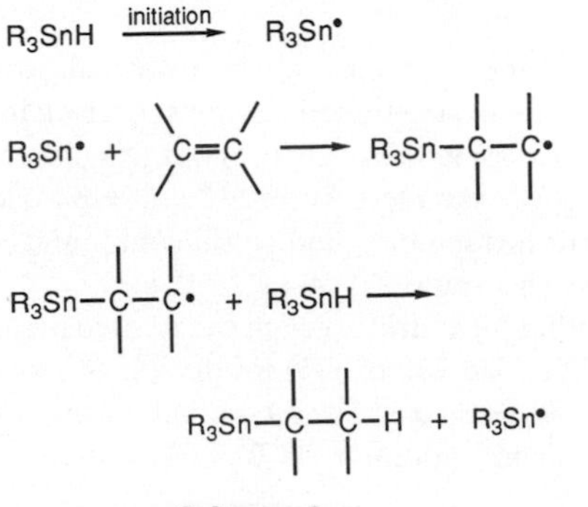

$$R_3SnH \xrightarrow{\text{initiation}} R_3Sn^{\bullet}$$

$$R_3Sn^{\bullet} + \gt C{=}C\lt \longrightarrow R_3Sn{-}\overset{|}{\underset{|}{C}}{-}\overset{|}{\underset{|}{C}}{\bullet}$$

$$R_3Sn{-}\overset{|}{\underset{|}{C}}{-}\overset{|}{\underset{|}{C}}{\bullet} + R_3SnH \longrightarrow R_3Sn{-}\overset{|}{\underset{|}{C}}{-}\overset{|}{\underset{|}{C}}{-}H + R_3Sn^{\bullet}$$

Scheme 2

irradiation or in the presence of a radical initiator, such as AIBN. For non-terminal species, the reactions are best carried out on UV irradiation unless these are activated by strain[68] or have substituents able to stabilize the intermediate organyl radical.

Usually for 1-alkenes, tin adds predominantly to the terminal carbon, e.g. eqns (5.45)[69] and (5.46), although for (**8**; R = H or Me_3Sn), appreciable amounts of the

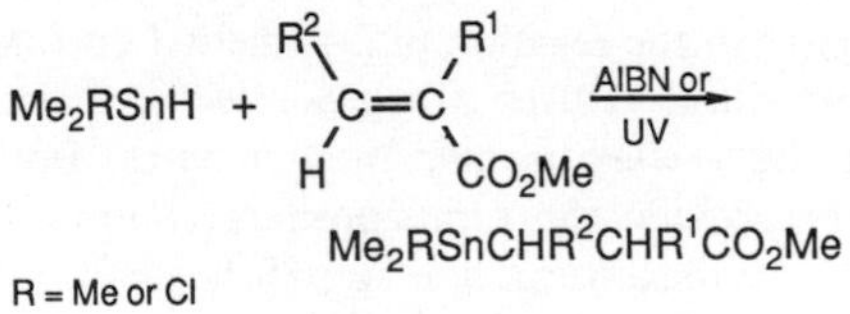

$$Me_2RSnH + (R^2)(H)C{=}C(R^1)(CO_2Me) \xrightarrow[\text{UV}]{\text{AIBN or}} Me_2RSnCHR^2CHR^1CO_2Me \qquad (5.45)$$

R = Me or Cl

$$Me_3SnH + \underset{(\mathbf{8})}{Me_3SnCR{=}CH_2} \xrightarrow[\text{UV}]{\text{AIBN or}} \underset{\substack{(\mathbf{9})\\ \text{major}}}{Me_3SnCHRCH_2SnMe_3} + \underset{\substack{(\mathbf{10})\\ \text{minor}}}{(Me_3Sn)_2CRMe} \qquad (5.46)$$

R = H, Ph, Bu, Me_3Si or Me_3Sn

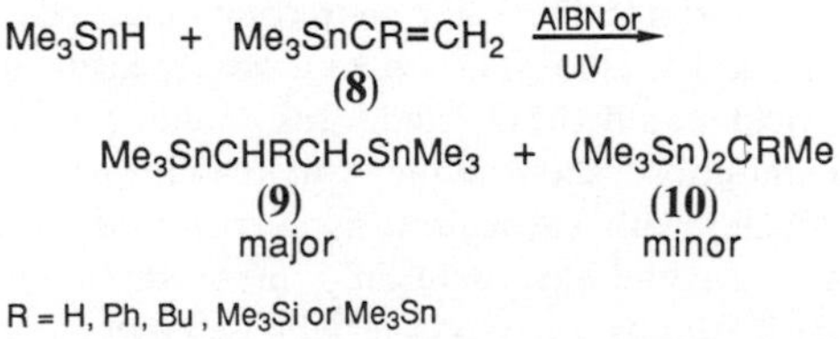

non-terminal adduct (**10**), are formed (in *c.* 40% yield compared to $< 5\%$ for the other R groups).[70] The regioselectivity of addition to acrylonitrile is dependent on the conditions and whether a nucleophilic (*a*) or radical (*b*) mechanism applies (eqn (5.47)).[65]

$$\begin{array}{c} Bu_3CHCNMe \\ \uparrow \text{(a) } CH_2{=}CHCN \text{ excess} \\ Bu_3SnH \\ \downarrow \text{(b) } CH_2{=}CHCN\text{, AIBN} \\ Bu_3SnCH_2CH_2CN \end{array} \qquad (5.47)$$

Variously, other functional groups can be tolerated, and hydrostannation is a particularly valuable route to substituted alkyltins, bearing functional groups in β-, γ- or more remote positions, using $CH_2{=}CH(CH_2)_nX$ ($X = SR$, SO_2R, OR, PR_2, $P(OR)_2$ etc.) (see Tables 5.2–5.5 for examples). Amines[65] and sulphoxides[71] catalyse decomposition of tin hydrides to distannanes, and so tin compounds containing such groups cannot be prepared by this route.

As well as R_3SnH, other tin hydrides which can be used include R_2SnH_2, $RSnH_3$ and R_2SnHCl (see eqn 5.45). The use of UV irradiation is recommended for R_2SnHCl reactions. Additions to R_2SnH_2 occur stepwise, and the mono-adducts can be trapped. For non-conjugated dienes, addition of R_2SnH_2 can lead to heterocycles,[77] e.g. eqn (5.48).

$$Ph_2SnH_2 + (H_2C{=}CH)_2SiPh_2 \longrightarrow Ph_2Sn\langle CH_2CH_2\rangle_2SiPh_2 \qquad (5.48)$$

Hydrostannation of bifluoroenylidene with chiral methylphenylneophyltin hydride takes place with some retention of configuration, implying that the intermediate tin radical is captured before it can become planar.[73] Some asymmetric induction (10–20% ee) has been obtained[74] in the reaction of (−)-methyl crotonate and Bu_3SnH.

In eqn (5.45), two chiral centres are generated[69]; diastereomeric mixtures are obtained, in which the *erythro* isomer predominates. Both (Z)- and (E)-$R^2CH{=}CR^1CO_2Me$ isomers provide the same product composition, resulting from the reversibility of the free-radical formation step (Scheme 3).[75]

Hydrostannation of terminal alkynes[65], containing a strongly electron-withdrawing group, e.g. CO_2R or CN, produces mainly, and possibly exclusively, the α-adduct *via* an ionic mechanism, e.g. eqn (5.49). Addition to compounds bearing an electron-releasing group, such as an alkyl or alkoxyl group, or a weakly electron-withdrawing group, e.g. CH_2OH or Ph, yield mainly the *cis*- and *trans*-β-adduct (*via* a free radical mechanism) as well as small amounts of the α-adduct. Initial additions occur in a *trans*-manner (to give the *cis*-adducts) with subsequent isomerizations of the *cis*-$R_3SnCH{=}CHR^1$ resulting to give the thermodynamically more stable *trans*-isomers, *via* radical intermediates. Hydrostannation of non-terminal alkynes by trialkyltin hydrides affords

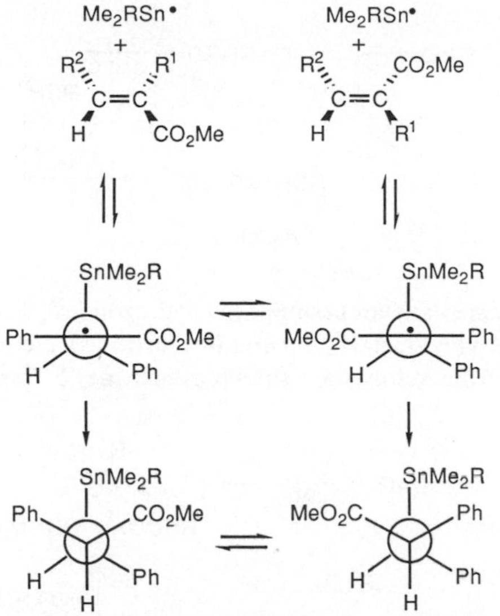

Scheme 3

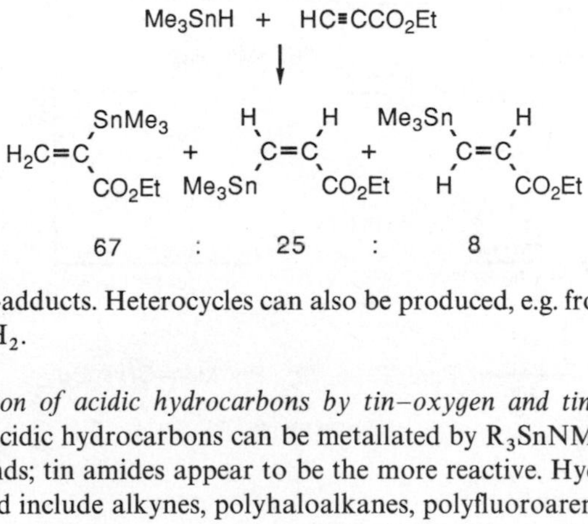

$$Me_3SnH + HC{\equiv}CCO_2Et \rightarrow H_2C{=}C(SnMe_3)CO_2Et + Me_3SnCH{=}CHCO_2Et + Me_3SnCH{=}CHCO_2Et \quad 67 : 25 : 8 \tag{5.49}$$

mainly the *trans*-adducts. Heterocycles can also be produced, e.g. from $HC{\equiv}CCH_2C{\equiv}CH$ and Bu_2SnH_2.

5.1.1.6 *Metallation of acidic hydrocarbons by tin–oxygen and tin–nitrogen bonded compounds.*[1,3] Acidic hydrocarbons can be metallated by R_3SnNMe_2, $(R_3Sn)_2O$ and related compounds; tin amides appear to be the more reactive. Hydrocarbons which can be metallated include alkynes, polyhaloalkanes, polyfluoroarenes, HN_2CCO_2Et, cyclopentadiene and fluorene as well as HCR_2Y ($R = H$ or alkyl; $Y = CN$, NO_2, $CONR_2$, COR or SO_2Ph), e.g. eqn (5.50).[76]

$$Me_3SnNEt_2 + HCN_2CO_2Et \rightarrow Me_3SnCN_2CO_2Et \tag{5.50}$$

5.1.2 *Formation of specific types of organotin compounds*

5.1.2.1 *Vinylstannanes.* Vinylstannanes are generated from the corresponding vinyl-Grignards and vinyl-lithium compounds with retention of configuration.[77] Vinyl

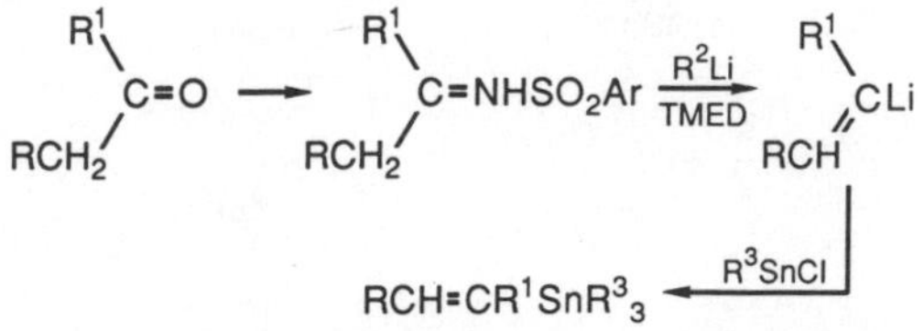

$$R^1(RCH_2)C{=}O \longrightarrow R^1(RCH_2)C{=}NHSO_2Ar \xrightarrow[TMED]{R^2Li} R^1C(Li){=}CHR \xrightarrow{R^3SnCl} RCH{=}CR^1SnR^3{}_3$$

Scheme 4

halides, substituted alkenes and arenesulphonyl hydrazones[78], (Scheme 4 and eqn 5.9) are useful sources of the vinyl-metal.[22] Scheme (4) illustrates a method of synthesis from ketones; other routes from ketones are shown in Schemes 5[79] and 6.[80]

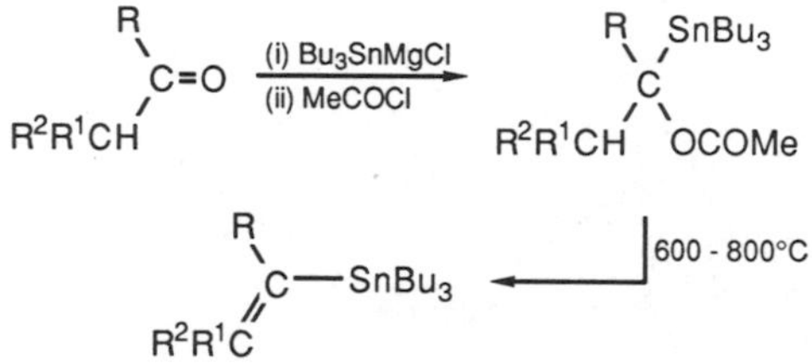

$$R(R^2R^1CH)C{=}O \xrightarrow[(ii)\ MeCOCl]{(i)\ Bu_3SnMgCl} R(R^2R^1CH)C(SnBu_3)(OCOMe) \xrightarrow{600-800°C} R^2R^1C{=}C(R)\text{—}SnBu_3$$

Scheme 5

$$PhCHMe(H)C{=}O \xrightarrow[(ii)\ H_2O]{(i)\ Bu_3SnLi} PhCHMe(H)C(OH)(SnBu_3) \xrightarrow[EtO_2CN{=}NCO_2Et]{Ph_3P,\ CH_2Cl_2} PhCHMe(H)C(Br)(SnBu_3)$$

separate diastereomers

$$\xrightarrow[-110°C]{DBu} \text{Pure } (Z)\text{- or } (E)\text{- } PhCMe{=}CHSnBu_3$$

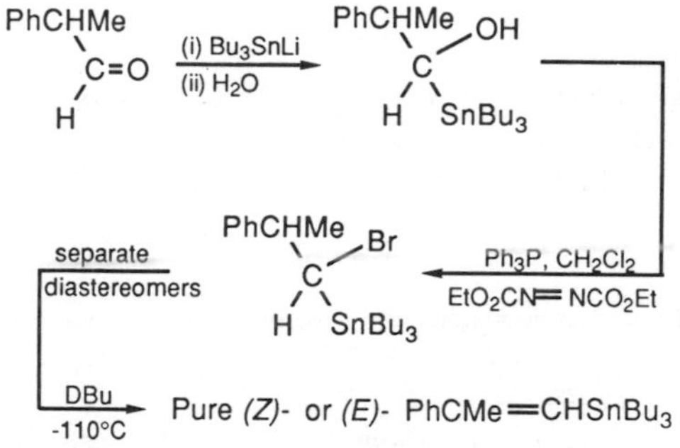

Scheme 6

Reactions of vinyl halides and triflates with triorganostannyl anions produce vinylstannanes.[30,81] Reactions of $Me_3SnMgMe/CuCN$ with vinyl triflates, but not iodides, are stereospecific (eqn 5.51)[58]

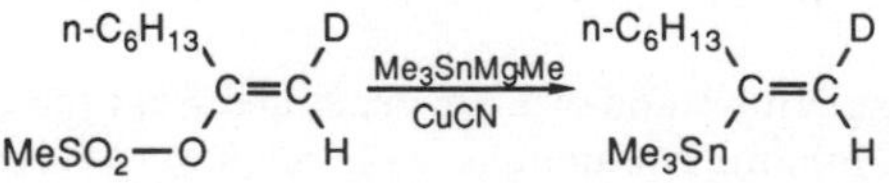

$$n\text{-}C_6H_{13}(MeSO_2\text{—}O)C{=}C(D)H \xrightarrow[CuCN]{Me_3SnMgMe} n\text{-}C_6H_{13}(Me_3Sn)C{=}C(D)H \quad (5.51)$$

Hydrostannation of alkynes provides vinylstannanes (section 5.1.1.5); however, these reactions are not always stereospecific.[82,83] In addition, certain isomers dominate and, as these may not be the ones required, use has been made of other methods involving alkynes.

Alkynylborates can be used in very stereoselective (> 98%) syntheses, (Scheme 7)[84].

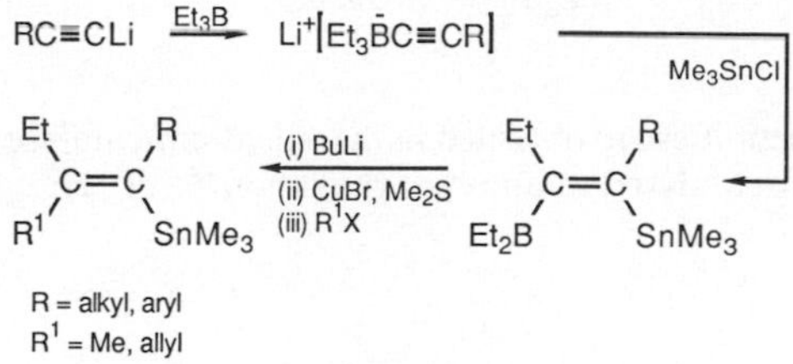

$$RC{\equiv}CLi \xrightarrow{Et_3B} Li^+[Et_3\bar{B}C{\equiv}CR] \xrightarrow{Me_3SnCl} (Et)(Et_2B)C{=}C(R)(SnMe_3) \xrightarrow{(i)\ BuLi;\ (ii)\ CuBr,\ Me_2S;\ (iii)\ R^1X} (Et)(R^1)C{=}C(R)(SnMe_3)$$

R = alkyl, aryl
R^1 = Me, allyl

Scheme 7

Triorganotin-metal compounds, including R_3SnSnR_3, R_3SnSiR_3, $Bu_3SnAlEt_2$, $(Bu_3Sn)_2Zn$ and triorganotin copper reagents, add in a *cis*-manner to alkynes.

Triorganotin copper(I) species add to terminal alkynes (eqn 5.52 and Scheme 8).[85,86]

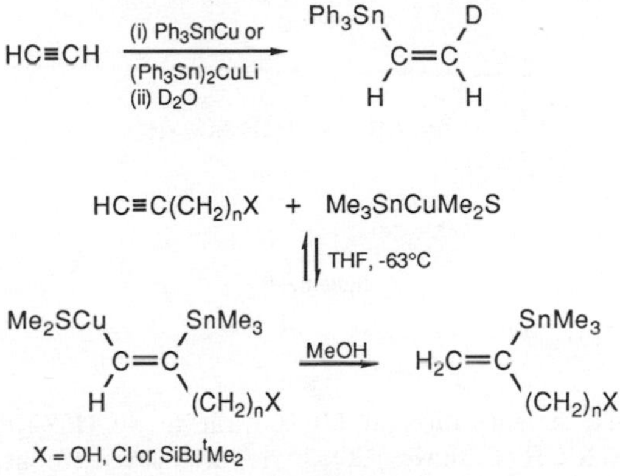

$$HC{\equiv}CH \xrightarrow{(i)\ Ph_3SnCu\ or\ (Ph_3Sn)_2CuLi;\ (ii)\ D_2O} (Ph_3Sn)(H)C{=}C(D)(H)$$

$$HC{\equiv}C(CH_2)_nX + Me_3SnCuMe_2S \underset{}{\overset{THF,\ -63°C}{\rightleftharpoons}} (Me_2SCu)(H)C{=}C(SnMe_3)(CH_2)_nX \xrightarrow{MeOH} H_2C{=}C(SnMe_3)(CH_2)_nX \qquad (5.52)$$

X = OH, Cl or $SiBu^tMe_2$

Scheme 8

Additions also occur to $HOCH_2C{\equiv}CCH_2OH$ using $Bu_3SnCu{\cdot}SMe_2{\cdot}LiBr$ providing (*E*)-$HOCH_2CH{=}C(SnBu_3)CH_2OH$.[87] An interesting ditin product, (*E*)-$RC(SnMe_3){=}C(SnMe_3)CO_2R'$, was isolated from $RC{\equiv}CCO_2R'$ and excess $(Me_3Sn)_2CuLi$ at −48°C in THF.[88]

As shown in eqns (5.39) and (5.40), ditins add to alkynes[59,60]. The initial (*Z*)-$Me_3SnCR{=}CXSnMe_3$ products (X = H, CO_2R' or $CONMe_2$) readily isomerize to the (*E*)-alkenes on UV irradiation (for X = H) or thermally (for X = CO_2R' or $CONMe_2$).

The addition of stannylsilanes to terminal alkynes[61,62] are regio- and stereo-specific, e.g. eqn (5.53).

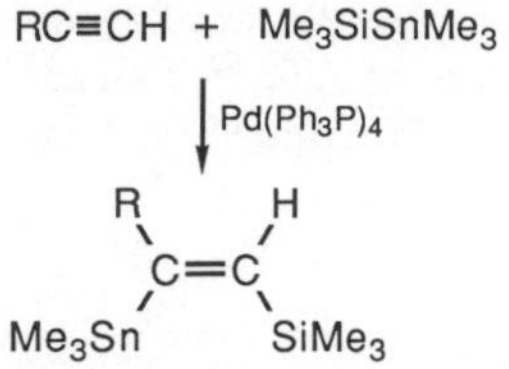

$$RC{\equiv}CH + Me_3SiSnMe_3 \xrightarrow{Pd(Ph_3P)_4} (R)(Me_3Sn)C{=}C(H)(SiMe_3) \quad (5.53)$$

Different regioselectivities are obtained in the palladium-catalysed stannylzincation and in the copper-catalysed stannylmagnesiation[58] of $HC{\equiv}C(CH_2)_2OCH_2Ph$ (Scheme 9).

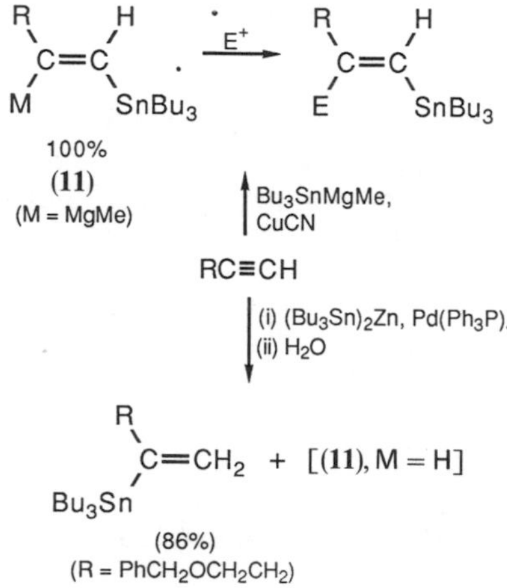

$$(R)(M)C{=}C(H)(SnBu_3) \xrightarrow{E^+} (R)(E)C{=}C(H)(SnBu_3)$$

100%
(11)
(M = MgMe)

$$RC{\equiv}CH \xrightarrow{Bu_3SnMgMe,\ CuCN} \mathbf{(11)}$$

$$RC{\equiv}CH \xrightarrow{(i)\ (Bu_3Sn)_2Zn,\ Pd(Ph_3P)_4;\ (ii)\ H_2O} (R)(Bu_3Sn)C{=}CH_2 + [\mathbf{(11)}, M = H]$$

(86%)
($R = PhCH_2OCH_2CH_2$)

Scheme 9

The stereospecific isomerization of $R'CH_2C(SnMe_3){=}CHCO_2R^2$ (R' = alkyl or $R_3SiOCH_2CH_2$) to $R'CH{=}C(SnMe_3)CO_2R^2$ has been devised (eqns 5.54 and 5.55).[89]

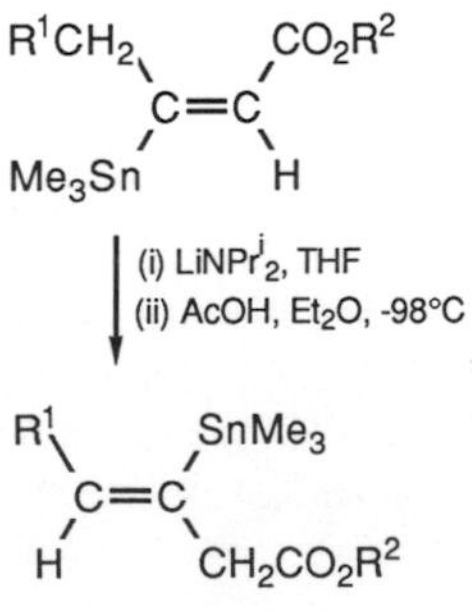

$$(R^1CH_2)(Me_3Sn)C{=}C(CO_2R^2)(H) \xrightarrow{(i)\ LiNPr^i_2,\ THF;\ (ii)\ AcOH,\ Et_2O,\ -98^\circ C} (R^1)(H)C{=}C(SnMe_3)(CH_2CO_2R^2) \quad (5.54)$$

$$\mathrm{(Me_3Sn)(R^1CH_2)C{=}C(H)(CO_2R^2)} \xrightarrow[\text{(ii) AcOH, Et}_2\text{O, -98°C}]{\text{(i) LiNPr}^i_2\text{, THF}} \mathrm{(R^1)(H)C{=}C(CH_2CO_2R^2)(SnMe_3)} \quad (5.55)$$

5.1.2.2 *Allylstannanes.* Allylstannanes have been obtained both from reactions of allylmagnesium or allyl-lithium compounds with organotin halides or even oxides,[90] and from triorganotin anionoid compounds (R_3SnM) with alkenyl halides or esters,[30,46,47] (see eqns 5.24, 5.25 and 5.56). Reaction (5.56) occurs with high regio- and stereo-selectivity.[91] An 1,7-homo-conjugate addition of $(Ph_3Sn)_2Zn$ produces an allylstannane (eq 5.57).[92]

$$\mathrm{RCH{=}CHCH_2OAc} \xrightarrow[\text{(ii) H}_2\text{O}]{\text{(i) Bu}_3\text{SnLiEt}_2\text{, Pd cat.}} \mathrm{RCH{=}CHCH_2SnBu_3} \quad (5.56)$$

$$\text{2-(prop-1-enyl)cyclopropane-1,1-}(CO_2Et)_2 \xrightarrow[\text{(ii) H}_2\text{O}]{\text{(i) (Ph}_3\text{Sn)}_2\text{Zn}} \mathrm{Ph_3SnCH_2CH{=}CHCH_2CH(CO_2Et)_2} \quad (E):(Z) = 82:18 \quad (5.57)$$

Various allylic derivatives (e.g. sulphones, xanthates, sulphides and tosylates)[93] react with organotin hydrides to give allylstannanes, e.g. eqns (5.58) and (5.59).

$$\left(\mathrm{RCH{=}CR^1CH_2OH}\right) \longrightarrow \mathrm{RCH(SC({=}O)SMe)CR^1{=}CH_2} \xrightarrow{\text{Bu}_3\text{SnH}} \mathrm{RCH{=}CR^1CH_2SnBu_3} \quad (5.58)$$

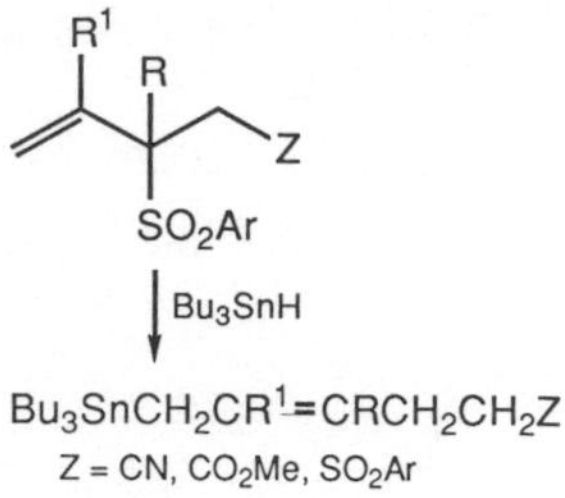

$$\text{(allylic sulfone: } CH_2{=}C(R^1){-}C(R)(SO_2Ar){-}CH_2CH_2Z) \xrightarrow{Bu_3SnH} Bu_3SnCH_2CR^1{=}CRCH_2CH_2Z$$

Z = CN, CO_2Me, SO_2Ar (5.59)

These reactions are, however, not usually stereospecific. Hydrostannation of 1,3-dienes also produce allylstannanes.[94]

Modifications of existing functionally-substituted organotins have been used, e.g. eqn (5.60) and Scheme 10.

$$Me_3SnCH_2CH{=}PPh_3 \xrightarrow{RR'C{=}O} Me_3SnCH_2CH{=}CRR' \quad (5.60)^{95}$$

$$Bu_3SnCRR'CH_2CO_2Me \xrightarrow{LiAlH_4} Bu_3SnCRR'CH_2CH_2OH \xrightarrow{ArSeCN,\ Bu_3P} Bu_3SnCRR'CH_2CH_2SeAr \xrightarrow[\text{Ref. 96}]{Ar'CO.O.O.H} Bu_3SnCRR^1CH{=}CH_2$$

Scheme 10

As shown in Scheme 10, α-substituted allyl compounds are obtained, in contrast to other synthetic methods which generate γ- or α/γ mixtures.

Addition of allylstannanes to ketones is a reversible reaction, and hence pyrolysis of homoalkyl enolates can be utilized in the synthesis of allylstannanes (eqn 5.61).[97]

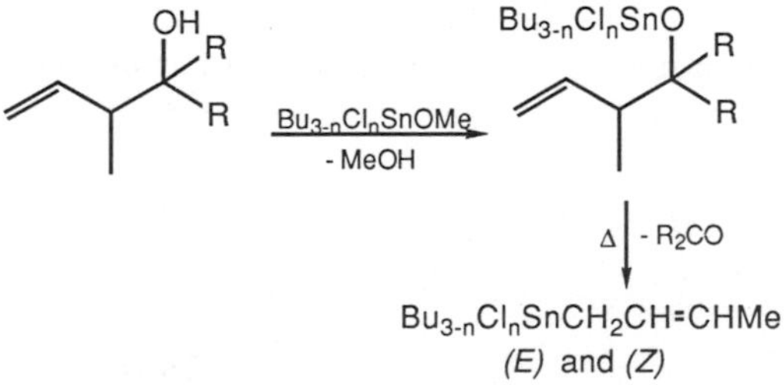

$$CH_2{=}CHCH(Me)C(OH)R_2 \xrightarrow[-\,MeOH]{Bu_{3-n}Cl_nSnOMe} CH_2{=}CHCH(Me)C(OSnCl_nBu_{3-n})R_2 \xrightarrow[\Delta]{-\,R_2CO} Bu_{3-n}Cl_nSnCH_2CH{=}CHMe$$

(*E*) and (*Z*) (5.61)

5.1.2.3 *Functionally-substituted alkylstannanes.* Functionally-substituted alkylstannyl species constitutes a useful but wide and disparate range of compounds. All the general routes discussed earlier can be used to advantage, e.g. (i) the use of organo-tin halides plus YCR^1R^2Li reactions (e.g. Y = nitrogen, phosphorus, sulphur, selenium, halo-, cyano- or carbonyl group) for α-substituted derivatives; (ii) hydrostannation of vinyl compounds, $CH_2{=}CH(CH_2)_nY$ for β, γ, or δ derivatives (Y = oxygen, sulphur, phosphorus, carbonyl groups, etc.); and (iii) reactions of R_3SnM and substituted alkyl halide $[Y\text{-}(\overset{|}{\underset{|}{C}})_nX]$. Clearly specific routes also have to be devised, and modifications of existing functional groups have been used to great effect, as follows:

(i) By conversion of hydroxyl groups to halogen groups with the possibility of further substitutions (e.g. Scheme 11) (related reactions occur for $R_3Sn(CH_2)_nOH$, $n > 3$).

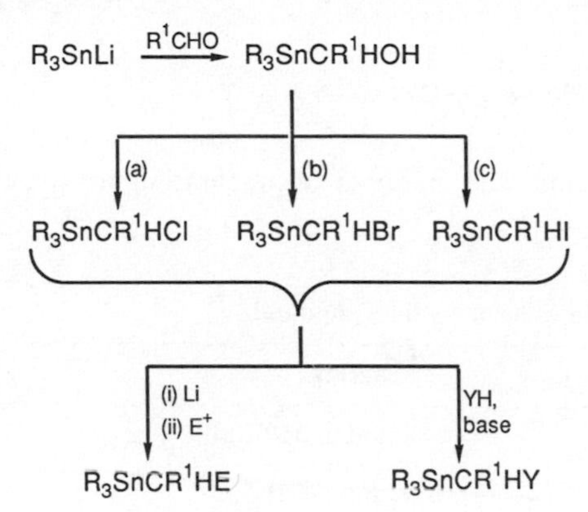

(*a*) CCl_4, Ph_3P or TosCl, py
(*b*) CBr_4, Ph_3P; or Ph_3P, $EtO_2CN{=}NCO_2Et$, CH_2Br_2
(*c*) MeI, Ph_3P, $EtO_2CN{=}NCO_2Et$

Scheme 11

E′ = CH_2CH_2OH or CR^1R^2OH; Y = OR^2, SR^2, SO_2R^2, SeR^2, NR^2_2, etc.

(ii) By conversion of cyano and carbonyl derivatives, e.g. Scheme 12:

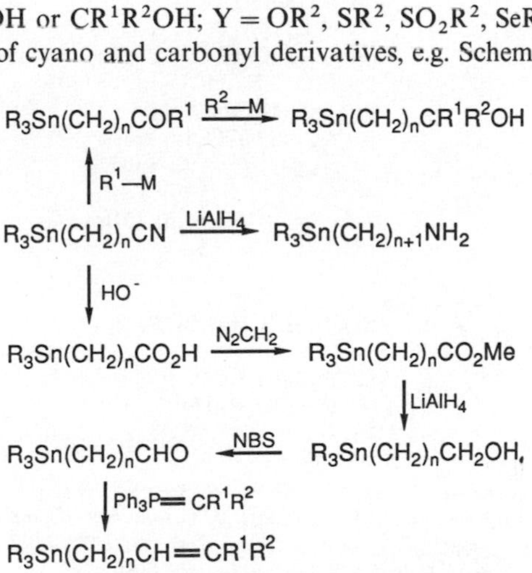

$$R_3Sn(CH_2)_nCOR^1 \xrightarrow{R^2\text{—}M} R_3Sn(CH_2)_nCR^1R^2OH$$

$$R_3Sn(CH_2)_nCN \xrightarrow{R^1\text{—}M} R_3Sn(CH_2)_nCOR^1$$

$$R_3Sn(CH_2)_nCN \xrightarrow{LiAlH_4} R_3Sn(CH_2)_{n+1}NH_2$$

$$R_3Sn(CH_2)_nCN \xrightarrow{HO^-} R_3Sn(CH_2)_nCO_2H \xrightarrow{N_2CH_2} R_3Sn(CH_2)_nCO_2Me$$

$$R_3Sn(CH_2)_nCO_2Me \xrightarrow{LiAlH_4} R_3Sn(CH_2)_nCH_2OH$$

$$R_3Sn(CH_2)_nCH_2OH \xrightarrow{NBS} R_3Sn(CH_2)_nCHO$$

$$R_3Sn(CH_2)_nCHO \xrightarrow{Ph_3P{=}CR^1R^2} R_3Sn(CH_2)_nCH{=}CR^1R^2$$

Scheme 12

(iii) By oxidations, e.g. of sulphur and phosphorus groups, as well as carbon groups, e.g. Scheme 12
(iv) On addition to vinyl- and 3-butenyl-stannanes by electrophiles, free radicals and carbenes (eqn 5.62).

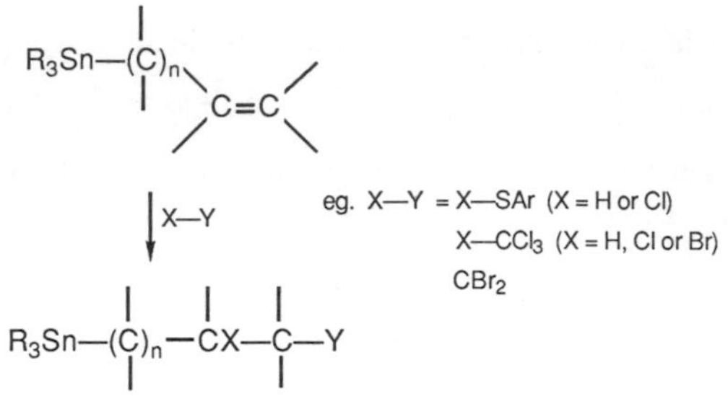

$$R_3Sn{-}(C)_n{-}C{=}C \xrightarrow{X-Y} R_3Sn{-}(C)_n{-}CX{-}C{-}Y \quad (5.62)$$

eg. X—Y = X—SAr (X = H or Cl)
X—CCl_3 (X = H, Cl or Br)
CBr_2

Examples of these and other methods of preparation are given in Tables 5.1–5.7.

Table 5.1 Preparation of haloalkyltin compounds.

Compound	Reagents	Ref.
$R_3SnCHR'I$ $R' = H$ or Me_3Sn	R_3SnI, $ICHR'ZnI$	1
$R_2Sn(CH_2I)_2$	R_2SnI_2, ICH_2ZnI	1
R_3SnCH_2X X = Cl, Br or I	R_3SnX, N_2CH_2	2
$R_3Sn(CH_2)_nCl$ (n = 1, > 3)	$R_3Sn(CH_2)_nOH$, CCl_4, PPh_3	3
$Bu_3SnCHRCl$	(i) Bu_3SnLi, RCHO	4
	(ii) TosCl, py	7
$Bu_3SnCHRBr$	(*a*) $Bu_3SnCHROH$, CBr_4, Ph_3P	4
	(*b*) $Bu_3SnCHROH$, Ph_3P $EtO_2CN{=}NCO_2Et$, CH_2Br_2	5
$Bu_3SnCHRI$	$Bu_3SnCHROH$, Ph_3P $EtO_2CN{=}NCO_2Et$, MeI	4
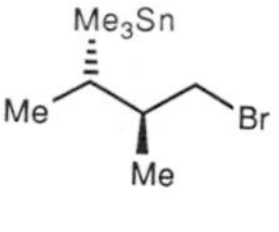 Me_3Sn, Me, Br, Me (structure)	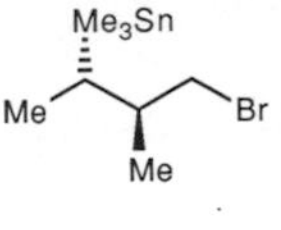 Me_3Sn, Me, Br, Me (structure) (i) $LiAlH_4$ (ii) TosCl (iii) LiBr	6
$Br_3Sn(CH_2)_nBr$ ($n > 3$)	$SnCl_2$, $Br(CH_2)_nBr$, R_3Sb	7
R_3SnCX_2X' (X = X' = halide)	(*a*) R_3SnCl, $LiCX_2X'$ (*b*) R_3SnBr, $PhHgCX_2X'$	8
$Et_3SnCHXCH_2CCl_3$ (X = H, Cl or Br)	$Et_3SnCH{=}CH_2$, $XCCl_3$ $(PhCO_2)_2$	9

1. D. Seyferth, S.B. Andrews and R.L. Lambert, *J. Organomet. Chem.*, 1972, **37**, 69.
2. D. Seyferth and E.G. Rochow, *J. Amer. Chem. Soc.*, 1955, **77**, 1302.
3. M. Gielen and J. Topart, *Bull. Soc. Chim. Belg.*, 1971, **80**, 655.
4. Y. Torisawa, M. Shibasaki and S. Ikegami, *Tetrahedron Lett.*, 1981, **22**, 2397.

5. M. Shibasaki, Y. Torisawa and S. Ikegami, *Tetrahedron Lett.*, 1982, **23**, 4607.
6. I. Fleming and M. Rowley, *Tetrahedron Lett.*, 1985, **26**, 3857.
7. E.J. Bulten, H.F.M. Gruter and H.F. Martens, *J. Organomet. Chem.*, 1976, **117**, 329.
8. D. Seyferth, F.M. Armbrecht, B. Prokai and R.J.C. Ross, *J. Organomet. Chem.*, 1966, **6**, 573.
9. D. Seyferth, *J. Org. Chem.*, 1957, **22**, 1252.

Table 5.2 Hydroxy- and alkoxy-alkyltin compounds.

Compound	Reagents	Ref.
$R_3SnCHR'OH$	R_3SnLi, R′CHO	1
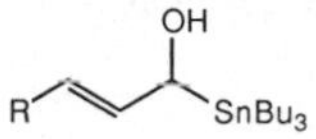 R–CH=CH–CH(OH)–$SnBu_3$	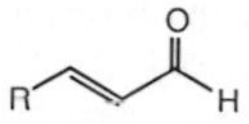 R–CH=CH–CHO, Bu_3SnLi	2
Ph_3SnCH_2OR	Ph_3SnCH_2I, NaOAr	3
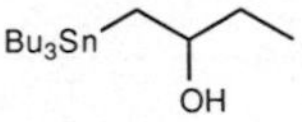 Bu_3Sn–CH_2–CH(OH)–CH_2CH_3	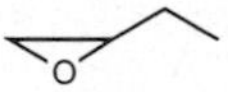 1,2-epoxybutane, $Bu_3SnMgCl$	4
$R_3SnCH_2CH_2OR'$	R_3SnH, $CH_2{=}CHOR'$	5
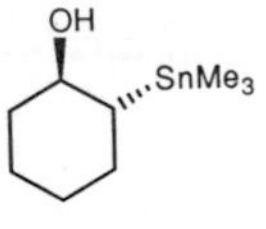 trans-2-(trimethylstannyl)cyclohexanol	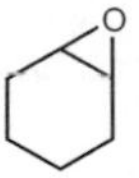 cyclohexene oxide, Me_3SnLi	6
$R_3Sn(CH_2)_nOH$	R = Me, n = 3–6 (*a*) Me_3SnH, $CH_2{=}CH(CH_2)_{n-2}OH$ (*b*) R = Ph, n = 3–5; $Ph_3Sn(CH_2)_{n-1}CO_2Me$, $LiAlH_4$	7 5(a)
$R_3SnCH_2CH_2CH_2OR'$	R_3SnH, $CH_2{=}CHCH_2OR'$	5
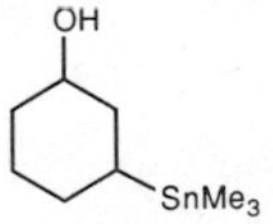 3-(trimethylstannyl)cyclohexanol, *cis : trans*	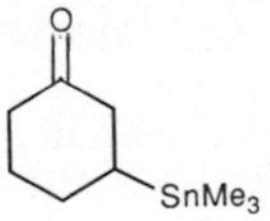 3-(trimethylstannyl)cyclohexanone, $LiAlH_4$	6
$Bu_3SnCH_2CHPhCMe_2OH$	$Bu_3SnCR_2CH_2C(O)R^2$, R^1Li or R^1MgX	8, 9
$Bu_3SnCH_2CHPhCMe_2OH$	$Bu_3SnCH_2CHPhCO_2Me$, MeLi	8
$Me_3Sn(CH_2)_3CHPhOH$	$Me_3Sn(CH_2)_3Br$ (i) Mg (ii) PhCHO (iii) H_2O	10
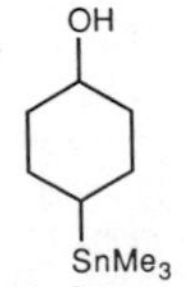 4-(trimethylstannyl)cyclohexanol, *cis & trans*	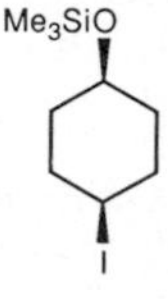 4-iodo-1-(Me_3SiO)cyclohexane, (i) Me_3SnLi (ii) H_2O	6

Table 5.2 (*Contd.*)

1. M. Shibasaki, Y. Torisawa and S. Ikegami, *Tetrahedron Lett.*, 1982, **23**, 4607.
2. M. Shibasaki, H. Suzuki, Y. Torisawa and S. Ikegami, *Chem. Lett.*, 1983, 1303.
3. R.D. Taylor and J.L. Wardell, *J. Organomet. Chem.*, 1974, **77**, 311.
4. J.C. Lahournere and J. Valade, *J. Organomet. Chem.*, 1971, **33**, C4.
5. (*a*) G.J.M. van der Kerk and J.G. Noltes, *J. Appl. Chem.*, 1959, **9**, 176. (*b*) W.P. Neumann, *Annalen*, 1962, **659**, 27.
6. A. Ricci, M. Taddei and G. Seconi, *J. Organomet. Chem.*, 1986, **306**, 23; see also M. Ochiai, T. Ukita, Y. Nago and E. Fujita, *J. Chem. Soc., Chem. Commun.*, 1985, 637.
7. D.G.J. Kingston, H.P. Tannenbaum and H.G. Kuivila, *Org. Mass. Spect.*, 1974, **9**, 31.
8. I. Fleming and C.J. Urch, *J. Organomet. Chem.*, 1985, **285**, 173.
9. J.F. Kadow and C.R. Johnson, *Tetrahedron Lett.*, 1984, **25**, 5255; K. Nakatani and S. Isoe, *Tetrahedron Lett.*, 1984, **25**, 5335; I. Fleming and M. Rowley, *Tetrahedron Lett.*, 1985, **26**, 3857; Y. Ueno, M. Ohta and M. Okawara, *Tetrahedron Lett.*, 1982, **23**, 2577.
10. E.J. Bulten, H.F.M. Gruter and H.F. Martens, *J. Organomet. Chem.*, 1976, **117**, 329.

Table 5.3 Sulphur-containing alkyltin compounds.

Compound	Reagents	Ref.
R_3SnCH_2SR'	(*a*) $R_3SnLi, ClCH_2SR'$	1
	(*b*) R_3SnCH_2I, HSR'	1, 2
$Bu_3SnCH(SiMe_3)SPh$	$LiCH(SiMe_3)SPh, Bu_3SnCl$	3
R S S $SnMe_3$	R S S Li, Me_3SnCl	4
Bu_3SnCH_2SOR	$Bu_3SnNMe_2, MeSOR$	5
($R = Me, Ph$)		
$Bu_3SnCH_2SO_2R$	(*a*) MCH_2SO_2R, ($M = Li, BrMg$)	6
(R = alkyl, Ph, NMe_2)	Bu_3SnCl	
	(*b*) $Bu_3SnNEt_2, MeSO_2R$	
$Bu_3SnCH_2\overset{+}{S}Me_2X^-$	Bu_3SnCH_2SMe, MeX	3
$Bu_3SnCH_2CH_2SH$	$Bu_3SnCH{=}CH_2, H_2S$	7
$Me_3SnCH_2CHMeSH$	Me_3SnLi, $MeCHCH_2S$ (ring)	8
$R_3SnCH_2CH_2SR'$	(*a*) $R_3SnCH{=}CH_2, HSR'$	1, 9
	(*b*) $R_3SnH, CH_2{=}CHSR'$	
$R^1R^2C{=}CR^3CH(SAr)CH_2SnBu_3$	$R^1R^2C{=}CR^3CH(SAr)Li, Bu_3SnCH_2I$	10
(Ar = 2-pyridyl)		
$Ph_3SnCH_2CH_2SO_2R$	$Ph_3SnH, CH_2{=}CHSO_2R$	11
$Bu_3SnCHR^2CHR^1SO_2Ph$	$Bu_3SnLi, R^2CH{=}CR'SO_2Ph$	12
$R_3Sn(CH_2)_3SR'$	$R_3Sn(CH_2)_3Cl, HSR'$	13
$Ph_3Sn(CH_2)_nSO_mR'$	$Ph_3Sn(CH_2)_nSR'$, m-$ClC_6H_4CO{\cdot}OOH$	11
($n = 3, 4; m = 1, 2$)		
$Ph_3Sn(CH_2)_4SR'$	(*a*) $Ph_3Sn(CH_2)_2CH{=}CH_2, HSR'$	13
	(*b*) $Ph_3SnH, H_2C{=}CH(CH_2)_2SR'$	
$Bu_3SnCH_2CH(CH_2R')S(O)Ph$	$Bu_3SnCH_2I, R^1CH_2CHLiS(O)Ph$	14
Ph_3SnCH_2SeR'	$Ph_3SnCH_2I, HSeAr$	15

1. R.D. Taylor and J.L. Wardell, *J. Organomet. Chem.*, 1974, **77**, 311; see also D.J. Peterson, *J. Organomet. Chem.*, 1971, **26**, 215; R.D. Brasington and R.C. Poller, *J. Organomet. Chem.*, 1973, **40**, 115.
2. J. McM. Wigzell, R.D. Taylor and J.L. Wardell, *J. Organomet. Chem.*, 1982, **235**, 29.
3. D.I. Ager, *J. Chem. Soc., Perkin Trans. 1*, 1986, 195.

4. G.M. Drew and W. Kitching, *J. Org. Chem.*, 1981, **46**, 558; D.J. Cane, W.A.G. Graham and L. Vancea, *Can. J. Chem.*, 1978, **56**, 1538.
5. D.J. Peterson and J.F. Ward, *Ger. Offen.*, 2410131 (*Chem. Abs.*, 1974, **81**, 169628).
6. D.J. Peterson, *J. Organomet. Chem.*, 1971, **26**, 215; US Patent 3794670 (*Chem. Abs.*, 1974, **80**, 108671).
7. W. Stamm, *J. Org. Chem.*, 1963, **28**, 3264.
8. A. Mordini, M. Taddei and G. Seconi, *Gazz. Chim. Ital.*, 1986, **116**, 239.
9. M.G. Voronkov, V.I. Rakhlin and R.G. Mirskov, *Dokl. Chem. (Engl. Trans.)* 1973, **209**, 261; A.A. Dzhafarov, I.A. Aslanov and D.A. Kochkin, *J. Gen. Chem., USSR (Engl. Trans.)*, 1975, **45**, 1986.
10. M. Ochiai, T. Ukita, E. Fujita and S.I. Tada, *Chem. Pharm. Bull.*, 1984, **32**, 1829.
11. J.L. Wardell and J. McM. Wigzell, *J. Organomet. Chem.*, 1983, **244**, 225; see also A. Hosomi, M. Inaba and H. Sakurai, *Chem. Lett.*, 1983, 1763.
12. M. Ochiai, T. Ukita and E. Fujita, *J. Chem. Soc. Chem. Commun.*, 1983, 619.
13. J.L. Wardell and J. McM. Wigzell, *J. Chem. Soc., Dalton Trans.*, 1982, 2321; see also G. Ayrey, R.D. Brasington and R.C. Poller, *J. Organomet. Chem.*, 1972, **35**, 105.
14. M. Ochiai, S.I. Tada, K. Sumi and E. Fujita, *Tetrahedron Lett.*, 1982, **23**, 2205.
15. R.D. Taylor and J.L. Wardell, *J. Chem. Soc., Dalton Trans.*, 1976, 1345.

Table 5.4 Amino group substituted alkyltin compounds.

Compound	Reagents	Ref.
$R_3SnCH_2NR'_2$	(*a*) R_3SnCH_2I, R'_2NH	1
	(*b*) R_3SnLi, R_2NCH_2X ($X = SR^2, SO_2Na$)	2
	(*c*) R_3SnCH_2Cl, MNR'_2 ($M = H$ or K)	3
$(Bu_3SnCH_2)_3N$	$(PhSCH_2)_3N$, Bu_3SnLi	1
$R_3SnCH_2CH_2NR'_2$ ($R' = H$, alkyl or aryl)	R_3SnM ($M = Na$ or Li), $ClCH_2CH_2NR'_2$	4
$R_3SnCH_2CH_2CH_2NH_2$	$R_3SnCH_2CH_2CN$, $LiAlH_4$	5
$R_3SnCH_2CH_2CH_2NMe_2$	(*a*) $R_3SnCH_2CH_2CH_2NH_2$, HCO_2H, H_2CO	5
	(*b*) R_3SnLi, $ClCH_2CH_2CH_2NMe_2$	6
$MeSn(CH_2CH_2CH_2)_3N$	$McSnCl_3$, $(ClMgCH_2CH_2CH_2)_3N$	7
$Bu_3Sn(CH_2)_nNMe_2$ ($n = 3, 4$)	$Bu_3Sn(CH_2)_nBr$, Me_2NH	8
$R_3Sn(CH_2)_n\overset{+}{N}R'_2R^2$, I^-	$R_3Sn(CH_2)_nNR'_2$, R^2I	9
$Ph_3SnCH_2CH_2CH_2NHCOMe$	Ph_3SnH, $CH_2{=}CHCH_2NHCOMe$	10
$Ph_2ClSnCH_2CH_2CH_2NHCOMe$	$Ph_2ClSnCH_2CH_2CH_2CMe{=}NOH$	11

1. E.W. Abel and R.J. Rowley, *J. Organomet. Chem.*, 1975, **97**, 159.
2. D.J. Peterson, *J. Amer. Chem. Soc.*, 1971, **93**, 4027.
3. R.G. Kostyanovskii and A.K. Prokofer, *Bull. Acad. Sci. USSR Div. Chem. Sci.*, 1965, 159.
4. H. Weichmann and A. Tzschach, *Z. Anorg. Allg. Chem.*, 1979, **458**, 291; Y. Sato, Y. Ban and H. Shirai, *J. Org. Chem.*, 1973, **38**, 4373; J.L. Wardell, *Inorg. Chim. Acta*, 1978, **26**, L18.
5. J.W. Suggs and K.S. Lee, *J. Organomet. Chem.*, 1986, **299**, 297.
6. J.G. Noltes and G.J.M. van der Kerk, *Functionally Substituted Organotin Compounds*, Tin Research Institute, Greenford, UK, 1958.
7. K. Jurkschat, A. Tzschach and J. Meunier-Piret, *J. Organomet. Chem.*, 1986, **315**, 45.
8. E.J. Bulten, H.F.M. Gruter and H.F. Martens, *J. Organomet. Chem.*, 1976, **117**, 329.
9. H. Gilman and T.C. Wu, *J. Amer. Chem. Soc.*, 1955, **77**, 3228; J.L. Wardell, unpublished observations.
10. G.J.M. van der Kerk and J.G. Noltes, *J. Appl. Chem.*, 1959, **9**, 106, 176.
11. S.Z. Abbas and R.C. Poller, *J. Organomet. Chem.*, 1976, **104**, 187.

Table 5.5 Phosphorus-containing alkyl tin compounds.

Compound	Reagents	Ref.
$Me_nSn(CH_2PMe_2)_{4-n}$	Me_nSnCl_{4-n}, $LiCH_2PMe_2$	1
$Ph_3SnCH_2P(S)Ph_2$	$Ph_2P(S)CH_2Li$, Ph_3SnCl	2
$R_3SnCH_2P(O)(OR')R^2$	(*a*) R_3SnCH_2I, $P(OR')_2R^2$	3
($R' = Me, Et; R^2 = OR', Ph$)	(*b*) R_3SnH, $N_2CHP(O)(OR')R^2$	
	(*c*) R_3SnNa, $ClCH_2P(O)(OR')R^2$	
	(*d*) R_3SnNMe_2, $CH_3P(O)(OR')R^2$	
$Me_3SnCHRP(Y)(OR')_2$	Me_3SnCl, $(R'O)_2P(Y)CHRCu$	4
($Y = O, S$)		
$Ph_3SnCHR\overset{+}{P}Ph_3, Cl^-$	Ph_3SnCl, Ph_3PCHR	5
($R = H, Me, Pr^i$)		
$Me_3SnCX{=}PC_6H_2Bu^t_3$	Me_3SnCl, $LiCX{=}P{-}C_6H_2Bu^t_3$	6
($X = Cl, Br$)		
$R_3SnCH_2CH_2PR'_2$	R_3SnH, $R'_2PCH{=}CH_2$	7
$Ph_3SnCH_2CHPhPPh_2$	Ph_3SnPPh_2, $CH_2{=}CHPh$	8
$Et_2Sn(CH_2CH_2PPhR')_2$	Et_2SnH_2, $CH_2{=}CHPPhR'$	9
($R' = Ph, Bu^t$)		
$Me_3Sn(CH_2)_{n+2}PR'R^2$	(*a*) Me_3SnH, $CH_2{=}CH(CH_2)_nPR'R^2$	10
($n = 0, 1$)	(*b*) Me_3SnNa, $Cl(CH_2)_{n+2}PR'R^2$	
$R_2R^1SnCH_2CH_2P(O)R^2Ph$	(*a*) R_2R^1SnH, $CH_2{=}CHP(O)(OEt)Ph$	11
($R^1 = Me, Ph; R = Me, Bu^t$)	(*b*) (i) $CH_2{=}CHPPhBu^t$, $R_2R'SnH$	
($R^2 = Bu^t, OEt$)	(ii) $KMnO_4$	
$Et_2Sn[CH_2CH_2P(S)PhR']_2$	$Et_2Sn(CH_2CH_2PPhR')_2$, S	9
$[Et_2Sn(CH_2CH_2PPhR'Me)_2]^{2+}2I^-$	$Et_2Sn(CH_2CH_2PPhR')_2$, MeI	9
$Et_3Sn(CH_2)_{n+2}P(O)(OEt)R$	Et_3SnH, $CH_2{=}CH(CH_2)_nPO(OEt)R$	12
($n = 0, 1; R = EtO, Ph$)		
$R_3SnCHMeCH{=}PPh_3$	R_3SnLi, $MeCH{=}CH^+PPh_3$, Br^-	13

1. H.H. Karsch and A. Appett, *Z. Naturforsch., Teil B*, 1983, **38**, 1399.
2. D. Seyferth, D.E. Welch and J.K. Heeren, *J. Amer. Chem. Soc.*, 1963, **85**, 642.
3. H. Weichmann, B. Ochsler, I. Duchek and A. Tzschach, *J. Organomet. Chem.*, 1979, **182**, 465.
4. P. Savignac and F. Mathey, *Synthesis*, 1982, 725.
5. Y. Yamamoto, *Bull. Chem. Soc. Jpn*, 1982, **55**, 3025.
6. R. Appel, C. Casser and M. Immenkeppel, *Tetrahedron Lett.*, 1985, **26**, 3551.
7. H. Weichmann, G. Quell and A. Tzschach, *Z. Anorg. Allg. Chem.*, 1980, **462**, 7.
8. H. Schumann, P. Jutzi and M. Schmidt, *Angew. Chem. Int. Edn. Engl.*, 1965, **4**, 869.
9. H. Weichmann and B. Rensch, *Z. Anorg. Allg. Chem.*, 1983, **503**, 106.
10. H. Weichmann, *J. Organomet. Chem.*, 1984, **262**, 279.
11. H. Weichmann, C. Mügge, A. Grand and J.B. Robert, *J. Organomet. Chem.*, 1982, **238**, 343.
12. H. Weichmann and A. Tzschach, *J. Prakt. Chem.*, 1976, **318**, 87.
13. M. Tsukamoto, H. Ito and T. Tokoroyama, *Tetrahedron Lett.*, 1987, **28**, 4561.

Table 5.6 Cyanoalkyltin compounds.

Compound	Reagents	Ref.
R_3SnCH_2CN	(*a*) R_3SnCl, $NaCH_2CN$	1
	(*b*) $Ph_3SnOCOCH_2CN$, Δ	2
$Ph_3SnCHCNMe$	$Ph_3SnOC(O)CHCNMe$, Δ	3
$R_3SnCH_2CH_2CN$	R_3SnH, $CH_2{=}CHCN$	3
$Ph_3SnCH_2CH_2CH_2CN$	Ph_3SnH, $CH_2{=}CHCH_2CN$	3, 4
$Me_3Sn(CH_2)_{n+2}CN$	Me_3SnH, $CH_2{=}CH(CH_2)_nCN$	5

1. A.A. Prishchenko, Z.S. Novikova and I.F. Lutsenko, *Zh. Obshch. Khim.*, 1981, **51**, 485

2. G.J.M. van der Kerk and J.G. Noltes, *J. Appl. Chem.*, 1956, **6**, 93.
3. J.G. Noltes and G.J.M. van der Kerk, *Functionally Substituted Organotin Compounds*, Tin Research Institute, Greenford, UK, 1958.
4. G.J.M. van der Kerk and J.G. Noltes, *J. Appl. Chem.*, 1959, **9**, 106.
5. H.G. Kuivila, J.E. Dixon, P.L. Maxfield, N.M. Scarpa, T.M. Topka, K.-H. Tsai and K.R. Wursthorn, *J. Organomet. Chem.*, 1975, **86**, 89.

5.2 Cleavages of carbon–tin bonds[1–4, 98–100]

5.2.1 *General aspects*

Tin–carbon bonds may be cleaved in both heterolytic and homolytic processes. For heterolytic processes, electrophilic attack at carbon is usually the more significant, although nucleophilic assistance at tin can also aid the cleavage reaction, and can even dominate (eqn 5.63).

$$R_3Sn{-}R^1 + E{-}Y \rightarrow R_3SnY + R^1{-}E \qquad (5.63)$$

Cleavages have been reported to occur with basic nucleophilic reagents, such as alcoholic alkali, alkoxide or alkali (e.g. with pentafluorophenyl-,[101] aryl-,[102] benzyl-,[103] allyl-,[104] and alkynyl-tin[105] compounds) and alkyl- or phenyl-lithium (with allyl-, vinyl-, and α-functionally-substituted alkyl-tin compounds (eqn 5.64)).[22,98]

$$R_3Sn{-}R^1 + R^2{-}Li \rightarrow R_3Sn{-}R^2 + R^1{-}Li \qquad (5.64)$$

R^1 = allyl, vinyl or XCH_2 ($X = R^3O$, R^3_2N, R^3S, etc.).

5.2.1.1 *Cleavage by electrophiles.* Among the various electrophilic reagents to react with a wide range of tin–carbon bonds are halogens, proton acids (e.g. hydrogen halides and carboxylic acids), metal(loid) halides (e.g. SnX_4, HgX_2, BX_3, PX_5, platinum and palladium halides, etc.) and sulphur dioxide. These reactions are used both for the synthesis of the organotin product, R_nSnY_{4-n} (especially Y = halide or carboxylate) and the organic product, R^1–E. Cleavage of more than one carbon–tin bond can be brought about.

The disproportionation reaction of tin tetrahalides and organotin compounds, the Kocheskov reaction, has had extensive preparative use for both alkyl and aryl derivatives, e.g. eqn (5.65).

$$(4-n)SnX_4 + nR_4Sn \rightarrow 4R_nSnX_{4-n} \qquad (5.65)$$

The mercury halide reactions, on the other hand, have been studied primarily for mechanistic purposes. These metal halide (MX) reactions show organotins acting as alkylating/arylating agents. (eqn 5.66).

$$R_3Sn{-}R^1 + M{-}X \rightarrow R_3Sn{-}X + R^1{-}M \qquad (5.66)$$

Organotins can also transfer organic groups to electrophilic carbon in transition metal-catalysed cross-coupling reactions with organic halides or esters, e.g. eqn (5.67).

$$R_3Sn{-}R^1 + R^2{-}X \xrightarrow{Pd^0 \text{ or } Pd^{II}} R^1{-}R^2 \qquad (5.67)$$

Halogens can react *via* homolytic reactions (e.g. in light and in non-polar solvents) as well as by heterolytic routes (i.e. in the dark and in polar solvents). The reaction type may have little consequence to the products of the reaction, but would have an impact on the stereochemistry of the reaction and on relative reactivities.

Many of the reactive electrophiles have only limited use and can be classed as being of more exotic than of either mechanistic or synthetic importance. Reactivity sequence for tin–carbon bonds in electrophilic cleavage reactions is generally allyl > phenyl > benzyl > vinyl > methyl > higher alkyl. The sequences for alkyl-tin bonds depend on the solvent and electrophile, and considerable variations have been realized.

Aryl-tin cleavages by electrophilic reagents are typical electrophilic aromatic substitutions[106], with greater electron releasing substituents, (*Z*), leading to faster reaction. Cleavage of Ar–Sn bonds provide Ar–E regiospecifically (eqn 5.68),[98] e.g. Ar–F (using $FOSO_4^-$, Cs^+,[107] F–OAc or F_2)[108] Ar–NO, from[109] NO–Cl, and Ar–D, from[98,110] D–Cl, D–O_2CCH_3, D–O_2CCF_3, etc. Other isotopically labelled groups *E* have been incorporated into aromatic groups *via* aryl-tin substrates.[98] Brominations, iodinations and protonolysis are the best studied reactions in regards to kinetic and mechanism of aryl-tin bond cleavages.

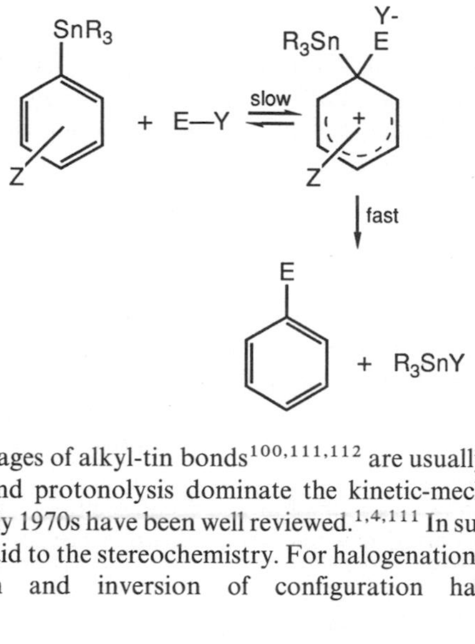

SnR3 + E—Y ⇌(slow) [R3Sn, E, Y-, Z, +] →(fast) E + R3SnY

(5.68)

Electrophilic cleavages of alkyl-tin bonds[100,111,112] are usually S_E2 processes. Here too, brominations and protonolysis dominate the kinetic-mechanistic studies. The findings up to the early 1970s have been well reviewed.[1,4,111] In subsequent work, more attention has been paid to the stereochemistry. For halogenations (I_2 and Br_2),[113–115] both net retention and inversion of configuration have been observed (eqn 5.69)[114]:

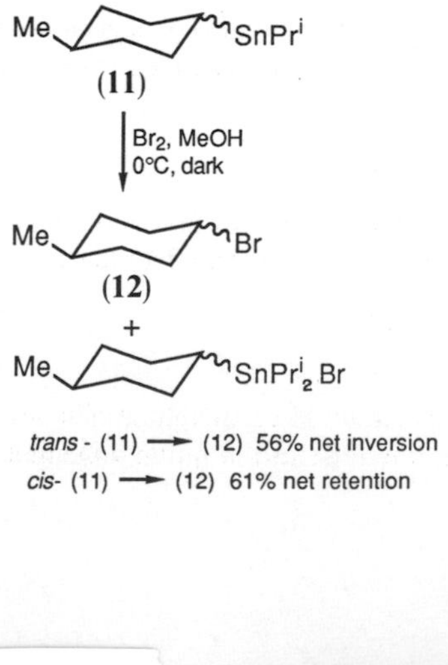

Me–C6H10–SnPr^i **(11)** →(Br2, MeOH, 0°C, dark) Me–C6H10–Br **(12)** + Me–C6H10–SnPr$^i_2$Br

trans- (11) → (12) 56% net inversion
cis- (11) → (12) 61% net retention

(5.69)

Three transition states have been considered[112]: (**13**), (**14**) and (**15**). Transition states (**13**) and (**14**) will lead to retention of configuration, with inversion arising from (**15**). Solvent, electronic effects and steric factors all play significant roles in determining the relative importance of these transition states and hence the overall stereochemistry.

In contrast to the electrophilic halogenations, shown in eqn (5.69), homolytic reactions of (**11**) with Br_2 in PhCl at 20°C in light gave (**12**) with no preferred stereochemistry. The stereochemical results for organotin compounds (e.g. eqn 5.69) contrast with the retention normally obtained for halogenations of dialkylmercurials.

An alternative view of halogenations of tetra-alkyltins has been presented by Kochi.[116] In this approach, the rate-determining step is considered to be an electron-transfer from a tetra-alkyltin-halogen charge-transfer complex (Scheme 13).

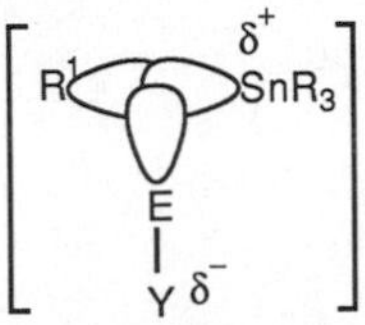

(**13**) 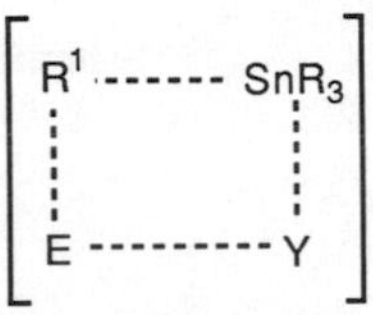 (**14**)

(**15**)

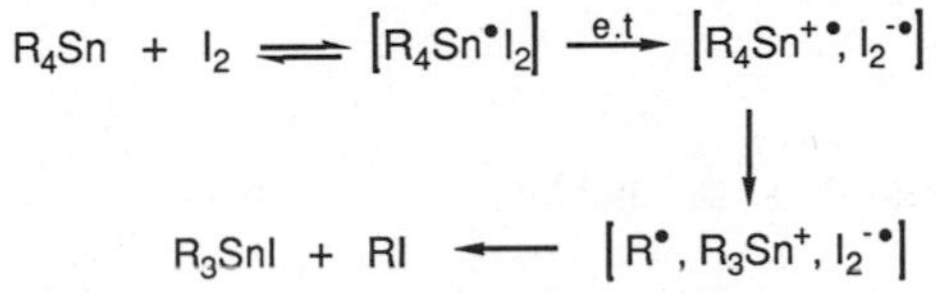

$$R_4Sn + I_2 \rightleftharpoons [R_4Sn^{\bullet}I_2] \xrightarrow{e.t.} [R_4Sn^{+\bullet}, I_2^{-\bullet}]$$

$$R_3SnI + RI \leftarrow [R^{\bullet}, R_3Sn^{+}, I_2^{-\bullet}]$$

Scheme 13

For protonolysis of alkyltins, retention of configuration is frequently obtained, e.g. reactions of trifluoroacetic acid with cyclohexyl(triisopropyl)stannanes occur stereospecifically with retention of configuration.

5.2.1.2 *Homolytic cleavages.*[3,117,118] Homolytic cleavage of a tin–carbon bond in a tetraorganotin is less common than is heterolytic cleavage. Examples of free radicals which have been shown to take part in S_H2 displacement at the tin centre are $Br^{\bullet}$, $I^{\bullet}$ and the succinimidyl radical (Scheme 14). Apparently organotin halides react more readily by this mechanism. The reactivity found for R_4Sn in Me_2CO towards the succinimidyl radicals (*viz* Me > Et > Pr ≈ Bu > Bu^s)[119] is the reverse of that found for free-radical reactions of I_2 or Br_2. The sequence obtained[120] for free-radical iodination of Me_nSnR_{4-n} is $Bu^t > Pr^i > Et > Pr^n$.

An electron transfer mechanism has also been proposed for free-radical iodination[120] (Scheme 15), and a similar mechanism was suggested[121] for the photochemical induced reaction between I_2 and R_3SnI.

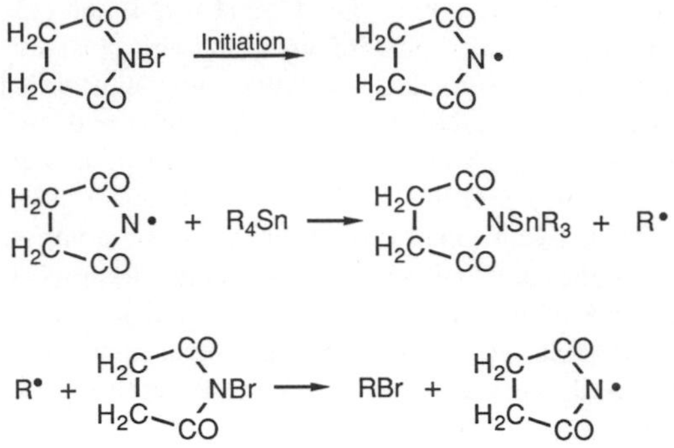

$$H_2C(CO)_2CH_2NBr \xrightarrow{\text{Initiation}} H_2C(CO)_2CH_2N^{\bullet}$$

$$H_2C(CO)_2CH_2N^{\bullet} + R_4Sn \longrightarrow H_2C(CO)_2CH_2NSnR_3 + R^{\bullet}$$

$$R^{\bullet} + H_2C(CO)_2CH_2NBr \longrightarrow RBr + H_2C(CO)_2CH_2N^{\bullet}$$

Scheme 14

$$R_4Sn + I^{\cdot} \xrightarrow{\text{slow}} [R_4Sn^{+\cdot}, I^-] \xrightarrow{\text{fast}} R_3SnI + R^{\cdot} \text{ etc.}$$

Scheme 15

Alkoxy radicals attack the α- and β-carbon atoms of tetraalkyltins.[112,123] In contrast, S_H2 reactions occur[124] at tin for trialkyltin compounds (Scheme 16).

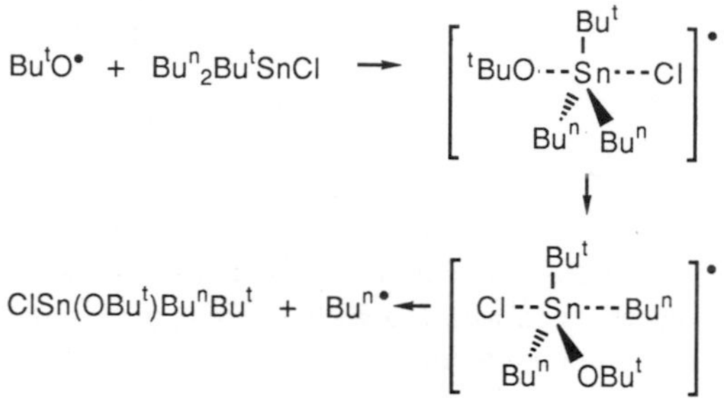

$$Bu^tO^{\bullet} + Bu^n{}_2Bu^tSnCl \longrightarrow [{}^tBuO\cdots Sn(Bu^t)(Bu^n)_2\cdots Cl]^{\bullet}$$

$$\downarrow$$

$$ClSn(OBu^t)Bu^nBu^t + Bu^{n\bullet} \longleftarrow [Cl\cdots Sn(Bu^t)(Bu^n)(OBu^t)\cdots Bu^n]^{\bullet}$$

Scheme 16

5.2.2 *Reactions of specific groups*[98]

5.2.2.1 *Vinylstannanes.* Vinyl-tin bonds are cleaved, with retention of configuration, by a variety of electrophilic reagents including proton acids,[125,126] halogens[126,127] (e.g. I_2 in MeOH), mercury(II) halides,[128] sulphur dioxide[129] and phenylselenyl bromide[127] (eqn 5.70).

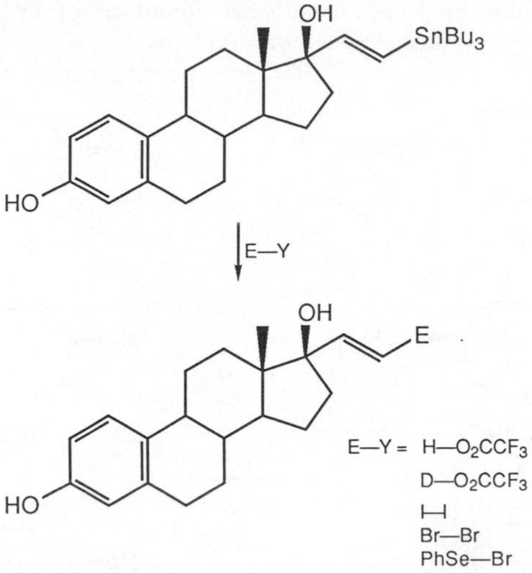

OH
$SnBu_3$
HO

E—Y

OH
E
HO

E—Y = H—O_2CCF_3
D—O_2CCF_3
I—I
Br—Br
PhSe—Br (5.70)

Reactions of $Ph_3SnCH{=}CH_2$ with electrophiles, such as I_2, usually result in Ph–Sn bond cleavage. However, sulphenyl halides react at the vinyl group, either resulting in addition to the vinyl group and/or cleavage of the vinyl-tin bond, e.g. eqn 5.71. Only the Markownikov adduct was isolated, which slowly decomposed in solution to Ph_3SnCl and $CH_2{=}CHSC_6H_4NO_2$-*o*. The relative proportions of cleavage to Markownikov addition varied with the sulphenyl halides.[130a] Other reagents, e.g. thiols, hydrogen sulphides, Cl_3C–X (X = Cl or Br) and perfluoroalkyl iodides, add to the vinyl groups of vinyl-tins in homolytic reactions.[1,4] Epoxidation[130b] and polymerization of vinyl-tins also results.

$$Ph_3SnCH{=}CH_2 + o\text{-}O_2NC_6H_4SCl \rightarrow Ph_3SnCHClCH_2SC_6H_4NO_2\text{-}o + Ph_3SnCl + o\text{-}O_2NC_6H_4SCH{=}CH_2 \quad (5.71)$$

Transmetallations between vinyl-tins and organo-lithiums occur with retention of configuration.

5.2.2.2 *Allylstannanes.* Allyl-tin bonds are among the most reactive carbon–tin bonds; cleavage reactions occur with even mild electrophiles, such as carbonyl compounds, (eqn 5.72).

$$R_3SnCH_2CH{=}CHR' + R^2CHO \rightarrow R_3SnOCHR^2CH_2CH{=}CHR' + R_3SnOCHR^2CHR'CH{=}CH_2 \quad (5.72)$$

Both unrearranged and rearranged allylic products arise (eqns 5.72 and 5.73)[131].

$$\underset{trans}{Ph_3SnCH_2CH{=}CHPh} + o\text{-}O_2NC_6H_4SCl \xrightarrow{-Ph_3SnCl}$$

$$o\text{-}O_2NC_6H_4SCH_2\text{–}\underset{trans}{CH{=}CHPh} + o\text{-}O_2NC_6H_4SCHPh\text{–}CH{=}CH_2 \quad (5.73)$$

Protonolyses of allyl-tin bonds have been found to occur γ-regio- and anti-stereospecifically,[98,132–134] e.g. Scheme 17.

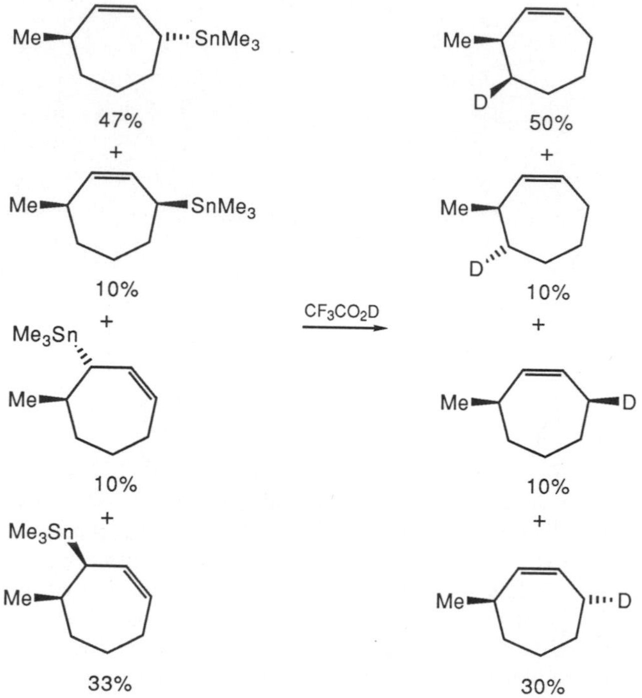

47% + 10% + 10% + 33% $\xrightarrow{CF_3CO_2D}$ 50% + 10% + 10% + 30%

Scheme 17

Sulphur dioxide reactions with allyl-tin compounds are also γ-regiospecific but exhibit *syn* stereospecificity (eqn 5.74).[133,135]

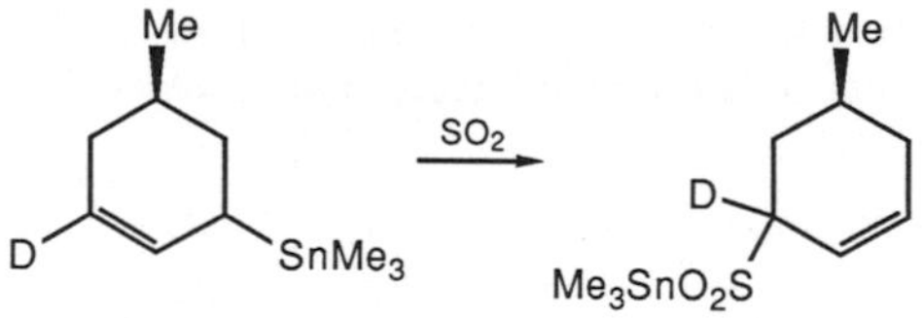

$\xrightarrow{SO_2}$ Me_3SnO_2S

cis : trans 29 : 71 *cis : trans* 25 : 75 (5.74)

Hydrogen sulphide adds to the double bond in allyl-tin derivatives on UV irradiation at -78°C.[136] Other homolytic reactions of allyl-stannanes lead to cleavage, e.g. by RS–SR, PhSe–SePh, RSO_2–Cl and R–X (Q–Y) (eqn 5.75).[98,137]

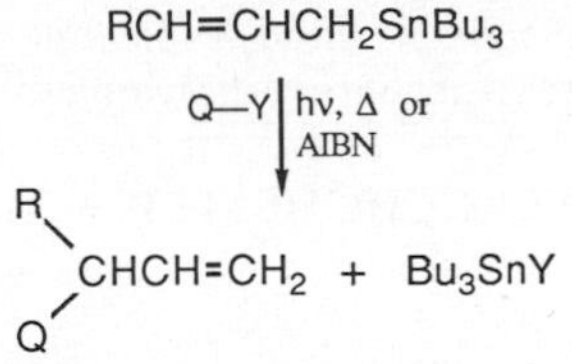

$$RCH{=}CHCH_2SnBu_3 \xrightarrow[\text{Q–Y}]{h\nu,\ \Delta \text{ or AIBN}} R(Q)CHCH{=}CH_2 + Bu_3SnY \quad (5.75)$$

5.2.2.3 *3-Butenylstannanes.* Reactions of 3-butenylstannanes with many electrophilic reagents result in carbon–tin bond cleavage and formation of cyclopropylmethyl derivatives (Scheme 18).[138,139] In contrast, an adduct, $Ph_3SnCH_2CHCHClCH_2SC_6H_4NO_2$-*o*, was isolated from the interaction of $Ph_3SnCH_2CH_2CH{=}CH_2$ with *o*-$O_2NC_6H_4SCl$ at room temperature.[140] More vigorous conditions can, however, lead to elimination in nitrobenzenesulphenyl chloride reactions,[139] as shown in eqn 5.76.

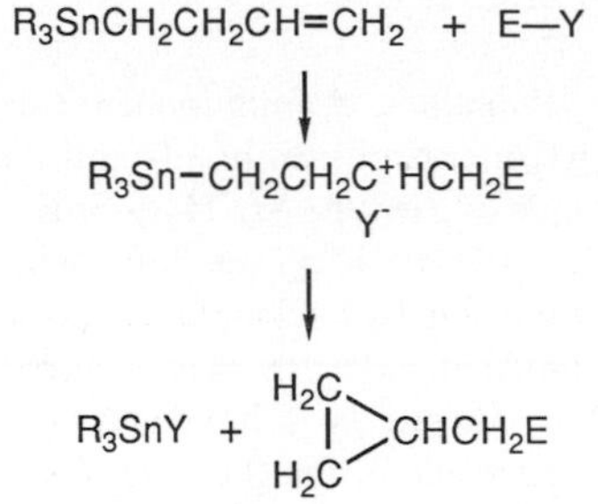

$$R_3SnCH_2CH_2CH{=}CH_2 + E{-}Y$$
$$\downarrow$$
$$R_3Sn{-}CH_2CH_2C^{+}HCH_2E \quad Y^{-}$$
$$\downarrow$$
$$R_3SnY + (H_2C)_2{>}CHCH_2E \text{ (cyclopropyl)}$$

E–Y = Cl–Cl; Br–Br, I–I, Ar–S–Cl etc.

Scheme 18

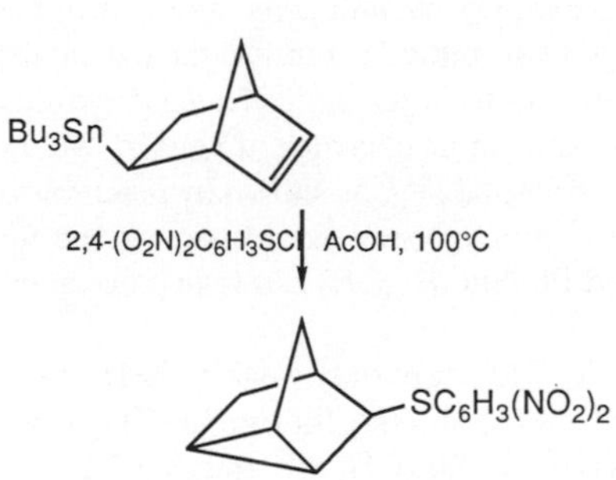

$$Bu_3Sn\text{-norbornenyl} \xrightarrow[\text{AcOH, 100°C}]{2,4\text{-}(O_2N)_2C_6H_3SCl} \text{nortricyclyl-}SC_6H_3(NO_2)_2 \quad (5.76)$$

5.2.2.4 *Propargyl- and allenyl-stannanes.* Propargyl- and allenyl-tin bonds are reactive bonds, and cleavage reactions occur with many electrophilic species, including carbonyl compounds (see eqn 5.77)[98,141]. Reactions proceed with complete rearrangement for substituted derivatives (e.g. eqns 5.77 and 5.78) while unsubstituted species provide equilibrium mixtures of allenyl-propargyl derivatives.

$$Me_3SnCH{=}C{=}CHMe \xrightarrow[\text{(ii) } H_2O]{\text{(i) } Cl_3CCHO} Cl_3CCHOHCMeHC{\equiv}CH \quad (5.77)$$

$$(Me_3Sn)(Me)C{=}C{=}CHMe \xrightarrow[\text{Ref. 142}]{Br_2} MeCHBrC{\equiv}CMe \quad (5.78)$$

Propargyltriphenylstannanes undergo S_H2' substitution reactions with sulphonyl chlorides or polychloromethanes, alkyl bromides or iodides under UV irradiation or AIBN initiation, e.g. eqn 5.79.[137,143]

$$RC{\equiv}CCH_2SnPh_3 + Q{-}Cl \xrightarrow{h\nu} QCR{=}C{=}\dot{C}H_2 + Ph_3SnCl \quad (5.79)$$

$R = H$, or Me; $Q = RSO_2$, CCl_3 or $CHCl_2$

5.2.2.5 *Functionally-substituted alkylstannanes.* A functional substituent in an alkyltin derivative can affect the reactivity of carbon–tin bonds and the type of reaction undergone. For example, in compounds of the type $X(CH_2)_n{-}SnR_3$, the substituent X can not only affect the reactivity of the $X(CH_2)_n{-}Sn$ bond (e.g. by its electronic influence) but also that of the Sn–R bond (e.g. by nucleophilic assistance by X at the tin centre). Reaction can also occur at X and lead to such processes as reduction, oxidation, 'onium salt formation and substitution without Sn–C bond cleavage, as well as leading to cleavage (*via* β or γ elimination) reactions of the $Sn{-}(CH_2)_nX$ bond.

When a strongly electron-withdrawing group X is at the α-carbon, cleavage of that tin–carbon bond is greatly facilitated. Cleavages of $Sn{-}CH_2X$ ($X = COR$, CO_2R, CH, etc.) occur readily with both basic nucleophiles and various electrophiles,[66] including aldehydes.[144] Reactions of such $Sn{-}CH_2X$ bonds occur much more readily than those of simple alkyl-tin bonds, e.g. as shown by halogens and proton acids.

α-Substituents such as COR, CO_2R and CN stabilize the partial negative charge on carbon in the transition state leading to heterolytic cleavage. α-Sulphido groups also stabilize carbonic centres, and electrophilic cleavage of $Sn{-}CH_2SR'$ bonds occur more readily than that of simple alkyl-tin bonds[145], as shown by reactions of Me_3SnCH_2SR with HCl, $HgCl_2$ and halogens, which occur exclusively at the $Sn{-}CH_2SR$ bond. Reactions of Ph_3SnCH_2SR (and Ph_3SnCH_2SeR) can take place at either or both tin–carbon bonds.[145]

Generally for α-substituents (X), which are only weakly electron-withdrawing or are electron-releasing groups, there is less tendency for the $Sn{-}CH_2Y$ bond to be cleaved, although $Sn{-}CH_2OR$, $Sn{-}CH_2NR_2$, $Sn{-}CH_2SR$ and $Sn{-}CH_2SeR$ bonds are all readily cleaved by organolithium reagents. Substitutions can occur at the Y group, e.g. eqns (5.80), (5.81). Ammonium salts have also been generated[142,149] (eqn 5.82).

$$R_3SnCH_2Y \xrightarrow[\text{HX, base}]{X^- \text{ or}} R_3SnCH_2X + Y^- \quad (5.80)$$

$Y = $ halide, $X^- = $ halide[146], NR^1_2, SR^1, OR^1 or SeR^1 [147]

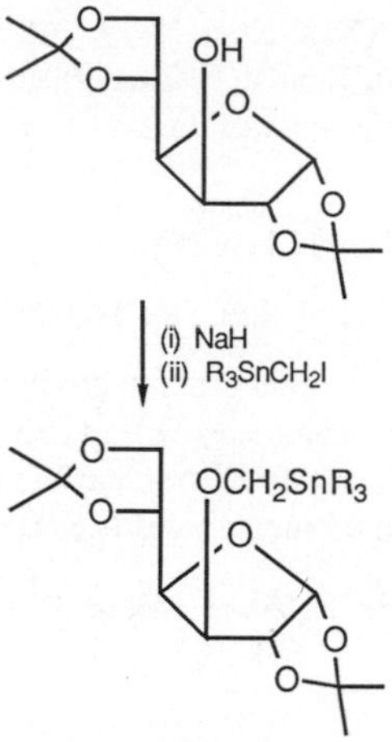

(i) NaH
(ii) R_3SnCH_2I

OCH_2SnR_3

(5.81)

$$R_3SnCH_2NMe_2 \xrightarrow{MeI} R_3SnCH_2\overset{+}{N}Me_3, I^- \qquad (5.82)$$

α-Haloalkyltins also can be used to form the corresponding Grignard and lithium reagents.

Initial attack at the substituent, Y, can lead to subsequent cleavage of the Sn–CH_2X bond. Two examples from α-sulphide substituents[150] are shown in Schemes 19 and 20, in which intermediate sulphonium salts undergo reaction at the tin centre by the halide ions.

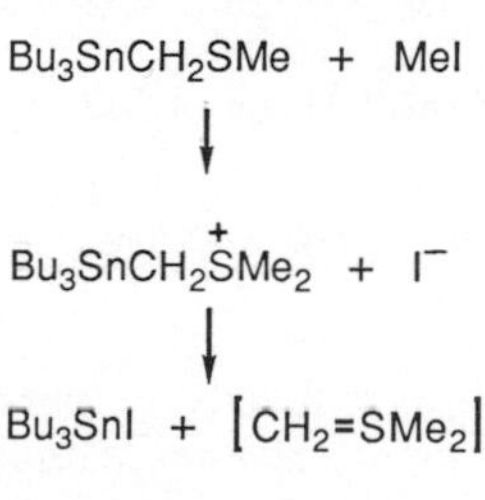

$$Bu_3SnCH_2SMe + MeI$$
$$\downarrow$$
$$Bu_3SnCH_2\overset{+}{S}Me_2 + I^-$$
$$\downarrow$$
$$Bu_3SnI + [CH_2{=}SMe_2]$$

Scheme 19

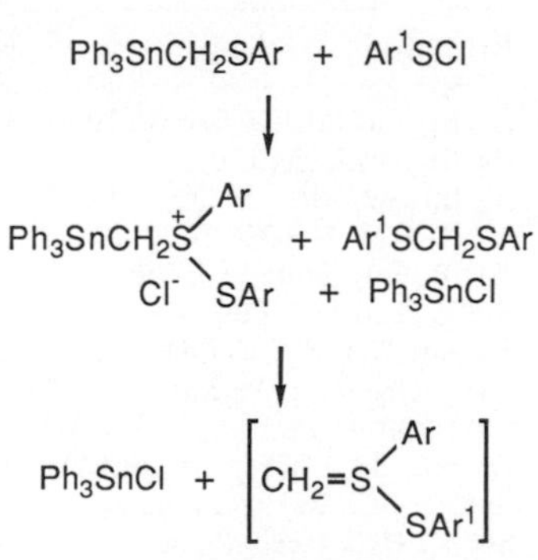

$$Ph_3SnCH_2SAr + Ar^1SCl$$
$$\downarrow$$
$$Ph_3SnCH_2\overset{+}{S}(Ar)(SAr)\ Cl^- + Ar^1SCH_2SAr + Ph_3SnCl$$
$$\downarrow$$
$$Ph_3SnCl + [CH_2{=}S(Ar)(SAr^1)]$$

Scheme 20

Electron-withdrawing groups (X) such as X = COR, CO_2R and CN, in β- or more remote sites, as a consequence of their decreased electronic effects at the carbon–tin bond, have no influence on the reactivity of the $Sn{-}(CH_2)_nX$ bond; compare eqns (5.83) and (5.84).[66,145]

$$Bu_3SnCH_2CN + I_2 \rightarrow Bu_3SnI + [ICH_2CN] \quad (5.83)$$

$$Bu_3SnCH_2CH_2CN + Br_2 \rightarrow Bu_2Sn(Br)CH_2CH_2CN + BuBr \quad (5.84)$$

However, such β-substituents can affect the reactivity of the other carbon–tin bonds as a consequence of nucleophilic assistance (e.g. eqn (5.85)). The nucleophilic assistance of the C=O group in the transition state of the cleavage reaction renders it possible for the methyl-tin bond to be cleaved by such a weakly electrophilic reagent as Me_3SnCl.[69]

$$Me_3SnCH_2CH_2CO_2Me + Me_3SnCl \rightarrow Me_2Sn(Cl)\text{—}CH_2CH_2C(=O\rightarrow Sn)OMe + Me_4Sn \quad (5.85)$$

Reactions of $R_3Sn(CH_2)_nY$ ($n > 2$; Y = COR, CO_2R or CN) do occur with nucleophiles or basic reagents, e.g. reduction using $LiAlH_4$, hydrolysis by NaOH and alkylations using Grignard reagents, e.g. eqns 5.86–5.89[66] (see also Tables 5.2, 5.7).

$$R_3Sn(CH_2)_nCO_2R \xrightarrow{LiAlH_4} R_3Sn(CH_2)_nCH_2OH \quad (5.86)$$

$$R_3Sn(CH_2)_nCN \xrightarrow[\text{(ii) MeI}]{\text{(i) NaOH}} R_3Sn(CH_2)_nCO_2Me \quad (5.87)$$

$$R_3Sn(CH_2)_nCN \xrightarrow[\text{(ii) } H_2O]{\text{(i) MeMgI}} R_3Sn(CH_2)_nCOMe \quad (5.88)$$

$$R_3Sn(CH_2)_nCOMe \xrightarrow[\text{(ii) } H_2O]{\text{(i) PhMgBr}} R_3Sn(CH_2)_nCPhMeOH \quad (5.89)$$

Table 5.7 Carbonyl- and carboxyl-substituted alkyltins.

Compound	Reagents	Ref.
Bu_3SnCOR	(*a*) $Bu_3SnMgCl$, excess RCHO	1
	(*b*) Bu_3SnLi, EtCOBr	
	(*c*) $Bu_3SnSnBu_3$, EtCOCl, Pd^0	
R_3SnCH_2COR'	(*a*) R_3SnH, N_2CHCOR'	2
	(*b*) $(R_3Sn)_2$, $Hg(CH_2COR')_2$	3
$Cl_3SnCHMeCOEt$	$SnCl_4$, $Et(Me_3SiO)C{=}CHMe$	4
$Bu_3SnCH_2CH_2CHO$	$Bu_3SnCH_2CH_2CH_2OH$, NBS, Me_2S	5
$Pr_3SnCH_2CH_2COMe$	$CH_2{=}CHCN$ (i) Pr_3SnH	6
	(ii) MeMgI	
$Cl_3SnCH_2CH_2COBu^t$	$Bu^t(Me_3SiO)\overline{CCH_2CH_2}$, $SnCl_4$	7
$Cl_3SnCMe_2CH_2COMe$	$SnCl_2$, HCl, Me_2CO (excess)	8
	$SnCl_2$, HCl, mesityl oxide	9
$Me_3Sn(CH_2)_{n+2}COR$	(*a*) $n = 0, 1, 2$; (i) Me_3SnH, $CH_2{=}CH(CH_2)_nCHOHMe$	
	R = Me (ii) CrO_3, Me_2CO	10

Table 5.7 (*Contd.*)

Compound	Reagents	Ref.
	(*b*) $n = 0, 1$; (i) Me_3SnH, $CH_2{=}CH(CH_2)_nCN$ $R = Me$ or Ph (ii) RMgX	10
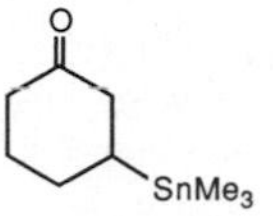 O, $SnMe_3$	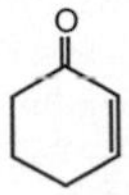 O, , Me_3SnLi	11
$Ph_3Sn(CH_2)_4COMe$	Ph_3SnH, $CH_2{=}CH(CH_2)_2COMe$	12
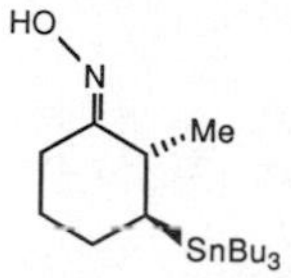 HO, N, Me, $SnBu_3$	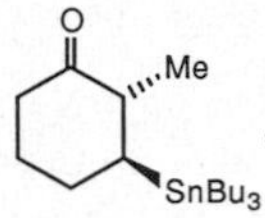 O, Me, $SnBu_3$, $HONH_2.HCl$	13
$R_3Sn(CH_2)_3CMe{=}NOH$	$R_3Sn(CH_2)_3COMe$, $HONH_2{\cdot}HCl$	14
$R_3SnCHR'CO_2R^2$	R_3SnCl, $NaCHR'CO_2R^2$	15
$R_2Sn(CHR'CO_2R^2)_2$	R_2SnCl_2, $NaCHR'CO_2R^2$	15
$X_2Sn(CH_2CO_2R)_2$ ($X = Br$ or I)	XCH_2CO_2R, Sn/Cu	16, 17
$Cl_3SnCH_2CH_2CO_2R$ ($R = H$, alkyl, aryl)	$SnCl_2$, HCl, $CH_2{=}CHCO_2R$	9
$Cl_2Sn(CH_2CH_2CO_2R)_2$	Sn, HCl, $CH_2{=}CHCO_2R$	9
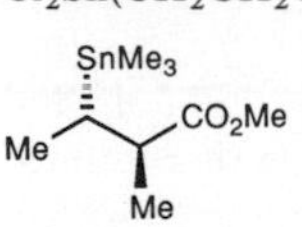 $SnMe_3$, CO_2Me, Me, Me	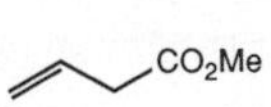 CO_2Me (i) Me_3SnLi (ii) MeI	18
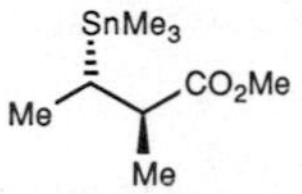 $SnMe_3$, CO_2Me, Me, Me	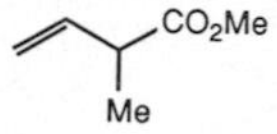 CO_2Me, Me (i) Me_3SnLi (ii) H_3O^+	18
$Me_2RSnCHR^2CHR'CO_2Me$ ($R = Cl$, Me)	Me_2RSnH, $CHR^2{=}CR'CO_2Me$	19
$R_3Sn(CH_2)_2COR'$	R_3SnH, $CH_2{=}CHCO_2R$	20, 21
$Ph_3Sn(CH_2)_3COR'$	(*a*) Ph_3SnH, $CH_2{=}CHCH_2CO_2R'$	6
	(*b*) Ph_3SnNa, $Br(CH_2)_3CO_2R'$	18, 22
$Ph_3Sn(CH_2)_4COR'$	(*a*) Ph_3SnH, $CH_2{=}CH(CH_2)_2CO_2R'$	23
	(*b*) Ph_3SnNa, $Br(CH_2)_4CO_2Et$	18
$Me_3Sn(CH_2)_3CO_2H$	$Me_3Sn(CH_2)_3Br$, (i) Mg (ii) CO_2	24
$Ph_3Sn(CH_2)_3CO_2H$	$Ph_3Sn(CH_2)_3CN$, NaOH	23
$Ph_3Sn(CH_2)_4CO_2H$	Ph_3SnH (i) $CH_2{=}CH(CH_2)_2CO_2Me$ (ii) NaOH	6
$Ph_3SnCH_2CONEt_2$	Ph_3SnCl, $BrZnCH_2CONEt_2$	25
$Ph_3SnCH_2CH_2CONH_2$	Ph_3SnH, $CH_2{=}CHCONH_2$	12
$Ph_3SnCH(COR)_2$ ($R = Me$, OEt)	$Hg[CH(COR)_2]_2$, $(Ph_3Sn)_2S$	17

1. M. Kosugi, N. Naka, H. Sano and T. Migita, *Bull. Chem. Soc. Jpn*, 1987, **60**, 3462.
2. M. Lesbre and R. Buisson, *Bull. Soc. Chim. France*, 1957, 1204.
3. D.H. Nguyen, V.S. Fainberg, Yu.I. Baukov and I.F. Lutsenko, *Zh. Obshch. Khim.*, 1968, **38**, 191.

Table 5.7 (*Contd.*)

4. E. Nakamura and J. Kuwajima, *Chem. Lett.*, 1983, 59.
5. Y. Ueno, M. Ohta and M. Okawara, *Tetrahedron Lett.*, 1982, 2577.
6. J.G. Noltes and G.J.M. van der Kerk, *Functionally Substituted Organotin Compounds*, Tin Research Institute, Greenford, UK, 1958.
7. I. Ryu, S. Murai and N. Sonoda, *J. Org. Chem.*, 1986, **51**, 2389.
8. J.W. Burley, P. Hope and A.G. Mack, *J. Organomet. Chem.*, 1984, **272**, 37.
9. R.E. Hutton, J.W. Burley and V. Oakes, *J. Organomet. Chem.*, 1978, **156**, 369.
10. H.G. Kuivila, P.L. Maxfield, K.-H. Tsai and J.E. Dixon, *J. Amer. Chem. Soc.*, 1976, **98**, 104; H.G. Kuivila, J.E. Dixon, P.L. Maxfield, N.M. Scarpa, T.M. Topka, K.-H. Tsai and K.R. Wursthorn, *J. Organomet. Chem.*, 1975, **86**, 89.
11. K.R. Wursthorn and H.G. Kuivila, *J. Organomet. Chem.*, 1977, **140**, 29.
12. G.J.M. van der Kerk and J.G. Noltes, *J. Appl. Chem.*, 1959, **9**, 106.
13. A. Nishiyama, H. Arai, T. Ohki and K. Itoh, *J. Amer. Chem. Soc.*, 1985, **107**, 5310.
14. S.Z. Abbas and R.C. Poller, *J. Organomet. Chem.*, 1976, **104**, 187.
15. A.N. Kashin, M.L. Tul'chinskii, I.P. Beletskaya and O.A. Reutov, *J. Org. Chem. USSR*, 1984, **20**, 1467; A.A. Prishcenko, Z.S. Novikova and I.F. Lutsenko, *Zh. Obshch. Khim.*, 1981, **51**, 485.
16. B. Emmert and W. Eller, *Ber. Bunsenges. Phys. Chem.*, 1911, **44**, 2328.
17. E.S. Paterson, PhD thesis, University of Aberdeen, 1983.
18. I. Fleming and M. Rowley, *Tetrahedron Lett.*, 1985, **26**, 3857.
19. A.B. Chopa, L.C. Koll, M.C. Savini, J.C. Podesta and W.P. Neumann, *Organometallics.*, 1985, **4**, 1036.
20. W.P. Neumann, H. Niemann and R. Sommer, *Annalen*, 1962, **659**, 27.
21. G.J.M. van der Kerk, J.G. Noltes and J.G.A. Luijten, *J. Appl. Chem.*, 1957, **7**, 356.
22. R.A. Howie, E.S. Paterson, J.L. Wardell and J.W. Burley, *J. Organomet. Chem.*, 1983, **259**, 71.
23. G.J.M. van der Kerk and J.G. Noltes, *J. Appl. Chem.*, 1959, **9**, 113.
24. E.J. Bulten, H.F.M. Gruter and H.F. Martens, *J. Organomet. Chem.*, 1976, **117**, 329.
25. R.C. Poller and D. Silver, *J. Organomet. Chem.*, 1978, **157**, 247.

Organotin compounds containing such β-substituents as OH, OR, SR, NR_2 or PR_2, undergo ready β-eliminations on reaction with certain electrophiles or on thermolysis, and examples are shown in eqns (5.90)–(5.93). 2-Acetoxyvinyl compounds eliminate acetylene on heating, with the triphenyltin derivative being more thermally stable than is the methyl analogue.[66]

$$Ph_3SnCH_2CH_2OH \xrightarrow{H_3O^+, H_2O} Ph_3SnOH + CH_2{=}CH_2 \qquad (5.90)^{57}$$

$$Ph_3SnCH_2CH_2SAr + E{-}Y \rightarrow Ph_3SnY + E{-}SAr + CH_2{=}CH_2 \qquad (5.91)^{152}$$

E–Y: Br–Br; I–I; Ar′S–Cl

$$Ph_3SnCH_2CH_2NMe_2 + E{-}Y \rightarrow Ph_3SnY + E{-}NMe_2 + CH_2{=}CH_2$$

$E{-}Y = ArSO_2{-}Cl$, ArS–Cl, RCO–Cl, Me–I (5.92)[149,153,154]

$$Et_2Sn(CH_2CH_2P(=S)(S^-)Ph)(CH_2CH_2NH_3^+) \xrightarrow{\Delta} Et_2Sn\overset{CH_2CH_2}{\underset{S}{\frown}}P(=S)Ph \qquad (5.93)^{155}$$

Alkene elimination can also occur on treatment of alkyltins, bearing β-hydrogens, for example on reaction with the hydride abstracting, agent Ph_3C^+, BF_4^-,[156] (eqn (5.94)) which proceeds by *trans*-elimination.

$$Me_3SnCHRCH_2R' + Ph_3C^+BF_4^- \rightarrow Me_3SnBF_4 + Ph_3CH + RCH{=}CHR' \qquad (5.94)$$

Substituents in positions more remote than beta from tin in $R_3Sn–(CH_2)_nY$ do not influence the reactivity of the $Sn–(CH_2)_nY$ bond unless γ-eliminations are possible. γ-Hydroxyalkyltin compounds undergo interesting stereospecific eliminations to give cyclopropanes on treatment with acids[157], e.g. eqn (5.95), or with thionyl chloride.[158]

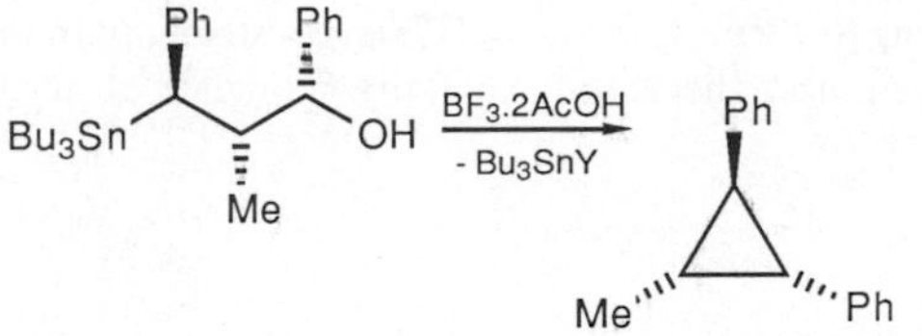

$$Bu_3Sn\text{–}CH(Ph)\text{–}CH(Me)\text{–}CH(Ph)\text{–}OH \xrightarrow[-\ Bu_3SnY]{BF_3.2AcOH} \text{Me, Ph, Ph-cyclopropane} \quad (5.95)$$

As already pointed out (e.g. eqn 5.85) substituents in one bond can increase the reactivity of the remaining R–Sn bonds (by nucleophilic assistance). The remote substituents Y can of course undergo their usual substitution, addition, oxidation or reduction chemistry, providing the reagents necessary for such reactions do not also react at tin–carbon bonds. Nucleophilic assistance has been found in a number of reactions, including reactions of $Ph_3Sn(CH_2)_nS(O)_mAr$ ($n = 2$, 3 or 4; $m = 0$, 1 or 2) with iodine. In particular, strong nucleophilic assistance was established for sulphoxide groups,[159] eqn 5.96.

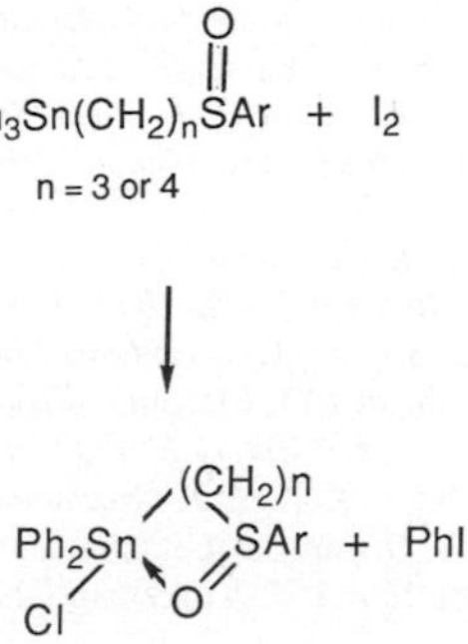

$$Ph_3Sn(CH_2)_nS(=O)Ar + I_2 \quad (n = 3 \text{ or } 4) \longrightarrow Ph_2Sn(Cl)(CH_2)_nS(Ar)=O\rightarrow Sn + PhI \quad (5.96)$$

Quite dramatic changes in reactivities have been observed[160] due to intramolecular nucleophilic assistance, including examples where alkyltin bonds are cleaved in preference to aryl- or vinyl-tin bonds, e.g. eqn 5.97 (cf. eqn 5.98).

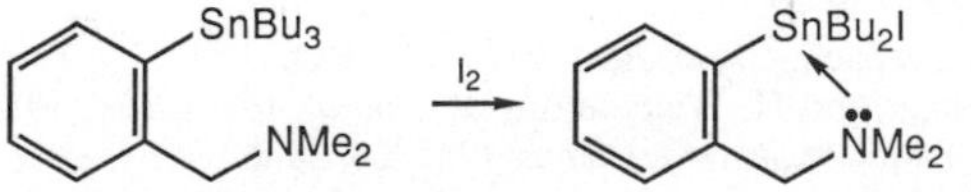

$$2\text{-}(SnBu_3)C_6H_4CH_2NMe_2\ (\mathbf{16}) \xrightarrow{I_2} 2\text{-}(SnBu_2I)C_6H_4CH_2\ddot{N}Me_2 \quad (5.97)$$

$SnBu_3$ $\xrightarrow{I_2}$ I

NMe_2 NMe_2

(**17**) (5.98)

The potential chelate ring (6-membered) from (**17**) is less strong than that formed from (**16**) (a 5-membered ring), and the usual reactivity sequence of aryltin > alkyltin is found.

References

1. R.C. Poller, *Chemistry of Organotin Compounds*, Logos, London, 1970.
2. W.P. Neumann, *Organic Chemistry of Tin*, Wiley, New York, 1970.
3. A.G. Davies and P.J. Smith, in *Comprehensive Organometallic Chemistry*, eds. G. Wilkinson, F.G.A. Stone and E.W. Abel, Pergamon, Oxford, 1982, Chap. 11.
4. J.G.A. Luijen and G.J.M. van der Kerk, in *Organometallic Compounds of the Group IV Elements*, Marcel Dekker, New York, 1968, Vol. 1, Part 11, p. 91; M. Gielen and J. Nasielski, in *Organotin Compounds*, ed. A.K. Sawyer, Marcel Dekker, New York, 1972, Vol. 3, Chap. 9.
5. M. Gielen, *Rev. Si, Ge, Sn, and Pb Cpds.*, 1981, **5**, 5.
6. J. Murphy and R.C. Poller, *J. Organomet. Chem. Lib.*, 1979, **9**, 189.
7. J. Kizlink, *Chem. Lisky*, 1984, **78**, 134.
8. T.W. Lapp, *The Manufacture and Use of Selected Alkyltin Compounds*, US. Environ. Protect. Agency Rept, No. EPA 560/6-6-76-011, March, 1976, 21.
9. F.S. Holland, *Appl. Organomet. Chem.*, 1987, **1**, 185.
10. E. Emmert and W. Eller, *Ber. Bunsenges. Phys. Chem.*, 1911, **44**, 2328.
11. I. Omae, S. Matsuda, A. Kirkkawa and R. Sato, *Kogyo Kagaku Zasshi.*, 1967, **70**, 705; M.S. Nomura, S. Matsuda, and S. Kirkkawa, *Kogyo Kagaku Zasshi.*, 1967, **70**, 710; T. Hayashi, J. Uchimura, S. Matsuda and S. Kirkkawa, *Kogyo Kagaku Zasshi*, 1967, **70**, 714.
12. K. Sisido, Y. Takeda and Z. Kinugawa, *J. Amer. Chem. Soc.*, 1961, **83**, 538; K. Sisido, S. Kozima and T. Hanada, *J. Organomet. Chem.*, 1967, **9**, 99; K. Sisido and S. Kozima, *J. Organomet. Chem.*, 1968, **11**, 503.
13. J.J. Habeeb and D.G. Tuck, *J. Organomet. Chem.*, 1977, **134**, 363.
14. G. Mengoli, *Rev. Si, Ge, Sn, and Pb Cpds.*, 1979, **4**, 59; H. Ulery, *J. Electrochem. Soc.*, 1972, **119**, 1474; M. Devaud, *Rev. Silicon, Germanium, Tin and Lead Cpds.*, 1976, **11**, 87.
15. M. Fleischmann, G. Mengoli and D. Fletcher, *Electrochim. Acta*, 1973, **18**, 231.
16. T.G. Jakobsen and V.P. Petrov, *Izv. Sib. otd. Akad. Nauk, SSR, Ser. Khim. Nauk.*, 1965, **2**, 75.
17. R.E. Hutton, J.W. Burley and V. Oakes, *J. Organomet. Chem.*, 1978, **156**, 369; R.E. Hutton, V. Hutton and V. Oakes, in *Organotin Compounds: New Chemistry and Applications*, ed. J.J. Zuckerman, Adv. Chem. Ser. **157**, American Chemical Society, Washington DC, 1976, 123.
18. E.J. Bulten, *J. Organomet. Chem.*, 1975, **97**, 167; E.J. Bulton, H.F.M. Cruter and H.F. Martens, *J. Organomet. Chem.*, 1976, **117**, 329.
19. G. van Koten, J.T.B.H. Jastrzebeski, J.G. Noltes, G.J. Verhoeckx, A.L. Spek and J. Kroon, *J. Chem. Soc. Dalton Trans.*, 1980, 1352.
20. E.-I. Negishi, *Organometallics in Organic Synthesis*, Wiley, New York, Vol. 1, 1980, 404.
21. Netherlands Patent, 6505, 767 (*Chem. Abs.*, 1966, **64**, 12723).
22. J.L. Wardell, Preparation and Use in Organic Synthesis of Organolithium and Group IA Organometallics, in *The Chemistry of the Metal–Carbon Bond*, ed. F.R. Hartley, Vol. 4, Part 1, (1987) Chap. 1.
23. B.T. Grobel and D. Seebach, *Chem. Ber.*, 1977, **119**, 852.
24. H.P. Abricht, C. Mugge and H. Weichmann, *Z. Anorg. Allg. Chem.*, 1980, **467**, 203.
25. W. Barth and L.A. Paquette, *J. Org. Chem.*, 1985, **50**, 2438.
26. D. Seyferth and S.B. Andrews, *J. Organomet. Chem.*, 1971, **30**, 151.
27. R.C. Poller and D. Silver, *J. Organomet. Chem.*, 1978, **157**, 247.

28. D.H. Nguyen, V.S. Fainberg, Yu.I. Baukov and I.F. Lutsenko, *Zh. Obshch. Khim.*, 1968, **38**, 191.
29. E.S. Patterson, PhD thesis, University of Aberdeen, 1983.
30. J.-P. Quintard and M. Pereye, *Rev. Si, Ge, Sn, and Pb Cpds.*, 1980, **4**, 153.
31. H.G. Kuivila, in *Organotin Compounds: New Chemistry and Applications*, ed. J.J. Zuckermann, Adv. Chem. Ser. **157**, American Chemical Society, Washington DC, 1976, 41.
32. M. Newcomb and M.G. Smith, *J. Organomet. Chem.*, 1982, **228**, 61.
33. R.M.G. Roberts, *J. Organomet. Chem.*, 1973, **63**, 159.
34. E.C. Ashby and T.N. Pham, *Tetrahedron Lett.*, 1987, **28**, 3183; see also M.S. Alnajjar and H.G. Kuivila, *J. Amer. Chem. Soc.*, 1985, **107**, 416; M.S. Alnajjar, C.F. Smith and H.G. Kuivila, *J. Org. Chem.*, 1984, **49**, 1271; W. Adcock, V.S. Iyer, W. Kitching and D. Young, *J. Org. Chem.*, 1985, **50**, 5706; M. Newcombe and M.G. Smith, *J. Organomet. Chem.*, 1982, **228**, 61; E.C. Ashby, W.Y. Su and T.N. Pham, *Organometallics*, 1985, **4**, 1493.
35. K. Sisido, K. Ban, T. Tsida and S. Kozima, *J. Organomet. Chem.*, 1971, **29**, C7.
36. R.D. Taylor and J.L. Wardell, *J. Organomet. Chem.*, 1974, **77**, 311.
37. E.S. Paterson, PhD thesis, University of Aberdeen, 1983.
38. J. San Filippo, J. Silbermann and P.J. Fagan, *J. Amer. Chem. Soc.*, 1978, **100**, 4834.
39. W. Kitching, H. Olszouwy, J. Waugh and D. Dodarel, *J. Org. Chem.*, 1978, **43**, 898.
40. G.S. Koermer, M.L. Hall and T.G. Taylor, *J. Amer. Chem. Soc.*, 1972, **94**, 7205.
41. J.L. Wardell and C. Welsh, unpublished observations.
42. D.J. Peterson and J.F. Ward, *J. Organomet. Chem.*, 1974, **66**, 209; E.W. Abel and R.J. Rowley, *J. Organomet. Chem.*, 1975, **97**, 159.
43. K.R. Wursthorn and H.G. Kuivila, *J. Organomet. Chem.*, 1977, **140**, 29.
44. J.P. Quintard, S. Havvette and M. Pereyre, *J. Organomet. Chem.*, 1978, **159**, 147.
45. E.S. Sivenkov, V.S. Zavgorodini and A.A. Petrov, *Zh. Obshch. Khim.*, 1969, **39**, 2673; M. LeQuan and P. Cadiot, *Bull. Soc. Chim. France*, 1965, 45.
46. G. Dumardin, J.-P. Quintard and M. Pereyre, unpublished observations, quoted in ref. [30].
47. D. Young and W. Kitching, *J. Org. Chem.*, 1985, **50**, 4098; see for related systems, D. Young, M. Jones and W. Kitching, *Austr. J. Chem.*, 1986, **39**, 563.
48. G.J.D. Peddle, *J. Organomet. Chem.*, 1968, **14**, 139; see also E. Linder and V. Kunge, *J. Organomet. Chem.*, 1970, **21**, 19.
49. W.C. Still, *J. Amer. Chem. Soc.*, 1978, **100**, 1481.
50. A. Ricci, M. Taddei and G. Seconi, *J. Organomet. Chem.*, 1986, **306**, 23.
51. H.G. Kuivila and F.V. di Stefano, *J. Organomet. Chem.*, 1976, **122**, 171.
52. I. Fleming and C.J. Urch, *J. Organomet. Chem.*, 1985, **285**, 173; see also J.F. Kakow and C.R. Johnson, *Tetrahedron Lett.*, 1984, **25**, 5255; M. Ochiai, T. Ukita, Y. Nagao and E. Fujita, *J. Chem. Soc., Chem. Commun.*, 1985, 637; G.J. McGarvey and J.M. Williams, *J. Amer. Chem. Soc.*, 1985, **107**, 1435; E. Piers and H.E. Morton, *J. Chem. Soc., Chem. Commun.*, 1971, 1034.
53. O.J. Taylor and J.L. Wardell, *Recl. Trav. Chim. Pays-Bas*, 1988, see also L.D. Hall, D.C. Miller and P.R. Steiner, *Carbohydr. Res.*, 1976, **52**, C1.
54. H. Gilman and S.D. Rosenberg, *J. Org. Chem.*, 1953, **18**, 1554.
55. A. Mordini, M. Taddei and G. Seconi, *Gazz. Chem. Ital.*, 1986, **116**, 238.
56. A. Ricci, M. Taddei and G. Seconi, *J. Organomet. Chem.*, 1986, **306**, 23; see also R.H. Fish and B.M. Broline, *J. Organomet. Chem.*, 1978, **159**, 255.
57. D.D. Davis and C.E. Gray, *J. Org. Chem.*, 1970, **35**, 1303.
58. S. Matsubara, J.-J. Hibino, Y. Morizawa, K. Oshima and H. Nozaki, *J. Organomet. Chem.*, 1985, **285**, 163.
59. T.N. Mitchell, A. Amamria, H. Killing and D. Putschow, *J. Organomet. Chem.*, 1986, **304**, 257, and refs. cited therein.
60. E. Piers and R.T. Skerlj, *J. Chem. Soc., Chem. Commun.*, 1986, 626
61. T.N. Mitchell, H. Killing, R. Dicke and R. Wickenkamp, *J. Chem. Soc., Chem. Commun.*, 1985, 354.
62. B.L. Chenard, E.D. Laganis, F. Davidson and T.V. RajenBabu, *J. Org. Chem.*, 1985, **50**, 3666.
63. H. Killing and T.N. Mitchell, *Organometallics*, 1984, **3**, 1163.
64. B.L. Chenard, *Tetrahedron Lett.*, 1986, **27**, 2805.
65. A.J. Lensink, Hydrostannation. A Mechanistic Study. PhD thesis, University of Utrecht, 1966.

66. J.G. Noltes and G.J.M. van der Kerk, *Functionally Substituted Organotin Compounds*, Tin Research Institute, Greenford, UK, 1958.
67. H.G. Kuivila, *Adv. Organomet. Chem.*, 1964, **1**, 47.
68. H.G. Kuivila, J.D. Kennedy, R.Y. Tien, I.J. Tyminiski, F.L. Pelczar and D.R. Khan, *J. Org. Chem.*, 1971, **36**, 2083.
69. A.B. Chopa, L.C. Koll, M.C. Savini, J.C. Podesta and W.P. Neumann, *Organometallics*, 1985, **4**, 1036.
70. (*a*) T.N. Mitchell, W. Reimann and C. Nettelbeck, *Organometallics*, 1985, **4**, 1044; (*b*) E.J. Bulten and H.A. Budding, *J. Organomet. Chem.*, 1976, **111**, C33.
71. J.L. Wardell and J. McM. Wigzell, *J. Organomet. Chem.*, 1983, **244**, 225.
72. M.C. Henry and J.G. Noltes, *J. Amer. Chem. Soc.*, 1960, **82**, 561; see also A.J. Leusink, J.G. Noltes, H.A. Budding and G.J.M. van der Kerk, *Recl. Trav. Chim.*, 1964, **83**, 1036.
73. M. Gielen and Y. Tondeur, *J. Organomet. Chem.*, 1977, **128**, C25.
74. A. Rahm, M. Degeuil-Castaing and M. Pereyre, *Tetrahedron Lett.*, 1980, **21**, 4649.
75. W.P. Neumann, H.J. Albert and W. Kaiser, *Tetrahedron Lett.*, 1967, 2041; H.G. Kuivila and R. Sommer, *J. Amer. Chem. Soc.*, 1967, **80**, 5616; R. Somer and H.G. Kuivila, *J. Org. Chem.*, 1968, **33**, 802.
76. J. Lorberth, *J. Organomet. Chem.*, 1968, **15**, 251.
77. D. Seyferth and L.G. Vaughan, *J. Organomet. Chem.*, 1963, **1**, 138.
78. A.R. Chamberlin and F.T. Bond, *Synthesis*, 1979, **44**,
79. J.G. Duboudin, M. Petrand, M. Ratier and B. Trouve, *J. Organomet. Chem.*, 1985, **288**, C6; see also G.J. McGarvey and J.S. Bajiva, *J. Org. Chem.*, 1984, **49**, 4091; R. Bloch, J. Abecassis and D. Hassan, *Can. J. Chem.*, 1984, **62**, 2019.
80. M. Shibasaki, Y. Torisawa and S. Ikegami, *Tetrahedron Lett.*, 1982, **23**, 4607.
81. E. Piers, J.M. Chong, K. Gustafson, R.J. Anderson, *Can J. Chem.*, 1984, **62**, 1; H. Westmijze, K. Ruitenberg, J. Meijer and P. Vemeer, *Tetrahedron Lett.*, 1982, 2797; J.-I. Hibino, S. Matsubara, Y. Morizawa, K. Oshima and H. Nozaki, *Tetrahedron Lett.*, 1984, 2151.
82. E. Negishi, *Organometallics in Organic Synthesis*, Vol. 1, Wiley, New York, 1980, 410.
83. E.J. Corey, P. Ulrich and J.M. Fitzpatrick, *J. Amer. Chem. Soc.*, 1978, **98**, 222; P.W. Collins, C.J. Jung, A. Gasiecki and R. Pappo, *Tetrahedron Lett.*, 1978, 3187.
84. K.-H. Chu and K.K. Wang, *J. Org. Chem.*, 1986, **51**, 767; B. Wrackmeyer, *Rev. Si, Ge, Sn, and Pb Cpds.*, 1982, **6**, 75.
85. H. Westmijze, K. Ruitenberg, J. Meijer and P. Vermeer, *Tetrahedron Lett.*, 1982, **23**, 2797.
86. E. Piers and J.M. Chong, *J. Chem. Soc., Chem. Commun.*, 1983, 934.
87. I. Fleming and M. Tadder, *Synthesis*, 1985, 898.
88. E. Piers and J.M. Chong, *J. Org. Chem.*, 1982, **47**, 1602.
89. E. Piers and A.V. Gavai, *J. Chem. Soc., Chem. Commun.*, 1985, 1241.
90. J. Grignon, C. Servens and M. Pereyre, *J. Organomet. Chem.*, 1975, **96**, 225; H. Yatagai, Y. Yamamoto and K. Maruyama, *J. Amer. Chem. Soc.*, 1980, **102**, 4548; E.W. Abel and R.J. Rowley, *J. Organomet. Chem.*, 1975, **84**, 199.
91. B.M. Trost and J.W. Herndon, *J. Amer. Chem. Soc.*, 1984, **106**, 6835.
92. K. Fugami, K. Oshima, K. Utimoto and H. Nozaki, *Bull. Chem. Soc. Jpn*, 1987, **60**, 2509.
93. Y. Ueno, T. Miyano and M. Okawara, *Tetrahedron Lett.*, 1982, **23**, 443; Y. Ueno, H. Sano and M. Okawara, *Tetrahedron Lett.*, 1980, **21**, 1767; Y. Ueno, M. Ohta and M. Okawara, *J. Organomet. Chem.*, 1980, **197**, C1; D.N. Jones and M.R. Peel, *J. Chem. Soc., Chem. Commun.*, 1986, 216; G.E. Keck and E.J. Ebholm, *Tetrahedron Lett.*, 1985, **26**, 3311.
94. U. Schroer and W.P. Neumann, *J. Organomet. Chem.*, 1976, **105**, 183.
95. D. Seyferth, K.R. Wursthorn and R.E. Mammarella, *J. Organomet. Chem.*, 1979, **19**, 25; S.J. Hannon and T.G. Taylor, *J. Chem. Soc., Chem. Commun.*, 1975, 630.
96. V.J. Jephede and E.J. Thomas, *Tetrahedron Lett.*, 1985, **26**, 5327.
97. A. Gambaro, D. Marton, V. Peruzzo and G. Tagliavini, *J. Organomet. Chem.*, 1981, **204**, 191.
98. M. Pereyre, J.-P. Quintard and A. Rahn, *Tin in Organic Synthesis*, Butterworth, London, 1987.
99. O.A. Reutov, *J. Organomet. Chem.*, 1983, **250**, 145.
100. M.H. Abraham, in *Electrophilic Substitution at a Saturated Carbon Atom*, eds. C.H. Bamford and C.F.H. Tipper, Elsevier, Amsterdam, 1973.
101. C. Tamborski, E.J. Soloski and S.M. Dee, *J. Organomet. Chem.*, 1965, **4**, 446.
102. C. Eaborn, H.L. Hornfield and D.R.M. Walton, *J. Chem. Soc., D*, 1967, 1036

103. C. Eaborn and G. Seconi, *J. Chem. Soc. Perkin Trans.*, **2**, 1979, 203.
104. R.M.G. Roberts and F. El Kaissi, *J. Organomet. Chem.*, 1968, **12**, 79.
105. C. Eaborn and D.R.M. Walton, *J. Organomet. Chem.*, 1965, **4**, 217.
106. C. Eaborn, A.R. Thompson, and D.R.M. Walton, *J. Organomet. Chem.*, 1971, **29**, 251; C. Eaborn, *Pure & Appl. Chem.*, 1969, **19**, 375.
107. A.R. Bryce, R.D. Chambers, S.T. Mullins and A. Perkin, *J. Chem. Soc., Chem. Commun.*, 1986, 1623.
108. M.J. Adam, J.J. Ruth, S. Jiran and B.D. Pate, *J. Fluorine Chem.*, 1984, **25**, 329.
109. E.H. Bartlett, C. Eaborn and D.R.M. Walton, *J. Chem. Soc. C*, 1970, 1717.
110. W.A. Asomaning, C. Eaborn and D.R. Walton, *J. Chem. Soc. Perkin Trans.* **1**, 1973, 137.
111. M. Gielen, *Acc. Chem. Res.*, 1973, **6**, 198.
112. D.S. Matteson, *Organometallic Reaction Mechanisms*, Academic Press, New York, 1974.
113. J.M. Fukuto and F.R. Jensen, *Acc. Chem. Res.*, 1983, **16**, 177 and refs. therein.
114. H.A. Olszowy and W. Kitching, *Organometallics*, 1984, **3**, 1676; *J. Org. Chem.*, 1982, **99**, 1672.
115. A. Rahm, J. Grimeau and M. Pereyre, *J. Organomet. Chem.*, 1985, **286**, 305.
116. S. Fukuzumi and J.K. Kochi, *J. Amer. Chem. Soc.*, 1980, **102**, 4397.
117. A.G. Davies, in *Organotin Compounds: New Chemistry and Applications*, ed. J.J. Zuckermann, Adv. Chem. Ser. **157**, American Chemical Society, Washington DC, 1976, 26.
118. R.C. Poller, *Rev. Si, Ge, Sn, and Pb Cpds.*, 1978, **3**, 243.
119. A.G. Davies, B.P. Roberts and J.M. Smith, *J. Chem. Soc. Perkin Trans.* **2**, 1972, 2221.
120. S. Fukuzumi and J.K. Kochi, *J. Org. Soc.*, 1980, **45**, 2654.
121. L. Verdonck, P.H. de Ryck, S. Hoste and G.P. van der Kelen, *J. Organomet. Chem.*, 1985, **288**, 289; P.H. de Ryck, L. Verdonck and G.P. van der Kelen, *Inst. J. Chem. Kinet.*, 1985, **17**, 95.
122. A.G. Davies, B.P. Roberts and M.-W. Tse, *J. Chem. Soc. Perkin Trans. 2*, 1978, 145.
123. R.A. Jackson, K.U. Ingold, D. Griller and A.S. Naznan, *J. Amer. Chem. Soc.*, 1985, **107**, 208.
124. A.G. Davies, B. Muggleton, B.P. Roberts, M.-W. Tse and J.N. Winter, *J. Organomet. Chem.*, 1976, **118**, 289.
125. J.C. Cochran, S.C. Bayer, J.T. Bilbo, M.S. Brown, L.B. Cohen, F.J. Gasparini, D.W. Goldsmith, M.D. Jamin, K.A. Nealy, C.T. Resnick, G.J. Schwartz, W.M. Short, K.R. Skarda, J.P. Spring and W.L. Strauss, *Organometallics*, 1982, **1**, 586.
126. R.N. Hanson and H. El-Wakil, *J. Org. Chem.*, 1987, **52**, 3687.
127. P. Baekelmans, M. Gielen, P. Malfro'd and C. Nasielski, *Bull. Soc. Chim. Belg.* 1968, **77**, 85.
128. A.N. Nesmeyanov and A.E. Borisov, *Tetrahedron*, 1957, **1**, 158.
129. C.W. Fong and W. Kitching, *J. Organomet. Chem.*, 1973, **59**, 213.
130. (*a*) J.L. Wardell, *J. Chem. Soc. Dalton Trans.*, 1975, 1786; (*b*) G.G. Barbieri and F. Taddei, *J. Chem. Soc. Dalton Trans.*, 1974, **71**, 387.
131. J.L. Wardell and S. Ahmed, *J. Organomet. Chem.*, 1974, **78**, 395.
132. W. Kitching, B. Laycock, I. Maynard and K. Penman, *J. Chem. Soc., Chem. Commun.*, 1986, 954; D. Young, W. Kitching and G. Wickham, *Tetrahedron Lett.*, 1983, **24**, 5789.
133. D. Young and W. Kitching, *J. Org. Chem.*, 1983, **48**, 614.
134. J.A. Mangravite, *J. Organomet. Chem.*, 1979, **7**, 45.
135. D. Young, M. Jones and W. Kitching, *Austr. J. Chem.*, 1986, **39**, 563.
136. W. Stamm, *J. Org. Chem.*, 1963, **28**, 3264.
137. G.A. Russell and L.L. Herold, *J. Org. Chem.*, 1985, **50**, 1037.
138. D.J. Peterson, M.D. Robins and J.R. Hansen, *J. Organomet. Chem.*, 1974, **73**, 237.
139. K.C. Nicolaou, D.A. Claremon, W.E. Barnette and S.P. Seitz, *J. Amer. Chem. Soc.*, 1979, **101**, 3704.
140. J. McWigzell and J.L. Wardell, *J. Organomet. Chem.*, 1982, **235**, 37.
141. M. Le Quan and G. Guillerm, *J. Organomet. Chem.*, 1973, **54**, 153.
142. J.C. Cochran and H.G. Kuivila, *Organometallics*, 1982, **1**, 97.
143. J.E. Baldwin, R.M. Adlington and A. Basak, *J. Chem. Soc., Chem. Commun.*, 1984, 1284.
144. F. Rijkens, M.J. Janssen, W. Drenth and G.J.M. van der Kerk, *J. Organomet. Chem.*, 1964, **2**, 347.
145. J.L. Wardell, in *Organotin Compounds: New Chemistry and Applications*, ed. J.J. Zuckermann, Adv. Chem. Ser. **157**, American Chemical Society, Washington DC, 1976, 113, and refs. therein.
146. R.W. Bott, C. Eaborn and T.W. Swaddle, *J. Organomet. Chem.*, 1966, **5**, 233.
147. R.D. Taylor and J.L. Wardell, *J. Chem. Soc. Dalton Trans.*, 1976, 1345.

148. O.J. Taylor, PhD thesis, University of Aberdeen, 1988.
149. J.L. Wardell, *Inorg. Chim. Acta*, 1978, **26**, L18.
150. D.J. Peterson, *J. Organomet. Chem.*, 1971, **26**, 215; R.D. Taylor and J.L. Wardell, *Tetrahedron Lett.*, 1982, 1735.
151. H.G. Kuivila, J.E. Dixon, P.L. Maxfield, N.M. Scarpa, T.M. Topka, K.-H. Tsai and K.R. Wursthorn, *J. Organomet. Chem.*, 1975, **86**, 89.
152. R.D. Taylor and J.L. Wardell, *J. Organomet. Chem.*, 1975, **94**, 15.
153. H. Weichmann and A. Tzschach, *Z. Anorg. Allg. Chem.*, 1979, **458**, 291.
154. Y. Sato, Y. Ban and H. Shirar, *J. Org. Chem.*, 1973, **38**, 4373.
155. H. Weichmann and A. Tzschach, *J. Organomet. Chem.*, 1975, **99**, 61; see also H. Weichmann, G. Quell and A. Tzschach, *Z. Anorg. Allg. Chem.*, 1980, **462**, 7; H. Weichmann and B. Rensett, *Z. Anorg. Allg. Chem.*, 1983, **503**, 106.
156. J.M. Jerkunica and T.G. Traylor, *J. Amer. Chem. Soc.*, 1971, **93**, 6278; S.J. Hannon and T.G. Taylor, *J. Org. Chem.*, 1981, **46**, 3645, 3651.
157. I. Fleming and C.J. Urch, *J. Organomet. Chem.*, 1985, **285**, 173.
158. J.F. Kadow and C.R. Johnson, *Tetrahedron Lett.*, 1984, **25**, 5255.
159. J. McM. Wigzell and J.L. Wardell, *J. Organomet. Chem.*,
160. B. Jousseaume and P. Villeneuve, *J. Chem. Soc., Chem. Commun.*, 1987, 513.

6 Organometallic compounds of tetravalent tin

KIERAN C. MOLLOY

6.1 Introduction

Over the last 30 years, research into the chemistry of organometallic compounds of tin in the +4 oxidation state has represented one of the most prolific areas of chemical activity. As a result of the serendipitous coupling of an arsenal of sophisticated analytical techniques (Chapter 3) and widespread industrial applications (Chapter 13), the many facets of this field have attracted the attention of chemists of all persuasions, as well as that of biochemists, biologists, pharmacologists, toxicologists and physicists, to name but a few. The reader's attention is drawn to existing books[1–3], review articles[4,5] and periodic surveys[6] covering this field. Invaluable databases relating to studies of organotin compounds by nmr[7–10] and Mössbauer spectroscopies,[11,12] and single crystal X-ray diffraction[13,14] also provide an entry into both structural and synthetic aspects of organotin chemistry.

6.2 Introducing reactivity: synthetic precursors to functionalized organotin (IV) compounds

Tetraorganotin compounds are usually colourless solids or oils, thermally stable at <200°C, and are stable in moist air, although some recently prepared *C*-organostannyl heterocycles are unusually aerobically sensitive[15]. Preparative and physical data for both symmetrical and unsymmetrical tetraorganotin compounds are contained in three volumes of the Gmelin series, covering literature up to 1975.[16–18] The formation and cleavage reactions of the Sn–C bond are described in Chapter 5.

Unlike most metal–carbon bonds, the Sn–C bond is of low polarity and hence is relatively unreactive. The stability of the Sn–C bond is largely responsible for both the extent and nature of the commercial utility of organotin compounds, but while R_3Al compounds for example are active olefin polymerization catalysts due to the highly polar Al–C bonds, R_4Sn compounds are too unreactive to be of significant importance save as precursors for less alkyl- or arylated species. Reactivity is introduced into an organotin moiety usually via tetraorganotins, by replacing Sn–C by more polar (e.g. Sn–Cl or Sn–O), or thermodynamically weaker (e.g. Sn–Sn) linkages. These compounds provide the synthetic base on which all organotin (IV) chemistry rests.

6.2.1 *Organotin (IV) halides*

The single most important class of organotin compounds embraces those species containing one or more tin–halogen bonds. These compounds have been comprehensively surveyed in four volumes of Gmelin.[19–22] Physical data for these compounds are given in Table 6.1. Organotin fluorides are generally only sparingly soluble in organic

Table 6.1 Melting points (°C) or boiling points (°C/mm Hg) for organotin halides, R_nSnX_{4-n}.[5]

R	n	X = F	Cl	Br	I
Me	3	375d	37–38	26–27	67–68/15
Me	2	> 300	107–108	75–77	43–44
Me	1	321–327	45–46	55	85
Et	3	302	210/760	224/760	234/760
Et	2	310–320	84	63	44
Et	1	269–272	86/12	46/0.1	181–184.5/19
Bu	3	248–252	152–156/14	120–122/2	108/0.07
Bu	2	156–157	40–41	20(96/0.1)	145/6
Bu	1	337–338	93/10	77–79/0.2d	154/5
Ph	3	357d	105	124–125	122–124
Ph	2	> 300	40–42	37	72–73
Ph	1	220	128/15	182–183/29	31–32

solvents due to their associated nature, while the other halides are readily soluble in organic solvents, and the methyltin compounds are also soluble in water. Organotin iodides are the least thermally stable of the series.

The classical preparation for these compounds is the Kocheshkov redistribution reaction between a tetraorganotin and a tin (IV) halide SnX_4.[23,24] Tin (IV) chloride is most commonly used, and, with the appropriate choice of reaction stoichiometry, mono-, di- or tri-organotin halides can be made the dominant products.

$$3R_4Sn + SnCl_4 \rightarrow 4R_3SnCl \tag{6.1}$$

$$R_4Sn + SnCl_4 \rightarrow 2R_2SnCl_2 \tag{6.2}$$

$$R_4Sn + 3SnCl_4 \rightarrow 4RSnCl_3 \tag{6.3}$$

Similar reactions proceed with unsymmetrical tetraorganotins:

$$Ph_3SnC_{18}H_{37} + 3SnCl_4 \rightarrow 3PhSnCl_3 + C_{18}H_{37}SnCl_3 \tag{6.4}$$

In addition to SnX_4, a variety of other inorganic halides participate in this type of redistribution reaction, namely BCl_3, $AlCl_3$, $SiCl_4$, $GeCl_4$, $BiCl_3$, $TiCl_4$ and $HgCl_2$, but are less commonly employed.[25]

Both halogens and hydrogen halides cleave Sn–C bonds, and provide a simple, economical way to form organotin halides. Unsymmetrical organotin halides are conveniently prepared in this way, since the relative ease of halide for hydrocarbon substitution broadly follows the stability of the leaving carbanion, viz: Ph > $PhCH_2$ > vinyl > Me > Et > Pr > Bu.[25] Iodine is the reagent of choice for cleavage of only one Sn–C bond, while Br_2 can substitute up to two organic groups depending upon stoichiometry and substrate. Typical examples are:[26–28]

$$Et_3OctSn + I_2 \rightarrow Et_2OctSnI + EtI \tag{6.5}$$

$$^iPr_2SnPh_2 + 2Br_2 \rightarrow Pr^i_2SnBr_2 + 2PhBr \tag{6.6}$$

$$Me_4Sn + 2HF \rightarrow Me_2SnF_2 + 2MeH \tag{6.7}$$

The above methods, in conjunction with the direct synthesis discussed in Chapter 5, are the most versatile and practical syntheses of organotin halides. Other Sn–halogen bond forming reactions are shown in Scheme 1, but since the reagents are themselves

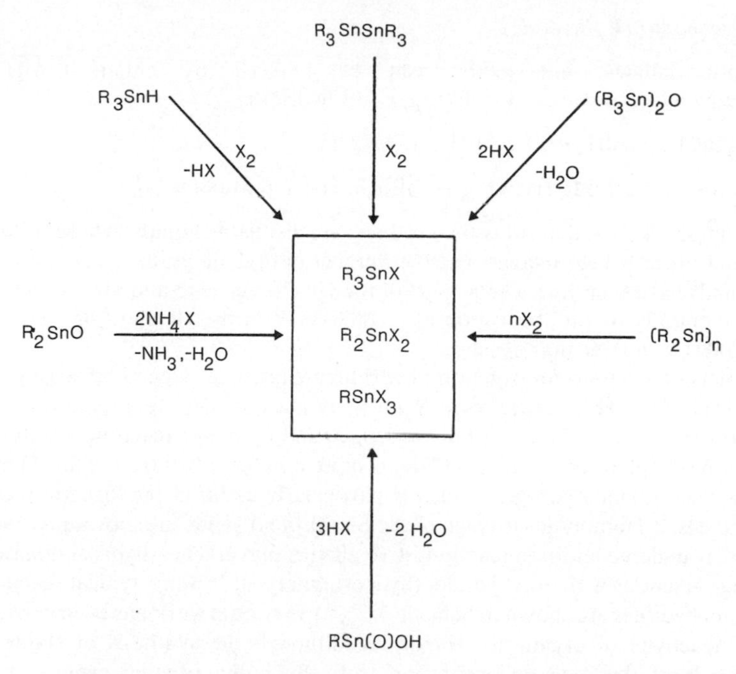

Scheme 1 Formation of organotin–halogen bonds (X = F, Cl, Br, I).

usually derived from organotin halides they offer no advantages over conventional methods. Once formed, the Sn–halogen bond will undergo a wide range of nucleophilic substitution reactions, some of which are shown in Scheme 2 for the triorganotin series.

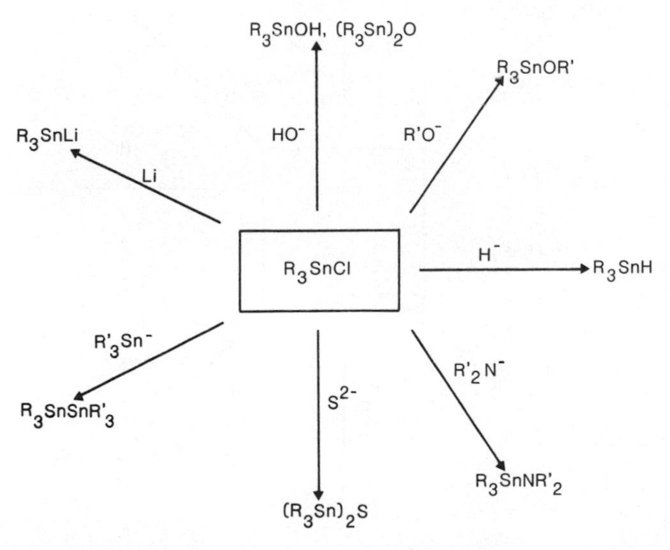

Scheme 2 Nucleophilic substitution reactions of triorganotin halides.

6.2.2 *Organotin (IV) hydrides*

Organotin halides and oxides can be reduced by either $LiAlH_4$ or poly(methylhydrosiloxane) to yield organotin hydrides:

$$2Ph_3SnCl + LiAlH_4 \rightarrow 2Ph_3SnH + LiAlCl_2H_2 \quad (6.8)$$

$$n(Bu_3Sn)_2O + 2\text{-}[Me(H)SiO]_n\text{-} \rightarrow 2nBu_3SnH + 2\text{-}[MeSiO_{1.5}]_n\text{-} \quad (6.9)$$

For common alkyl- and aryltins the products are distillable liquids, whose instability in air and towards heat increases as the number of organic groups decreases. These compounds have been reviewed as part of the Gmelin series,[29] and are characterized spectroscopically by an absorption at *c.* 1800 cm^{-1} in the IR and $^1J(^{119}Sn-^1H) \sim$ 1500–2000 Hz in their nmr spectra.

A variety of reactions for organotin hydrides are possible, depending largely on the mechanism of Sn–H bond cleavage. With weak nucleophiles, the organotin hydride will provide a source of nucleophilic hydrogen and exchange reactions result.[30] If a stronger nucleophilic centre such as NR_2 is present in the substrate, the Sn–H moiety releases electrophilic hydrogen, which is particularly useful in the formation of Sn–metal bonds.[31] Homolytic cleavage of the Sn–H bond yields organostannyl radicals which may undergo addition reactions with alkynes and alkenes (hydrostannation) or exchange reactions with alkyl halides (hydrostannolysis).[32] Some typical reactions of organotin hydrides are shown in Scheme 3.[30–37] Very little work has been carried out on the reactivity of organotin trihydrides, although the synthesis of stable solid compounds of this type incorporating sterically bulky organic groups such as $(Me_3Si)_3CSnH_3$[38] may promote interest in this area.

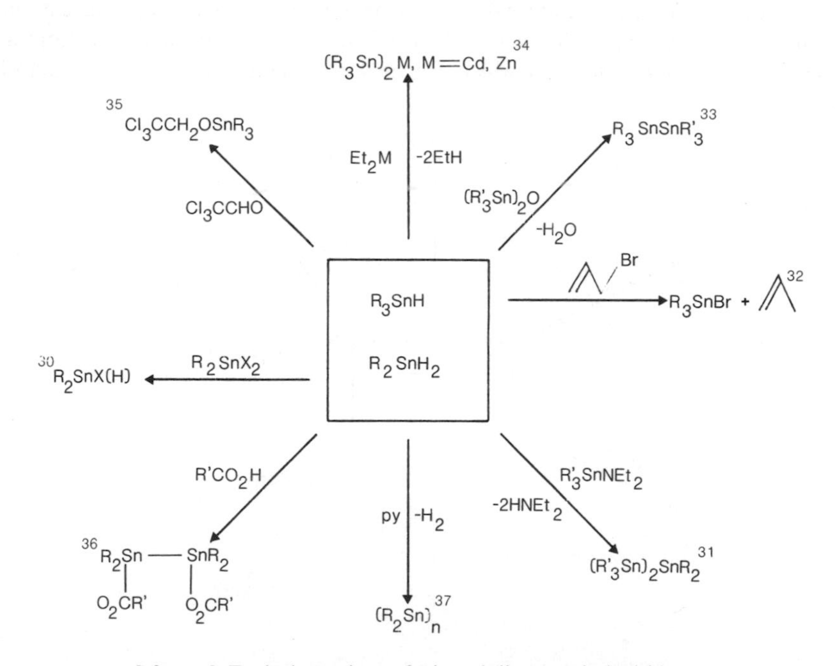

Scheme 3 Typical reactions of tri- and diorganotin hydrides.

6.2.3 *Organotin (IV) oxides, hydroxides and alkoxides*

Base hydrolysis of organotin halides leads to a variety of organotin oxygen compounds (Scheme 4) which have a rich structural and synthetic chemistry. In addition, $(Bu_3Sn)_2O$, Ph_3SnOH, Cy_3SnOH and $[(Neophyl)_3Sn]_2O$ are commercial biocides.

The nature of the hydrolysis products increases in complexity with increasing halogen content of the precursor. Triorganotin halides yield initially hydroxides, which are stable where R = Me, aryl, or bulky alkyl (e.g. cyclo-C_6H_{11}). Other triorganotin hydroxides spontaneously condense to the corresponding bis-(triorganotin)oxides. In the diorganotin series, the 1, 1, 3, 3-tetraorgano-1, 3-dihalodistannoxanes and diorganotin oxides are well characterized, but other partial hydrolysis products have only been isolated in cases where steric bulk inhibits condensation reactions, e.g. $Bu^t_2Sn(OH)Cl$,[39] $ClPr^i_2SnOSnPr^i_2OH$[40], $[(OH)Bu^tMe_3SiCH_2Sn]_2O$.[41] Similarly, while most diorganotin oxides are amorphous, polymeric materials, several oligomeric oxides are now known, e.g. $[Bu^t_2SnO]_3$.[42] Less attention has been paid to the hydrolysis products of organotin trihalides, although $EtSn(OH)Cl_2 \cdot H_2O$ has been isolated as a hydroxy-bridged dimer.[43] The stannoic acids, RSn(O)OH, are usually polymeric powders, but again bulky hydrocarbons yield soluble, presumably oligomeric, materials, e.g. $(Me_3Si)_3CSn(O)OH$.[44]

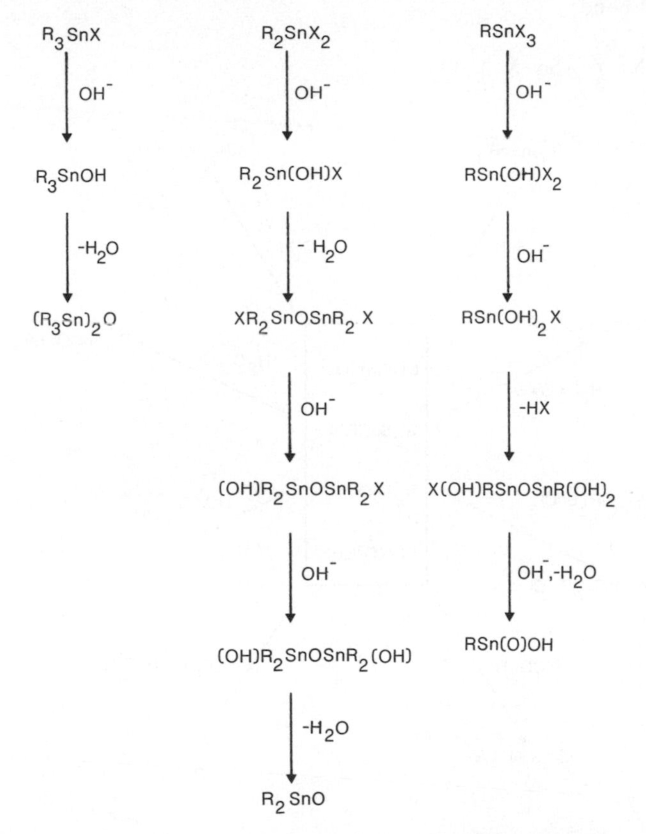

Scheme 4 Stepwise hydrolysis products of organotin halides (X = halogen).

Closely related to these systems from a synthetic viewpoint are the organotin alkoxides. These are prepared by nucleophilic substitution of Cl^- by RO^- within an organotin halide, and, unlike the oxides/hydroxides, are often aerobically sensitive.

$$Bu_3SnCl + MeONa \rightarrow Bu_3SnOMe + NaCl \quad (6.10)$$

As synthetic precursors to other organotin compounds, synthons containing Sn–O bonds may be utilized in one of two reaction classes. The predominant class is heterolytic substitution, usually involving protic ligands. Reactions of the following general form are typical:

$$(R_3Sn)_2O + 2HL \rightarrow 2R_3SnL + H_2O \quad (6.11)$$

$$R_2Sn(OR')_2 + 2HL \rightarrow R_2SnL_2 + 2HOR' \quad (6.12)$$

Reactions in which H_2O is the by-product are driven by its azeotropic removal using toluene as solvent. Carboxylic[45], oxy-sulphur,[46–48] oxy- and thiophosphorus,[49,50] and nitric acids[51] make an arbitrary but representative list of ligands which form organotin esters by this route. Other examples of substitution reactions involving Sn–O bonded compounds are shown in Scheme 5[33,52–56].

Also included in Scheme 5 are examples of a second reaction type, insertion into the polar Sn–O bond.

$$Sn{-}O + \overset{\delta-}{X}{=}\overset{\delta+}{Y} \rightarrow Sn{-}X{-}Y{-}O \quad (6.13)$$

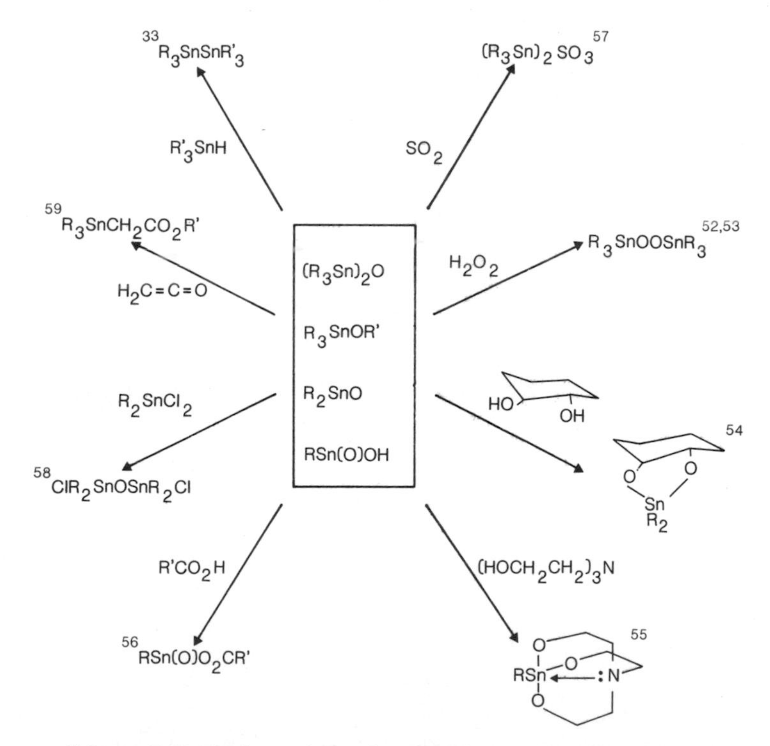

Scheme 5 Typical reactions of organotin–oxygen compounds.

One of the more environmentally significant consequences of this type of reactivity is the conversion of the wood preservative $(Bu_3Sn)_2O$ to $(Bu_3Sn)_2CO_3$ within the cellulose structure.[60]

6.2.4 *Organotin (IV) amines*

Amino-nucleophiles, usually as lithium but occasionally as other alkali metal, magnesium or organosilicon derivatives, displace Cl^- from organotin chlorides to yield organostannyl amines. For example:[61]

$$RSnCl_3 + 3LiNMe_2 \rightarrow RSn(NMe_2)_3 + 3LiCl \quad (6.14)$$

Derivatives of less volatile amines can then be synthesized by a transamination route:[62]

$$4Me_3SnNEt_2 + N_2H_4 \rightarrow (Me_3Sn)_2NN(SnMe_3)_2 + 4HNEt_2 \quad (6.15)$$

The Sn–N bond can be cleaved by even mildly protic ligands to form other organotin compounds, and the evolution of gaseous amine used to promote reaction. Insertion reactions are also known, and in each of these reaction types the Sn–N compounds are more reactive than their Sn–O counterparts.

6.2.5 *Organostannyl anions*

For practical purposes, organostannylanions can be considered simply as their lithium or sodium salts, although magnesium derivatives are also of value. Organotin-copper reagents[63] R_3SnCuX have also found application in organic synthesis. Triorganostannylanions have been considerably more widely exploited than R_2SnM_2 or $RSnM_3$ type species. The former class of compound can be prepared by the routes of Scheme 6, and their synthetic chemistry has been reviewed.[64,65]

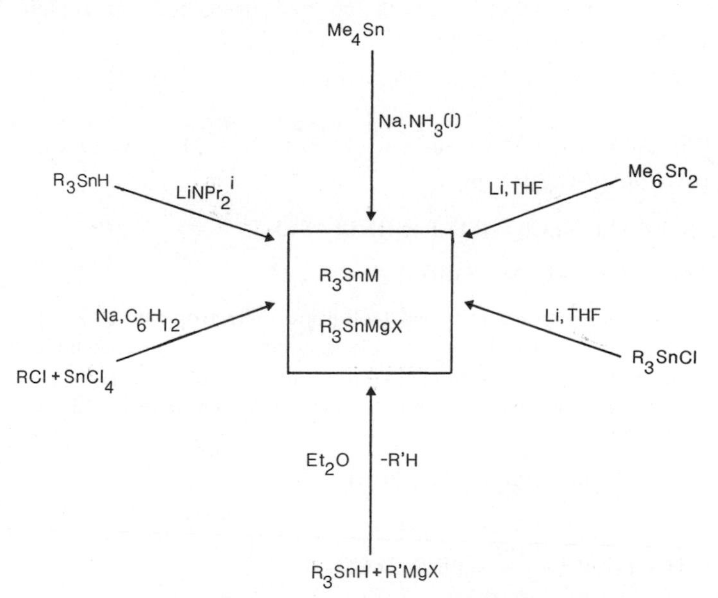

Scheme 6 Formation of organostannyl anions.

Organotin nucleophiles will react with a variety of carbonium centres, such as alkyl halides,[66] aldehydes, ketones and oxetanes[67,68] to form Sn–C bonds. In the case of reactions with alkyl halides, the stereochemistry of the reaction is dependent on the nature of R, X, M and the solvent, and mechanisms such as S_N2 at C, S_N2 at X, and radical-pair have been suggested.[65,69]

Alternatively, organostannylanions react with metal halides to form new Sn–metal bonds and this reaction type has been widely exploited, as exemplified by[70,71]

$$4Ph_3SnLi + SnCl_4 \rightarrow (Ph_3Sn)_4Sn + 4LiCl \quad (6.16)$$

$$2Ph_3SnLi + Ph_2MCl_2 \rightarrow (Ph_3Sn)_2MPh_2 + 2LiCl \quad (6.17)$$

M = Si, Ge

6.3 Construction and reactivity of Sn–element bonds

The synthesis of R_nSn–element bonds is covered systematically by Group, except for the halogens whose chemistry has already been covered in preceding sections.

6.3.1 *Groups 1, 2, 12, 13*

Apart from the Li, Na and Mg organostannylanions, very little attention has been paid to the derivatives of Groups 1 and 2. The solvent dependence of R_3SnM reactivity suggests that solvent coordination is an important structural component of these systems, but this is yet to be confirmed crystallographically.[212] R_2SnM_2 species have not been developed to the same extent as their triorganotin analogues, nor have systems based on the more electropositive of the Group 1, 2 elements (Rb, Cs, Ca, Sr, Ba).

Organotin–boron compounds are better established, but no structural data is yet available. Four approaches have been reported, using both nucleophilic[72,73] and electrophilic[74–76] boron reagents:

$$Me_3SnCl + LiB_5H_8 \rightarrow \mu\text{-}Me_3SnB_5H_8 + LiCl \quad (6.18)$$

$$\frac{n}{2}(TPE)_2Co(BPh_2)_2 + Me_{4-n}SnX_n \rightarrow \frac{n}{2}(TPE)_2CoX_2 + Me_{4-n}Sn(BPh_2)_n \quad (6.19)$$

$(TPE = Ph_2PCH_2CH_2PPh_2)$

$$2Me_3SnLi + Et_2NBCl_2 \rightarrow (Me_3Sn)_2BNEt_2 + 2LiCl \quad (6.20)$$

$$Me_3SnLi + BH_3 \rightarrow Li[Me_3SnBH_3] \quad (6.21)$$

Equation 6.21 has also been applied to the more electropositive Group 3 elements using R_3M as Lewis acids, but the products appear to be quite unstable.[76]

Several examples of Sn–M bonded compounds (M = Zn, Cd, Hg) are known, and their structures have been assigned, somewhat tentatively, on spectroscopic evidence.[34,77–80]

$$2Ph_3SnH + Et_2M \rightarrow (Ph_3Sn)_2M + 2EtH \quad (6.22)$$

$$Ph_3SnH + EtZnCl \xrightarrow{-EtH} [Ph_3SnZnCl] \rightarrow \quad (6.23)$$

Ph₂Sn–Cl:–ZnPh / PhZn–:Cl–SnPh₂ (cyclic dimer: Ph_2Sn bridged via :Cl to ZnPh; PhZn bridged via :Cl to $SnPh_2$)

$$2R_3SnCl + (Me_3Si)_2Hg \rightarrow 2Me_3SiCl + (R_3Sn)_2Hg \xrightarrow{>-10^\circ} R_6Sn_2 + Hg \quad (6.24)$$

$(R_3Sn)_2Hg$ species can also be synthesized by electrochemical methods.[81] The direct Sn–Hg bond is corroborated by the large one-bond $^1J(^{119}Sn-^{1199}Hg)$ coupling constant of ~6000 Hz.[82]

6.3.2 *Group 14*

Although synthetic aspects of the Sn–C linkage have already been covered, one recent example from what is likely to be an expanding area of activity serves to underscore several features of the chemistry of this bond type. The Lewis acidity of tin makes it suitable for incorporation into macrocycles designed for anion binding, to complement the well-established area of cation binding by basic host macrocycles, e.g. crown ethers. Scheme 7 outlines routes developed by Newcombe *et al.*, and elegantly demonstrates the use of lithium and Grignard reagents in Sn–C bond formation, and the selective cleavage of Sn–C(aryl) over Sn–C(alkyl) bonds to introduce reactivity. Macrocycles of architecture $n = 8, 10, 12$ all bind Cl^- ions strongly, and a small macrocycle selectivity effect exists in favour of the 18-ring ($n = 8$) species.[83,84]

Synthesis of the basic skeletons of organic molecules which contain tin and other Group 14 elements in place of carbon continues to attract interest. Unfunctionalized

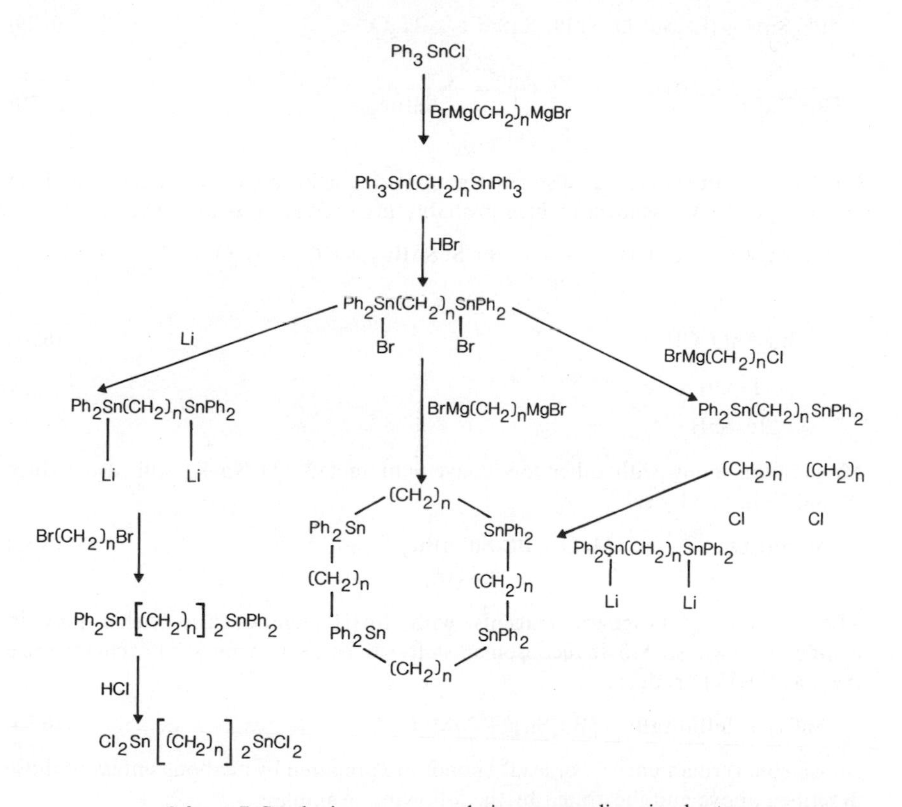

Scheme 7 Synthetic routes to polytin macrocyclic anion hosts.

polytin compounds can be prepared by coupling of triorganotin halides in the presence of metal, both chemically[85,86] or electrochemically:[87]

$$2(R_2SnX)CR'_2 \xrightarrow[\text{or Hg/DMF/2e}^-]{\text{Na/NH}_3} \begin{matrix} & R' \quad R' & \\ R_2Sn & \diagdown\!\!\diagup & SnR_2 \\ | & & | \\ R_2Sn & \diagup\!\!\diagdown & SnR_2 \\ & R' \quad R' & \end{matrix} + 4NaBr \tag{6.25}$$

$$2Bu_3SnCl \xrightarrow{\text{Mg/THF}} Bu_3SnSnBu_3 + MgCl_2 \tag{6.26}$$

Preformed organostannylanions may also be used:

$$Ph_3SnLi + Ph_3SnBr \xrightarrow{\text{Et}_2\text{O}} Ph_3SnSnPh_3 + LiBr \tag{6.27}$$

Alternatively, Sn–N,[89] Sn–O[33] and Sn–S[90] bonds may be reduced by tin hydrides to yield Sn–Sn bonds:

$$3Me_2SnH_2 + 3Me_2Sn(NEt_2)_2 \xrightarrow[\text{dark}]{0^\circ} \text{cyclo-}(Me_2Sn)_6 + 6HNEt_2 \tag{6.28}$$

$$2Ph_3SnH + (Et_3Sn)_2O \rightarrow 2Ph_3SnSnEt_3 + H_2O \tag{6.29}$$

$$[Bu^t_2SnS]_2 \xrightarrow{\text{LiAlH}_4} \begin{matrix} Bu^t_2Sn - S & \\ | & \diagdown \\ & SnBu^t_2 \\ | & \diagup \\ Bu^t_2Sn - S & \end{matrix} \tag{6.30}$$

Bis-(triorganotin)oxides will also react with formic acid to yield ditin compounds in 40–70% yield, by a sequence which probably involves an organotin hydride:[91]

$$\begin{matrix} (Bu_3Sn)_2O + HCO_2H & \longrightarrow & Bu_3SnSnBu_3 + CO_2 + H_2O \\ \downarrow -H_2O & & \nearrow \\ Bu_3SnO_2CH & & \\ \downarrow -CO_2 & \nearrow_{(Bu_3Sn)_2O/-H_2O} & \\ Bu_3SnH & & \end{matrix} \tag{6.31}$$

Magnesium, along with other less convenient metals (Ti, Na, K) will also reduce $(R_3Sn)_2O$ in good yield:[92]

$$(Bu_3Sn)_2O \xrightarrow{\text{Mg/THF}} MgO + \underset{(>80\%)}{Bu_3SnSnBu_3} \tag{6.32}$$

The reaction of Grignard reagents with tin(II) halides also yields poly-tin compounds, whose Mössbauer isomer shift values ($\sim 1.55\ mm\ s^{-1}$)[93] characterize them as tin(IV) products:

$$nSnCl_2 + 2nBuMgBr \rightarrow [Bu_2Sn]_n + 2nMgClBr \tag{6.33}$$

Tin-silicon, germanium[94,95] or lead[96] bonds are prepared by methods similar to those described above and illustrated by the following examples:

$$Me_3SnLi + Me_3SiCl \rightarrow Me_3SnSiMe_3 + LiCl \qquad (6.34)$$

$$Bu_3GeNEt_2 + Ph_3SnH \rightarrow Bu_3GeSnPh_3 + Et_2NH \qquad (6.35)$$

$$2PbCl_2 + 4Ph_3SnLi \rightarrow (Ph_3Sn)_4Pb + Pb + 4LiCl \qquad (6.36)$$

^{119}Sn chemical shifts are sensitive to the degree of chain branching in these systems, moving to higher field as tin is increasingly metallated from a primary to a quaternary centre.[97]

The reactive centre in compounds of type $R_3SnMR'_3$ (M = Si, Ge, Sn, Pb) and their cyclic or longer chain relatives is always the Sn–M bond. Some typical reactions are shown in Scheme 8.[5] Of particular interest is the recent demonstration that cyclic polytin compounds can be photochemically converted to Sn=Sn compounds.[98]

$$2\ \text{cyclo-}(Ar_2Sn)_3 \xrightarrow{h\nu} 3Ar_2Sn{=}SnAr_2 \qquad (6.37)$$

Ar = 2, 4, 6-triisopropylphenyl

Polytin compounds which incorporate other reactive centres, such as Sn–halogen bonds, have been less well investigated than the simple, unfunctionalized polyorganotins, although these compounds are precursors to a wide area of unexplored structural chemistry. Such compounds can be made by coupling of diorganotin(halogen)hydrides (eqn 6.38),[99] controlled opening of cyclic polytins by halogens (Scheme 8),[100] or electrochemical methods:[101]

$$R_2SnH_2 + R_2SnCl_2 \rightarrow 2R_2Sn(H)Cl \xrightarrow{\text{pyridine}} ClR_2SnSnR_2Cl + H_2 \qquad (6.38)$$

$ClMe_2SnSnMe_2Cl$ has been shown to exist as a chlorine-bridged, double-chain polymer.[102] Some reactions of this type of compound are shown in Scheme 8.[103–105]

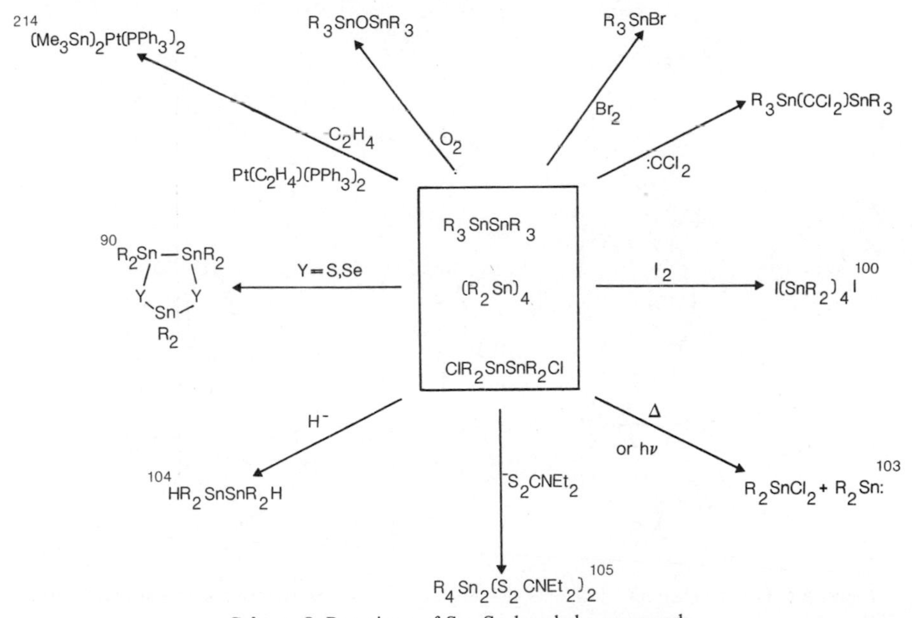

Scheme 8 Reactions of Sn–Sn bonded compounds.

The most thoroughly investigated reaction type is with bidentate ligands such as carboxylates, which span the Sn–Sn bond.[105–108]

6.3.3 *Group 15*

Organotin amines of formulae $R_nSn(NR'_2)_{4-n}$, $(R_3Sn)_nNR'_{3-n}$ $(n = 1, 2, 3)$ and $(R_2SnNR')_n$ $(n = 2, 3)$ are known, and their chemistry has been reviewed.[109,110] Preparation of these species by lithioamination and transamination methods has been described previously. The synthetic value of these compounds is the ease with which the Sn–N bond is cleaved by protic reagents, although this also renders these compounds sensitive to atmospheric moisture. The triorganotin amines are usually monomeric, but trimethyltin aziridine has been ascribed a polymeric structure on the basis of its Mössbauer quadrupole splitting ($2.24\,\mathrm{mm\,s^{-1}}$) and spectral area temperature-dependence data.[111] Triorganotin derivatives of imidazoles and triazoles[112,113] are more stable than the simple amines, and these species form N-bridged polymeric structures. The structure of the acaricidally active Cy_3Sn-1, 2, 4-triazole is shown in Figure 6.1.[114] In contrast, electron diffraction data show $(Me_3Sn)_3N$ to have a planar $[Sn_3N]$ skeleton with an Sn–N bond length of 203.8 pm.[115]

The bond between tin and the heavier Group 15 elements may be synthesized by similar means to the Sn–N bond. For phosphorus in particular, the use of polyfunctional reagents has led to a variety of ring and cage structures,[116–118] although the products of eqns (6.39)–(6.41) await crystallographic verification.

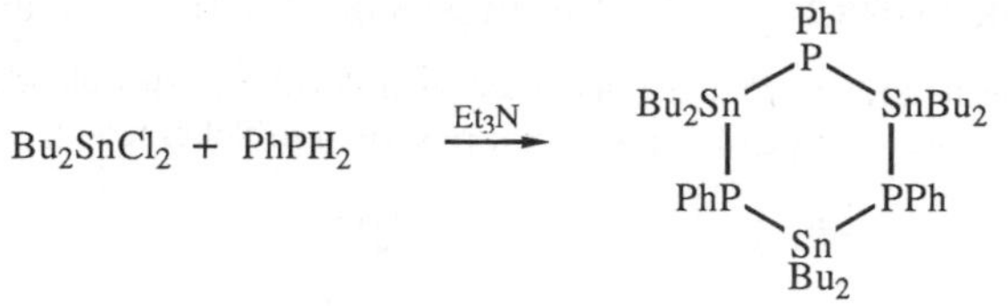

$$Bu_2SnCl_2 + PhPH_2 \xrightarrow{Et_3N} \text{cyclo-}[(Bu_2Sn)(PPh)]_3 \quad (6.39)$$

Ph P Bu₂Sn SnBu₂ PhP PPh Sn Bu₂

(1)

Sn (1) N (1) N (3) Sn (1') N (2)

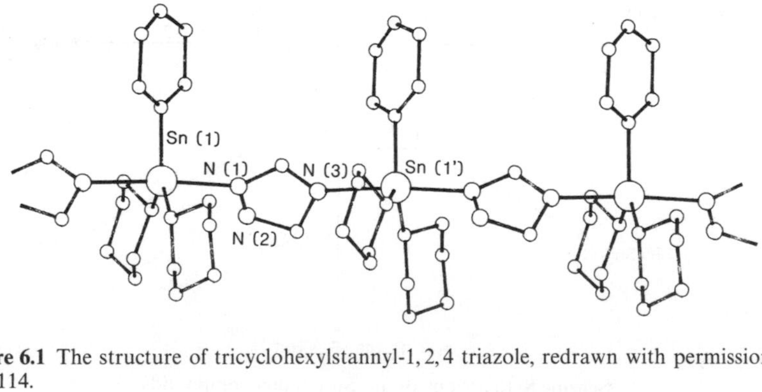

Figure 6.1 The structure of tricyclohexylstannyl-1, 2, 4 triazole, redrawn with permission from Ref. 114.

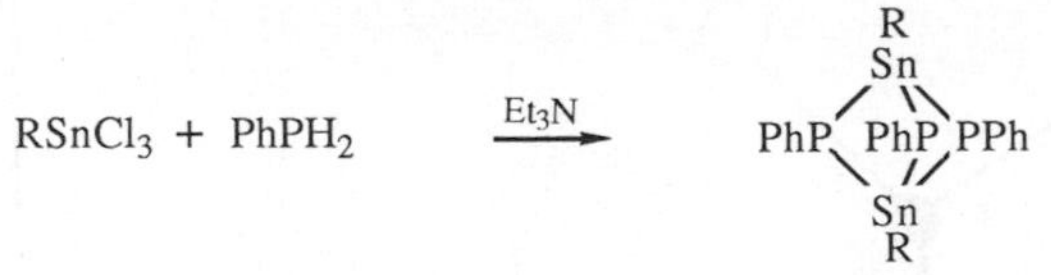

$$RSnCl_3 + PhPH_2 \xrightarrow{Et_3N} (RSn)_2(PPh)_3 \quad \textbf{(2)} \tag{6.40}$$

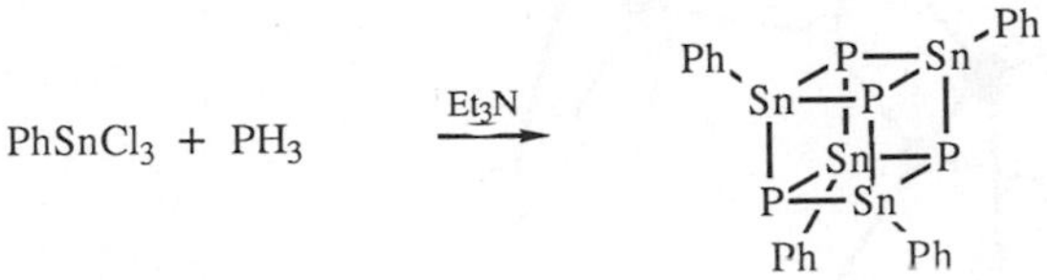

$$PhSnCl_3 + PH_3 \xrightarrow{Et_3N} (PhSn)_4P_4 \quad \textbf{(3)} \tag{6.41}$$

When elemental phosphorus is reduced by the ditin dihydride $Me_4Sn_2H_2$, an unusual cage structure (Figure 6.2*a*)[119] results which is stable in the dark, but in sunlight eliminates dimethylstannylene, Me_2Sni, to yield a norbornane type structure (Figure 6.2*b*).[120]

The first example of a Sn–P double bond, kinetically stabilized by bulky ligands, has recently been reported:[121]

$$R_2SnF_2 + ArP(H)Li \rightarrow R_2Sn(F){-}PAr(H) \xrightarrow{Bu^tLi} R_2Sn{=}PAr + Bu^tH + LiF \tag{6.42}$$

$R = (Me_3Si)_2CH \quad Ar = 2,4,6\text{-}t\text{-butylphenyl}$

One crystallographically authenticated organotin–antimony compound, $(Me_3Sn)_4Sb_2$, produced by photolytic coupling of $(Me_3Sn)_3Sb$, has an Sn–Sb bond length of 279.7 pm.[122]

Further examples of Sn–Group 15 bonds arise in the organotin pseudohalides R_nSnX_{4-n} ($X = NCS, NCO, N_3$ etc; $n = 2, 3$),[22] and the numerous donor–acceptor complexes these and organotin halides form with N, P, As ligands.[14]

6.3.4 *Group 16*

Ease of preparation, high reactivity and diversity of compound type has made Sn–O bonded compounds amongst the most intensively investigated of all organotin systems. Whereas mono- and diorganotin oxides often form amorphous polymers, the corresponding compounds of S, Se and Te tend to form smaller aggregates, and reports of a number of interesting heterocyclic derivatives of these elements with tin have encouraged a resurgence of activity in this area. Simple organotin sulphides and alkyl thiolates have been known for a number of years,[123,124] but a proposal that the biocidal activity of diorganotin compounds arises from binding to enzymatic

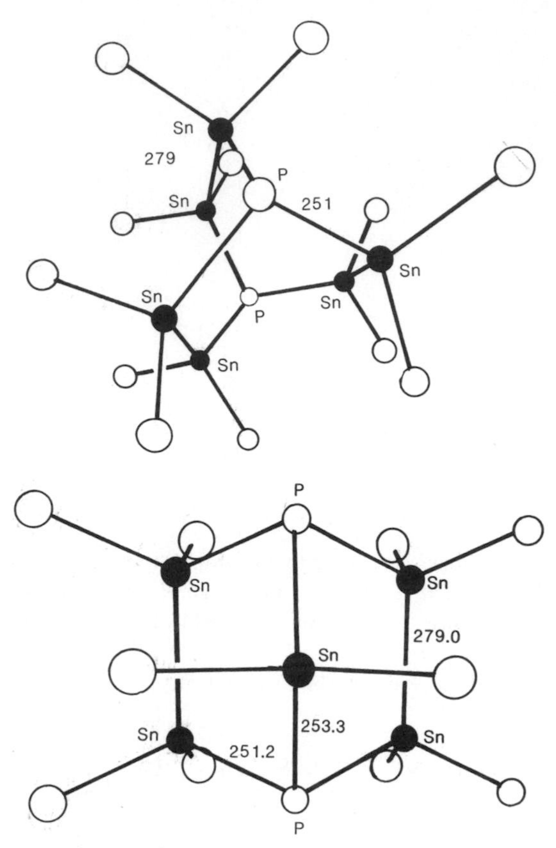

Figure 6.2 The structures of (*a*) $(Me_2Sn)_6P_2$ and (*b*) $(Me_2Sn)_5P_2$, redrawn with permission from Refs. 119 and 120 respectively. Distances in pm.

dithiolates, such as reduced lipoic acid,[125] has renewed interest in this area of chemistry also, and several new thiolates have recently been reported:[126]

$$\text{4-Me-}C_6H_3(SH)_2\text{-1,2} \xrightarrow[\text{(ii) } R_2SnCl_2]{\text{(i) KOH}} \text{4-Me-}C_6H_3(S)_2SnR_2 \qquad (6.43)$$

R = Me, Ph

Like many of the corresponding hydroxides, the hydrochalcogenides R_3SnXH (X = S, Se, Te) readily disproportionate to give bis-(triorganotin)chalcogenides. Some preparative methods and reaction chemistry are shown in Scheme 9.[127-134] The Sn–chalcogen bond is less reactive than Sn–O, but Sn–Te bonds are moisture-sensitive.

Several novel tin–chalcogen heterocycles have recently been synthesized (eqns 6.44–6.46) and their structures determined by crystallographic methods.[90,135,136]

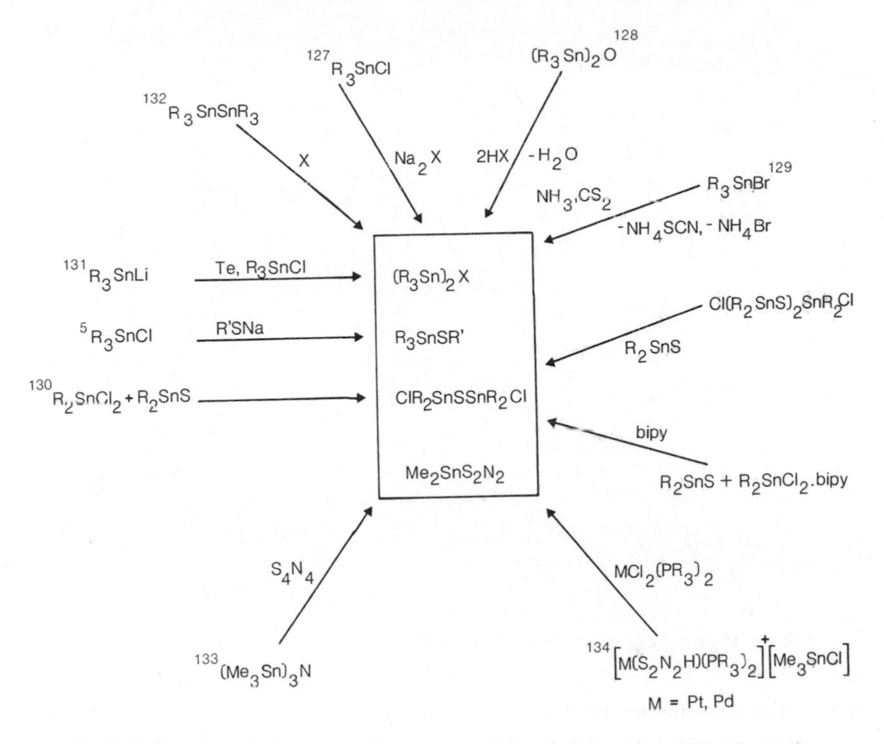

Scheme 9 Synthesis and reactions of the organotin–chalcogen bond (X = S, Se, Te).

$$I(Bu^t_2Sn)_4I + H_2X + Et_3N \longrightarrow \text{cyclo-}[(Bu^t_2Sn)_4X] \tag{6.44}$$

$$3(Bu^t_2Sn)_4 + 8X \longrightarrow \text{cyclo-}[Bu^t_2Sn{-}SnBu^t_2{-}X{-}SnBu^t_2{-}X] \tag{6.45}$$

$$5Bu^t_2SnS + 2S \xrightarrow{-Bu^t_2S}$$
$$4Bu^t_2SnCl_2 + 2Na_4SnS_4 \xrightarrow{-SnS,\ 8NaCl} \text{spiro-}Sn[(X)_2(R_2SnXSnR_2)]_2 \tag{6.46}$$
$$2(Pr^i_2SnCl)_2X + Na_4SnSe_4 \xrightarrow{-4NaX}$$

Typical $^1J(^{119}Sn-^{77}Se)$ couplings in these systems are ~1500 Hz.[136] Organotin chalcogenides show less tendency to associate via coordinate bonds, thereby expanding

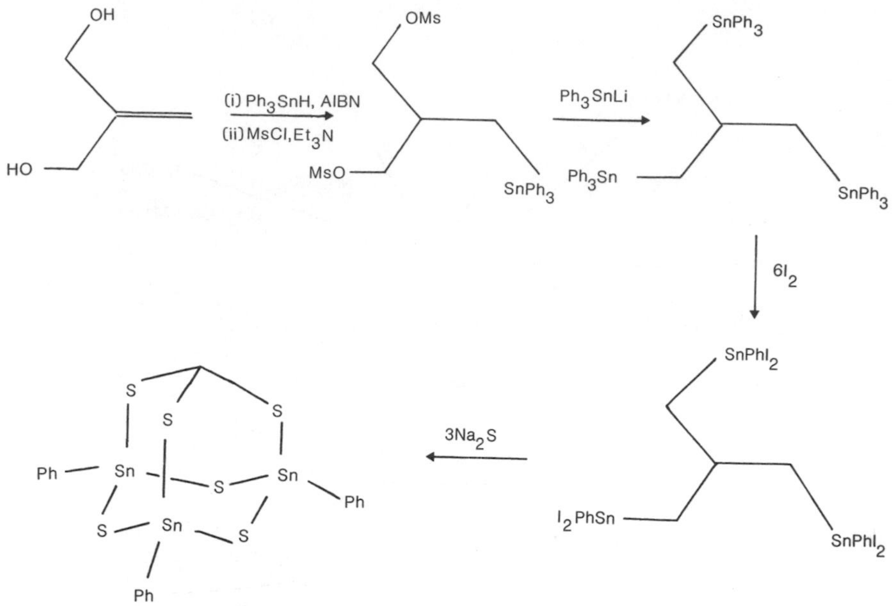

Scheme 10 Synthesis of 1, 3, 5-triphenyl-2, 4, 6-trithia-1, 3, 5-tristannaadamantane.

the coordination number at tin to 5 or 6, than corresponding compounds containing more electronegative oxygen atoms. For example, $(ClR_2Sn)_2X$ compounds are monomers, except where X = O when they are dimers.[137] However, di- and mono-organotin chalcogenides form ring and cage structures respectively[138–143] (see Chapter 2). A polymeric modification of Pr^i_2SnS is also known.[144] 1, 3, 5-Triphenyl-2, 4, 6-trithia-1, 3, 5-tristannaadamantane, whose synthesis is shown in Scheme 10, also adopts a cage structure.[145] The apical proton is very labile due to strain at the bridgehead carbon, and this compound is a useful reducing agent in organic synthesis.[146]

6.3.5 *Groups 3–11, lanthanides and actinides*

The formation of bonds between these elements and tin is covered in Chapter 8.

6.4 Structural variations

Arguably the single most important discovery in organotin chemistry was the establishment of $Me_3SnCl \cdot py$ as containing five-coordinated tin. Hulme's X-ray diffraction study in 1963[147] demonstrated that the coordination number at tin is not confined to four as is normally the case with carbon, and initiated a quarter-century of study which has revealed a structural chemistry as diverse as that of any element of the Periodic Table. Organotin (IV) compounds are now known to contain tin in coordination numbers three through seven, in varying isomeric forms and lattice arrays. A second legacy of Hulme's work has been the importance of X-ray diffraction studies to the structural tin chemist, and despite the contribution of ^{119}Sn nmr and

Mössbauer spectroscopies, no new structural form of tin can be firmly established without crystallographic confirmation. Two compendia of tin crystal structures are currently available for reference.[13,14]

6.4.1 *Molecular systems*

Organotin (IV) compounds are known with coordination numbers (CN) of 3, 4, 5, 6 and 7 at tin, within geometries which are predictable by VSEPR methods. However, it should be noted that distortions from regular polyhedra are common and give rise to some of the more debated structural analyses so far reported.

The coordination number adopted by tin in any particular molecule is essentially dictated by two opposing effects: increasing Lewis acidity at the metal, brought about by increasing the number of electronegative substituents on tin, will promote high coordination numbers; conversely, the coordination number at tin will be limited by the steric demands of the ligands, and bulky moieties will favour low coordination numbers. These effects will be evident in the following examples.

Only one example of CN = 3, trigonal planar organotin (IV) is known, that being the compound containing a Sn=C double bond [see structure (**1**), Chapter 2] stabilized by bulky organic groups. Four-coordinate tin is always of tetrahedral geometry and is typified by a plethora of R_4Sn structures of both symmetric (e.g. R = Ph,[148] $PhCH_2$,[149] $C_6H_4Me_2$[150]) and asymmetric (e.g. Ph_3SnCH_2I[151]) persuasions. When R is replaced by a more electronegative group X, CN > 4 usually occurs, although this tendency is checked if X is of relatively low electronegativity (e.g. Br, I *v.* Cl; S *v.* O) or if one or more of the ligands is bulky. The tetrahedral nature of $Me_3SnS_2CNMe_2$,[152] in comparison to $Me_3SnO_2CCH_3$[153] which is five-coordinated, is a manifestation of the first of these phenomena, while $[(Me_3Si)_2CH]_3SnCl$ (CN = 4)[154] compared with Me_3SnCl (CN = 5)[155] exemplifies the second.

Five-coordinated structures are generally trigonal bipyramidal in shape, and this is commonly the structure adopted by R_3SnX compounds and, to a lesser extent, R_2SnX_2 species. Three isomeric forms are possible for this geometry, of which the *trans*-isomer (**4**) (e.g. $(Bu_3SnCl_2)^-(BzPPh_3)^+$,[156] $Me_3SnCl\cdot py$,[147] or (3-thienyl)$_3SnBr\cdot Ph_3PO$,[157] is more prevalent than the *cis*-analogue (**5**) [e.g. $Ph_3SnON(Ph)CO\cdot Ph$].[158] The distribution of structures between these two isomers, and indeed the lack of any *mer*-$[R_3SnX_2]$ examples (**6**), can be rationalized by a qualitative MO analysis.[159] For

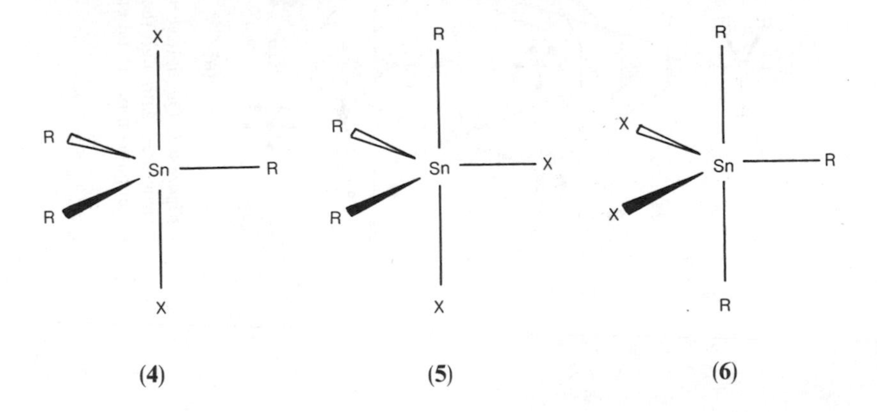

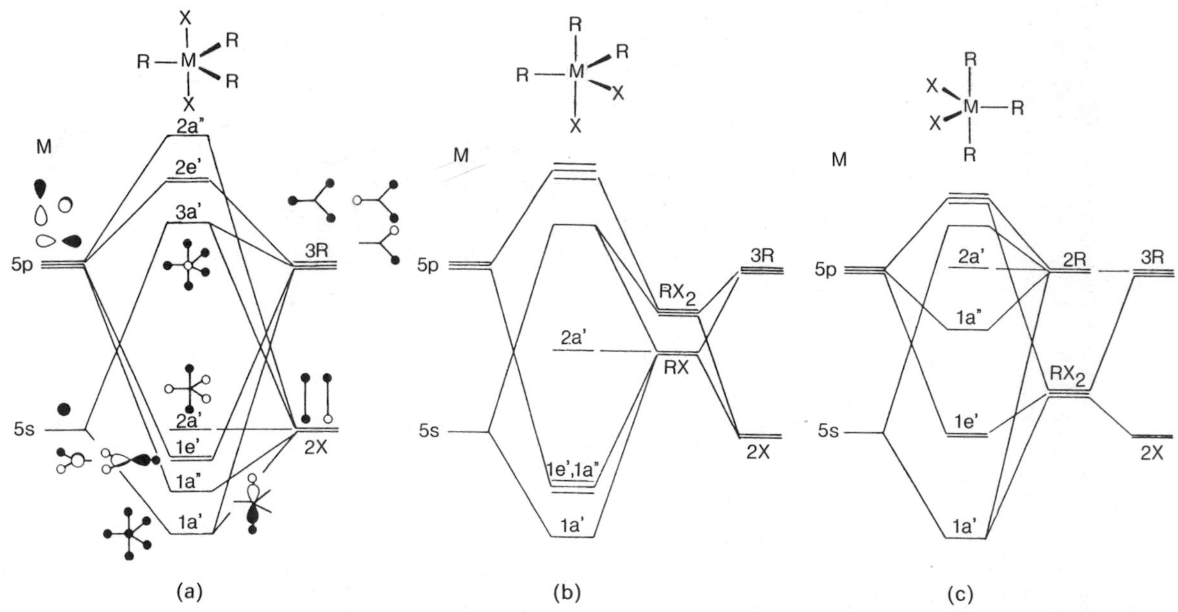

Figure 6.3. Qualitative molecular orbital energy level diagrams for (*a*) *trans*-$[X_2SnR_3]$ (*b*) *cis*-$[X_2SnR_3]$ and (*c*) *mer*-$[R_3SnX_2]$ structures. Although the symmetry labels for MOs are correct only for (*a*), they have been retained throughout to allow comparison of orbital energies between the three isomers.

the *trans*-isomer (Figure 6.3*a*) the HOMO is a non-bonding MO centred on the two axial ligands X (2*a'*). This is lowest in energy when the energies of the axial ligands are as low as possible, i.e. X is of high electronegativity. This analysis parallels the 'axially most electronegative' valence-bond terminology of Bent.[160] The D_{3h} structure, or distorted versions of it, occur when X (e.g. O, N, halogen) is more electronegative than R, which is usually the case.

For the *cis*-[R_3SnX_2] isomer (Figure 6.3*b*), the effect on the energy of the non-bonding HOMO, brought about by forming symmetry-adapted axial R and X orbital combinations, is to raise its energy. However, this is minimized, and thus favoured, when the R-AOs are similar in energy to the X-AOs. This geometry is thus more common when X (e.g. S) and R (e.g. C_6H_5) are of similar electronegativity, and bidentate sulphur ligands usually chelate, while for oxygen bridging between axial sites is more usual. For the *mer*-isomer (Figure 6.3*c*) the non-bonding HOMO is at its highest energy, arising as it does from the symmetry adapted combination of R-AOs, and it is not surprising that no crystallographic examples of this arrangement have been reported. A combination of high-electronegativity R (e.g. C_6F_5) and low-electronegativity X (e.g. S) will most likely be required to realize this goal. It should also be appreciated that five-coordinated diorganotin compounds always adopt the *cis*-[R_2SnX_3] geometry (**7**), *mer* with respect to the X ligands, which follows from an adaptation of Figure 6.3*c* and provided X is more electronegative than R which is usual, as exemplified by the structure of diphenyltin glycylglyinate, $Ph_2SnGlyGly$, which adopts this geometry, as shown in Figure 6.4.[161]

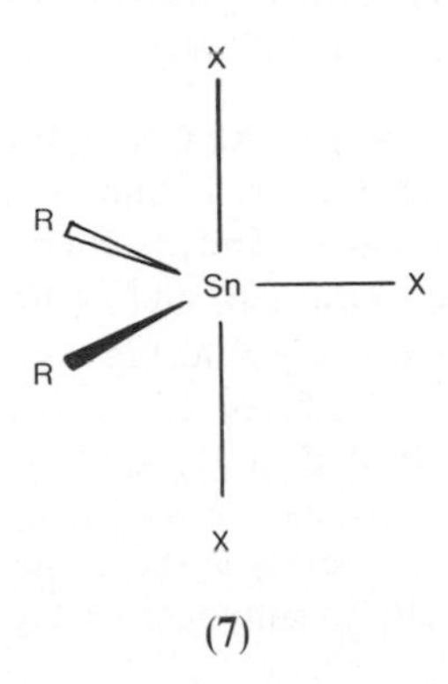

(**7**)

The onset of five-coordination (or higher) is readily evident from up-field ^{119}Sn nmr shifts (see Chapter 3). Mössbauer QS values are also capable of distinguishing the three isomers (**4**; 3.00–4.00 mm s^{-1}), (**5**; 1.70–2.40 mm s^{-1}) and (**6**; 3.50–4.10 mm s^{-1}), but tetrahedral R_3SnX species (1.00–2.40 mm s^{-1}) can be confused with the *cis*-configuration, particularly when the QS of the tetrahedral species is enhanced by deviations from ideal geometry (see also Chapter 3).[5]

Six-coordinate systems are common for the highly Lewis-acidic R_2SnX_2, $RSnX_3$ classes, but rare for tetra- and triorganotins. Bis[3-(2-pyridyl)-2-thienyl-*C,N*]diphenyltin (IV)[162] is unique in being the only six-coordinated tetraorganotin compound, although a five-coordinated stannatrane $MeSn(CH_2CH_2CH_2)_3N$ has been reported,[163] as have unsubstantiated claims for the adduct

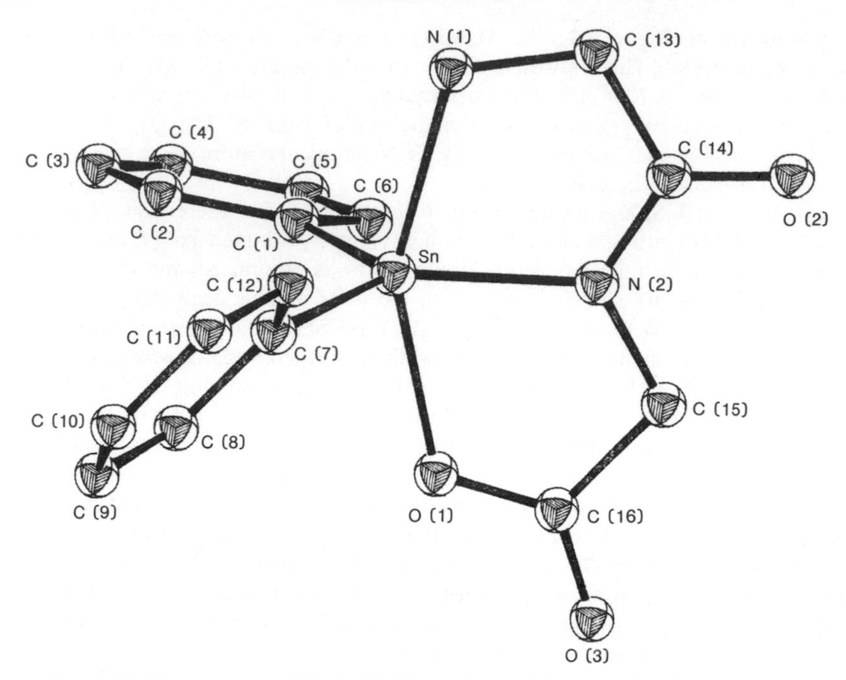

Figure 6.4 The structure of $Ph_2SnGlyGly$, redrawn with permission from Ref. 161.

$Me_3SnCF_3 \cdot HMPA$.[164] R_3SnX species with CN = 6 are equally rare, but one example each of both *mer*-(**8**) and *fac*-(**9**) isomers is known. The latter is formed using the tridentate tris(pyrazolyl)borate ligand, and its complex with Me_3Sn- is shown in (2.52).[165] The *mer*-isomer is less clear cut, and occurs in the coordination polymer $Ph_3SnO_2CH_3$, but the sixth bond is very weak (Sn–O, 320.6 pm)[166]. Again, Mössbauer QS values are predicted to be quite different for the two forms [(**8**), QS ~ 3.50 mm s^{-1}; (**9**), QS ~ 0 mm s^{-1}]. Data for $Ph_3SnO_2CH_3$ (QS = 3.36 mm s^{-1}) are consistent with the *mer*-isomer value, but five-coordinate *trans*-O_2SnR_3 systems give too similar a QS value for this experiment to conclusively authenticate CN = 6. Both Mössbauer and ^{119}Sn nmr data are, unfortunately, still outstanding for the six-coordinated *fac*-compound.

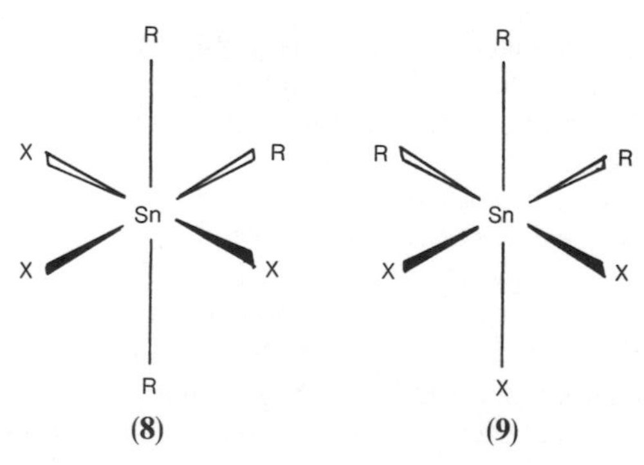

Six is the common coordination number for diorganotin systems. Both *trans*-(**10**) and *cis*-(**11**) isomers are known, the former being markedly more common. Adducts of $(4\text{-ClC}_6H_4)_2SnCl_2$ with $4,4'\text{-Me}_2$bipy(2, 2′) occur in both *cis*- and *trans*- forms, which

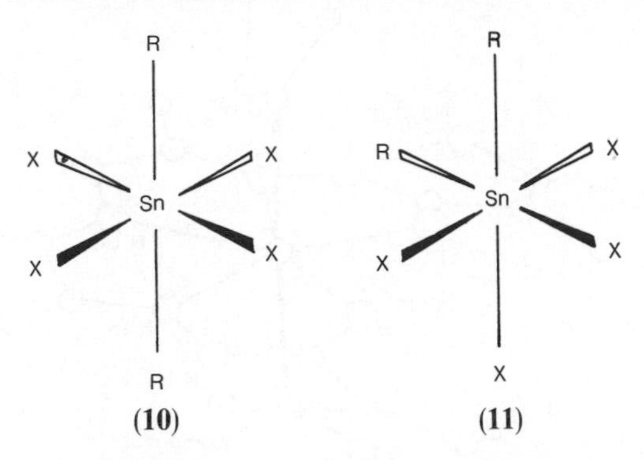

(10) **(11)**

isomerize in solution.[167] This is the only confirmed example in which both isomers of the same compound have been isolated, so clearly a fine balance of as yet undetermined factors has fortuitously been achieved in this instance. $Ph_2Sn[S_2CNEt_2]_2$ exemplifies the less common *cis*-isomer,[168] which, empirically, seems to be more favoured by the presence of two bidentate, rather than four monodentate, ligands. This latter compound is, however, distorted, and this is a common feature of this geometry.

The structure of $Me_2Sn[S_2CN(CH_2)_4]_2$[169] highlights two other structural features common to distorted $[R_2SnX_4]$ geometries. Firstly, the C–Sn–C bond angle is a function of the strength of the fifth and sixth bonds to tin, arising from the chelating ligands, and weak, anisobidentate chelation results in only slight opening of the C–Sn–C skeleton (137.3°) from 109° towards 180°. When chelation is symmetrical or isobidendente, a linear C–Sn–C arrangement accrues, e.g. $Ph_2Sn[S_2P(OPr^i)_2]_2$.[170] Secondly, bond lengths within the ligand should reflect its coordinating strength to tin. In the dithiocarbamate, anisobidentate chelation results in C–S and C=S which have clearly distinguishable bond lengths. Isobidentate chelation, as in the thiophosphate, induces a totally delocalized S–P–S system, with S–P bonds of equal length. These ligand bond length patterns therefore provide important ancillary evidence concerning the validity or otherwise of weak, coordinate tin-donor bonds.[213]

Monoorganotin compounds are also strongly Lewis-acidic, and readily form six-coordinated structures. Mono-organotin trihalide complexes with O, N, S donor

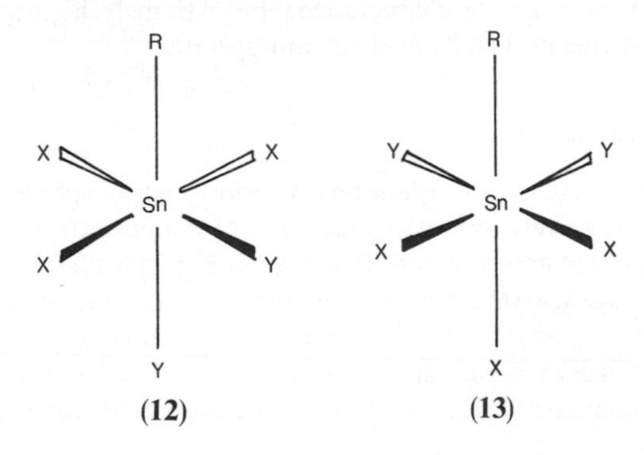

(12) **(13)**

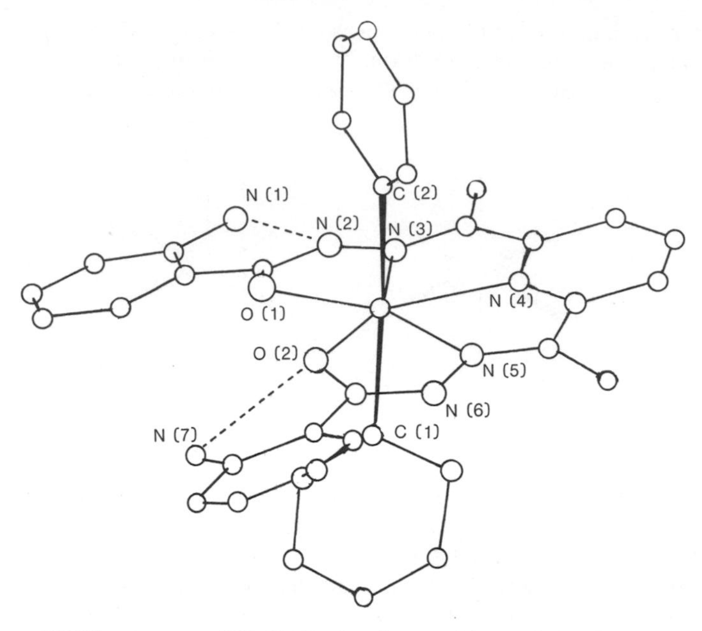

Figure 6.5 The structure of Ph_2Sn(dapa), redrawn with permission from Ref. 172.

ligands of stoichiometry $RSnX_3 \cdot 2L$ are common examples of this effect. Two isomeric forms, *mer*- (**12**) and *fac*- (**13**), are possible, and can be distinguished by Mössbauer QS values by comparison with the predicted values based upon a point charge model. The $BuSnCl_3$ adducts with either Ni(salphen) or Ni(salmphen) can produce either isomer depending on crystallization solvent, but have not yet been characterized crystallographically.[171]

Seven-coordinated organotin compounds have been less extensively studied, and considerable scope for further study exists in this area. Both mono- and diorganotin systems can achieve this coordination number, and universally adopt a pentagonal bipyramidal geometry. An example involving a pentadentate ligand, Ph_2Sn(dapa),[172] is shown in Figure 6.5, but these are less common than monoorganotins bound to three bidentate ligands, e.g. $MeSn(NO_3)_3$[173], or adducts incorporating ligands of lower denticity, e.g. $Me_2Sn(NCS)_2 \cdot$terpy,[174] or $Ph_2Sn(NO_3)_2 \cdot Ph_3PO$.[175] All compounds of this coordination number are characterized by extremely high-field (*ca.* -450 to -700 ppm) ^{119}Sn chemical shifts in their nmr spectra.[176]

6.4.2 *Rings, ladders and chains*

The examples chosen to depict organotin(IV) coordination spheres in the preceding section centred exclusively, with the exception of triphenyltin acetate, on discrete, molecular systems. It is, however, one of the fascinating aspects of this area of chemistry that molecules will associate through their ligands to produce clusters of increasing complexity, from dimeric units to three-dimensional networks.

Molecular association in one dimension to form linear coordination polymers is one of the more common lattice structures in organotin chemistry. Me_3SnX [X = Cl

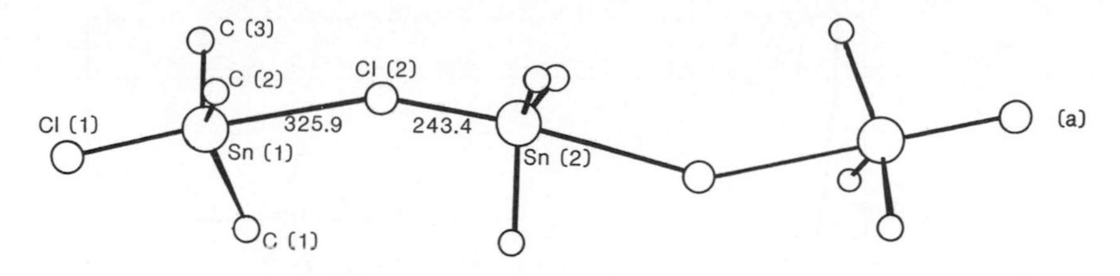

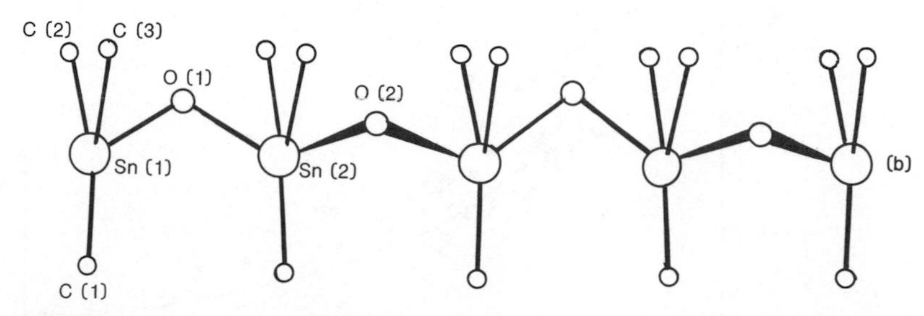

Figure 6.6 The structures of (*a*) Me_3SnCl and (*b*) a provisional analysis of the crystallographic data for Me_3SnOH, the latter redrawn with permission from Ref. 180. Distances in pm.

(Figure 6.6*a*),[115] OMe,[177] NCS,[178] N_3[179]] are typical, and all are based upon a *trans*-$[X_2SnR_3]$ repeat geometric unit. Uniquely, and without obvious rationale, Me_3SnOH adopts a structure which eclipses CH_3- groups, and places all the oxygen atoms on the same side of the polymer spine (Figure 6.6*a*).[180] This structure is made more surprising by the fact that Ph_3SnOH[181] crystallizes in the conventional array of Figure 6.6*a*.

Diorganotin dihalides (Me_2SnCl_2,[182] Et_2SnX_2, X = Cl, Br, I[183]) form similar one-

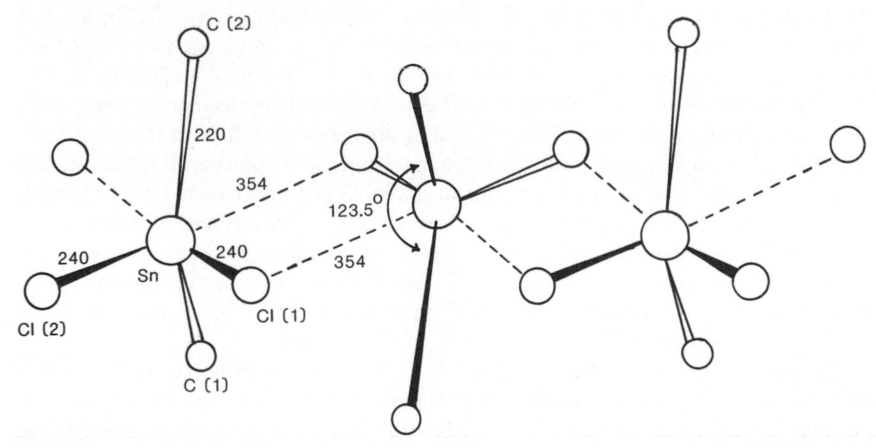

Figure 6.7 Part of the Me_2SnCl_2 coordination polymer, redrawn with permission from Ref. 182. Distances in pm.

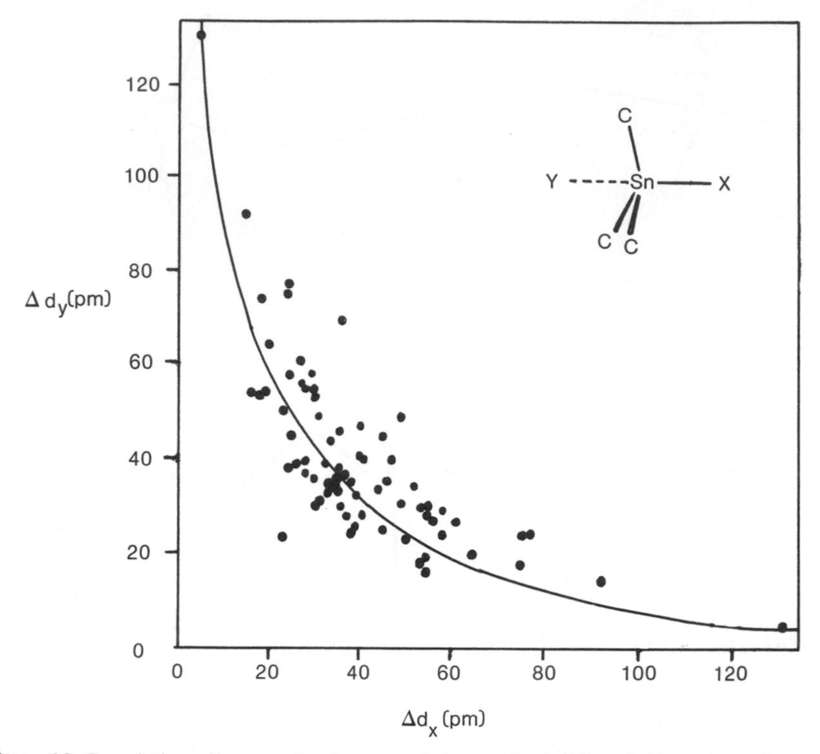

Figure 6.8 Correlation diagram for intra- and intermolecular bonds in polymeric *trans*-$[R_3SnXY]$ systems indicating the analogy with the S_{N2} reaction profile. $\Delta dy = d(Sn{-}Y)_{obs} - d(Sn{-}Y)_{single}$; Δdx is defined analogously. Redrawn with permission from Ref. 184.

dimensional chains, now with double halide bridges. Figure 6.7 shows the arrangement for Me_2SnCl_2. It is clear from this pictorial representation that the octahedral $[Me_2SnCl_4]$ unit is not regular and the non-linear C–Sn–C unit arises in this case from the weak, intermolecular Cl····Sn bridging bonds in a manner analogous to that described for molecular, chelated species. These deviations in molecular geometry at tin for varying strengths of intermolecular bonding are nicely visualized as frozen pictures of points along a reaction pathway for a nucleophilic displacement reaction. S_N2 substitution with inversion at a tetrahedral centre is thus depicted by polymeric *trans*-$[R_3SnX_2]$ geometries, whose structures map the reaction profile for this reaction (Figure 6.8). As the intermolecular Y····Sn bond becomes stronger, so the intramolecular Sn–X bond weakens and the C–Sn–C bond angle closes towards 90°. A regular $[R_3SnX_2]$ trigonal bipyramid with equal Sn–X, Sn–Y bonds corresponds to the S_N2 transition state.[184]

The manner in which the coordination polymer propagates through space is determined by the steric demands of both the hydrocarbon and ligand residues. The polymer spines of Ph_3SnO_2CR (R = H, Me) are shown in Figure 6.9. Bending of the acetate chain to minimize steric clashes between CH_3 and C_6H_5 is visually apparent.[185] As the bulk of substituents on tin or the acid increases, polymer bending becomes

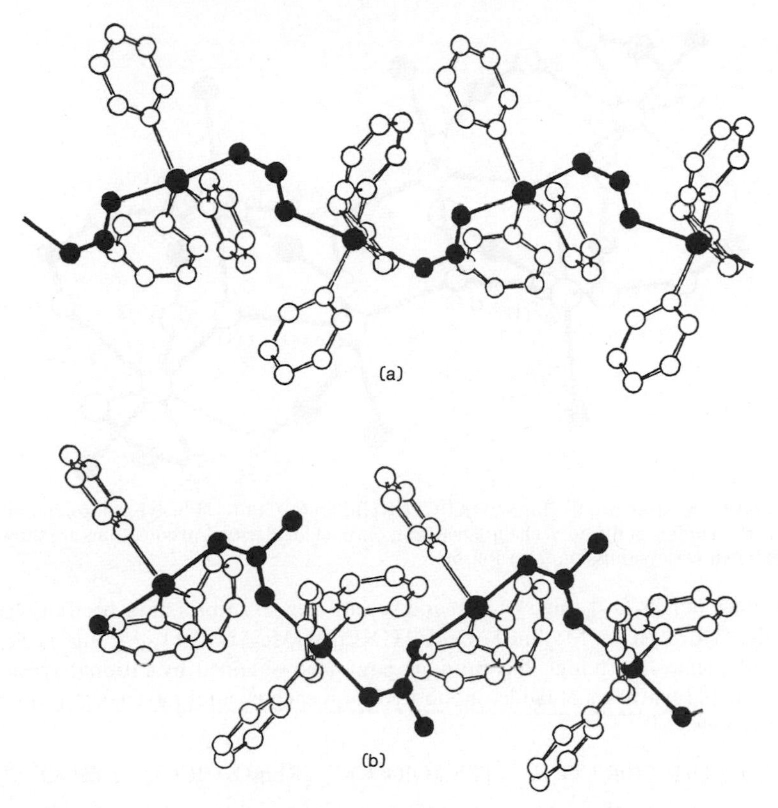

Figure 6.9 Polymer propagation in Ph_3SnO_2CR, (*a*) R = H, (*b*) R = CH_3. Data from Ref. 185.

accompanied by a weakening of the intermolecular bond, e.g. $(C_6H_{11})_3SnO_2CMe$[186], until finally only tetrahedral monomers are possible, e.g. $Ph_3SnO_2CCMe_3$.[187]

Coordination polymers which have CN > 4 at tin can be distinguished from discrete molecules with the same metal coordination number by comparison of solution and solid-state spectra. Intramolecular coordination is retained in solution, whereas intermolecular bonds are concentration dependent with complete polymer fragmentation accompanied by lowering of the tin coordination occurring in dilute solutions, and both ir and nmr methods can be used to investigate such phenomena (see Chapter 3).

Lattice association into two dimensions begins with the formation of cyclic oligomers, which then continue to link together into small sheets or 'ladder' arrays. Of the cyclic oligomers, dimers and trimers are well characterized, and exemplified by the auto-dimerization of organotin alkoxides. The enhanced coordination number at tin in the dimer is reflected in changes in ^{119}Sn nmr shifts.[188,189] Steric bulk on the OR′ (R′ = Pr^i, Bu^s, Bu^t) ligand inhibits dimer formation, and this effect similarly influences the degree of nuclearity of diorganotin chalcogenide oligomers, viz. $(Me_2SnX)_3$ *v.* $(Bu^t_2SnX)_2$ (X = S, Se).[139–141,190] The largest ring system known that is formed in cyclic coordination is hexameric $[Ph_3SnO_2P(OPh)_2]_6$[191] (see Figure 2.6).

Ladder structures, which are essentially a fusion in one direction of small rings, are particularly a feature of tin–oxygen compounds (probably due to the widespread

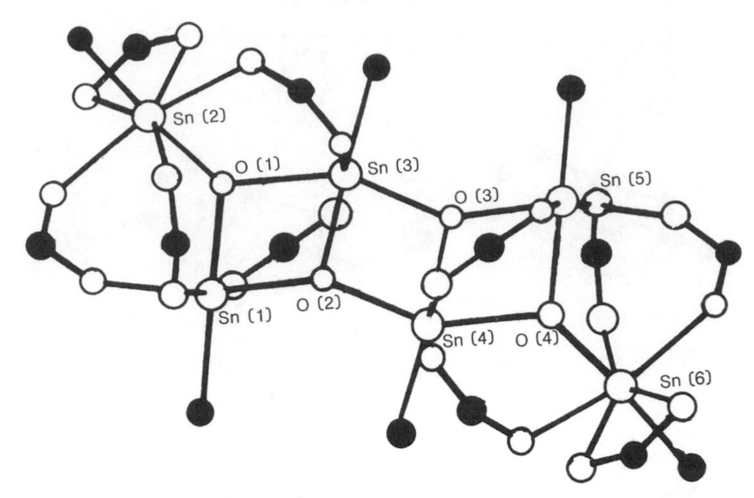

Figure 6.10 The structure of $[Bu^nSn(O)O_2CPh]_2[Bu^nSn(O_2CPh)_3]$. Phenyl groups and all but the α-carbon atoms of the butyl chains have been omitted for clarity. Carbon atoms are shown as ●. Redrawn with permission from Ref. 56.

occurrence of $[Sn_2O_2]$ ring systems) and several new examples have been published recently. Hydrolysis of Me_2SnCl_2 yields $[HNEt_3][(Me_2SnCl)_5O_3]$[192] and is clearly part of Scheme 4, although this product cannot be obtained by rational synthesis. Reproducible syntheses of ladder monoorganotin carboxylates have been devised by two methods:[56,193]

$$6RSn(O)OH + 10R'CO_2H \rightarrow \{[RSn(O)O_2CR']_2[RSn(O_2CR')_3]\}_2 + 8H_2O \quad (6.47)$$

$$6RSnCl_3 + 10AgO_2CR' + 4H_2O \rightarrow \{[RSn(O)O_2CR']_2[RSn(O_2CR')_3]\}_2 + 10AgCl + 8HCl \quad (6.48)$$

The structure of $[Bu^nSn(O)O_2CPh]_2[Bu^nSn(O_2CPh)_3]$ is illustrated in Figure 6.10.[56] Longer oligomeric[194] and also polymeric[195] ladder structures have been characterized for 1, 3, 2-dioxastannolanes and related systems.

6.4.3 *Layer compounds*

When intermolecular association takes place in two dimensions, layer structures result. Examples of this structural type are relatively rare, particularly in comparison with one-dimensional chain compounds, probably since increasing association reduces solubility and inhibits the growing of crystals. Most of the available examples occur with diorganotin halides or pseudohalides, e.g. Me_2SnF_2,[196] $Me_2Sn(CN)_2$,[197] $Me_2Sn[N(CN)_2]$.[198] Figure 6.11 shows a further example, $Me_2Sn(SO_3F)_2$.[199]

6.4.4 *Cages and networks*

Monoorganotins naturally incorporate three ligands, and are thus potentially able to form three-dimensional arrays. Where small cages, rather than extended networks, are

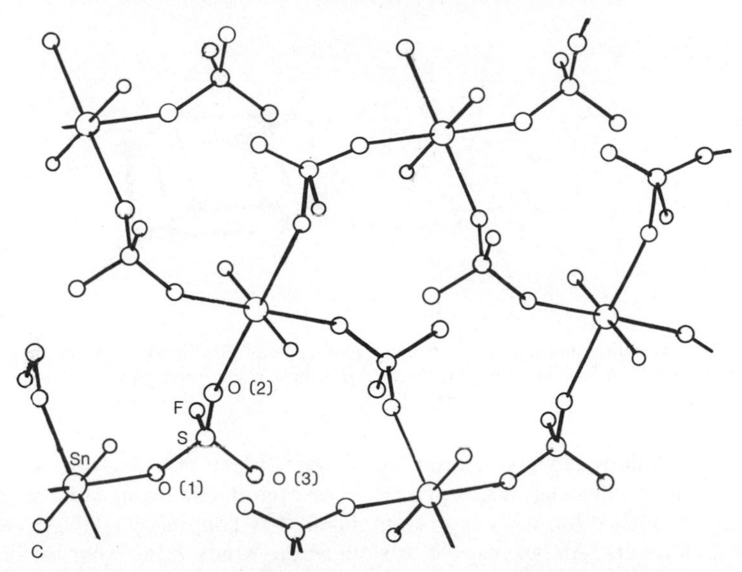

Figure 6.11 The layer structure of $Me_2Sn(SO_3F)_2$, redrawn with permission from Ref. 199.

formed, solubility is likely to be more favourable, and among the organotin derivatives of Groups 15 and 16 trigonal bipyramidal, cubic and adamantane cages have already been described. Of these, only the latter has been established crystallographically. To add to this range, a rather more exotic class of monoorganotin–oxygen compounds has recently appeared, which can be related to the ladder stannoxane of Figure 6.10. Scheme 11 shows how the ladder can be considered as an 'open drum', and by hydrolysis converted to a 'drum' structure. Synthesis of the drum, as with the ladder, originates with either a stannoic acid or organotin trihalide:[56,193,200]

$$6RSn(O)OH + 6R'CO_2H \rightarrow [RSn(O)O_2CR']_6 + 6H_2O \quad (6.49)$$

$$6RSnCl_3 + 6NaO_2CR' \rightarrow 6[RSnCl_2(O_2CR')] \xrightarrow{H_2O} [RSn(O)(O_2CR')]_6 + 12HCl \quad (6.50)$$

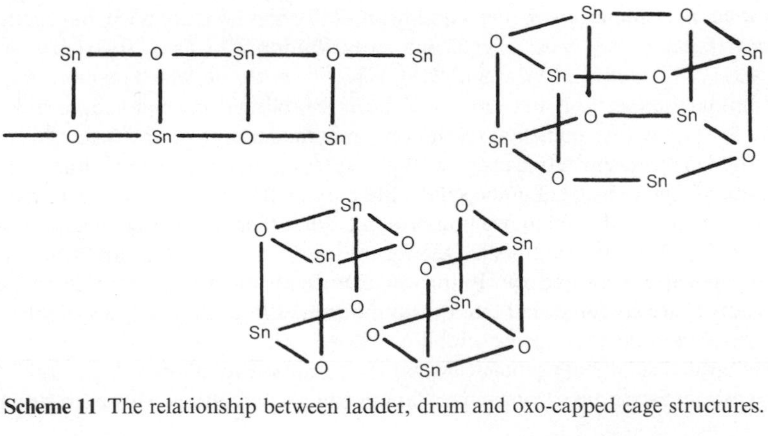

Scheme 11 The relationship between ladder, drum and oxo-capped cage structures.

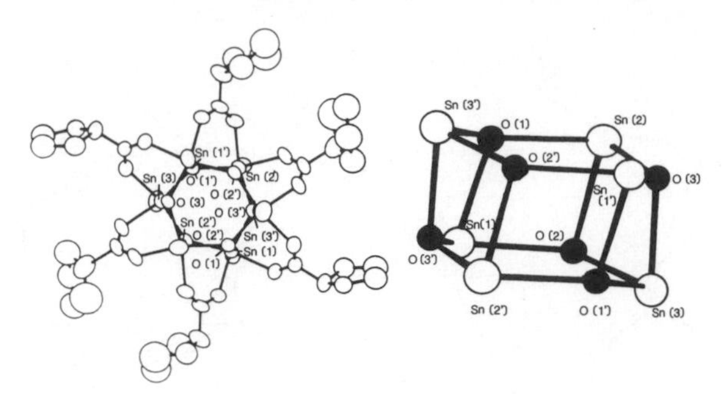

Figure 6.12 Two representations of the drum architecture of $[Bu^nSn(O)O_2CC_5H_9]_6$, redrawn with permission from Ref. 193. Only the α-carbon atoms of the butyl groups are included for clarity.

Two representations of a typical 'drum' structure, $[Bu^nSn(O)O_2CC_5H_9]_6$, are shown in Figure 6.12. If bulkier ligands are present, a cubic tetrameric analogue of the drum of identical empirical formula and exemplified by $[Bu^nSn(O)O_2P(C_6H_{11})_2]_4$[201] (Figure 2.7) results. An oxo-capped tristannoxane arises from what is formally hydrolysis of the drum arrangement (Scheme 11), an example of which is $[Bu^nSn(OH)O_2PPh_2)_3O]^+[Ph_2PO_2]^-$[202] (Figure 2.8). All of these structures, including the ladder, derive from the SnOSnO moiety, so cages with even numbers of tin atoms appear to predominate over odd-tin arrays which are subsequently derived from them.

Crystalline network structures are relatively rare, but occur when an organotin is bonded to a polyfunctional ligand. $(Me_3Sn)_3[Co(CN)_6]$ is a recent example.[203]

6.4.5 *The solution state*

Two topics are important when considering structure in solution. Firstly, what species are present, are there equilibria operating and are these solvent-dependent? Secondly, to what extent can data from the solid state–X-ray diffraction, Mössbauer spectroscopy–be correlated with species in solution? Studies relating to the first of these questions are remarkably limited, and even for aqueous media and the attendant environmental implications, the equilibria of Tobias (Figure 6.13) for methyltin systems remain the most significant contribution.[204] Two solvated cations, $[Bu_3Sn(H_2O)_2]^+$[205] and $[Cy_3Sn(NCMe)_2]^+$,[206] both of which exhibit trigonal bipyramidal geometry at tin, have now been established crystallographically. The former is of clear environmental relevance, while the latter type of structure has been implicated in the racemization of optically active triorganotin halides.[207] Furthermore, products which are isolated in the solid state may be the least-soluble rather than the thermodynamically-favoured reaction product, and other species may predominate in solution. Equilibrium constants for the formation of adducts of di- and triorganotin halides have been measured by ^{31}P-nmr and calorimetry, and suggest that in both cases 1:1 products are favoured, and that the dominance of $R_2SnCl_2 \cdot L_2$ species in the solid state merely reflects their lower solubility.[208,209]

Structurally, solid-state geometries and coordination numbers will generally hold for the solution state only if intermolecular coordination is absent. However, structural

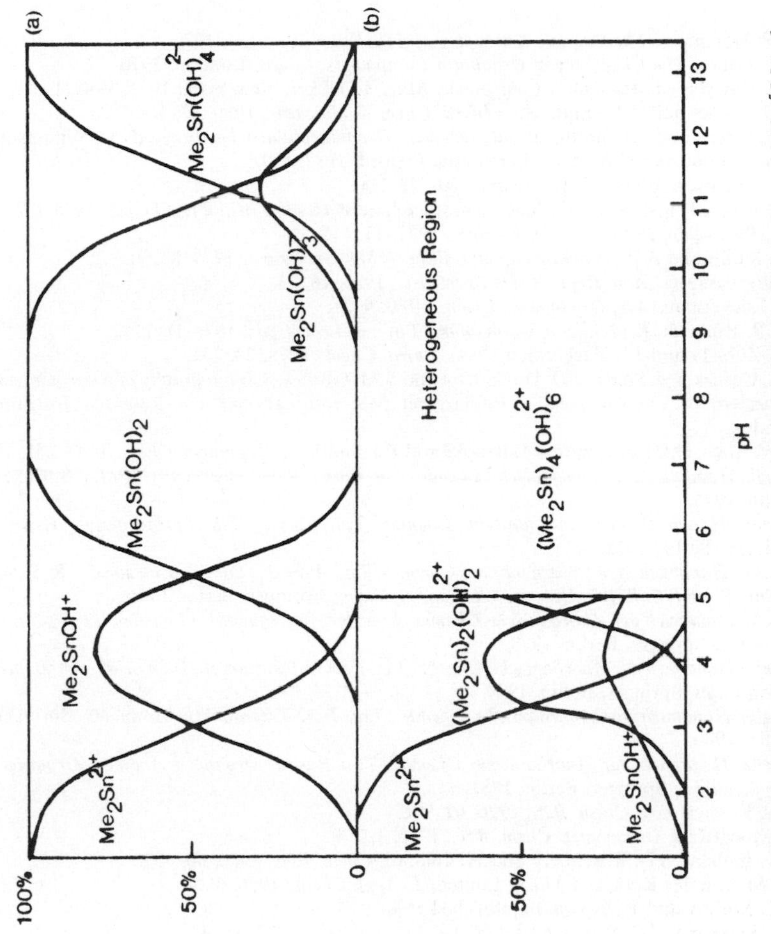

Figure 6.13 Species distribution diagrams for aqueous solutions of dimethyltin at (*a*) 10^{-5} M and (*b*) 10^{-2} M. Redrawn from ref. 204.

dynamics must also be considered in solution, and many fluxional organotin compounds are known. Of these, derivatives of cyclopentadiene and related carbocycles are the most celebrated, and have been widely studied by nmr methods.[210] Two-dimensional nmr methods have also been used to investigate structural equilibria,[211] and no doubt more use will be made of this technique in coming years.

References

1. W.P. Neumann, *The Organic Chemistry of Tin*, Wiley, London, 1970.
2. R.C. Poller, *The Chemistry of Organotin Compounds*, Logos, London, 1970.
3. A.K. Sawyer, ed., *Organotin Compounds*, Marcel Dekker, New York, 1971, Vols. 1–3.
4. A.G. Davies and P.J. Smith, *Adv. Inorg. Chem. Radiochem.*, 1980, **23**, 1.
5. A.G. Davies and P.J. Smith, in *Comprehensive Organometallic Chemistry*, eds. G. Wilkinson, F.G.A. Stone and E.W. Abel, Pergamon, Oxford, 1982, 519.
6. P.G. Harrison, *Coord. Chem. Revs.*, 1986, **75**, 200.
7. J.D. Kennedy and W. McFarlane, *Rev. Silicon, Germanium, Tin and Lead Cpds*, 1974, **I**, 235.
8. V.S. Petrosyan, *Prog. NMR Spectrosc.*, 1977, **11**, 115.
9. P.J. Smith and A.P. Tupčiauska, *Ann. Rept. NMR Spectrosc.*, 1978, **8**, 291.
10. B. Wrackmeyer, *Ann. Rept. NMR Spectrosc.*, 1985, **16**, 73.
11. J.J. Zuckerman, *Adv. Organomet. Chem.*, 1970, **9**, 21.
12. J.N.R. Ruddick, *Rev. Silicon, Germanium, Tin and Lead Cpds*, 1976, **II**, 115.
13. J.A. Zubieta and J.J. Zuckerman, *Prog. Inorg. Chem.*, 1978, **24**, 251.
14. P.A. Cusack, P.J. Smith, J.D. Donaldson and S.M. Grimes, *A Bibliography of X-ray Crystal Structures of Tin Compounds*, Publication 588, International Tin Research Institute, London.
15. D.W. Allen, D.J. Derbyshire, J.S. Brooks and P.J. Smith, *J. Organomet. Chem.*, 1983, **251**, 45.
16. *Gmelin Handbuch der Anorganische Chemie. Tin. Part 1: Tin Tetraorganyls* SnR_4, Springer, Berlin, 1975.
17. *Gmelin Handbuch der Anorganische Chemie. Tin. Part 2: Tin Tetraorganyls* R_3SnR, Springer, Berlin, 1975.
18. *Gmelin Handbuch der Anorganische Chemie. Tin. Part 3: Tin Tetraorganyls* $R_2SnR'_2$, $R_2SnR'R''$, $RR'SnR''R'''$. *Heterocyclics and Spiranes*, Springer, Berlin, 1976.
19. *Gmelin Handbuch der Anorganische Chemie. Tin. Part 5: Organotin Fluorides, Triorganotin Chlorides*, Springer, Berlin, 1978.
20. *Gmelin Handbuch der Anorganische Chemie. Tin. Part 6: Diorganotin Dichlorides, Organotin Trichlorides*, Springer, Berlin, 1979.
21. *Gmelin Handbuch der Anorganische Chemie. Tin. Part 7: Organotin Bromides*, Springer, Berlin, 1980.
22. *Gmelin Handbuch der Anorganische Chemie. Tin. Part 8: Organotin Iodides, Organotin Pseudohalides*, Springer, Berlin, 1981.
23. K.A. Kocheshkov, *Chem. Ber.*, 1926, **62**, 996.
24. K. Moedritzer, *Organomet. Chem. Rev.*, 1966, **1**, 179.
25. R.K. Ingham, S.D. Rosenberg and H. Gilman, *Chem. Rev.*, 1960, **60**, 459.
26. G.J.M. van der Kerk and J.G.A. Luijten, *J. Appl. Chem.*, 1956, **6**, 56.
27. K.C. Molloy and P. Brown, unpublished results.
28. L.E. Levchuck, J.R. Sams and F. Aubke, *Inorg. Chem.*, 1972, **11**, 43.
29. *Gmelin Handbuch der Anorganische Chemie. Tin. Part 4: Organotin Hydrides*, Springer, Berlin, 1976.
30. A.K. Sawyer and H.G. Kuivila, *Chem. Ind. (London)*, 1961, 260.
31. R. Somer, B. Schneider and W.P. Neumann, *Liebigs Ann. Chem.*, 1966, **692**, 12.
32. G.J.M. van der Kerk, J.G. Noltes and J.G.A. Luijten, *J. Appl. Chem.*, 1957, **7**, 356.
33. W.P. Neumann and B. Schneider, *Angew. Chem., Int. Edn. Engl.*, 1964, **3**, 751.
34. F.J.A. des Tombe, G.J.M. van der Kerk, H.M.J.C. Creemers, N.A.D. Corey and J.G. Noltes, *J. Organomet. Chem.*, 1972, **44**, 247.
35. A.J. Levsink, H.A. Budding and W. Drenth, *J. Organomet. Chem.*, 1968, **11**, 541.
36. A.K. Sawyer and H.G. Kiuvila, *J. Org. Chem.*, 1962, **27**, 610.

37. W.P. Neumann, *Liebigs Ann. Chem.*, 1964, **667**, 1.
38. F. Glockling, P. Harriott and W.-K. Ng, *J. Chem. Res.*, 1979, (S) 12.
39. C.K. Chu and J.D. Murray, *J. Chem. Soc. (A)*, 1971, 360.
40. H. Puff, I. Bung, E. Friedrichs and A. Jansen, *J. Organomet. Chem.*, 1983, **254**, 23.
41. H. Puff, E. Friedrichs and F. Visel, *Z. Anorg. Allg. Chem.*, 1981, **477**, 50.
42. H. Puff, W. Schuh, R. Sievers and R. Zimmer, *Angew. Chem., Int. Edn. Engl.*, 1981, **20**, 591.
43. C. Le Compte, J. Protas and M. Devaud, *Acta Crystallogr., Sect. B*, 1976, **32**, 923.
44. F. Glockling and W.-K. Ng, *J. Chem. Res.*, 1980, (S) 230.
45. R.C. Mehrotra and R. Bohra, *Metal Carboxylates*, Academic Press, London, 1983.
46. H.H. Anderson, *Inorg. Chem.*, 1964, **3**, 108.
47. R. Okawara and M. Wada, in Ref. 3., Vol. 2, 253.
48. J. Kizlink and V. Rattay, *Coll. Czech. Chem. Comm.*, 1987, **52**, 1514.
49. K.C. Molloy and J.J. Zuckerman, *Acc. Chem. Res.*, 1983, **16**, 386.
50. C. Silvestru, F. Ilies, I. Haiduc, M. Gielen and J.J. Zuckerman, *J. Organomet. Chem.*, 1987, **330**, 315.
51. K. Yasuda and R. Okawara, *J. Organomet. Chem.*, 1965, **3**, 76.
52. A. Rieche and J. Dahlmann, *Liebigs Ann. Chem.*, 1964, **675**, 19.
53. M.F. Salomon and R.G. Salomon, *J. Amer. Chem. Soc.*, 1977, **99**, 3500.
54. S.J. Blunden, P.A. Cusack and P.J. Smith, *J. Organomet. Chem.*, 1987, **325**, 141.
55. A.G. Davies, L. Smith and P.J. Smith, *J. Organomet. Chem.*, 1972, **39**, 279.
56. R.R. Holmes, C.G. Schmid, V. Chandrasekhar, R.O. Day and J.M. Holmes, *J. Amer. Chem. Soc.*, 1987, **109**, 1408.
57. A.J. Bloodworth, A.G. Davies and S.C. Vasishtha, *J. Chem. Soc. (C)*, 1967, 1309.
58. A.G. Davies, P.G. Harrison and P.R. Palan, *J. Organomet. Chem.*, 1967, **10**, 33.
59. A.G. Davies, *Synthesis*, 1969, 56.
60. A.J. Crowe, R. Hill and P.J. Smith, Publication No. 599, International Tin Research Institute, London.
61. K. Jones and M.F. Lappert, *J. Chem. Soc.*, 1965, 1944.
62. T. Gasparis-Ebeling, H. Nöth and B. Wrackmeyer, *J. Chem. Soc., Dalton Trans.*, 1983, 97.
63. M. Pereyre, J.-P. Quintard and A. Rahm, *Tin in Organic Synthesis*, Butterworth, London, 1987, 17.
64. D.D. Davis and C.L. Gray, *Organomet. Chem. Rev.*, 1970, **6**, 283.
65. H.G. Kuivila, in *Organotin Compounds: New Chemistry and Applications*, ed. J.J. Zuckerman, Adv. Chem. Ser. **157**, American Chemical Society, Washington DC, 1976, 227.
66. E.C. Juenge, T.E. Snider and Y.-C. Lee, *J. Organomet. Chem.*, 1970, **22**, 403.
67. J.-C. Lahournère and J. Valade, *J. Organomet. Chem.*, 1970, **22**, C3.
68. J.-C. Lahournère and J. Valade, *J. Organomet. Chem.*, 1971, **33**, C4.
69. H. Kuivila, J.L. Considine and J.D. Kennedy, *J. Amer. Chem. Soc.*, 1972, **94**, 7206.
70. L.C. Willemsens and G.J.M. van der Kerk, *J. Organomet. Chem.*, 1964, **2**, 260.
71. S. Adams and M. Dräger, *J. Organomet. Chem.*, 1987, **323**, 11.
72. D.F. Gaines and T.V. Iorns, *J. Amer. Chem. Soc.*, 1968, **90**, 6617.
73. H. Nöth, H. Schafer and G. Schmid, *Z. Naturforsch., Teil B*, 1971, **26**, 497.
74. H. Nöth and R. Schwerthoffer, *Chem. Ber.*, 1981, **114**, 3056.
75. W. Biffar, H. Nöth, H. Pommerening, R. Schwerthoffer, W. Storch and B. Wrackmeyer, *Chem. Ber.*, 1981, **114**, 49.
76. A.T. Weibel and J.P. Oliver, *J. Amer. Chem. Soc.*, 1972, **94**, 8590.
77. H. Blaukat and W.P. Neumann, *J. Organomet. Chem.*, 1973, **63**, 27.
78. F.J.A. des Tombe, G.J.M. van der Kerk and J.G. Noltes, *J. Organomet. Chem.*, 1972, **43**, 323.
79. T.N. Mitchell, *J. Organomet. Chem.*, 1975, **92**, 311.
80. D.C. McWilliam and P.R. Wells, *J. Organomet. Chem.*, 1975, **85**, 347.
81. C. Feasson and M. Devaud, *Bull. Soc. Chim. Fr.*, 1983, 40.
82. Yu. K. Grishin, V.A. Roznyatovskii, V.A. Ustynyuk, M.N. Bochkarev, G.S. Kalinina and G.A. Razuvaev, *Izv. Akad. Nauk SSSR, Ser. Khim.*, 1980, 2190.
83. Y. Azuma and M. Newcomb, *Organometallics*, 1984, **3**, 9.
84. M. Newcomb, A.M. Madonik, M.T. Blanda and J.K. Judice, *Organometallics*, 1987, **6**, 145.
85. T.N. Mitchell, B. Fabisch, R. Wickenkamp, H.G. Kuivila and T.J. Karol, *Silicon, Germanium, Tin and Lead Cpds*, 1986, **IX**, 57.
86. H. Shirai, Y. Sato and M. Niwa, *Yakugaku Zasshi*, 1970, **90**, 59; *Chem. Abstr.*, 1972, **72**, 90593.

87. M. Engel and M. Devaud, *J. Chem. Res.*, 1984, (S) 152.
88. G. Wittig, E.J. Meyer and G. Lange, *Liebigs Ann. Chem.*, 1951, **571**, 167.
89. B. Watta, W.P. Neumann and J. Sauer, *Organometallics*, 1985, **4**, 1954.
90. H. Puff, B. Breuer, W. Schuh, R. Sievers and R. Zimmer, *J. Organomet. Chem.*, 1987, **332**, 279.
91. B. Jousseame, E. Chanson, M. Bevilacqua, A. Saux, M. Pereyre, B. Barbe and M. Petraud, *J. Organomet. Chem.*, 1985, **294**, C41.
92. B. Jousseame, E. Chanson and M. Pereyre, *Organometallics*, 1983, **5**, 1271.
93. V.I. Goldanskii, V. Ya. Rocher and V.V. Kharpov, *Proc. Acad. Sci. USSR, Phys. Chem. Sect. (Engl. Transl.)*, 1964, **156**, 571.
94. H. Schumann and S. Ronecker, *Z. Naturforsch., Teil B*, 1967, **22**, 452.
95. F. Glockling, *The Chemistry of Germanium*, Academic Press, London, 1969, 160.
96. P.G. Harrison, in Ref. 5, Chapter 12, 665.
97. S. Adams and M. Dräger, *Abstr. 5th Int. Conf. Chem. Si, Ge, Sn and Pb Cpds*, Padua, Italy, 1986, 9.
98. S. Masamune and L.R. Sita, *J. Amer. Chem. Soc.*, 1985, **107**, 6390.
99. W.P. Neumann and J. Pedain, *Tetrahedron Lett.*, 1964, 2461.
100. S. Adams and M. Dräger, *J. Organomet. Chem.*, 1985, **288**, 295.
101. M. Devaud, M. Engel, C. Feasson and J.-L. Lecat, *J. Organomet. Chem.*, 1985, **281**, 181.
102. S. Adams, M. Dräger and B. Mathiasch, *Z. Anorg. Allg. Chem.*, 1986, **532**, 81.
103. U. Schröer and W.P. Neumann, *Angew. Chem., Int. Edn. Engl.*, 1975, **14**, 246.
104. B. Mathiasch, *J. Organomet. Chem.*, 1977, **141**, 295.
105. B. Mathiasch and T.N. Mitchell, *J. Organomet. Chem.*, 1980, **185**, 351.
106. A.K. Sawyer, in Ref. 3, Vol. 3, 845.
107. T. Birchall and J.P. Johnson, *Can. J. Chem.*, 1979, **60**, 934.
108. M. Dräger, *J. Organomet. Chem.*, 1987, **326**, 173.
109. K. Jones and M.F. Lappert, *Organomet. Chem. Rev.*, 1966, **1**, 67.
110. M.F. Lappert, P.P. Power, A.R. Sanger and R.C. Srivastava, *Metal and Metalloid Amides*, Ellis Horwood, Chichester, 1980, 235.
111. K.C. Molloy, M.P. Bigwood, R.H. Herber and J.J. Zuckerman, *Inorg. Chem.*, 1982, **21**, 3709.
112. J.G.A. Luijten, M.J. Janssen and G.J.M. van der Kerk, *Rec. Trav. Chim. Pays-Bas*, 1962, **81**, 202.
113. S. Kozima, T. Itano, N. Mihara, K. Sisido and T. Isida, *J. Organomet. Chem.*, 1972, **44**, 117.
114. I. Hammann, K.H. Büchel, K. Bungarz and L. Born, *Pflanzenschutz-Nachr.*, 1978, **31**, 61.
115. L.S. Khaikin, A.V. Belyakov, G.S. Koptev, A.V. Golubinskii, L.V. Vilkov, N.V. Girbasova, E.T. Bogoradovskii and V.S. Zavgorodnii, *J. Mol. Struct.*, 1980, **66**, 191.
116. H. Schumann and H. Benda, *Angew. Chem., Int. Edn. Engl.*, 1968, **7**, 812.
117. H. Schumann and H. Benda, *Angew. Chem., Int. Edn. Engl.*, 1969, **8**, 989.
118. H. Schumann and H. Benda, *Angew. Chem., Int. Edn. Engl.*, 1968, **7**, 813.
119. M. Dräger and B. Mathiasch, *Angew. Chem., Int. Edn. Engl.*, 1981, **20**, 1029.
120. B. Mathiasch, *J. Organomet. Chem.*, 1979, **165**, 295.
121. C. Couret, J. Escudie, J. Satgé, A. Raharinirina and J.D. Andriamizaka, *J. Amer. Chem. Soc.*, 1985, **107**, 8280.
122. S. Roller, M. Dräger, H.-J. Breunig, M. Ates and S. Gülec, *J. Organomet. Chem.*, 1987, **329**, 319.
123. *Gmelin Handbuch der Anorganische Chemie. Tin. Part 9: Triorganotin Sulphur Compounds*, Springer, Berlin, 1982.
124. *Gmelin Handbuch der Anorganische Chemie. Tin. Part 10: Mono- and Diorganotin Sulphur Compounds. Organotin-Selenium and Tellurium Compounds*, Springer, Berlin, 1983.
125. W.N. Aldridge, in Ref. 65, 186.
126. K. Gratz, F. Huber, A. Silvestri, G. Alonzo and R. Barbieri, *J. Organomet. Chem.*, 1985, **290**, 41.
127. E.W. Abel and D.A. Armitage, *Adv. Organomet. Chem.*, 1967, **5**, 1.
128. G.S. Sasin, *J. Org. Chem.*, 1953, **18**, 1142.
129. W.T. Reichle, *Inorg. Chem.*, 1962, **1**, 650.
130. A.G. Davies and P.G. Harrison, *J. Chem. Soc. (C)*, 1970, 2035.
131. H. Schumann, K.F. Thom and M. Schmidt, *J. Organomet. Chem.*, 1964, **2**, 361.
132. C.A. Kraus and W.V. Sessions, *J. Amer. Chem. Soc.*, 1925, **47**, 2361.
133. H. Roesky and H. Weizer, *Angew. Chem., Int. Edn. Engl.*, 1973, **12**, 674.

134. R. Jones, C.P. Warrens, D.J. Williams and J.D. Woolins, *J. Chem. Soc., Dalton Trans.*, 1987, 907.
135. H. Puff, A. Bongartz, W. Schuh and R. Zimmer, *J. Organomet. Chem.*, 1983, **248**, 61.
136. H. Puff, E. Friedrichs, R. Hundt and R. Zimmer, *J. Organomet. Chem.*, 1983, **259**, 79.
137. P.G. Harrison, M.J. Begley and K.C. Molloy, *J. Organomet. Chem.*, 1980, **186**, 213.
138. H. Puff, R. Gattermeyer, R. Hundt and R. Zimmer, *Ang. Chem., Int. Edn. Engl.*, 1977, **16**, 547.
139. B. Menzebach and P. Bleckmann, *J. Organomet. Chem.*, 1975, **91**, 291.
140. H.-J. Jacobsen and B. Krebs, *J. Organomet. Chem.*, 1977, **136**, 333.
141. M. Dräger, A. Blecher, H.-J. Jacobsen and B. Krebs, *J. Organomet. Chem.*, 1978, **161**, 319.
142. D. Kobelt, E.F. Paulus and H. Scherer, *Acta Crystallogr., Sect. B*, 1972, **28**, 2323.
143. A. Blecher, M. Dräger and B. Mathiasch, *Z. Naturforsch., Teil B*, 1981, **36**, 1361.
144. H. Puff, A. Bongartz, R. Sievers and R. Zimmer, *Angew. Chem. Int. Edn. Engl.*, 1978, **17**, 939.
145. A.L. Beauchamp, S. Latour, M.J. Oliver and J.D. Wuest, *J. Amer. Chem. Soc.*, 1983, **105**, 7778.
146. Y. Ducharme, S. Latour and J.D. Wuest, *J. Amer. Chem. Soc.*, 1984, **106**, 1499.
147. R. Hulme, *J. Chem. Soc.*, 1963, 1524.
148. N.A. Akhmed and G.G. Aleksandrov, *J. Struct. Chem.*, 1970, **11**, 824.
149. G.R. Davies, J.A. Jarvis and B.T. Kilbourn, *J. Chem. Soc., Chem. Commun.*, 1971, 1511.
150. L. Prasad and F.E. Smith, *Acta Crystallogr., Sect. B*, 1982, **38**, 1815.
151. P.G. Harrison and K.C. Molloy, *J. Organomet. Chem.*, 1978, **152**, 53.
152. G.M. Sheldrick and W.S. Sheldrick, *J. Chem. Soc. (A)*, 1970, 490.
153. M. Chih and B.R. Penfold, *J. Cryst. Mol. Struct.*, 1973, **3**, 285.
154. M.J.S. Gynane, M.F. Lappert, S.J. Miles, A.J. Carty and N.J. Taylor, *J. Chem. Soc., Dalton Trans.*, 1977, 2009.
155. J.L. Lefferts, K.C. Molloy, M.B. Hussain, D. van der Helm and J.J. Zuckerman, *J. Organomet. Chem.*, 1982, **240**, 349.
156. P.G. Harrison, K.C. Molloy, R.C. Phillips, P.J. Smith and A.J. Crowe, *J. Organomet. Chem.*, 1978, **160**, 421.
157. D.W. Allen, D.J. Derbyshire, I.W. Nowell and J.S. Brooks, *J. Organomet. Chem.*, 1984, **260**, 263.
158. T.J. King and P.G. Harrison, *J. Chem. Soc., Dalton Trans.*, 1974, 2298.
159. K.C. Molloy, S.J. Blunden and R. Hill, *J. Chem. Soc., Dalton Trans.*, 1988, 1259.
160. H.A. Bent, *Chem. Rev.*, 1961, **61**, 275.
161. F. Huber, H.-J. Preut, H. Preut, R. Barbieri and M.T. Lo Guidice, *Z. Anorg. Allg. Chem.*, 1977, **432**, 51.
162. V.G. Kumar Das, L.K. Mun, C. Wei and T.C.W. Mak, *Organometallics*, 1987, **6**, 10.
163. K. Jurkschat and A. Tzschach, *Pure Appl. Chem.*, 1986, **58**, 639.
164. V.S. Petrosyan and O.A. Reutov, *Pure Appl. Chem.*, 1974, **37**, 147.
165. B.K. Nicholson, *J. Organomet. Chem.*, 1984, **265**, 153.
166. K.C. Molloy, T.G. Purcell, K. Quill and I.W. Nowell, *J. Organomet. Chem.*, 1984, **267**, 237.
167. V.G. Kumar Das, Y.C. Keong, C. Wei, P.J. Smith and T.W.C. Mak, *J. Chem. Soc., Dalton Trans.*, 1987, 129.
168. P.F. Lindley and P. Carr, *J. Cryst. Mol. Struct.*, 1974, **4**, 173.
169. T.P. Lockhart, W.F. Manders and E.O. Schlemper, *J. Amer. Chem. Soc.*, 1985, **107**, 7451.
170. K.C. Molloy, M.B. Hossain, D. van der Helm, J.J. Zuckerman and I. Haiduc, *Inorg. Chem.*, 1980, **19**, 2041.
171. D. Cunningham, J. Fitzgerald and M. Little, *J. Chem. Soc., Dalton Trans.*, 1987, 2261.
172. C. Pelizzi, G. Pelizzi and G. Predieri, *J. Organomet. Chem.*, 1984, **263**, 9.
173. G.S. Brownlee, A. Walker, S.C. Nyburg and J.R. Szynaski, *J. Chem. Soc., Chem. Commun.*, 1971, 1073.
174. D.V. Naik and W. Scheidt, *Inorg. Chem.*, 1973, **12**, 272.
175. M. Nardelli, C. Pelizzi and G. Pelizzi, *J. Chem. Soc., Dalton Trans.*, 1978, 131.
176. J. Otera, T. Hinoishi and R. Okawara, *J. Organomet. Chem.*, 1980, **202**, C93.
177. A.M. Domingos and G.M. Sheldrick, *Acta Crystallogr., Sect. B*, 1974, **30**, 519.
178. R.A. Forder and G.M. Sheldrick, *J. Organomet. Chem.*, 1970, **21**, 115.
179. R. Allmann, R. Hohlfeld, A. Waskowska and J. Lorberth, *J. Organomet. Chem.*, 1980, **192**, 353.
180. A.G. Davies, S.D. Slater, D.C. Povey and G.W. Smith, unpublished work; cited in *Revs. Si, Ge, Sn, and Pb Cmpds.*, 1986, **IX**, 87.

181. C. Glidewell and D.C. Liles, *Acta Crystallogr., Sect. B*, 1978, **34**, 129.
182. A.G. Davies, J. Milledge, D.C. Puxley and P.J. Smith, *J. Chem. Soc. (A)*, 1970, 2862.
183. N.W. Alcock and J.F. Sawyer, *J. Chem. Soc., Dalton Trans.*, 1977, 1090.
184. D. Britton and J.D. Dunitz, *J. Amer. Chem. Soc.*, 1981, **102**, 2971.
185. K.C. Molloy, K. Quill and I.W. Nowell, *J. Chem. Soc., Dalton Trans.*, 1987, 101.
186. N.W. Alcock and R.E. Timms, *J. Chem. Soc. (A)*, 1968, 1876.
187. B.F.E. Ford and J.R. Sams, *J. Organomet. Chem.*, 1970, **21**, 345.
188. P.J. Smith, R.F.M. White and L. Smith, *J. Organomet. Chem.*, 1972, **40**, 341.
189. J.D. Kennedy, W. McFarlane, P.J. Smith, R.F.M. White and L. Smith, *J. Chem. Soc., Perkin Trans., 2*, 1973, 1785.
190. H. Puff, R. Gattermeyer, R. Hundt and R. Zimmer, *Angew. Chem., Int. Edn. Engl.*, 1977, **16**, 547.
191. K.C. Molloy, F.A.K. Nasser, C.L. Barnes, D. van der Helm and J.J. Zuckerman, *Inorg. Chem.*, 1982, **21**, 960.
192. N.W. Alcock, M. Pennington and G.R. Willey, *J. Chem. Soc., Dalton Trans.*, 1985, 2683.
193. V. Chandrasekhar, C.G. Schmid, S.D. Burton, J.M. Holmes, R.O. Day and R.R. Holmes, *Inorg. Chem.*, 1987, **26**, 1050.
194. C.W. Holzapfel, J.M. Kochmoer, C.M. Marais, G.J. Kruger and J.A. Pretorius, *S. Afr. J. Chem.*, 1982, **35**, 81.
195. A.G. Davies, A.J. Price, H.M. Davies and M.B. Hursthouse, *J. Chem. Soc., Dalton Trans.*, 1986, 297.
196. E.O. Schlemper and W.C. Hamilton, *Inorg. Chem.*, 1966, **5**, 995.
197. J. Konnert, D. Britton and Y.M. Chow, *Acta Crystallogr.*, 1972, **B28**, 180.
198. Y.M. Chow, *Inorg. Chem.*, 1971, **10**, 1938.
199. F.H. Allen, J.A. Lerbscher and J. Trotter, *J. Chem. Soc. (A)*, 1971, 2507.
200. V. Chandrasekhar, R.O. Day and R.R. Holmes, *Inorg. Chem.*, 1985, **24**, 1970.
201. K.C. Kumara Swamy, R.O. Day and R.R. Holmes, *J. Amer. Chem. Soc.*, 1987, **109**, 5546.
202. R.O. Day, J.M. Holmes, V. Chandrasekhar and R.R. Holmes, *J. Amer. Chem. Soc.*, 1987, **109**, 940.
203. K. Yünlii, N. Hock and R.D. Fischer, *Angew. Chem., Int. Edn. Engl.*, 1985, **24**, 879.
204. R.S. Tobias, Adv. Chem. Ser., **82**, American Chemical Society, Washington DC, 1978, 130.
205. A.G. Davies, J.P. Goddard and M.B. Hursthouse, *J. Chem. Soc., Chem. Commun.*, 1983, 597.
206. W.A. Nugent, R.J. McKinney and R.L. Harlow, *Organometallics*, 1984, **3**, 1315.
207. M. Gielen, *Pure Appl. Chem.*, 1980, **52**, 657.
208. J.N. Spencer, R.B. Belser, S.R. Moyer, R.E. Haines, M.A. DiStravalo and C.H. Yoder, *Organometallics*, 1986, **5**, 118.
209. C.H. Yoder, D. Mokrynka, S.M. Coley, J.C. Otter, R.E. Haines, A. Grushaw, L.J. Ansel, J.W. Hovick, J. Mikus, M.A. Shermak and J.N. Spencer, *Organometallics*, 1987, **6**, 1679.
210. A.D. McMaster and S.R. Stobart, *J. Amer. Chem. Soc.*, 1982, **104**, 2109.
211. C. Wyants, G. van Binst, C. Mügge, K. Jurkschat, A. Tzschach, H. Pepermans, M. Gielen and R. Willem, *Organometallics*, 1985, **4**, 1906.
212. The structure of Ph_3SnK-18-crown-6 has been reported in which a pyramidal Ph_3Sn^- anion is >600 pm from the nearest 18-crown-6 complexed K^+ cation. T. Birchall and J.A. Ventrone, *J. Chem. Soc., Chem. Commun.*, 1988, 877.
213. A discussion of distorted octahedral complexes in terms of a skew-trapezoidal bipyramid can be found in: S.W. Ng, V.G. Kumar Das and T.C.W. Mak, *J. Organomet. Chem.*, 1987, **334**, 295, and references therein.
214. C.S. Cundy, B.M. Kingston and M.F. Lappert, *Adv. Organomet. Chem.*, 1975, **11**, 253.

7 Organometallic compounds of bivalent tin

P.D. LICKISS

7.1 Review articles

Early work on the nature of stannylenes has been reviewed by Neumann.[1] An article by Connolly and Hoff[2] in 1981 gives a good account of work up to about 1979. Transition metal complexes of silylenes, germylenes, and plumbylenes as well as stannylenes have been reviewed recently,[3] as has the chemistry of stannocenes and other Group IV π complexes.[4]

7.2 σ-Bonded compounds

7.2.1 *Synthesis*

Simple dialkyl- and diaryltins, R_2Sn ($R = Me, Ph, etc.$) are thought to be reactive intermediates in many reactions, but in the absence of trapping reagents cyclic, linear or branched oligomers are formed. A recent example is the reaction[5] of 9-phenanthrylmagnesium bromide with $SnCl_2$ which yields a mixture of cyclic, open-chain and polymeric stannanes rather than di(9-phenanthryl)tin as earlier work had suggested.[6] However, simple R_2Sn species may be trapped either at low temperature in a matrix or chemically.

For example, dimethylstannylene, Me_2Sn, is formed when the 1,4-ditellura heterocycle (**1**) is photolysed and can be trapped with CH_3I giving Me_3SnI (eqn 7.1).[7]

$$\text{cyclo-}(Me_2Sn\text{-}Te\text{-}SnMe_2\text{-}SnMe_2\text{-}Te\text{-}SnMe_2)\ (\mathbf{1}) \xrightarrow{h\nu} [Me_2Sn] \xrightarrow{MeI} [Me_3SnI] \tag{7.1}$$

Neumann[8] found that thermolysis of cyclo-hexastannanes and subsequent trapping in an argon matrix at 5 K was a good clean source of monomeric Me_2Sn and $(CD_3)_2Sn$, which enabled a detailed IR spectroscopic study to be carried out (eqn 7.2).

$$(R_2Sn)_6 \xrightarrow{130^\circ C} R_2Sn \text{ (trapped in Ar matrix)} \tag{7.2}$$

$$R = CH_3 \text{ or } CD_3$$

Treatment of $SnCl_2$ with $LiCHCl_2$ at $-78^\circ C$ in THF solution gives rise to a product which was formulated as $(CHCl_2)_2Sn \cdot x THF$ (eqn 7.3), but attempts to remove the non-coordinated THF resulted in the formation of oligomers, making difficult the characterization and proof of the nature of the compound.[9]

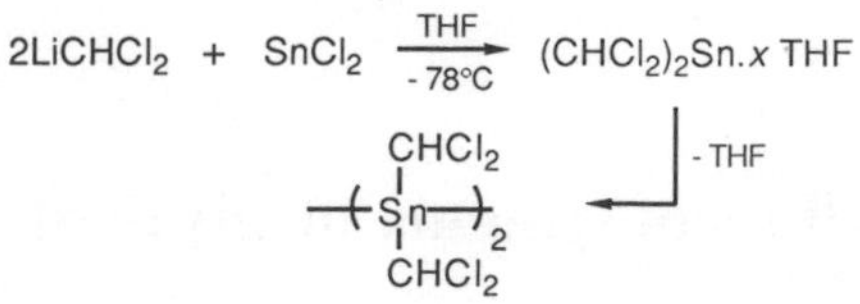

$$2LiCHCl_2 + SnCl_2 \xrightarrow[-78^{\circ}C]{THF} (CHCl_2)_2Sn.x\,THF \xrightarrow{-THF} {-}\!\!\left[Sn(CHCl_2)_2\right]\!\!{-}_2 \qquad (7.3)$$

Evidence for the formation of $(CF_3)_2Sn$ and CF_3SnI in the reaction between SnI_2 and $(CF_3)_2Cd$ (eqn 7.4) was seen in the ^{19}F nmr and Mössbauer spectra.[10]

$$(CF_3)_2Cd + SnI_2 \rightleftharpoons CF_3SnI + CF_3CdI \rightleftharpoons CdI_2 + (CF_3)_2Sn \qquad (7.4)$$

Neumann *et al.*[11,12] have developed methods for preparing a variety of R_2Sn species in solution at 20°C from the thermal decomposition of 7-stannanorbornenes (eqn 7.5) or at -30°C by reduction of R_2SnCl_2 via $R_2SnClLi$ (eqn 7.6).

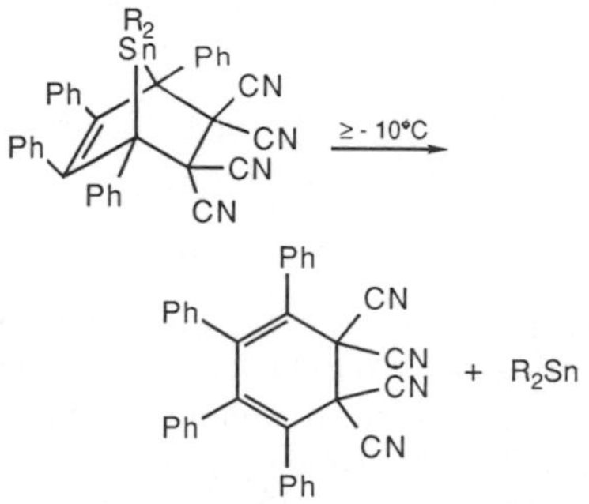

(7.5)

$$[\text{anthracene}\cdot 2Li] + R_2SnCl_2 \xrightarrow{\geqslant -80^{\circ}} R_2SnClLi \xrightarrow{-30^{\circ}} R_2Sn \qquad (7.6)$$

$$R = Me,\ CD_3,\ Et,\ Bu^t$$

Although transient alkylSn (II) species were thought to be present in many reaction systems and evidence for them could be gained from trapping experiments, the synthesis of a stable dialkyltin (II) compound remained a considerable challenge. The problem was eventually overcome by Lappert, who used the sterically demanding bis(trimethylsilyl)methyl ligand to prepare bis[bis(trimethylsilyl)methyl]tin (II), which is kinetically stable with respect to oligomerization[13,14] (eqn 7.7). Several synthetic methods are available: substitution of $SnCl_2$, $Sn(C_5Me_5)_2$ or $Sn[N(SiMe_3)_2]_2$ using the alkyllithium reagent, or by reduction of the corresponding dialkyl (IV) dichloride with Li_2COT:[13–16]

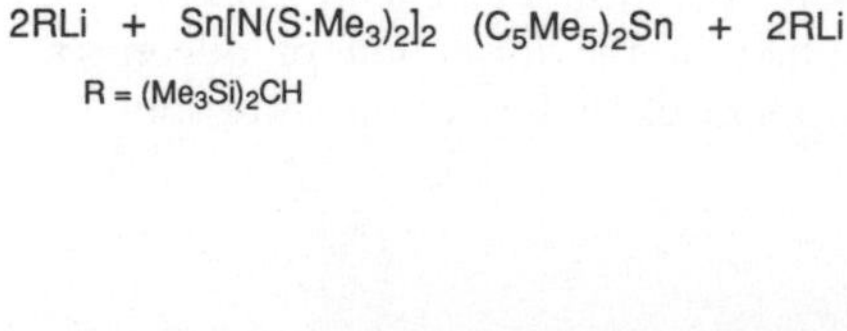

(7.7)

The cyclic stannaindene (**2**) has been obtained from a dilithium reagent and either a tin(IV) or (in better yield) a tin(II) precursor[17] (eqn 7.8).

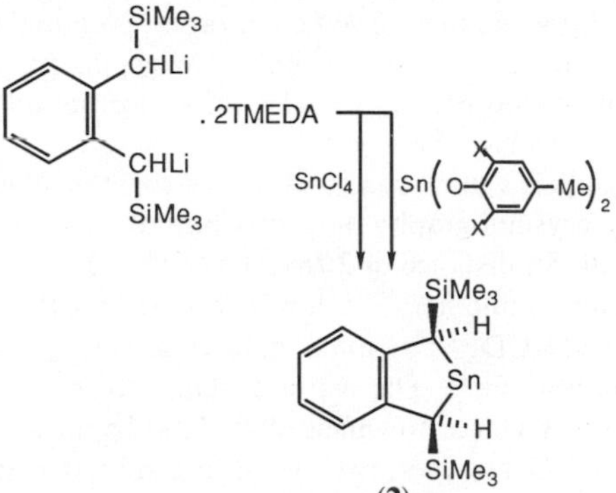

$$C_6H_4[CH(SiMe_3)Li]_2 \cdot 2TMEDA \xrightarrow{SnCl_4 \text{ or } Sn(OAr)_2} (\mathbf{2}) \qquad (7.8)$$

The sterically crowded distannene $[(2,4,6\text{-}(Pr^i)_3C_6H_2)_2Sn]_2$ has been shown to be in thermal equilibrium with a cyclotristannane rather than a free monomeric stannylene species.[18] However, stable diaryl tin(II) compounds with sterically demanding groups on the tin have been reported (eqn 7.9) as products in the reactions of $SnCl_2$ with aryllithium reagents bearing *ortho* and *para* alkyl substituents.[19] Aryl groups bearing alkylamino substituents gave aryltin(II) chlorides coordinated to LiCl. Unfortunately, none of these aryl derivatives have been characterized by X-ray diffraction, ^{119}Sn Mössbauer data being the main means of identification.

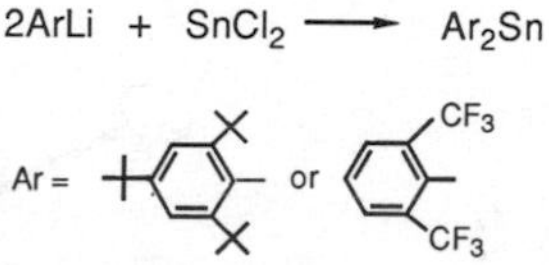

$$2ArLi + SnCl_2 \longrightarrow Ar_2Sn \qquad (7.9)$$

Ar = 2,4,6-tri-*tert*-butylphenyl or 2,6-bis(trifluoromethyl)phenyl (CF_3)

Treatment of $SnCl_2$ with $(PPh_2)_2CHLi$ leads to formation of a three-coordinate Sn(II) species, $[(PPh_2)_2CH]_2Sn$, in which one ligand bonds through carbon and the other acts as a chelating diphosphinomethanide group. When $(Me_2P)_2CH$ was used a ligand, the product, $[(Me_2P)_2CH]_2Sn$, has four-coordinated tin with two chelating diphosphinomethanide ligands and no tin–carbon bonds. ^{31}P-nmr spectroscopy suggests that at 50°C the structure of $[(PPh_2)_2CH]_2Sn$ is similar to that of $\{[(Me_3Si)_2CH]_2Sn\}$, i.e. with the tin coordinated to the two central carbon atoms of the ligands.[20–22]

7.2.2 *Structure*

Although many alkyl- and aryltin(II) compounds have been written as R_2Sn, their actual structures have subsequently been found to be polymeric in nature. Matrix isolation techniques have enabled the ir spectra of both Me_2Sn and $(CD_3)_2Sn$ to be recorded at 5 K. The trapped stannylenes had C_{2v} symmetry and Sn–C stretching frequencies in Me_2Sn of 518 and 504 cm^{-1}.[8] *Ab initio* SCF calculations using a pseudopotential method gave an Sn–C bond length of 2.203 Å, a C–Sn–C bond angle of 95.3° and a dipole moment of 0.43 D for Me_2Sn with a singlet ground state[8]. Dewar *et al.*[23] have also calculated parameters for Me_2Sn as a singlet ground state: ΔH_f 0.25 kcal mol^{-1}, dipole moment 2.94 D, ionization potential 9.63 eV, Sn–C bond length 2.03 Å, and C–Sn–C bond angle 99.1°. No explanation for the significant differences between the results of the two calculations for bond length and angle and dipole moment appears to have been given.

The structure of Lappert's brick-red solid bis(bistrimethylsilylmethyl)tin has been determined by X-ray crystallography, and is a non-planar centrosymmetric dimer (Figure 7.1) with an Sn–Sn distance of 2.768(1) Å.[13,14,24] In cyclohexane or benzene solution it is, however, monomeric at low concentrations[13], although at higher concentrations in $C_6D_5CD_3$ ^{13}C-nmr data indicate some evidence for a monomer–dimer equilibrium.[25] The variable-temperature ^{119}Sn-nmr spectra of $\{[(Me_3Si)_2CH]_2Sn\}$ show a large downfield shift of 2315 ppm at 100°C attributable to the monomer. At -108°C, however, two signals are seen at higher field which have been attributed to different conformations of the dimeric species.[26]

The solid-state variable-temperature ^{119}Sn- and ^{13}C-nmr spectra of $\{[(Me_3Si)_2CH]_2Sn\}$, together with the solution spectra, are all consistent with a monomer–dimer equilibirum, the energy of dissociation being calculated as 12.8 kcal mol^{-1}. The small dissociation energy and the low $^1J(^{119}Sn{-}^{117}Sn)$ of 1340 $\pm$ 10 Hz both suggest that in the dimer the bonding between the tin atoms is best described as a weak dative interaction (of the sort Lappert originally proposed[14]), and not a covalent double bond.[26]

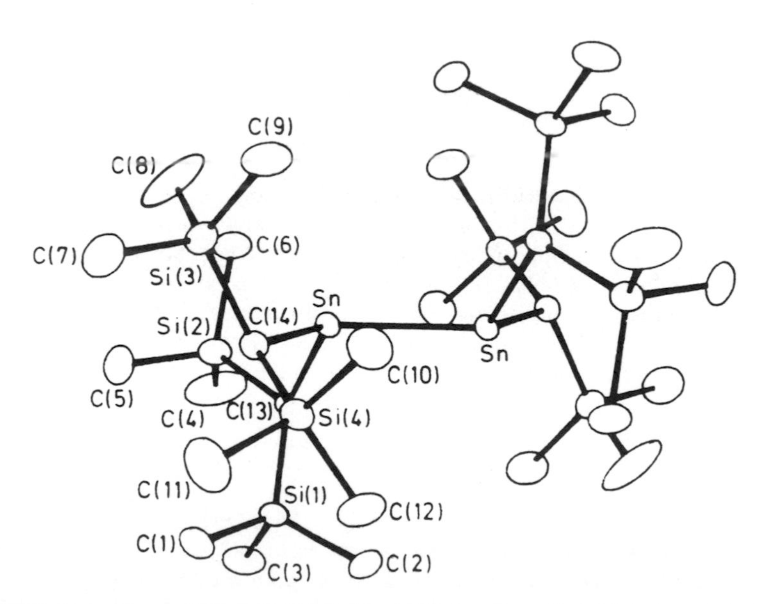

Figure 7.1 Molecular structure of $\{[(Me_3Si)_2CH]_2Sn\}_2$. Reproduced with permission from Ref. 24.

Electron diffraction data show $\{[(Me_3Si)_2CH]_2Sn\}$ to be a 'V' shaped monomer in the gas phase with a C–Sn–C angle of 97(2)° and a Sn–C bond length of 2.22(2) Å.[15,27] He(I) photoelectron spectra of R_2Ge, R_2Sn and R_2Pb ($R = (Me_3Si)_2CH$) have been recorded,[28] and the first ionization potential was found to be close to that for the atomic metal (7.42 eV for R_2Sn) and attributed to the lone pair orbital on the metal. *Ab initio* MO calculations on H_2SnSnH_2 show that a *trans*-folded structure (as is found experimentally for $\{[(Me_3Si)_2CH]_2Sn\}$) is more stable than a planar structure. These

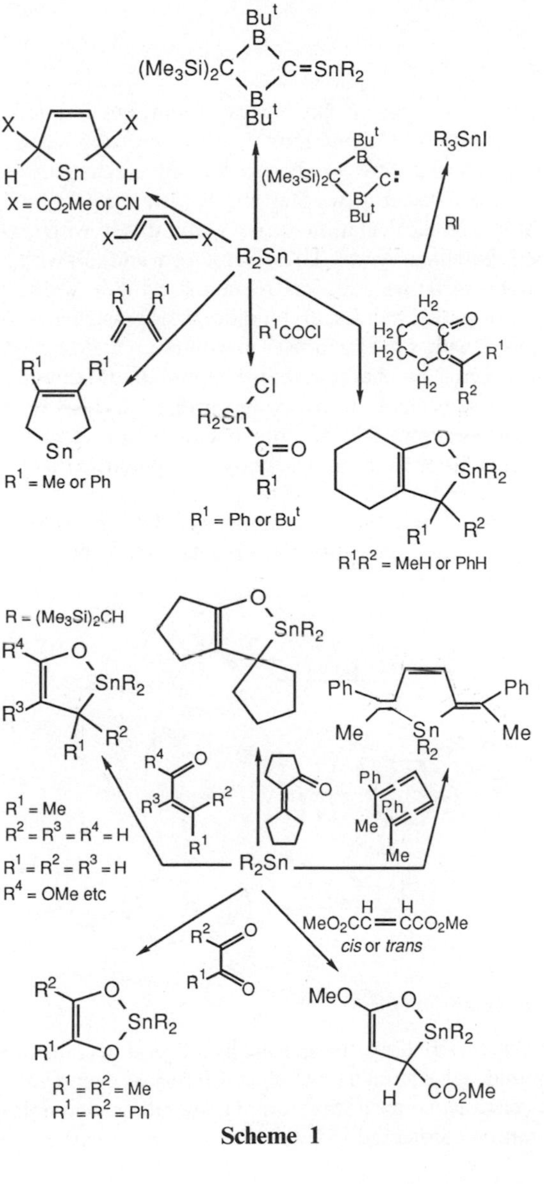

Scheme 1

calculations also give[24,29] an Sn–Sn distance of 2.71 Å and an Sn–Sn dissociation energy of 90 kJ mol^{-1}. MNDO calculations by Dewar *et al.* also predict that the *trans*-folded structure is preferred.[30]

The crystal structure of $[(PPh_2)_2CH]_2Sn$ shows the compound to be monomeric with a pyramidal $[SnCP_2]$ core.[22] However, the structure of the stannaindene [*o*-$(Me_3SiCH)_2C_6H_4$]Sn is tetrameric in the solid, with an $[Sn_4]$ ring with the four tin atoms in a distorted tetrahedral arrangement and an Sn–Sn distance of 2.852(3) Å, slightly longer than that in $\{[(Me_3Si)_2CH]_2Sn\}$. In the gas phase the compound is thought to be monomeric.[17]

7.2.3 *Reactions*

Unless the R group in R_2Sn of alkyl or arylstannylenes is bulky, their dominant reaction is polymerization.[1] Calculations[23] on the insertion reactions of Me_2Sn with Me_2SnCl_2, Me_3SnCl and MeI predict concerted mechanisms for reaction with Me_2SnCl_2 and Me_3SnCl, but a two-step mechanism via $CH_3^{\cdot}$ and $Me_2SnI^{\cdot}$ for the reaction with MeI. These calculations agree qualitatively with experimental results.

Bis[bis(trimethylsilyl)methyl]tin (II) does not form adducts with Lewis acids such as BF_3, but undergo oxidative-addition reactions with a wide variety of organic compounds including alkyl halides, aryl halides, halogens, dienes, ketones, aldehydes, acid chlorides, and with the organoboron compound, $(Me_3Si)_2C(BBu^t)_2\ddot{C}$, which affords the first structurally characterized compound containing a Sn=C bond[31–34] (Scheme 1). A radical mechanism involving $[(Me_3Si)_2CH]_2XSn^{\cdot}$ as an intermediate (X = halogen) has been proposed for the reactions with alkyl and aryl halides.[32] 1, 2-Diketones react faster than vinylcarbonyl compounds which in turn react faster than 1, 3-dienes.

Little has been reported about the chemistry of the substituted diaryl stannylenes. Bis(2, 6-bis(trifluoromethyl)phenyl)tin (II) reacts with both methanol and *p*-toluenethiol[19] (eqn 7.10).

$$[2,6\text{-}(CF_3)_2C_6H_3]_2Sn \xrightarrow{MeOH} 2,6\text{-}(CF_3)_2C_6H_3SnOMe$$

$$[2,6\text{-}(CF_3)_2C_6H_3]_2Sn \xrightarrow{Me\text{-}C_6H_4\text{-}SH} (Me\text{-}C_6H_4\text{-}S)_2Sn \qquad (7.10)$$

7.2.4 *Transition metal complexes*

An organometallic tin (II) derivative can act as a ligand in transition metal complexes either as a terminal or bridging ligand, or as a terminal ligand with one or more base molecules also coordinated to the tin atom. The bonding in complexes with a terminal tin (II) ligand can be considered to be similar to that in carbene complexes, with σdonation from the tin to the transition metal and back donation from the transition metal to the empty *p* or *d* orbitals of the tin. In the base-complexed terminal complexes, the tin also functions as a Lewis acid, and there is additional coordination of a base by the vacant *p* orbital of the tin. In the bridged complexes the tin(II) ligand bridges between two transition metals.

Terminal complexes containing alkyl- or aryltin(II) derivatives appear to be restricted to those complexes of $\{[(Me_3Si)_2CH]_2Sn\}$ (R_2Sn). The stannylene $\{[(Me_3Si)_2CH]_2Sn\}$ can displace CO or R_3P from transition metal complexes and also inserts into M–M, M–alkyl and M–H bonds. The mono-stannylene complexes $[M(CO)_5SnR_2]$ (M = Cr or Mo) were obtained by uv irradiation of hexane solutions of SnR_2 and $M(CO)_6$.[31] The crystal structure of $[Cr(CO)_5SnR_2]$[35] showed the $(Me_3Si)_2\mathit{C}H$ carbons, and the chromium and tin atoms to be coplanar, with an Sn–Cr bond length of 2.562 Å. Recently, a compound containing a terminally-bound $\{[(Me_3Si)_2CH]_2Sn\}$ ligand on a triosmium cluster has been prepared which has a Sn–Os distance (2.573 Å), rather shorter than is found in terminally-bound trialkyltin–osmium complexes.[36] Other reactions of SnR_2 with transition metal complexes are summarized in Scheme 2.

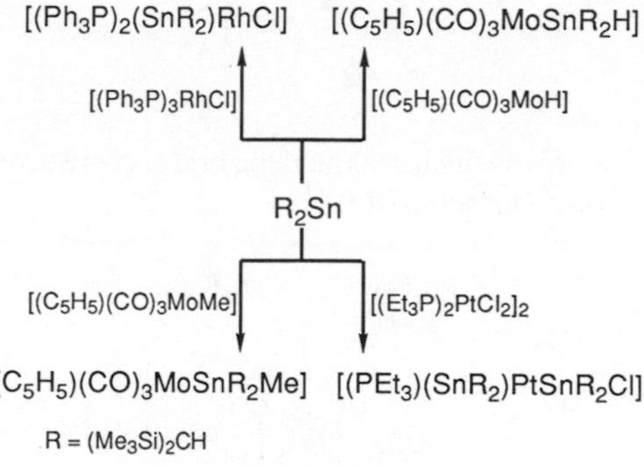

$$[(Ph_3P)_2(SnR_2)RhCl] \xleftarrow{[(Ph_3P)_3RhCl]} R_2Sn \xrightarrow{[(C_5H_5)(CO)_3MoH]} [(C_5H_5)(CO)_3MoSnR_2H]$$

$$[(C_5H_5)(CO)_3MoSnR_2Me] \xleftarrow{[(C_5H_5)(CO)_3MoMe]} R_2Sn \xrightarrow{[(Et_3P)_2PtCl_2]_2} [(PEt_3)(SnR_2)PtSnR_2Cl]$$

R = $(Me_3Si)_2CH$

Scheme 2

Terminal complexes of Me_2Sn, Bu^t_2Sn and Ph_2Sn with a base such as THF or pyridine coordinated to the tin can be prepared in solution either by dissolving a bridged dimer in a coordinating solvent according to the equilibrium shown in eqn 7.11, or by treating dialkyldichlorostannanes with $Na_2Cr_2(CO)_{10}$ in the presence of a Lewis base.[37,38]

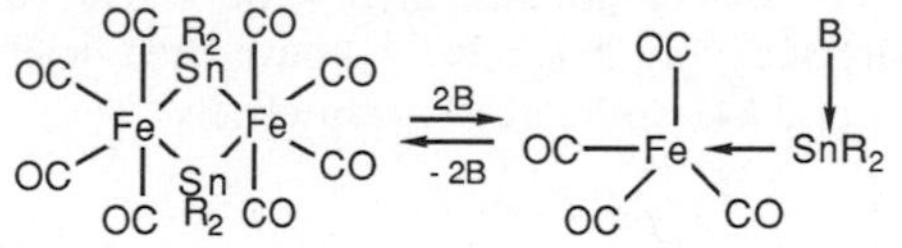

$$[(CO)_4Fe(\mu\text{-}SnR_2)]_2 \underset{-2B}{\overset{2B}{\rightleftharpoons}} (CO)_4Fe \leftarrow SnR_2 \leftarrow B \quad \textbf{(3)} \qquad (7.11)$$

R = Me, Bu^t, or Ph   B = THF or pyridine

On removal of solvent, only the bridged dimers were recovered using THF as base, but with the more basic pyridine the complex $[(CO)_4Fe(SnBu^t_2)\cdot pyridine]$ (**3**, R = Bu^t, B = pyridine) could be isolated.[39] The crystal structure of the relatively stable chromium complex $[(CO)_5Cr(SnBr^t_2)\cdot pyridine]$[40] showed that the tin was tetracoordinated with an Sn–Cr distance of 2.654 Å i.e. 0.092 Å shorter than that in the terminal complex $[(CO)_5Cr\{Sn[CH(SiMe_3]_2\}]$.

Complexes containing base-stabilized arylstannylene ligands have been prepared in which the base donor atoms are substituents on the aryl rings and intramolecular coordination makes the tin penta-coordinated. Examples are shown in eqn 7.12. The W–Sn distances in the two complexes are 2.749 and 2.762 Å for E = NMe_2 and PPh_2, respectively.[41,42]

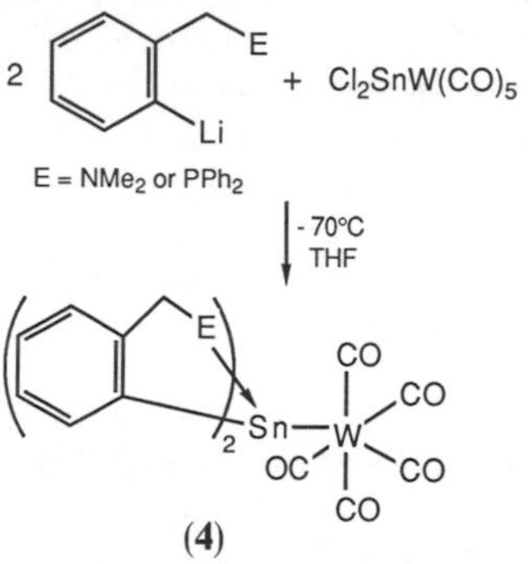

2 (2-$ECH_2C_6H_4Li$) + $Cl_2SnW(CO)_5$ → (−70°C, THF) → $(2\text{-}ECH_2C_6H_4)_2Sn{-}W(CO)_5$ **(4)**

E = NMe_2 or PPh_2 (7.12)

Compounds such as (**5**) in which a stannylene bridges between two metals have been prepared by several methods (eqn 7.13):[43]

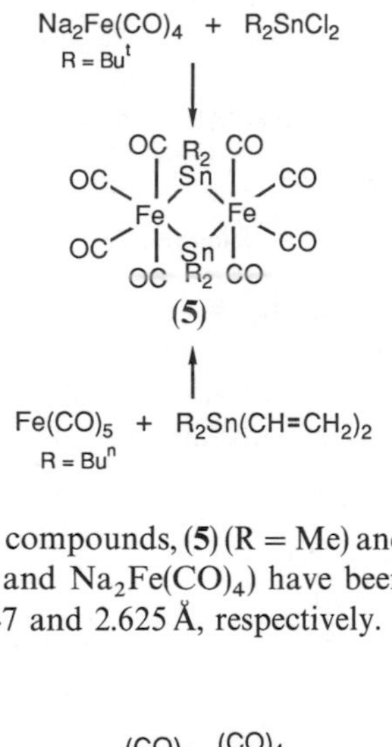

$Na_2Fe(CO)_4$ + R_2SnCl_2 (R = Bu^t) → $[(CO)_4Fe(\mu\text{-}SnR_2)]_2$ **(5)** ← $Fe(CO)_5$ + $R_2Sn(CH{=}CH_2)_2$ (R = Bu^n) (7.13)

The crystal structures of two compounds, (**5**) (R = Me) and (**6**), (both products from the reaction between $MeSnCl_3$ and $Na_2Fe(CO)_4$) have been determined,[44,45] and have Sn–Fe bond lengths of 2.647 and 2.625 Å, respectively.

$Me_2Sn[\mu\text{-}Fe(CO)_4]_2Sn[\mu\text{-}Fe(CO)_4]_2SnMe_2$

(6)

7.3 π-Bonded complexes

7.3.1 *Synthesis*

The chemistry of π-bonded organometallic complexes of bivalent tin is more extensive than that of σ-bonded compounds and dates back to the first preparation of bis(cyclopentadienyl)tin or stannocene* in 1956 according to eqn 7.14.[46]

$$SnCl_2 + 2NaC_5H_5 \rightarrow (C_5H_5)_2Sn + 2NaCl \tag{7.14}$$

The same salt elimination method using $SnCl_2$ and an alkali metal salt of a cyclopentadiene derivative has been used extensively to prepare substituted stannocenes as illustrated in Scheme 3.[47–56]

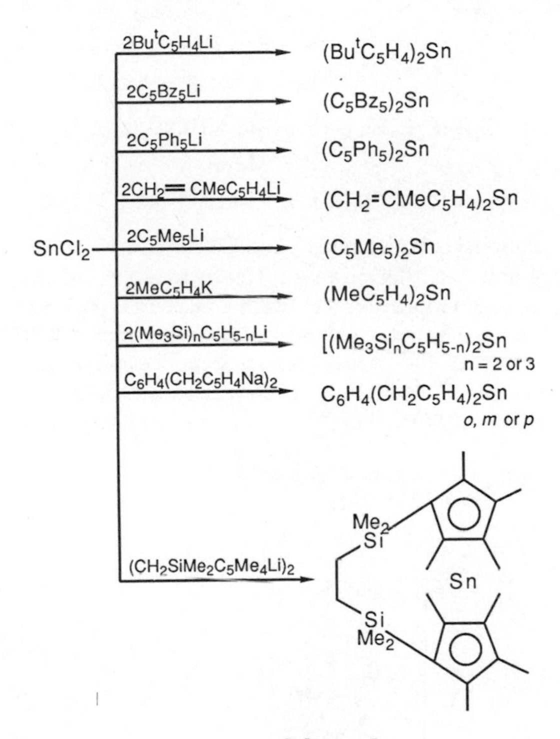

Scheme 3

Decamethylstannocene has also been prepared by reduction of the corresponding diorganotin (IV) dihalides, $(C_5Me_5)_2SnX_2$ (X = Cl or Br), with either lithium naphthalenide or dilithium cyclooctatetraenide. The best yield (70%) is obtained by using the dichloride and lithium naphthalenide.[57]

Treatment of stannocene with one equivalent of $SnCl_2$ or $SnBr_2$ leads to an exchange reaction and the formation of cyclopentadienyltin (II) chloride or bromide

*Although stannocene is not a systematic name, it has been universally accepted by workers in the field.

respectively.[58] Many other stannocene derivatives have been obtained by substitution at the coordinated cyclopentadienyl rings, and these are discussed in section 7.3.3.

A few arene–tin(II) complexes have been prepared as colourless, very air-sensitive crystals by reaction of $SnCl_2$ and $AlCl_3$ in the presence of the arene as solvent (eqn 7.15), although they have been much less well studied.[59–61]

$$(Arene)SnCl(AlCl_4) \xleftarrow[AlCl_3]{Arene} SnCl_2 \xrightarrow[2AlCl_3]{Arene} (Arene)_2Sn(AlCl_4)_2 \qquad (7.15)$$

Arene = benzene or *p*-xylene

Several compounds in which a bivalent tin is incorporated into a carborane cage have been prepared using the salt elimination route[62–66] (eqns 7.16–7.18).

$$SnCl_2 + Na[C,C'\text{-}R,R'C_2B_4H_5] \rightarrow Sn[R,R'C_2B_4H_4] \qquad (7.16)$$

$R = R' = H, Me, Me_3Si$

$R = H, Me\ R' = Me_3Si$

$$SnCl_4 + Na[(Me_3Si)(R)C_2B_4H_5] \rightarrow Sn[(Me_3Si)(R)C_2B_4H_4] \qquad (7.17)$$

$R = Me_3Si$, Me or H

$$SnCl_2 + Na_2[7,8\text{-}Me_2\text{-}7,8\text{-}C_2B_9H_9] \rightarrow 2,3\text{-}Me_2\text{-}1\text{-}Sn\text{-}2,3\text{-}C_2B_9H_9 \qquad (7.18)$$

The isolobal relationship between Cp^+ and $CpCo(C_3B_2)^-$ has prompted the use of the 1,3-diborenyl unit as a π-ligand to tin. Hence, treatment of $SnCl_2$ with anion (**7**) leads to a dark orange tetradecker sandwich compound, (**8**),[67] (eqn 7.19). The 1,2-azaborolinyl ligand has also been used as a π-ligand to tin, and treatment of $SnCl_2$ with the lithium salt (**9**) gives the orange–yellow stannocene-like compound (**10**) as a mixture of diastereomers[68] (eqn 7.20). The tetradecker compound (**8**) is quite thermally stable and decomposes at 140°C, but (**10**) decomposes with formation of tin at −20°C.

$SnCl_2$ + 2 $[CpCoC_2Et_2B_2Me_2CH]^-K^+$

(**7**)

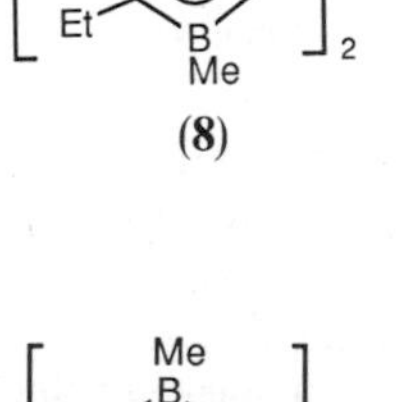

↓

$[$Co, Et, Et, BMe, B, Me$]_2$Sn

(**8**) (7.19)

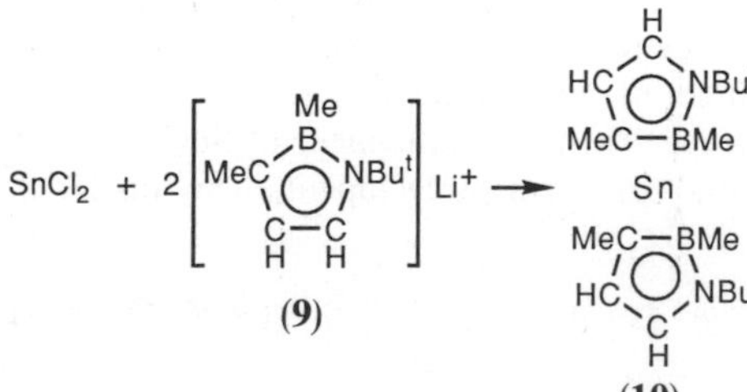

$SnCl_2$ + 2 [MeC, B, Me, NBu^t, C–C, H H] Li^+ →

(**9**)

HC, C, H, NBu^t, MeC—BMe

Sn

MeC—BMe, HC, C, H, NBu^t

(**10**) (7.20)

7.3.2 *Structure and bonding*

The first observations[46] on the structure of stannocene, that it was monomeric in benzene solution and that it had a non-zero dipole moment, were thought to imply a σ-bonded structure. However, the true pentahapto π-bonded sandwich structure of stannocene and its derivatives has since been demonstrated in numerous ways including X-ray diffraction studies. Infrared and Raman spectra of stannocene and its derivatives have been recorded,[48,69] ir spectroscopy being used to distinguish mono- from penta-hapto bonding.

The solid-state structure of stannocene has been determined by X-ray crystallography[70] and, unlike plumbocene, was found to be monomeric with two crystallographically independent molecules in the unit cell. In contrast to ferrocene, the cyclopentadienide rings are not parallel in either molecule and have ring centroid–metal–ring centroid angles (V) of 148.0° and 143.7° for the two different molecules. The gas phase electron diffraction structure[71] showed a smaller angle V of about 125°, but the data was poor and no detailed comparisons between the structure in the solid and gas phases can be drawn. The structure of several substituted stannocenes have also been determined and details are summarized in Table 7.1.

The 'bent sandwich' structure of most of the stannocene derivatives is attributed to the tin having sp^2 hybridization, with two of the hybrid orbitals interacting with the π-system of the rings and the other containing an unshared pair of electrons which, in all but the decaphenyl derivative, overcomes any steric repulsion caused by the rings. Steric factors do, however, play a rôle, and the data in Table 7.1 show that the larger the substituents on the rings the nearer to being parallel they become. Care must be exercised when comparing V angles, as the ring centroid–tin–ring centroid angle is different to that subtended by perpendiculars to the rings crossing at tin. This has led to the erroneous comment that V is larger in $(C_5H_5)_2Sn$ than $(Me_5C_5)_2Sn$,[70] but this has since been corrected and reversed.[72]

Table 7.1 Structural data for stannocene and its derivatives.

Compound	Method	α* (deg)	V* (deg)	Reference
$(C_5H_5)_2Sn$	X-ray	48.4, 45.9	148.0, 143.7	70, 72
$(C_5H_5)_2Sn$	ED	~55	~125	71
$(MeC_5H_4)_2Sn$	ED	34 ± 7		72
$(Me_5C_5)_2Sn$	X-ray	36.4, 35.4		52
$(Bz_5C_5)_2Sn$	X-ray	32.8(2)	155.89(9)	48
$[(1,2,4\text{-}Me_3Si)_3C_5H_2]_2Sn$	X-ray		162 ± 2	54
$(Ph_5C_5)_2Sn$	X-ray	0	180	49
$[(Pr^i_2N)_2PC_5H_4]_2Sn$	X-ray		150.2	73
$\{[BF_4]^-(C_5H_5)_2Sn[C_5H_5Sn]^+thf\}$	X-ray		138.7[a]	74
$[Bu^tNB(Me)(CH)_2CMe]_2Sn$	X-ray	46.5		68
$[(C_5H_5)CoBMe(CEt)_2BMeCH]_2Sn$	X-ray	68.66		67

[a]in the $(C_5H_5)_2Sn$ moiety

*

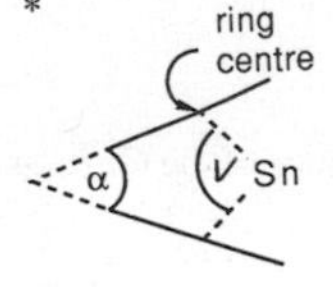

The ^{1}H and ^{13}C nmr spectra of $(C_5H_4Me)_2Sn$, $(C_5H_4Me)_2Ge$ and $(C_5H_4Me)_2Pb$ are found to be quite similar,[53] suggesting similar sandwich structures. Coupling between Sn and C or H in $(C_5H_4Me)_2Sn$ was not observed at ambient temperature, consistent with rapid intermolecular exchange. Sn–H coupling is observed at temperatures $\leqslant -40°C$. Ring-substitution of cyclopentadienes seems to slow down intermolecular exchange, and coupling to ^{119}Sn is observed for $(Me_5C_5)_2Sn$ and $(Me_5C_5)SnBF_4$ at ambient temperatures. The ^{119}Sn nmr chemical shifts for stannocene derivatives are shifted more than 2100 ppm upfield from Sn(IV) species, which can be attributed to shielding caused by increased *s* electron density at the tin nucleus. The He(I)[75,76] and He(II)[76] photoelectron spectra of $(Me_5C_5)_2Sn$ and the He(I) and He(II)[75,77] spectra of $(C_5H_5)_2Sn$ have been recorded. The lowest ionization potential for $(C_5H_5)_2Sn$ is at 7.57 eV and has been attributed to electron loss from the C_5H_5 ring.[75]

In $(Ph_5C_5)_2Sn$ (Figure 7.2), which is bright yellow and has S_{10} molecular symmetry, the lone pair is stereochemically inert.

The two sets of phenyl rings in $(Ph_5C_5)_2Sn$ are almost perpendicular to each other, a conformation shown to be the lowest-energy one by molecular modelling.[78] Fenske–Hall MO calculations for $(Ph_5C_5)_2Sn$ predict that the tin lone pair is not delocalized into the ligand ring systems but is in a tin 5*s*-like orbital. Figure 7.3 shows the molecular orbital diagram for $(Ph_5C_5)_2Sn$ as calculated by this method.[78]

The crystal structures of the salts $(Bu^tC_5H_4)Sn^+BF_4^-$,[79] $C_5Me_5Sn^+BF_4^-$ [52,80]) and $C_5H_5Sn^+Cl^-$ [81] show that the tin atom sits above the centre of the ring with an average $Sn-C_{ring}$ distance of 2.462 Å, somewhat shorter than that found in $(C_5Me_5)_2Sn$ (2.585–2.770 and 2.567–2.767 Å in the two crystallographically distinct molecules)[52] indicat-

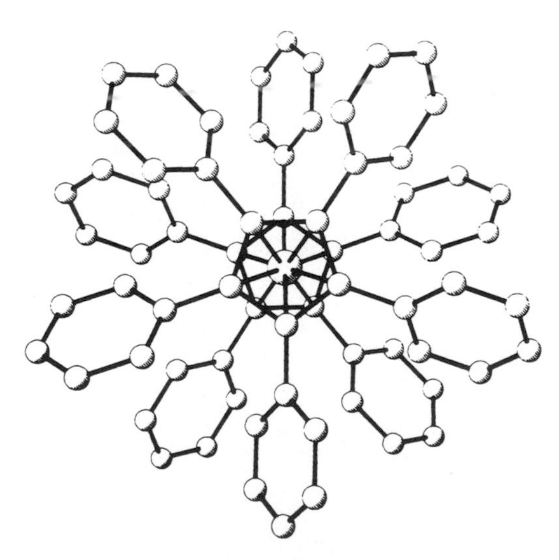

Figure 7.2 Molecular structure of $(C_5Ph_5)_2Sn$ viewed perpendicular to the planes of the C_5 rings. Reproduced with permission from Ref. 49.

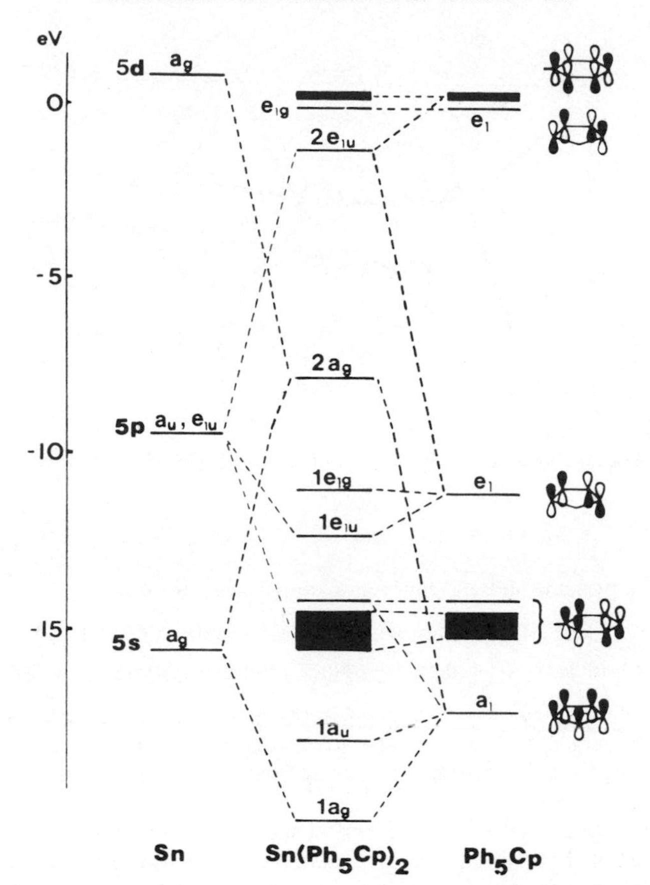

Figure 7.3 M.O. diagram for $(C_5Ph_5)_2Sn$ as calculated by the Fenske–Hall method. Reproduced with permission from Ref. 78.

ing a stronger interaction between the ring and the tin in the $[C_5Me_5Sn^+]$ cation than in the neutral $(C_5Me_5)_2Sn$. In both $(Bu^tC_5H_4)Sn^+BF_4^-$ and in $C_5H_5Sn^+Cl^-$ the tin atom is not above the ring centre, with Sn–C_{ring} distances of 2.443–2.559 and 2.45–2.74 Å, respectively, $C_5H_5Sn^+Cl^-$ having a ring centroid–Sn–Cl angle of 117.4°. Extended Huckel calculations predicted the HOMO for $C_5H_5Sn^+BF_4^-$ to be a tin lone pair[52]. However, MNDO calculations, which are probably more reliable for calculations of ionization energies, predict a pair of nearly degenerate HOMOs localized on the ring to lie above the tin lone pair.[30]

Jutzi *et al.* have also prepared *N*-donor complexes of $Me_5C_5Sn^+$ cations by treating $Me_5C_5Sn^+CF_3SO_3^-$ with various amines (eqn 7.21).[82,83] The X-ray crystal structures of $[Me_5C_5Sn{\cdot}amine]^+[CF_3SO_3]^-$ (amine = pyridine or 2,2′-bipyridine) have been determined and show that amine coordination distorts the bonding between the Me_5C_5 and the tin from η^5 towards η^2 or η^3 with Sn–N distances of about 2.62 and

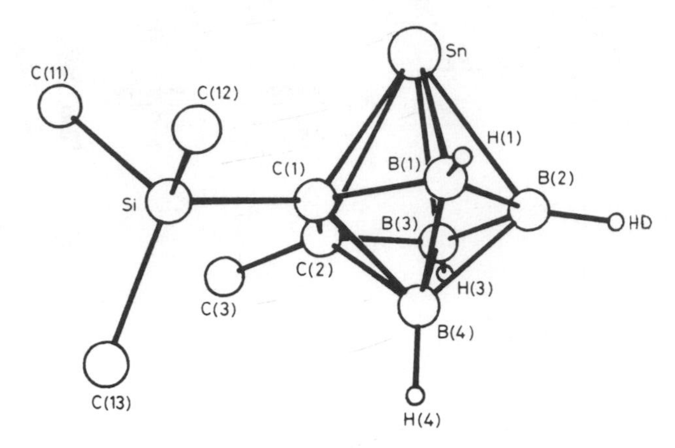

Figure 7.4 Molecular structure of $Sn(Me_3Si)(Me)C_2B_4H_4$. Reproduced with permission from Ref. 63.

2.48 Å in the pyridine and 2,2′-bipyridine complexes, respectively.

$$[C_5Me_5Sn]^+[CF_3SO_3]^- + \text{amine} \rightarrow [C_5Me_5Sn{\cdot}\text{amine}]^+[CF_3SO_3]^- \qquad (7.21)$$

amine = pyridine, 2-, or 4-methylpyridine, 3,5-dimethylpyridine or 2,2′-bipyridine

Although the 1,3-diborolenyl ligands in (**8**) (eqn 7.19) interact with the tin in an essentially pentahapto fashion, in (**10**) (eqn 7.20) the tin atom is displaced in the direction of the CCB ring fragment and away from the nitrogen, suggesting a trihapto interaction.

The X-ray crystal structures of several of the stannacarboranes mentioned in the previous section have recently been determined by X-ray crystallography. Both $Sn(Me_3Si)(Me)C_2B_4H_4$[63] (Figure 7.4) and $Sn(Me_3Si)C_2B_4H_5$[84] show the tin atom to be at an apex of a distorted pentagonal bipyramid and can be regarded as being η^5 bonded to the $[C_2B_3]$ face, with slightly longer bonds to carbon than to boron. The ^1H, ^{11}B, ^{13}C and ^{29}Si nmr spectra for the stannacarboranes $Sn(Me_3Si)(R)C_2B_4H_4$ (R = Me_3Si, Me or H) have been recorded and are all consistent with the proposed structures.[64]

Coordination of the tin atoms in these compounds by 2,2′-bipyridine results in ‘slippage’ of the tin away from the carbon atoms in the $[C_2B_3]$ ring, and becomes essentially η^3 bonded to the three borons and interacts only weakly with the carbons.[63,84,85] A similar although less pronounced distortion also occurs when $Sn(Me_3Si)_2C_2B_4H_4$ is treated with 2,2′-bipyrimidine to give a red complex in which the bipyrimidine bridges between two $[Sn(Me_3Si)_2C_2B_4H_4]$ units.[86]

All three known arenetin(II) complexes have been structurally characterized. In both (η^6-arene)$SnCl(AlCl_4)$ (arene = C_6H_6 or p-$Me_2C_6H_4$, (Figure 7.5) complexes the ring is asymmetrically bound to the tin atom with $Sn{-}C_{ring}$ distances ranging from 2.92–3.27 and 3.05–3.339 Å when the arene is C_6H_6 or p-$Me_2C_6H_4$, respectively, i.e. much longer than those found in $C_5H_5Sn^+Cl^-$ (2.45–2.74 Å). Each tin is also part of a planar $[Sn_2Cl_2^{2+}]$ dimeric unit and is chelated by two chlorines of one $AlCl_4^-$ unit and also one chlorine from a second $AlCl_4^-$ unit, leading to the formation of a chain

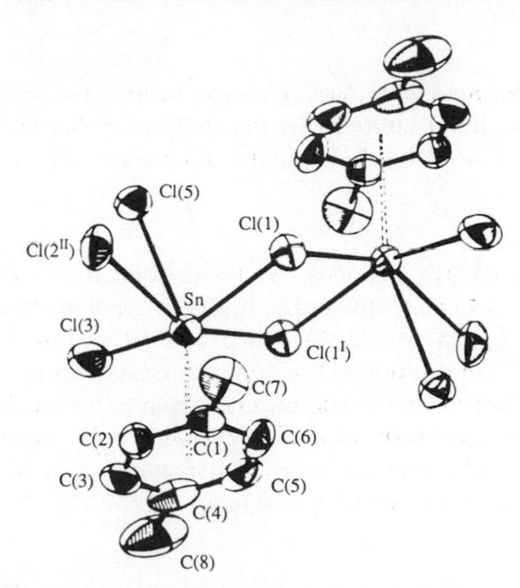

Figure 7.5 Molecular structure of the environment about the $[Sn_2Cl_2^{2+}]$ dimer in *p*-$Me_2C_6H_4ClSn(AlCl_4)$. Reproduced with permission from Ref. 61.

structure.[60,61] In the structure of $(\eta^6\text{-}C_6H_6)Sn(AlCl_4)_2 \cdot C_6H_6$,[59,87] (Figure 7.6) the tin is coordinated by three pairs of chlorines from three different $[AlCl_4^-]$ units and interacts fairly symmetrically with one benzene ring, with an average $Sn-C_{ring}$ distance of 3.06 Å again forming a chain structure. The second molecule of benzene lies in a cleft position between the chains and is not bound to the tin.

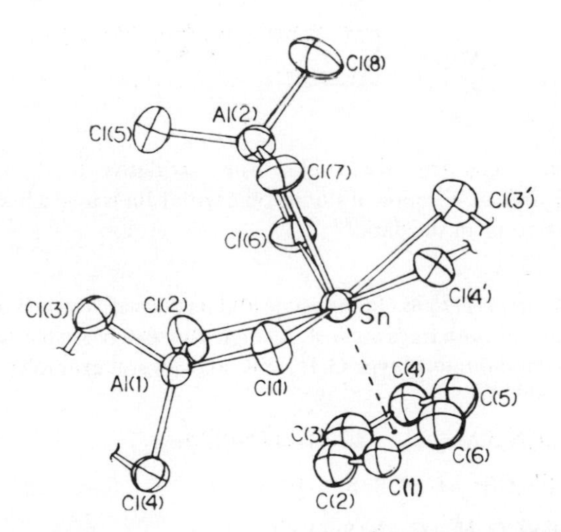

Figure 7.6 Structure of the environment around tin in $\eta^6\text{-}C_6H_6Sn(AlCl_4)_2 \cdot C_6H_6$. Reproduced with permission from Ref. 59.

7.3.3 *Reactions*

The reactions of stannocene and its derivatives can be broadly divided into two types: those of an oxidative-addition nature where the Sn (II) is oxidized to Sn (IV), and those where the oxidation state of the tin remains unchanged. These will be discussed separately below.

7.3.3.1 *Oxidative-addition type reactions.* Although solid stannocene does not polymerize spontaneously at room temperature, it is slowly polymerized by DMF to the tetravalent $\{Sn(C_5H_5)_2\}_n$, a reaction[88] which can be regarded as an oxidative-addition. The degree of substitution of the cyclopentadienyl rings plays an important rôle in the reactivity of substituted stannocenes. Hence, although stannocene itself reacts rapidly with atmospheric oxygen, $(C_5Me_5)_2Sn$ is stable in air for several hours and $(Bz_5C_5)_2Sn$ is air-stable. Several other oxidative-addition reactions of stannocenes are known and some are summarized below (eqns 7.22, 7.23):[51,88–90]

$$(C_5H_5)_2Sn \xrightarrow{RX} (C_6H_5)_2Sn(R)X \quad RX = MeI,\ CH_2{=}CHCH_2Br,\ C_6H_5CH_2Br$$
$$(C_5H_5)_2Sn \xrightarrow{I_2} (C_5H_5)_2SnI_2$$
$$(C_5H_5)_2Sn \xrightarrow{O_2} [(C_5H_5)_2SnO]_n$$
$$(C_5H_5)_2Sn \xrightarrow{PhSSPh} (C_5H_5)_2Sn(SPh)_2 \rightleftharpoons (C_5H_5)_3SnSPh + (C_5H_5)Sn(SPh)_3 \quad (7.22)$$

$$(C_5Me_5)_2Sn \xrightarrow{X_2} (C_5Me_5)_2SnX_2 \quad X = Br\ or\ I$$
$$(C_5Me_5)_2Sn \xrightarrow{SnCl_4\ or\ HgCl_2} (C_5Me_5)_2SnCl_2 \quad (7.23)$$

It has been suggested that, like the oxidative-addition reactions of $\{[(Me_3Si)_2CH]_2Sn\}$, the reaction of stannocene with MeI is also a free radical process since it does not occur in the dark.[89]

7.3.3.2 *Non-oxidative reactions.* Stannocene and its derivatives can undergo reactions in which either one or both rings are lost. A single ring can be displaced by a variety of reagents to give compounds of type C_5H_5SnX, and several examples are given below (eqns 7.24–7.26):[89,91,92]

$$(C_5H_5)Sn + Sn[N(SiMe_3)_2]_2 \rightarrow 2C_5H_5Sn[N(SiMe_3)_2] \quad (7.24)$$

$$(C_5H_5)_2Sn + Ph_3CBr \rightarrow C_5H_5SnBr + Ph_3CC_5H_5 \quad (7.25)$$

$$(C_5Me_5)_2Sn + RCO_2H \rightarrow C_5Me_5SnO_2CR \quad (7.26)$$
$$R = CF_3,\ CCl_3$$

Ionic salts are formed if a relatively weakly nucleophilic anion is available, and examples are shown in eqns 7.27–7.31.[52,79,80,92,93] A similar reaction is undergone by the tetradecker sandwich compound (**8**) formed in eqn 7.19, which forms the salt $[CpCoBMe(CEt)_2BMeCHSn]^+BF_4^-$ on treatment with HBF_4.[67]

$$(C_5Me_5)_2Sn \xrightarrow{CF_3SO_3R} [C_5Me_5Sn]^+CF_3SO_3^- \quad R = H \text{ or } Me \qquad (7.27)$$

$$(C_5Me_5)_2Sn \xrightarrow{2AlCl_3} [C_5Me_5Sn]^+AlCl_4^- \qquad (7.28)$$

$$(C_5Me_5)_2Sn \xrightarrow{HBF_4} [C_5Me_5Sn]^+BF_4^- \qquad (7.29)$$

$$(C_5Me_5)_2Sn \xrightarrow{HC_5(CO_2Me)_5} [C_5Me_5Sn]^+C_5(CO_2Me)_5^- \qquad (7.30)$$

$$(Bu^tC_5H_4)_2Sn + BF_3 \longrightarrow [Bu^tC_5H_4Sn]^+BF_4^- \qquad (7.31)$$

Stannocene was originally thought to form a donor–acceptor complex, $(C_5H_5)_2Sn \cdot BF_3$, when treated with $BF_3 \cdot THF$[94]. However, a recent crystal structure determination[74] has shown that this is not correct. Rather, the product has a complicated polymeric structure represented by $\{[BF_4]^-\{\mu\text{-}\eta^5C_5H_5)_2Sn\text{-}[\mu\text{-}\eta^5C_5H_5Sn]^+THF\}_n$. Stannocene does, however, form 1:1 adducts with BBr_3, $AlCl_3$ and $AlBr_3$.[95] An anionic tetrakis(cyclopentadienyl)tin(II) species has been prepared by treating stannocene with Me_5Sb. The structure of the spontaneously flammable complex $[Me_4Sb]_2^+[(C_5H_5)_4Sn]^{2-}$, is however, unknown.[96]

The chlorine in C_5H_5SnCl can be displaced using sodium or lithium reagents to prepare asymmetrical stannocenes. The mono-substituted acetyl- and alkoxycarbonyl-stannocenes $C_5H_5SnC_5H_4C(O)R$ (R = Me, OMe or OEt) can be prepared from C_5H_5SnCl and $RC(O)C_5H_5Na$ in very good yields.[97] Attempts to make symmetrically-substituted $[RC(O)C_5H_5]_2Sn$ compounds from $SnCl_2$ and the appropriate sodium reagent failed. Other asymmetrical stannocenes are prepared according to Scheme 4:[49,55,97]

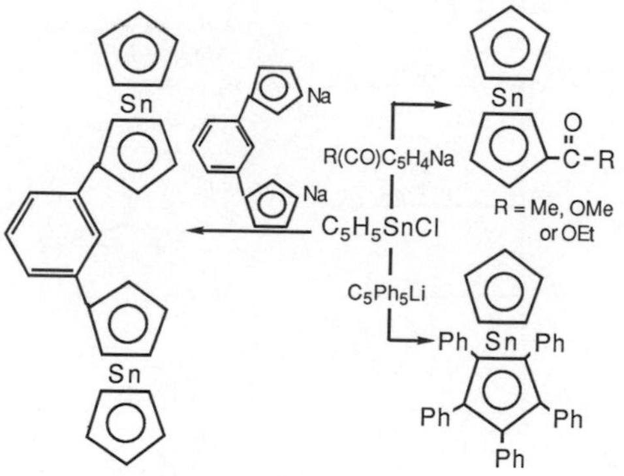

Sn Na Na $R(CO)C_5H_4Na$ C_5H_5SnCl C_5Ph_5Li Sn Sn O C–R R = Me, OMe or OEt Ph Sn Ph Ph Ph Ph

Scheme 4

Stannocene forms a pale brown 1:1 and a brown 1:2 adduct with tetracyanoethylene, and a pale lime-green 1:1 adduct with 7, 7, 8, 8-tetracyanoquinodimethane. These complexes are thermally quite stable and decompose only above 160°C. Their structures are unknown, but it is thought that they are charge-transfer complexes involving the cyclopentadiene π-electron system.[98]

A route to synthetically useful tin(II) compounds without Sn–C bonds is the cleavage of stannocenes (usually $(C_5H_4Me)_2Sn$) by protic compounds such as alcohols,[99,100] phenols,[101] thiols,[100] hydroxylamines,[99] oximes,[99] cyclic amines,[99] carboxylic acids,[99,102] sulphonic acids,[102] acetoacetals,[103,104] and the Schiff base *N*, *N'*-ethylenebis(acetylidenemine).[105] These reactions usually proceed rapidly at room temperature in good yield to give products that are readily isolated. Such reactions have now been used to generate tin(II) species as intermediates for the organic synthesis of a variety of compounds such as optically active glycerol derivatives, esters and amides, N-alkyl imides, esters and imides, and α,β-dihydroxyketones.[106–110] Examples are shown in Scheme 5.

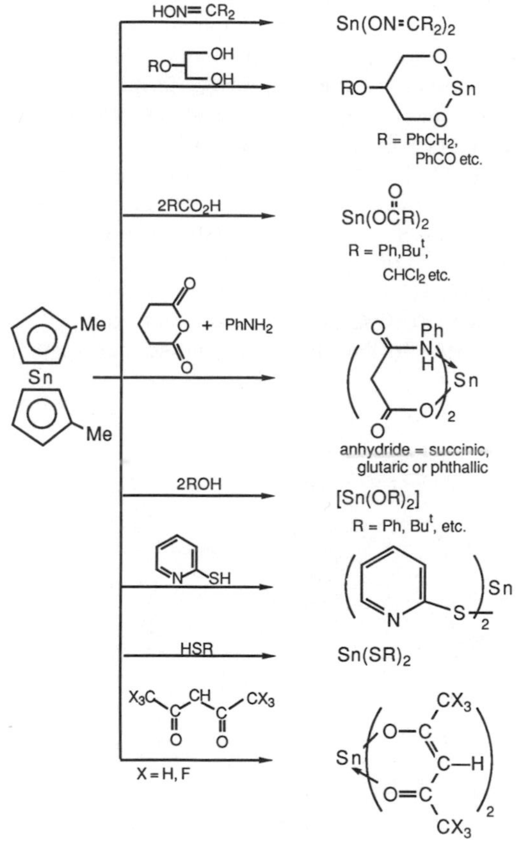

Scheme 5

Stannocene itself has also been found to react with Bu^tSH to give $(Bu^tS)_2Sn$[111]. Treatment of $(C_5Me_5)_2Sn$ with $HC_5(CO_2Me)_5$ was originally thought to give the stannocene derivative $[C_5(CO_2Me)_5]_2$[112], but X-ray crystallography has since shown that the product is in fact a covalent tin(II) product containing Sn–O bonds.[93]

Transition metal hydrides react with stannocene with loss of both cyclopentadienyl rings from tin. Reaction with $HMn(CO)_5$ gives the ditin dihydride, $\{H[Mn(CO)_5]_2Sn\}_2$,[113] whilst $HMo(CO)_3(C_5H_5)$ affords $HSn[Mo(CO)_3(C_5H_5)]_3$.[114] The nature of the product formed in the reaction of 1, 1′-dimethylstannocene with $HW(CO)_2(C_5H_5)$ has been the subject of several reports. The reaction was originally reported to yield the tin(IV) polymer $\{Sn[W(CO)_3C_5H_5]_2]_n\}$,[99] but later the material was reformulated as an unusual transition metal derivative of tin(II) $Sn[W(CO)_3C_5H_5]_2$.[115] The reaction is now thought to proceed like that of the molybdenum analogue to give initially the hydride $HSn[W(CO)_3C_5H_5]_3$, which can react further with a chlorinated solvent to give the chloride $ClSn[W(CO)_3C_5H_5]_3$.[116,117]

Ring substitution reactions of stannocenes may be carried out without their displacement from the tin. A recent and potentially very useful reaction is the metallation of the cyclopentadienide rings by butyllithium giving a bis-lithium derivative (**11**) which can subsequently be derivatized as shown in eqn (7.32). Repeated metallation and derivatization of $(Me_3SiC_5H_4)_2Sn$ leads to formation of $[(Me_3Si)_2C_5H_3]_2Sn$ and $[(Me_3Si)_3C_5H_2]_2Sn$.[54]

$(C_5H_5)_2Sn \xrightarrow[-78°C]{2BuLi}$ $(LiC_5H_4)_2Sn$ (**11**); (**11**) $\xrightarrow{2Me_3SiCl}$ $(Me_3SiC_5H_4)_2Sn$; (**11**) $\xrightarrow{2(Pr^i_2N)_2PCl}$ $[(Pr^i_2N)_2PC_5H_4]_2Sn$ (7.32)

Other reactions in which ring substitutions occur are the oxidative addition of a C–H bond to the phosphenium salt $[(Pr^i_2N)_2P]^+[AlCl_4]^-$[118] affording (**12**), and the reaction of stannocene with Me_3SnNEt_2 to give a low yield of the mixed Sn (II)–Sn (IV) product (**13**) (eqn 7.33).[119]

$(C_5H_5)_2Sn$ $\xrightarrow{[Pr^i_2N_2P]^+[AlCl_4]^-}$ $[(C_5H_4PH(NPr^i_2)_2)Sn(C_5H_4Me)]^+[AlCl_4]^-$ (**12**)

$(C_5H_5)_2Sn$ $\xrightarrow{2Me_3SnNEt_2}$ $(Me_3SnC_5H_4)_2Sn$ (**13**) (7.33)

The isopropenyl stannocene $(CH_2{=}C(Me)C_5H_4)_2Sn$ does not undergo homopolymerization with $SnCl_4$ as initiator, but does undergo free radical copolymerization with styrene initiated by AIBN.[50]

The stannacarboranes $Sn(Me_3Si)_2C_2B_4H_4$, $Sn(Me_3Si)(Me)C_2B_4H_4$ and $Sn(Me_3Si)C_2B_4H_5$ do not react with either BF_3 or $BH_3 \cdot THF$ at room temperature,[64] but do form donor–acceptor complexes with 2, 2′-bipyridine with the tin atom acting as a Lewis acid.[63,84] The tin atoms in these stannacarboranes can be displaced by treatment with half an equivalent of $GeCl_4$ in the absence of solvent at 150–160°C, giving $SnCl_2$ and the germanium (IV) sandwich compounds, $Ge[2,3\text{-}(Me_3Si)(R)\text{-}2,3\text{-}C_2B_4H_4]_2$ ($R = Me_3Si$, Me or H).[65] The stannadicarbadodecaborane, $2,3\text{-}Me_2\text{-}1\text{-}Sn\text{-}2,3\text{-}C_2B_9H_9$, is unreactive towards CF_3SO_3Me, MeI and $SnCl_4$, but it does form adducts with, for example, 2, 2′-bipyridine, tetrahydrofuran and triphenylphosphine.[66]

7.3.4 *Transition metal complexes*

The cyclopentadienyl tin (II) derivatives form a variety of complexes with transition metals. Treatment of $(C_5H_5)_2Sn$ or $(MeC_5H_4)_2Sn$ with some pentacarbonylmetal–THF complexes gives rise to simple donor–acceptor complexes (eqns 7.34, 7.35):[115]

$$(RC_5H_4)_2Sn \xrightarrow{M(CO)_5 \cdot THF} (RC_5H_4)_2Sn{:}M(CO)_5 \qquad (7.34)$$

$$R = H,\ CH_3; \quad M = Cr,\ Mo,\ W.$$

However, reaction with $Fe_2(CO)_9$ gives the bridged dinuclear compound (**14**) in which the mode of attachment of the rings to tin has changed from penta- to monohapto. On dissolving these complexes in pyridine or THF, monomeric base-stabilized species (**15**) are formed (eqn 7.35):[115,120]

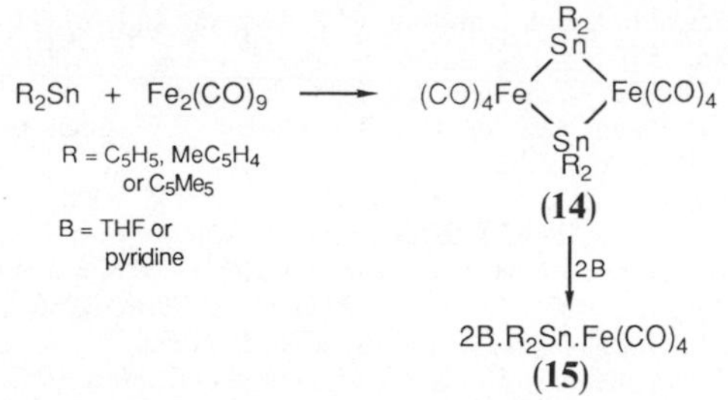

$$R_2Sn + Fe_2(CO)_9 \longrightarrow (CO)_4Fe(\mu\text{-}SnR_2)_2Fe(CO)_4 \ (\mathbf{14}) \xrightarrow{2B} 2B.R_2Sn.Fe(CO)_4 \ (\mathbf{15}) \quad (7.35)$$

R = C_5H_5, MeC_5H_4 or C_5Me_5

B = THF or pyridine

Transition metal complexes containing the CpSnX (Cp = C_5H_5 or C_5Me_5, X = Cl or Br) unit have been prepared similarly (eqn 7.36),[120] or by alkylation of an $SnCl_2$-complex (eqn 7.37).[102]

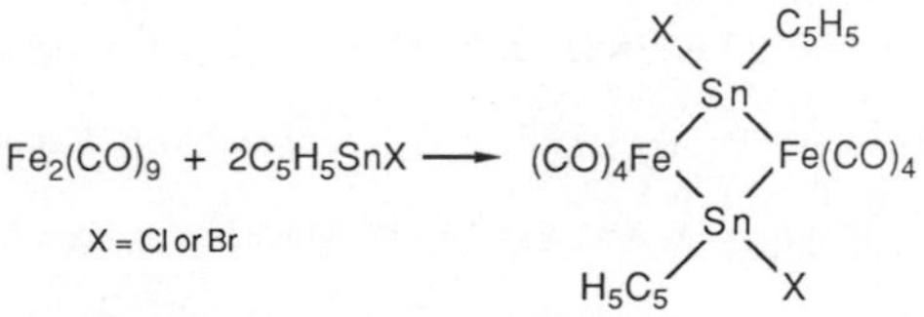

$$Fe_2(CO)_9 + 2C_5H_5SnX \longrightarrow (CO)_4Fe(\mu\text{-}Sn(X)C_5H_5)_2Fe(CO)_4 \quad (7.36)$$

X = Cl or Br

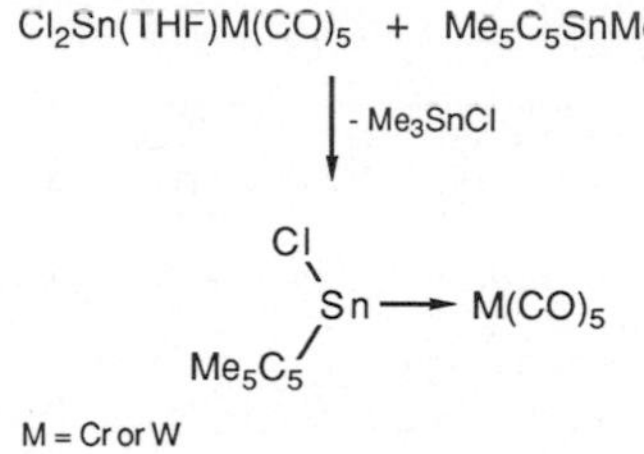

$$Cl_2Sn(THF)M(CO)_5 + Me_5C_5SnMe_3 \xrightarrow{-Me_3SnCl} (Me_5C_5)(Cl)Sn \longrightarrow M(CO)_5 \quad (7.37)$$

M = Cr or W

References

1. W.P. Neumann, in *The Organometallic and Coordination Chemistry of Germanium, Tin and Lead*, eds. M. Gielen and P.G. Harrison, Freund, Tel Aviv, 1978, 51.
2. J.W. Connolly and C. Hoff, *Adv. Organomet. Chem.*, 1981, **19**, 123.
3. W. Petz, *Chem. Rev.*, 1986, **86**, 1019.
4. P. Jutzi, *Adv. Organomet. Chem.*, 1986, **26**, 217.
5. W.P. Neumann and J. Fu, *J. Organomet. Chem.*, 1984, **273**, 295.
6. G. Bähr and R. Gelius, *Chem. Ber.*, 1958, **91**, 829.
7. B. Mathiasch, *J. Organomet. Chem.*, 1980, **194**, 37.
8. P. Bleckmann, H. Maly, R. Minkwitz, W.P. Neumann and B. Watta, *Tetrahedron Lett.*, 1982, **23**, 4655.
9. R. Hani and R.A. Geanangel, *Inorg. Chim. Acta*, 1985, **96**, 225.
10. R. Hani and R.A. Geanangel, *Polyhedron*, 1982, **1**, 826.
11. L.-W. Gross, R. Moser, W.P. Neumann and K.-H. Scherping, *Tetrahedron Lett.*, 1982, **23**, 635.
12. C. Grugel, W.P. Neumann and M. Schiewer, *Angew. Chem., Int. Edn. Engl.*, 1979, **18**, 543.
13. P.J. Davidson, D.H. Harris and M.F. Lappert, *J. Chem. Soc., Dalton Trans.*, 1976, 2268.
14. D.E. Goldberg, D.H. Harris, M.F. Lappert and K.M. Thomas, *J. Chem. Soc., Chem. Commun.*, 1976, 261.

15. T. Fjeldberg, A. Haaland, B.E.R. Schilling, M.F. Lappert and A.J. Thorne, *J. Chem. Soc., Dalton Trans.*, 1986, 1551.
16. P. Jutzi and B. Hielscher, *Organometallics*, 1986, **5**, 2511.
17. M.F. Lappert, W.-P. Leung, C.L. Raston, A.J. Thorne, B.W. Skelton and A.H. White, *J. Organomet. Chem.*, 1982, **233**, C28.
18. S. Masamune and L.R. Sita, *J. Amer. Chem. Soc.*, 1985, **107**, 6390.
19. M.P. Bigwood, P.J. Corvan and J.J. Zuckerman, *J. Amer. Chem. Soc.*, 1981, **103**, 7643.
20. H.H. Karsch, A. Appelt and G. Müller, *Organometallics*, 1986, **5**, 1664.
21. H.H. Karsch, A. Appelt and G. Hanika, *J. Organomet. Chem.*, 1986, **312**, C1.
22. A.L. Balch and D.E. Oram, *Organometallics*, 1986, **5**, 2159.
23. M.J.S. Dewar, J.E. Friedheim and G.L. Grady, *Organometallics*, 1985, **4**, 1784.
24. D.E. Goldberg, P.B. Hitchcock, M.F. Lappert, K.M. Thomas, A.J. Thorne, T. Fjeldberg, A. Haaland and B.E.R. Schilling, *J. Chem. Soc., Dalton Trans.*, 1986, 2387.
25. M.F. Lappert, *Rev. Silicon, Germanium, Tin, Lead Cpds*, 1986, **9**, 129.
26. K.W. Zilm, G.A. Lawless, R.M. Merrill, J.M. Millar and G.W. Webb, *J. Amer. Chem. Soc.*, 1987, **109**, 7236.
27. T. Fjeldberg, A. Haaland, M.F. Lappert, B.E.R. Schilling, R. Seip and A.J. Thorne, *J. Chem. Soc., Chem. Commun.*, 1982, 1407.
28. D.H. Harris, M.F. Lappert, J.B. Pedley and G.J. Sharp, *J. Chem. Soc., Dalton Trans.*, 1976, 945.
29. T. Fjeldberg, A. Haaland, B.E.R. Schilling, H.V. Volden, M.F. Lappert and A.J. Thorne, *J. Organomet. Chem.*, 1984, **276**, C1.
30. M.J.S. Dewar, G.L. Grady, D.R. Kuhn and K.M. Merz Jr., *J. Amer. Chem. Soc.*, 1986, **106**, 6773.
31. J.D. Cotton, P.J. Davidson and M.F. Lappert, *J. Chem. Soc., Dalton Trans.*, 1976, 2275.
32. M.J.S. Gynane, M.F. Lappert, S.J. Miles, A.J. Carty and N.J. Taylor, *J. Chem. Soc., Dalton Trans.*, 1977, 2009.
33. H. Meyer, G. Baum, W. Massa, S. Berger and A. Berndt, *Angew. Chem., Int. Edn. Engl.*, 1987, **26**, 546.
34. K. Hillner and W.P. Neumann, *Tetrahedron Lett.*, 1986, **27**, 5347.
35. J.D. Cotton, P.J. Davidson, D.E. Goldberg, M.F. Lappert and K.M. Thomas, *J. Chem. Soc., Chem. Commun.*, 1974, 893.
36. R.A. Bartlett, C.J. Cardin, D.J. Cardin, M.B. Hursthouse, G.A. Lawless, J.M. Power and P.P. Power, *J. Chem. Soc., Chem. Comm.*, 1988, 312.
37. T.J. Marks, *J. Amer. Chem. Soc.*, 1971, **93**, 7090.
38. G.W. Grynkewich, B.Y.K. Ho, T.J. Marks, D.L. Tomaja and J.J. Zuckerman, *Inorg. Chem.*, 1973, **12**, 2522.
39. T.J. Marks and A.R. Newman, *J. Amer. Chem. Soc.*, 1973, **95**, 769.
40. M.D. Brice and F.A. Cotton, *J. Amer. Chem. Soc.*, 1973, **95**, 4529.
41. K. Jurkschat, H.-P. Abicht, A. Tzschach and B. Mahieu, *J. Organomet. Chem.*, 1986, **309**, C47.
42. H.-P. Abicht, K. Jurkschat, A. Tzschach, K. Peters, E.-M. Peters and H.G. von Schnering, *J. Organomet. Chem.*, 1987, **326**, 357.
43. R.B. King and F.G.A. Stone, *J. Amer. Chem. Soc.*, 1960, **82**, 3833.
44. R.M. Sweet, C.J. Fritchie and R.A. Schunn, *Inorg. Chem.*, 1967, **6**, 749.
45. C.J. Gilmore and P. Woodward, *J. Chem. Soc., Dalton Trans.*, 1972, 1387.
46. E.O. Fischer and H. Grubert, *Z. Naturforsch.*, 1956, **11B**, 423.
47. R. Hani and R.A. Geanangel, *J. Organomet. Chem.*, 1985, **293**, 197.
48. H. Schumann, C. Janiak, E. Hahn, C. Kolax, J. Loebel, M.D. Rausch, J.J. Zuckerman and M.J. Heeg, *Chem. Ber.*, 1986, **119**, 2656.
49. M.J. Heeg, C. Janiak and J.J. Zuckerman, *J. Amer. Chem. Soc.*, 1984, **106**, 4259.
50. K. Gonsalves, L. Zhan-Ru, R.W. Lenz and M.D. Rausch, *J. Polym. Sci. Polym. Chem. Ed.*, 1985, **23**, 1707.
51. P. Jutzi and F. Kohl, *J. Organomet. Chem.*, 1979, **164**, 141.
52. P. Jutzi, F. Kohl, P. Hofmann, C. Krüger and Y.-H. Tsay, *Chem. Ber.*, 1980, **113**, 757.
53. A. Bonny, A.D. McMaster and S.R. Stobart, *Inorg. Chem.*, 1978, **17**, 935.
54. A.H. Cowley, P. Jutzi, F.X. Kohl, J.G. Lasch, N.C. Norman and E. Schlüter, *Angew. Chem., Int. Edn. Engl.*, 1984, **23**, 616.

55. T.S. Dory and J.J. Zuckerman, *J. Organomet. Chem.*, 1984, **264**, 295.
56. P. Jutzi and R. Dickbreder, *Chem. Ber.*, 1986, **119**, 1750.
57. P. Jutzi and B. Hielscher, *Organometallics*, 1986, **5**, 1201.
58. K.D. Bos, E.J. Bulten and J.G. Noltes, *J. Organomet. Chem.*, 1972, **39**, C52.
59. P.F. Rodesiler, Th. Auel and E.L. Amma, *J. Amer. Chem. Soc.*, 1975, **97**, 7405.
60. M.S. Weininger, P.F. Rodesiler, A.G. Gash and E.L. Amma, *J. Amer. Chem. Soc.*, 1972, **94**, 2135.
61. M.S. Weininger, P.F. Rodesiler and E.L. Amma, *Inorg. Chem.*, 1979, **18**, 751.
62. K.-S. Wong and R.N. Grimes, *Inorg. Chem.*, 1977, **16**, 2053.
63. A.H. Cowley, P. Galow, N.S. Hosmane, P. Jutzi and N.C. Norman, *J. Chem. Soc., Chem. Commun.*, 1984, 1564.
64. N.S. Hosmane, N.N. Sirmokadam and R.H. Herber, *Organometallics*, 1984, **3**, 1665.
65. M.S. Islam, U. Siriwardane, N.S. Hosmane, J.A. Maguire, P. de Meester and S.S.C. Chu, *Organometallics*, 1987, **6**, 1936.
66. P. Jutzi, P. Galow, S. Abu-Orabi, A.M. Arif, A.H. Cowley and N.C. Norman, *Organometallics*, 1987, **6**, 1024.
67. H. Wadepohl, H. Pritzkow and W. Siebert, *Organometallics*, 1983, **2**, 1899.
68. G. Schmid, D. Zaika and R. Boese, *Angew. Chem., Int. Edn. Engl.*, 1985, **24**, 602.
69. P.G. Harrison and M.A. Healy, *J. Organomet. Chem.*, 1973, **51**, 153.
70. J.L. Atwood, W.E. Hunter, A.H. Cowley, R.A. Jones and C.A. Stewart, *J. Chem. Soc., Chem. Commun.*, 1981, 925.
71. A. Almenningen, A. Haaland and T. Motzfeldt, *J. Organomet. Chem.*, 1967, **7**, 97.
72. J. Almlöf, L. Fernholt, K. Fægri, A. Haaland, B.E.R. Schilling, R. Seip and K. Taugbøl, *Acta Chem. Scand.*, 1983, **37A**, 131.
73. A.H. Cowley, J.G. Lasch, N.C. Norman, C.A. Stewart and T.C. Wright, *Organometallics*, 1983, **2**, 1691.
74. T.S. Dory, J.J. Zuckerman and C.L. Barnes, *J. Organomet. Chem.*, 1985, **281**, C1.
75. S.G. Baxter, A.H. Cowley, J.G. Lasch, M. Lattman, W.P. Sharum and C.A. Stewart, *J. Amer. Chem. Soc.*, 1982, **104**, 4064.
76. G. Bruno, E. Ciliberto, I.L. Fragalà and P. Jutzi, *J. Organomet. Chem.*, 1985, **289**, 263.
77. S. Cradock and W. Duncan, *J. Chem. Soc., Faraday Trans. 2*, 1978, **74**, 194.
78. R.L. Williamson and M.B. Hall, *Organometallics*, 1986, **5**, 2142.
79. R. Hani and R.A. Geanangel, *J. Organomet. Chem.*, 1985, **293**, 197.
80. P. Jutzi, F. Kohl and C. Krüger, *Angew. Chem., Int. Edn. Engl.*, 1979, **18**, 59.
81. K.D. Bos, E.J. Bulten, J.G. Noltes and A.L. Spek, *J. Organomet. Chem.*, 1975, **99**, 71.
82. P. Jutzi, F. Kohl, C. Krüger, G. Wolmershäuser, P. Hofmann and P. Stauffert, *Angew. Chem., Int. Edn. Engl.*, 1982, **21**, 70.
83. F.X. Kohl, E. Schlüter, P. Jutzi, C. Krüger, G. Wolmershäuser, P. Hofmann and P. Stauffert, *Chem. Ber.*, 1984, **117**, 1178.
84. N.S. Hosmane, P. de Meester, N.N. Maldar, S.B. Potts, S.S.C. Chu and R.H. Herber, *Organometallics*, 1986, **5**, 772.
85. U. Siriwardane, N.S. Hosmane and S.S.C. Chu, *Acta Cryst.*, 1987, **C43**, 1067.
86. N.S. Hosmane, M.S. Islam, U. Siriwardane and J.A. Maguire, *Organometallics*, 1987, **6**, 2447.
87. H. Lüth and E.L. Amma, *J. Amer. Chem. Soc.*, 1969, **91**, 7515.
88. P.G. Harrison and J.J. Zuckerman, *J. Amer. Chem. Soc.*, 1969, **91**, 6885.
89. K.D. Bos, E.J. Bulten and J.G. Noltes, *J. Organomet. Chem.*, 1975, **99**, 397.
90. K.D. Bos, E.J. Bulten and J.G. Noltes, *J. Organomet. Chem.*, 1974, **67**, C13.
91. D.H. Harris and M.F. Lappert, *J. Chem. Soc., Chem. Commun.*, 1974, 895.
92. F.X. Kohl and P. Jutzi, *Chem. Ber.*, 1981, **114**, 488.
93. P. Jutzi, F.-X. Kohl, E. Schlüter, M.B. Hursthouse and N.P.C. Walker, *J. Organomet. Chem.*, 1984, **271**, 393.
94. P.G. Harrison and J.J. Zuckerman, *J. Amer. Chem. Soc.*, 1970, **92**, 2577.
95. P.G. Harrison and J.A. Richards, *J. Organomet. Chem.*, 1976, **108**, 35.
96. K.D. Bos, E.J. Bulten, H.A. Meinema and J.G. Noltes, *J. Organomet. Chem.*, 1979, **168**, 159.
97. T.S. Dory, J.J. Zuckerman and M.D. Rausch, *J. Organomet. Chem.*, 1985, **281**, C8.
98. A.B. Cornwell, C.-A. Cornwell and P.G. Harrison, *J. Chem. Soc., Dalton Trans.*, 1976, 1612.
99. P.G. Harrison and S.R. Stobart, *J. Chem. Soc., Dalton Trans.*, 1973, 940.
100. P.G. Harrison and S.R. Stobart, *Inorg. Chim. Acta*, 1973, **7**, 306.

101. P.F.R. Ewings and P.G. Harrison, *J. Chem. Soc., Dalton Trans.*, 1975, 2015.
102. P.F.R. Ewings and P.G. Harrison, *J. Chem. Soc., Dalton Trans.*, 1975, 1717.
103. P.F.R. Ewings, P.G. Harrison and D.E. Fenton, *J. Chem. Soc., Dalton Trans.*, 1975, 821.
104. A.B. Cornwell and P.G. Harrison, *J. Chem. Soc., Dalton Trans.*, 1975, 1722.
105. P.F.R. Ewings, P.G. Harrison and A. Mangia, *J. Organomet. Chem.*, 1976, **114**, 35.
106. J. Ichikawa, M. Asami and T. Mukaiyama, *Chem. Lett.*, 1984, 949.
107. T. Mukaiyama, J. Ichikawa and M. Asami, *Chem. Lett.*, 1983, 683.
108. T. Mukaiyama, J. Ichikawa, M. Toba and M. Asami, *Chem. Lett.*, 1983, 879.
109. T. Mukaiyama, J. Ichikawa and M. Asami, *Chem. Lett.*, 1983, 293.
110. J. Ichikawa and T. Mukaiyama, *Chem. Lett.*, 1985, 1009.
111. W.-W. du Mont and M. Grenz, *Chem. Ber.*, 1985, **118**, 1045.
112. F.X. Kohl, E. Schlüter and P. Jutzi, *J. Organomet. Chem.*, 1983, **243**, C37.
113. K.D. Bos, E.J. Bulten, J.G. Noltes and A.L. Spek, *J. Organomet. Chem.*, 1975, **92**, 33.
114. C.D. Hoff and J.W. Connolly, *J. Organomet. Chem.*, 1978, **148**, 127.
115. A.B. Cornwell, P.G. Harrison and J.A. Richards, *J. Organomet. Chem.*, 1976, **108**, 47.
116. P.G. Harrison, *J. Organomet. Chem.*, 1981, **212**, 183.
117. T.S. Dory, J.J. Zuckerman, C.D. Hoff and J.W. Connolly, *J. Chem. Soc., Chem. Commun.*, 1981, 521.
118. A.H. Cowley, R.A. Kemp and C.A. Stewart, *J. Amer. Chem. Soc.*, 1982, **104**, 3239.
119. E.J. Bulten and H.A. Budding, *J. Organomet. Chem.*, 1978, **157**, C3.
120. V. Sriyunyongwat, R. Hani, T.A. Albright and R. Geanangel, *Inorg. Chim. Acta*, 1986, **122**, 91.

8 Tin-metal bonded compounds

F. GLOCKLING

This chapter is devoted to the formation, structure, reactivity and physical properties of molecules containing one or more bonds between tin and another metal, including boron and silicon. Attention has been concentrated on work published since 1970. Earlier work is covered in a number of review articles, some of which include comprehensive lists of compounds.[1-12]

8.1 Tin–alkali metal compounds

Compounds of the type R_3SnM are known for all of the alkali metals M, and those of the type R_2SnM_2 for lithium and sodium. Although only one structure has been reported, it is apparent from chemical behaviour, conductivity, Mössbauer and nmr studies that these compounds are essentially ionic with strong ion-pairing.[13,14]

The main preparative methods involve the cleavage of Sn–halide, Sn–H, Sn–C or Sn–Sn bonds by an alkali metal or an organo-derivative of an alkali metal. In general, these reactions proceed most satisfactorily in Lewis basic solvents such as ether, THF, 1,2-dimethoxyethane, liquid ammonia or HMPT, although it is as well to bear in mind that as strong nucleophiles they react slowly with some of these solvents. Illustrative examples are shown in eqns (8.1)–(8.4).[15-21]

$$Ph_3SnCl + 2Li \xrightarrow[\text{or dioxan}]{\text{THF}} Ph_3SnLi + LiCl \tag{8.1}$$

$$Me_3SnH + Na/NH_3 \text{ or } NaNH_2 \longrightarrow Me_3SnNa \tag{8.2}$$

$$Me_2SnH_2 + 2M/NH_3 \rightarrow Me_2SnM_2 + H_2 \quad (M = Li, Na) \tag{8.3}$$

$$Bu_6Sn_2 + 2Li \xrightarrow{\text{THF or HMPT}} 2Bu_3SnLi \tag{8.4}$$

Bu_3SnM (M = K, Cs) have been obtained[22] from Bu_3SnH and $MNPr^i_2/LiOBu^t$. Selective cleavage of Sn–C bonds can result in unsymmetrical derivatives (eqn 8.5)[23]:

$$(C_6H_{11})_3SnPh + K \xrightarrow{NH_3} (C_6H_{11})_2PhSnK \tag{8.5}$$

Stannane, SnH_4, may be converted into either H_3SnNa or H_2SnNa_2 by the action of sodium in liquid ammonia.[19,24] Organohydrido tin anions $[R_nSnH_{3-n}]^-$ are similarly formed.[25] Cleavage of Sn–C, Sn–Sn or Sn–halide bonds by sodium naphthalide in 1,2-dimethoxyethane may also be used to form R_3SnNa compounds.[26]

Cyclic or polymeric organotin compounds have been converted into R_3SnM or R_2SnM_2 by the types of reactions shown in eqn (8.6).[17,26-29]

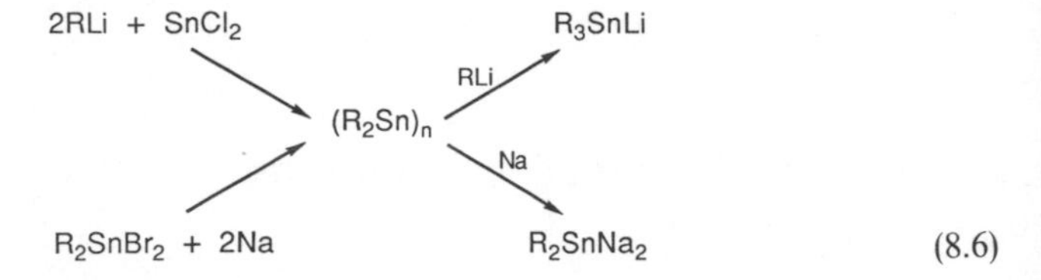

(8.6)

In addition to organo-derivatives, Sn–O bonded analogues have been obtained[30] according to eqn (8.7):

$$Sn(OBu^t)_2 + MOBu^t \rightarrow (Bu^tO)_3SnM \quad (8.7)$$

(M = an alkali metal)

Both R_3SnM and R_2SnM_2 compounds are strong nucleophiles and undergo a wide range of reactions analogous to those of organolithium and Grignard reagents. The utility of these reactions as synthetic reagents is discussed elsewhere (Chapter 10). Some of their less familiar reactions are the following:[31,32]

$R_3SnM \xrightarrow{[(OC)_5Cr{=}CNEt_2]BF_4} [(OC)_5Cr{=}C(NEt_2)](SnR_3)$ (8.8)

$R_3SnM \xrightarrow{>CBr\text{--}CBr<} >C{=}C< + R_3SnBr \xrightarrow{R_3SnM} R_6Sn_2$ (8.9)

Irradiation of Ph_3SnM compounds in the presence of polycyclic aromatic hydrocarbons gives the radical anion of the hydrocarbon.[33] R_3SnLi compounds decompose rapidly with first-order kinetics in THF in the presence of Sn_2R_6, giving mainly SnR_4, together with Sn_2R_5Li, elemental tin, and polymeric material.[34,35]

Although halogen–metal exchange and subsequent coupling reactions are frequently observed in reactions with organic or metal halides, there are many examples, such as eqn (8.10),[36] of the nucleophilic substitution product being formed in quite high yield:

$$Me_3SnLi + CCl_4 \rightarrow (Me_3Sn)_4C + Me_6Sn_2 \quad (8.10)$$

Until recently, little information on the structure of these compounds has been available. The crystal structure of Li[Sn(2-furyl)$_3$], as a dioxan solvate, reveals the ionic structure $[Li(dioxane)_4]^+[Sn(2\text{-furyl})_3Li(2\text{-furyl})_3Sn]^-$ in which lithium appears in both the anion and cation.[30] In solution R_3SnM compounds (R = Et, Bu) appear, on the basis of their reaction with CO_2, to be in equilibrium with RLi[26] (eqn 8.11).

$$R_3SnLi \rightleftharpoons R_2Sn + RLi \xrightarrow{CO_2} RCO_2Li \quad (8.11)$$

When R = Ph, this equilibrium is said to be well to the left, and on reaction with CO_2 it yields Ph_6Sn_2 and oxalate, consistent with a charge transfer reaction and subsequent coupling of the resulting radicals. Similarly, oxidation of Ph_3SnNa with O_2 gives the products of radical coupling (Ph_6Sn_2 and Na_2O_2).[26] R_3SnM compounds form charge transfer complexes with electron acceptors like 1, 3, 5-trinitrobenzene.[37] Solutions of $SnMe_4$ and MeLi in THF/HMPA form an *ate* complex $Li[SnMe_5]$, probably with a trigonal bipyramidal arrangement of methyl groups about tin.[38]

8.2 Tin–Group II metal compounds

No attempts have apparently been made to form tin–beryllium compounds, even though their structures would be of considerable interest in relation to corresponding lithium and Grignard derivatives. Nor is there any report of tin compounds with calcium, strontium or barium.

In the reaction between Grignard reagents and $SiCl_4$ or $GeCl_4$, appreciable yields of R_6M_2 products result, especially in the presence of free finely-divided magnesium or with sterically demanding R groups (eqn 8.12). Similar observations have been made for the analogous reaction of tin tetrahalides, and these can be rationalized in terms of the formation of an intermediate magnesium halide derivative R_3SnMgX.[39,40]

$$\begin{array}{l} R_3SnX + Mg \searrow \\ \qquad\qquad R_3SnMgX \xrightarrow{R_3SnX} R_6Sn_2 \\ R_3SnX + RMgX \nearrow \end{array} \tag{8.12}$$

Similarly, sterically hindered Grignard reagents react with Bu_3SnH with the formation of $Bu_3SnMgBr$:[41]

$$Bu_3SnH + Bu^tMgCl \rightarrow C_4H_{10} + Bu_3SnMgCl \xrightarrow{D_2O} Bu_3SnD \tag{8.13}$$

Ph_3SnH and $EtMgBr{\cdot}NEt_3$ in ether at $-15°C$ yield an oil formulated as $Ph_3SnMgBr{\cdot}NEt_3$ which loses Et_3N *in vacuo* giving the dimer $(Ph_3SnMgBr)_2$.[42] Further evidence for R_3SnMgX derivatives comes from the reaction between Ph_3SnCl or Ph_6Sn_2 and magnesium in THF containing a small amount of ethyl bromide since, on hydrolysis, Ph_3SnH is formed.[43] As with R_3SnLi compounds there is some evidence for dissociation in solution (eqn 8.14).[42]

$$Ph_3SnMgBr \rightleftharpoons PhMgBr + Ph_2Sn \xrightarrow{MeI} Ph_2MeSnI \tag{8.14}$$

8.3 Scandium, yttrium, lanthanides and actinides

The Sc–Sn (II) complex $(C_5H_5)_2Sc(Me)Sn[N(SiMe_3)_2]_2$ has been obtained[44] as a pale yellow oil by reacting $(C_5H_5)_2Sc(\mu\text{-}Me_2)AlMe_2$ with $Sn[N(SiMe_3)_2]_2$.

Three types of lanthanide–tin bonded compounds have been briefly described. Reaction between MCl_3 and Ph_3SnLi in THF at $-80°$ has so far yielded $M(SnPh_3)_3$ for M = Pr and Nd; which with excess Ph_3SnLi are converted into the complex anions $Li(thf)_n[M(SnPh_3)_4]$. Reaction between Ph_3SnLi and $(C_5H_5)_2MCl$ (M = Er or Yb) gives the coloured complexes $(C_5H_5)_2MSnPh_3$. Thermal decomposition of the complexes is observed above 200°.[45]

The lanthanides praseodymium, neodymium, holmium and lutetium react with

$Hg[Sn(C_6F_5)_3]_2$ to give air- and water-sensitive products of the type $(C_6F_5)_3SnMHgSn(C_6F_5)_3$[46] like the germanium analogues.[47] Reaction of $(Me_3SiCH_2)_3SnH$ with $M[N(SiMe_3)_2]_3$ in 1,2-dimethoxyethane yields, for praseodymium and neodymium, the complexes $(R_3Sn)_3M$(1,2-dimethoxyethane). Similar reactions with Et_3SnH and praseodymium or neodymium are more complex due to cleavage of a methyl group from the solvent. In these cases the products have been formulated as $Et_3SnM(OCH_2CH_2OMe)_2$.[48]

The only U–Sn complex so far described has been obtained by the room-temperature reaction of $(C_5H_5)_3UX$ ($X = CH_3$, NEt_2) in toluene with Ph_3SnH, forming $(C_5H_5)_3USnPh_3$ as a bright green solid. It decomposes at 60–70°C, is unstable in Lewis base solvents, and is readily oxidized and hydrolysed. Unexpectedly, Bu_3SnH apparently does not undergo this reaction.[49]

8.4 Titanium, zirconium and hafnium

All three metals are known to form tin complexes in their +2, +3 and +4 oxidation states. Ti(III)– and Ti(IV)–tin compounds have been obtained by classical nucleophilic displacement reactions (eqns 8.15 and 8.16):

$$(C_5H_5)_2TiCl + Ph_3SnM \xrightarrow{THF} (C_5H_5)_2TiSnPh_3(thf) \qquad (8.15)$$

$$(C_5H_5)_2TiCl_2 + Ph_3SnM \xrightarrow{THF} (C_5H_5)_2Ti(SnPh_3)_2 \qquad (8.16)$$

With an excess of Ph_3SnK in monoglyme, the anion $[(C_5H_5)_2Ti(SnPh_3)_2]^-$ is obtained.[50,51] Titanium (IV) complexes have also been formed from the amide and tin hydride by hydrostannolysis (eqn 8.17).

$$Ti(NMe_2)_4 + 4Ph_3SnH \rightarrow Ti(SnPh_3)_4 \qquad (8.17)$$

In the presence of added methanol, this reaction can yield $Ti(SnPh_3)_2(OMe)_2$.[52]

The dark green paramagnetic Ti(III)–Sn complex $(C_5H_5)_2TiSnPh_3(thf)$ is oxidized in $CHCl_3$ solution to $(C_5H_5)_2Ti(Cl)SnPh_3$, a diamagnetic green solid.[50]

For all three metals this type of compound can be made directly from the M (IV) complex (eqn 8.18).[53]

$$(C_5H_5)_2MCl_2 + R_3SnLi \rightarrow (C_5H_5)_2M(Cl)SnR_3 \qquad (8.18)$$

These chloro-complexes are coloured yellow or orange and are thermally stable solids. The M–Sn bond is cleaved by hydrogen halides to give Ph_3SnH and $(C_5H_5)_2MX_2$, consistent with bond polarity in the sense $M^{\delta+}–Sn^{\delta-}$.[54]

The zirconium complex, $Zr(SnPh_3)_4$, results from hydrostannolysis of the zirconium amide,[52] and both the zirconium and hafnium complexes (**1**) have been obtained by an oxidative addition reaction (eqn 8.19).[55]

$(C_6H_5CH_3)_2MPMe_3$ + $Me_3SnSnMe_3$

M = Zr, Hf

↓

$(Me_3Sn)_2M(\eta^6\text{-toluene})_2$ (8.19)

(**1**)

The hafnium complex (**1**, M = Hf) forms black crystals decomposing in air over several hours, and very rapidly oxidized in solution. The Sn–Hf–Sn bond angle is $<90°$.[55]

Hydrostannolysis of $Ti(NMe_2)_3$ or $[(C_5H_5)_2TiNMe_2]_2$ using Ph_3SnH results in the formation of titanium(III) compounds, believed to be $Ti(SnPh_3)_3$ and $(C_5H_5)_2TiSnPh_3$, respectively.[56] The far less thermally stable triethyltin complexes $(C_5H_5)_2MSnEt_3$ (M = Ti, Zr) have been formed from $(C_5H_5)_2M$ and Et_3SnLi in THF or toluene at $-40°$. Decomposition occurs at $-20°C$ to afford Sn_2Et_6.[57]

8.5 Vanadium, niobium and tantalum

All of these metals form bonds to tin, with a similar chemistry for each metal. The formal oxidation states involved are -1, $+1$, $+3$, $+4$ and $+5$.

The high reactivity of the Sn–H bond has been used in forming V–Sn compounds (eqn 8.20).[58]

$$(C_5H_5)_2VCH_2SiMe_3 + Et_3SnH \rightarrow (C_5H_5)_2VSnEt_3 + SiMe_4 \quad (8.20)$$

$(C_5H_5)_2VSnEt_3$, is a dark blue solid, stable at room temperature but very readily oxidized by air. It reacts with 1,2-dibromethane forming ethene, Et_3SnBr and $(C_5H_5)_2VBr_2$.[58] The V–Sn bond is cleaved by acids and alcohols and thermal decomposition gives primarily $(C_5H_5)_2V$ and Sn_2Et_6. The reverse type of reaction appears to take place with Sn_2Bu_6 in THF (eqn 8.21).[59]

$$(C_5H_5)_2V + Sn_2Bu_6 \rightarrow (C_5H_5)_2V(SnBu_3)_2 \quad (8.21)$$

However, the product has not been isolated. A further preparative method makes use of the high reactivity of Sn–N bonds[60] (eqn 8.22), although Me_nSnCl_{4-n} compounds also show similar reactivity towards tantalum hydrides.[61]

$$(C_5H_5)_2TaH_3 + Me_3SnNMe_2 \rightarrow Me_3SnTa(H)_2(C_5H_5)_2 + Me_2NH \quad (8.22)$$

Mössbauer, ir and nmr spectroscopy reveal that compounds of this type have a symmetrical structure with two terminal Ta–H and a central Ta–Sn bond, a structure which has been confirmed by X-ray analysis in the case of $(C_5H_5)_2TaH_2(SnMe_2Cl)$. Other $+3$ oxidation state complexes which have been examined crystallographically include $Ph_3SnNb(CO)(C_5H_5)_2$ and its $SnCl_3$ analogue.[62]

Seven-coordinate V(I)–Sn(IV) complexes have been obtained from the 18-electron six-coordinate hexacarbonyl vanadium anion (eqn 8.23):

$$\begin{array}{ccc} V(CO)_6^- + Ph_3SnCl & \longrightarrow & Ph_3SnV(CO)_6 \\ & & \downarrow PPh_3 \\ [V(CO)_5PPh_3]^- + Ph_3SnCl & \longrightarrow & Ph_3SnV(CO)_5PPh_3 \end{array} \quad (8.23)$$

These coloured complexes are of low thermal stability and readily oxidized[63]. Other compounds of vanadium, niobium and tantalum in their $+1$ oxidation state include $Ph_3SnM(CO)_5[P(OPh)_3]$ and $Ph_3SnM(CO)_4[PPh_2CH_2]_2$.[64]

Anionic complexes have been made from reactions of the type:

$$Na[Nb(CO)_3(C_5H_5)H] + Ph_3SnCl + Et_4NCl \xrightarrow{CH_3CN} Et_4N[Ph_3SnNb(CO)_3(C_5H_5)] \quad (8.24)$$

In this complex, tin is in a highly distorted tetrahedral environment, the overall geometry at niobium being that of a four-legged 'piano stool'.[65,66] Another 7-coordinate ion $[(Ph_3Sn)_2V(CO)_5]^-$ (formed from $[(CO)_5VSnPh_3]^{2-}$ and Ph_3SnCl), does not have the expected pentagonal bipyramidal structure. Rather various descriptions of the geometry are possible: monocapped trigonal prism, trigonal base-tetragonal base or a mono-capped octahedron.[67]

Doubly charged anionic complexes have been made by two routes (eqn 8.25):[67]

$$\begin{array}{c} Et_4N[(Ph_3Sn)_2V(CO)_5] + Ph_3SnLi \\ \downarrow \\ (Et_4N)_2[Ph_3SnV(CO)_5] \\ \uparrow \\ Et_4N[(Ph_3SnV(CO)_6] + Na/Hg \end{array} \qquad (8.25)$$

This reaction has also been applied to the synthesis of phosphine-substituted derivatives. Reaction of the $[(Ph_3Sn)_2V(CO)_5]^-$ ion with acetic acid in THF gives what is possibly the hydride $[H(Ph_3Sn)V(CO)_5]^-$ as an unstable orange–red complex. Mixed organotin anions have been obtained by the reaction

$$[Ph_3SnV(CO)_5]^{2-} + Me_3SnCl \rightarrow [Ph_3Sn(Me_3Sn)V(CO)_5]^-$$

Trichlorotin complexes may be made by similar reactions (eqn 8.26).[68]

$$C_5H_5V(CO)_4 + Et_4NSnCl_3 \xrightarrow{h\nu} (Et_4N)_2[(Cl_3Sn)_2V(CO)_2C_5H_5] \qquad (8.26)$$

8.6 Chromium, molybdenum and tungsten

All three metals form Sn–M complexes with M in oxidation states of −1, zero and +2 as neutral and anionic complexes, and with tin in its +2 and +4 oxidation states. Coordination numbers of 2, 4 and 6 are encountered for tin, and with one exception Lewis-base ligands are bonded to the transition metal. Although interesting structures have been found, the chemistry has so far been limited because of their low reactivity.

The reaction of SnX_2 with M–M bonded complexes of the type $[M(C_5H_5)(CO)_3]_2$ produces two products according to the experimental conditions: e.g. $X_2Sn[Mo(C_5H_5)(CO)_3]_2$ and $X_3SnMo(C_5H_5)(CO)_3$.[69] Insertion of SnX_2 into M–X bonds has also been extensively used. Subsequent oxidation by halogen cleaves the M–Sn bond (eqns 8.27 and 8.28):[70]

$$(C_5H_5)(CO)_2(Ph_3P)MoCl + SnCl_2 \rightarrow (C_5H_5)(CO)_2(Ph_3P)MoSnCl_3 \qquad (8.27)$$

$$(C_5H_5)_2MCl_2 + SnCl_2 + HCl \rightarrow (C_5H_5)_2ClMSnCl_3 \xrightarrow{Cl_2} [(C_5H_5)_2MCl_2]SnCl_3 \qquad (8.28)$$

The crystal structure of $Br_3SnMo(C_5H_5)_2Br$ confirms the presence of the Sn–Mo bond.[71]

A versatile method of forming Sn–M bonds for these metals is the reaction between an organotin amide and a metal hydride complex (eqn 8.29). The order of reactivity is CrH > MoH > WH for related complexes.[60]

$$Me_3SnNMe_2 + (C_5H_5)_2MH_2 \rightarrow Me_3SnM(H)(C_5H_5)_2 \quad (M = Mo, W) \tag{8.29}$$

Anionic transition metal complexes may also be used in synthesis (eqns 8.30–8.34).[72–76]

$$SnCl_4 + Na_2[M_2(CO)_{10}] \xrightarrow{thf} (thf)Cl_2SnM(CO)_5 \quad (M = Cr, Mo, W) \tag{8.30}$$

$$Me_3SnCl + Na[(C_5H_5)Mo(CO)_2(CH_3CN) \rightarrow Me_3SnMo(CO)_2(C_5H_5)(CH_3CN) \tag{8.31}$$

$$Ph_3SnCl + Na[(C_7H_7)Mo(CO)_2] \rightarrow Ph_3SnMo(CO)_2(C_7H_7) \tag{8.32}$$

$$Ph_3SnCl + Cs_2Cr(CO)_5 \rightarrow Cs[Ph_3SnCr(CO)_5] \tag{8.33}$$

$$Ph_3SnCo(CO)_4 + Na[Mo(CO)_3)C_5H_5] \rightarrow Ph_3SnMo(CO)_3C_5H_5 \tag{8.34}$$

Reaction of $Na_2[W_2(CO)_{10}]$ with $SnCl_4$ gives the planar metal cluster complex (**2**) in which the W–Sn ring bond lengths are 278 pm, the exo W–Sn length being significantly shorter at 270 pm.[77]

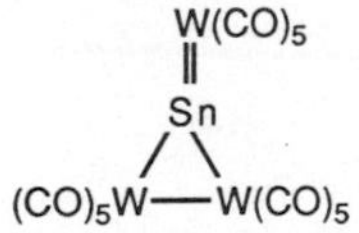

W(CO)5
||
Sn
(CO)5W—W(CO)5

(**2**)

Reduction of complexes of the type $W(CO)_3L_3$ (L = 1, 1, 4, 7, 7-pentamethyldiethylenetriamine) with K/NH_3 gives tungsten carbonyl anions of unknown structure which react with Ph_3SnCl in ethanol to give 7-coordinate complexes such as $(Me_4N)_2[HW(CO)_3(SnPh_3)_3]$ or $Ph_4P[(Ph_3Sn)_2\{(Ph_2Sn)_2OEt\}W(CO)_3]$. In the latter compound, the $(Ph_2Sn)_2OEt$ unit functions as a bidentate ligand as in (**3**).[78,79]

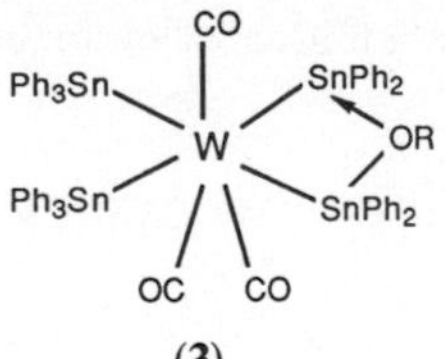

CO
Ph3Sn SnPh2
W OR
Ph3Sn SnPh2
OC CO

(**3**)

When $SnCl_4$ reacts with neutral mölybdenum or tungsten complexes, cationic products with an $M–SnCl_3$ bond are formed (eqn 8.35).[80]

$$SnCl_4 + [M(CO)_3C_5H_5]_2 \rightarrow [Cl_3SnM(CO)_3C_5H_5]Cl \xrightarrow{SnCl_4} [Cl_3SnM(CO)_3C_5H_5]SnCl_5 \tag{8.35}$$

Sn–M bonds may be formed using anionic tin species (eqns 8.36, 8.37).[68]

$$Me_4NSnCl_3 \xrightarrow{M(CO)_6} Me_4N[Cl_3SnM(CO)_5]$$
$$Me_4NSnCl_3 \xrightarrow{C_6H_6M(CO)_3} (Me_4N)_3[(Cl_3Sn_3M(CO)_3]$$
$$Me_4NSnCl_3 \xrightarrow{M(C_6H_6)_2} (Me_4N)_6[(Cl_3Sn)_6M] \quad (8.36)$$

$$Ph_3SnLi + M(CO)_6 \rightarrow Li[Ph_3SnM(CO)_5] \quad (8.37)$$
$(M = Cr, Mo, W)$

Some of these anionic complexes are stable to air and water.[75] The complex $Et_4N[(Ph_3Sn)_3Cr(CO)_4]$, in which chromium is 7-coordinate with approximately C_{3v} symmetry, reacts with Ph_3SnLi in HMPA forming the 6-coordinate anion $[(Ph_3Sn)_2Cr(CO)_4]^{2-}$.[81] Starting from an Mo(III) compound with a Mo≡Mo triple bond, this type of reaction has yielded the triply-bonded Mo–Mo complex (**4**), with a Mo≡Mo bond length of 220.1 pm[82].

$$(Me_2N)_4Mo_2Cl_2 + (THF)_3Li[Sn(SnMe_3)_3] \longrightarrow$$
$$(Me_3Sn)_3Sn(Me_2N)_2Mo \equiv Mo(NMe_2)_2Sn(SnMe_3)_3 \quad (8.38)$$

(**4**)

Oxidative addition of organotin hydrides to chromium, molybdenum and tungsten complexes represents a further method of establishing the tin–metal bond (eqns 8.39, 8.40).

$$2Me_3SnH + [Mo(C_5H_5)(CO)_3]_2 \rightarrow H_2 + 2Me_3SnMo(C_5H_5)(CO)_3 \quad (8.39)$$

$$Ph_3SnH + [Cr(C_5H_5)(CO)_3]_2 \rightarrow Ph_3SnCr(C_5H_5)(CO)_3 + CrH(C_5H_5)(CO)_3 \quad (8.40)$$

Closely related is the oxidative addition of cyclopentadienyltin compounds to the transition metal (eqn 8.41).[85]

$$Me_3SnC_5H_5 + M(CO)_3(CH_3CN)_3 \rightarrow Me_3SnM(CO)_3C_5H_5 + 3CH_3CN \quad (8.41)$$
$M = Cr, Mo, W$

The greater reactivity of M–Si or M–Ge bonds can be used in synthesis:[86]

$$Me_3SiM(CO)_3C_5H_5 + Me_3SnOSiMe_3 \rightarrow Me_3SnM(CO)_3C_5H_5 + Me_6Si_2O \quad (8.42)$$

Cleavage of the W–CH_3 bond in $C_5H_5(CH_3)W(CO)_3$ by $SnCl_2$ gives $(C_5H_5)(CO)_3WSnCl_3$.[87]

Reactions of stable stannylenes with transition metal complexes can take a variety of forms: they can behave as neutral Lewis-base donors, displacing, for example, CO (eqn 8.43)[44].

$$Sn(NR_2)_2 + M(CO)_6 \rightarrow (R_2N)_2SnM(CO)_5 \quad (M = Cr, Mo, W. \; R = SiMe_3) \qquad (8.43)$$

With substituted metal carbonyls $M(CO)_4L_2$, *trans*-disubstituted derivatives can be isolated $[(R_2N)_2Sn]_2M(CO)_4$. Reaction of $Sn(NR_2)_2$ with $M(CO)_5(THF)$ (M = Cr, Mo, W) gives the dimer $[(R_2N)_2SnM(CO)_5]_2$ (**5**) with a 4-membered $\overline{SnNSnN}$ ring. In THF this complex breaks down to the monomer $(R_2N)_2(thf)SnM(CO)_5$.[88]

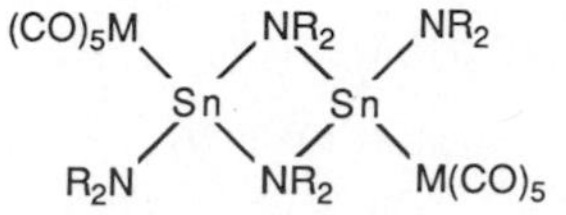

(CO)$_5$M, NR$_2$, NR$_2$, Sn, Sn, R$_2$N, NR$_2$, M(CO)$_5$

(**5**)

Irradiation of $M(CO)_6$ (M = Cr, Mo, W) and $Sn[N(SiMe_3)_2]_2$ in hexane gives both mono- and di-substitution products[89,90], the stereochemistry about tin being trigonal planar.[90] Similarly, using $M(CO)_4$(norbornadiene) gives exclusively *trans*-$(R_2Sn)_2M(CO)_4$.[91] However, with less bulky ligands the tin becomes tetrahedrally coordinated by incorporation of a solvent molecule, e.g. in $(thf)Bu^t_2SnCr(CO)_5$. The pyridine analogue has an Sn–Cr bond length of 265.4 pm with distorted tetrahedral geometry about tin.[90,92,93] This is also the case for the product of reaction (8.44)[94]:

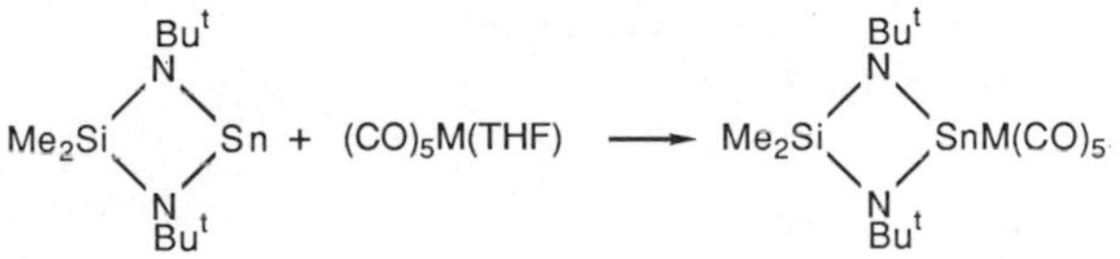

$$Me_2Si(NBu^t)_2Sn + (CO)_5M(THF) \longrightarrow Me_2Si(NBu^t)_2SnM(CO)_5 \qquad (8.44)$$

$$[Me_2Si(NBu^t)_2Sn]_2Mo(CO)_4$$

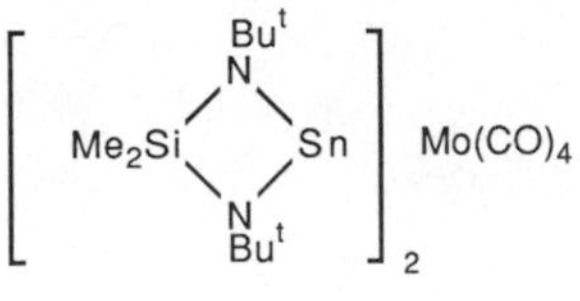

(**6**)

The similar molybdenum complex (**6**) has also been prepared.

Reaction of $Sn(NR_2)_2$ with hydride complexes of Mo and W has given Sn(IV) products by oxidative addition[95] (eqn 8.45).

$$Sn(NR_2)_2 + 3M(CO)_3(C_5H_5)H \rightarrow HSn[M(CO)_3C_5H_5]_3 \qquad (8.45)$$

A similar reaction leads to the formation of the dark red two-coordinated Sn(II) complex (7)[44].

$$Sn(NR_2)_2 + 2C_5H_5Mo(CO)_3H \rightarrow \underset{(7)}{Sn[Mo(C_5H_5)(CO)_3]_2} \qquad (8.46)$$

Biscyclopentadienyltin reacts with $HMo(CO)_3(C_5H_5)$ in the same way as the tin(II) amide to give $HSn[Mo(CO)_3(C_5H_5)]_3$. This reaction seems to proceed via $Sn[Mo(CO)_3C_5H_5]_2$, which then reacts oxidatively with $HMo(CO)C_5H_5$[96]. In its photochemically-induced reaction with $Cr(CO)_6$, $Sn(C_5H_5)_2$ behaves as a 2-electron donor giving $(C_5H_5)_2SnCr(CO)_5$.[60]

Bis[bis(trimethylsilyl)methyl]tin, R_2Sn ($R = CH(SiMe_3)_2$), reacts oxidatively with chromium and molybdenum hydrides and alkyls[91,97] (eqns 8.47, 8.48):

$$R_2Sn + Mo(C_5H_5)(CO)_3H \rightarrow R_2HSnMo(CO)_3C_5H_5 \qquad (8.47)$$

$$R_2Sn + Mo(CO)_5Me \rightarrow R_2MeSnMo(CO)_3C_5H_5 \qquad (8.48)$$

Other oxidative addition reactions of this type employ Sn–Sn and Sn–Cl bonded compounds[85,98–100] (eqns 8.49 and 8.50).

$$Sn_2Me_6 + [C_5H_5M(CO)_3]_2 \rightarrow Me_3SnM(CO)_3C_5H_5 \qquad (8.49)$$

$$R_3SnCl + (bipy)M(CO)_4 \rightarrow R_3SnMCl(CO)_3(bipy) \qquad (8.50)$$

The high reactivity of Sn–Hg bonded compounds can be used to form tin–transition metal bonds (eqn 8.51), although there are difficulties in handling these stannyl mercurials because of their toxicity and thermal instability.[101]

$$(R_3Sn)_2Hg + [Mo(CO)_3C_5H_5]_2 \rightarrow R_3SnMo(CO)_3C_5H_5 \quad (R = Me_3SiCH_2) \qquad (8.51)$$

The stannylcarbene–chromium complex (**8**) undergoes a carbene→carbyne rearrangement with migration of tin from carbon to chromium:

$$[(CO)_5Cr{\equiv}CNEt_2]BF_4 + Ph_3SnK \rightarrow \underset{(\mathbf{8})}{(CO)_5Cr{=}C(NEt_2)(SnPh_3)} \rightarrow Ph_3Sn{-}Cr(CO)_4{\equiv}CNEt_2 \qquad (8.52)$$

This reaction occurs spontaneously at room temperature in the solid state and in solution following first-order kinetics, and is essentially independent of solvent polarity.[31,102]

Some reactions of these complexes have already been referred to in discussing their formation. Neutral ligand exchange occurs readily and may result in subsequent disproportionation[103] (eqn 8.53):

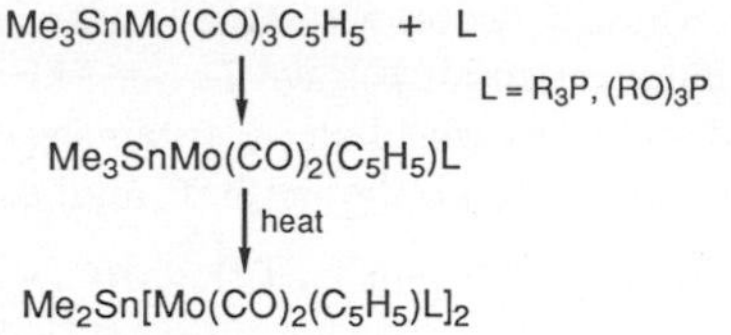

$$\begin{array}{c} Me_3SnMo(CO)_3C_5H_5 + L \\ \downarrow \quad L = R_3P, (RO)_3P \\ Me_3SnMo(CO)_2(C_5H_5)L \\ \downarrow \text{heat} \\ Me_2Sn[Mo(CO)_2(C_5H_5)L]_2 \end{array} \tag{8.53}$$

The Sn–W bonds in stannylene complexes can react with the carbonyl groups of $Fe(CO)_5$ affording the complexes (**9**):[88]

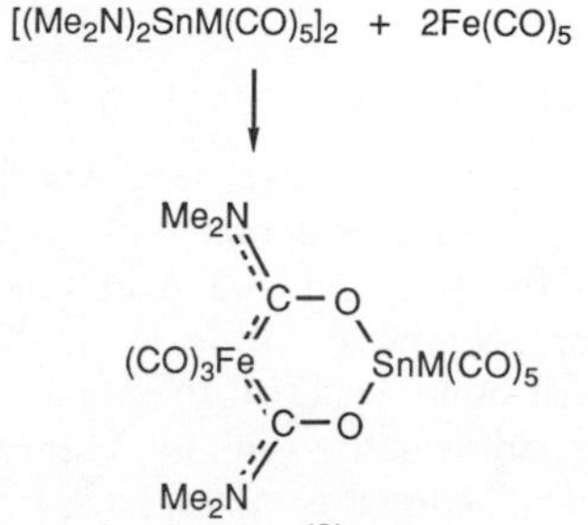

$$[(Me_2N)_2SnM(CO)_5]_2 + 2Fe(CO)_5 \longrightarrow$$

Me2N
C–O
(CO)3Fe SnM(CO)5
C–O
Me2N
(**9**) (8.54)

Substitution of anionic ligands on tin proceeds without Sn–M bond breaking, and in the case of 8-hydroxyquinoline (eqn 8.55) the product is 6-coordinate about tin.

$$Cl_3SnMo(CO)_3C_5H_5 + \text{oxine} \rightarrow Cl(\text{oxinato})_2SnMo(CO)_3C_5H_5 \tag{8.55}$$

Similarly, halide exchange occurs at tin:[105]

$$L_nM\text{–}SnCl_3 \xrightarrow{SnBr_2} L_nM\text{–}SnBr_3 \xrightarrow{SnI_2} L_nM\text{–}SnI_3 \tag{8.56}$$

Compounds described as $R_2Sn[Cr(CO)_3(\text{arene})]_2$ (R = Me, Ph) can be oxidized electrochemically to stable cations. However, the formulation of these compounds must be incorrect, and they probably contain R_2Sn groups bonded to the arene ligand.[106] Methyllithium reacts with a CO ligand of $Ph_3SnM(CO)_3C_5H_5$ forming a carbene without cleaving the Sn–Mo bond.[107] The complexes $Ph_3SnM(CO)_3C_5H_5$ (M = Mo, W) react with liquid sulphur dioxide at 25°C, mainly by Sn–C bond cleavage to give di-insertion products. By contrast, the chromium compound is unreactive, as are the complexes $Ph_2Sn[M(CO)_3C_5H_5]_2$ (M = Mo and W). The analogous trimethyltin derivatives (M = Mo, W) absorb 1 or 2 moles of SO_2 by cleavage of the Sn–M bond with formation of the bridged complexes (**10**) and (**11**).[108]

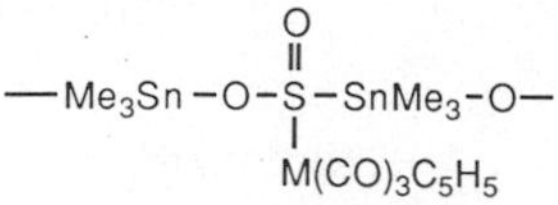

—Me3Sn–O–S(=O)(M(CO)3C5H5)–SnMe3–O—
(**10**)

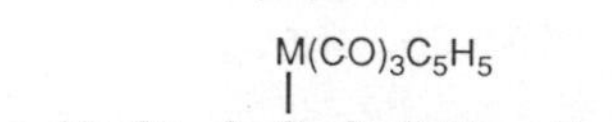

—Me3Sn–O–S(M(CO)3C5H5)–O–SnMe3–O—
(**11**)

Reaction of the $[Ph_3SnW(CO)_5]^-$ anion with HCl results in cleavage of the Sn–Ph bonds.[109] Reactions with halogens and ICl cleave the Sn–M bond. The kinetics suggest that the reactions proceed via an intermediate in a two step process.[110–113] RHgCl and $HgCl_2$ also cleave the Sn–M bond, though not in all cases:[114]

$$Me_3SnM(CO)_3C_5H_5 + HgCl_2 \rightarrow Me_3SnCl + ClHgM(CO)_3C_5H_5 \quad (8.57)$$

$$(allyl)_3SnM(CO)_3C_5H_5 + HgCl_2 \rightarrow (allyl)HgCl + (allyl)_2ClSnM(CO)_3C_5H_5 \quad (8.58)$$

Photolysis of $Me_2Sn[Mo(CO)_3C_5H_5]_2$ or $Me_2ClSnMo(CO)_3C_5H_5$ gives $Mo_2(CO)_6(C_5H_5)_2$.[115]

8.7 Manganese, technetium and rhenium

No technetium–tin bonded compounds have been described. For manganese, all examples involve the metal in its +1 oxidation state combined either to Sn(II) or Sn(IV). For rhenium, both the +1 and +3 oxidation states have yielded tin complexes. Neutral and cationic complexes feature in the chemistry. In most cases the transition metal has Lewis base donors (CO, R_3P) bonded to it, and these complexes are generally of high thermal and oxidative stability. Manganese and rhenium form a wide range of readily prepared anionic complexes (eqns 8.59, 8.60) and these are commonly used in synthesis.[116–118]

$$NaM(CO)_5 + R_3SnCl \rightarrow R_3SnM(CO)_5 \quad (8.59)$$

$$Na[M(CO)_3(phen)] + R_3SnCl \rightarrow R_3SnM(CO)_3(phen) \quad (8.60)$$

With tin halides possessing more than one Sn–Cl bond further substitution can occur, and with $SnCl_4$, three chlorines may be replaced giving $ClSn[Mn(CO)_5]_3$.[116] Complications can occur due to halogen–metal exchange and subsequent coupling reactions[116] e.g. eqns (8.61 and 8.62):

$$X_2Sn[Mn(CO)_5]_2 + 2NaMn(CO)_5 \rightarrow [(CO)_5Mn]_3SnSn[Mn(CO)_5]_3 \quad (8.61)$$

$$C_6F_5Li + Ph_2(Cl)SnMn(CO)_5 \rightarrow [(CO)_5MnSnPh_2]_2 + Ph_2(C_6F_5)SnMn(CO)_5 \quad (8.62)$$

The dimeric structure of $[(CO)_5MnSnPh_2]_2$ is based on a 4-membered Mn–Sn–Mn–Sn ring. In $XSn[Mn(CO)_4(PPh_3)]_3$ (X = Cl, Br, I), the Ph_3P ligand is *trans* to tin for all three halogens.[119–120] The crystal structure of $Br_2Sn[Re(CO)_4PPh_3]_2$ has also been reported.[121]

The reaction between SnX_2 and $M_2(CO)_8(PPh_3)_2$ gives two types of product: $ClSn[M(CO)_4PPh_3]_3$ and $M_2(CO)_8[\mu\text{-}Sn(Cl)M(CO)_4PPh_3]_2$. The latter structure of the compound (M = Re) (**12**) consists of a planar $[Re_2Sn_2]$ rhombus with a transannular Re–Re bond; with each tin(IV) atom also being bonded to a chlorine and $[Re(CO)_4(PPh_3)]$.[122]

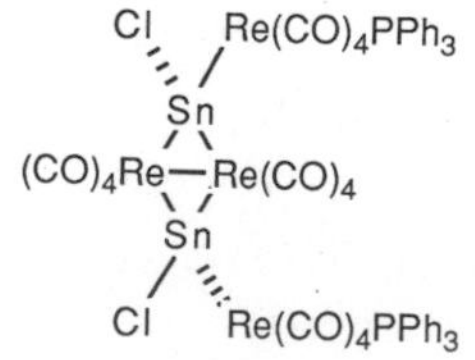

Cl Re(CO)4PPh3
Sn
(CO)4Re—Re(CO)4
Sn
Cl Re(CO)4PPh3

(**12**)

Reaction of SnX_2 (X = Br, I) and $Mn_2(CO)_{10}$ gives the metal cluster complexes $X_2Sn[Mn(CO)_5]_2$ and $Mn_2(CO)_8[\mu\text{-}(XSn)Mn(CO)_5]_2$.[123] Photochemical methods have been successful in forming M–Sn bonds[124] (eqns 8.63, 8.64):

$$C_5H_5Mn(CO)_3 + THF \xrightarrow{h\nu} C_5H_5Mn(CO)_2(THF) \xrightarrow{Ph_4AsSnCl_3} Ph_4As[C_5H_5(CO)_2MnSnCl_3] \quad (8.63)$$

$$Mn_2(CO)_{10} + Ph_4As[SnCl_3] \xrightarrow{h\nu} Ph_4As[Mn_2(CO)_9SnCl_3] + CO \quad (8.64)$$

In the latter reaction no disubstitution is observed. Reduction of $M(CO)_5^-$ [M = Mn or Re] by sodium in HMPT gives the metal in its lowest oxidation state, and subsequent reaction with organotin halides results in the formation of anionic complexes[125]:

$$Na_3[M(CO)_4] + Ph_3SnCl \rightarrow [(Ph_3Sn)_2M(CO)_4]^- \quad (8.65)$$

The reaction of the amidostannylene $[(Me_3Si)_2N]_2Sn$ with $BrMn(CO)_5$ at 20°C results in oxidation of the tin and formation of an Sn–Mn bond giving $Br(R_2N)_2SnMn(CO)_5$ (R = Me_3Si).[44] When the transition metal has more than one neutral ligand bonded to it, *cis*- and *trans*-isomers of the resulting octahedral complex are usually isolated, as in the reaction between Me_3SnBr and $Na[Mn(CO)_3(PPh_3)_2]$. Isomer interchange in the resulting complex, $Me_3SnMn(CO)_3(PPh_3)_2$, proceeds by an intramolecular process. This reaction also yields $Me_2BrSnMn(CO)_3(PPh_3)_2$ by a redistribution process.[126,127] Other unpredictable reactions include that between $Sn(C_5H_5)_2$ and $HMn(CO)_5$ where the product, $[(CO)_5Mn]_2Sn(H)Sn(H)[Mn(CO)_5]_2$, contains an Sn–Sn as well as Sn–H bonds.[128]

Rhenium hydrides have been used in the synthesis of Re(III) complexes:

$$(C_5H_5)_2ReH + Me_nSnCl_{4-n} \xrightarrow{Et_3N} (C_5H_5)_2ReSnMe_nCl_{3-n} \quad (8.66)$$

These coloured complexes have been examined crystallographically, which shows that substitution of chlorine by methyl on tin lengthens the Re–Sn bond by 2 pm per methyl group. The two cyclopentadienyl rings are non-parallel.[129] A number of Mn(I)–Sn(II) complexes have been reported in which tin is bonded to a porphyrinato or β-diketonato ligand[130,131] (eqns 8.67, 8.68).

$$Mn(CO)_2(C_5H_4Me)THF + Sn\left[\begin{matrix} O-CCF_3 \\ CH \\ O-CCF_3 \end{matrix}\right]_2 \longrightarrow (MeC_5H_4)(CO)_2MnSn\left[\begin{matrix} O-CCF_3 \\ CH \\ O-CCF_3 \end{matrix}\right]_2 \quad (8.67)$$

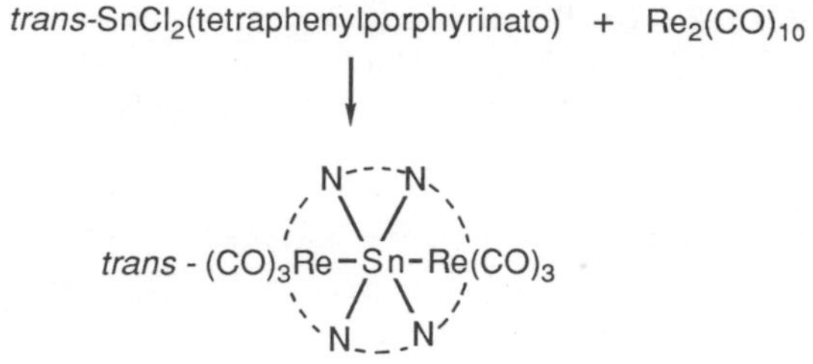

trans-$SnCl_2$(tetraphenylporphyrinato) + $Re_2(CO)_{10}$

↓

trans - $(CO)_3Re-Sn-Re(CO)_3$ (with four N atoms of the porphyrinato ligand about Sn)

(8.68)

A striking feature of the chemistry of these complexes is their low chemical reactivity. Electrophiles commonly, though not invariably, cleave the Sn–C rather than the Sn–M bond,[132] e.g.:

$$Ph_3SnMn(CO)_5 + 3Cl_2 \rightarrow Cl_3SnMn(CO)_5 \quad (8.69)$$

$$Ph_2Sn[Mn(CO)_5]_2 + HCl \rightarrow Cl_2Sn[Mn(CO)_5]_2 \quad (8.70)$$

Redistribution of groups on tin occurs readily:

$$2Ph_3SnMn(CO)_5 + Cl_3SnMn(CO)_5 \rightarrow 3Ph_2ClSnMn(CO)_5 \quad (8.71)$$

A further reflection of their low reactivity is the formation of anionic complexes[133]:

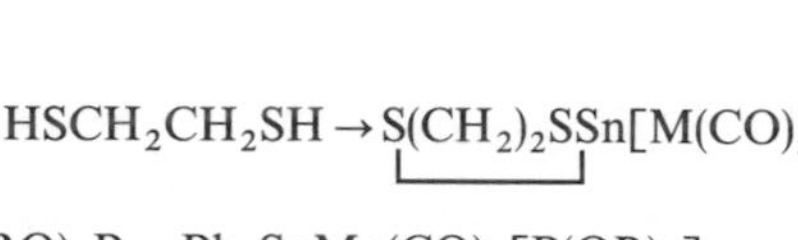

$Me_2ClSnMn(CO)_5$ + $AgBF_4$ ↘

[$Me_2SnMn(CO)_5$]BF_4 + AgCl

$Me_3SnMn(CO)_5$ + BF_3 ↗

(8.72)

In compounds with tin–halide bonds the halide may be exchanged without Sn–M bond cleavage, and ligand substitution reactions at the transition metal occur readily, e.g. eqns (8.73)–(8.75).[116,134,135]

$$Cl_2Sn[M(CO)_5]_2 + HSCH_2CH_2SH \rightarrow \underline{S(CH_2)_2S}Sn[M(CO)_5]_2 \quad (8.73)$$

$$Ph_3SnMn(CO)_5 + (RO)_3P \rightarrow Ph_3SnMn(CO)_4[P(OR)_3] \quad (8.74)$$

$$Me_3SnMn(CO)_5 + C_2H_4 \xrightarrow{h\nu} Me_3SnMn(CO)_4(C_2H_4) \quad (8.75)$$

A number of ligand substitution reactions involving tertiary phosphines have been examined kinetically. The rate law is consistent with slow dissociation of CO followed by nucleophilic attack by the phosphine.[136] Most Sn(IV)–M complexes are 4-coordinate about tin. However, in some acetonylacetonates the tin is 6-coordinate, and in $(acac)_2PhSnMn(CO)_5$ *cis*, *trans* and *fac* geometric isomers have been isolated[137]:

$$PhCl_2SnMn(CO)_5 + 2Tl(acac) \rightarrow \underset{\textbf{(13)}}{(acac)_2PhSnMn(CO)_5} \quad (8.76)$$

In the acetolysis of $Ph_3SnMn(CO)_5$ and $Ph_2Sn[Mn(CO)_5]_2$ only Sn–Ph bonds are cleaved,[138] and both $SiHCl_3$ and $SnCl_4$ react with $Me_3SnMn(CO)_5$ by chlorinative cleavage of one methyl group.[139] Similarly, reaction with Me_3SnCF_3 gives $Me_2(CF_3)SnMn(CO)_5$.[140] $Cl_2Sn[Mn(CO)_5]_2$ is reduced by $LiAlH_4$ to $H_2Sn[Mn(CO)_5]_2$.[141]

Many reactions of Sn–Mn and Sn–Re bonded complexes do result in Sn–M bond cleavage. Complexes of the type $Ph_3SnRe(CO)_3$(phen) have a low-lying charge transfer excited state. The HOMO has sigma Sn–Re (or Sn–Mn) bonding character whilst the LUMO is mainly delocalized on the phenanthroline. Thus, reduction is reversible since addition of an electron to the LUMO does not destabilize the Sn–Re bond:

$$Ph_3SnRe(CO)_3(phen) + e^- \xrightarrow{-1.4V\ (CH_3CN)} [Ph_3SnRe(CO)_3(phen)]^{-\cdot} \qquad (8.77)$$

In contrast, oxidation is irreversible since it removes an electron from the HOMO:[118]

$[(CH_3CN)Re(CO)_3(phen)]^+$

$Ph_3SnRe(CO)_3(phen) \xrightarrow{+0.85V\ (CH_3CN)} [(CH_3CN)Re(CO)_3(phen)]^+$

$Ph_3SnRe(CO)_3(phen) \xrightarrow{e} [Ph_3SnRe(CO)_3(phen)]^{+\bullet}$

$[Ph_3SnRe(CO)_3(phen)]^{+\bullet} \xrightarrow{fast} Ph_3Sn^{\bullet} + [Re(CO)_3(phen)]^+$

$[Re(CO)_3(phen)]^+ \xrightarrow{(CH_3CN)} [(CH_3CN)Re(CO)_3(phen)]^+ \qquad (8.78)$

The Sn–Mn bond in $Me_3SnMn(CO)_5$ is more readily cleaved than in the triphenyl analogue (due possibly to π-interaction of the phenyl groups strengthening the Sn–Mn bond,[133] and reagents which do so include RHgX, $HgCl_2$, $C_2H_4Br_2$,[139] halogens, HX, ICl,[133] Ph_2PCl, and Ph_3PAuCl.[142] Cleavage by bromine proceeds at a rate some 4000 times faster than for iodine. In other related reactions ICI forms a detectable charge transfer complex prior to bond cleavage. The rate of bond cleavage by iodine is mainly determined by the size of M (Mn, Re) and the effectiveness of shielding by other ligands.[111,112]

The reaction of $Ph_3SnMn(CO)_5$ with tetraphenylcyclopentadienone results in the formation of an Sn–O bond by transfer of Ph_3Sn from manganese to the ring with concomitant formation of the π-complex (**14**)[143] (eqn 8.79).

$Ph_3SnMn(CO)_5$ + O=C₅Ph₄ (tetraphenylcyclopentadienone)

↓

Ph_3SnO–C₅Ph₄–$Mn(CO)_3$ + 2CO

(14) (8.79)

Reactions of these compounds with liquid sulphur dioxide show variations according to the groups bonded to tin. $Ph_3SnM(CO)_5$ (M = Mn, Re) reacts by cleavage of Sn–C bonds giving a variety of structures in which Sn–OS(=O)Ph groups may be terminal or chelating. By contrast, the trimethyltin–manganese complex incorporates 1.5 mol SO_2 whereas the rhenium analogue gives $(CO)_5ReSnMe_2[OS(=O)Me]$, with the $MeSO_2$ ligand chelating the tin atom. With $Ph_2Sn[Mn(CO)_5]_2$ two mols of SO_2 insert into Sn–C bonds and one Sn–Mn bond is cleaved.[144] Photolysis of $Me_2ClSnMn(CO)_5$ gives the bridged dimer (**15**) as a sublimable solid.[115,145]

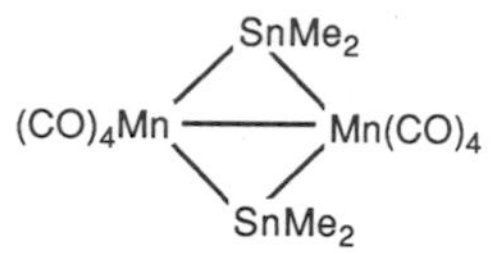

$SnMe_2$

$(CO)_4Mn$ — $Mn(CO)_4$

$SnMe_2$

(**15**)

Difluorocarbene fails to insert into the Sn–Mn bond,[140] but the reactions of perfluoroalkenes and alkynes are of some interest. UV irradiation facilitates the insertion of C_2F_4 into the Sn–Mn bond of $Me_3SnMn(CO)_5$ by what would seem to be a free radical mechanism, although the absence of products resulting from multiple insertions, i.e. $[Me_3Sn(C_2F_4)_nMn(CO)_5]$, is perhaps evidence against this. Other products can be accounted for by decomposition of the initially formed molecule:[135,146]

$$Me_3Sn(C_2F_4)Mn(CO)_5 \rightarrow Me_3SnF + CF_2{=}CFMn(CO)_5 + [CF_2{=}CFMn(CO)_4]_2 \quad (8.80)$$

A similar reaction with $CClF{=}CF_2$ gives evidence for non-selective insertion into the Sn–Mn bond (formation of Me_3SnCl and Me_3SnF). With $CF_3CF{=}CF_2$, Me_3SnF is obtained together with *cis*- and *trans*-isomers of the alkene complex, $CF_3CF{=}CFMn(CO)_5$. Perfluorocyclobutene reacts in a similar way, and irradiation of $CF_2{=}CFCF{=}CF_2$ gives the *cis* isomer of $CF_2CF{=}CFCF{=}CFMn(CO)_5$.[147] The carbene complex $Ph_3SnRe(CO)_4[C(OEt)NR_2]$ reacts with BI_3 by cleavage of Ph–Sn bonds, giving finally the I_3Sn–Re complex.[148]

8.8 Iron, ruthenium and osmium

Many papers have appeared on these metals bonded to tin and a large number of crystal structures have been reported. All of the metals bond to tin in either their 0 or +2 oxidation states with Lewis base ligands completing the coordination at the transition metal. Tin may be in its +2 or +4 oxidation states with coordination numbers of 2, 4, and less commonly 5 and 6. In octahedral complexes like $(Me_3Sn)_2Os(CO)_4$ both *cis*- and *trans*-isomers exist. Exchange of carbonyl groups between axial and equatorial positions is a non-dissociative process,[149] and in favourable cases geometrical isomers can be identified from their ir spectra in the carbonyl region.[150] Preparative methods are similar for the three metals, and all of the methods employed have already been referred to in earlier parts of this chapter.

Syntheses often make use of anionic complexes of the transition metal, as illustrated in the following examples:[132,151–153]

$$R_3SnCl + [Fe(CO)_2(C_5H_5)]^- \rightarrow R_3SnFe(CO)_2(C_5H_5) \quad (8.81)$$

$$R_3SnX + [Fe(CO)_4]^{2-} \rightarrow cis\text{- or } trans\text{-}(R_3Sn)_2Fe(CO)_4 \quad (8.82)$$

$$Ph_2SnCl_2 + [Fe(CO)_2(C_5H_5)]^- \rightarrow Ph_2Sn[Fe(CO)_2C_5H_5]_2 \quad (8.83)$$

The reaction of $[Fe(CO)_2(C_5H_5)]^-$ with $PhSnCl_3$ or $SnCl_4$ gives $PhSn[Fe(CO)_2(C_5H_5)]_3$ and $Sn[Fe(CO)_2(C_5H_5)]_4$, respectively.[153] However, with SnI_4 the reaction may be terminated at the stage $I_2Sn[Fe(CO)_2(C_5H_5)_2$.[154] This type of reaction often gives products such as (**16**) and (**17**) resulting from secondary redistribution of ligands bonded to tin[155]:

$$MeSnCl_3 + Na_2Fe(CO)_4 \longrightarrow Me_2Sn(\mu\text{-}Fe(CO)_4)_2SnMe_2\ (\mathbf{16}) + Me_2Sn(\mu\text{-}Fe(CO)_4)_2Sn(\mu\text{-}Fe(CO)_4)_2SnMe_2\ (\mathbf{17}) \quad (8.84)$$

Anionic tin derivatives have occasionally been used in synthesis[156].

$$Ph_3SnLi + Et_4N[HFe_3(CO)_{11}] \rightarrow Et_4N[Ph_3SnFe(CO)_4] \quad (8.85)$$

Insertion of SnX_2 into M X or M–M bonds (eqns 8.86, 8.87) gives X_3Sn–M or $M(SnX_2)M$ derivatives,[157] a type of reaction which has been examined kinetically.[158]

$$SnX_2 + M(CO)_2C_5H_5X \rightarrow X_3SnM(CO)_2C_5H_5 \quad (8.86)$$

(M = Fe, Ru, Os)

$$SnX_2 + [Ru(CO)_2C_5H_5]_2 \rightarrow X_2Sn[Ru(CO)_2C_5H_5]_2 \quad (8.87)$$

The reactions of $Fe(CO)_5$[159] or $Fe_2(CO)_9$[160] with tin(II) halides are more complex:

$$SnCl_2 + Fe(CO)_5 \rightarrow Sn[Fe(CO)_4]_4 \quad (8.88)$$

$$SnX_2 + Fe_2(CO)_9 \rightarrow X_2Sn[Fe(CO)_4]_2 \xrightarrow{\text{pyridine}} (py)X_2SnFe(CO)_4 \quad (8.89)$$

Insertion of SnX_2 into Fe–C bonds (eqn 8.90) occurs readily via a radical chain process.[161,162]

$$SnX_2 + Fe(CO)_2(C_5H_5)R \rightarrow RX_2SnFe(CO)_2C_5H_5 \quad (8.90)$$

If $RuCl_3{\cdot}xH_2O$ is dissolved in ethanol and CO passed into the solution followed by $SnCl_2$, Ph_3P, and acetone, the monomeric complex,

$Cl_3SnRuCl(CO)(Ph_3P)_2(Me_2CO)$, is produced in which the phosphines are mutually *trans* and $[SnCl_3]$ is *trans* to Cl.[163] Insertion of $SnCl_2$ into the Ru–Cl bond of a stereoisomer of $Ru[Ph_2PCH(CH_3)CH_2PPh_2](C_5H_5)Cl$ occurs with retention of the geometry at the ruthenium atom. This means that $SnCl_2$ could attack at either chlorine or ruthenium, or both to give a 3-centred transition state.[164] The complex, $RuCl(C_6H_6)(CH_3)[Ph_2PNHCH(CH_3)Ph]$, is optically active, and reaction with $SnCl_2$ gives separable isomers in which the chirality at the ruthenium is different. The diastereoisomer which is laevorotatory at 436 nm has the R configuration, which is also the configuration of the chiral phosphine ligand.[165]

The related cleavage of M–M bonds by SnX_4 (e.g. eqns 8.91, 8.92) has also been used in synthesis,[166] the products obtained often depending on the reaction conditions.[167]

$$SnCl_4 + [Fe(CO)_2C_5H_5]_2 \rightarrow Cl_3SnFe(CO)_2C_5H_5 + Fe(CO)_2(C_5H_5)Cl \qquad (8.91)$$

$$SnCl_4 + Ru_3(CO)_{12} \xrightarrow{CO} trans\text{-}(Cl_3Sn)_2Ru(CO)_4$$
$$SnCl_4 + Ru_3(CO)_{12} \xrightarrow{20^\circ C} ClRu(CO)_4Ru(CO)_4Ru(CO)_4SnCl_3$$
$$SnCl_4 + Ru_3(CO)_{12} \xrightarrow{135^\circ C} (CO)_3Ru(\mu\text{—}Cl)_3Ru(CO)_2SnCl_3 \qquad (8.92)$$

In $(Me_3Sn)_2Ru_2(CO)_8$, the Sn–Ru–Ru–Sn chain is almost linear.[168] Osmium carbonyls and carbonyl hydrides react with $SnCl_4$ forming Os–$SnCl_3$ bonds[169,170]:

$$SnCl_4 \xrightarrow{Os_3(CO)_{12}} Cl_3Sn[Os(CO)_4]_3Cl$$
$$SnCl_4 \xrightarrow{H_2Os(CO)_4} cis\text{-}Os(CO)_4(SnCl_3)H$$
$$SnCl_4 \xrightarrow{H_2Os_2(CO)_8} Os_2(CO)_8(SnCl_3)H \qquad (8.93)$$

Oxidative addition of $SnCl_4$ to $Fe(CO)_4PPh_3$ gives $Cl_3SnFe(CO)_3(PPh_3)Cl$.[171] SnX_4 and $Fe(CO)_5$ yield either $Fe(CO)_4SnX_3$ or *cis*-$Fe(CO)_3(SnX_3)_2$, according to the experimental conditions.[172] Similar oxidative addition products are formed in the reactions between organotin halides and neutral or anionic metal carbonyls and their derivatives[159]:

$$Bu_3SnCl + Fe(CO)_5 \rightarrow [Bu_2SnFe(CO)_4]_2 + Bu_4Sn_3[Fe(CO)_4]_4 + Sn[Fe(CO)_4]_4 \qquad (8.94)$$

These reactions have been applied to the formation of ruthenium and osmium complexes[173,174] (eqns 8.95, 8.96).

$$Na[Ru(CO)_4SiMe_3] \xrightarrow{Me_3SnCl} Me_3SnRu(CO)_4SiMe_3$$
$$Na[Ru(CO)_4SiMe_3] \xrightarrow{Me_2SnCl_2} [Me_2SnRu(CO)_4]_2 + Me_3SiCl \qquad (8.95)$$

$$Ph_3SnCl + Na_2Os(CO)_4 \rightarrow (Ph_3Sn)_2Os(CO)_4 + Ph_3SnOs(CO)_4H \quad \text{(after hydrolysis)} \tag{8.96}$$

Oxidative addition of organotin hydrides to the metal carbonyl (often with other ligands in place of some CO groups) occurs with displacement of CO and loss of hydrogen. With $M_3(CO)_{12}$ more than one product may result by redistribution of the ligands about tin (eqns 8.97–8.99).[173,175,176]

$$(R_3Sn)_2Ru(CO)_4 + R_3Sn(CO)_3Ru[\mu\text{—}(SnR_2)_2]Ru(CO)_3SnR_3 \xleftarrow{R_3SnH} Ru_3(CO)_{12} \xrightarrow{Me_2SnH_2} (Me_3Sn)_2Ru((CO)_4 + [Me_2SnRu(CO)_4]_2 \tag{8.97}$$

$$Me_2SnH_2 + Fe_3(CO)_{12} \rightarrow (Me_3Sn)_2Fe(CO)_4 + [Me_2SnFe(CO)_4]_2 \tag{8.98}$$

$$Me_3SnH + [R_3SiRu(CO)_4]_2 \rightarrow (Me_3Sn)_2Ru(CO)_4 + Me_3SiH \tag{8.99}$$

In the case of osmium, an Os–H bonded product can be isolated[170] (eqn 8.100):

$$Me_3SnH + Os_3(CO)_{12} \rightarrow Me_3SnOs(CO)_4H + (Me_3Sn)_2Os(CO)_4 \tag{8.100}$$

The products of the reaction of Me_3SnH with $Fe(CO)_5$ depend on the conditions employed: $(Me_3Sn)_2Fe(CO)_4$, $[Me_2SnFe(CO)_4]_2$, $Me_4Sn_3[Fe(CO)_4]_4$ and $Sn[Fe(CO)_4]_4$ have been isolated.[177] In its reaction with $Os_3(\mu_3\text{-}S)(\mu_3\text{-}\eta^2\text{-}SCH_2)(CO)_8PMe_2Ph$, Me_3SnH yields two Os_3-hydride products in which Me_3Sn is bonded to one exterior osmium atom of the Os_3–S cluster.[178]

Oxidative addition of a wide range of R_4Sn and Sn_2R_6 compounds to transition metal complexes results in Sn–M bond formation[179–184] (eqns 8.101–8.106):

$$R_2Sn(CH{=}CH_2)_2 + Fe(CO)_5 \rightarrow [R_2SnFe(CO)_4]_2 \leftarrow R_2Sn(C{\equiv}CR)_2 + Fe_3(CO)_{12} \tag{8.101}$$

$$Me_3SnC_5H_5 + M(CO)_3(CNMe)_3 \rightarrow Me_3SnM(CO)_3C_5H_5 \quad M = Cr, Mo, W \tag{8.102}$$

$$R_2SnO + Fe(CO)_5 \rightarrow [R_2SnFe(CO)_4]_2 + CO_2 \tag{8.103}$$

$$\text{Me}_2\text{Sn}(\text{CH}_2)_4 \text{ (stannacyclopentane)} + Fe_2(CO)_9 \rightarrow \text{Me}_2\text{Sn–Fe(CO)}_4 \text{ ring} + Fe(CO)_5 \tag{8.104}$$

$$Me_3SnC_5H_5 + Fe(CO)_5 \rightarrow [Me_2SnFe(CO)_4]_2 \quad (8.105)$$

$$Me_6Sn_2 + Fe(CO)_5 \rightarrow (Me_3Sn)_2Fe(CO)_4 \quad (8.106)$$

Tin–iron bonded compounds have been obtained from Hg–Fe complexes by reaction with tin halides, or by the action of Hg–Sn compounds on iron complexes[101,185] (eqns 8.107, 8.108):

$$SnCl_4 + Hg[Fe(CO)_3NO]_2 \rightarrow Cl_3SnFe(CO)_3NO \quad (8.107)$$

$$(R_3Sn)_2Hg + [Fe(CO)_2C_5H_5]_2 \xrightarrow{THF} R_3SnFe(CO)_2C_5H_5 \quad (8.108)$$

Stannylenes react in a variety of ways with complexes of these metals. Insertion into M–halide, M–C and M–M bonds can occur[44,91,97] (eqns 8.109, 8.110):

$$(R_2N)_2Sn + Fe(CO)_2(C_5H_5)X \rightarrow (R_2N)_2XSnFe(CO)_2C_5H_5 \quad (8.109)$$

$R = Me_3Si$

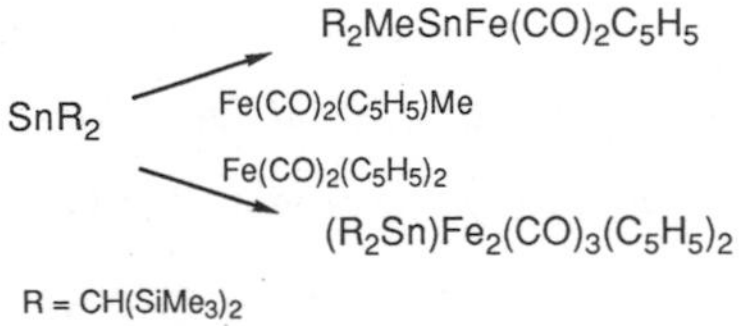

$$SnR_2 \xrightarrow{Fe(CO)_2(C_5H_5)Me} R_2MeSnFe(CO)_2C_5H_5$$

$$SnR_2 \xrightarrow{Fe(CO)_2(C_5H_5)_2} (R_2Sn)Fe_2(CO)_3(C_5H_5)_2$$

$R = CH(SiMe_3)_2$ (8.110)

The reaction between R_2Sn [$R = (Me_3Si)_2CH$] and $Os_3H_2(CO)_{10}$ gives the complex $Os_3Sn(\mu\text{-}H)_2R_2(CO)_{10}$ (**18**) in which the Os_3Sn atoms form a closed plane with one hydrogen bridging an Os–Sn bond and the other bridging an Os–Os bond.[186] This complex transforms thermally into $RSnOs_3(\mu\text{-}H)_2(\mu\text{-}OCR)(CO)_9$ (**19**) via migration of a bulky R group from Sn to a carbonyl, giving the acyl complex[186]:

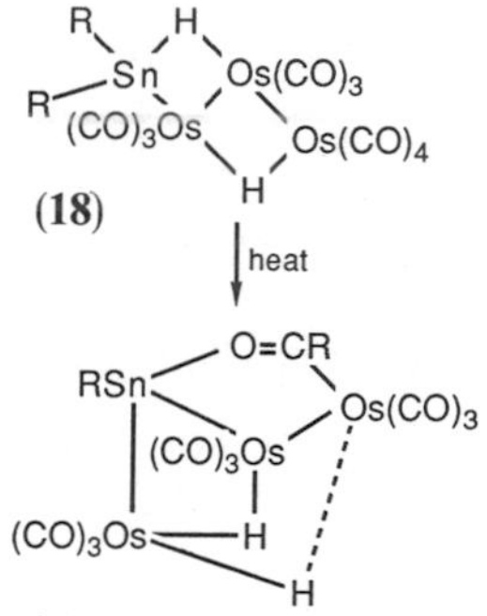

R, H, Sn, R, $Os(CO)_3$, $(CO)_3Os$, $Os(CO)_4$, H

(**18**)

heat

O=CR, RSn, $Os(CO)_3$, $(CO)_3Os$, $(CO)_3Os$, H, H

(**19**) (8.111)

The reaction between $(C_5H_5)_2Sn$ and $Fe_2(CO)_9$ yields $[(C_5H_5)_2SnFe(CO)_4]_2$ in which the cyclopentadienyl rings are monohapto.[187,188] The photochemically-induced decomposition of $PhMeSn[Fe(CO)_4]_2$ gives a mixture of two isomers, (**20**) and (**21**), which, from their nmr spectra, interconvert in solution by a combination of bridge deformation and iron–tin bond breaking.[189]

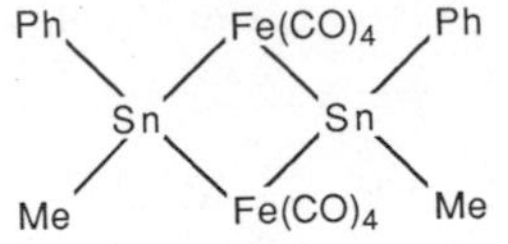

Ph Fe(CO)4 Ph
Sn Sn
Me Fe(CO)4 Me

(20)

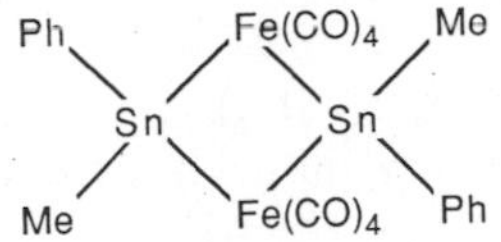

Ph Fe(CO)4 Me
Sn Sn
Me Fe(CO)4 Ph

(21)

The ^{13}C-nmr spectra of $(Me_3Sn)_2M(CO)_4$ ($M = Fe$, Ru, Os) show that, for $M = Fe$ and Ru, they exist in solution as the *cis* isomers, whereas for $M = Os$ there is an appreciable amount of the *trans* isomer present. Isomerization is non-dissociative since $^{119}Sn-^{13}C$ coupling is retained.[190] In *cis*-$(R_3Sn)_2Fe(CO)_4$ complexes, interchange of axial and equatorial CO groups occurs intramolecularly.[191]

The low reactivity of the Sn–M ($M = Fe$, Ru, Os) bonds is reflected in the large number of reactions in which the Sn–M bond remains intact. Phosphines, R_3P,[180,192] and butadiene[193] will displace CO, whilst other reagents attack RSn or SnX rather than Sn–M bonds (eqns 8.112–8.116)[105,132,157,166,194]

$$I_3SnFe(CO)_2C_5H_5 + SnCl_2 \rightarrow Cl_3SnFe(CO)_2C_5H_5 \quad (8.112)$$

$$Ph_3SnFe(CO)_2C_5H_5 + SO_2 \rightarrow Ph(PhSO_2)_2SnFe(CO)_2C_5H_5 \quad (8.113)$$

$$Ph_2Sn[Fe(CO)_2C_5H_5]_2 + SO_2 \rightarrow (PhSO_2)_2Sn[Fe(CO)_2C_5H_5]_2 \quad (8.114)$$

$$X_2Sn[Ru(CO)_2C_5H_5]_2 + RMgBr \rightarrow R_2Sn[Ru(CO)_2C_5H_5]_2 \quad (8.115)$$

$$Ph_3SnFe(CO)_2C_5H_5 + HX \rightarrow X_3SnFe(CO)_2C_5H_5 \quad (8.116)$$

However, Lewis bases cleave the Sn–Fe bonds of dimeric structures such as $(CO)_4Fe[(\mu\text{-}SnR_2)]_2Fe(CO)_4$, producing monomeric $R_2LSnFe(CO)_4$ reversibly.[195,196] In the osmium complex. $(Bu_2PhSn)_2Os(CO)_4$, only the Ph–Sn bonds are cleaved by HCl.[197] Reaction with chelating ligands such as 8-hydroxyquinoline and 2,2′-bipyridyl yields complexes in which tin is 6-coordinated[104] (e.g. eqn 8.117).

$$Cl_3SnFe(CO)_2C_5H_5 + bipy \rightarrow (bipy)Cl_3SnFe(CO)_2C_5H_5 \quad (8.117)$$

Tin (II) β-ketoenolates and $Fe_2(CO)_9$ also give complexes which are 6-coordinated about tin,[198] whereas in (porphyrin)$SnFe(CO)_4$ the tin is 5-coordinated.[199]

In $Ph_3MFe(CO)_2C_5H_5$ compounds ($M = Si$, Ge, Sn) the stability of the radical anion increases with the size of M.[200] Reduction of $(R_2Sn)_2Fe_2(CO)_8$ using NaH or Na/Hg in the presence of a Lewis base gives the stable dianion $[R_2SnFe(CO)_4]^{2-}$. This reduction can be considered as populating a non-bonding orbital on tin, with the possibility of some back-bonding to iron.[201] Electrochemical reduction of $Cl_3SnFe(CO)_2C_5H_5$ proceeds in two one-electron steps, only the first of which involves the X_3Sn groups, corresponding to the formation of X_3Sn^- and $[Fe(CO)_2C_5H_5]$.

Reactions with fluoro-alkenes and -alkynes are similar to those discussed under manganese, and usually result in cleavage of the Sn–M bond forming R_3SnF[147] (eqn 8.118).

$$Me_3SnFe(CO)_2C_5H_5 + CF_3C{\equiv}CCF_3 \xrightarrow{h\nu} Me_3SnF$$
$$+ Me_3SnC(CF_3){=}C(CF_3)Fe(CO)_2C_5H_5 + Fe(CO)_2(C_5H_5)[cyclo\text{-}C_5F_2(CF_3)_3] \quad (8.118)$$

The iron–tin bond of $R_3SnFe(CO)_2(C_5H_5)$ complexes is cleaved by iodine (eqn 8.119),[112] whilst only one ruthenium–tin bond of $(Me_3Sn)_2Ru(CO)_4$ is cleaved (eqn 8.120).

$$R_3SnFe(CO)(C_5H_5) + I_2 \rightarrow R_3SnI + Fe(CO)_2(C_5H_5)I \quad (8.119)$$

$$(Me_3Sn)_2Ru(CO)_6 + I_2 \rightarrow Me_3SnRu(CO)_4I \rightarrow [Me_3SnRu(CO)_3I]_2 \quad (8.120)$$

Photolysis of $Me_2Sn[Fe(CO)_2C_5H_5]_2$ results in loss of one CO, forming $C_5H_5(CO)$-$Fe[\mu$-(CO)-μ-$(Me_2Sn)]Fe(CO)C_5H_5$.[115] The thermolysis of $[(C_5H_5)_2SnFe(CO)_4]_2$ results in transfer of cyclopentadienyl groups from tin to iron producing the metal cluster (**22**).[204]

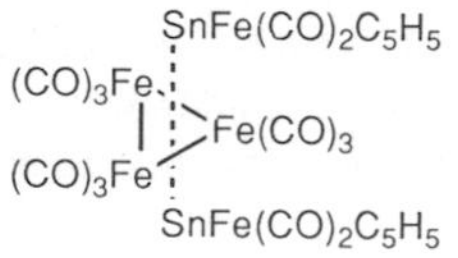

$$\begin{array}{c} SnFe(CO)_2C_5H_5 \\ (CO)_3Fe \quad \quad \\ \quad | \quad Fe(CO)_3 \\ (CO)_3Fe \quad \quad \\ SnFe(CO)_2C_5H_5 \end{array}$$

(22)

Photolysis of $[PhMeSnFe(CO)_4]_2$ gives $(PhMeSn)_2Fe_2(CO)_7$ (**23**); nmr evidence is consistent with three isomeric forms of this compound resulting from the relative positions of the methyl and phenyl groups.[205]

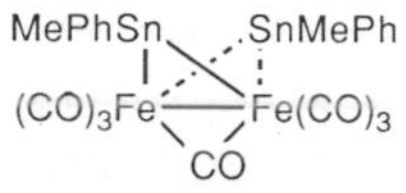

$$\begin{array}{ccc} MePhSn & & SnMePh \\ | & & \vdots \\ (CO)_3Fe & \text{—} & Fe(CO)_3 \\ & CO & \end{array}$$

(23)

8.9 Cobalt, rhodium and iridium

All three transition metals form complexes with tin. For cobalt, the metal is mostly in its +1 oxidation state, whereas for rhodium both the +1 and +3 states form Sn-bonded complexes, and with iridium there is in addition a Sn–Ir(V) complex. Coordination numbers of 4, 5, and 6 are found for the transition metals. Tin in these complexes may be in its +2 or +4 oxidation states with coordination numbers of 3 or 4, or 6 in the case of chelate complexes. The tendency of these metals to form cluster compounds makes for an interesting structural chemistry, though relatively few reactions have been studied beyond ligand exchange and electrophilic substitution at either metal. The question of $d\pi$–$d\pi$ bonding between tin and cobalt has been considered in relation to data from various physicochemical methods. The tentative conclusion, not surprisingly, is that the π interaction is non-zero, and that R_3Sn is a stronger σ-donor than Cl_3Sn[206].

Reactions of complex anions of the transition metal form the basis of many syntheses for example eqns (8.121)–(8.123).[207,208]

$$R_3SnBr + Na[Rh(PPh_3)_2(CO)_2] \rightarrow R_3SnRh(PPh_3)_2(CO)_2 \quad (8.121)$$

$$R_3SnBr + Na[Ir(PPh_3)(CO)_3] \rightarrow R_3SnIr(PPh_3)(CO)_3 \quad (8.122)$$

$$R_3SnCl + Li[(Me_5C_5)IrH_3] \rightarrow R_3SnIr(H)_3(Me_5C_5) \quad (8.123)$$

Although complexes with PF_3 ligands are poor nucleophiles, Sn–M (M = Rh, Ir) bonded complexes have been isolated (eqn 8.124; cf. the corresponding reactions with silicon or germanium) which are unproductive.[209]

$$Ph_3SnCl + NaM(PF_3)_4 \rightarrow Ph_3SnM(PF_3)_4 \quad M = Rh, Ir \quad (8.124)$$

Nucleophilic substitution reactions on the transition metal halides have been used to form Sn–M bonds:[210]

$$Ph_3SnNa + (Ph_3P)_3CoCl \xrightarrow{THF} Na(THF)_7[(Ph_3Sn)_2Co(PPh_3)_2] \quad (8.125)$$

With chelating ligands on tin, 6-coordinate Sn(IV) complexes result:[211]

$$Cl_2Sn(acac)_2 + Na[Co(CO)_4] \rightarrow (CO)_3Co[\mu\text{-}CO(\mu\text{-}Sn(acac)_2]Co(CO)_3 \quad (8.126)$$

Reactions of SnX_2 and SnX_4 compounds with carbonyl, substituted carbonyl, or alkene derivatives of these metals result in Sn–M bond formation[212–214] (eqns 8.127–8.129).

$$SnX_2 + [Co(CO)_3PPh_3]_2 \rightarrow X_2Sn[Co(CO)_3(PPh_3)]_2 \quad (8.127)$$

$$SnCl_2 + [(norbornadiene)RhCl]_2 + R_3P \rightarrow Cl_3SnRh(nbd)(PR_3)_2 \quad (8.128)$$

$$SnX_4 + Co(CO)_2(C_5H_5) \rightarrow X_3SnCo(CO)(C_5H_5)X + (X_3Sn)_2Co(CO)C_5H_5 \quad (8.129)$$

The reaction of $SnCl_2$ with Co–Co complexes occurs in two stages[215], and reactions of this type can result in unexpected products[216] (eqns 8.130, 8.131).

$$SnCl_2 + Co_2(CO)_6(PR_3)_2 \rightarrow Cl_2Sn[Co(CO)_3(PR_3)]_2 \xrightarrow{SnCl_2} ClSn[Co(CO)_3)(PR_3)]_3 \quad (8.130)$$

$$SnCl_2 + [Co(CO)_3(PBu_3)]_2 \longrightarrow Sn[Co(CO)_3(PBu_3)]_4 + HSn[Co(CO)_3PBu_3]_3 \quad (8.131)$$

$SnCl_2$ and Ir(III) complexes form a range of Cl_3Sn–Ir(III) hydride products; the hydrogen coming from the solvent [EtOH or $MeO(CH_2)_2OH$][217] (eqns 8.132, 8.133):

$$SnCl_2 + Na_3IrCl_6 + Ph_3P \rightarrow Cl_3SnIr(PPh_3)HCl + Cl_3SnIr(PPh_3)_3(H)_2 \quad (8.132)$$

$$SnCl_2 + Ir(PPh_3)_2(CO)Cl \xrightarrow{EtOH} Cl_3SnIr(PPh_3)_2(CO)HCl \quad (8.133)$$

Rhodium in its + 3 oxidation state is reduced to rhodium (I) by $SnCl_2$ in acid solution to give coloured anionic complexes[218,219] (eqn 8.134).

$$RhCl_3 + HCl + SnCl_2 \rightarrow [Rh(SnCl_3)_4]^{3-} \quad (8.134)$$

$SnCl_4$ reacts with $Co(PMe_3)_2C_5H_5$ to form a 1:1 adduct at $-70°$, which with excess $SnCl_4$ is converted into $[Cl_3SnCo(PMe_3)_2C_5H_5]SnCl_5$.[220] Reaction of Me_2SnCl_2 with $Co_2(CO)_8$ gives the tetrahedral metal cluster $MeSnCo_3(CO)_9$,[221] whereas $MeSnCl_3$ forms $MeSn[Co(CO)_4]_3$.[222]

Cleavage of tin–carbon bonds gives Sn–M complexes, as in the following examples:[222,223]

$$Sn(CH{=}CH_2)_4 + Co_2(CO)_8 \rightarrow (CH_2{=}CH)_2Sn[Co(CO)_4]_2 + CH_2{=}CHSn[Co(CO)_4]_3 \quad (8.135)$$

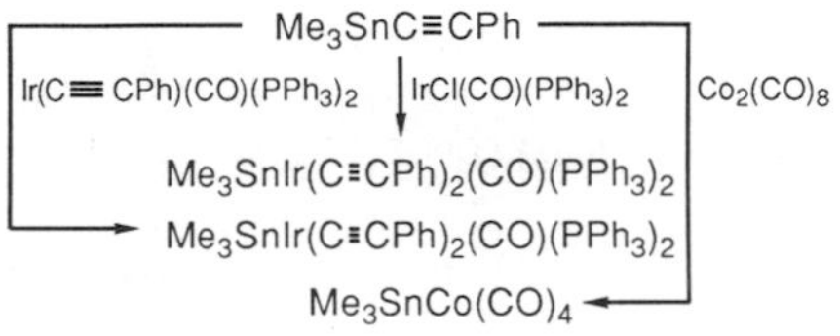

Me₃SnC≡CPh

- $Ir(C{\equiv}CPh)(CO)(PPh_3)_2$ → $Me_3SnIr(C{\equiv}CPh)_2(CO)(PPh_3)_2$
- $IrCl(CO)(PPh_3)_2$ → $Me_3SnIr(C{\equiv}CPh)_2(CO)(PPh_3)_2$
- $Co_2(CO)_8$ → $Me_3SnCo(CO)_4$

(8.136)

Likewise di–tin compounds undergo oxidative addition to M(I) complexes[224] (eqn 8.137):

$$Me_6Sn_2 + M(CO)_2C_5H_5 \rightarrow (Me_3Sn)_2M(CO)C_5H_5 + [Me_2SnM(CO)C_5H_5]_2 \quad (8.137)$$

Reaction between Me_3SnH and $Co_2(CO)_8$ gives[83] hydrogen and $Me_3SnCo(Co)_4$.

The reactivity of Sn–H compounds towards iridium(I) is markedly dependent on the groups bonded to both tin and iridium. Thus, the oxidative addition of trimethyltin hydride to *trans*-$Ir(PPh_3)_2(CO)Cl$ gives an Ir (III) product as a colourless crystalline solid whose ^1H nmr spectrum suggests the presence of isomers $Me_3SnIr(PPh_3)_2(CO)ClH$ and $Me_3SnIr(PPh_3)_2(H)_2(CO)$ in solution. Bu_3SnH also undergoes a complex reaction with the same iridium complex.

Trimethyltin complexes of iridium(III) are perhaps more readily obtained by exchange reactions[83,225] (eqns 8.138 and 8.139).

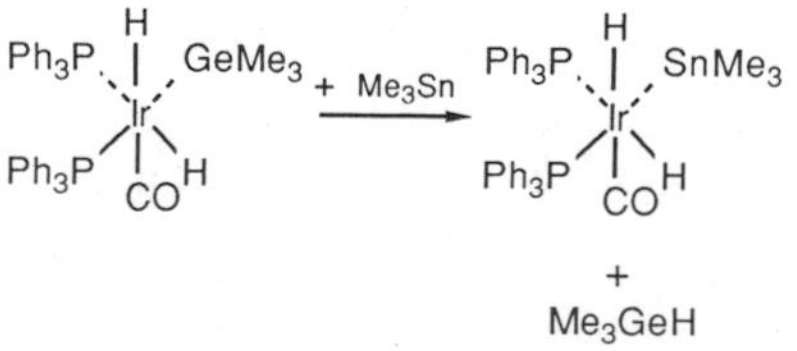

$$(Ph_3P)_2Ir(H)_2(CO)(GeMe_3) + Me_3Sn \rightarrow (Ph_3P)_2Ir(H)_2(CO)(SnMe_3) + Me_3GeH \quad (8.138)$$

$$Me_3SiCo(CO)_4 + Me_3SnH \rightarrow Me_3SnCo(CO)_4 + Me_3SiH \quad (8.139)$$

Other types of exchange reactions may be used to form tin–rhodium bonds[226]:

$$Ph_3SnC_5H_5 + Me_3GeRh(CO)(C_5H_5)I \rightarrow Ph_3SnRh(CO)(C_5H_5)I \quad (8.140)$$

$$Me_3SnX + (Me_3Ge)_2Rh(CO)C_5H_5 \xrightarrow{h\nu} (Me_3Sn)(Me_3Ge)Rh(CO)C_5H_5 \quad (8.141)$$

In the case of Et_3SnH, the complex $Et_3SnIr(PPh_3)_2(CO)HCl$ has been isolated.[227] The related reactions with Rh(I) are rapid and exothermic. Use of Me_3SnH only affords decomposition products, but Bu_3SnH and $Rh(PPh_3)_3Cl$ give the 5-coordinate complex, $Bu_3SnRh(PPh_3)_2HCl$.[228] With other rhodium(I) complexes, reactions of Me_3SnH (eqns 8.142, 8.143) can be more satisfactorily controlled:[226]

$$Me_3SnH + Rh(CO)_2C_5H_5 \xrightarrow{\text{heat or } h\nu} (Me_3Sn)_2Rh(CO)C_5H_5 + mer\text{-}(Me_3Sn)_3Rh(CO)_3 \quad (8.142)$$

$$Me_3SnH + Rh_2(CO)_3(C_5H_5)_2 \rightarrow Me_3SnRh(CO)(C_5H_5)H + (Me_3Sn)_2Rh(CO)C_5H_5 \quad (8.143)$$

Few reactions of organotin dihydrides have been examined. The reaction of Me_2SnH_2 with $Co_2(CO)_8$ gives the bridged product $(CO)_3Co[(\mu\text{-}SnMe_2)_2]Co(CO)_3$ (**24**) as a bright yellow solid:

$$Me_2SnH_2 + Co_2(CO)_8 \rightarrow \underset{\mathbf{(24)}}{(CO)_3Co[(\mu\text{-}SnMe_2)_2]Co(CO)_3} \quad (8.144)$$

Rapid exchange of axial and equatorial CO groups occurs in (**24**) even at low temperature, and exchange of methyl group positions takes place in a concerted fashion.[229] Intramolecular exchange of axial and equatorial CO groups also occurs readily in the five-coordinated complexes, $R_3SnCo(CO)_4$. In general, the barrier to exchange seems to correlate with the size of the R groups.[230] The oxidative addition of Ph_3SnH to $Ir(PPh_3)_3(CO)H$ proceeds by dissociation of phosphine and *cis* addition of Sn–H giving $Ph_3SnIr(PPh_3)_2(CO)(H)_2$.[231]

Stannylenes R_2Sn have been used in forming Sn–Co bonds. In the majority of cases they behave as Lewis bases, although products of oxidative addition are formed in some reactions[91,187,232,233] (eqns 8.145–8.148):

Me₂SiCH₂ / O / Me₂SiCH₂ ring–Sn + $Co_2(CO)_8$

↓

Me₂SiCH₂ / O / Me₂SiCH₂ ring–$Sn[Co(CO)_4]_2$ (8.145)

$$(MeC_5H_4)_2Sn + Co_2(CO)_8 \rightarrow Sn[Co(CO)_4]_4 \quad (8.146)$$

$$R_2Sn + Rh(PPh_3)_2Cl(C_2H_4) \rightarrow Rh(SnR_2)(PPh_3)_2Cl \qquad R = (Me_3Si)_2CH \quad (8.147)$$

$$R_2Sn + [Rh(C_2H_4)_2Cl]_2 \xrightarrow{\text{toluene}, C_8H_{14}} R_2ClSnRh(C_6H_5CH_3)(\text{cyclooctene}) \qquad R = (Me_3Si)_2N \quad (8.148)$$

Ligand exchange and alkylation reactions can be carried out without cleaving the Sn–M bond:[234,235]

$$R_3SnCo(CO)_4 + R_3P \rightarrow R_3SnCo(CO)_3PPh_3 \quad (8.149)$$

$$X_2Sn[Co(CO)_4]_2 + RMgX \rightarrow R_2Sn[Co(CO)_4]_2 \quad (8.150)$$

A number of the 16-electron Sn–M complexes will reversibly add a Lewis base, e.g. $Ph_3SnCo(PMe_3)_3$ adds a mole of Me_3P.[236] X_3Sn–Co complexes react with chelating ligands such as 8-hydroxyquinoline to give 6-coordination at tin[104] (eqn 8.151):

$$Cl_3SnCo(CO)_3PBu_3 + \text{Hoxine} \rightarrow (\text{oxinato})_2ClSnCo(CO)_3PBu_3 \quad (8.151)$$

Photolysis of the stannylene complex, $Me_2Sn[Co(CO)_4]_2$, results in the formation of the complexes (**24**) and (**25**)[115] (eqn 8.152):

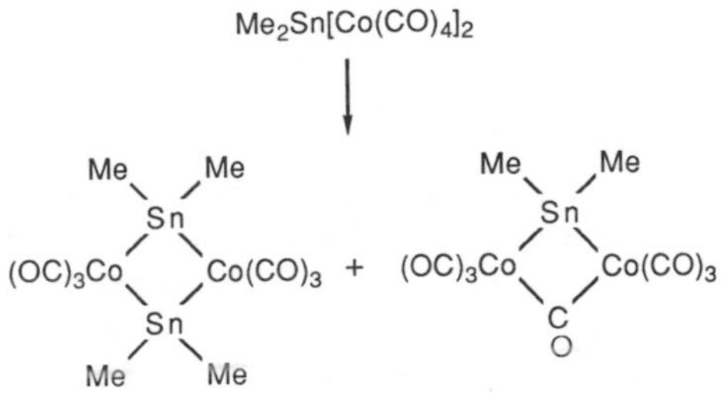

$Me_2Sn[Co(CO)_4]_2$

↓

(24): $(OC)_3Co$ bridged by two $SnMe_2$ groups to $Co(CO)_3$ + (25): $(OC)_3Co$ bridged by $SnMe_2$ and CO to $Co(CO)_3$

(**24**) (**25**)

(8.152)

The tin–cobalt bond in $Me_3SnCo(CO)_4$ is cleaved by halogens, Me_2AsCl, Ph_2PCl and Ph_3AuCl. In all cases Me_3SnX and a Co(I) complex are formed.[206] Fluoroalkenes insert into the Sn-Co bond,[237] and the Sn–Co bond in $[R_3SnCo(PMe_3)_2C_5H_5]Cl$ is cleaved by $HgCl_2$ to give Me_3SnCl.[220] The reaction of $Cl_3SnCo(CO)_4$ with tertiary phosphines and arsines (L) results in tin–cobalt bond cleavage and formation of an ionic complex $[Cl_3Sn][Co(CO)_3L_2]$, with good evidence that these reactions proceed by a radical chain mechanism, since they are sensitive to oxygen and radical scavengers. The kinetics have been interpreted in terms of a series of 14 reactions of which the first is shown in eqn (8.153).[238]

$$Cl_3SnCo(CO)_4 + L \rightarrow Cl_3LSnCo(CO)_4 \rightarrow Cl_3SnL + Co(CO)_4 \quad (8.153)$$

The tin–iridium bond in iridium(I) complexes is cleaved by electrophiles[207] (eqn 8.154):

$$\begin{array}{c} SnBr_4 + Ir(PPh_3)(CO)_2Br_3 + CO + PhBr \\ \uparrow Br_2 \\ Ph_3SnIr(CO)_3PPh_3 \\ \downarrow HgI_2 \\ Ir(HgI)_2(CO)_2PPh_3 + CO + Ph_3SnI \end{array} \quad (8.154)$$

8.10 Nickel, palladium and platinum

Nickel–tin complexes have been obtained in which nickel is formally in its zero, +2 and +4, and tin in its +2 and +4 oxidation states.

The dimeric stannylene $[Sn(OBu^t)_2]_2$ reacts with $Ni(CO)_4$ with displacement of one CO forming $[(OC)_3NiSn(OBu^t)_2]_2$ as a colourless dimeric crystalline solid, in which two Bu^tO groups bridge the two tin atoms. Presumably for steric reasons, the dimer $[Sn(OR)_2]_2$ ($R = SiMe_3$) reacts with only one mol of $Ni(CO)_4$ forming $(CO)_3Ni(OR)(\mu\text{-}OR)Sn(OR)_2$.[239] Anionic Ni(0) complexes result from reactions such as those illustrated in eqn (8.155).[210]

$$\begin{array}{c} (Ph_3P)_2NiC_2H_4 + Ph_3SnNa \\ \downarrow THF \\ Na(THF)_4[(Ph_3P)_3NiSnPh_3] + \\ Na_3(THF)_5[Ph_3PNi(SnPh_3)_3] \\ \uparrow \\ Ni(COD)_2 + Ph_3SnNa + Ph_3P \end{array} \quad (8.155)$$

Five-coordinate trigonal bipyramidal nickel(II) cations of the type $[LNiSnPh_3]BPh_4$ have been formed by similar nucleophilic substitution reactions (eqn 8.156).

$$[LNiCl]BPh_4 + Ph_3SnLi \rightarrow [LNiSnPh_3]BPh_4 \quad (8.156)$$
$$L = (Ph_2PCH_2CH_2)_3P$$

Cationic complexes of these types are diamagnetic and fairly air stable.[240] A further synthetic method for nickel(II) complexes which is of wide applicability involves the cleavage of Sn–Sn bonds[99] (eqn 8.157):

$$[(C_5H_5)Ni(CO)]_2 + Me_6Sn_2 \rightarrow 2(C_5H_5)(CO)NiSnMe_3 \quad (8.157)$$

Both nickel(0) and nickel(II) complexes have been isolated containing $Ni–SnCl_3$ and $Ni–SnCl_2–Ni$ groups[7] (eqns 8.158, 8.159).

$$Ni(CO)_4 + KSnCl_3 + Ph_4AsCl \rightarrow (Ph_4As)_4[Ni(SnCl_3)_4] \quad (8.158)$$

$$C_5H_5(R_3P)NiCl + SnCl_2 \rightarrow C_5H_5(R_3P)NiSnCl_3 \xrightarrow{PR_3} C_5H_5(R_3P)_2NiSnCl_3 \quad (8.159)$$

Ni(II)-Sn complexes are coloured (brown, gold, green) and mostly decompose between 100–150°C in the absence of air.[7]

The Ni(0)–$SnCl_3$ complexes are oxidized by air both in the solid state and in solution, and there is evidence for reversible dissociation in solution into $SnCl_2$

and the nickel chloride complex anion $[NiCl_4]^{4-}$. The high reactivity of the Sn–cyclopentadienyl bond may be used to form Ni–Sn complexes[241] (eqn 8.160).

$$[(C_5H_5)(CO)Ni]_2 + Me_3SnC_5H_5 \rightarrow Ni(C_5H_5)_2 + (C_5H_5)(CO)NiSnMe_3 \qquad (8.160)$$

Reaction between $Ni(Ph_3P)_3$ and Me_3SnCl or Ph_3SnCl in toluene leads to the isolation of low spin Ni(IV) complexes which are colourless, stable in air and inert towards substitution by $(Ph_2PCH_2)_2$, and probably have the configuration (**26**).[242]

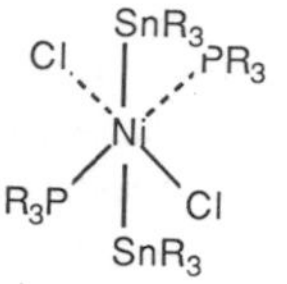

SnR_3
Cl　　PR_3
Ni
R_3P　　Cl
SnR_3

(**26**)

Both organotin–Pd and X_3Sn–Pd compounds have been described with palladium in its zero, +1 and +2 states and tin in its +2 and +4 oxidation states. The palladium (II) complex $Pd(COD)Cl_2$ reacts with the stannylene $Sn(NR_2)_2$ ($R = Me_3Si$) to give, surprisingly, the M(0) complex (**27**), probably via $Pd[(COD)\{Sn(NR_2)_2Cl\}_2]_2$ which undergoes reductive elimination in the presence of excess $Sn(NR_2)_2$.[243] The platinum analogue reacts similarly. In (**27**), palladium, tin and nitrogen all have trigonally planar coordination. Both the palladium and platinum complexes are extremely air-sensitive. On reaction with CO, (**27**) forms the cluster (**28**), in which the stannylene $Sn(NR_2)_2$ functions as a bridging group between adjacent palladium atoms.[243]

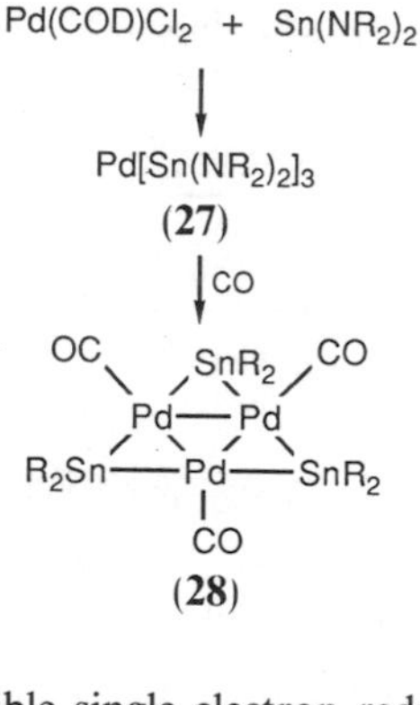

$Pd(COD)Cl_2 + Sn(NR_2)_2$
↓
$Pd[Sn(NR_2)_2]_3$
(**27**)
↓ CO

OC　SnR_2　CO
Pd—Pd
R_2Sn—Pd—SnR_2
CO

(**28**) (8.161)

The cluster undergoes a reversible single electron reduction in THF, and at more negative potentials two further irreversible reductions. The radical anion initially generated decomposes in THF at −40°C to yield a paramagnetic product that contains two equivalent tin nuclei.[244] $(\eta^3\text{-}C_3H_5)PdCl(SnR_2)_2$, a dark yellow solid decomposing at 80°C, is formed from $[\eta^3\text{-}C_3H_5PdCl]_2$ and $Sn(NR_2)_2$ ($R = Me_3Si$).[44]

In general, palladium clusters react with Sn(II) compounds to form new metal clusters, sometimes with displacement of palladium as well as neutral ligands, for example:

$$Sn(acac)_2 + Pd_4(CO)_5(PPh_3)_4 \rightarrow (acac)_4Sn_2Pd_3(CO)_2(PPh_3)_3 \quad (8.162)$$

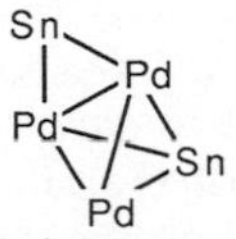

(29)

(30)

In (**29**), the $[Pd_3Sn_2]$ core has a propeller-shaped framework (**30**).[245] $SnCl_2$ will insert into either one or both of the Pd–Cl bonds in the complex (**31**)[246] (eqn 8.163).

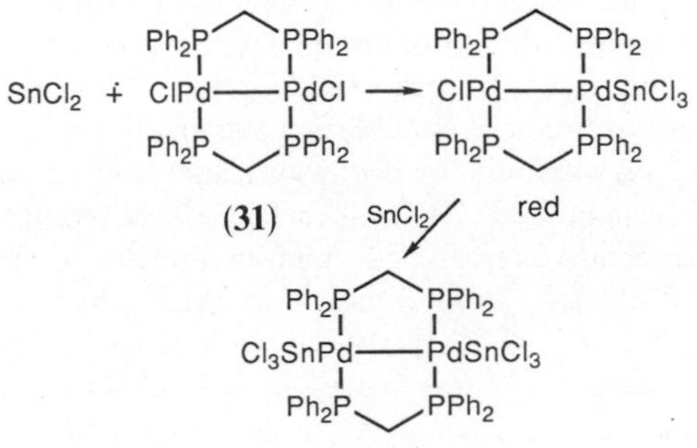

$SnCl_2$ + (**31**) → red; $\xrightarrow{SnCl_2}$ purple (8.163)

Palladium (II)–Sn complexes have been obtained by insertion of $SnCl_2$ into Pd–Cl bonds, yielding both neutral and anionic complexes ($L = Ph_3P$, etc)[7,247,248,249] (eqns 8.164–8.166)

$$L_2PdCl(SnCl_3) \xleftarrow{SnCl_2} L_2PdCl_2 \xrightarrow{2SnCl_2} L_2Pd(SnCl_3)_2 \quad (8.164)$$

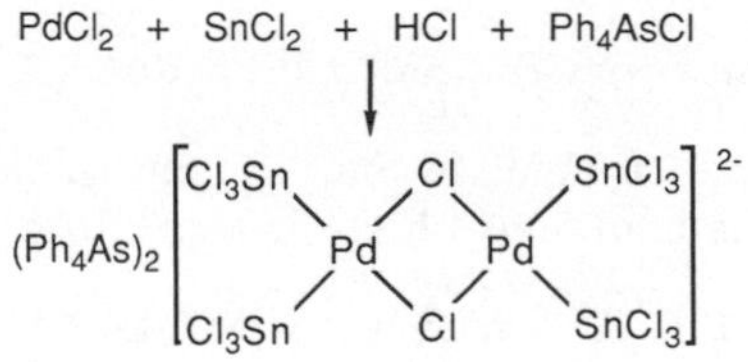

(*cis* and *trans*) (*cis* and *trans*)

$$PdCl_2 + SnCl_2 + HCl + Ph_4AsCl \rightarrow (Ph_4As)_2[(Cl_3Sn)_2Pd(\mu\text{-}Cl)_2Pd(SnCl_3)_2]^{2-} \quad (8.165)$$

$$(PdCl_4)^{2-} + SnCl_2 \rightarrow [Pd(SnCl_3)_4]^{2-} \quad (8.166)$$

However, an excess of $SnCl_2$ reacts with $(\eta^3\text{-}C_3H_5PdCl)_2$ in benzene or THF to give ultimately the decomposition products Pd, $SnCl_4$ and biallyl.[250]

The Pd–Sn bonds in compounds of these types are cleaved by iodine. Treatment of the anion $[(OC)Pd(SnCl_3)_2Cl]^-$ with Ph_3P yields $(Ph_3P)Pd_2Cl_2$, in contrast to the behaviour of the platinum analogue where dissociative cleavage of the platinum–tin bonds does not occur. In solution, these phosphine- and arsine--$PdSnCl_3$ complexes undergo rapid phosphine or arsine exchange at room temperature.[7,247] In η^3-$C_3H_5(Ph_3P)PdSnCl_3$, the Pd–Cl distance of 256 pm is about 10 pm less than the sum of the covalent radii.[7]

Platinum–tin chemistry has been quite extensively explored and has produced some extremely novel structures. Complexes have been studied with platinum in zero, +2 and +4 and tin in its +2 and +4 oxidation states. Neutral, anionic and cationic complexes have been identified, often in the form of metal cluster compounds. Pt(0)–Sn complexes are 4-coordinate (tetrahedral) about platinum; those of Pt(II) are 4-(planar) or 5-(trigonal bipyramidal) coordinate, whilst Pt(IV)–Sn complexes are octahedrally coordinated about platinum.

Cationic Pt(II)–$SnCl_3$ complexes were first prepared in 1835, although it was many years before they were recognized as such. The original paper ('On some combinations of protochloride of platina with protochloride of tin') describes how 'these bodies unite in two different proportions; that containing least tin is of an olive brown colour, crystalline and very deliquescent, decomposed by water... The second which contains most tin is of an intensely red colour, soluble in water giving a splendid red solution.... The colour of the solution was found by Professor Kane, on examination by a prism, to be an absolutely homogeneous red'.[251] The formation of a red complex in this way is the basis of a standard qualitative test for platinum in aqueous solution.[252]

Examples of Pt(0)–Sn compounds include the trigonal complex $Pt\{Sn[N(SiMe_3)_2]_2\}_3$, which is isomorphous with the corresponding palladium complex already described.[253,254] It is formed from $Pt(COD)_2$ and the stannylene in toluene. An orange Sn(II)–Pt(0) complex $[(acac)_2Sn]_2Pt(PPh_3)_2$, with tetrahedral geometry about platinum is formed from $Pt(PPh_3)_2$ and $Sn(acac)_2$. With excess acetylacetone a more complex metal cluster results.[225] In the reaction of SnR_2 $[R=(Me_3Si)_2CH]$ with $PtCl_2(PEt_3)_2$, or (better) $[PtCl_2(PEt_3)]_2$, one SnR_2 coordinates as a Lewis base whilst the other inserts into a Pt–Cl bond giving $(R_2Sn)(R_2ClSn)Pt(PEt_3)Cl$.[91] By contrast, $(R_2N)_2Sn$ and $(COD)PtCl_2$ give $(COD)Pt[SnCl(NR_2)_2]_2$.[44,91,243,253]

The oxidative addition of R_3SnX and R_2SnX_2 to Pt(0) was originally thought to involve the formation of a Pt–X bond, but in most cases this is not so. R_3SnCl reacts with $(Ph_3P)_2Pt(C_2H_4)$ by displacement of ethene and insertion of platinum into an Sn–C bond[256] (eqn 8.167).

$$(Ph_3P)_2PtC_2H_4 + R_3SnCl \rightarrow cis\text{-}PtR(SnR_2Cl)(Ph_3P)_2 \quad (8.167)$$

Ph_2SnCl_2 reacts in the same way, but Me_2SnCl_2, $PhSnCl_3$, $MeSnCl_3$ and $SnCl_4$ react by insertion of platinum into an Sn–Cl bond[257] (eqn 8.168).

$$Pt(PPh_3)_4 \text{ or } (Ph_3P)_2PtC_2H_4 + MeSnCl_3 \rightarrow cis\text{-}(MeCl_2Sn)PtCl(Ph_3P)_2 \quad (8.168)$$

Oxidative addition of $SnCl_4$ and $MeSnCl_3$ to chelated Pt(II) yields 6-coordinated complexes[258] (eqns 8.169, 8.170):

$$SnCl_4 + (bipy)PtMe_2 \rightarrow Cl_3SnPtMe_2Cl(bipy) \quad (8.169)$$

$$MeSnCl_3 + (bipy)PtMe_2 \rightarrow (MeCl_2Sn)_2PtMe_2(bipy) \quad (8.170)$$

The reaction of *cis*-$Pt(CO)LCl_2$ $(L=R_3P)$ with Me_3SnPh probably proceeds through an oxidative addition stage which is followed by phenyl migration and elimination of Me_3SnCl[259](eqn 8.171).

$$Me_3SnPtCl_2(CO)(Ph)L \rightarrow Me_3SnCl + cis\text{- and } trans\text{-}Pt_2(COPh)_2Cl_2L_2 \quad (8.171)$$

In reactions of platinum hydridohalides with $SnCl_4$, hydrogen chloride is eliminated and the initially formed platinum(IV) complexes (**32**) readily undergo reductive elimination[260] (eqn 8.172).

$$\begin{array}{c} trans\text{-}Pt(PPh_3)_2HCl + SnCl_4 \\ \downarrow \\ (Cl_3Sn)_2Pt(PPh_3)_2Cl_2 \\ (\mathbf{32}) \\ \downarrow \\ (Cl_3Sn)Pt(PPh_3)_2Cl \end{array} \quad (8.172)$$

Many reactions originally thought to be simple are seen to be far more complex when examined by ^{31}P-nmr. For example, the reaction of Me_3SnNMe_2 with PtL_2HCl gives at least three products rather than just that expected from the elimination of Me_2NH. Loss of HCl generates PtL_2 which then adds Me_3SnCl to give $Me_2ClSnPtL_2Me$. Even nucleophilic substitution reactions are often complex for similar reasons[256,259] (eqn 8.173).

$$\begin{array}{c} cis\text{-}[PtCl_2L_2] + Ph_3SnLi \\ \downarrow \\ PtL_2 + Ph_3SnCl \\ \downarrow \\ cis\text{-}Ph_2ClSnPtPhL_2 \end{array} \quad (8.173)$$

The report that $(Ph_3Sn)Pt(PPh_3)_2Cl$ decomposes in hot acetone to $Pt(PPh_3)_2PhCl$[260] is incorrect, and can also be ascribed to these effects[256].

Oxidative addition of organotin hydrides occurs readily both to platinum(0) and to platinum(II) complexes[261,262] (eqn 8.174).

$$R_3SnH + Pt(PPh_3)_4 \rightarrow trans\text{-}R_3SnPt(PPh_3)_2H + 2PPh_3 + (R_3Sn)_2Pt(PPh_3)_2 \quad (8.174)$$

The related reaction with $Pt(diphos)_2$ [diphos = $(Ph_2PCH_2)_2$] proceeds first to the platinum(II) complex (**33**) and in the presence of excess hydride to the platinum(IV) complex (**34**) (eqn 8.175).

$$\begin{array}{c} Me_3SnH + Pt(diphos)_2 \\ \downarrow \\ (Me_3Sn)_2Pt(diphos) + diphos + H_2 \\ (\mathbf{33}) \\ \rightleftharpoons Me_3SnH \\ (Me_3Sn)_3Pt(diphos)H \\ (\mathbf{34}) \end{array} \quad (8.175)$$

Similarly, platinum(II) complexes and Me_3SnH yield the platinum(IV) product (**35**) which dissociates reversibly in solution (eqn 8.176):

$$Me_3SnH + (diphos)PtCl_2 \rightarrow (Me_3Sn)_2Pt(diphos)HCl \rightleftharpoons (Me_3Sn)Pt(diphos)Cl \quad (8.176)$$

(**35**)

The stereochemistry of these platinum(IV) complexes with unidentate phosphine ligands is readily determined from their ^1H- and ^{31}P-nmr spectra.[256,257] The reaction between R_3SnH and $Pt(PMe_3)_2(CO_3)$ gives $(R_3Sn)_2Pt(PMe_3)_2(H)_2$ with a *cis*-[H–Pt–H] and *trans*-[P–Pt–P] configuration. In benzene this complex loses H_2 reversibly giving *trans*-$(R_3Sn)_2Pt(PMe_3)_2$.[263]

The extent and rate of these addition reactions can be controlled using bulky phosphines such as tricyclohexylphosphine. If a very bulky phosphine is used, no reaction occurs. In the case of a Pt-dihydride complex, hydrogen is eliminated[264] (eqn 8.177).

$$Ph_3SnH + Pt(PR_3)_2(H)_2 \rightarrow Ph_3SnPt(PR_3)_2H + H_2 \quad (8.177)$$

Platinum–tin bonds are formed by the exchange reactions of Pt–Si or Pt–Ge bonded complexes with Me_3SnH. These reactions are equilibria where the Pt–Sn product is strongly favoured[265] (eqns 8.178, 8.179).

$$Pt(diphos)(SiMe_3)_2 + Me_3SnH \rightarrow (Me_3Sn)_2Pt(diphos) + Me_3SiH \quad (8.178)$$

$$Pt(diphos)(SiMe_3)Cl + \text{excess } Me_3SnH \rightarrow (Me_3Sn)_2Pt(diphos)HCl \quad (8.179)$$

Most unexpectedly, the Sn–Pt(IV) complex (**36**) is formed in the reaction of (carbonato)$Pt(SEt_2)(PR_3)$ with (p-tolyl)$_3$SnH in methanol.

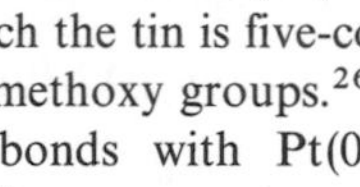

MeO
R_2Sn H SnR_2
Pt OMe
R_3Sn SnR_2
Et_3P

(**36**)

This highly unusual structure involves three Sn(IV) ligands (R_3Sn and two R_2SnOMe) and one stannylene (R_2Sn) in which the tin is five-coordinate due to coordination of lone pair electrons from the two methoxy groups.[266]

The reaction of tin–carbon bonds with Pt(0) complexes can take various forms[223,256,267] (eqns 8.180–8.182).

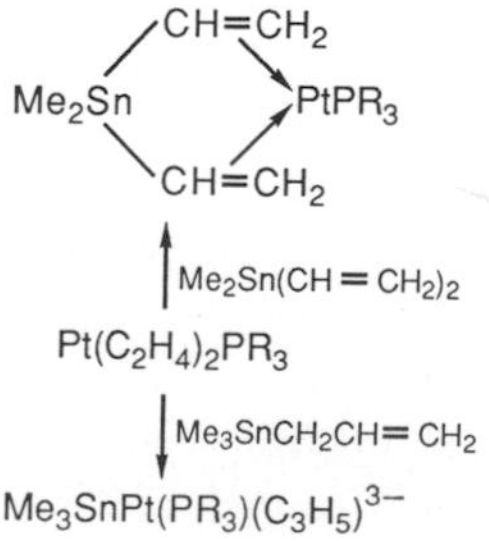

$Me_2Sn(CH{=}CH_2)_2 \rightarrow PtPR_3$ (chelate: Me_2Sn, CH=CH$_2$, CH=CH$_2$, $PtPR_3$)

↑ $Me_2Sn(CH=CH_2)_2$

$Pt(C_2H_4)_2PR_3$

↓ $Me_3SnCH_2CH=CH_2$

$$Me_3SnPt(PR_3)(C_3H_5)^{3-} \quad (8.180)$$

$$SnMe_4 + Pt(PEt_3)_4 \rightarrow cis\text{-}Me_3SnPt(PEt_3)_2Me \quad (8.181)$$

$$Me_3SnC{\equiv}CPh + Pt(PPh_3)_2 \rightarrow Me_3SnPt(C{\equiv}CPh)(PPh_3)_2 \quad (8.182)$$

Di-tin and mercury–tin compounds will also add to Pt(0)[262,268,269] (eqns 8.183–8.185).

$$Sn_2Me_6 + Pt(PPh_3)_2C_2H_4 \rightarrow (Me_3Sn)Pt(PPh_3)_2Cl \quad (8.183)$$

$$Ph_2P(CH_2)_2SnMe_2{-}SnMe_2(CH_2)_2PPh_2 + Pt(PPh_3)_4$$

$$\downarrow$$

Me₂Sn PPh₂ / Pt / Ph₂P SnMe₃ (chelate ring) + Me₂Sn SnMe₂ / Pt / Ph₂P PPh₂ (chelate rings) (8.184)

$$[(C_6F_5)_3Sn]_2Hg + Pt(PPh_3)_2 \rightarrow (C_6F_5)_3SnPt(PPh_3)_2HgSn(C_6F_5)_3 \quad (8.185)$$

The 'mixed' compound $R_3SnHgGeR_3$ reacts in the same way, selectively forming an Sn–Pt bond.

Insertion of $SnCl_2$ into Pt–Cl bonds has already been referred to. The chemistry of this reaction has been extensively studied because some of the products act as catalysts for the homogeneous hydrogenation and isomerization of alkenes. Neutral complexes with one or two Pt–$SnCl_3$ groups may be produced[260,270] (eqns 8.186, 8.187).

$$Pt(PPh_3)_2HCl + SnCl_2 \rightarrow (Cl_3Sn)Pt(PPh_3)_2H \quad (8.186)$$

$$Pt(AsPh_3)_2Cl_2 + SnCl_2 \rightarrow (Cl_3Sn)_2Pt(AsPh_3)_2 + (Cl_3Sn)Pt(AsPh_3)_2Cl \quad (8.187)$$

When these reactions are carried out in acid solution, anionic complexes are formed.[271] Five-coordinate cationic complexes such as $[L_4PtSnCl_3][BPh_4]$ are also known.[247] Similar anionic clusters are formed from $SnCl_2$ and $PtCl_2$ or Na_2PtCl_4[272,273] (eqn 8.188).

$$SnCl_2 + Na_2PtCl_4 \longrightarrow trans\text{-}[(Cl_3Sn)_2PtCl_2]^{2-} \xrightarrow{CO} trans\text{-}[(Cl_3Sn)_2Pt(CO)Cl]^{-} \quad (8.188)$$

If a platinum (IV) compound is used, reduction to platinum (II) occurs. Many of these anions are 5-coordinate and, if combined with large cations, may be obtained as crystals. Examples are $[(Cl_3Sn)_5Pt](PMePh_3)_3$ and $[(Cl_3Sn)_4PtH](NMe_4)_3$.[218] The reaction between K_2PtCl_4, Me_4NCl and $SnCl_2$ in acetone gives a cluster compound with the probable structure $(Me_4N)_4[(Cl_3Sn)_6Pt_3(\mu_3\text{-}SnCl)_2]$, i.e. with the [SnCl] units lying above and below a triangle of platinum atoms.[244] Similar anionic clusters are formed from $SnCl_2$ and $PtCl_2$.[272,273]

Ethyne and other alkynes react with $Pt(MeCN)_2Cl_2$ in the presence of $SnCl_2$, forming (**37**) and (**38**) in which the ethyne, dimerized to cyclobutadiene, remains coordinated to platinum:

(C₄H₄)Pt(μ-Cl)₃Pt(C₄H₄) — Pt, Cl, Cl, Cl, Pt

(**37**)

(C₄H₄)–Pt(SnCl₃)₃

(**38**)

Heating this ionic complex with HCl/acetone results in loss of all the tin from the anion as $SnCl_2$. With sterically hindered alkynes, dimerization does not occur and the reaction stops at the stage of mono- and dimeric products (**39**) and (**40**)[274]

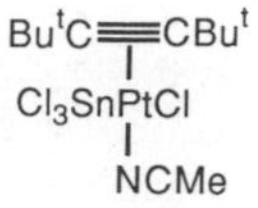

$Bu^tC{\equiv}CBu^t$
$Cl_3SnPtCl$
NCMe
(**39**)

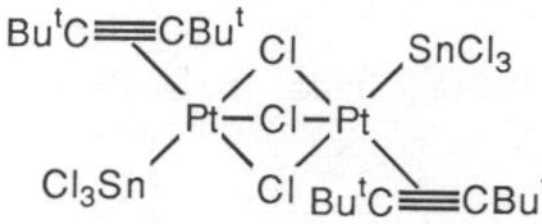

$Bu^tC{\equiv}CBu^t$ Cl $SnCl_3$
Pt—Cl—Pt
Cl_3Sn Cl $Bu^tC{\equiv}CBu^t$
(**40**)

$(COD)_3Pt_3(SnCl_3)_2$ has a trigonal bipyramidal structure with axial $SnCl_3$ groups and each equatorial platinum atom coordinated to a diene.[273] Mercury–tin compounds react oxidatively with Pt(0) complexes.[268]

Many electrophilic reagents e.g. $PhC{\equiv}CH$, CH_3CO_2H, HCl, $HOCH_2CH_2SH$, $SnCl_4$, I_2, CH_3I, CCl_4 and $HgCl_2$ cleave one or both of the tin–platinum bonds in the complex $(Me_3Sn)_2Pt(CH_2PPh_2)_2$. This complex is stable to 10% NaOH, and is thermally stable at 150°C *in vacuo* over three weeks. At 200°, it decomposes forming methane.[275]

Cleavage of the tin–platinum bond in platinum(IV) complexes has been observed[258] (eqn 8.189).

$$Ph_2ClSnPtMe_2Cl(bipy) + Ph_2PbCl_2 \rightarrow Ph_2SnCl_2 + Ph_2ClPbPtMe_2Cl(bipy) \quad (8.189)$$

8.11 Copper, silver and gold

Tin complexes of these metals all involve the transition metal in its + 1 oxidation state with coordination numbers of 2, 3 or 4.

Surprisingly, in view of its general chemistry, a two-coordinate copper complex (**41**) has been described together with the related salt $Li[Me_3SnCuSPh]$:[276]

$$Me_3SnLi + Me_2SCuCl \rightarrow \underset{(\mathbf{41})}{Me_3SnCuSMe_2} \quad (8.190)$$

The reaction between $CuBr{\cdot}SMe_2$ and Me_3SnLi in THF at −78°C gives an uncharacterized complex written as $Me_3SnCu{\cdot}LiBr{\cdot}SMe_2$.[277]

A versatile preparative method is the insertion of $SnCl_2$ into the M–Cl bond[278] (eqns 8.191, 8.192).

$$(Ph_3P)_3MCl + SnCl_2 \rightarrow (Ph_3P)_3MSnCl_3 \quad (M = Cu, Ag) \quad (8.191)$$

$$Ph_3PAuCl + SnCl_2 + 2Ph_3P \xrightarrow{Me_2CO} (Ph_3P)_3AuSnCl_3 \quad (8.192)$$

These four-coordinate complexes are white, crystalline and stable to water. The usual strong tendency of gold (I) to form two-coordinate complexes does not seem to apply. Heating $(Ph_3P)_3AuSnCl_3$ results in loss of 1 mole of Ph_3P, but further heating causes dissociation into Ph_3PAuCl and $SnCl_2$. The action of $SnCl_4$ on $Ph_3AuGePh_3$ in the presence of Ph_3P results in the following reaction:[279]

$$Ph_3PAuGePh_3 + Ph_3P + \text{excess } SnCl_4 \rightarrow (Ph_3P)_2AuSnCl_3 + Ph_3GeCl \quad (8.193)$$

The structure of the three-coordinate complex $(Me_2PhP)_2AuSnCl_3$, suggests that it is best formulated as the salt $[(Me_2PhP)_2Au](SnCl_3)$ since the tin–gold distance is large at 288 pm. The PAuP angle is 134°.[280] Use of the bulky phosphine $(Me_3SiCH_2)_3P$ gives the stable three-coordinate complex (**42**).[281]

$$(Me_3SiCH_2)_3PAuCl + SnCl_2 \rightarrow \underset{\mathbf{(42)}}{[(Me_3SiCH_2)_3P]_2AuSnCl_3} \quad (8.194)$$

Of the Group I metals, gold is especially prone to form metal cluster complexes, and two examples, neither of which has been characterized, have the compositions $(Ph_3P)_4Au_4SnCl_3$ and $(Ph_3P)_2Au_3SnI_2$.[282]

8.12 Zinc, cadmium and mercury

Organotin derivatives of all three metals have been reported. The thermal stability increases on going down the Group, but all compounds are sensitive to oxygen and water. Thermal stability is enhanced by having bulky ligands which cannot undergo β-elimination of alkene (such as Me_3SiCH_2, neopentyl) bonded to tin, or having strong Lewis bases coordinated to the otherwise vacant coordination sites on zinc, cadmium or mercury. Examples are illustrated by eqns 8.195–8.197.[283–286]

$$Bu_3SnH + Bu^t_2Hg \xrightarrow{-25^\circ} Hg[SnBu_3]_2 \quad (8.195)$$

$$(Me_3CCH_2)_3SnH + Et_2Cd \xrightarrow{20^\circ} Cd[Sn(CH_2CMe_3)_3]_2 \quad (8.196)$$

$$Ph_3SnH + R_2Zn + L \rightarrow Zn(SnPh_3)_2L \quad (8.197)$$

L = tetramethylethylenediamine or bipyridyl

For the tin–zinc complex in the absence of Lewis bases, phenyl transfer from tin to zinc occurs. Electrochemical oxidation of zinc, cadmium and mercury in non-aqueous solutions containing Ph_3SnCl yields Ph_3SnMCl, and these compounds can be isolated if strong donors such as bipyridyl are present.[287] The mercury compound, $Hg[Sn(C_6F_5)_3]_2$, may be formed by the tin hydride route or by irradiating a mixture of $(C_6F_5)_3SnBr$ and $(Et_3Ge)_2Hg$.[288] However, the hydride method does not always work. For example, Me_3SnH and Bu^t_2Hg give Bu_3SnH, Me_6Sn_2 and Hg at 20°.[289]

Unsymmetrical compounds have been made by exchange reactions[290] for example, eqn (8.198):

$$R_3SnH + Et_2Hg \rightarrow R_3SnHgEt \quad (8.198)$$

$R = Me_3CCH_2$

In other cases this type of exchange results in elimination of mercury.[291]

$$[(C_6F_5)_3Sn]_2Hg + [(Me_3SiCH_2)_3Sn]_2Hg \rightarrow Hg + (C_6F_5)_3SnSn(CH_2SiMe_3)_2 \quad (8.199)$$

Pyrolysis or irradiation of these compounds results in the formation of mercury and

Sn_2R_6,[292] although more profound decomposition reactions may occur giving tin and SnR_4.[290]

The reactivity of these compounds is high and various examples have been referred to in connection with the synthesis of tin–transition metal bonds. They are highly photosensitive, and react with most electrophiles and nucleophiles, often under mild conditions. Examples are I_2, MeI,[285,286] HgX_2, HgR_2, Se,[292] and Li.[290] $(Me_3Sn)_2Hg$ is unstable even at $-10°$, and is hydrolysed to Me_3SnH.[289] $Pt(PPh_3)_2$ inserts into the tin–mercury bond[267] eqn (8.200):

$$[(C_6F_5)_3Sn]_2Hg + Pt(PPh_3)_2 \rightarrow (C_6F_5)_3SnHgPt(PPh_3)_2Sn(C_6F_5)_3. \quad (8.200)$$

8.13 Boron, aluminium and thallium

Boron compounds have been made using standard methods[293,294]:

$$Me_3SnLi + ClB(NR_2)_2 \rightarrow Me_3SnB(NR_2)_2 \quad (8.201)$$

$$Et_3SnLi + Cl_2BNR_2 \rightarrow (Et_3Sn)_2BNR_2 \quad (8.202)$$

Alternatively, $Me_nSn(BPh_2)_{4-n}$ compounds may be formed by reacting methyltin halides with complexes such as $[(Ph_2PCH_2)_2]_2Co(BPh_2)_2$.[295] The carborane anion $(B_9C_2H_{11})^{2-}$ forms a tin complex by reaction with $SnCl_2$ or R_2SnCl_2.[296] The reaction between Me_3SnCl and $NaB_{10}H_{13}$ gives $Me_2SnB_{10}H_{12}$ as a crystalline solid melting at 123°, which is fairly stable in air but cleaved by HCl to Me_2SnCl_2.[297] In general, these compounds are thermally stable. The tin–boron bond is cleaved by H_2, X_2, sulphur and ROH.

The complex $(Me_3Sn)_3Al(THF)$ has been obtained from $(Me_3Sn)_2$ Hg and aluminium and is unstable above 20°C.[298] The cubane cage structure $(SnNBu^t)_4$ reacts with $AlCl_3$ forming an adduct which contains Al–Sn bonds.[299] $Bu_3SnAlEt_2$, made from Bu_3SnLi and Et_2AlCl but not isolated, finds use as an organic reagent for stannylation reactions.[300] The thallium–tin compound (**44**) has been made by hydrostannolysis:[301,302]

$$R_3SnH + TlEt_3 \rightarrow \underset{(\mathbf{44})}{Tl[(SnR_3)]_3} \quad (8.203)$$

$$(R = Me_3SiCH_2)$$

(**44**) reacts with Li, O_2, Br_2, Hg and EtOH with cleavage of the tin–thallium bond.

8.14 Antimony and bismuth

Tin–antimony and tin–bismuth bonded compounds have been little studied and again are prepared by hydrostannolysis, bismuth being more reactive than antimony:

$$R_3SnH + MR'_3 \rightarrow (R_3Sn)_3M \quad (8.204)$$

$$M = Sb, Bi$$

In the presence of $AlCl_3$ they decompose to SnR_4, Sn and M. With RBr they are converted into R_3SnBr and MR_3.[303] On photolysis $(Me_3Sn)_3Sb$ is converted into $(Me_3Sn)_4Sb_2$.[304]

8.15 Silicon, germanium and lead

Nucleophilic substitution reactions have been applied in various ways for the preparation of bonds to silicon, germanium and lead:[305–308]

$$\begin{array}{l} R_3SnLi + R'_3MCl \searrow \\ \qquad\qquad\qquad\qquad R_3SnMR'_3 \\ R'_3MLi + R_3SnCl \nearrow \end{array} \tag{8.205}$$

$$Et_3SnH + Et_3GeLi \rightarrow Et_3SnGeEt_3 + LiH \tag{8.206}$$

In general, these reactions give reasonable yields of the desired product, but side reactions occur due to halogen- or hydrogen-metal exchange and subsequent coupling to give the symmetrical products, R_6Sn_2 and R'_6M_2. A further complication is the possible dissociation of R_3SnLi to $(R_2Sn)_n$ and the more strongly nucleophilic RLi. Despite these limitations this remains a good synthetic method. Other alkali-metal derivatives may be used, and solvents range from ether, THF, 1,2-dimethoxyethane and HMPA to liquid ammonia.

One extremely useful method of synthesis, apparently not applicable to silicon, makes use of the high reactivity of Ge–N, Sn–N or Pb–N bonds in reactions where the amine is eliminated[293,309,310] (eqns 8.207, 8.208).

$$Bu_3GeNMe_2 + Bu_2SnH_2 \rightarrow Bu_3GeSnBu_2H \xrightarrow{\text{base}} Bu_3GeSnBu_2SnBu_2GeBu_3 \tag{8.207}$$

$$Me_3SnH + Me_3PbNEt_2 \rightarrow Me_3SnPbMe_3 \tag{8.208}$$

Synthesis by thermal or photochemical decomposition of the appropriate mercury compounds (eqns 8.209, 8.210) has already been discussed[292].

$$R_3SnHgSiR'_3 \rightarrow Hg + R_3SnSiR'_3 + Sn_2R_6 + Si_2R'_6 \tag{8.209}$$

$$\begin{array}{l} R_3SnCl + (Et_3Ge)_2Hg \searrow \\ \qquad\qquad\qquad\qquad R_3SnGeEt_3 \\ R_3SnH + Et_3GeHgEt \nearrow \end{array} \tag{8.210}$$

Branched chain compounds can be made by variations of these methods:[310,311]

$$Et_2SnH_2 + 2Ph_3GeSnEt_2NEt_2 \rightarrow Et_2Sn(SnEt_2GePh_3)_2 \tag{8.211}$$

$$2Ph_3SnLi + Ph_2SiCl_2 \rightarrow (Ph_3Sn)_2SiPh_2 \tag{8.212}$$

A further method of some interest is the insertion of germylenes into Sn–Sn bonds[312,313] (eqn 8.213):

$$R_6Sn_2 + GeF_2 \rightarrow R_3SnGeF_2SnR_3 \tag{8.213}$$

Electrophilic cleavage of the Sn–M bond (e.g. eqn 8.212) was reported as early as 1933.[314]

$$Ph_3SiSnMe_3 + Br_2 \rightarrow Ph_3SiBr + Me_3SnBr \tag{8.214}$$

Nucleophilic cleavage reactions have been reported for all combinations of elements[314,315] (eqns 8.215–8.217).

$$Me_3SnSiPh_3 + Na/NH_3 \rightarrow Me_3SnNa + Ph_3SiNa \quad (8.215)$$

$$(Me_3Si)_4Sn + H_2O/OH^- \rightarrow (Me_3Si)_2O + Sn + H_2 \quad (8.216)$$

$$Ph_3SnGePh_3 + PhLi \rightarrow Ph_4Sn + Ph_4Ge + Ph_3SnLi + Ph_3GeLi \quad (8.217)$$

Organic groups may be cleaved from either metal without breaking the Sn–M bond if reactions are carried out under sufficiently controlled conditions as in the following sequence:[316]

$$Ph_3SnMPh_3 \xrightarrow{HOAc} (AcO)_3SnMPh_3 \xrightarrow{HOAc} (AcO)_3SnM(OAc)_3 \quad (8.218)$$

$$(AcO)_3SnM(OAc)_3 \xrightarrow{HCl,\ -100^\circ C} Cl_3SnMCl_3 \xrightarrow{<0^\circ C} SnCl_2 + MCl_4$$

$$(AcO)_3SnM(OAc)_3 \xrightarrow{LiAlH_4,\ -90^\circ C} H_3SnMH_3 \xrightarrow{0^\circ C} Sn + H_2 + MH_4 \quad (8.219)$$

8.16 Tin–tin bonded compounds

A polymeric diethyl tin compound derived from EtI and Na/Sn was described in 1852[317]. Synthetic methods for establishing tin–tin bonds have much in common with those described earlier in this section, as in the following illustrative examples[288,292,309,318–320]:

$$Me_3SnBr + Na/NH_3 \rightarrow Me_6Sn_2 \quad (8.220)$$

$$Ph_3SnNa + Me_3SnBr \rightarrow Me_3SnSnPh_3 \quad (8.221)$$

$$R_3SnNEt_2 + R'_3SnH \rightarrow R_3SnSnR'_3 \quad (8.222)$$

$$R_3SnH + (R'_3Sn)_2O \text{ (or } R'_3SnOR) \rightarrow R_3SnSnR'_3 \quad (8.223)$$

$$(R_3Sn)_2Hg \xrightarrow{h\nu} R_6Sn_2 \quad (8.224)$$

$$(Bu_3Sn_2)O + Ti \text{ (dispersion in THF)} \rightarrow Bu_6Sn_2 \quad (8.225)$$

The last of these reactions works with dispersions of sodium, potassium and magnesium in place of titanium. Electrolytic methods involving reduction at a mercury cathode can be utilized in a number of ways, for example, eqn 8.226.[321]

$$R_2SnCl_2 + 2e \xrightarrow{CH_3Ca} ClR_2SnSnR_2Cl \quad (8.226)$$

Electrolytic reduction of Ph_3SnCl occurs by two different routes depending on the reduction potential:[322]

$$Ph_3Sn^- + Ph_3Sn^+ \rightarrow Ph_6Sn_2 \leftarrow 2Ph_3Sn^\bullet \quad (8.227)$$

Optically active tin–tin compounds have been isolated via chiral hydrides[323] (eqn 8.228):

$$PhMeRSnH + Pd/C \rightarrow PhMeRSn{-}SnRMePh \qquad (8.228)$$
$$R = PhCMe_2CH_2$$

The reaction of Ph_2SnH_2 with CH_3CO_2H gives $Ph_4Sn_2(OCOCH_3)_2$, in which the two acetato groups bridge the tin atoms so that each tin is five-coordinated.[324] These methods can be extended to yield unbranched tin–tin bonded chain compounds, as in the following examples:[27,320,325,326]

$$Ph_3SnNa + Ph_2SnCl_2 \rightarrow Ph_8Sn_3 \qquad (8.229)$$

$$R_3SnSnR_2H + Pt \downarrow R_{10}Sn_4 \uparrow R_3SnNEt_2 + HR_2SnSnR_2H \qquad (8.230)$$

$$Me_3SnBr + NaMe_2Sn(SnMe_2)SnMe_2Na \rightarrow Me_{12}Sn_5 \qquad (8.231)$$

$$Et_7Sn_3H + Et_2NH \rightarrow H_2 + Et_{14}Sn_6 \qquad (8.232)$$

Further adaptations of these methods result in the formation of branched compounds[320,327] (eqns 8.233, 8.234).

$$Ph_3SnLi + SnCl_2 \rightarrow (Ph_3Sn)_3SnLi \xrightarrow{Ph_3SnCl} (Ph_3Sn)_4Sn \qquad (8.233)$$

$$(Bu_3Sn)_2SnPhH + Et_2NH \rightarrow H_2 + (Bu_3Sn)_2[SnPh]_2(SnBu_3)_2 \qquad (8.234)$$

In addition, ring systems containing from four to nine tin atoms have been prepared, usually from organotin hydrides. The size of the ring depends on the particular experimental conditions employed, for example:[328]

$$Bu^t_2SnH_2 + Bu^t_2Sn(NEt_2)_2 \rightarrow cyclo\text{-}Bu^t_8Sn_4 \qquad (8.235)$$

(45)

The four-membered ring in (**45**) is planar.[329] Other reactions leading to cyclic tetramers include the thermolysis of $Bu^t_2SnH_2$ in pyridine[6] and the reaction of $Bu^t_2SnCl_2$ with excess Bu^tMgCl in THF.[330] Cyclic pentamers are obtained from related reactions. For example, thermolysis of Ph_2SnH_2 in boiling dimethylformamide yields $Ph_{10}Sn_5$, and this reaction also gives the hexamer, $Ph_{12}Sn_6$, which has a chair conformation.[331] This hexamer has also been prepared from Ph_2SnCl_2 and sodium naphthalide in THF.[331] The reaction between Me_2SnH_2 and $Me_2Sn(NEt_2)_2$, if carried out in the absence of air, oxygen and light, gives exclusively the hexamer, $Me_{12}Sn_6$, which in solution equilibrates to include cyclomers with 7, 8, and 9 tin atoms in the ring.[332] Similarly, Ph_2SnH_2 and Bu^t_2Hg at $-30°C$ form *cyclo*-$(Ph_2Sn)_n$ with $n = 4$, 5, 6 and 7,[289] and decomposition of Et_2SnH_2 with pyridine and Et_2SnCl_2 gives a 94% yield of $(Et_2Sn)_9$.[333]

In addition to these well-defined compounds, tin forms polymers of the type $(R_2Sn)_n$, usually along with linear and cyclic compounds in the reactions already referred to.

Chemical degradation of these polymers shows that they contain $[R_3Sn]$, $[R_2Sn]$, $[RSn]$ and $[Sn]$ units[1].

Chemical reactions of tin–tin bonded compounds mostly result in cleavage of the Sn–Sn bond. Reversible homolytic dissociation of the Sn–Sn bond in di–tin compounds occurs if bulky substitutents are attached to tin, e.g. eqn (8.236).

$$R_6Sn_2 \rightleftharpoons 2R_3Sn^{\cdot} \qquad R = 2,4,6\text{-}Me_3C_6H_2 \tag{8.236}$$

Esr signals of the radicals can be detected at elevated temperature but disappear on cooling.[334] Other di-tin compounds are thermally stable to 150–200°, the main decomposition products being R_4Sn, tin and hydrocarbons.[335] Polymeric $(R_2Sn)_n$ compounds decompose to R_4Sn, R_6Sn_2 and other products.[336]

In the cation $Me_6Sn_2^+$, generated by radiolysis of Me_6Sn_2 at 77 K in $CFCl_3$, the odd electron couples to the spin $\frac{1}{2}$ isotopes of tin. Analysis of the spectrum suggests that each Me_3Sn group is nearly planar, the Sn–Sn bonding MO being largely $4p_z + 4p_z$ in character.[337]

R_6Sn_2 compounds react with $R^{\cdot}$ radicals to give R_4Sn and R_3Sn, and diacetyl peroxide cleaves the tin–tin bond in a non-radical process forming $R_3SnOCOCH_3$.[338] Reaction with oxygen occurs readily, especially for alkyls, giving $(R_3Sn)_2O$.[339] Sulphur and selenium form analogous products,[340] and peroxides cleave the tin–tin bond[341] (eqn 8.237).

$$R_6Sn_2 + Bu^tOOBu^t \rightarrow R_3SnOBu^t \tag{8.237}$$

Electrophiles readily cleave the tin–tin bond. For iodine, kinetic studies suggest that a polar rather than a four-centred transition state is involved.[342] The product of halogen cleavage, R_3SnX, may undergo subsequent redistribution reactions, depending on the conditions used, and this consideration applies to cleavage of the tin–tin bond by $SnCl_4$, organotin halides and HX.[6,335] Cyclic tin compounds are more reactive, and with electrophiles can be limited to ring opening:[343]

$$(Bu^t_2Sn)_4 + I_2 \rightarrow I(Bu^t_2Sn)_4I \tag{8.238}$$

Solvolysis of Ph_6Sn_2 by RCO_2H leads to the formation of the mixed-valence complexes $[Sn^{II}Sn^{IV}O(O_2CR)_4O(OCR)_2]_2$, in which the four tin atoms are held together by bridging carboxylate groups and two μ_3-oxo bridges. Both Sn(IV) atoms are octahedrally coordinated, and both Sn(II) atoms have pentagonal bipyramidal geometry.[344] Me_6Sn_2 and $ClCH_2CO_2H$ react under mild conditions without cleavage of the tin–tin bond forming $Me_4Sn_2(OCOCH_2Cl)_2$ with bridging chloroacetato groups.[345] $HgCl_2$ gives mercury and Me_3SnCl,[346] and Me_3PbCl reacts in a related way forming Me_4Pb, $PbCl_2$ and Me_3SnCl.[347] Me_6Sn_2 and tetracyanoethylene form a charge transfer complex, the product being the highly reactive adduct, $Me_3SnC(CN)_2C(CN)_2SnMe_3$.[348,349]

Sulphur dioxide reacts with Me_6Sn_2, forming a polymeric product by insertion into the tin–tin bond. By contrast, Ph_6Sn_2 reacts by SO_2 insertion into the Ph–Sn bonds.[350] C_2F_4 and other fluoroalkenes insert into the tin–tin bond, as do halocarbenes[351,352] (eqns 8.239, 8.240):

$$R_6Sn_2 + C_2F_4 \rightarrow R_3SnC_2F_4SnR_3 \tag{8.239}$$

$$R_6Sn_2 + CX_2 \rightarrow R_3SnCX_2SnR_3 \tag{8.240}$$

CF_3I also cleaves the tin–tin bond in Me_6Sn_2 forming Me_3SnI and Me_3SnCF_3.[353]

Addition of Me_6Sn_2 to allene is catalysed by Pt(0) complexes and yields $Me_3SnCH_2C(SnMe_3){=}CH_2$. Other allenes give mixtures of tin-substituted monoalkenes.[354] Diphenylacetylene undergoes a similar reaction to give *trans*-$R_3SnC(Ph){=}C(Ph)SnR_3$.[355] Oligomeric $(Me_2Sn)_n$ compounds and $(Me_2SnS)_3$ combine to give $Me_2SnSnMe_2SSnMe_2S$.[356]

Most nucleophilic reactions have already featured in earlier sections. Ph_6Sn_2 is evidently unaffected by hot NaOH in ethanol.[6] For polymeric compounds reaction with alkali metals proceeds in stages[27] (eqn (241)).

$$(Me_2Sn)_n + Na/NH_3 \rightarrow [Me_2SnSnMe_2]2Na \xrightarrow{Na} Me_2SnNa_2 \qquad (8.241)$$

R_3SnLi and Me_6Sn_2 undergo rapid exchange in THF[357] (eqn (242)).

$$Me_3SnLi + Et_6Sn_2 \rightleftharpoons Et_3SnLi + Me_3SnSnEt_3 \rightleftharpoons Me_6Sn_2 \qquad (8.242)$$

8.17 Structure and bonding

Many tin–metal compounds have been examined by X-ray crystallography, and for transition metals it appears that the Sn–M bond length is significantly shorter than the sum of the covalent radii. In R_3Sn–M compounds, the Sn–M length decreases with increasing electronegativity of the R groups (Me_3Sn–M > Cl_3Sn–M for related structures).[129] For transition metal complexes the length of the Sn–M bond increases with increasing π-acceptor properties of, in particular, the ligand *trans* to tin. No systematic changes in Sn–M bond lengths can be correlated with changes in the formal oxidation state of the transition metal.[8] These effects can be accounted for in terms of variations in the extent of $d\pi$–$d\pi$ interaction between tin and the transition metal. In Cl_3Sn–M complexes the electronegative chlorine atoms lower the energy of the d orbitals on tin, making them more compatible with the energies of filled d orbitals of the transition metal. Similarly, it can be argued that, in the system *trans*-R_3Sn–M–L, changing L from say CO to a tertiary phosphine, which is a stronger σ-donor and a weaker π-acceptor than CO, makes a predictable change in the Sn–M bond length (Ph_3Sn–$Mn(CO)_5$, 267.4 pm, and *trans*-Ph_3Sn–$Mn(CO)_4(PPh_3)$, 262.7 pm). Although these arguments seem reasonably self-consistent, definitive proof is lacking and bond length variations can be accounted for in terms of purely σ-bond effects based on variations in mixing of the metal bonding orbitals: a high s character to the Sn–M bond resulting in bond shortening. This in general seems to be substantiated by nmr studies of coupling constants between tin and transition metals with a spin $\frac{1}{2}$ isotope, and between tin and hydrogen in methyltin compounds.[358] In complexes of the type $(R_3P)_2Pt(Ph)(SnR'_3)$, trends in Pt–P and Sn–P coupling constants lead to the conclusion that Sn–Pt bonds are predominantly σ in character.[359] In platinum, rhodium and ruthenium complexes containing more than one $SnCl_3$ group bonded to the transition metal, ^2J(Sn–Sn) couplings are extremely large, especially for *trans*-$Cl_3SnMSnCl_3$.[360]

In *cis*-$[PtCl_2(SnCl_3)_2]^{2-}$, the platinum–tin length of 235.6 pm is shorter than in $Pt[(SnCl_3)_5]^{3-}$, (255.3 pm), and this correlates with the greater ^1J(SnPt) coupling in the former (27 640 Hz), implying a high s-character to the tin–platinum bonds.[257,359,361] It also appears that π-bonding is greater in four- than in five-coordinate complexes.[362] Mössbauer and photoelectron spectroscopic studies on these types of compounds have led to the conclusion that π-bonding is minimal.[10,137] Distortion of the bond angles at tin from tetrahedral in R_3Sn-M compounds correlate with Sn–M coupling constants, which in turn may be correlated with the *s*-character of the Sn–M bond.[65] In tin–iron complexes, both tin and iron Mössbauer spectra have been examined. For β-diketone derivatives of the type $[(chelate)_2SnFe(CO)_4]_2$, the quadrupole splittings confirm that tin is octahedrally coordinated and iron isomer shifts suggest that tin is a stronger σ donor than CO or R_3P.[198] Examination of Mössbauer data for a range of Cl_3Sn–M compounds suggests that the Mössbauer centre shift does not satisfactorily distinguish between the formal +2 and +4 oxidation states of tin.[363] Various other methods (^{59}Co and ^{35}Cl NQR, ^{119}Sn γ-resonance, dipole moments, ^{59}Co wide-line nmr and vibrational spectroscopy) of assessing $d\pi$–$d\pi$ contributions to bonding in tin–cobalt complexes have been considered, with the conclusion that π-bonding is non-zero and that R_3Sn is a stronger σ-donor than Cl_3Sn.[206] Electronic and ^{31}P-nmr spectra have been reported for $C_5H_5(Ph_3P)NiSnCl_3$ and the $SnPh_3$ analogue; these suggest that $SnCl_3$ and $SnPh_3$ lie high in the spectrochemical series.[364]

Stannylene-transition metal bond lengths are considerably shorter than those of Sn(IV)–M bonds. For example, the tin–chromium distance in $R_2SnCr(CO)_5$ (R = $(Me_3Si)_2N$) is 256.2 pm, whereas for a range of Sn(IV)–Cr complexes it is about 285 pm.[91] This change is probably due to extensive π interaction giving double bond character to the stannylene–Cr bond, as in carbene complexes of transition metals. In the ylide complex $Bu^t_2(py)SnCr(CO)_5$ the Sn–Cr bond length is 265.2 pm, which may be interpreted as a decrease in π interaction, relative to the stannylene, due to coordination of a Lewis base to tin.[93,365] Similar correlations can be made for Pt–Sn(II) and Pt–Sn(IV) complexes.[251]

In polystannanes such as $Me_2Sn(SnPh_3)_2$ and $(Ph_3Sn)_4Sn$, the ^{119}Sn-nmr chemical shifts are strongly influenced by chain branching and the two bond tin–tin coupling correlates reasonably well with bond lengths and bond angles.[366]

References

1. W.P. Neumann, *The Organic Chemistry of Tin*, Wiley, London, 1970.
2. W. Petz, *Chem. Rev.*, 1986, **86**, 1019.
4. R.C. Poller, *Chemistry of Organotin Compounds*, Logos, London, 1970.
4. E.H. Brooks and R.J. Cross, *Organometal. Chem. Rev.*, 1970, **A6**, 227.
5. D.D. Davis and C.E. Gray, *Organometal. Chem. Rev.*, 1970, **A6**, 283.
6. A.K. Sawyer and M.J. Newlands, (1972) in *Organotin Compounds*, Vol. 3, ed. A.K. Sawyer, Marcel Dekker, New York.
7. F. Glockling and S.R. Stobart, *MTP Int. Rev. Sci.*, 1972, **6**, 63.
8. A.T.T. Hsieh, *MTP Int. Rev. Sci. Ser. 2*, 1974, **6**, 109.
9. K.M. Mackay and B.K. Nicholson, *Comp. Organomet. Chem.*, 1982, **6**, 1043.
10. J.A. Zubieta and J.J. Zuckermann, *Progr. Inorg. Chem.*, 1978, **24**, 336.
11. H. Schumann and I. Schumann, *Gmelins Handbuch der Anorganischen Chemie*, **35**, Part 4, 1976.
12. E.H. Brooks and R.J. Cross, *Organometal. Chem. Rev.*, 1970, **A6**, 227.
13. V.I. Gol'danskii, B.V. Borshagovskii, E.F. Makarov, R.A. Stukan, K.A. Anisimov, K. Kolobova and V.V. Skripkin, *Teor. Eksperim. Khim.*, 1967, **3**, 478.
14. W.L. Wells and T.L. Brown, *J. Organometal. Chem.*, 1968, **11**, 271.
15. H. Gilman, O.L. Mars and S.Y. Sim, *J. Org. Chem.*, 1962, **27**, 4232.
16. C. Tamborski, F.E. Ford and E.J. Soloski, *J. Org. Chem.*, 1963, **28**, 181.
17. G. Wittig, *Angew. Chem.*, 1959, **63**, 231.
18. C.A. Kraus and W.N. Greer, *J. Amer. Chem. Soc.*, 1922, **44**, 2629.

19. S.F.A. Kettle, *J. Chem. Soc.*, 1959, 2936.
20. J.-P. Quintard, S. Hauvette-Frey and M. Pereyre, *J. Organometal. Chem.*, 1978, **159**, 147.
21. C.A. Kraus and W.V. Sessims, *J. Amer. Chem. Soc.*, 1925, **47**, 2361.
22. M. Newcomb and M.G. Smith, *J. Organometal. Chem.*, 1982, **225**, 61.
23. H. Weichmann and A. Tzschach, *Z. anorg. allg. Chem.*, 1979, **458**, 291.
24. J.J. Eméleus and S.F.A. Kettle, *J. Chem. Soc.*, 1958, 2444.
25. T. Birchall and J. Vetrone, *Proc. Vth Int. Conf. Ge, Sn, Pb (Padova)*, 1986, P15.
26. D. Blake, G.E. Coates and J.M. Tate, *J. Chem. Soc.*, 1961, 618.
27. C.A. Kraus and W.N. Greer, *J. Amer. Chem. Soc.*, 1925, **47**, 2568.
28. G. Wittig, F.J. Meyer and G. Lange, *Annalen*, 1951, **571**, 167.
29. H. Gilman and S.D. Rosenberg, 1953, **75**, 3592.
30. M. Veith, R. Roesler, D. Kaefer, C. Ruloff and V. Huch, *Proc. Vth Int. Conf. Ge, Sn, Pb, (Padova)*, 1986, C1.
31. E.O. Fischer, R.B.A. Pardy and U. Schubert, *J. Organometal. Chem.*, 1979, **181**, 37.
32. H.G. Kuivila and Y.M. Choi, *J. Org. Chem.*, 1979, **44**, 4774.
33. T. Aruga, O. Ito and M. Matsuda, *J. Chem. Phys.*, 1982, **228**, 61.
34. W. Kitching, H.A. Olszowy and G.M. Drew, *Organometallics*, 1982, **1**, 1244.
35. K. Kobayashi, M. Kawanisi, T. Hitomi and S. Kozima, *J. Organometal. Chem.*, 1982, **233**, 299.
36. T.N. Mitchell and M. El-Behairy, *J. Organometal. Chem.*, 1979, **172**, 293.
37. G.A. Arkamkina, M.P. Egorov, I.P. Beletskaya and O.A. Reutov, *J. Organometal. Chem.*, 1979, **182**, 185.
38. H.J. Reich and N.H. Phillips, *J. Amer. Chem. Soc.*, 1986, **108**, 2102.
39. F. Glockling and K.A. Hooton, *J. Chem. Soc.*, 1962, 3509.
40. A. Carrick and F. Glockling, *J. Chem. Soc. A*, 1966, 623.
41. J.-C. Lahournere and J. Valade, *J. Organometal. Chem.*, 1970, **15**, C3.
42. H.M.J.C. Creemers, J.G. Noltes and G.J.M. van der Kerk, *J. Organometal. Chem.*, 1968, **14**, 217.
43. C. Tamborski and E.J. Soloski, *J. Amer. Chem. Soc.*, 1961, **83**, 3734.
44. M.F. Lappert and P.P. Power, *J. Chem. Soc., Dalton Trans.*, 1985, 51.
45. H. Schumann and M. Cygon, *J. Organometal. Chem.*, 1978, **144**, C43.
46. G.S. Kalinina, L.N. Bochkarev, G.A. Razuvaev and M.N. Bochkarev, *Proc. XIX Int. Conf. Coord. Chem. (Prague)*, 1978, 65.
47. G.A. Razuvaev, L.N. Bochkarev, G.S. Kalinina and M.N. Bochkarev, *Inorg. Chim. Acta*, 1977, **24**, L40.
48. G.A. Razuvaev, G.S. Kalinina and E.A. Fedorova, *J. Organometal. Chem.*, 1980, **190**, 157.
49. M. Porchia, U. Casellato, F. Ossolo, G. Rossetto, P. Zanella and R. Graziani, *Proc. Vth Int. Conf. Ge, Sn, Pb (Padova)*, 1986, P17.
50. R.S.P. Coutts and P.C. Wailes, *Chem. Commun.*, 1968, 260.
51. J.G. Kenworthy and J. Myatt, *J. Chem. Soc. D*, 1970, 447.
52. H.M.J.C. Creemers, F. Verbeek and J.G. Noltes, *J. Organometal. Chem.*, 1968, **15**, 125.
53. B.M. Kingston and M.F. Lappert, *J. Chem. Soc., Dalton Trans.*, 1972, 69.
54. B.M. Kingston and M.F. Lappert, *Inorg. Nucl. Chem. Lett.*, 1968, **4**, 371.
55. F.G.N. Cloke, K.P. Cox, M.L.H. Green, J. Bashkin and K. Prout, *J. Chem. Soc., Chem. Commun.*, 1981, 117.
56. M.F. Lappert and A.R. Sanger, *J. Chem. Soc. A*, 1971, 1314.
57. G.A. Razuvaev, L.I. Vyshinskaya, G.A. Vasil'eva, V.L. Latyaeva, S.Y. Timoshenko and N.L. Ermolaev, *Izv. Akad. Nauk SSSR Ser. Khim.*, 1978, 2584.
58. G.A. Razuvaev, V.N. Latyaeva, E.N. Gladyshev, E.V. Krasilnikova and G. Lineva, *Inorg. Chim. Acta*, 1978, **31**, L357.
59. G.A. Razuvaev, V.N. Latyaeva, V.P. Mar'in, L.I. Vyshinskaya, S.P. Korneva, Yu.A. Andrianov and E.V. Krasil'nikova, *J. Organometal. Chem.*, 1982, **225**, 233.
60. D.H. Harris, S.A. Keppie and M.F. Lappert, *J. Chem. Soc., Dalton Trans.*, 1973, 1653.
61. T.M. Arkhireeva, B.M. Bulychev, A.N. Protsky, G.L. Soloveichik and V.K. Belsky, *J. Organometal. Chem.*, 1986, **317**, 33.
62. A.A. Pasynskii, Yu.V. Skripkin, O.G. Volkov, V.T. Kalinnikov, M.A. Porai-Koshits, A.S. Antsyshkina, L.M. Dikareva and K. Ostrikova, *Bull. Acad. Sci. USSR Div. Chem. Sci.*, 1983, **32**, 1093.
63. A. Davidson and J.E. Ellis, *J. Organometal. Chem.*, 1970, **23**, C1.

64. A. Davidson and J.E. Ellis, *J. Organometal. Chem.*, 1972, **36**, 113.
65. F. Neumann, J. Kopf and D. Rehder, *J. Organometal. Chem.*, 1984, **267**, 249.
66. F. Pforr, F. Neumann and D. Rehder, *J. Organometal. Chem.*, 1983, **258**, 189.
67. J.E. Ellis, J.G. Hayter and R.E. Stevens, *J. Organometal. Chem.*, 1981, **216**, 191.
68. T. Kruck and H. Breuer, *Chem. Ber.*, 1974, **107**, 263.
69. P. Hackett and A.R. Manning, *J. Chem. Soc., Dalton Trans.*, 1972, 2434.
70. M.L.H. Green, A.H. Lynch and M.G. Swanwick, *J. Chem. Soc., Dalton Trans.*, 1972, 1445.
71. T.S. Cameron and C.K. Prout, *J. Chem. Soc., Dalton Trans.*, 1972, 1447.
72. H. Behrens, M. Moll and E. Sixtus, *Z. Naturforsch.*, 1977, **32B**, 1105.
73. R.D. Adams, *J. Organometal. Chem.*, 1975, **88**, C38.
74. E.E. Isaacs and W.A.G. Graham, *Can. J. Chem.*, 1975, **53**, 975.
75. J.E. Ellis, S.G. Hentges, D.G. Kalina and G.P. Hagen, *J. Organometal. Chem.*, 1975, **73**.
76. A.N. Nesmeyanov, N.E. Kolobova, V.N. Khandozhko and K.N. Anisimov, *J. Gen. Chem. USSR*, 1974, **44**, 298.
77. G. Huttner, U. Weber, B. Sigwarth, O. Scheidsteger, H. Lang and L. Zsolnai, *J. Organometal. Chem.*, 1985, **282**, 331.
78. G.L. Rochfort and J.E. Ellis, *J. Organometal. Chem.*, 1983, **250**, 265.
79. G.L. Rochfort and J.E. Ellis, *J. Organometal. Chem.*, 1983, **250**, 277.
80. A.G. Ginzberg and A.N. Nesmeyanov, *Proc. V Int. Conf. Ge, Sn, Pb (Padova)*, 1986, P12.
81. J.T. Lin, G.P. Hagen and J.E. Ellis, *Organometallics*, 1984, **3**, 1288.
82. M.H. Chisholm, H.T. Chiu, K. Folting and J.C. Huffman, *Inorg. Chem.*, 1984, **23**, 4097.
83. G.F. Bradley and S.R. Stobart, *J. Chem. Soc., Dalton Trans.*, 1974, 264.
84. A. Mikaye, H. Kondo and A. Aoyama, *Angew. Chem. Int. Edn. Engl.*, 1969, **8**, 520.
85. S.A. Kellie and M.F. Lappert, *J. Chem. Soc. A*, 1971, 3216.
86. W. Malisch, H. Schmidbaur and M. Kuhn, *Angew. Chem. Int. Edn. Engl.*, 1972, **11**, 516.
87. F. Bonati and G. Wilkinson, *J. Chem. Soc.*, 1964, 179.
88. W. Petz, *J. Organometal. Chem.*, 1979, **165**, 199.
89. J.E. Shade, B.F. Johnson, D.H. Gibson, W.L. Hsu and C.D. Schaeffer, *Inorg. Chim. Acta*, 1985, **99**, 99.
90. J.D. Cotton, P.J. Davison, D.E. Goldberg, M.F. Lappert and K.M. Thomas, *Chem. Commun.*, 1974, 893.
91. J.D. Cotton, P.J. Davidson and M.F. Lappert, *J. Chem. Soc., Dalton Trans.*, 1976, 2275.
92. P.J. Davidson and M.F. Lappert, *Chem. Commun.*, 1973, 317.
93. M.D. Brice and F.A. Cotton, *J. Amer. Chem. Soc.*, 1973, **95**, 4529.
94. M. Veith, H. Lange, K. Braeuer and R. Bachmann, *J. Organometal. Chem.*, 1981, **216**, 377.
95. M.F. Lappert and M.J. Michalczyk, *Proc. Vth Int. Conf. Ge, Sn, Pb (Padova)*, 1986, C7.
96. C.D. Hoff and J.W. Connolly, *J. Organometal. Chem.*, 1978, **148**, 127.
97. M.J.S. Gynane, M.F. Lappert, S.J. Miles and P.P. Power, *Chem. Commun.*, 1976, 256.
98. K. Edgar, B.F.G. Johnson, J. Lewis and S.B. Wild, *J. Chem. Soc. A*, 1968, 2851.
99. E.W. Abel and S. Moorhouse, *J. Organometal. Chem.*, 1970, **24**, 687.
100. R. Kummer and W.A.G. Graham, *Inorg. Chem.*, 1968, **7**, 310.
101. B.I. Petrov, G.S. Kalinina and Y.A. Sorokin, *J. Gen. Chem. USSR*, 1975, **45**, 1873.
102. E.O. Fischer, H. Fischer, U. Schubert and R.B.A. Pardy, *Angew. Chem. Int. Edn. Engl.*, 1979, **18**, 872.
103. T.A. George, *Inorg. Chem.*, 1972, **11**, 77.
104. F. Bonati and G. Minghetti, *J. Organometal. Chem.*, 1969, **16**, 332.
105. M.J. Mays and S.M. Pearson, *J. Chem. Soc. A*, 1969, 136.
106. R.D. Rieke, S.N. Milligan, I. Tucket, K.A. Dowler and B.R. Willeford, *J. Organometal. Chem.*, 1981, **218**, C25.
107. W.K. Dean and W.A.G. Graham, *Inorg. Chem.*, 1977, **16**, 1061.
108. U. Kunze and S.B. Sastrawan, *Chem. Ber.*, 1979, **112**, 3149.
109. E.E. Isaacs and W.A.G. Graham, *Can. J. Chem.*, 1975, **53**, 467.
110. J.R. Chipperfield, *J. Organometal. Chem.*, 1977, **137**, 355.
111. J.R. Chipperfield, J. Ford, A.C. Hayter, D.J. Lee and D.E. Webster, *J. Chem. Soc., Dalton Trans.*, 1976, 1024.
112. J.R. Chipperfield, J. Ford, A.C. Hayter and D.E. Webster, *J. Chem. Soc., Dalton Trans.*, 1976, 360.
113. J.R. Chipperfield, A.C. Hayter and D.E. Webster, *J. Organometal. Chem.*, 1976, **121**, 185.

114. R.M.G. Roberts, *J. Organometal. Chem.*, 1972, **40**, 359.
115. K. Triplett and M.D. Curtis, *Inorg. Chem.*, 1976, **15**, 431.
116. W.A.G. Graham and J.A.J Thompson, *Inorg. Chem.*, 1967, **6**, 1365.
117. R.A. Burnham and S.R. Stobart, *J. Chem. Soc. Dalton Trans.*, 1977, 1489.
118. J.C. Luong, R.A. Faltynek and M.S. Wrighton, *J. Amer. Chem. Soc.*, 1980, **102**, 7892.
119. H. Preut and H.J. Haupt, *Acta Crystallogr.*, 1982, **B28**, 1290.
120. H. Preut and H.J. Haupt, *Acta Crystallogr.*, 1981, **B37**, 688.
121. H. Preut and H.J. Haupt, *Acta Crystallogr.*, 1983, **C39**, 981.
122. H.J. Haupt, P. Balsaa, B. Schwab and U. Floerke, *Z. anorg. allg. Chem.*, 1984, **513**, 22.
123. H.J. Haupt and J. Hoffmann, *Z. anorg. allg. Chem.*, 1977, **429**, 162.
124. J.K. Ruff, *Inorg. Chem.*, 1971, **10**, 409.
125. J.E. Ellis and R.A. Faltynek, *J. Amer. Chem. Soc.*, 1977, **99**, 1801.
126. S. Onaka, Y. Yoshikawa and H. Yamatera, *J. Organometal. Chem.*, 1978, **157**, 187.
127. S. Onaka, *Chem. Lett.*, 1978, 1163.
128. A.L. Spek, K.D. Bos, E.J. Bulton and J.G. Noltes, *Inorg. Chem.*, 1976, **15**, 339.
129. V.K. Belsky, A.N. Protsky, I.V. Molodnitskaya, B.M. Bulychev and G.L. Soloveichik, *J. Organometal. Chem.*, 1985, **293**, 69.
130. A.B. Cornwell and P.G. Harrison, *J. Chem. Soc., Dalton Trans.*, 1976, 1054.
131. S. Kato, I. Noda, M. Mizuta and Y. Itoh, *Angew. Chem. Int. Edn. Engl.*, 1979, **18**, 82.
132. R.D. Gorsich, *J. Amer. Chem. Soc.*, 1962, **84**, 2486.
133. M.R. Booth, D.J. Cardin, N.A.D. Carey, H.C. Clark and B.R. Sreenathan, *J. Organometal. Chem.*, 1970, **21**, 171.
134. L.M. Bower and M.H.B. Stiddard, *J. Chem. Soc. A*, 1968, 706.
135. H.C. Clark and J.H. Tsai, *Inorg. Chem.*, 1966, **5**, 1407.
136. G.R. Dobson and E.P. Ross, *Inorg. Chim. Acta*, 1971, **5**, 199.
137. G.M. Bancroft and T.K. Sham, *J. Chem. Soc., Dalton Trans.*, 1976, 467.
138. W. Schubert, H.J. Haupt and F. Huber, *Z. anorg. allg. Chem.*, 1982, **485**, 190.
139. R.A. Burnham, F. Glockling and S.R. Stobart, *J. Chem. Soc., Dalton Trans.*, 1972, 1991.
140. H.C. Clark and B.K. Hunter, *J. Organometal. Chem.* 1971, **31**, 227.
141. J.P. Collman, J.K. Hoyano and D.W. Murphy, *J. Amer. Chem. Soc.*, 1973, 3434.
142. E.W. Abel and G.V. Hutson, *J. Inorg. Nucl. Chem.*, 1968, **30**, 2339.
143. R.D. Gorsich, *J. Organometal. Chem.*, 1966, **5**, 5105.
144. U. Kunze and S.B. Sastrawan, *J. Organometal. Chem.*, 1978, **154**, 233.
145. N.A.D Carey and H.C. Clark, *Can J. Chem.*, 1968, **46**, 643.
146. H.C. Clark, J.D. Cotton and J.H. Tsai, *Inorg. Chem.*, 1966, **5**, 1582.
147. M. Green, N. Mayne and F.G.A. Stone, *J. Chem. Soc. A*, 1968, 902.
148. A.C. Filippou, E.O. Fischer and H.G. Alt, *J. Organometal. Chem.*, 1987, **330**, 325.
149. R.K. Pomeroy and W.A.G. Graham, *J. Amer. Chem. Soc.*, 1972, **94**, 272.
150. J. Dalton, I. Paul and F.G.A. Stone, *J. Chem. Soc. A*, 1968, 1215.
151. J.G.A. Reuvers and J. Takats, *J. Organometal. Chem.*, 1979, **175**, C13.
152. J.M. Burtlitch and R.C. Winterton, *Inorg. Chem.*, 1979, **18**, 2309.
153. A.N. Nesmeyanov, K.N. Anisimov, N.E. Kolobova and V.V. Skripkin, *Izv. Akad. Nauk SSSR Ser. Khim.*, 1966, 1292.
154. N. Flitcroft, D.A. Harbourne, I. Paul, P.M. Tucker and F.G.A. Stone, *J. Chem. Soc. A*, 1966, 1130.
155. R.M. Sweet, C.J. Fritchie and R.A. Schunn, *Inorg. Chem.*, 1967, **6**, 749.
156. E.E. Isaacs and W.A.G. Graham, *J. Organometal. Chem.* 1975, **85**, 237.
157. T. Blackmoore, J.D. Cotton, M.I. Bruce and F.G.A. Stone, *J. Chem. Soc. A*, 1968, 2931.
158. P.F. Barrett and K.K.W. Sun, *Can. J. Chem.*, 1970, **48**, 3300.
159. J.D. Cotton, J. Duckworth, S.A.R. Knox, P. Lindley, I. Paul, F.G.A. Stone and P. Woodward, *Chem. Commun.*, 1966, 253.
160. A.B. Cornwell, P.G. Harrison and J.A. Richards, *J. Organometal. Chem.*, 1974, **76**, C26.
161. J.D. Cotton and G.A. Morris, *J. Organometal. Chem.*, 1978, **145**, 245.
162. C.V. Magatti and W.P. Giering, *J. Organometal. Chem.*, 1974, **73**, 85.
163. R.O. Gould, W.J. Sime and T.A. Stephenson, *J. Chem. Soc., Dalton Trans.*, 1978, 76.
164. G. Consiglio, F. Morandini, G. Ciani, A. Sironi and M. Kretschmer, *J. Amer. Chem. Soc.*, 1983, **105**, 1391.
165. J.D. Korp and I. Bernal, *Inorg. Chem.*, 1981, **20**, 4065.

166. R.C. Edmondson and M.J. Newlands, *Chem. Commun.*, 1968, 1219.
167. R.K. Pomeroy, M. Elder, D. Hall and W.A.G. Graham, *Chem. Commun.*, 1969, 381.
168. J.A.K. Howard, S.C. Kellett and P. Woodward, *J. Chem. Soc., Dalton Trans.*, 1975. 2332.
169. J.R. Moss and W.A.G. Graham, *J. Organometal. Chem.*, 1969, **18**, P24.
170. S.A.R. Knox and F.G.A. Stone, *J. Chem. Soc. A*, 1970, 3147.
171. T. Takano, *Bull. Chem. Soc. Jpn*, 1973, **46**, 522.
172. W.A.G. Graham and R. Kummer, *Inorg. Chem.*, 1968, **7**, 1208.
173. S.A.R. Knox and F.G.A. Stone, *J. Chem. Soc. A*, 1969, 2559.
174. R.D. George, S.A.R. Knox, and F.G.A. Stone, *J. Chem. Soc., Dalton Trans.*, 1973, 972.
175. J.D. Cotton, S.A.R. Knox and F.G.A. Stone, *J. Chem. Soc. A*, 1968, 2758.
176. S.F. Watkins, *J. Chem. Soc. A*, 1969, 1552.
177. J.D. Cotton, S.A.R. Knox, I. Paul and F.G.A. Stone, *J. Chem. Soc. A*, 1967, 264.
178. R.D. Adams and D.A. Katahira, *Organometallics*, 1982, **1**, 460.
179. R.B. King and F.G.A. Stone, *J. Amer. Chem. Soc.*, 1960, **83**, 3833.
180. S.D. Ibekwe and M.J. Newlands, *J. Chem. Soc. A*, 1967, 1783.
181. S.A. Keppie and M.F. Lappert, *J. Organometal. Chem.*, 1969, **19**, P5.
182. R.A. Burnham, M.A. Lyle and S.R. Stobart, *J. Organometal. Chem.*, 1977, **125**, 179.
183. E.J. Bulten and H.A. Budding, *J. Organometal. Chem.*, 1979, **166**, 339.
184. E.W. Abel and S. Moorehouse, *Inorg. Nucl. Chem. Lett.*, 1971, **7**, 905.
185. M. Casey and A.R. Manning, *Chem. Commun.*, 1970, 674; *J. Chem. Soc., A*, 1971, 256.
186. C.J. Cardin, D.J. Cardin, J.M. Power and M.B. Hursthouse, *J. Amer. Chem. Soc.*, 1985, **107**, 505.
187. A.B. Cornwell, P.G. Harrison and J.A. Richards, *J. Organometal. Chem.* 1976, **108**, 47.
188. P.G. Harrison, T.J. King and J.A. Richards, *J. Chem. Soc., Dalton Trans.*, 1975. 2097.
189. G.W. Grynkewich and T.J. Marks, *J. Chem. Soc., Dalton Trans.*, 1976, 972.
190. L. Vancea, R.K. Pomeroy and W.A.G. Graham, *J. Amer. Chem. Soc.*, 1976, **98**, 1407.
191. R.K. Pomeroy, L. Vancea, H.P. Calhoun and W.A.G. Graham, *Inorg. Chem.*, 1977, **16**, 1508.
192. W.R. Cullen, J.R. Sams and J.A.J. Thompson, *Inorg. Chem.*, 1971, **10**, 843.
193. A.N. Nesmeyanov, N.E. Kolobova, V.V. Skripkin, K.N. Anisimov and L.A. Fedorov, *Dokl. Akad. Nauk SSSR*, 1970, **195**, 368.
194. R.C. Edmondson, D.S. Field and M.J. Newlands, *Can J. Chem.*, 1971, **49**, 618.
195. C.J. Gilmore and P. Woodward, *J. Chem. Soc., Dalton Trans.*, 1972, 1387.
196. T.J. Marks and A.R. Newman, *J. Amer. Chem. Soc.*, 1973, **95**, 769.
197. J.P. Collman, D.W. Murphy, E.B. Fleischer and D. Swift, *Inorg. Chem.*, 1974, **13**, 1.
198. A.B. Cornwell and P.G. Harrison, *J. Chem. Soc., Dalton Trans.*, 1975, 2017.
199. J.-M. Barbe, R. Guilard, C. Lecompte and R. Gerdin, *Polyhedron*, 1984, **3**, 889.
200. D. Miholova and A.A. Vlcek, *Inorg. Chim. Acta*, 1983, **73**, 249.
201. B.A. Sosinsky, J. Shelly and R. Shong, *Inorg. Chem.*, 1981, **20**, 1370.
202. D. Miholova and A.A. Vlcek, *Inorg. Chim. Acta*, 1980, **43**, 43.
203. M.J. Ash, A. Brooks, S.A.R. Knox and F.G.A. Stone, *J. Chem. Soc. A*, 1971, 458.
204. T.J. McNeese, S.S. Wreford, D.L. Tipton and R. Bau, *Chem. Commun.*, 1977, 390.
205. T.J. Marks and G.W. Grynkewich, *J. Organometal. Chem.*, 1975, **91**, C9.
206. L.F. Wuyts, G.P. van der Kelen and Z. Eeckhaut, *J. Mol. Struct.* 1976, **33**, 107.
207. J.P. Collman, F.D. Vastine and W.R. Roper, *J. Amer. Chem. Soc.*, 1968, **90**, 2282.
208. T.M. Gilbert, F.J. Hollander and R.G. Bergman, *J. Amer. Chem. Soc.*, 1985, **107**, 3508.
209. M.A. Bennett and D.J. Patmore, *Inorg. Chem.*, 1971, **10**, 2387.
210. E. Uhlig, B. Hipler and P. Mueller, *Z. anorg. allg. Chem.*, 1978, **447**, 18.
211. W.A.G. Graham and D.J. Patmore, *Inorg. Chem.*, 1967, **6**, 981, 1879.
212. D.J. Patmore and W.A.G. Graham, *Inorg. Chem.*, 1966, **5**, 1405.
213. J.N. Crosby and R.D.W. Kemmitt, *J. Organometal. Chem.*, 1971, **26**, 277.
214. R. Kummer and W.A.G. Graham, *Inorg. Chem.*, 1968, **7**, 523.
215. P.F. Barrett, *Can. J. Chem.*, 1974, **52**, 3773.
216. P. Hackett and A.R. Manning, *J. Organometal. Chem.*, 1974, **66**, C17.
217. R.C. Taylor, J.F. Young and G. Wilkinson, *Inorg. Chem.*, 1966, **5**, 20.
218. J.F. Young, R.D. Gillard and G. Wilkinson, *J. Chem. Soc.*, 1964, 5176.
219. V. Garcia, M.A. Garralda and E. Zugasti, *J. Organometal. Chem.*, 1987, **322**, 249.
220. K. Dey and H. Werner, *J. Organometal. Chem.*, 1977, **137**, C28.
221. K.E. Schwarzhans, *Z. Naturforsch.*, 1979, **34B**, 1456.

222. D.J. Patmore and W.A.G Graham, *Inorg. Chem.*, 1966, **5**, 2222.
223. B. Cetinkaya, M.F. Lappert, J. McMeeking and D.E. Palmer, *J. Chem. Soc., Dalton Trans.*, 1973, 1202.
224. T. Blackmore, M.I. Bruce and F.G.A. Stone, *J. Chem. Soc. A*, 1971, 2376.
225. F. Glockling and J.G. Irwin, *Inorg. Chim. Acta*, 1972, **6**, 355.
226. R. Hill and S.A.R Knox, *J. Chem. Soc., Dalton Trans.*, 1975, 2622.
227. M.F. Lappert and N.F. Travers, *J. Chem. Soc. A*, 1970, 3303.
228. F. Glockling and G.C. Hill, *J. Chem. Soc. A*, 1971, 2137.
229. R.D. Adams, F.A. Cotton, W.R. Cullen, D.L. Hunter and L. Mihichuk, *Inorg. Chem.*, 1975, **14**, 1395.
230. D.L. Lichtenberger and T.L. Brown, *J. Amer. Chem. Soc.*, 1977, **99**, 8197.
231. J.P. Fawcett and J.F. Harrod, *Can. J. Chem.*, 1976, **54**, 3102.
232. G.K.I. Magomedov, G.V. Druzhkova, T.G. Basanina and V.I. Shiryaev, *J. Gen. Chem. USSR*, 1981, **51**, 2054.
233. M. Hawkins, P.B. Hitchcock and M.F. Lappert, *J. Chem. Soc., Chem. Commun.*, 1985, 1592.
234. H. Schumann and W. Feldt, *Z. anorg. allg. Chem.*, 1979, **458**, 257.
235. F. Bonati, S. Cenini, D. Morelli and R. Ugo, *J. Chem. Soc. A*, 1966, 1052.
236. H.F. Klein, K. Ellrich, D. Neugebauer, O. Orama and A. Krueger, *Z. Naturforsch.*, 1983, **B38**, 303.
237. A.D. Beveridge and H.C. Clark, *J. Organometal. Chem.*, 1968, **11**, 601.
238. M.A. Halabi and T.L. Brown, *J. Amer. Chem. Soc.*, 1977, **99**, 2982.
239. M. Grenz and W.W D́u Mont, *J. Organometal. Chem.*, 1983, **241**, C5.
240. S. Midollini, A. Orlandini and L. Sacconi, *J. Organometal. Chem.* 1978, **162**, 109.
241. E.W. Abel, S.A. Keppie, M.F. Lappert and S. Moorhouse, *J. Organometal. Chem.*, 1970, **22**, C31.
242. P.E. Garrou and G.E. Hartwell, *Chem. Commun.*, 1972, 881.
243. P.B. Hitchcock, M.F. Lappert and M.C. Misra, *J. Chem. Soc., Chem. Commun.*, 1985, 863.
244. G.K. Campbell, P.B. Hithcock, M.F. Lappert and M.C. Misra, *J. Organometal. Chem.*, 1985, **289**, C1.
245. V.V. Bashilov, V.I. Sokolov, Yu.L. Slovokhotov, Yu.T. Struchkov, E.G. Mednikov and N.K. Eremenko, *J. Organometal. Chem.*, 1987, **327**, 285.
246. M.M. Olmstead, L.S. Benner, H. Hope and A.L. Balch, *Inorg. Chim. Acta*, 1979, **32**, 193.
247. K.H.A. Ostoja-Starzewski, P.S. Pregosin and H. Ruegger, *Helv. Chim. Acta*, 1982, **63**, 785.
248. C. Arz, I.R. Herbert and P.S. Pregosin, *J. Organometal. Chem.*, 1986, **308**, 373.
249. G.E. Batley and J.C. Bailar, *Inorg. Nucl. Chem. Lett.*, 1968, **4**, 577.
250. S.P. Gubin, A.Z. Rubezhov, L.I. Voronchikhina and A.N. Nesmeyanov, *Bull. Acad. Sci. USSR Div. Chem. Sci.*, 1972, **21**, 1317.
251. R.J. Kane, *Phil. Mag.*, 1835, **7**, 399.
252. C.L. Wilson and D.W. Wilson, *Compr. Anal. Chem.*, 1959, **1A**, 338.
253. T.A.K. Al-Allaf, C. Eaborn, P.B. Hitchcock, M.F. Lappert and A. Pidcock, *J. Chem. Soc., Chem. Commun.*, 1985, 548.
254. M.F. Lappert and P.P. Power, *J. Chem. Soc., Dalton Trans.*, 1985, 51.
255. G.W. Bushnell, D.T. Eadie, A. Pidcock, A.R. Sam, R.D. Holmes-Smith and S.R. Stobart, *J. Amer. Chem. Soc.*, 1982, **104**, 5837.
256. C. Eaborn, A. Pidcock and B.R. Steele, *J. Chem. Soc., Dalton Trans.*, 1976, 767.
257. G. Butler, C. Eaborn and A. Pidcock, *J. Organometal. Chem.*, 1979, **181**, 47.
258. J. Kuyper, *Inorg. Chem.*, 1977, **16**, 2171.
259. C. Eaborn, K.J. Odell and A. Pidcock, *J. Chem. Soc., Dalton Trans.*, 1978, 1288.
260. M.C. Baird, *J. Inorg. Nucl. Chem.*, 1967, **29**, 367.
261. A.F. Clemmit and F. Glockling, *J. Chem. Soc. A*, 1971, 1164.
262. M. Akhtar and H.C. Clark, *J. Organometal. Chem.*, 1970, **22**, 233.
263. C. Eaborn, A. Pidcock and B.R. Steele, *J. Chem. Soc., Dalton Trans.*, 1985, 809.
264. H.C. Clark, A.B. Goel and C. Billard, *J. Organometal. Chem.*, 1979, **182**, 431.
265. F. Glockling and R.J.I. Pollock, *J. Chem. Soc., Dalton Trans.*, 1975, 497.
266. J.F. Almeida, K.R. Dixon, C. Eaborn, P.B. Hitchcock, A. Pidcock and J. Vinaixa, *J. Chem. Soc., Chem. Commun.*, 1982, 1315.
267. A. Christofides, M. Ciriano, J.L. Spencer and F.G.A. Stone, *J. Organometal. Chem.*, 1979, **178**, 273.

268. V.I. Sokolov, V.V. Bashilov, O.A. Reutov, M.M. Bochkarev, L.P. Maijorova and G.A. Razuvaev, *J. Organometal. Chem.*, 1976, **112**, C47.
269. H. Weichmann, *J. Organometal. Chem.*, 1982, **238**, C49.
270. J.C. Bailar and H. Itatani, *Inorg. Chem.*, 1965, **4**, 1618.
271. J.V. Kingston and G.R. Scollary, *J. Chem. Soc. A*, 1971, 3765.
272. R.V. Lindsey, G.W. Parshall and U.G. Stollberg, *Inorg. Chem.*, 1966, **5**, 109.
273. L.J. Guggenberger, *Chem. Commun.*, 1968, 512.
274. S. Moreto and P.M. Maitlis, *J. Chem. Soc., Dalton Trans.*, 1980, 1368.
275. F. Glockling and P.J.M.L. Ssebuwufu, *Inorg. Chim. Acta*, 1978, **31**, 105.
276. E. Piers and J.M. Chong, *J. Chem. Soc., Chem. Commun.*, 1983, 934.
277. E. Piers, J.M. Chong and H.E. Morton, *Tetrahedron Lett.* 1981, **22**, 4905.
278. D.D Dilts and M.P. Johnson, *Inorg. Chem.*, 1966, **5**, 2079.
279. F. Glockling and M.D. Wilby, *J. Chem. Soc. A*, 1968, 2168.
280. W. Clegg, *Acta Crystallogr.*, 1978, **B34**, 278.
281. A.T.T. Hsieh, J.D. Ruddick and G. Wilkinson, *J. Chem. Soc., Dalton Trans.*, 1972, 1966.
282. L. Malatesta, L. Naldini, G. Simonetta and F. Cariati, *Coord. Chem. Rev.*, 1966, **1**, 255.
283. W.P. Neumann and A. Blaukat, *Angew. Chem. Int. Edn. Engl.*, 1969, **8**, 610.
284. B.V. Fedot'ev, O.A. Kruglaya and N.S. Vyazankin, *Bull. Acad. Sci. USSR Div. Chem. Sci.*, 1974, 713.
285. F.J.A.D. Tombe, G.J.M. van der Kerk and J.G. Noltes, *J. Organometal. Chem.*, 1972, **43**, 323.
286. F.J.A.D. Tombe, G.J.M. van der Kerk and J.G. Noltes, *J. Organometal. Chem.*, 1972, **44**, 247.
287. J.J. Habeeb, A. Osnan and D.G. Tuck, *Inorg. Chim. Acta*, 1979, **35**, 105.
288. M.N. Bochkarev, S.P. Korneva, L.P. Maiorova, V.A. Kuzetsov and N.S. Vyazankin, *J. Gen. Chem. USSR*, 1974, **44**, 293.
289. U. Blaukat and W.P. Neumann, *J. Organometal. Chem.*, 1973, **63**, 27.
290. O.A. Kruglaya, G.S. Kalinina, B.I. Petrov and N.S. Vyazankin, *J. Organometal. Chem.*, 1972, **46**, 51.
291. M.N. Bochkarev, N.S. Vyazankin, L.P. Maiorova and G.A. Razuvaev, *J. Gen. Chem. USSR*, 1978, **48**, 2454.
292. O.A. Kruglaya, B.V. Fedot'ev, I.B. Fedot'eva and N.S. Vyazankin, *J. Gen. Chem. USSR*, 1976, **46**, 1482.
293. J.D. Kennedy, W. McFarlane and B. Wrackmeyer, *Inorg. Chem.*, 1976, **15**, 1299.
294. H. Noth and R. Schwerthoeffer, *Chem. Ber.*, 1981, **114**, 3056.
295. H. Noth, H. Schaeffer and G. Schmidt, *Angew. Chem. Int. Edn. Engl.*, 1969, **8**, 515.
296. J.J. Eméleus and S.F.A. Kettle, *J. Chem. Soc.*, 1958, 2189.
297. R.E. Loffredo and A.D. Norman, *J. Amer. Chem. Soc.*, 1971, **93**, 5587.
298. L. Roesch and W. Erb, *Angew. Chem. Int. Edn. Engl.*, 1978, **17**, 604.
299. M. Veith and W. Franks, *Angew. Chem. Int. Edn. Engl.*, 1985, **24**, 223.
300. M. Trost and J.W. Herndon, *J. Amer. Chem. Soc.*, 1984, **106**, 6835.
301. G.S. Kalinina, E.A. Shchupak, O.A. Kruglaya, N.S. Vyazankin. *Bull. Acad. Sci. USSR Div. Chem. Sci.*, 1973, 1154.
302. G.S. Kalinina, E.A. Shchupak, N.S. Vyazankin and G.A. Razuvaev, *Bull. Acad. Sci. USSR Div. Chem. Sci.*, 1976, 1289.
303. N.S. Vyazankin, O.A. Kruglaya, G.A. Razuvaev and G.S. Semchikova, *Dokl. Chem. Engl. Trans.*, 1966, **166**, 8.
304. S. Roller, M. Draeger, H.J. Breunig, M. Ates and S. Gulec, *J. Organometal. Chem.*, 1987, **329**, 319.
305. N.S. Vyazankin, G.A. Razuvaev, E.N. Gladyshev and S.P. Korneva, *J. Organometal. Chem.*, 1967, **7**, 353.
306. E. Amberger and E. Muelhofer, *J. Organometal. Chem.*, 1968, **12**, 55.
307. N.S. Vyazankin, E.N. Gladyshev, S.P. Korneva, G.A. Razuvaev and E.A. Archangelskaya, *J. Gen. Chem. USSR*, 1968, **38**, 1757.
308. C. Tamborski, F.E. Ford and E.J. Sokolski, *J. Org. Chem.*, 1963, **28**, 181.
309. W.P. Neumann, B. Schneider and R. Sommer, *Annalen*, 1966, **692**, 1.
310. H.M.J.C. Creemers and J.G. Noltes, *J. Organometal. Chem.*, 1967, **7**, 237.
311. S. Adams, M. Draeger, *J. Organometal. Chem.*, 1987, **323**, 11.
312. P. Rivière, J. Satgé and A. Boy, *J. Organometal Chem.*, 1975, **96**, 25.
313. J. Barrau and J. Satgé, *J. Organometal. Chem.*, 1978, **148**, C9.

314. C.A. Kraus and H. Eatough, *J. Amer. Chem. Soc.*, 1933, **55**, 5008.
315. H. Gilman and C.W. Gerow, *J. Org. Chem.*, 1957, **22**, 334.
316. E. Wiberg, E. Amberger and H. Cambensi, *Z. anorg. allg. Chem.*, 1966, **351**, 164.
317. C. Lowig, *Annalen*, 1852, **84**, 308.
318. C.A. Kraus and R.H. Bullard, *J. Amer. Chem. Soc.*, 1926, **48**, 2131.
319. B. Jousseaume, E. Chanson and M. Pereyre, *Organometallics*, 1986, **5**, 1271.
320. H.M.J.C. Creemers and J.G. Noltes, *Rec. Trav. Chim.*, 1965, **84**, 382.
321. M. Devaud, M. Engel, C. Feasson and J.L. Lecat, *J. Organometal. Chem.*, 1985, **281**, 181.
322. G.A. Mazzocchim, R. Seeber and G. Bontempelli, *J. Organometal. Chem.*, 1976, **121**, 55.
323. M. Gielen and Y. Tondeur, *Chem. Commun.*, 1978, 81.
324. S. Adams, M. Draeger and B. Mathiasch, *J. Organometal. Chem.*, 1987, **326**, 173.
325. W.P. Neumann, K. Konig and G. Burkhardt, *Annalen*, 1964, **677**, 18.
326. R. Sommer, B. Schneider and W.P. Neumann, *Annalen*, 1966, **692**, 12.
327. L.C. Williamsens and G.J.M. van der Kerk, *J. Organometal. Chem.*, 1964, **2**, 260.
328. W.P. Neumann, J. Pedain and R. Sommer, *Annalen*, 1966, **694**, 9.
329. H. Puff, C. Bach, W. Schuh and R. Zimmer, *J. Organometal. Chem.*, 1986, **312**, 313.
330. W.V. Farrer and H.A. Skinner, *J. Organometal. Chem.*, 1964, **1**, 434.
331. W.P. Neumann and K. Konig, *Annalen*, 1964, **667**, 1.
332. B. Watta, W.P. Neumann and J. Sauer, *Organometallics*, 1985, **4**, 1954.
333. W.P. Neumann and J. Pedain, *Annalen*, 1964, **672**, 34.
334. H.U. Buschaus and W.P. Neumann, *Angew. Chem. Int. Edn. Engl.*, 1978, **17**, 59.
335. G.A. Razuvaev, N.S. Vyazankin and O.A. Shchepetkova, *J. Gen. Chem. USSR*, 1961, **31**, 3515.
336. K. Sisido, S. Kozima and T. Isibasi, *J. Organometal. Chem.*, 1967, **10**, 439.
337. M.C.R. Symons, *J. Chem. Soc., Chem. Commun.*, 1981, 1251.
338. M. Lehnig, W.P. Neumann and P. Seifert, *J. Organometal. Chem.*, 1978, **162**, 145.
339. Yu.A. Aleksandrov and B.A. Radbil, *J. Gen. Chem. USSR*, 1966, **36**, 562.
340. C.A. Kraus and W.V. Sessions, *J. Amer. Chem. Soc.*, 1925, **47**, 2361.
341. N.S. Vyazankin and V.T. Bychkov, *J. Gen. Chem. USSR*, 1966, **36**, 1681.
342. G. Tagliavini, S. Faleschini, G. Pilloni and G. Plazzogna, *J. Organometal. Chem.*, 1966, **5**, 136.
343. S. Adams and M. Draeger, *J. Organometal. Chem.*, 1985, **288**, 295.
344. T. Birchall and J.P. Johnson, *Inorg. Chem.*, 1982, **21**, 3724.
345. T. Birchall and J.P. Johnson, *Can. J. Chem.*, 1979, **57**, 160.
346. D.C. McWilliam and P.R. Wells, *J. Organometal. Chem.*, 1975, **85**, 335, 347.
347. D.P. Arnold and P.R. Wells, *J. Organometal. Chem.*, 1976, **108**, 345.
348. O.A. Reutov, V.I. Rozenberg, V.A. Nikanovov and G.V. Gavrilov, *Dokl. Chem.*, 1977, **237**, 690.
349. A.B. Cornwell, P.G. Harrison and J.A. Richards, *J. Organometal, Chem.*, 1977, **140**, 273.
350. U. Kunze and L. Steinmann, *J. Organometal. Chem.*, 1978, **150**, 39.
351. M.A.A. Beg and H.C. Clark, *Chem. Ind.*, 1962, 140.
352. D. Seyferth and K.Y. Darragh, *J. Organometal. Chem.*, 1968, **11**, P9.
353. H.C. Clark and C.J. Willis, *J. Amer. Chem. Soc.*, 1960, **82**, 1888.
354. H. Killing and T.N. Mitchell, *Organometallics*, 1984, **3**, 1318.
355. E.J. Bulten, H.A. Budding and J.G. Noltes, *J. Organometal. Chem.*, 1970, **22**, C5.
356. B. Mathiasch, *J. Organometal. Chem.*, 1976, **122**, 345.
357. K. Kobayashi, M. Kawanisi, S. Kozima, T. Hitomi, H. Iwamura and T. Sugawara, *J. Organometal. Chem.*, 1981, **217**, 315.
358. R.H. Summerville and R. Hoffman, *J. Amer. Chem. Soc.*, 1979, **101**, 3821.
359. S. Carr, R. Colton and D. Dakternieks, *J. Organometal. Chem.*, 1983, **249**, 327.
360. H. Moriyama, T. Aoki, S. Shinoda and Y. Saito, *Chem. Commun.*, 1982, **500**,
361. G. Butler, C. Eaborn and A. Pidcock, *J. Organometal. Chem.*, 1980, **185**, 367.
362. N.W. Alcock and J.H. Nelson, *J. Chem. Soc., Dalton Trans.*, 1982, 2415.
363. M.J. Mays and P.L. Sears, *J. Chem. Soc., Dalton Trans.*, 1974, 2254.
364. J. Thompson and M.C. Baird, *Can. J. Chem.*, 1973, **51**, 1179.
365. A. Tzschach, K. Jurkschat, M. Sheer, J. Meunier-Piret and M. van Meerssche, *J. Organometal. Chem.*, 1983, **259**, 165.
366. S. Adams and M. Draeger, *Proc. Vth Int. Conf. on Ge, Sn, Pb (Padova)*, 1986, P5.

9 Radical chemistry of tin

A.G. DAVIES

9.1 Introduction

Free radical reactions of organotin compounds, particularly those which involve intermediate neutral trialkyltin radicals, $R_3Sn^{\cdot}$, have attracted a lot of attention in recent years. The important types of homolytic reactions are:

(i) The formation and reactions of tin-centred radicals, $R_n\dot{Sn}X_{3-n}$
(ii) Homolytic reactions which occur at a ligand in an organotin compound
(iii) Bimolecular homolytic substitution at the tin centre in tin(IV) compounds, R_nSnX_{4-n}
(iv) Electron transfer reactions to generate the radical cations, $R_nSnX_{4-n}^{\cdot +}$, or anions, $R_nSnX_{4-n}^{\cdot -}$.

A brief review of the field was published in 1976,[1] and specific aspects, which are referred to later, have been reviewed elsewhere. A major development has involved the use of trialkyltin radicals as reactive intermediates in organic synthesis. This chemistry has been thoroughly covered in several reviews,[2–6] and will not be considered in detail here.

9.2 Organotin radicals

9.2.1 *Formation of organotin radicals*

The principal routes by which organotin radicals can be generated are summarized in Scheme 1.

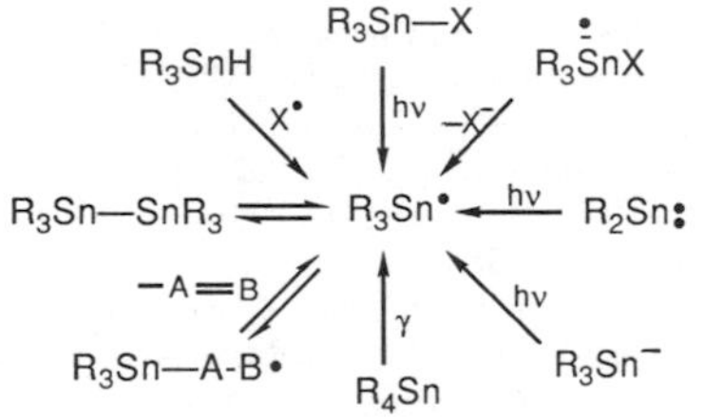

$$R_3SnH \xrightarrow{X^{\bullet}} R_3Sn^{\bullet}$$
$$R_3Sn{-}X \xrightarrow{h\nu} R_3Sn^{\bullet}$$
$$R_3\dot{\bar{S}}nX \xrightarrow{-X^-} R_3Sn^{\bullet}$$
$$R_3Sn{-}SnR_3 \rightleftharpoons R_3Sn^{\bullet}$$
$$R_2Sn{:} \xrightarrow{h\nu} R_3Sn^{\bullet}$$
$$R_3Sn{-}A{-}B^{\bullet} \underset{}{\overset{-A=B}{\rightleftharpoons}} R_3Sn^{\bullet}$$
$$R_4Sn \xrightarrow{\gamma} R_3Sn^{\bullet}$$
$$R_3Sn^- \xrightarrow{h\nu} R_3Sn^{\bullet}$$

Scheme 1

For both physical studies and synthetic applications, trialkyltin hydrides provide the most common source of $R_3Sn^{\cdot}$ radicals. For physical studies, the hydrogen can conveniently be abstracted by *t*-butoxyl radicals generated by photolysis of di-*t*-butyl peroxide. For example, if a solution of tri-*n*-butyltin hydride and di-*t*-butyl peroxide in pentane is photolysed in the cavity of an esr spectrometer, the spectrum of the $Bu_3Sn^{\cdot}$ radical can be observed (eqns 9.1, 9.2)[7]:

$$Bu^tOOBu^t \xrightarrow{h\nu} 2Bu^tO^{\cdot} \tag{9.1}$$

$$Bu^tO^{\cdot} + HSnBu_3 \longrightarrow Bu^tOH + {}^{\cdot}SnBu_3 \tag{9.2}$$

If the *t*-butoxyl radicals are generated by laser flash photolysis, the formation of the $R_3Sn^{\cdot}$ radical can be monitored by optical spectroscopy to give values for the rate constants for reaction (9.2) in the narrow range of $2–4 \times 10^8\,l\,mol^{-1}\,s^{-1}$, depending on the nature of R, at 22°C in di-*t*-butyl peroxide as solvent.[8,9] Alternatively, the hydrogen can be abstracted by a ketone which is photoexcited to the triplet state.[10] For example, reaction (9.3) takes place[8] with a rate constant of $2.9 \times 10^8\,l\,mol^{-1}\,s^{-1}$.

$$Ph_2CO^T + HSnBu_3 \rightarrow Ph_2\dot{C}OH + {}^{\cdot}SnBu_3 \tag{9.3}$$

When organotin hydrides are used in organic synthesis, the reactions may occur spontaneously, or may be initiated by light or by a reagent such as azoisobutyronitrile. The chain carriers[4] are the organotin radical and frequently an alkyl radical $R'^{\cdot}$. The propagation step (eqn 9.4) then has a rate constant of about $1 \times 10^6\,l\,mol^{-1}\,s^{-1}$ at room temperature, and shows an isotope effect[11–13] of k_H/k_D *c.* 2.

$$R'^{\cdot} + HSnBu_3 \rightarrow R'H + {}^{\cdot}SnBu_3 \tag{9.4}$$

Phenyl radicals are much more reactive, with a rate constant of $5.9 \times 10^8\,l\,mol^{-1}\,s^{-1}$.

Trialkyl- or triaryl-tin radicals can also be formed by the homolysis of the Sn–Sn bond in a distannane (eqn 9.5). For simple hexa-alkyl- or hexa-aryl-ditins, the reaction can be brought about photolytically; the quantum yield can be improved if di-*t*-butyl peroxide is added to the system, when the bimolecular reaction (eqn 9.6) is superimposed on the unimolecular homolysis (eqn 9.5).[14]

$$R_3Sn\text{–}SnR_3 \rightarrow 2R_3Sn^{\cdot} \tag{9.5}$$

$$Bu^tO^{\cdot} + R_3Sn\text{–}SnR_3 \rightarrow Bu^tOSnR_3 + {}^{\cdot}SnR_3 \tag{9.6}$$

However, hexa-alkylditins are normally much less sensitive to homolytic attack than are the trialkyltin hydrides. Their reactions involve little chain character, and they are used more in physical studies than in organic synthesis.

If the distannanes carry bulky ligands, these can sterically weaken the Sn–Sn bond so that reversible thermal dissociation may occur without other decomposition.[15] Table 9.1 shows a correlation between the structure of the distannane and the Sn–Sn bond vibration frequency and dissociation energy, and the lowest temperature at which the esr spectrum of the $R_3Sn^{\cdot}$ radical can be detected.[16] Some other types of organotin

Table 9.1 Reversible dissociation of R_3SnSnR_3

R	ν(Sn–Sn) (cm^{-1})	ΔH_{diss}(Sn–Sn) (kJ mol^{-1})	T (°C)
Me	200	276 ± 16	—
1,3,5-$Me_3C_6H_2$	102	205 ± 8	180
1,3,5-$Et_3C_6H_2$	92	111 ± 8	100
1,3,5-$Pr^i_3C_6H_2$		35.5 ± 4	20

compounds $R_3Sn–X$, where the Sn–X bond is weakened sterically or electronically, can similarly yield the radicals $R_3Sn^{\cdot}$ on photolysis. Examples are given by $X = NPr^i_2$, OAr, CH_2Ph,[17] and particularly C_5H_5.[18–20] The ease of preparation and the general photosensitivity of the cyclopentadienyltin compounds makes them convenient sources of a series of radicals with the composition R_nSnX_{3-n} ($X = Cl$ or C_5H_5) (eqn 9.7), so that the effect of ligands R and X on the properties of the radicals can be studied.

$$R_nX_{3-n}Sn{-}C_5H_5 \xrightarrow{h\nu} R_nX_{3-n}Sn^{\cdot} + C_5H_5^{\cdot} \quad (9.7)$$

Sonolysis provides an attractive alternative to photolysis. Janzen has studied the ultrasonic decomposition of organotin compounds in benzene, trapping organic radicals with nitrosodurene.[21] For example, hexabutylditin in benzene yielded both butyl radicals and tributyltin radicals which were detected as shown in eqns (9.8) and (9.9):

$$Bu^{\cdot} + Me_4C_6H{-}N{=}O \longrightarrow Me_4C_6H{-}N(Bu){-}O^{\cdot} \quad (9.8)$$

$$Bu_3Sn^{\cdot} + C_6H_6 \longrightarrow Bu_3Sn{-}C_6H_6^{\cdot} \quad (9.9)$$

γ-Irradiation has been used for generating organotin radicals (and radical anions and cations) in either a host matrix or a self-matrix. For example, irradiation of tetramethyltin in an adamantane matrix at 77 K gave an isotropic ten-line spectrum with binomial intensity ratios,[22] and esr parameters in agreement with those obtained for the $Me_3Sn^{\cdot}$ radical in fluid solution.[23]

Photoinduced electron transfer from a triphenylstannyl anion to a solvent of furan or methylfuran, or a hydrocarbon such as anthracene has been used to generate triphenyltin radicals (e.g. eqn 9.10):[24–26]

$$Ph_3Sn^-M^+ + ArH \rightarrow Ph_3Sn^{\cdot} + M^+ + ArH^{\cdot -} \quad (9.10)$$

Similar radicals can be prepared by photoinduced electron transfer from an electron-rich alkene to a triaryltin chloride in toluene at low temperature (eqn 9.11).[27] A similar mechanism is presumably involved in the reaction of triaryltin halides with alkali metals to give the corresponding hexa-arylditins.

$$Ar_3SnCl + \text{(MeN)}_2C_2H_4C{=}CC_2H_4\text{(NMe)}_2 \xrightarrow{h\nu} Ar_3SnCl^{\cdot -} + [\text{(MeN)}_2C_2H_4C{=}CC_2H_4\text{(NMe)}_2]^{\cdot +} \quad (9.11)$$

$$Ar_3SnCl^{\cdot -} \rightarrow Ar_3Sn^{\cdot} + Cl^-$$

($Ar = 2,4,6\text{-}Me_3C_6H_2$)

Most organotin(II) compounds, R_2Sn:, except biscyclopentadienyltin(II), rapidly form oligomers $(R_2Sn)_n$, but where R is a very bulky alkyl or aminyl group, a few examples are known where the stannylenes exist in solution as stable monomers. Photolysis of these solutions then generates the corresponding persistent radicals, $R_3Sn^\cdot$; one example is shown in eqn (9.12).[28,29]

$$[(Me_3Si)_2CH]_2Sn: \xrightarrow{h\nu} [(Me_3Si)_2CH]_3Sn^\cdot \qquad (9.12)$$

The final route to organotin (III) radicals, listed in Scheme 1, involves β-scission of a radical carrying an organotin substituent at the β-position. For example, the bistrimethyltin derivative of benzpinacol (**1**) is in equilibrium at room temperature with the ketyl radicals (**2**) which at 60° dissociate into benzophenone and trimethyltin radicals (eqn 9.13).[30]

$$\underset{(\mathbf{1})}{Me_3SnOCPh_2CPh_2OSnMe_3} \rightleftharpoons \underset{(\mathbf{2})}{2Me_3SnO\dot{C}Ph_2} \rightleftharpoons 2Me_3Sn^\cdot + O{=}CPh_2 \qquad (9.13)$$

A second example is shown in eqn (9.14).[31] When di-*t*-butyl peroxide is photolysed in the presence of trimethylisobutyltin, hydrogen is abstracted preferentially from the β-position of the isobutyl group, and the β-stannylalkyl radical (**3**) then undergoes β-scission to yield the trimethyltin radical, which can be observed by esr spectroscopy directly, or through its reaction with ethene or an alkyl bromide.

$$Me_2CHCH_2SnMe_3 \xrightarrow{Me_3CO} \underset{(\mathbf{3})}{Me_2\dot{C}CH_2SnMe_3} \rightarrow Me_2C{=}CH_2 + {}^\cdot SnMe_3 \qquad (9.14)$$

Table 9.2 Esr spectra of tin (III) radicals.

Radical	T(K)	a(H)/mT	$a(^{119}Sn)$/mT	g	Ref.
$H_3Sn^\cdot$	4	2.6(3H)		2.017	33
$Me_3Sn^\cdot$	193	0.31(9H)	161.1	2.0163	7, 23
$Me_2\dot{S}nCl$	77			2.0113	22
$Me\dot{S}nCl_2$	77			2.0009	22
$Cl_3Sn^\cdot$	77			1.9974	22
$Me_3Sn\dot{S}nMe_2$			117.1 (and 25)		34
$Et_3Sn^\cdot$	193			2.015	7
$Pr_3Sn^\cdot$	193	0.30(6H)		2.0160	7
$Bu_3Sn^\cdot$	193	0.31(6H)		2.0160	7
$Ph_3Sn^\cdot$	223		186.6	2.0023	7
$Ph_2MeSn^\cdot$	203	0.30(3H)		2.0082	7
$PhMe_2Sn^\cdot$	198	0.30(6H)		2.0124	7
$Ph_2EtSn^\cdot$	213			2.0125	7
$(PhCMe_2CH_2)_3Sn^\cdot$	293	0.31(6H)	138.0	2.0150	35
$(2,3,5\text{-}Me_3C_6H_2)_3Sn^\cdot$	453			2.0073	29, 35
$(2,3,5\text{-}Et_3C_6H_2)_3Sn^\cdot$	373			2.0076	16, 35
$(2,3,5\text{-}Pr^i_3C_6H_2)_3Sn^\cdot$	483		167.8	2.0078	7, 35
$[(Me_3Si)_2CH]_3Sn^\cdot$		0.21(3H)	177.6	2.0094	
$[(Me_3Si)_2N]_3Sn^\cdot$		1.09(3H)	342.6	1.9912	29
$[(Me_3C)(Me_3Si)N]_3Sn^\cdot$		1.27(3H)		1.9928	28
$[(Me_3Ge)_2N]_3Sn^\cdot$		1.07(3H)		1.9924	28
$[(Et_3Ge)_2N]_3Sn^\cdot$		1.19(3H)		1.9939	28

9.2.2 *Physical properties of organotin radicals*

The first observations of tin-centred radicals by esr spectroscopy were on samples in the solid state, but more recently many spectra have been recorded on fluid solutions. It is an advantage to use high microwave power, when the spectra of interfering organic radicals are saturated, but those of the tin-centred radicals are enhanced. Spectral parameters are listed in Table 9.2.[32] The high values of the hyperfine coupling to $^{117/119}Sn$ imply that the tin-centred radicals have a pyramidal structure (**4**), with the unpaired electron in an orbital with a substantial degree of *s* character and a direct interaction with the tin nucleus.[22] Application of Kaptein's rules in the ^{1}H-CIDNP spectra of $PhCH_2SnMe_3$ formed from freely diffusing benzyl and trimethyltin radicals, leads to the conclusion that the sign of $a(^{117,119}Sn)$ in $Me_3\dot{S}n$ is negative.[36] Radicals centred on silicon or germanium have a similar pyramidal structure, but carbon-centred radicals have a planar or near-planar structure (**5**). MNDO calculations parameterized for tin lead to a similar pyramidal structure for $Me_3Sn^{\cdot}$, and it has been suggested that the controlling factor is the σ-conjugative interactions which are greatest when all the atomic orbitals have the same sp^3 hybridization.[37]

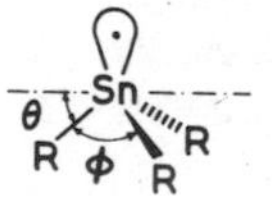

(**4**) 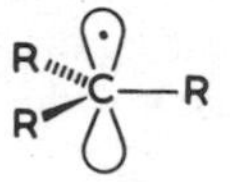 (**5**)

The ^{119}Sn hyperfine coupling constants for the other radicals in Table 9.2 imply that these radicals, even when the ligands are bulky substituted aryl groups, are similarly non-planar. This structure is confirmed by the spectrum of the trineophyltin radical $(PhCMe_2CH_2)_3Sn^{\cdot}$, which shows an alternating line width effect indicating hindered rotation about the Sn–C bonds.[35] It is concluded that the radical is configurationally stable, with an angle θ of 14.0°. Again, reactions of optically active tin compounds which proceed through an intermediate tin radical can retain asymmetry at the tin (e.g. eqn 9.15).[38,39]

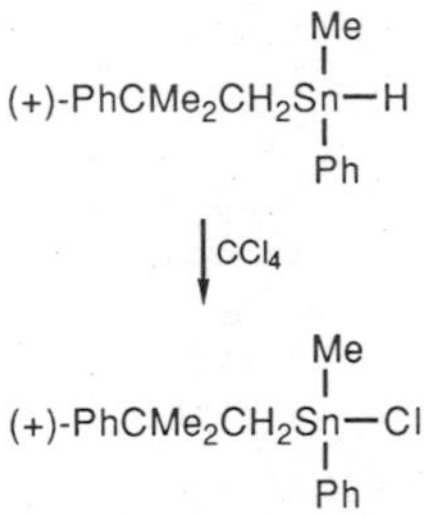

$$(+)\text{-}PhCMe_2CH_2Sn(Me)(Ph)\text{-}H \xrightarrow{CCl_4} (+)\text{-}PhCMe_2CH_2Sn(Me)(Ph)\text{-}Cl \qquad (9.15)$$

Hyperfine coupling to β-CH in the radicals $(CH_3)_3M^{\cdot}$ decreases in the sequence M = C (2.27 mT) > Si (0.63 mT) > Ge (0.55 mT) > Sn (0.31 mT). Hyperfine coupling to aryl protons decreases in the same sequence: M = C [a(Hp) 0.57 mT] > Si (0.12 mT) > Ge (0.093 mT) > Sn(< 0.05 mT), implying that the unpaired electron remains centred largely on the tin atom, and this is confirmed by the similar (0.30–0.31 mT) values of $a(CH_3)$ in the radicals $Me_3Sn^{\cdot}$, $PhMe_2Sn^{\cdot}$ and $Ph_2MeSn^{\cdot}$. In contrast, in the radicals $Me_3C^{\cdot}$, $PhMe_2C^{\cdot}$. In contrast, the values of $a(CH_3)$ are 2.27, 1.65 and 1.56 mT, respectively.

Optical spectroscopy has been used for characterizing organotin radicals and monitoring their reactions in solution.[40] Thus the radical $Bu_3Sn^{\cdot}$ shows a maximum absorption at 400 nm and the radical $Ph_3Sn^{\cdot}$ shows a maximum at 325 nm.

9.2.3 *Reactions of organotin radicals*

The principal reactions of organotin radicals are summarized in Scheme 2.

$R_3\overset{*}{Sn}:$ $R_3SnOO^{\bullet}$ $R'^{\bullet}$ (X = halogen, SR", SeR", TeR", NO_2, etc.)

$R_3\overset{*}{Sn}H$ O_2 R'X

$R_3Sn{-}SnR_3 \rightleftharpoons R_3Sn^{\bullet} \xrightarrow{C=C} R_3SnC{-}C^{\bullet}$

X=C< R₃Sn-allyl

$R_3Sn{-}X{-}C^{\bullet}$ R_3Sn-allyl

(X = O, S, Ne)

Scheme 2

We will consider here the principles underlying these reactions, rather than their extensive applications in organic synthesis.

The reversible association of sterically hindered triaryltin radicals to give hexa-arylditins has been referred to above (Table 9.1). Simple trialkyltin radicals undergo self-reaction to give the corresponding ditins at rates close to diffusion control ($2k_t = 1{-}3 \times 10^9\,l\,mol^{-1}\,s^{-1}$).[11,23] A ^{119}Sn CIDNP effect has been observed in the $Me_3SnSnMe_3$ which is formed.[36]

Simple $R_3Sn^{\cdot}$ radicals react rapidly with their parent hydrides by the identity reaction (9.16).

$$R_3Sn^{\cdot} + H{-}SnR_3 \rightleftharpoons R_3SnH + {}^{\cdot}SnR_3 \qquad (9.16)$$

which can lead to line broadening in the nmr spectrum of R_3SnH, and line broadening and exchange narrowing in the esr spectrum of $R_3Sn^{\cdot}$, and the rate constant for reaction has been estimated[41] to be $10^6{-}10^8\,l\,mol^{-1}\,s^{-1}$. The high rate of the reaction has been ascribed to the length of the Sn–H bond, which reduces the repulsion between the R_3Sn groups in the transition state $[R_3Sn\cdots H\cdots SnR_3]^{\cdot}$.

Triorganotin radicals react rapidly[42] with oxygen to give the corresponding peroxyl radicals, $R_3SnO_2^{\cdot}$. For $Bu_3Sn^{\cdot}$, the rate constant is $(7.5 \pm 1.4) \times 10^9\,l\,mol^{-1}\,s^{-1}$, and it has been suggested that the spin selection factor of 1/3 for the reaction doublet + triplet → doublet, may be overcome by the effect of the heavy tin atom. Whereas *t*-alkylperoxyl radicals ($ROO^{\cdot}$) are in equilibrium at low temperature with the corresponding tetroxides (ROOOOR), the stannylperoxyl radicals are long-lived at low temperature. In the radicals $R_3SiOO^{\cdot}$ and $R_3GeOO^{\cdot}$, ^{17}O studies show that the two oxygen atoms are non-equivalent but, remarkably, ^{17}O labelled (I 5/2) $Bu_3SnOO^{\cdot}$ shows only a single sextet with $a(^{17}O)$ 2.5 mT, g 2.0265.[43,44] This implies that the two oxygen atoms are equivalently bound to the tin in a structure such as (**6**).

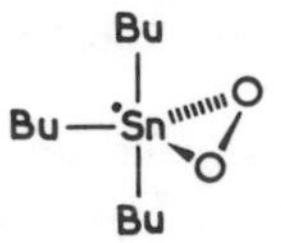

Bu
|
Bu—$\dot{Sn}$(O—O)
|
Bu
(6)

The observation in 1957[45,46] that alkyl halides were reduced to propene by trialkyltin hydrides was the first example of a reaction which has since achieved great importance in synthetic and mechanistic organic chemistry. The reaction shows all the usual characteristics of a radical chain process, and is well-established to involve an S_H2 reaction by a tin radical at the halogen centre to displace an organic radical (eqn 9.17), which abstracts hydrogen from the tin hydride to regenerate the tin radical (eqn 9.18).[4,47–50]

$$R_3Sn^{\cdot} + XR' \rightarrow R_3SnX + R'^{\cdot} \qquad (9.17)$$

$$R'^{\cdot} + HSnR_3 \rightarrow R'H + {}^{\cdot}SnR_3 \qquad (9.18)$$

Reaction (9.18) has been discussed above. The most direct evidence for reaction (9.17) is provided by the fact that esr spectra of the radicals $R'^{\cdot}$ can be observed when a distannane such as hexabutylditin is photolysed in the presence of an alkyl halide (eqn 9.5, 9.17).[14,23,51] Stronger spectra are obtained when di-*t*-butyl peroxide is present, because of the occurrence of reaction (9.6). Some typical reactions which emphasise the homolytic mechanism are shown in eqns (9.19)–(9.21).

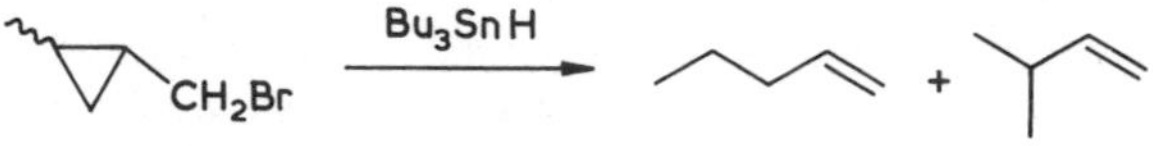

CH_2Br $\xrightarrow{Bu_3SnH}$ + (9.19)[52]

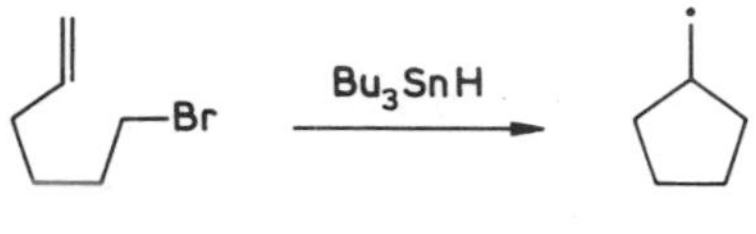

Br $\xrightarrow{Bu_3SnH}$ (9.20)[53]

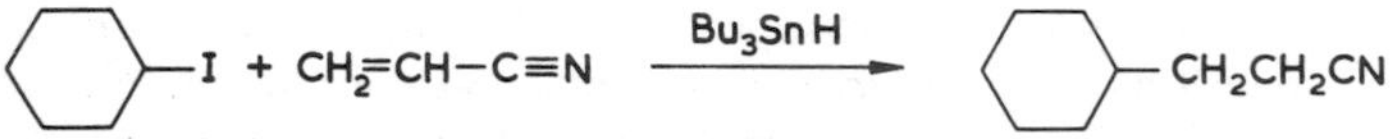

I + $CH_2{=}CH{-}C{\equiv}N$ $\xrightarrow{Bu_3SnH}$ CH_2CH_2CN (9.21)[54]

The presence of chloro or cyclopentadienyl ligands on the tin reduces the reactivity of the tin-centred radicals in reaction (9.17).[19,20] A few measurements have been made of the absolute rate constants for the halogen abstraction reaction (9.17), and many of the relative values. A selection of the data is given in Table 9.3.[55] Reactivity follows the sequences RCl < RBr < RI and *p*-RX < *s*-RX < *t*-RX, reflecting the relative stabilities of the alkyl radicals.

Thiols and thioethers are cleaved by trialkyltin hydrides by a mechanism involving bimolecular substitution at the heteroatom X (eqn 9.22):[56,57]

$$Bu_3Sn^{\bullet} + R'{-}X{-}R \longrightarrow Bu_3SnXR' + R^{\bullet} \qquad (9.22)$$

Table 9.3 Rate constants at 298 K for the reaction of trialkyltin radicals with alkyl halides (R′X).

$R_3Sn^·$	R′	X	$k(l\,mol^{-1}\,s^{-1})$
$Bu_3Sn^·$	$^cC_5H_{11}$	Cl	2.0×10^3
$Bu_3Sn^·$	$^cC_5H_{11}$	Bu	2.2×10^7
$Bu_3Sn^·$	CH_3	I	2.5×10^9
$Bu_3Sn^·$	$^nC_5H_{11}$	Cl	8.5×10^2
$Bu_3Sn^·$	$^cC_6H_{11}$	Cl	2.0×10^3
$Bu_3Sn^·$	Me_3C	Cl	1.6×10^4
$Bu_3Sn^·$	$PhCH_2$	Cl	6.4×10^5
$Me_3Sn^·$	Me_3C	Cl	5.9×10^3
$Bu_3Sn^·$	Me_3C	Cl	1.6×10^4
$Ph_3Sn^·$	Me_3C	Cl	2.0×10^4

Table 9.4 Rate constants $(l\,mol^{-1}\,s^{-1})$ for the reaction of tributyltin radicals with compounds R_2X and R_2X_2 (X = O, S, Se, Te).

	$k_2{}^a$		$k_2{}^b$
MeSMe	< 10	EtOOEt	2.5×10^4
		MeCOOOCOMe	1.4×10^5
MeSeMe	3.7×10^5	Bu^tOOBu^t	10^2
MeTeMe	1.4×10^7	BuSSBu	1.1×10^6
		Bu^tSSBu^t	7.9×10^4

[a] At 298 K in hydrocarbons.
[b] At 283 K in benzene.

The reactivity increases in the sequence $R = Me < R^p < R^s < R^t <$ allyl or benzyl. Selenides and tellurides are more reactive in the same sense. In contrast, ethers (or alcohols) cannot be reduced by the $Bu_3Sn^·$ radical, but both peroxides and disulphides are cleaved,[58,59] the latter 10 to 100 times more readily. Some rate constants are given in Table 9.4.

Organotin radicals add reversibly to alkenes and alkynes to give β-stannylalkyl radicals (eqn 9.23). If the source of the tin radicals is an organotin hydride, a chain reaction is set up by reaction (9.24), leading to overall hydrostannation of the alkene or alkyne.[60,61]

$$R_3Sn^\bullet + \gt C{=}C\lt \longrightarrow R_3Sn{-}\overset{|}{\underset{|}{C}}{-}\overset{|}{\underset{|}{C}}^\bullet \tag{9.23}$$

$$R_3Sn\overset{|}{\underset{|}{C}}\overset{|}{\underset{|}{C}}^\bullet + R_3SnH \longrightarrow R_3Sn\overset{|}{\underset{|}{C}}\overset{|}{\underset{|}{C}}H + R_3Sn^\bullet \tag{9.24}$$

The reactions may occur spontaneously or may be initiated by light, by a palladium catalyst,[62] or by a free-radical source such as azoisobutyronitrile.[63] Examples of the synthesis of organotin compounds by hydrostannation are given in eqns (9.25)[63] and (9.26).[64]

$$\text{(cyclohexenyl)}\text{-CH=CH}_2 \xrightarrow{\text{Et}_3\text{SnH}} \text{(cyclohexenyl)}\text{-CH}_2\text{CH}_2\text{SnEt}_3 \quad (9.25)$$

If tributyltin radicals are generated from hexabutylditin in the presence of an aldehyde or ketone (but not ester), the esr spectrum of the corresponding stannyloxyalkyl radical can be observed (e.g. eqn 9.27).[65]

$$\text{HC}{\equiv}\text{C-CH}_2\text{-C}{\equiv}\text{CH} \xrightarrow{\text{Bu}_2\text{SnH}_2} \text{(stannacyclohexadiene, SnBu}_2\text{)} \quad (9.26)$$

$$\text{Bu}_3\text{Sn}^{\cdot} + \text{O=CMe}_2 \rightarrow \text{Bu}_3\text{SnO}\dot{\text{C}}\text{Me}_2 \quad (9.27)$$

If a tin hydride is used as the source of the tin radicals, a chain reaction is set up leading to overall hydrostannation of the carbonyl group, and trialkyltin hydrides, or dialkyltin dihydrides (particularly in the presence of a small amount of dialkyltin dihalide) can be used for reducing aldehydes or ketones to alcohols.[4] The rate constants for the reaction of $Bu_3Sn^{\cdot}$ radicals with cyclohexanone and with 9-fluorenone have been determined by the laser flash photolysis technique to be $<5 \times 10^4$ and $(3.8 \pm 0.5) \times 10^8 \, l \, mol^{-1} \, s^{-1}$, respectively, at 300 K.[66] The stannyl radicals are thus less reactive than the corresponding silyl radicals, but similar in reactivity to the germyl radicals.

More work has been carried out on the adducts which are formed with *para*- or *ortho*-quinones, or with 1,2-diones, when the products may adopt a variety of structures.[67] The rate constant for the addition of the $Bu_3Sn^{\cdot}$ radical to duroquinone has been shown to be $(1.4 \pm 0.1) \times 10^9 \, l \, mol^{-1} \, s^{-1}$.[66] Tetracyanobenzoquinone reacts with hexa-alkylditins to give stable free radicals which presumably have the structure (**7**) (R = Me, Pr or Bu).[68] 2, 5-Dimethoxybenzoquinone gives a 1:5 mixture of the isomers (**8**) and (**9**),[69] but 2, 5-di-*t*-butylbenzoquinone gives only the adduct (**10**).[70]

$$\text{(NC)}_4\text{C}_6\text{O}_2 \xrightarrow{\text{R}_3\text{Sn}^{\cdot}} \text{R}_3\text{SnO-C}_6\text{(CN)}_4\text{-O}^{\cdot} \quad (9.28)$$

(**7**)

$$\text{2,5-(MeO)}_2\text{C}_6\text{H}_2\text{O}_2 \xrightarrow{\text{Bu}_3\text{Sn}^{\cdot}} \text{Bu}_3\text{SnO-C}_6\text{H}_2\text{(OMe)}_2\text{-O}^{\cdot} + {}^{\cdot}\text{O-C}_6\text{H}_2\text{(OMe)}_2\text{-OSnBu}_3 \quad (9.29)$$

(**8**) (**9**)

$$\text{2,5-Bu}^t_2\text{C}_6\text{H}_2\text{O}_2 \xrightarrow{\text{Bu}_3\text{Sn}^{\cdot}} \text{Bu}_3\text{SnO-C}_6\text{H}_2\text{Bu}^t_2\text{-O}^{\cdot} \quad (9.30)$$

(**10**)

A similar reaction occurs between 3,5- and 3,6-di-*t*-butylbenzo-1,2-quinone (Scheme 3), and the same adducts can also be made by a variety of other processes such as the reaction of the semidione radical ion with the appropriate alkyltin halide,[71,72] or by generating the $R_3Sn^{\cdot}$ radical by photolysis of a cyclopentadienyltin compound.[20]

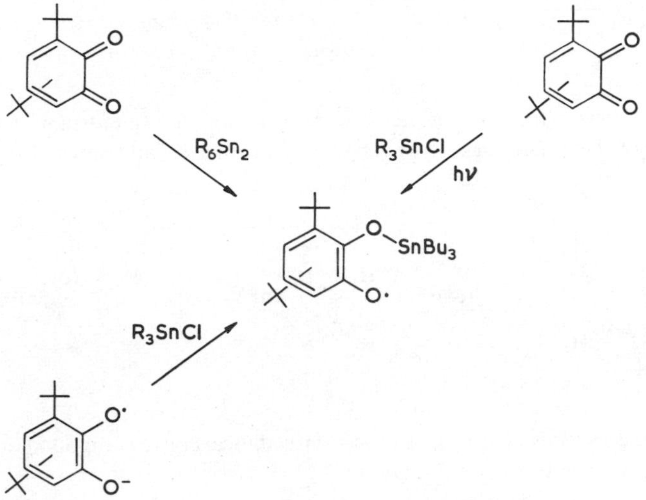

R_6Sn_2; R_3SnCl, $h\nu$; $SnBu_3$; R_3SnCl

Scheme 3

By variable-temperature esr spectroscopy, the rate constant for the metallotropic shift (eqn 9.31) has been shown to be *c.* $2.5 \times 10^6\ s^{-1}$ at 333 K when $R_3Sn = Me_2SnCl$ or Bu_2SnCl, but $> 10^9\ s^{-1}$ when $R_3Sn = Me_3Sn$ or Bu_3Sn.[20,73] For comparison, the rate constant for the corresponding shift of Me_3Si is *c.* $2.5 \times 10^6\ s^{-1}$.

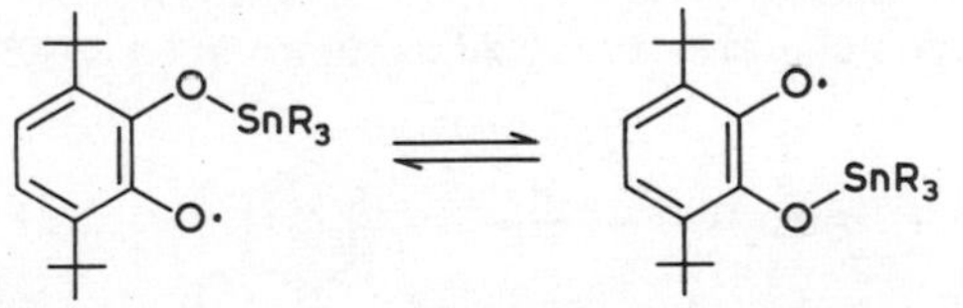

SnR_3 ⇌ SnR_3 (9.31)

A larger variety of adduct structures can be observed with biacetyl.[19] The esr spectra indicate that adducts formed from the $Cl_3Sn^{\cdot}$ and $BuCl_2Sn^{\cdot}$ radicals have the 5-coordinate trigonal bipyramidal structures (**11**) and (**12**). The radical $Bu_2ClSn^{\cdot}$ forms the 4-coordinate adduct (**13**) which is not fluxional on the esr time scale up to 0°C, but $Bu_3Sn^{\cdot}$ gives the rapidly fluxional *cis*-adduct (**14**), and the more slowly fluxional *trans*-adduct (**15**).[19]

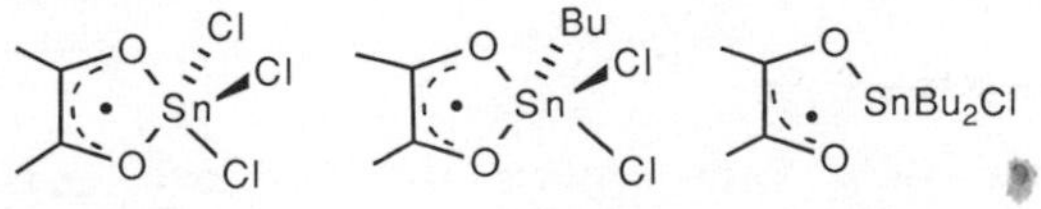

(**11**) (**12**) (**13**)

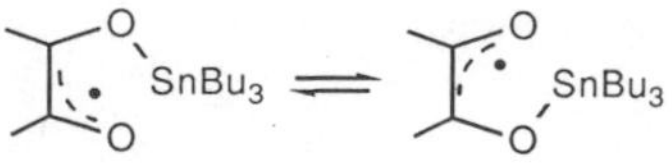

(14)

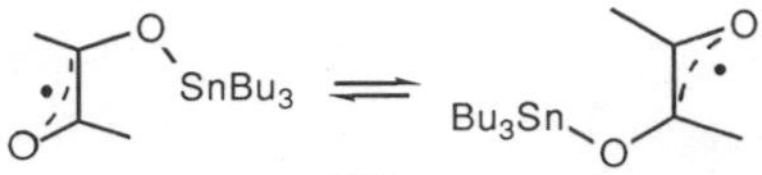

(15)

Similarly, triphenyltin radicals react with 1,4-diaza-1,3-butadienes (the nitrogen analogues of the 1,2-diones) to give rapidly fluxional *cis* and slowly fluxional *trans* adducts (eqn 9.32).[74]

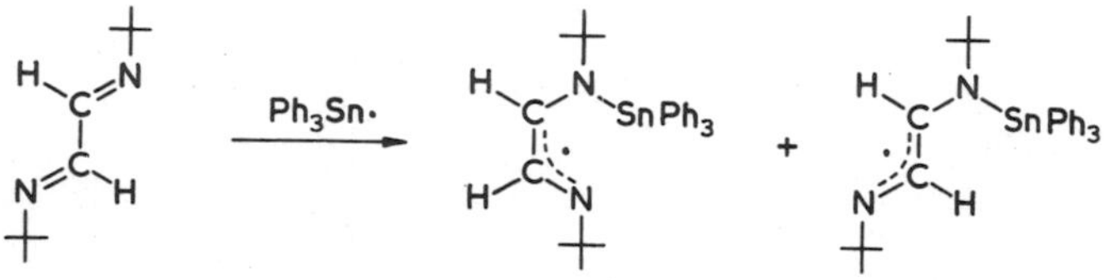

$Ph_3Sn\cdot$ (9.32)

Tin–centered radicals also form adducts with thiocarbonyl compounds.[75–77] Some examples are given in eqns (9.33) and (9.34):

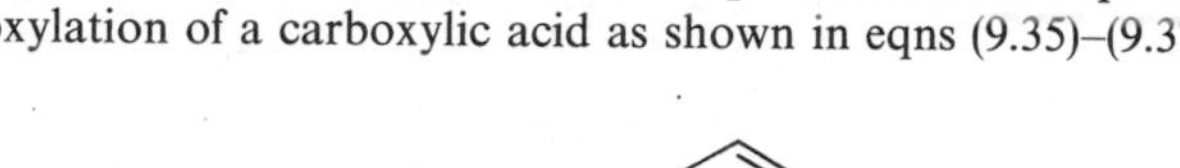

$+ Ph_3SnH \xrightarrow{h\nu}$ $\cdot SSnPh_3$ (9.33)

$$(MeOC_6H_4)_2C{=}S + Me_6Sn_2 \rightarrow (MeOC_6H_4)_2(\ SnMe_3 \qquad (9.34)$$

The attack of tin-centred radicals at the sulphur of a thiocarbonyl group is a component step in a number of organic synthetic procedures. An important example is the decarboxylation of a carboxylic acid as shown in eqns (9.35)–(9.37).[78]

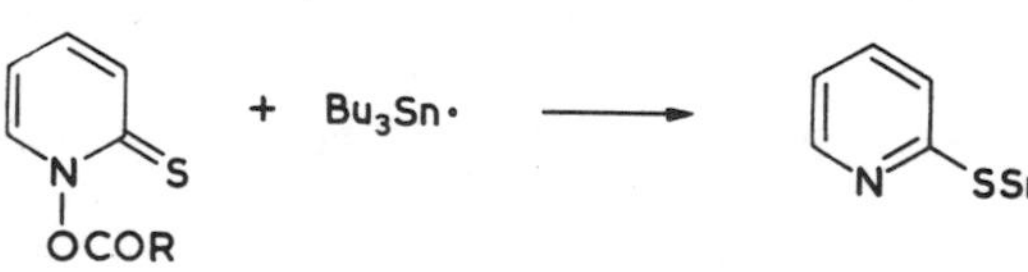

$+ RCO_2H \longrightarrow$ (9.35)

$+ Bu_3Sn\cdot \longrightarrow$ $SSnBu_3$ $+ R\cdot + CO_2$ (9.36)

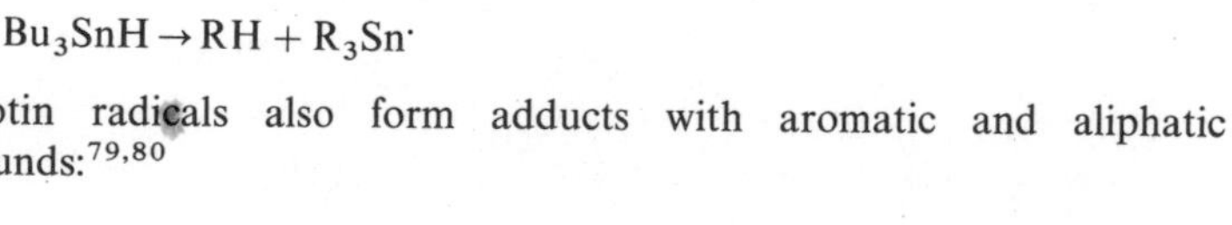

$$R^{\cdot} + Bu_3SnH \rightarrow RH + R_3Sn^{\cdot} \qquad (9.37)$$

Organotin radicals also form adducts with aromatic and aliphatic nitro-compounds:[79,80]

$$\mathrm{R'NO_2 + R_3Sn^\bullet \longrightarrow R'N(O^\bullet)OSnR_3} \tag{9.38}$$

When R′ = alkyl, a(N) = c. 3 mT, and the radical is probably pyramidal at nitrogen, but when R′ = aryl, a(N) = c. 1.5 mT, and the structure is probably planar. With an excess of the metallating agent, these stannyloxynitroxyl radicals are reduced further to stannyloxyaminyl radicals, $\mathrm{R'\dot{N}OSnR_3}$, probably by the reactions shown in eqns (9.39) and (9.40).[79]

$$\mathrm{R'N(O^\cdot)OSnR_3 \xrightarrow{R_3Sn^\cdot} R'N(OSnR_3)_2 \rightarrow R'N{=}O + (R_3Sn)_2O} \tag{9.39}$$

$$\mathrm{R'N{=}O + R_3Sn^\cdot \rightarrow R'\dot{N}OSnR_3} \tag{9.40}$$

When R′ is a tertiary or secondary alkyl group, the stannyloxynitroxyl radicals undergo β-scission to give alkyl radicals, and tributyltin hydride in refluxing benzene will reduce s- or t-alkyl compounds to the corresponding alkanes:[81,82]

$$\mathrm{R'N(O^\cdot)OSnBu_3 \rightarrow R'^\cdot + O{=}NOSnBu_3} \tag{9.41}$$

$$\mathrm{R'^\cdot + Bu_3SnH \rightarrow R'H + Bu_3Sn^\cdot} \tag{9.42}$$

9.3 Homolytic reactions at the ligands in organotin compounds

The introduction of an R_3Sn substituent into an alkane enhances the ease of abstraction of hydrogen on the α- or β-positions by t-butoxyl radicals.[1,31,83,84] Relative reactivities (per hydrogen atom) are given as follows:

$Et_3C{-}CH_2{-}CH_3$		$Et_3Sn{-}CH_2{-}CH_3$	
4.2	1.0	47	31

A similar but smaller effect is found for R_3Si and R_3Ge substituents. The enhanced reactivity at the α-position may be ascribed to the contribution which the polar canonical form (**16**) makes to the transition state, the positive charge being stabilized by inductive electron release by the Et_3Sn group.

$$\mathrm{Et_3Sn{-}\overset{CH_3}{\underset{H}{C^+}}\quad H^\bullet\ {}^-OBu^t}$$

(**16**)

Anchimeric assistance of the abstraction of β-hydrogen implies incipient stabilization of the developing carbon radical as illustrated in (**17**).[84]

This interaction is apparent in the fully developed β-trialkylstannylalkyl radicals, which have the preferred eclipsed conformation (**18**). This confers a low value and a positive temperature coefficient on a(Hβ) (1.584 mT at 173 K for $Me_3SnCH_2\dot{C}H_2$ compared with 2.471 mT for $Me_3CCH_2\dot{C}H_2$),[51] and is usually ascribed to hyperconjugation involving the singly occupied $2p$ orbital and the Cβ-Sn σ^* orbital, with perhaps some contribution from homoconjugation involving a vacant $5d$ orbital on

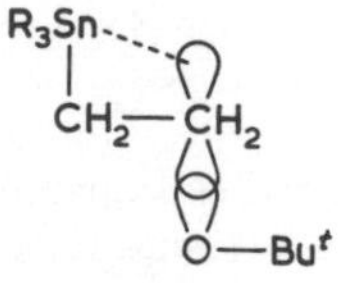

(17) 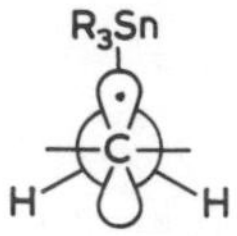 (18)

tin.[1,51,84–86] This ready formation of the β-stannylalkyl radicals, coupled with the β-scission reaction, has been exploited in the use of trimethylisobutyltin as a source of $Me_3Sn^{\cdot}$ radicals (eqn 9.14).[31] It is also the basis of the conjugate homolytic displacement (S_H2') reactions which occur with allytin compounds (eqn 9.43).[5,87]

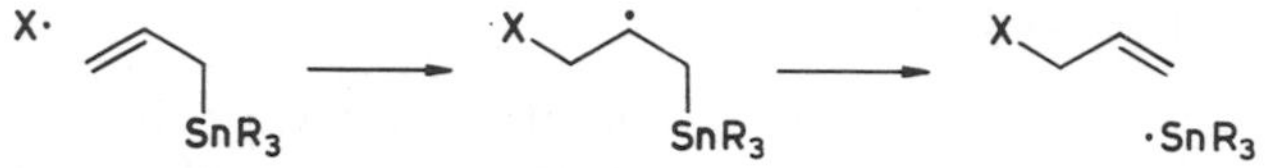

(9.43)

Russell[88] has shown that the radicals $PhS^{\cdot}$, $RS^{\cdot}$, $PhSe^{\cdot}$ and $PhSO_2^{\cdot}$ ($Q^{\cdot}$) will react with 1-alkenyl- or 1-alkynyl-tributyltin compounds by an addition-elimination process (eqns 9.44, 9.45) involving the formation and fission of an intermediate β-stannylalkyl radical.

$$Q^{\cdot} + RR'C{=}CHSnBu_3 \rightarrow RR'\dot{C}{-}CH(Q)SnBu_3 \quad (9.44)$$

$$RR'\dot{C}{-}CH(Q)SnBu_3 \rightarrow RR'C{=}CHQ + {}^{\cdot}SnBu_3 \quad (9.45)$$

If the radical $Q^{\cdot}$ is derived from a reagent QY (PhS–SPh, RS–SR, $PhSe{-}SO_2Ar$, or $PhSO_2{-}Cl$) or from a mercury(II) derivative Q_2Hg $\{Hg(SPh)_2$, $Hg(SePh)_2$, $Hg(O_2SPh)_2$, or $Hg[PO(OEt)_2]\}$, the displaced $Bu_3Sn^{\cdot}$ radical can regenerate the radical $Q^{\cdot}$ by reactions (9.46) and (9.47), and a chain reaction can be set up.

$$Bu_3Sn^{\cdot} + Q{-}Y \rightarrow Bu_3SnY + Q^{\cdot} \quad (9.46)$$

$$Bu_3Sn^{\cdot} + Q_2Hg \rightarrow Bu_3SnQ + QHg^{\cdot} \rightarrow Q^{\cdot} + Hg^{\circ} \quad (9.47)$$

Two examples of such reactions are shown in eqns (9.48) and (9.49).

$$CH_2{=}CHSnBu_3 + PhSSPh \xrightarrow[4h]{h\nu} CH_2{=}CHSPh + Bu_3SnPh \quad (9.48)$$

$$2Ph_2C{=}CHSnBu_3 + Hg(O_2SPh)_2 \xrightarrow[4h]{h\nu} Ph_2C{=}CHO_2SPh + Hg^{\circ} \quad (9.49)$$

Towards the radical $PhS^{\cdot}$, $PhC{\equiv}CSnBu_3$ is 300 times less reactive than $PhCH{=}CHSnBu_3$.

The trimethylstannylcyclopentadienyl radical has been generated by photolysis of bis(trimethylstannyl)cyclopentadiene (eqn 9.50):

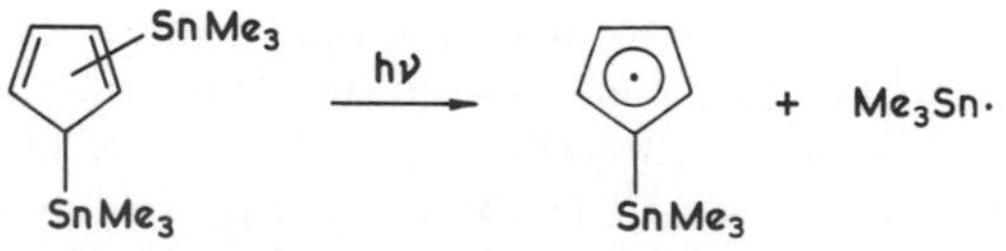

(9.50)

The esr spectrum shows that the Me_3Sn substituent stabilizes the ψ_S orbital so that the unpaired electron resides principally in the ψ_A orbital (Scheme 4).[89] Similarly, the trimethylstannylbenzene radical anion can be generated by reduction of trimethylphenyltin, and the esr spectrum shows that the ψ_S orbital is stabilized by the substituent (Scheme 5).[90]

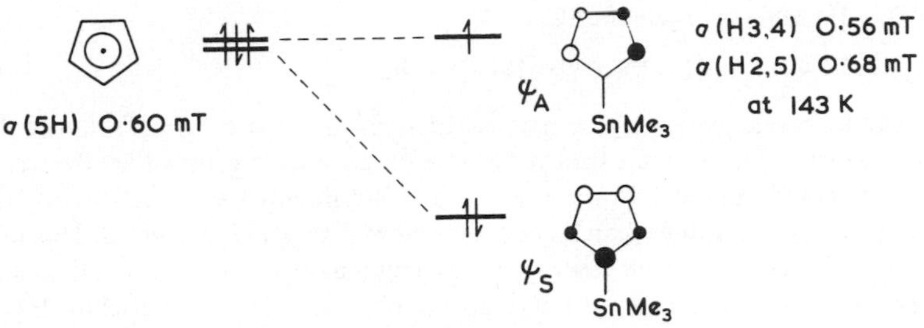

a(5H) 0·60 mT

ψ_A SnMe$_3$ a(H3,4) 0·56 mT a(H2,5) 0·68 mT at 143 K

ψ_S SnMe$_3$

Scheme 4

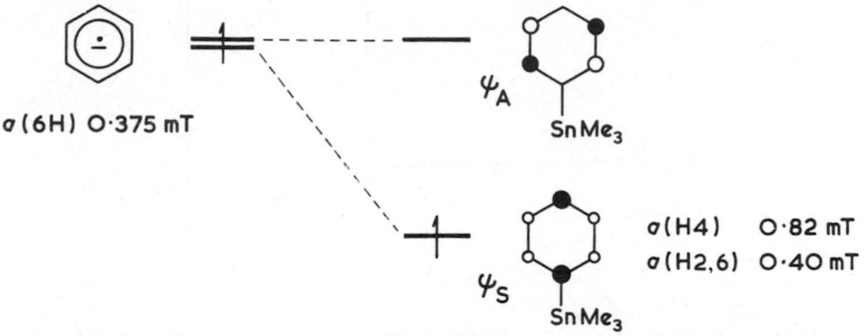

a(6H) 0·375 mT

ψ_A SnMe$_3$

ψ_S SnMe$_3$ a(H4) 0·82 mT a(H2,6) 0·40 mT

Scheme 5

ortho-Aminophenols react with organotin compounds R_3SnX or R_2SnX_2 to give stable paramagnetic complexes.[91] The structure of one such complex (**19**) has been determined by single-crystal X-ray diffraction.[92] The compound may be regarded as being formed by loss of a hydrogen atom from α-nitrogen, and the configuration about the tin is that of a distorted trigonal bipyramid, with equatorial aryl groups and nitrogen, and the oxygen ligands distorted away from the apical positions to give an OSnO angle of 153.5°.

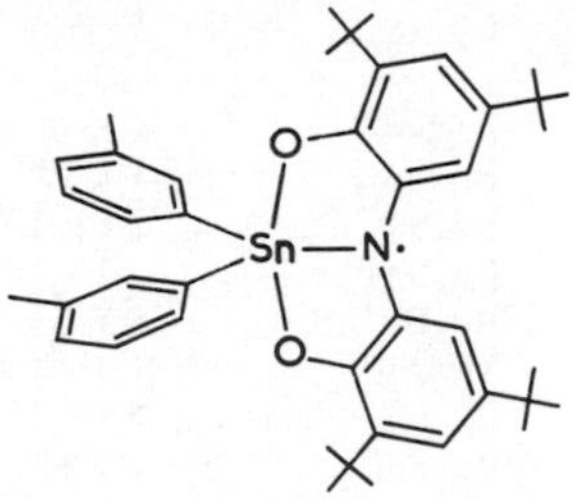

Sn—N·

(**19**)

9.4 Homolytic reactions at the tin centre

Whereas alkoxyl radicals and ketone triplets react with tetraalkyltin compounds at hydrogen centres in the alkyl groups, the presence of electronegative ligands on the tin in compounds R_nSnX_{4-n} directs the attack to the tin centre, and an alkyl radical is displaced.[1,83,93–96] Two examples are shown in eqns (9.51) and (9.52).

$$Bu^tO^{\cdot} + Pr_3SnCl \rightarrow Pr_2SnCl(OBu^t) + Pr^{\cdot} \qquad (9.51)$$

$$Ph_2C{=}O^T + Bu_3SnCl \rightarrow Ph_2\dot{C}OSnBu_2Cl + Bu^{\cdot} \qquad (9.52)$$

This change in reactivity may be associated with the greater availability of the *5d* orbitals to form a 5-coordinate transition state or intermediate, for which the structure (**19**) might provide a model. For the reaction of $Bu^n_2Bu^tSnCl$, esr spectroscopy shows that the primary $Bu^{n\cdot}$ rather than the more stable $Bu^{t\cdot}$ radical is displaced. This may be rationalized in terms of apical cleavage in the trigonal bipyramidal transition state or intermediate where the bulky *t*-butyl group occupies the less sterically hindered equatorial position (eqn 9.53).[97]

$$Bu^tO^{\bullet} + Bu^n_2Bu^tSnCl \rightarrow \left[{}^tBuO\cdots Sn(Bu^t)(Bu^n)(Bu^n)\cdots Cl\right]^{\bullet} \rightarrow \left[Cl\cdots Sn(Bu^t)(Bu^n)(OBu^t)\cdots Bu^n\right]^{\bullet} \rightarrow ClSn(OBu^t)Bu^nBu^t + Bu^{n\bullet} \qquad (9.53)$$

Rate constants for these reactions are collected in Table 9.5. With certain organotin compounds, or certain reagents, S_H2 reaction at tin can occur even in tetra-alkyltin compounds. Thus the stannacyclopentanes show an enhanced Lewis acidity, probably because complex formation leads to relief of strain in the ring, and these compounds react with $Bu^tO^{\cdot}$, $Me_3SiO^{\cdot}$, $PhCO_2^{\cdot}$, and $PhS^{\cdot}$ radicals with ring-opening (eqn 9.54)[97]:

Table 9.5 Rate constants ($l\,mol^{-1}\,s^{-1}$) for S_H2 reactions at tin.

Reagent	R_nSnX_{4-n}	k	T(K)	Ref.
$Bu^tO^{\cdot}$	Pr_3SnCl	1×10^6	283	93
	Et_2SnCl_2	1×10^6	283	93
	Bu_3SnCl	2.1×10^6	213	96
	Bu_2SnCl_2	3.0×10^6	213	96
	Bu_2Bu^tSnCl	4.5×10^6	213	96
	Me_2Sn	1.0×10^6	213	97
	Bu_2Sn	5.5×10^6	213	97
$PhMeCO^T$	Bu_3SnCl	4×10^8	283	93
$(4\text{-}MeC_6H_4)MeCO^T$	Pr_3SnCl	2.2×10^8	283	93

$$X\cdot + R_2Sn(CH_2)_4 \longrightarrow [X\text{—}SnR_2(CH_2)_4]^{\cdot} \longrightarrow X\text{—}SnR_2(CH_2)_3CH_2\cdot \quad (9.54)$$

Thiols can therefore bring about ring-opening in a radical chain process as shown in eqns (9.55) and (9.56).[97]

$$PhS\cdot + Bu_2Sn(CH_2)_4 \longrightarrow PhSBu_2Sn(CH_2)_3CH_2\cdot \quad (9.55)$$

$$PhSSnBu_2(CH_2)_3CH_2^{\cdot} + PhSH \rightarrow PhSSnBu_3 + PhS^{\cdot} \quad (9.56)$$

If the alkyl groups are bulky (e.g. Bu^t), or if the ring is large, some reaction also occurs at hydrogen atoms within the ring.

There are three important reactions in which bimolecular homolytic substitution has been suggested to occur at the tin centre in acyclic tetralkyltin compounds. The reaction with *N*-halogenosuccinimides (eqn 9.57) follows a steric sequence in R,[98] the reaction with bromine follows the opposite sequence (eqn 9.58),[99,100] and the reaction with oxygen has as yet been established only where R = methyl (eqn 9.59).[101]

$$(CH_2CO)_2N^{\cdot} + R_4Sn \rightarrow R_3SnN(COCH_2)_2 + R^{\cdot} \quad (9.57)$$

$$Br^{\cdot} + R_4Sn \rightarrow R_3SnBr + R^{\cdot} \quad (9.58)$$

$$MeOO^{\cdot} + Me_4Sn \rightarrow Me_3SnOOMe + Me^{\cdot} \quad (9.59)$$

In all of these reactions, there may be some contribution of an electron-transfer complex to the transition state, as discussed below.

9.5 Electron transfer reactions

Fukuzumi and Kochi have shown that the selectivity (R/Me) in the reaction of photolytically generated $I^{\cdot}$ atoms with organotin compounds, Me_nSnR_{4-n}, follows the same sequence ($Bu^t > Pr^i > Et > Pr^n$) as that which is observed in the reaction of hexachloroiridate(IV), which involves an electron transfer process.[102] They suggest that the S_H2 reaction involves electron transfer to give the tetraalkyltin radical cation (eqn 9.60).

$$R_4Sn + I^{\cdot} \xrightleftharpoons{\text{slow}} [R_4Sn^{\cdot +}I^-] \xrightarrow{\text{fast}} R_3SnI + R^{\cdot} \quad (9.60)$$

Electron transfer was also proposed to be involved in the reaction between R_4Sn and I_2 (eqn 9.61)[103].

$$R_4Sn + I_2 \rightleftharpoons [R_4Sn, I_2] \xrightarrow{\text{slow}} [R_4Sn^{\cdot +}I_2^{\cdot -}] \xrightarrow{\text{fast}} [R^{\cdot}R_3Sn^+I_2^{\cdot -}] \downarrow R^{\cdot} + R_3SnI + I^{\cdot} \quad (9.61)$$

Organotin radical cations can also be formed by photoinduced electron transfer from a benzyltrialkyltin compound ($4\text{-}XC_6H_4CH_2SnR_3$; X = Cl, F, H, Me, OMe; R = Me, Bu)

Table 9.6 Esr spectra of tin radical cations.

Radical	Structure	Nucleus	$A_{\perp}$(mT)	$A_{\parallel}$(mT)
$SnH_4^{+\cdot}$	C_{2v}	^{119}Sn	−237	−310
		2H	8.5	8.5
$SnH_4^{+\cdot}$	C_{3v}	^{119}Sn	−318	−365
		1H	17.5	17.5
$MeSnH_3^{+\cdot}$	C_{2v}	^{119}Sn	−238	−315
		2H	8.5	8.5
$Me_2SnH_2^{+\cdot}$	C_{2v}	^{119}Sn	−238	−322
		2H	8.5	8.5
$Me_3SnH^{+\cdot}$	C_{2v} or C_{3v}	^{119}Sn	−167	−242
$Me_4Sn^{+\cdot}$	C_{3v}	^{119}Sn	−7.8	−21
		3H	−1.35	−1.35
		^{13}C	5.3	12
$Me_3SnCMe_3^{+\cdot}$	C_{3v}	^{119}Sn	8.8	
		9H	0.76	
		^{13}C	18.7	
$Me_3SnSnMe_3^{+\cdot}$		^{119}Sn	11.5	

to methylene blue, Rose Bengal or 9, 10-dicyanoanthracene (Sens). The radical cation then fragments to give the trialkyltin cation and the benzyl radical, which can initiate polymerization.[104]

$$Sens^* + R_3SnCH_2Ar \rightarrow Sens^{\cdot -} + R_3SnCH_2Ar^{\cdot +} \rightarrow R_3Sn^+ + \dot{C}H_2Ar \qquad (9.62)$$

Flavin analogues and their Mg^{2+} complexes have similarly been used for the photosensitized formation of simple tetra-alkyltin radical cations. The rates of reaction follow the sequence $Me_4Sn < Et_4Sn$, $Bu_4Sn < Pr^i_4Sn$, reflecting the relative donor ability, and the radical cations fragment to form R_3Sn^+ and $R^{\cdot}$. If oxygen is present, tetramethyltin is thereby converted into the peroxide $Me_3SnOOMe$ by reaction (9.59).[101]

The most direct evidence for the formation of organotin(IV) radical cations comes from esr studies of the γ-radiolysis of solid solutions in Freon matrices.[105,106] Spectral data are listed in Table 9.6.

The salient feature is that in $Me_4Sn^{\cdot +}$, the hyperfine coupling to tin is small and there is a unique methyl ligand. The Me_3Sn unit is almost planar, and the bond to the remaining methyl group is elongated and weakened, as shown in (**20**). The compounds Me_3SnR (R = Et and Pr^i) showed only the spectrum of the radical $R^{\cdot}$, but when R = Bu^t, the radical cation was observed with a structure similar to (**20**). The low value of $a(^{119}Sn)$ in Me_6Sn_2 similarly implies that the Me_3Sn units are nearly planar, as shown in (**21**).

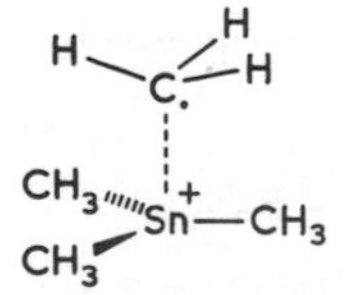

(**20**) 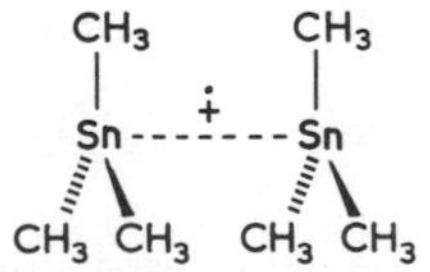 (**21**)

Table 9.7 Esr spectra of tin radical anions.

Radical	T(K)	$g_\perp$	$g_\parallel$	Nucleus	$A_\perp$(mT)	$A_\parallel$(mT)	Ref.
$SnH_4^{\cdot-}$	77	2.000	2.010	$2H_{ax}$	13.8	13.7	112
				$2H_{eq}$	0.77	0.77	
$SnH_4^{\cdot-}$	100	$g = 2.000\,37$		$2H_{ax}$	$a = 14.35$		111
				$2H_{eq}$	$a = 0.80$		
				^{117}Sn	$a = 212.9$		
$MeSnH_3^{\cdot-}$	77	1.997	2.012	$2H_{ax}$	13.2	13.1	112
	103			$1H_{eq}$	$a = 0.80$		
$Me_2SnH_2^{\cdot-}$	77	1.996	2.014	$2H_{ax}$	12.7	12.6	112
$Me_3SnH^{\cdot-}$	77	1.995	2.014	$1H_{ax}$	14.0	13.9	112
$Me_4Sn^{\cdot-}$	77	2.0		Sn	210.1	167.2	110

Calculations by the UHF-MNDO method are broadly in accord with these results, except that they find the Me_3Sn groups in $Me_6Sn_2^{\cdot+}$ to be nonplanar.[107,108]

The observation of the esr spectrum of the radical anion $Me_3SnPh^{\cdot-}$ has been referred to earlier.[89] Photoelectron spectroscopy confirms that the first ionization potential of bis(1,4-trimethylstannyl) benzene (8.5 eV) is less than that of benzene (9.25 eV).[109] The esr spectrum of the radical anion of 1,4-bis(trimethylstannyl)naphthalene has also been studied. The hyperfine coupling constants (mT) are shown in **(22)**.[109]

(22) **(23)**

The radical anions of the methylstannanes, $Me_nSnH_{4-n}^{\cdot-}$, have been generated by γ-irradiation of frozen matrices such as tetramethylsilane. Data on the esr spectra are given in Table 9.7.[110–112] It is concluded that all these radical cations have a trigonal bipyramidal configuration, with the unpaired electron acting as an equatorial 'phantom ligand', and the methyl groups preferring to occupy an equatorial position. This is exemplified for $Me_2SnH_2^{\cdot-}$ in **(23)**.

References

1. A.G. Davies, in *Organotin Compounds: New Chemistry and Applications*', ed. J.J. Zuckermann, Adv. Chem. Ser., **157**, American Chemical Society, Washington, DC, 1976, 26.
2. M. Pereyre, J.P. Quintard and A. Rahm, *Tin in Organic Synthesis*, Pergamon, London, 1986.
3. B. Giese, *Radicals in Organic Synthesis: Formation of Carbon–Carbon Bonds*, Pergamon, London, 1986.
4. W.P. Neumann, *Synthesis*, 1987, 665.
5. Y. Yamamoto, *Aldrichim. Acta*, 1987, **20**, 45.

6. M. Ramaiah, *Tetrahedron*, 1987, **43**, 3541.
7. M. Lehnig and K. Dören, *J. Organomet. Chem.*, 1981, **210**, 331.
8. J.C. Scaiano, *J. Amer. Chem. Soc.*, 1980, **102**, 5399.
9. C. Chatgilialoglu, K.U. Ingold, J. Lusztyk, A.S. Nazran, and J.C. Scaiano, *Organometallics*, 1983, **2**, 1332.
10. P.J. Wagner, P.A. Kelso, and R.G. Zepp, *J. Amer. Chem. Soc.*, 1972, **94**, 7480.
11. D.J. Carlsson and K.U. Ingold, *J. Amer. Chem. Soc.*, 1968, **90**, 7047.
12. L.J. Johnston, J. Lusztyk, D.D.M. Wayner, A.N. Abeywickreyma, A.L.J. Beckwith, J.C. Scaiano and K.U. Ingold, *J. Amer. Chem. Soc.*, 1986, **107**, 4594.
13. Landolt-Börnstein, *Radical Reaction Rates in Liquids*, Vol. II, 1983, 13b, 5; Vol. II, 1983, 13c, 323, Springer, Berlin.
14. J. Cooper, A. Hudson and R.A. Jackson, *J. Chem. Soc., Perkin Trans. 2*, 1973, 1056.
15. M.J. Gynane, M.F. Lappert, P.I. Riley, P. Riviere and M. Riviere-Baudet, *J. Organomet. Chem.*, 1980, **202**, 5.
16. H.U. Buschhaus, W.P. Neumann and T. Apoussidis, *Annalen*, 1981, 1190.
17. M. Lehnig, *Tetrahedron Lett.*, 1974, 3323.
18. P.J. Barker, A.G. Davies and M.-W. Tse, *J. Chem. Soc., Perkin Trans. 2*, 1980, 941.
19. P.J. Barker, A.G. Davies, J.A.-A. Hawari and M.-W. Tse, *J. Chem. Soc., Perkin Trans. 2*, 1980, 1488.
20. A.G. Davies and J.A.A. Hawari, *J. Organomet. Chem.*, 1980, **201**, 221.
21. D. Rehorek and E.G. Janzen, *J. Organomet. Chem.*, 1984, **268**, 135.
22. R.V. Lloyd and M.T. Rogers, *J. Amer. Chem. Soc.*, 1973, **95**, 2459.
23. G.B. Watts and K.U. Ingold, *J. Amer. Chem. Soc.*, 1972, **94**, 491.
24. B.A. King and F.B. Bramwell, *J. Inorg. Nucl. Chem.*, 1981, **43**, 1479.
25. A. Tamotsu, O. Ito and M. Matsuda, *J. Phys. Chem.*, 1982, **86**, 2950.
26. K. Mochida, M. Wakasa, Y. Sakaguchi and H. Hayashi, *Chem. Lett.*, 1986, 773.
27. M.J.S. Gynane, M.F. Lappert, P.I. Riley, P. Riviere and M. Riviere-Baudet, *J. Organomet. Chem.*, 1980, **202**, 5.
28. M.J.S. Gynane, D.H. Harris, M.F. Lappert, P.P. Power, P. Riviere and M. Riviere-Baudet, *J. Chem. Soc., Dalton Trans.*, 1977, 2004.
29. A. Hudson, M.F. Lappert and P.W. Lednor, *J. Chem. Soc. Dalton Trans.*, 1986, 2369.
30. H. Hillgärtner, W.P. Neumann and B. Schroedner, *Annalen*, 1975, 586.
31. A.G. Davies, B.P. Roberts and M.-W. Tse, *J. Chem. Soc., Perkin Trans. 2*, 1978, 145.
32. Landolt-Börnstein, *Magnetic Properties of Free Radicals*, Vol. II, 1977, 9a, 245; Vol. II, 1979, 9c2, 312, Springer, Berlin.
33. R.L. Morehouse, J.J. Christiansen and W. Gordy, *J. Chem. Phys.*, 1965, **45**, 1751.
34. S.A. Fieldhouse, A.R. Lyons, H.C. Starkie and M.C.R. Symons, *J. Chem. Soc. Dalton Trans.*, 1974, 1966.
35. M. Lehnig, H.U. Buschhaus, W.P. Neumann and T. Apoussidis, *Bull. Soc. Chem. Belg.*, 1980, **89**, 907.
36. M. Lehnig, *Chem. Phys.*, 1975, **8**, 419.
37. M.J.S. Dewar, G.L. Grady, D.R. Kuhn and K.M. Merz, *J. Amer. Chem. Soc.*, 1984, **106**, 6773.
38. M. Gielen and Y. Tondeur, *Nouv. J. Chim.*, 1978, **2**, 117.
39. M. Gielen and Y. Tondeur, *J. Organomet. Chem.*, 1979, **169**, 265.
40. C. Chatgilialoglu, K.U. Ingold, J. Lusztyk, A.S. Nazran and J.C. Scaiano, *Organometallics*, 1983, **2**, 1332.
41. M. Lehnig, *Tetrahedron Lett.*, 1977, 3663.
42. B. Maillard, K.U. Ingold and J.C. Scaiano, *J. Amer. Chem. Soc.*, 1983, **105**, 5095.
43. J.A. Howard and J.C. Tait, *J. Amer. Chem. Soc.*, 1977, **99**, 8349.
44. J.A. Howard, J.C. Tait and S.B. Tong, *Can. J. Chem.*, 1979, **57**, 2761.
45. G.J.M. van der Kerk, J.G. Noltes and J.G.A. Luijten, *J. Appl. Chem.*, 1957, **7**, 356.
46. J.G. Noltes and G.J.M. van der Kerk, *Chem. Ind. (London)*, 1959, 294.
47. H.G. Kuivila, *Adv. Organometallic Chem.*, 1964, **1**, 47.
48. H.G. Kuivila, L.W. Menapace and C.R. Warner, *J. Amer. Chem. Soc.*, 1962, **84**, 3584.
49. H.G. Kuivila, *Acc. Chem. Res.*, 1968, **1**, 299.
50. H.G. Kuivila, *Synthesis*, 1970, 499.
51. P. Krusic and J.K. Kochi, *J. Amer. Chem. Soc.*, 1971, **93**, 846.
52. M. Castaing, M. Pereyre, M. Ratier, P.M. Blum and A.G. Davies, *J. Chem. Soc., Perkin Trans. 2*, 1979, 287.

53. A.L.J. Beckwith, *Tetrahedron*, 1981, **37**, 3073.
54. B. Giese, *Angew. Chem. Int. Edn. Engl.*, 1985, **24**, 161.
55. Landolt-Börnstein, *Radical Reaction Rates in Liquids*, Vol. II, 1983, 13c, 325, Springer, Berlin.
56. J.C. Scaiano, P. Schmid and K.U. Ingold, *J. Organomet. Chem.*, 1976, **121**, C4.
57. C.G. Gutierrez and L.R. Summerhayes, *J. Org. Chem.*, 1984, **49**, 5206.
58. J.L. Brockenshire and K.U. Ingold, *Int. J. Chem. Kinetics*, 1970, **2**, 157.
59. J.L. Brockenshire and K.U. Ingold, *Int. J. Chem. Kinetics*, 1971, **3**, 343.
60. G.J.M. van der Kerk, J.G. Noltes and J.G.A. Luijten, *Chem. Ind. (London)*, 1956, 352.
61. W.P. Neumann, H. Nierman and R. Sommer, *Angew. Chem.*, 1961, **73**, 768.
62. T.N. Mitchell, *J. Organomet. Chem.*, 1986, **304**, 1.
63. W.P. Neumann, H. Niermann and B. Schneider, *Annalen*, 1967, **707**, 15.
64. A.J. Ashe and P. Shu, *J. Amer. Chem. Soc.*, 1971, **93**, 1804.
65. J. Cooper, A. Hudson and R.A. Jackson, *J. Chem. Soc., Perkin Trans. 2*, 1973, 1933.
66. K.U. Ingold, J. Lusztyk and J.C. Scaiano, *J. Amer. Chem. Soc.*, 1984, **106**, 343.
67. Landolt-Börnstein, *Magnetic Properties of Free Radicals*, Vol. II, 1977, *9a*, 298; Vol. II, 1987, *17b*, 451, Springer, Berlin.
68. A.B. Cornwell, P.G. Harrison and J.A. Richards, *J. Organomet. Chem.*, 1977, **140**, 273.
69. A. Alberti and G.F. Pedulli, *J. Organomet. Chem.*, 1983, **248**, 261.
70. K.S. Chen, T. Foster and J.K.S. Wan, *J. Chem. Soc., Perkin Trans. 2*, 1979, 1288.
71. G.A. Razuvaev, V.A. Tsarjapkin, L.V. Gorbunova, V.K. Cherkasov, G.A. Abakumov and E.S. Klimov, *J. Organomet. Chem.*, 1979, **174**, 47.
72. A.I. Prokof'ev, T.I. Prokof'eva, N.N. Bubnov, S.P. Solodovnikov, I.S. Belostatskaya, V.V. Ershov and M.I. Kabachnik, *Doklady Akad. Nauk*, 1978, **239**, 1367.
73. S.G. Kukes, A.I. Prokof'ev, N.N. Bubnov, S.P. Solodovnikov, E.D. Korniets, D.N. Kravtsov and M.I. Kabachnik, *Doklady Akad. Nauk*, 1976, **229**, 877.
74. A. Alberti and A. Hudson, *J. Organomet. Chem.*, 1983, **24**, 313.
75. J.C. Scaiano and K.U. Ingold, *J. Amer. Chem. Soc.*, 1976, **98**, 4727.
76. A. Alberti, F.P. Colonna and G.F. Pedulli, *Tetrahedron*, 1980, **36**, 3043.
77. B.B. Adeleka, K.S. Chen and J.K.S. Wan, *J. Organomet. Chem.*, 1981, **208**, 317.
78. D.H.R. Barton, D. Crich and W.B. Motherwell, *J. Chem. Soc., Chem. Commun.*, 1983, 939.
79. K. Reuter and W.P. Neumann, *Tetrahedron Lett.*, 1978, 5235.
80. A.G. Davies and J.A.-A. Hawari, *J. Organomet. Chem.*, 1980, **201**, 221.
81. N. Ono, H. Miyake, R. Tamura and A. Kaji, *Tetrahedron Lett.*, 1981, 1705.
82. N. Ono and A. Kaji, *Synthesis*, 1986, 693.
83. A.G. Davies, B.P. Roberts and M.W. Tse, *J. Chem. Soc. Perkin Trans. 2*, 1977, 1499.
84. R.A. Jackson, K.U. Ingold, D. Griller and A.S. Nazran, *J. Amer. Chem. Soc.*, 1985, **107**, 208.
85. M.C.R. Symons, *Chem. Phys. Lett.*, 1973, **19**, 61.
86. D. Griller and K.U. Ingold, *J. Amer. Chem. Soc.*, 1973, **95**, 6459.
87. G.E. Keck and J.B. Yates, *J. Amer. Chem. Soc.*, 1982, **104**, 5829.
88. G.A. Russell, P. Ngoviwatchai, H. Tashkoush and J. Hershberger, *Organometallics*, 1987, **6**, 1414.
89. P.J. Barker, A.G. Davies, R. Henriquez and J.-Y. Nedelec, *J. Chem. Soc., Perkin Trans. 2*, 1982, 745.
90. S.P. Solodovinkov and A.K. Prokof'ev, *Russ. Chem. Rev.*, 1970, **39**, 591.
91. H.B. Stegmann, K. Scheffler and F. Stöcker, *Chem. Ber.*, 1970, **103**, 1279.
92. W. Uber, H.B. Stegmann and K. Scheffler, *Z. Naturforsch.*, 1977, **32b**, 355.
93. A.G. Davies, B.P. Roberts and J.C. Scaiano, *J. Organomet. Chem.*, 1972, **39**, C55.
94. A.G. Davies and J.C. Scaiano, *J. Chem. Soc., Perkin Trans. 2*, 1973, 1777.
95. K.U. Ingold and B.P. Roberts, *Free Radical Substitution Reactions*, Wiley-Interscience, New York, 1970.
96. A.G. Davies and B.P. Roberts, in *Free Radicals*, ed. J.K. Kochi, Wiley, New York, 1973, Chap. 10.
97. A.G. Davies, B. Muggleton, B.P. Roberts, M.-W. Tse and J.N. Winter, *J. Organomet. Chem.*, 1976, **118**, 289.
98. A.G. Davies, B.P. Roberts and J.M. Smith, *J. Chem. Soc., Perkin Trans. 2*, 1972, 2221.
99. S. Boué, M. Gielen and J. Nasielski, *J. Organomet. Chem.*, 1969, **9**, 461.
100. S. Boué, M. Gielen and J. Nasielski, *J. Organomet. Chem.*, 1969, **9**, 491.
101. S. Fukuzumi, S. Kuroda and T. Tanaka, *J. Chem. Soc., Perkin Trans. 2*, 1986, 25.

102. S. Fukuzumi and J.K. Kochi, *J. Org. Chem.*, 1980, **45**, 2654.
103. S. Fukuzumi and J.K. Kochi, *J. Amer. Chem. Soc.*, 1980, **102**, 2141.
104. D.F. Eaton, *Pure Appl. Chem.*, 1984, **54**, 1191.
105. A. Hasegawa, S. Kaminaka, T. Wakabayashi, H. Michiro, M.C.R. Symons and J. Ridout, *J. Chem. Soc., Dalton Trans.*, 1984, 1667.
106. B.W. Walther, F. Williams, W. Lau and J.K. Kochi, *Organometallics*, 1983, **2**, 688.
107. C. Glidewell, *J. Organomet. Chem.*, 1985, **294**, 173.
108. M.J.S. Dewar, G.L. Grady and D.R. Kuhn, *Organometallics*, 1985, **4**, 1041.
109. W. Kaim, H. Tesmann and H. Bock, *Chem. Ber.*, 1980, **113**, 3221.
110. S.A. Fieldhouse, H.C. Starkie and M.C.R. Symons, *Chem. Phys. Lett.*, 1973, **23**, 508.
111. J.R. Morton and K.F. Preston, *Mol. Phys.*, 1975, **30**, 1213.
112. A. Hasegawa, T. Yamaguchi and M. Michiro, *Chem. Lett.*, 1980, 611.

10 The uses of organotin compounds in organic synthesis

J.L. WARDELL

10.1 Introduction

Over the last decade or so, organotin compounds have been shown to be extremely valuable and versatile reagents in organic synthesis, capable of allowing various transformations to occur with excellent regio- and stereoselective control. Not only have reactions of tin–carbon bonds found application, but so have those of compounds containing bonds to other elements, particularly hydrogen and oxygen. A most valuable and timely book on the uses of organotin compounds in synthesis has recently been published.[1]

10.2 Use of organotin hydrides[1,2]

Organotin hydrides have been extensively used in transforming a variety of $\equiv$C–Y bonds to $\equiv$C–H bonds (eqn 10.1). Of particular importance are reductions of carbon–

$$\equiv C\text{–}Y \xrightarrow{R_3SnH} \equiv C\text{–}H \tag{10.1}$$

halogen bonds. The most frequently used reagent is Bu_3SnH; Ph_3SnH and other organotin hydrides have had more limited employment. Other variations include polymer-bound (immobilized) organotin hydrides, organotin hydrides formed *in situ*, and reductions using $LiAlH_4$ (or $NaBH_4$) with catalytic quantities of Bu_3SnCl. The corresponding deuterides react similarly to give $\equiv$C–D.

10.2.1 *Reductions of halides*

For reactions involving alkyl halides (R′–X) a free-radical chain mechanism occurs (eqns 10.2–10.4). Heat, AIBN or uv light may be used to initiate the reaction, which may be carried out either in the absence of solvent or in such solvents as benzene, toluene, THF, Et_2O or cyclohexane.

$$R_3SnH \xrightarrow{\text{initiation}} R_3Sn^\bullet \tag{10.2}$$

$$R_3Sn^\bullet + R'\text{–}X \xrightarrow{k_3} R_3SnX + R'^\bullet \tag{10.3}$$

$$R'^\bullet + R_3SnH \xrightarrow{k_4} R'\text{–}H + R_3Sn^\bullet. \tag{10.4}$$

Values of the rate constants, k_3, are of the order of 10^2, 10^7 and $10^9\ M^{-1}s^{-1}$ for X = Cl, Br or I respectively. Alkyl fluorides do not normally react.

Rate constants, k_4, for the second propagating step are *c.* $10^6 M^{-1} s^{-1}$. Other organic halides to react include acyl chlorides, cycloalkyl, vinyl, allyl and propargyl chlorides, bromides and iodides, while in the aromatic series, successful reaction is restricted to bromides and iodides. Organic halides can be reduced in the presence of such functional groups as alkenyl, alkynyl, alcohol, ether, peroxide, ketal epoxide, ketone, ester, lactone, lactam, amine, amide, carbamate, sulphide, thioester, sulphone, sulphonate, sulphonamide, aziridine, nitro and cyano groups. The compatibility of organotin hydrides with such a range of groups renders it a particularly useful reagent for this reaction. Work-up procedures are well established and present few problems for small-scale preparations. For larger-scale syntheses, the complete removal of the organotin halide product has been reported not to be as convenient a procedure.

The intermediate free radical, $R'^{\bullet}$, may be trapped by additional substrates, including alkenes (eqn 10.5), or may undergo cyclization or another rearrangement (eqn 10.6) before reaction with R_3SnH. These alternative reactions have been well established in synthetic procedures. The relative importances of reactions (10.3)–(10.8) depend on the values of the rate constants and reagent concentrations.

R'• + >C=C<(Z) $\xrightarrow{k_5}$ R'–C–C•(Z) (10.5)

$$R'^{\bullet} \xrightarrow{k_6} R^{*\bullet} \quad (10.6)$$

$$R^{*\bullet} + R_3SnH \xrightarrow{k_7} R^{*}\text{—H} + R_3Sn^{\bullet} \quad (10.7)$$

R*• + >C=C<(Z) $\xrightarrow{k_8}$ R*–C–C•(Z) (10.8)

Some examples of uncomplicated reductions of organic halides are given in eqns (10.9)–(10.13).[3–7]

$$C_4F_9CH_2CHICH_2OAc \xrightarrow{Bu_3SnH, AIBN} C_4F_9(CH_2)_3OAc \quad (10.9)$$

(93%)

Gem di- or poly-halides can undergo more than one replacement, e.g. eqns (10.14)[8], (10.5)[9] and (10.6).[10] The silylation of the carboxylic acid group in eqn (10.15) is to prevent destruction of the tin hydride reagent by the free acid.

HO(O=)C-alkenoic acid $\xrightarrow{I_2}$ iodomethyl lactone $\xrightarrow{Bu_3SnH}$ methyl lactone (10.10)

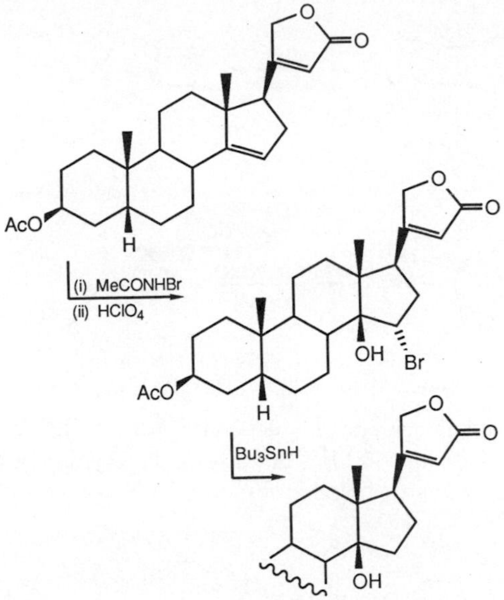

AcO H

(i) MeCONHBr
(ii) $HClO_4$

AcO H OH Br

Bu_3SnH

OH

(10.11)

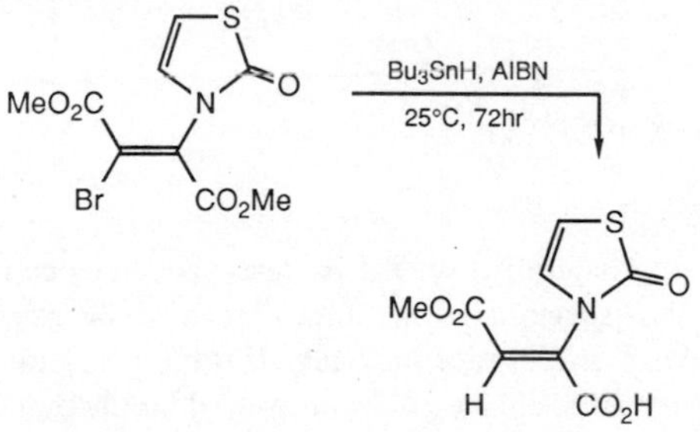

MeO_2C N S O Br CO_2Me

Bu_3SnH, AIBN
25°C, 72hr

MeO_2C N S O H CO_2H

(10.12)

X Br

Bu_3SnD, hν
25°C, ½ - 1hr

X D

X = CN, F or Cl

(>82%)

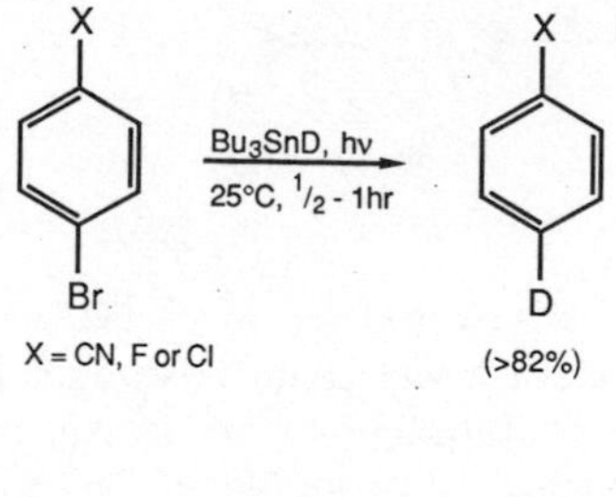

(10.13)

$$CCl_4 \xrightarrow{Bu_3SnH} \underset{(85\%)}{HCCl_3} \xrightarrow{Bu_3SnH} \underset{(94\%)}{H_2CCl_2} \qquad (10.14)$$

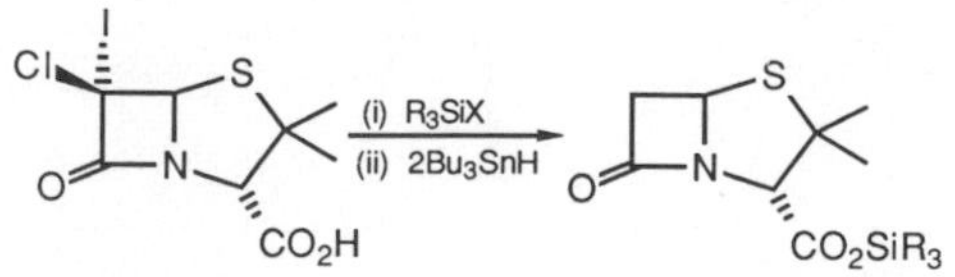

(i) R_3SiX (ii) $2Bu_3SnH$ — CO_2H → CO_2SiR_3 (10.15)

$2Bu_3SnH$, AIBN, C_4H_6, 80°C, 12h (10.16)

The reactivity of the halogen not only depends on the particular halogen (see for example eqns 10.9[3], 10.13[7] and 10.17[11]) but also on its position in the molecule, e.g. eqn (9.16)[10]. Chlorines α to a carbonyl group (or to a benzyl group) are particularly reactive.

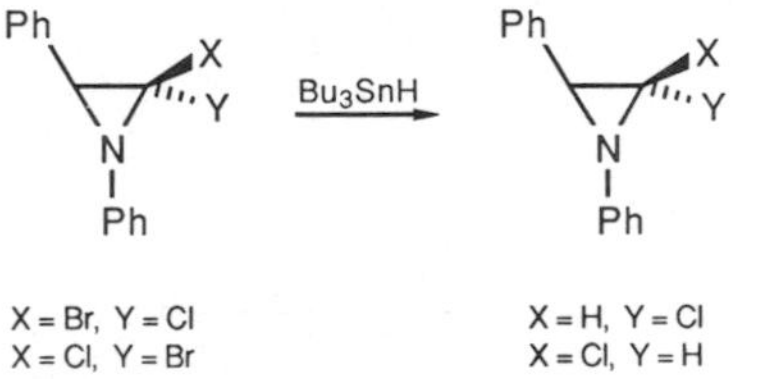

Bu_3SnH

X = Br, Y = Cl → X = H, Y = Cl
X = Cl, Y = Br → X = Cl, Y = H (10.17)

10.2.1.1 *Stereochemistry.* Generally little if any stereospecificity can be expected in this free-radical reaction unless special circumstances prevail. The benzylic[12] halide, (+)-MeCHClPh, for example, provides racemic MeCHDPh on reaction with Ph_3SnD. An example of the lack of stereospecificity in a cyclopropyl halide reaction is given in eqn (10.18).[13]

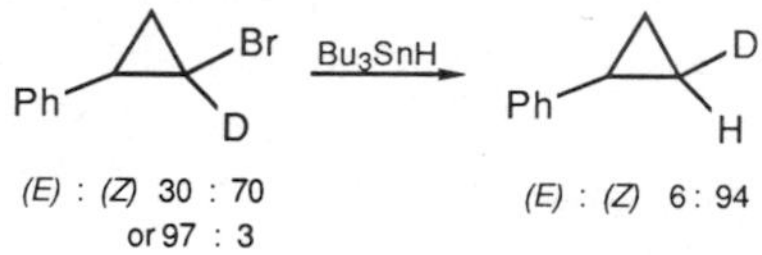

Bu_3SnH

(*E*) : (*Z*) 30 : 70 or 97 : 3 → (*E*) : (*Z*) 6 : 94 (10.18)

The intermediate radicals are usually considered to be planar or rapidly interconverting pyramidal species. Some selectivity could arise, after H transfer, if the radical populations at the conformational equilibrium are not equal. In extreme cases, radicals with complete configurational stability are found. This must be the case in eqn (10.17) for the intermediate 2-halo-2-aziridinyl radical. The strongly electron-withdrawing α-F group in (**1**) stabilizes the pyramidal structure of the intermediate radicals, which lead to highly stereospecific reactions (with retention of configuration) (eqn 10.19).[14]

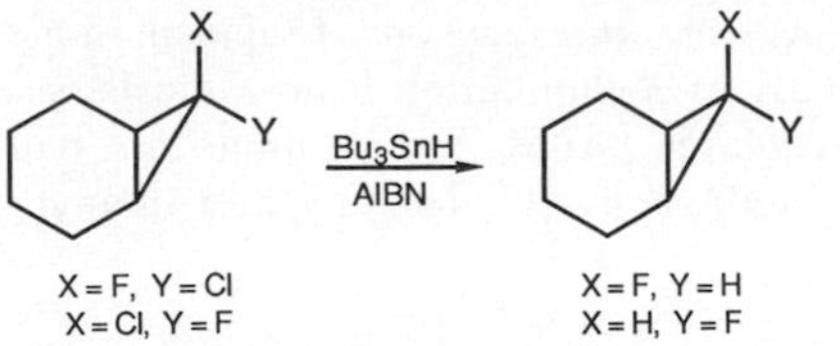

Bu_3SnH, AIBN

X = F, Y = Cl; X = Cl, Y = F **(1)** → X = F, Y = H; X = H, Y = F (10.19)

Higher temperatures and lower concentrations of Bu_3SnH lead to lower stereospecificity. Differences in the degree of steric hindrance to the approach of the tin reagent can lead to preferential attack either from one side of a planar radical or on one of the equilibrating radicals. This must happen in the reaction of 7-halo-7- substituted norcaranes, e.g. (**2**) in eqn (10.20).[15] There is a rapid inversion of the intermediate radical which is then attacked by the organotin hydride from the least hindered side.

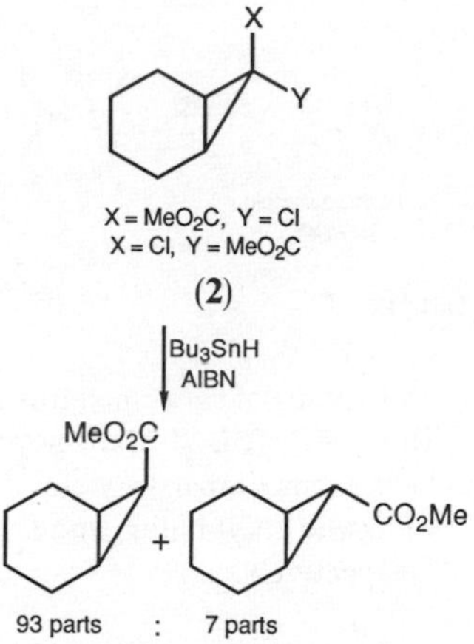

X = MeO_2C, Y = Cl; X = Cl, Y = MeO_2C **(2)**

Bu_3SnH, AIBN

MeO_2C ... + ... CO_2Me

93 parts : 7 parts (10.20)

Steric control[16] is also realized in eqn (10.21).

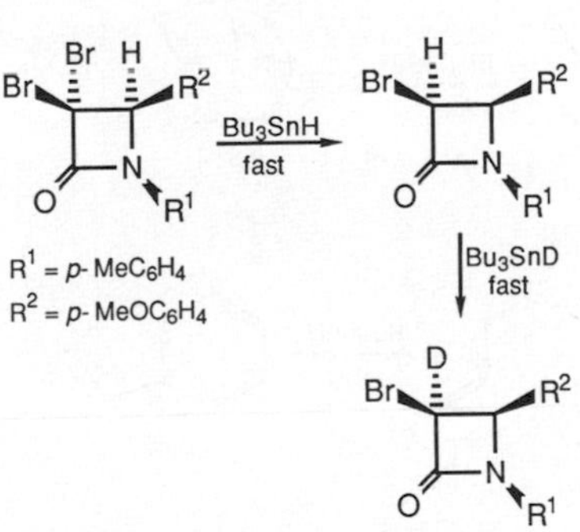

R^1 = p-MeC_6H_4
R^2 = p-$MeOC_6H_4$

Bu_3SnH, fast; Bu_3SnD, fast (10.21)

Vinyl halide reductions are reported not to be very stereospecific.[1] Moreover, stereospecific reductions are *not* found with 2-halonorbornanes (major attack on the intermediate radical by R_3SnH occurring from the less hindered exo side) and 7-norborenyl halides (predominant attack of R_3SnH from the side opposite the double bond).

10.2.1.2 *Radical rearrangements and other reactions.* Fragmentation of the intermediate radical may arise, particularly by β-elimination from radicals bearing good leaving groups (e.g. halogen or arylthio) in β-sites. A most important reaction of the initial radical is a ring-closure in suitably substituted alkenyl and alkynyl species (Scheme 1).

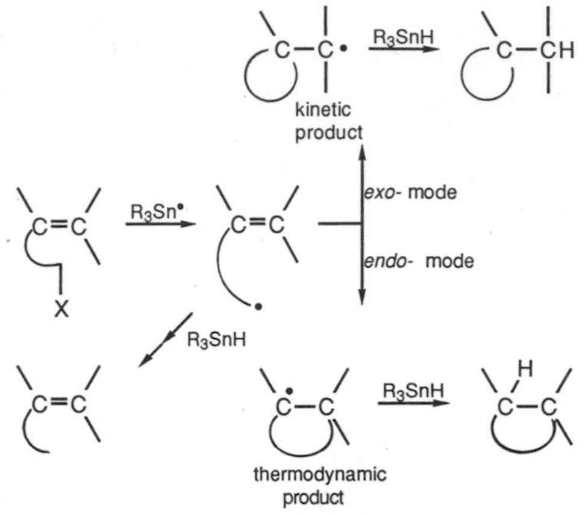

$R_3Sn^\bullet$; R_3SnH; kinetic product; *exo-* mode; *endo-* mode; R_3SnH; thermodynamic product; R_3SnH

Scheme 1

Relative concentrations of reagents and values of rate constants are clearly of great significance in deciding the relative extents of cyclized and uncyclized products. In many cases conditions have been found which maximize the yields of desirable cyclized products. Examples of cyclization for a bicyclic alkyl halide and for an aryl halide are shown in eqns (10.22)[17] and (10.23)[18], respectively.

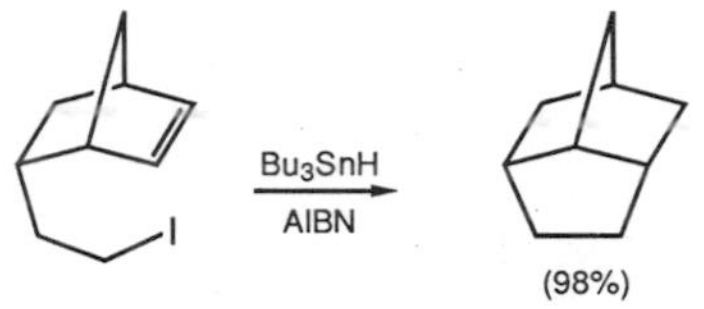

Bu_3SnH, AIBN (98%) (10.22)

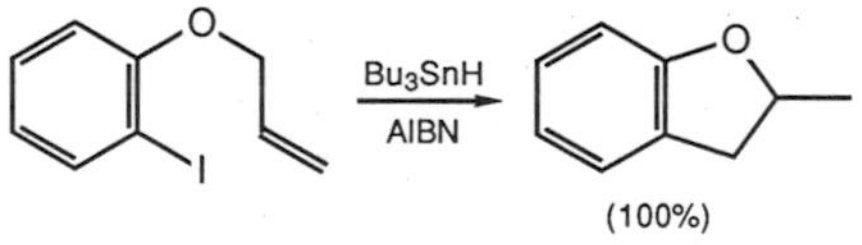

Bu_3SnH, AIBN (100%) (10.23)

Examples of other heterocyclic formations are provided in eqns (10.24)–(10.26).[19–21]

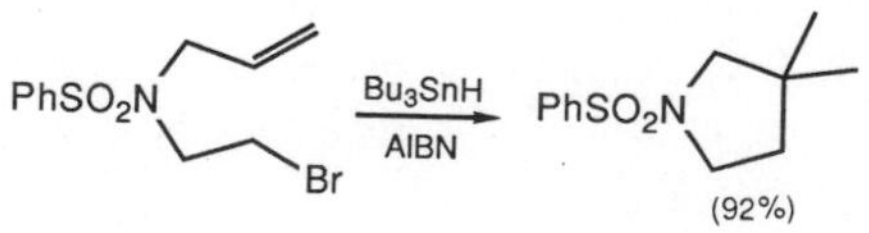

$PhSO_2N$ ... Br $\xrightarrow[AIBN]{Bu_3SnH}$ $PhSO_2N$ ... (92%) (10.24)

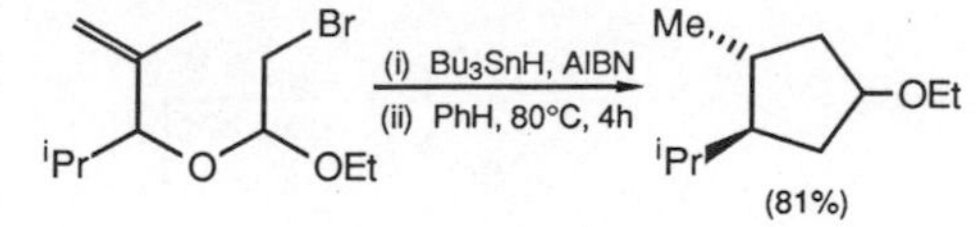

(i) Bu_3SnH, AIBN
(ii) PhH, 80°C, 4h
(81%) (10.25)

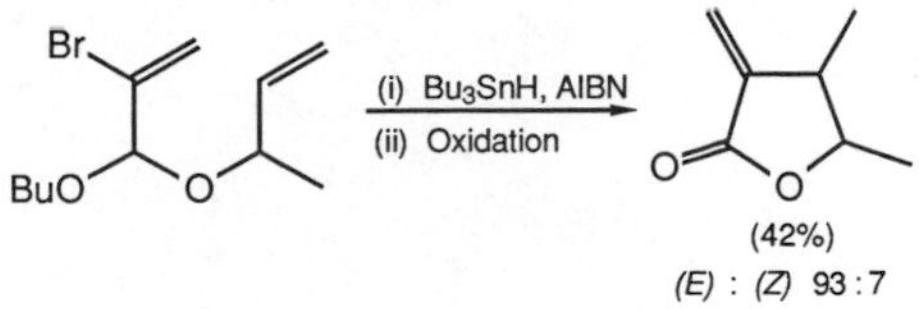

(i) Bu_3SnH, AIBN
(ii) Oxidation
(42%)
(E) : (Z) 93 : 7 (10.26)

These examples illustrate the preference for the formation of a five-membered over a six-membered ring, i.e. the *exo*-mode. However, *endo*-cyclization can be favoured whenever there is a radical-stabilizing group, such as phenyl, on the non-terminal carbon of the unsaturated group.[22]

Frequently *cis*-cyclization occurs, for example as shown in eqn (10.27).[23]

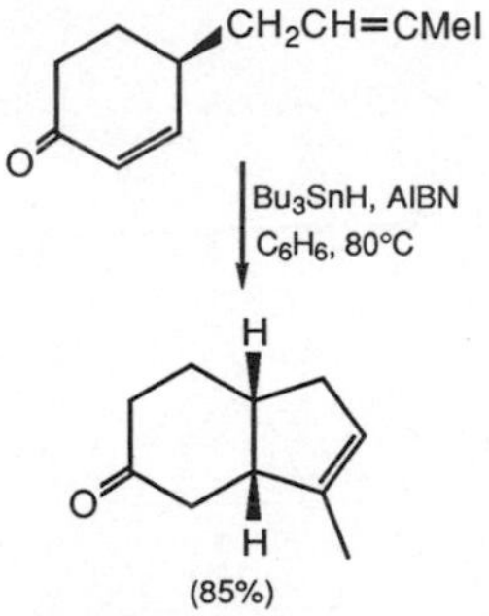

$CH_2CH{=}CMeI$

Bu_3SnH, AIBN
C_6H_6, 80°C

(85%) (10.27)

Formation of two rings (a tandem cyclization) can result with suitably substituted precursors, e.g. Scheme 2[24] and eqn (10.28).[25]

Six-membered rings can be obtained (by *exo*-mode) from 6-hept-enyl and -ynyl related radicals as illustrated in eqns (10.29)–(10.31):[20,26,27]

In eqn (10.30), *cis*-cyclization clearly occurs. The *exo*-cyclization of (**3**) shown in eqn (10.31) is in contrast to the *endo*-cyclization exhibited by (**4**) in eqn (10.32), the difference being attributed to the presence of a terminal radical stabilizing group (phenyl) in (**3**) but not in (**4**).[27]

10.2.1.3. *Intermolecular C–C coupling.*[28] Trapping of the initial radical R′• by an external substrate, e.g. an alkene, has been to provide interesting and valuable products. For successful trapping, the alkene has to win the competition with the tin hydride, R_3SnH, for the radical R′•, e.g. eqns (10.3) and (10.5). Not only are the relative concentrations of R_3SnH and alkene of importance, but clearly so are the values of the rate constants, k_3 and k_5. Success becomes more probable by choosing an electron-poor alkene to react with a nucleophilic alkyl radical (Scheme 3[28] and (10.33)[29]), or

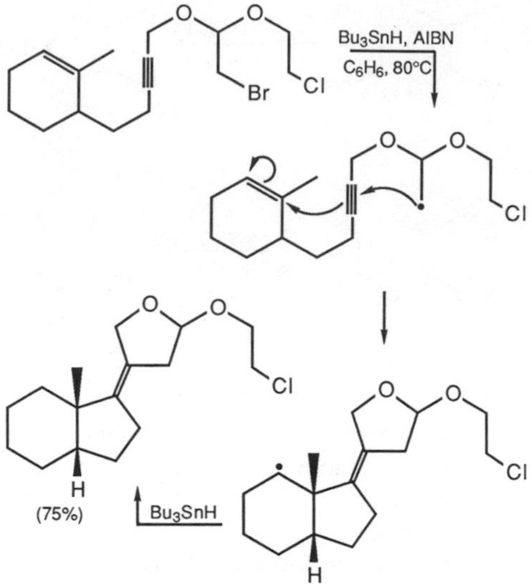

Bu$_3$SnH, AIBN
C_6H_6, 80°C

(75%)

Bu$_3$SnH

Scheme 2

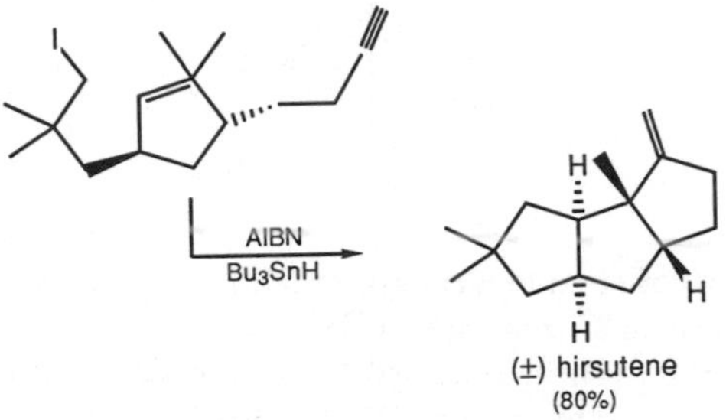

AIBN
Bu$_3$SnH

(±) hirsutene
(80%)

(10.28)

Bu$_3$SnH, AIBN
C_6H_6

(75%)

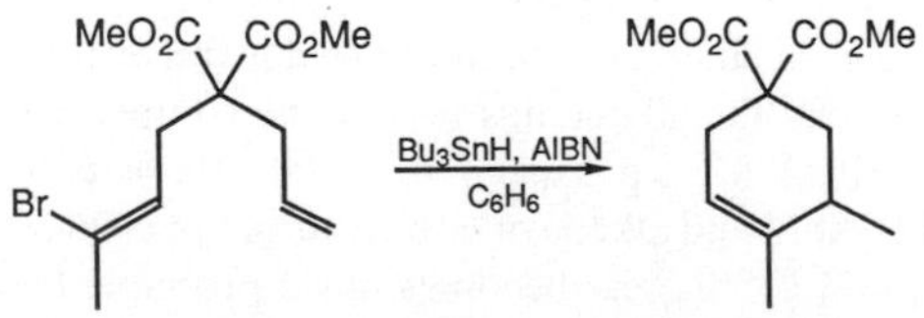

(10.29)

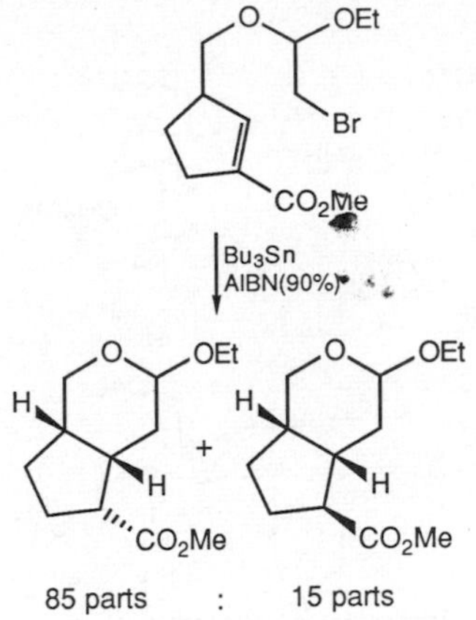

OEt, Br, CO_2Me

Bu_3Sn AIBN(90%)

OEt, H, H, CO_2Me + OEt, H, H, CO_2Me

85 parts : 15 parts

(10.30)

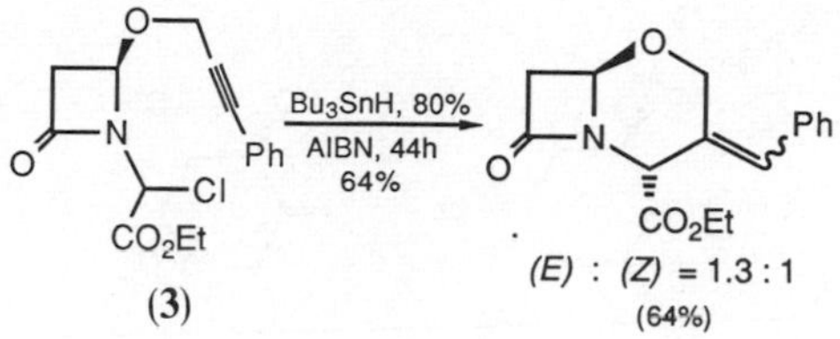

Bu_3SnH, 80% AIBN, 44h 64%

(3) → CO_2Et, Ph; (E) : (Z) = 1.3 : 1 (64%)

(10.31)

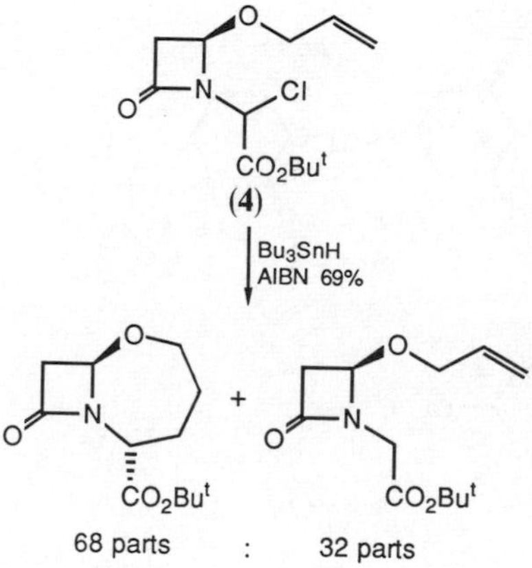

(4) CO_2Bu^t, Cl

Bu_3SnH AIBN 69%

CO_2Bu^t + CO_2Bu^t

68 parts : 32 parts

(10.32)

conversely an electron-rich alkene to trap an electrophilic radical, such as $(EtO_2C)_2CH\cdot$, (eqn (10.34)).

The high stereospecificity in Scheme 3 is due to the stabilization of the intermediate sugar radical (**5**) by the neighbouring oxygen. This then leads to the thermodynamically less stable α-anomer product (**6**).

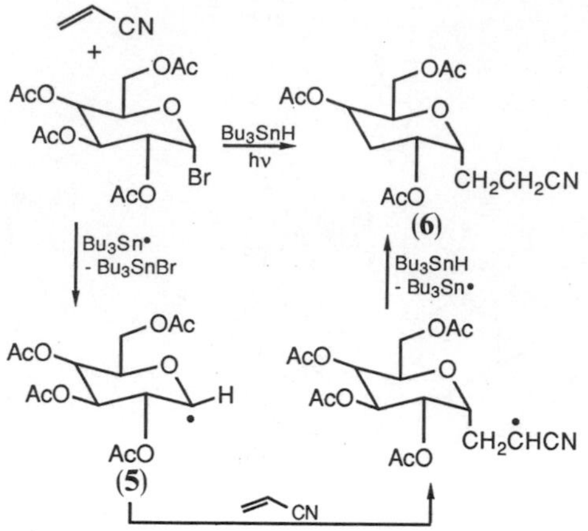

Scheme 3

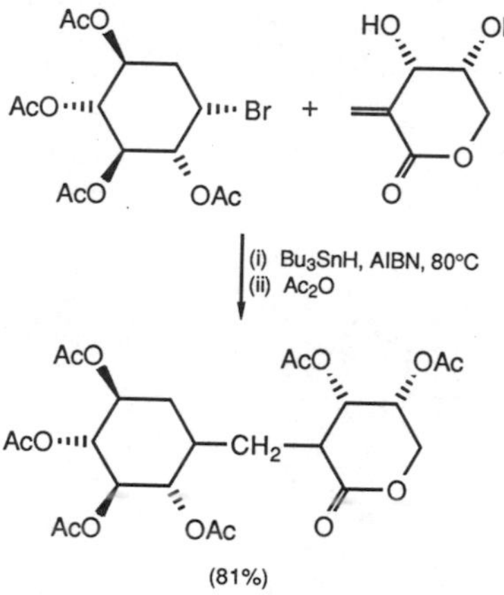

(10.33)

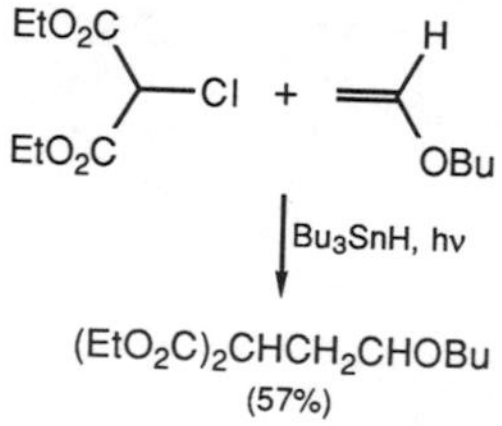

$(EtO_2C)_2CHCH_2CHOBu$ (57%) (10.34)

A further twist in these reactions is the intermolecular trapping after the rearrangement of the initial radical (e.g. Scheme 4, which illustrates a stereo and regio-selective sequence of an intramolecular C–C coupling followed by an intermolecular C–C coupling).[30]

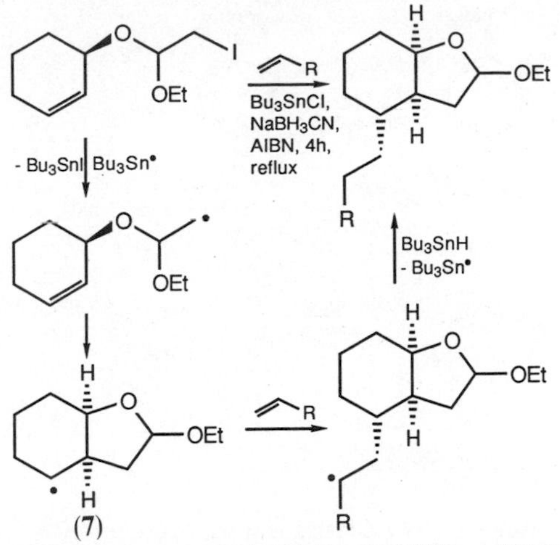

(7)

Scheme 4; R = CN, COEt, CO_2Me, $P(O)(OEt)_2$ or SO_2R'

Scheme 4

Trapping of ($7 \equiv R^*\cdot$) by Bu^tNC provides R^*–CN in a yield of 60%.

Aryl radicals have been trapped by electron-poor alkenes, as in the reaction of ArI, CH_2=CHCN and Bu_3SnH, which gives $ArCH_2CH_2CN$.[28]

10.2.2 *Reductions of carbonyl compounds*

10.2.2.1 *Reduction of aldehydes and ketones*[2]. Tin hydrides add to carbonyl compounds to give tin alkoxides, subsequent hydrolysis providing alcohols (eqn 10.35). Aldehydes are more reactive than ketones.

$$R^1R^2C{=}O \xrightarrow[\text{(ii) } H^+]{\text{(i) } R_3SnH} R^1R^2CHOH \qquad (10.35)$$

Reactions can proceed both via a free-radical mechanism (using AIBN or uv initiation) or preferably via a Lewis-acid-catalysed polar mechanism. A particularly useful Lewis acid catalyst is silica gel. Other functional groups such as sulphoxide, esters and cyanides are unaffected.[31] Tributyltin triflate has also proved to be a valuable catalyst.[32] Conjugate reduction of α,β-unsaturated carbonyl compounds[33] is achieved using Bu_3SnH in the presence of $Pd(PPh_3)_4$, e.g. eqn (10.36).

$$PhCH{=}C{-}CHO \xrightarrow[Pd(PPh_3)_4]{Bu_3SnH} PhCH_2CH_2CHO \quad (>99\%) \qquad (10.36)$$

10.2.2.2 *Reduction of acyl halides*[1]. Acyl halides, R^1COX, are reduced by R_3SnH to aldehydes, R^1CHO. However, the competitive formation of esters, $R^1CO_2R^1$, reduces the yields of aldehydes.[34] Better yields of aldehydes are obtained using Bu_3SnH with $Pd(PPh_3)_4$ as catalyst (e.g. eqns 10.37, 10.38)[35].

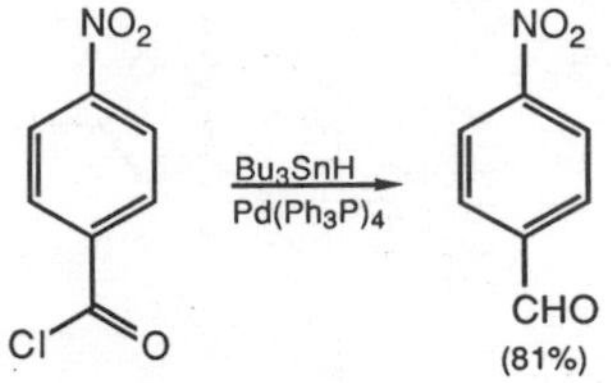

(10.37)

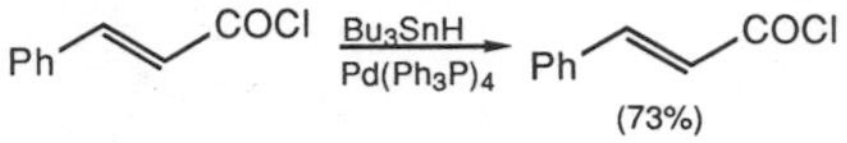

(10.38)

10.2.3 *Reductions of other bonds*

10.2.3.1 *Alcohols.* While a C–OH bond is not directly reduced to C–H by organotin hydrides, simple conversions to other groups open up routes. One method is via the conversion to a halide as illustrated in Scheme 5.[36,37]

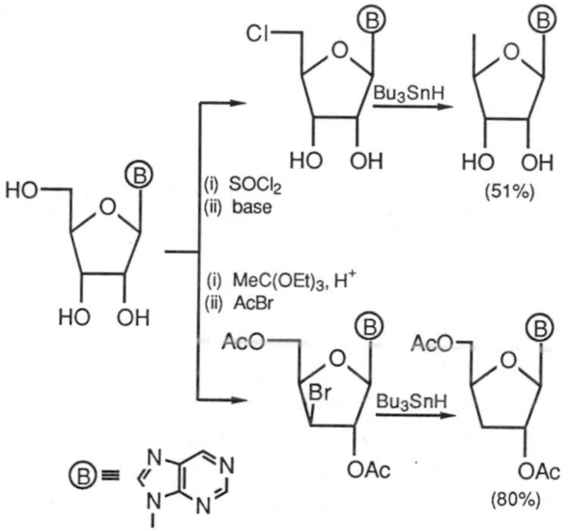

Scheme 5

Particularly useful conversions of the alcohol, R^1–OH, are to thiocarbonyl derivatives,[38] R^1–O–C(S)R^2, including R^2=SMe,[38] H,[38] 1-imidazoyl[39] and OPh[40] (Scheme 6).

S-Methyl thiocarbonates, R^1OC(S)SMe, are convenient intermediates (e.g. eqn 10.39), being converted to R^1–H at 130–150°C for R^1 = primary alkyl and at 80–110°C for secondary groups.[41] Many groups do not usually interfere, including tosylate, mesylate, carbonate, alkenyl, alkynyl, carbonyl, epoxide, amine, alcohol and ester groups; however, nitro, halogen, and cyano on the other hand can compete.

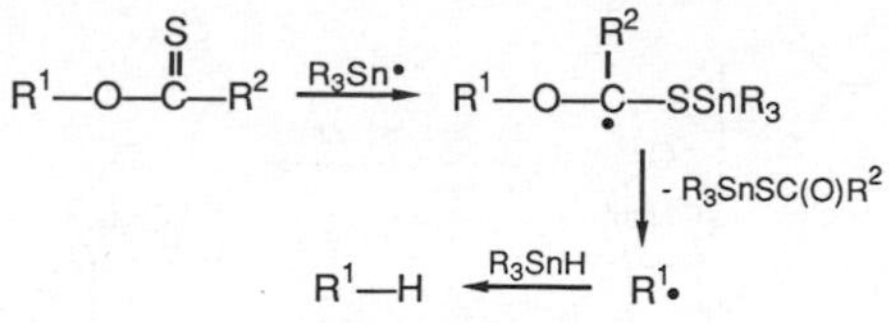

$$R^1{-}O{-}C(=S){-}R^2 \xrightarrow{R_3Sn\bullet} R^1{-}O{-}\dot{C}(R^2){-}SSnR_3 \xrightarrow{-R_3SnSC(O)R^2} R^1\bullet \xrightarrow{R_3SnH} R^1{-}H$$

Scheme 6

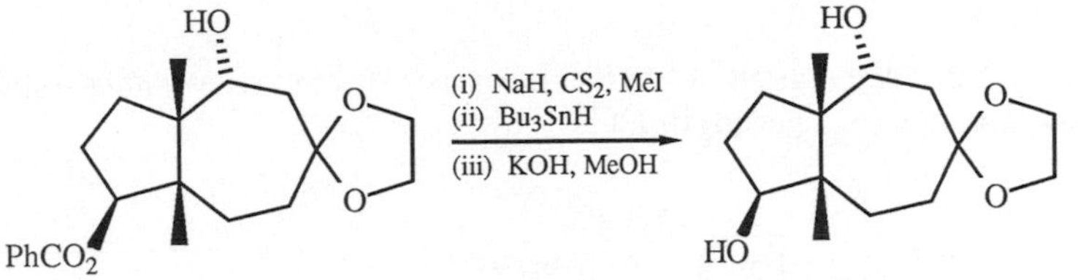

(i) NaH, CS_2, MeI
(ii) Bu_3SnH
(iii) KOH, MeOH

28% (10.39)

The stereochemistry of the product depends on the stability of the intermediate radical(s) and on steric hindrance to the approach of the organotin hydride. The attack of the Bu_3SnD reagent on the intermediate radicals in eqn (10.40) is directed by steric factors.[42]

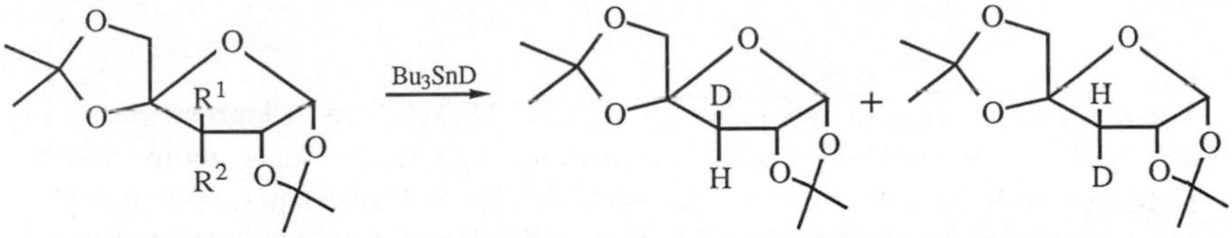

Bu_3SnD

85 parts : 15 parts (10.40)

Cyclic thiocarbonates, obtained from 1,2-diols,[1,2,38,43] react with Bu_3SnH to furnish a monohydroxy compound. Ease of deoxygenation of hydroxy groups is in the sequence tertiary > secondary > primary, i.e. the sequence for stabilities of the intermediate radicals, such as (**8**) in Scheme 7. A particular example is shown in eqn (10.41).[43]

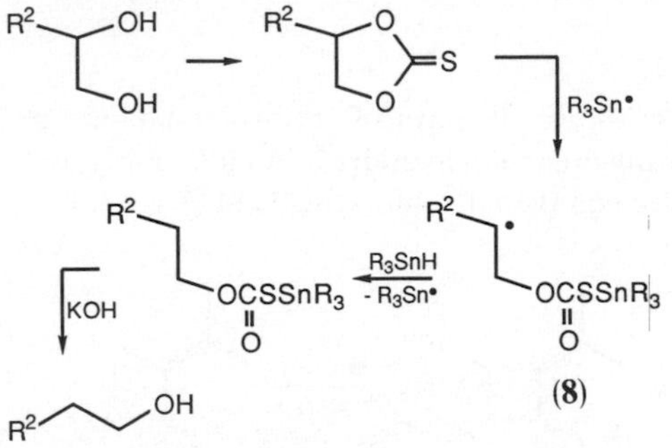

$R_3Sn\bullet$; R_3SnH, $-R_3Sn\bullet$; KOH

(**8**)

Scheme 7

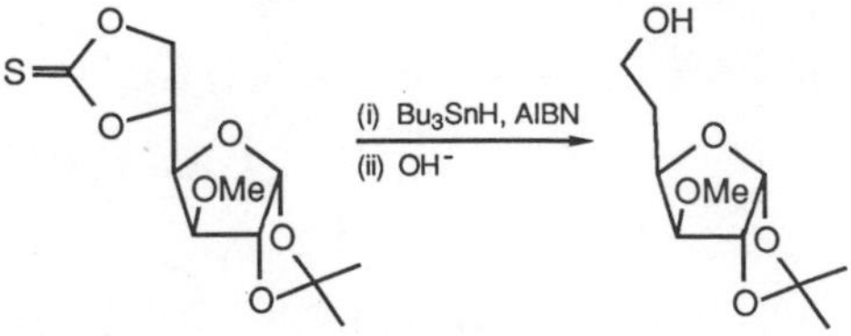

(10.41)

The intermediate alkyl radicals, R′•, in the thiocarbonate reductions can be trapped by electron-poor alkenes, e.g. eqn (10.42).[28]

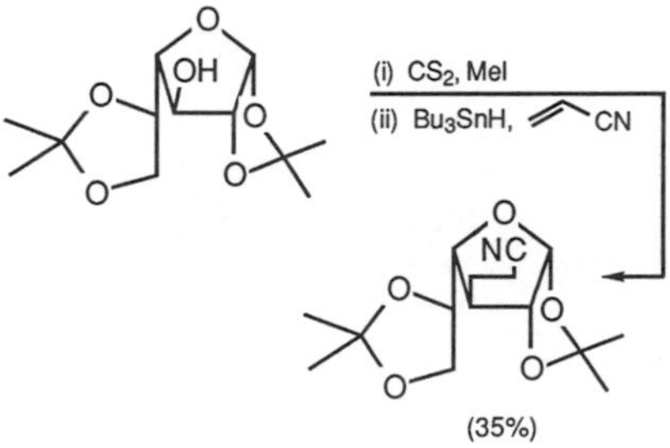

(10.42)

Benzoates of reactive alcohols (e.g. allylic, benzylic, or α-keto-alcohols) are reduced[38] by Bu_3SnH, but p-cyanobenzoates apparently react more readily.[2] Tosylates, in the presence of NaI, also react with Bu_3SnH, e.g. eqn (10.43), reduction occurring via the iodides (eqn 10.43).[44] Vinyl triflates are readily reduced by Bu_3SnH in the presence of $Pd(Ph_3P)_4$ and LiCl.[45]

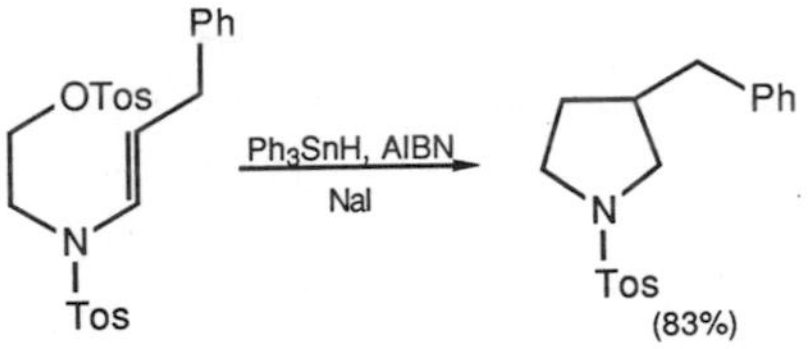

(10.43)

10.2.3.2 *Sulphur derivatives.* Removal of a thiol or a sulphido group, R^2S, from carbon can be achieved using organotin hydrides.[46] A particularly useful leaving group is the PhS group[47,48] (see eqn (10.44)[47] and Scheme 8).[49]

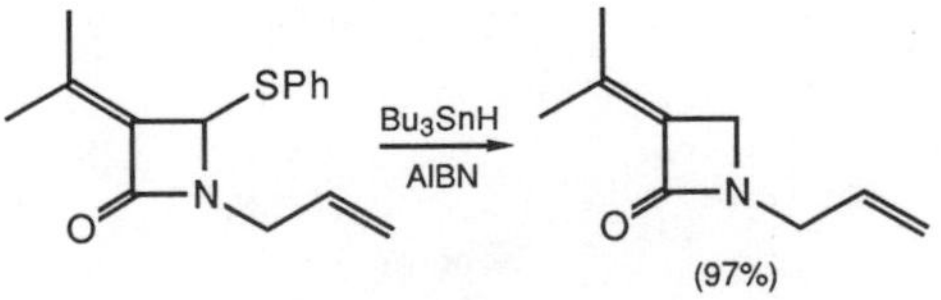

(10.44)

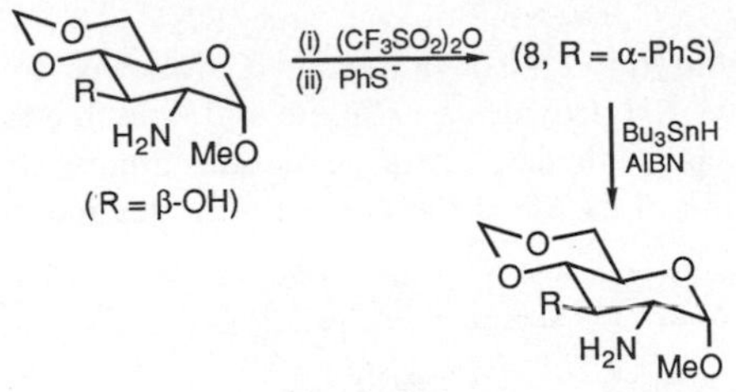

Scheme 8

Scheme 8 indicates an overall reduction of a hydroxy compound, via its phenylthio derivative.

Thioketals, $R^1{}_2C(SR^2)_2$, are reduced to the hydrocarbon[50] $R^1{}_2CH_2$.

10.2.3.3 *Selenium and tellurium derivatives.* Phenyl-selenides[51] and -tellurides[52] are reduced by organotin hydrides. Examples of the phenylselenide[53–55] reductions are shown in eqns (10.45)–(10.47).

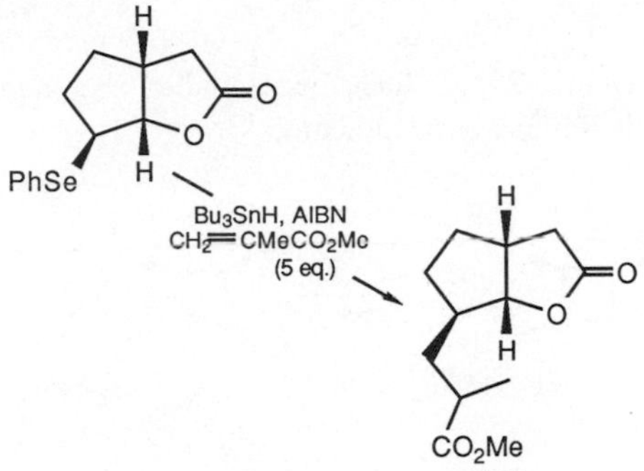

(10.45)

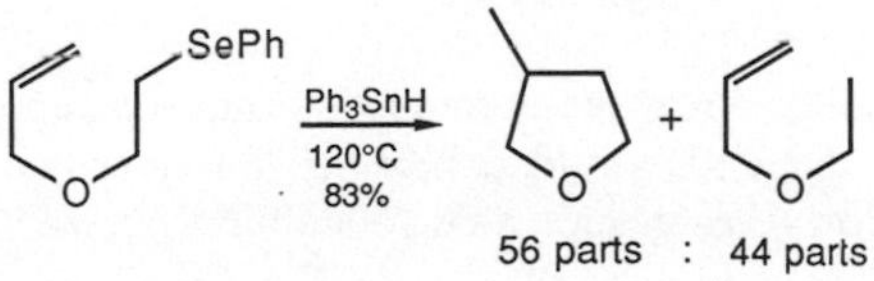

(10.46)

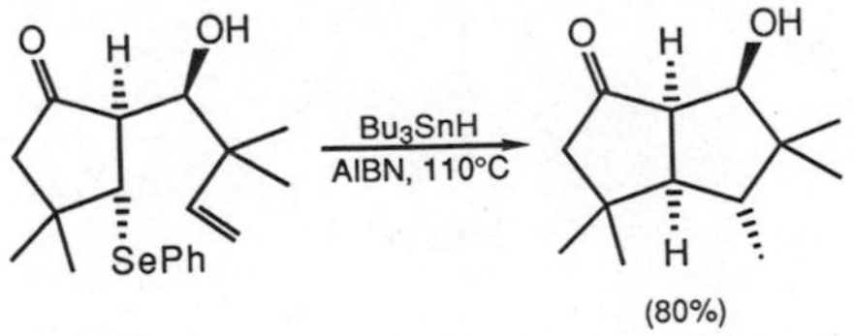

(10.47)

Like thioketals, selenoketals, $R^1{}_2C(SeR^2)_2$, are also reduced[53] by organotin hydrides to $R^1{}_2CH_2$.

10.2.3.4 *Deamination*[56,57]. Conversion of an amine, RNH_2 (R = primary, secondary or tertiary alkyl but not aryl) to an isocyanide, RNC, followed by treatment with an organotin hydride to give RH provides a valuable and mild free-radical deamination sequence. Groups such as aldehydes, esters, secondary amines and sulphonates are unaffected. An illustration of its use is the overall tetra-deamination of neamine (**9**), shown in eqn (10.48).[58]

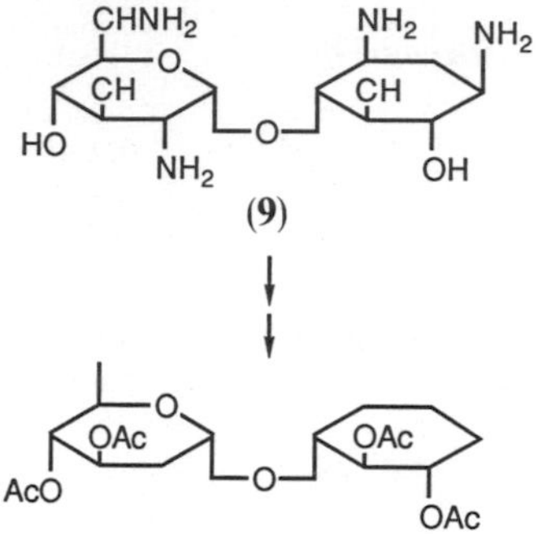

(**9**)

(10.48)

Reductions of isocyanides, R^1NC, have been studied, (e.g. eqn 10.49)[59], and the intermediacy of radicals $R^1\cdot$ has been indicated.[1]

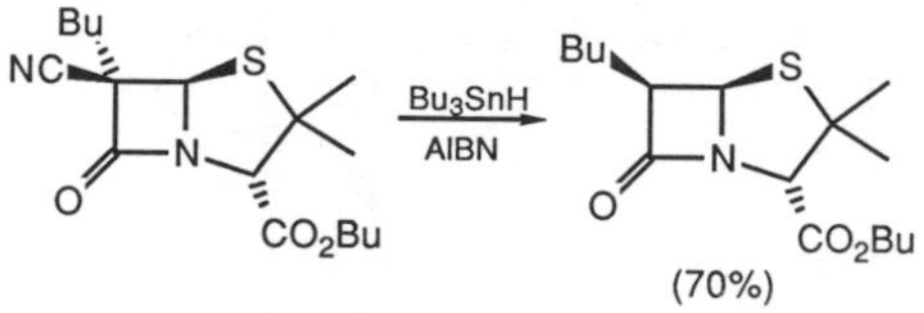

Bu_3SnH, AIBN

(70%) (10.49)

Isothiocyanates and iso-selenocyanates can be similarly reduced.[1,58]

10.2.3.5 *Denitration*[1,60]. The direct replacement of a nitro group by hydrogen using R_3SnH occurs readily and under mild conditions for *tert*-alkyl, benzyl and allyl compounds as well as for other groups able to stabilize free radicals. More forcing

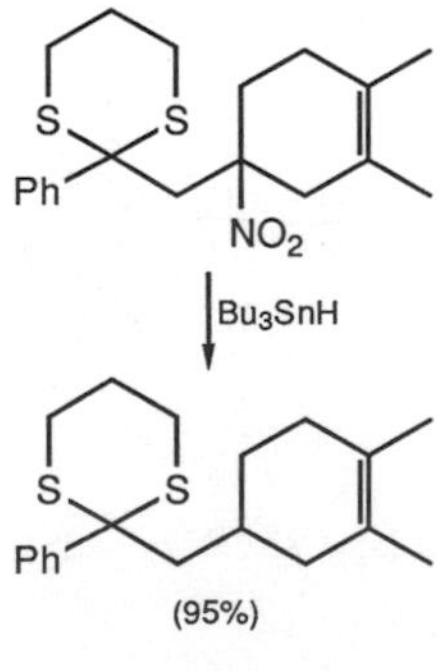

Bu_3SnH

(95%) (10.50)

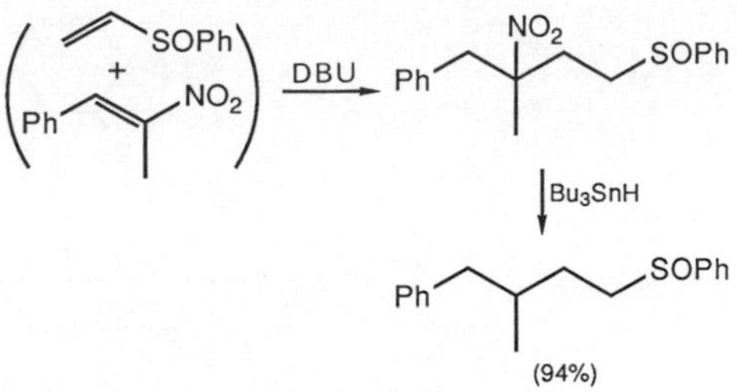

(10.51)

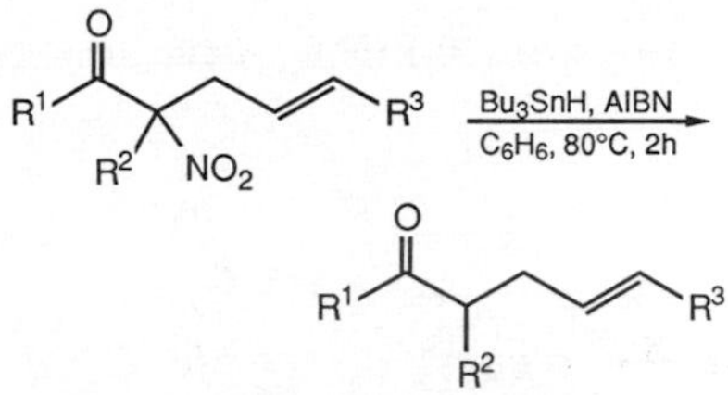

(10.52)

conditions are required for secondary nitroalkanes, while nitroarenes and primary nitroalkanes are unreactive. An s.e.t. mechanism has been proposed. Such groups as Cl, HO, SOR, COR, CO_2R, SO_2R and OMe are tolerated. Examples are shown in eqns (10.50)–(10.52).[60–62]

In contrast to eqn (10.51), α,β-unsaturated nitro compounds react with Bu_3SnH to give either nitroalkanes or carbonyl compounds (Scheme 9).[63]

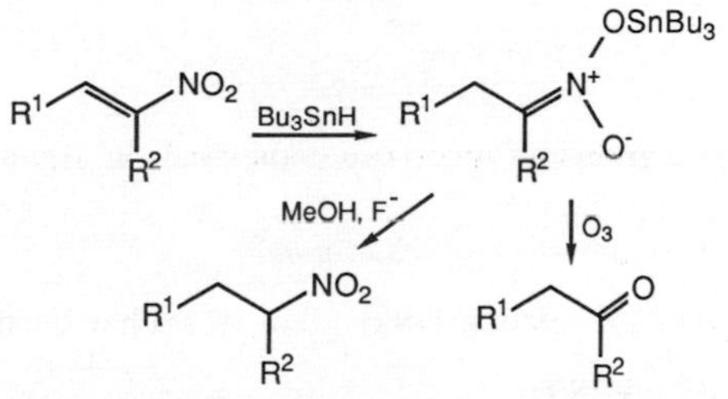

Scheme 9

1,2-Dinitro compounds and other β-substituted nitroalkanes (e.g. β-nitrosulphones and -sulphides) can lead to alkene formation. Elimination from a β-nitrosulphone, such as $EtMeCNO_2C(CN)MeSO_2Ph$, occurs with high *anti*-stereospecificity.[64]

10.2.3.6 *Decarboxylations.* Decarboxylation of $R^1{-}CO_2H$ to $R^1{-}H$ can be brought about by triorganotin hydrides in two ways:

(i) via N-hydroxypyridine-2-thione esters (e.g. Scheme 10).[65]

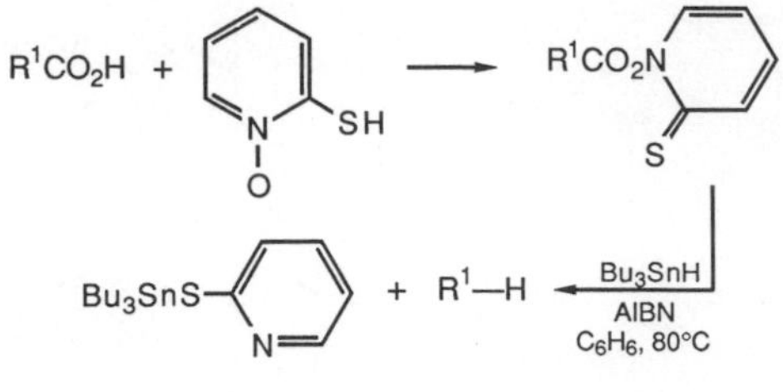

Scheme 10

and (ii) via the ester of 9-HO-10-Cl- or -10-PhS-9, 10-dihydrophenanthrene (10, R*–OH) (Scheme 11).[56]

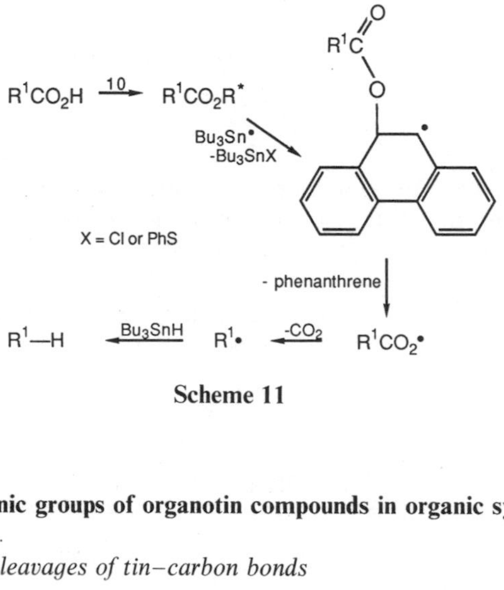

X = Cl or PhS

Scheme 11

10.3 Use of the organic groups of organotin compounds in organic synthesis[1]

10.3.1 *Electrophilic cleavages of tin–carbon bonds*

Electrophilic cleavage of tin–carbon bonds (eqn 10.53) has been variously used in

$$RSn\equiv + E{-}Y \rightarrow R{-}E + Y{-}Sn\equiv \qquad (10.53)$$

organic synthesis; Aryl-, vinyl- and allyl-tin compounds are especially useful, and protonolysis, halogenation and nitrosation can all be accomplished readily (see eqns (10.54)[66] and (10.55)[67] and Schemes 12[68,69] and 13[70]). Included in this area, particularly for allyltins and other reactive organotin species, are additions to aldehydes and ketones (see Section 10.3.2).

AcOD, D_2O, 60 - 80%; isotopic purity ≥ 98% (10.54)

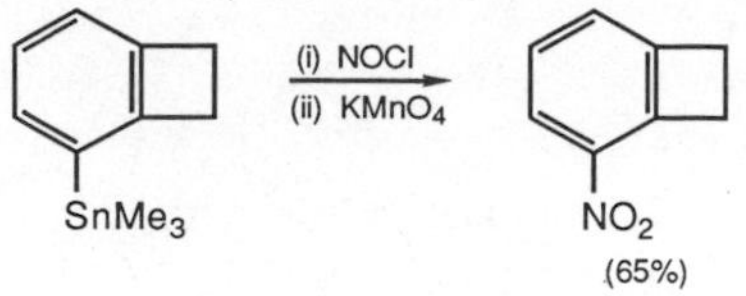

(10.55)

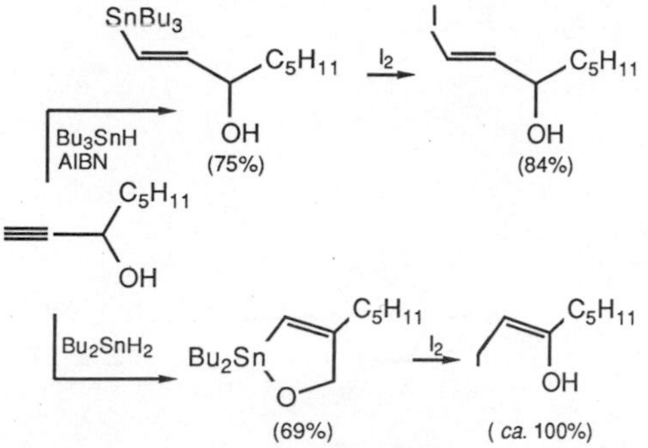

Scheme 12

Electrophilic cleavages of tin–carbon bonds have been used to introduce isotopic labelling, including deuterium (e.g. eqn 10.54), tritium (e.g. using[71] $(CF_3CO_2)_2O + T_2O$), ^{82}Br (e.g. using[72] $NH_4{}^{82}Br$, chloroamine-T and HCl; as well as other bromine isotopes), and ^{125}I (using[73] $^{125}I_2$ or ^{125}ICl).

There are particular advantages in using electrophilic aryltin cleavages over direct substitution of arenes, in that the cleavage product, which may not even be obtainable directly from the arene, is obtained regioselectively. Cleavages of vinyltin bonds occur, in the main, with retention of configuration while those of allyltin bonds usually proceed with allylic rearrangement (Scheme 13). The stereoselectivity is also high with proton acids, an anti S'_E mechanism has been established (see eqn 10.56).[74] Induced isomerization of the allyl-tin prior to reaction will limit the stereoselectivity. Protonolysis of allyltins provides a useful route to alkenes (e.g. eqn (10.56) and Scheme 13).

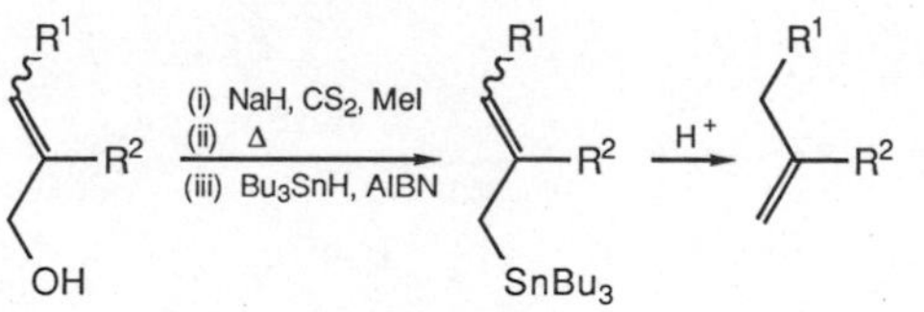

Scheme 13

10.3.2 *Reaction with carbonyl compounds*[1]

Allyltin compounds add (reversibly) to carbonyl compounds to provide, after hydrolysis, homoallylic alcohols (Scheme 14).[75]

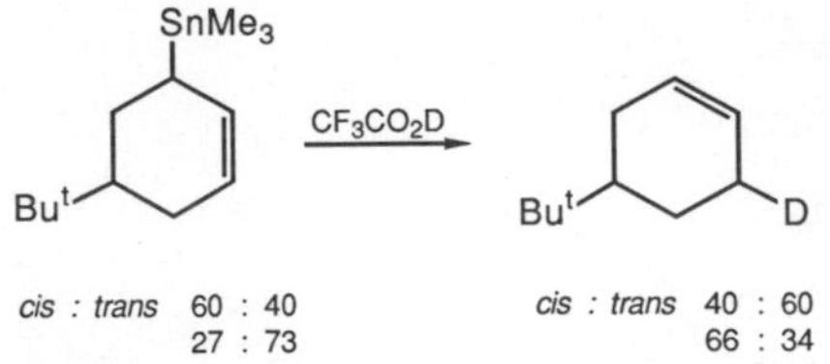

$SnMe_3$ ... CF_3CO_2D ... Bu^t ... D

cis : *trans* 60 : 40 27 : 73 — *cis* : *trans* 40 : 60 66 : 34 (10.56)

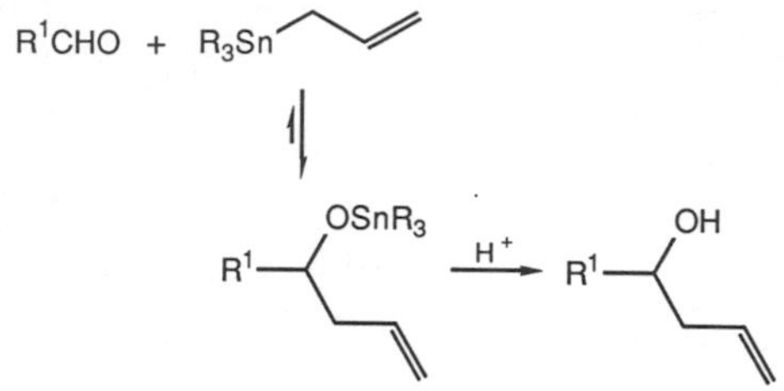

$R^1CHO + R_3Sn$ ⇌ $OSnR_3$, R^1 → (H^+) OH, R^1

Scheme 14

The presence of a Lewis acid (such as BF_3 or $TiCl_4$), or use of allyltin halides (e.g. $ClBu_2SnCH_2CH{=}CH_2$ rather than $Bu_3SnCH_2CH{=}CH_2$) increases the reactivity, and thus the range of carbonyl compounds which can be allylated under reasonable conditions. High-pressure reactions have also been used.[76]

The allyltin halide approach, pioneered by Tagliavini,[77] is successful for many aldehydes and ketones (e.g. eqn 10.57). The reactivity of the allylic tin species has been established as $BrCl_2SnAll > Cl_2BuSnAll > ClBu_2SnAll > Bu_3SnAll$ (All = allyl or crotyl)[78,79]. Such mixed chlorides can be made *in situ* (e.g. Scheme 15).

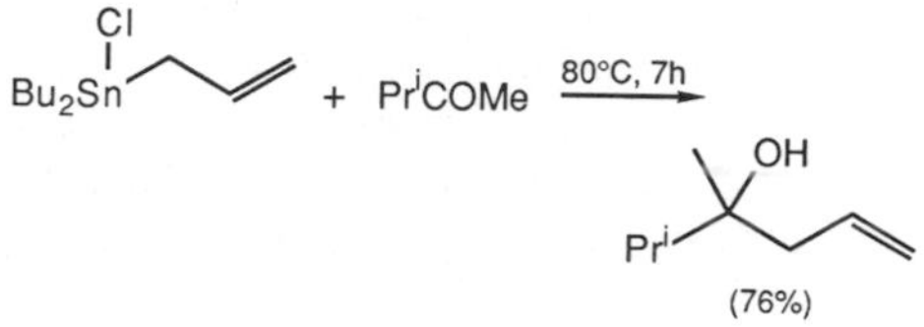

Cl, Bu_2Sn + Pr^iCOMe → (80°C, 7h) OH, Pr^i (76%) (10.57)

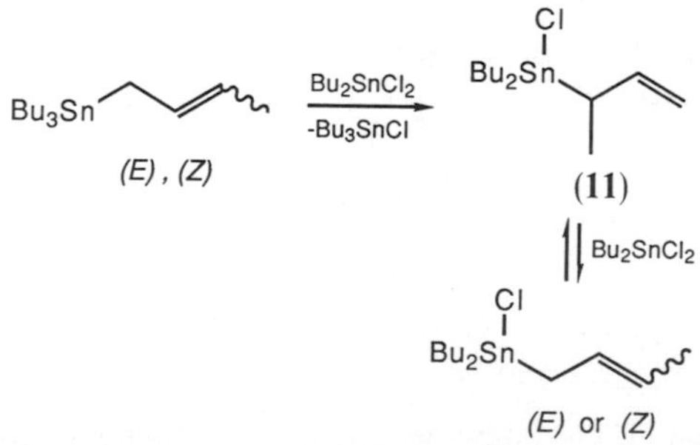

Bu_3Sn (*E*), (*Z*) → (Bu_2SnCl_2, $-Bu_3SnCl$) Cl, Bu_2Sn (**11**) ⇌ (Bu_2SnCl_2) Cl, Bu_2Sn (*E*) or (*Z*)

Scheme 15

The initially-formed species (**11**) in Scheme 15 can react with RCHO as it is generated to give as the major product (*Z*)-$RCHOHCH_2CH{=}CHMe$ as well as *threo*- and *erythro*-$RCHOHCHMeCH{=}CH_2$, in proportions depending upon R (eqn 10.58).[80]

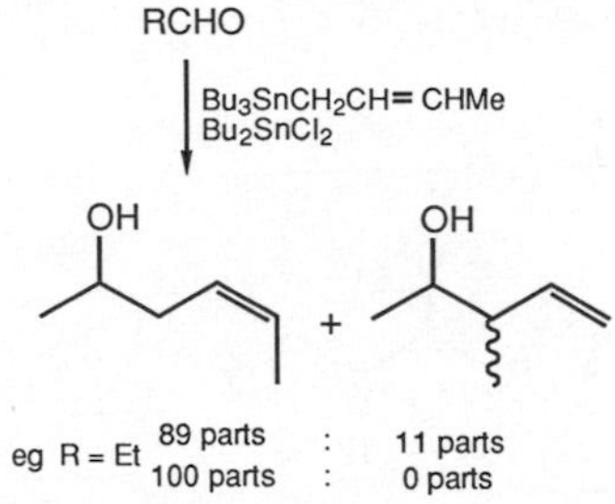

RCHO

$Bu_3SnCH_2CH{=}CHMe$
Bu_2SnCl_2

OH + OH

eg R = Et 89 parts : 11 parts
100 parts : 0 parts

(10.58)

The reactions can be carried out neat, in solution, or in the presence of water.[81] Other variations include the *in-situ* formation of the allylic tin from an allylic substrate with either tin or $Sn^{II}X_2$ species,[82,83] (e.g. eqn 10.59).

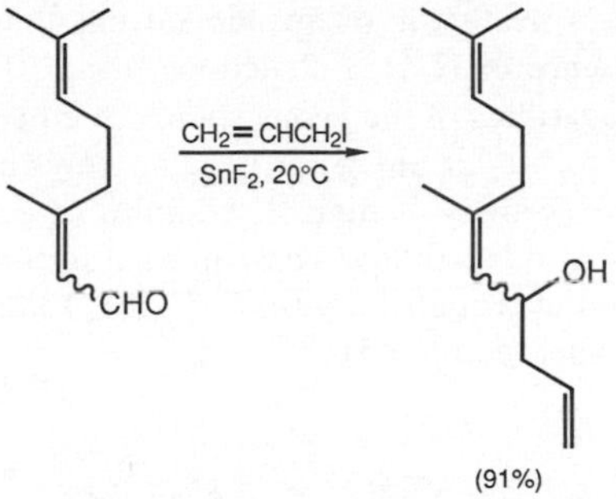

CHO

$CH_2{=}CHCH_2I$
SnF_2, 20°C

OH

(91%)

(10.59)

Lewis-acid-catalysed reactions can occur under very mild conditions (eqns 10.60[84] and 10.61[85]).

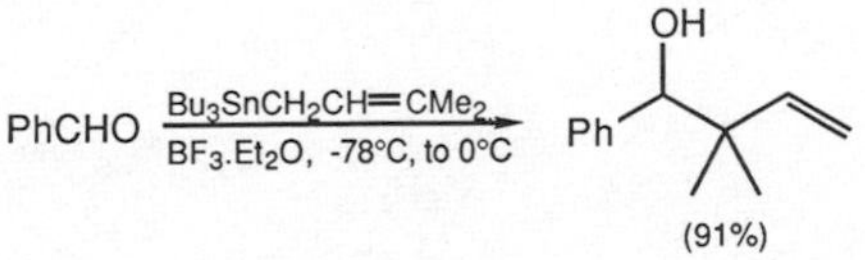

PhCHO $\xrightarrow[BF_3.Et_2O,\ -78°C,\ to\ 0°C]{Bu_3SnCH_2CH{=}CMe_2}$ Ph OH

(91%)

(10.60)

$SnBu_3$

$(H_2CO)_n$, $BF_3.Et_2O$
CH_2Cl_2, 25°C

OH

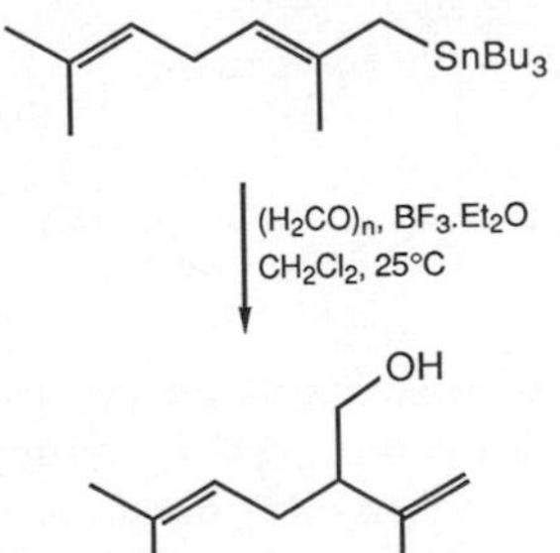

(10.61)

10.3.2.1 *Stereochemistry.* The stereochemistry of the reaction has been frequently and variously studied for many substituted allyltin reagents, under different conditions. *A priori*, both *erythro* and *threo* isomers are possible (eqn 10.62).

$$R^1CHO + R_2CH{=}CHCH_2SnR_3 \longrightarrow \textit{erythro} + \textit{threo} \qquad (10.62)$$

For crotyl-systems, high stereospecificity results [(*E*)-allyl → *threo*; (*Z*)-allyl → *erythro*] when thermal uncatalysed[86] or high-pressure reactions[76] are used. In contrast, crotylstannations using crotyltin halides show poor stereoselectivity, and the *threo*-isomer is slightly preferred in neat[87] or in aprotic solvent,[88] while *erythro* is the favoured isomer in the presence of H_2O.[89] Reactions using $BF_3 \cdot Et_2O$ catalysis[90] provide *erythro* selectivity regardless of the geometry of the crotyl unit. Other Lewis acid catalysts, $MgBr_2$, $ZnBr_2$, ZnI_2 or $SnCl_4$, provide mixtures of stereo- and regio-isomers.[91] With $TiCl_4$, the selectivity is high.[91] Addition of crotyltributyltin to a solution of cyclohexane carboxyaldehyde and $TiCl_4$ (normal addition) gives mainly the *erythro*-isomer, while addition of crotyltributyltin and $TiCl_4$ to the aldehyde (inverse addition) yields the *threo*-isomer (eqn 10.63).

$$c\text{-}C_6H_{11}CHO + Bu_3SnCH_2CH{=}CHCH_3 \xrightarrow{TiCl_4} c\text{-}C_6H_{11}CH(OH)CH(CH_3)CH{=}CH_2 + c\text{-}C_6H_{11}CH(OH)CH_2CH{=}CHCH_3 \qquad (63)$$

Normal addition	1.05 eq.	1.05 eq.	>97% *erythro* : *threo* 93 : 7	<3%
Inverse addition	2 eq.	2.1 eq.	>95% *erythro* : *threo* 5 : 95	<5%

(10.63)

In the inverse addition, it is probable that crotyl-titanium species are formed, which are known to give *threo*-selectivity, and it is possible that transmetallations also arise in other cases. Allyltins add to quinones,[92] and other organotin compounds which are reactive towards aldehydes include propargyl-, allenyl- and alkynyltin compounds[1].

10.3.3 *Palladium-catalysed cross-coupling reactions*[1,93,94]

The palladium-catalysed cross-coupling reaction of an organotin with an organic halide or tosylate has become a most valuable and versatile route to new carbon–carbon bonds (eqn 10.64).

$$RSn{\equiv}R^1X \xrightarrow{[Pd]} R{-}R^1 \quad (10.64)$$

Yields are usually high under mild and neutral conditions and virtually all functional groups (including aldehydes) are tolerated. The reaction is relatively insensitive to steric hindrance and high catalytic turnovers are observed. Only one group from the tetraorganotin is transferred at an appreciable rate, and use can be made of R_4Sn (for readily available groups), $RSnMe_3$ or $RSnBu_3$ for the transfer of R groups. Various catalysts have been employed, including $(PhCH_2)PdCl(PPh_3)_2$ and $Pd(PPh_3)_4$. For reactions involving triflates, the presence of LiCl is also required.

Conclusions regarding stereospecificity are: (i) inversion of configuration occurs at sp^3 hybridized carbon centres (either bonded to tin or in the organic electrophile); (ii) for vinyl groups, retention of configuration is normally found; and (iii) in allyltin reactions, regio-selective couplings result, usually with allylic rearrangement.

A limitation in the reaction is the β-eliminations which result from organic halides or tosylates having a β-hydrogen to the halide or tosylate. However, the use of [Pd(bipy)(E–NCCH=CHCN)] has been recommended for reactions in which β-elimination could be a problem.[95] A general catalytic cycle for the reactions is shown in Scheme 16.[93]

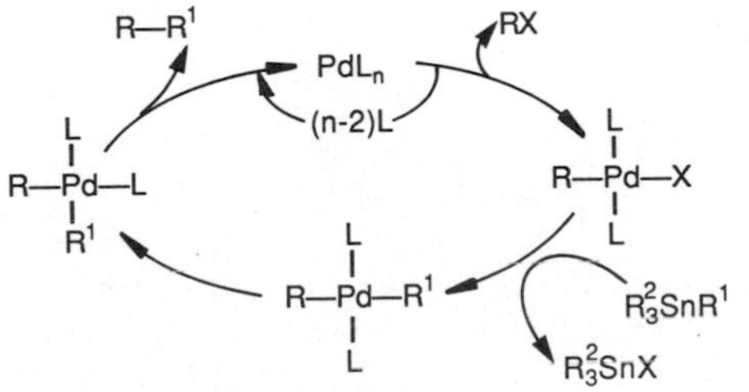

Scheme 16

Coupling systems studied[1,93] include:

(i) Benzylic halides with allyl- or alkyltins (reaction of Me_4Sn leads to inversion of configuration at the benzylic carbon).
(ii) Allyl halides or acetates with alkyl-, heteroaryl-, aryl- or allyltin reagents (e.g. eqns 10.65, 10.66). In eqn (10.65). inversion of configuration occurs at the allylic carbon[96,97].
(iii) Vinyl halides and triflates, with alkyl-, vinyl-, allyl-, alkynyl- or aryltin compounds (Scheme 17).[98]
(iv) Aryl halides with alkyl-, vinyl-,[99] allyl-,[100] alkynyl-,[101] XCH_2–(X = HO, MeO, CN or MeCO), aryl-[94], and heteroaryltin[102] species (eqn 10.67 and Scheme 18).

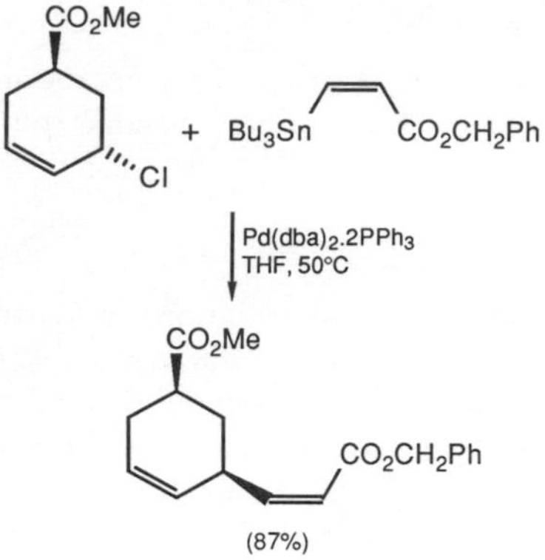

CO_2Me

Cl + Bu_3Sn CO_2CH_2Ph

$Pd(dba)_2.2PPh_3$
THF, 50°C

CO_2Me

CO_2CH_2Ph

(87%) (10.65)

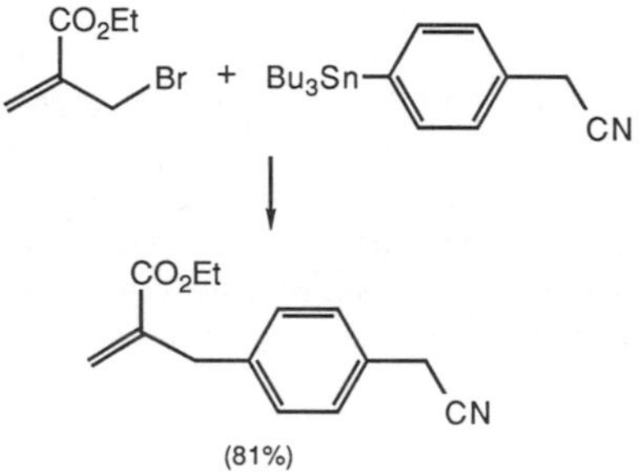

CO_2Et

Br + Bu_3Sn CN

CO_2Et

CN

(81%) (10.66)

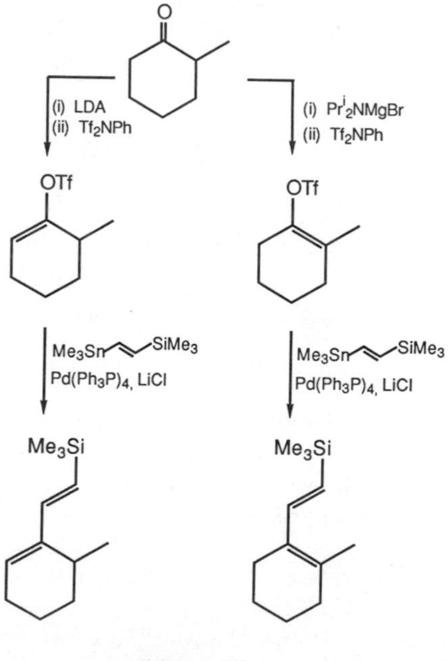

O

(i) LDA
(ii) Tf_2NPh

(i) Pr^i_2NMgBr
(ii) Tf_2NPh

OTf

OTf

Me_3Sn $SiMe_3$
$Pd(Ph_3P)_4$, LiCl

Me_3Sn $SiMe_3$
$Pd(Ph_3P)_4$, LiCl

Me_3Si

Me_3Si

Scheme 17

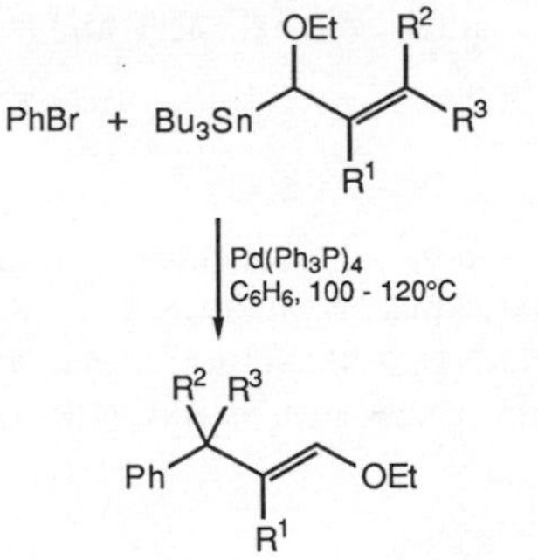

PhBr + Bu3Sn–CH(OEt)–C(R1)=C(R2)R3 → [Pd(Ph3P)4, C6H6, 100 - 120°C] → Ph–C(R2)(R3)–C(R1)=CH–OEt

(10.67)

PhC≡CSnMe3, PhPdI(PPh3)2, HMPT, 20°C → 4-NO2C6H4C≡CPh (100%) Ref. 101

Bu3SnCF=CF2, PhPdI(PPh3)2, HMPT, 70°C → 4-NO2C6H4CF=CF2 (87%) Ref. 102

YC6H4SnMe3, PhPdI(PPh3)2, HMPT, 70°C → 4-NO2C6H4–C6H4Y (75 - 100%) Ref. 101

Y = H, Me, OMe, Cl, NO2

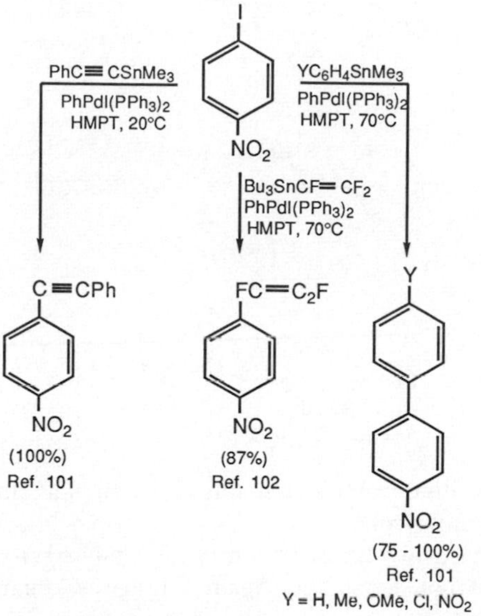

Scheme 18

(v) Heteroaryl halides with heteroaryl-[103] or vinyltin[104] compounds.
(vi) Aryl triflates[105] with alkyl-, vinyl-, alkynyl- or aryltins (eqn 10.68)

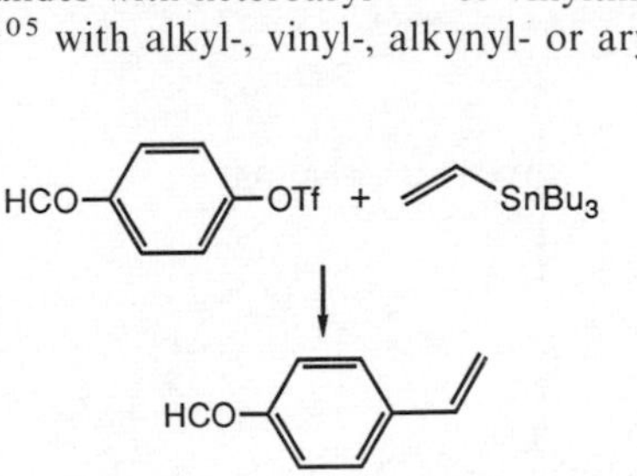

HCO–C6H4–OTf + CH2=CH–SnBu3 → HCO–C6H4–CH=CH2

(10.68)

(vii) Polyfluoroalkyl iodides with alkenyl-, allyl- or alkynyl-tin compounds.[106]

(viii) Acyl chlorides with alkyl-, vinyl-, allyl-, alkynyl-, aryl- and heteroaryl-tin species (eqn 10.69).

$$R^1COCl + R\text{–}Sn\equiv \xrightarrow{[Pd]} R^1COR \qquad (10.69)$$

The latter reaction leads to ketones.[1,93] Particular examples are shown in eqns (10.70) and (10.71). A reactivity sequence established for R groups is PhC≡C > PhCH=CH > Ph > $PhCH_2$ > $MeOCH_2$ > Me > Bu.[107] As well as palladium catalysts (e.g. eqn (10.70)),[108] rhodium species such as $RhCl(PPh_3)_3$ may also be used (e.g. eqn (10.71)).

Me3Sn–C(=CH2)–OMe + R1COCl → (PhCH2PdCl(PPh3)2, C6H6, reflux) → R1CO–C(=CH2)–OMe (77 - 100%)

R1 = Me, alkyl, Ph, 2-furyl etc. (10.70)

Pri–C6H4–CH2SnMe3 + Me2C=CHCOCl → (CH2Cl2, RhCl(PPh3)3) → Pri–C6H4–CH2–CO–CH=CMe2 (53%) (10.71)

To avoid decarbonylation in this coupling reaction, an atmosphere of carbon monoxide has been employed.[107]

Other ketone syntheses involve transition-metal-catalysed cross-coupling reactions in the presence of carbon monoxide. Again a range of organic electrophiles and organotin species successfully react[1,93,109,110]; (e.g. eqns 10.72 and 10.73).

3-iodo-5,5-dimethylcyclohex-2-enone + MeCH=CH–SnBu3 → (PhCH2PdCl(PPh3)2, THF, 50°C, CO (3 atm.)) → 3-(but-2-enoyl)-5,5-dimethylcyclohex-2-enone (71%) (10.72)

$$CH_3CH(Br)CO_2Et + Me_4Sn \xrightarrow[120^\circ C]{PdCl_2(AsPh_3)_2,\ HMPT,\ CO\ (20\ atm.)} CH_3COCH(CH_3)CO_2Et\ (62\%) \qquad (10.73)$$

Another ketone synthesis involves reactions of $Bu_3SnC(OEt){=}CH_2$ (eqn 10.74).[111]

$$Bu_3SnC(OEt){=}CH_2 + RX \xrightarrow[(ii)\ H^+]{(i)\ Pd\ catalyst} RCOMe \qquad (10.74)$$

R = vinyl, allyl, aryl, benzyl

Related routes to 1,2-diketones include the palladium-catalysed reactions of acyl chlorides with either α-methoxyvinyltin[108] and acyltin reagents[112] (for asymmetric species) or hexa-alkylditins (for symmetrical ones)[112,113]. A route to aldehydes, using acyl halides and Bu_3SnH in the presence of palladium catalyst, has already been mentioned. An alternative route to aldehydes is to use organic halides, R′–X (R′ = alkyl, aryl or vinyl), CO and Bu_3SnH (eqn (10.75)).[93]

$$R'{-}X + CO + Bu_3SnH \xrightarrow[50^\circ C,\ 1-3\ atm]{Pd(PPh_3)_4} R^1{-}CHO \qquad (10.75)$$

In the presence of Bu^tOOH and $Pd(OAc)_2$, oxidative coupling of 1-alkenyl- and 2-alkenyl-tin compounds can arise (eqns 10.76, 10.77).[114]

$$2\,R^1CH{=}CHSnR^2_3 \xrightarrow[Bu^tOOH]{Pd(OAc)_2} R^1CH{=}CH{-}CH{=}CHR^1 \qquad (10.76)$$

$$R^1CH{=}CHSnR^2_3 + CH_2{=}CHCH_2SnR^2_3 \xrightarrow[Bu^tOOH]{Pd(OMe)_2} R^1CH{=}CHCH_2CH{=}CH_2 \qquad (10.77)$$

10.3.4 *Other coupling reactions*

10.3.4.1 *Free-radical couplings.*[115] Allyltin compounds can take part in free-radical coupling reactions with a variety of organic halides, including polyhaloalkanes, allylic halides and XCR_2Y (where X = halide, Y = ester, alkoxy, sulphido, carbonyl or cyano group), as well as with phenylsulphide and phenylselenide derivatives. These reactions, which occur thermally, upon uv irradiation or AIBN initiation, are particularly useful for γ-unsubstituted allyltins. Some examples are shown in eqns (10.78)–(10.80).[1,116,117]

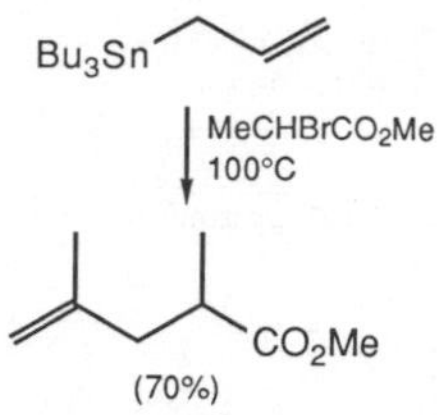

Bu_3Sn

$MeCHBrCO_2Me$, 100°C

CO_2Me (70%) (10.78)

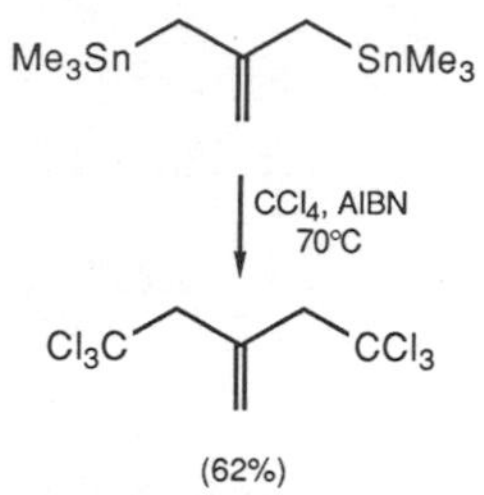

Me_3Sn $SnMe_3$

CCl_4, AIBN, 70°C

Cl_3C CCl_3 (62%) (10.79)

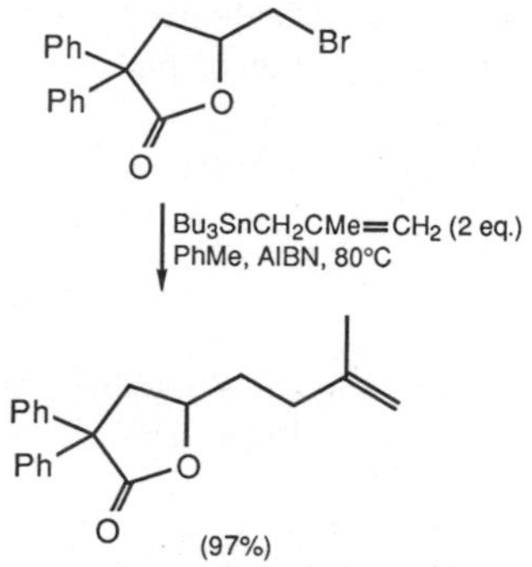

Ph Ph Br O O

$Bu_3SnCH_2CMe{=}CH_2$ (2 eq.), PhMe, AIBN, 80°C

Ph Ph O O (97%) (10.80)

Allenylation can occur using propargylic-tin derivatives (eqn 10.81).[118,119]

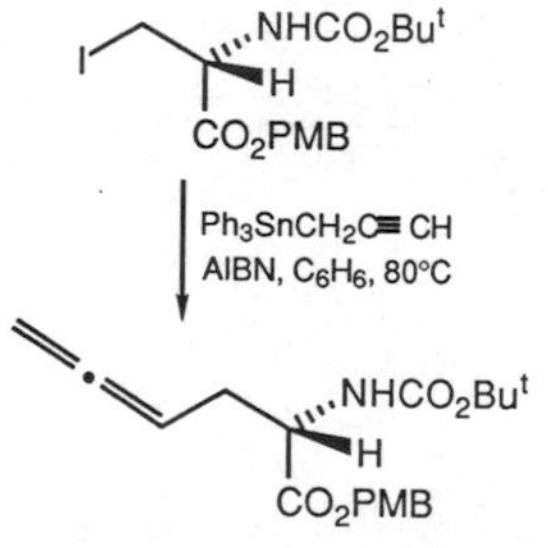

I $NHCO_2Bu^t$ H CO_2PMB

$Ph_3SnCH_2C{\equiv}CH$, AIBN, C_6H_6, 80°C

$NHCO_2Bu^t$ H CO_2PMB (10.81)

Free-radical coupling reactions of vinyltins have also been achieved, for example by using uv irradiation for coupling with CCl_4 or alkyl iodides.[120] Other interesting reactions include the photostimulated[120] (or AIBN-initiated) reactions of vinyltins, $R^1CH{=}CHSnR_3$, with alkylmercury halides, R^2HgX (which provide $R^1CH{=}CHR^2$)

and the reactions of organic halides with vinyltin reagents[121] in the presence of hexa-alkylditins (Scheme 19).

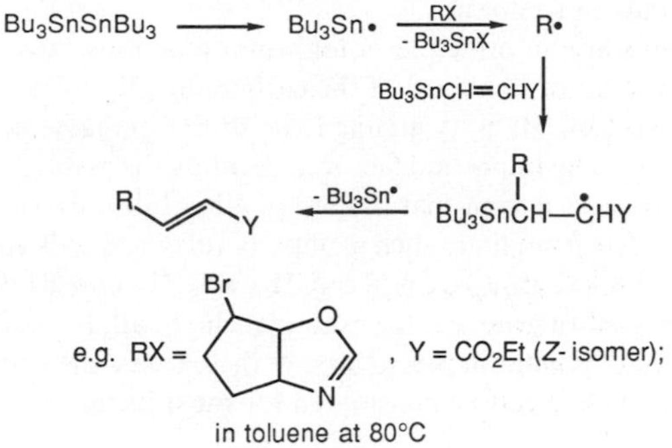

$$Bu_3SnSnBu_3 \longrightarrow Bu_3Sn\bullet \xrightarrow[-Bu_3SnX]{RX} R\bullet$$

$$R\bullet \xrightarrow{Bu_3SnCH{=}CHY} Bu_3SnCH(R){-}\dot{C}HY \xrightarrow{-Bu_3Sn^\bullet} RCH{=}CHY$$

e.g. RX = (bromo-substituted bicyclic oxazoline), $Y = CO_2Et$ (*Z*-isomer); in toluene at 80°C

Scheme 19

10.3.4.2 *Lewis-acid catalysed couplings.*[1] Allyltins couple under Lewis-acid catalysis with various organic halides (and sulphides), acetals and thioketals, as well as with allylic substrates. A wide range of catalysts have been used, including $ZnCl_2$, BF_3, $(Et_2Al)_2SO_4$, Me_3SiOTf and $Me_2(MeS)S^+BF_4^-$ but the choice of catalyst is of some importance. Two examples are shown in eqns (10.82)[122] and (10.83).[123]

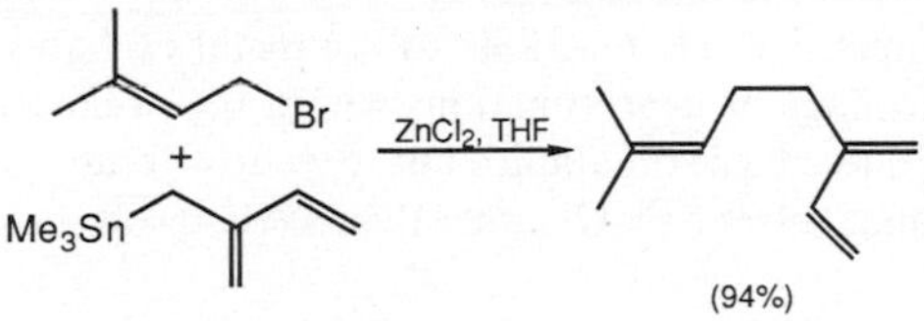

$$Me_2C{=}CHCH_2Br + Me_3SnCH_2C(=CH_2)CH{=}CH_2 \xrightarrow{ZnCl_2,\ THF} \text{product } (94\%) \qquad (10.82)$$

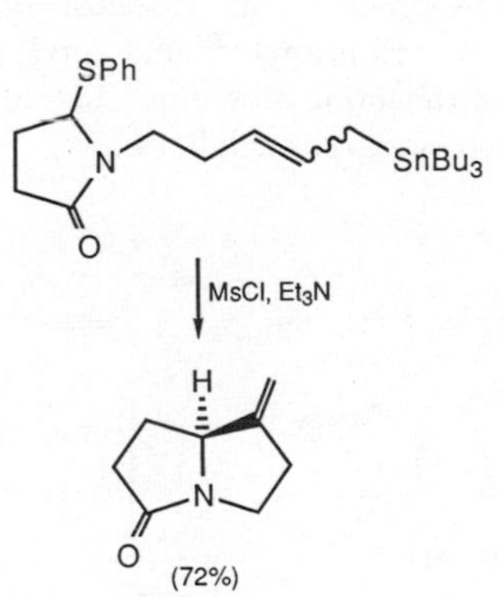

$$\text{SPh-pyrrolidinone–}SnBu_3 \xrightarrow{MsCl,\ Et_3N} \text{pyrrolizidinone } (72\%) \qquad (10.83)$$

10.3.5 *Tin–lithium exchange: transmetallation*[1,124]

The transmetallation reaction (eqn 10.84) is a most important route

$$m(R^1Li)_n + mn{\cdot}RSnR^2_3 \rightleftharpoons n(R{-}Li)_m + nmR^1SnR^2_3 \qquad (10.84)$$

to organolithiums, R–Li, especially for those which are either difficult or impossible to prepare by other routes or required free of lithium halides. Since the extent of organolithium reactions is vast,[124,125] this reaction substantially increases the scope of organotin compounds in synthesis.

The transmetallations are, in principle, equilibrium reactions. At equilibrium, it is generally found that the more stabilized of the carbanions, R^- or R^{1-}, provides the major organolithium species. Stability arising from intra-aggregate co-ordination, as in $(ROCH_2Li)_n$, will also be an important factor in deciding the position of equilibrium. From such considerations, it is seen that the use of alkyl-lithiums, such as BuLi, will result in extensive transfers from tin of such groups as vinyl, allyl, alkynyl, cyclopropyl and some α-substituted alkyl groups, e.g. R_2NCH_2, $RSCH_2$ and $ROCH_2$. Similarly, phenyl-lithium can be used to generate, for example, vinyl-, allyl-, α-substituted alkyl-lithiums from appropriate organotin precursors. In these cases, the equilibrium lies far to the right and the exchange can be considered for most purposes to be practically complete.

In addition to thermodynamic considerations, kinetic effects are also important. The choice of solvent and added donors have influences on rates of exchange, e.g. no reaction occurred between $Bu_3SnCH_2SiMe_3$ and BuLi in hexane at 20°C even after 24 h, while BuLi in THF produced a good yield of $LiCH_2SiMe_3$ at 0°C within 30 minutes.[126] The passive groups, R^2, attached to tin also effect the reactivity of R^1–SnR^2_3 compounds, the rates of cleavage of the Sn–R^1 bond increasing in the sequence, R^2=Me > Bu > cyclohexyl.

The presence of the new tetraorganotin product, $R^1SnR^2_3$, usually Me_4Sn, Bu_4Sn or Ph_4Sn, provides no problem, since these are passive and can be readily removed at the work-up stage. Competitive side reactions to the metal exchange reactions, such as lithium–halogen exchange or deprotonations within the organotin reagent, can also arise. The choice of a less basic organolithium reagent will limit deprotonation while that of the poorer donor solvent, Et_2O, rather than THF, results in less lithium–halogen exchange.

These transmetallations proceed with retention of configuration for secondary alkyl,[127] α-alkoxyalkyl,[128] cyclopropyl,[129] and vinyl[130] groups.

Scheme 20 illustrates the retention of configuration about the double bond of a vinyl system in a sequence involving transmetallation.[131]

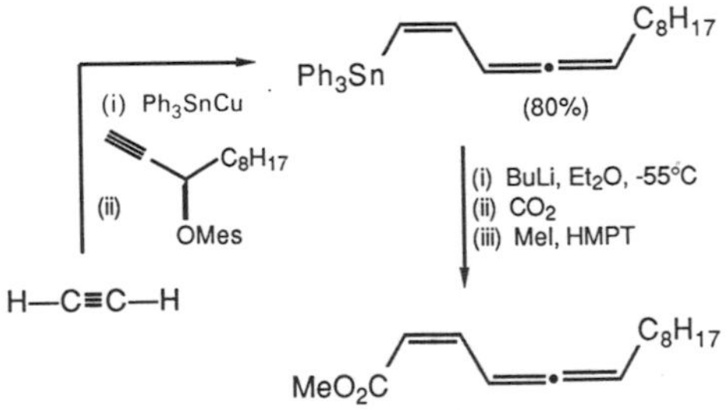

(i) Ph_3SnCu
(ii) [alkyne with C_8H_{17} and OMes]
H—C≡C—H → Ph_3Sn–CH=CH–CH=C=CH–C_8H_{17} (80%)
(i) BuLi, Et_2O, -55°C
(ii) CO_2
(iii) MeI, HMPT
→ MeO_2C–CH=CH–CH=C=CH–C_8H_{17}

Scheme 20

A use of methylenecyclohexyltributyltin is shown in Scheme 21.[132]

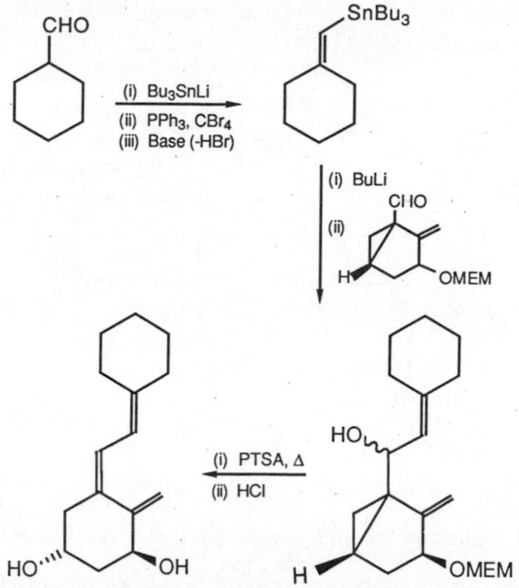

CHO
(i) Bu_3SnLi
(ii) PPh_3, CBr_4
(iii) Base (-HBr)
$SnBu_3$
(i) BuLi
(ii) CHO, H, OMEM
(i) PTSA, Δ
(ii) HCl
HO, OH
HO, H, OMEM

Scheme 21

Scheme 22 illustrates a sequence in the synthesis of the California red scale pheromone and which involves a [2, 3]-sigmatropic rearrangement of an α-alkoxymethyl-lithium reagent, obtained from the corresponding tin derivative.[133]

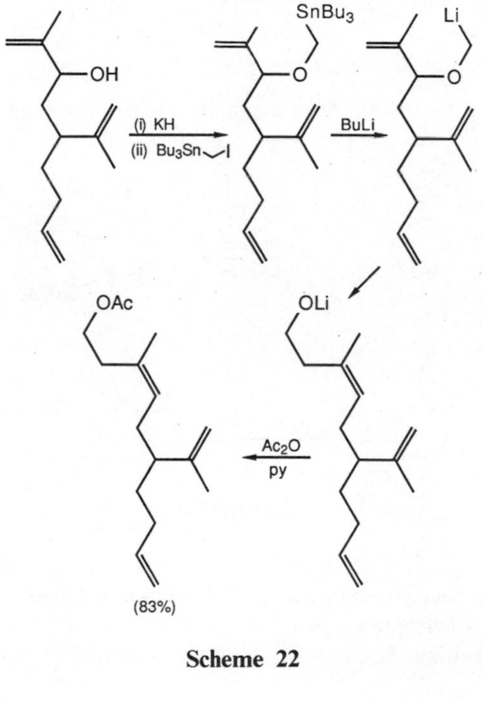

OH
(i) KH
(ii) Bu_3Sn⌒I
$SnBu_3$
O
BuLi
Li
O
OLi
Ac_2O
py
OAc
(83%)

Scheme 22

As part of a synthetic sequence to hybridolactone, a chiral tributyltin cyclopropylcarbinol derivative was transmetallated (Scheme 23).[129,134]

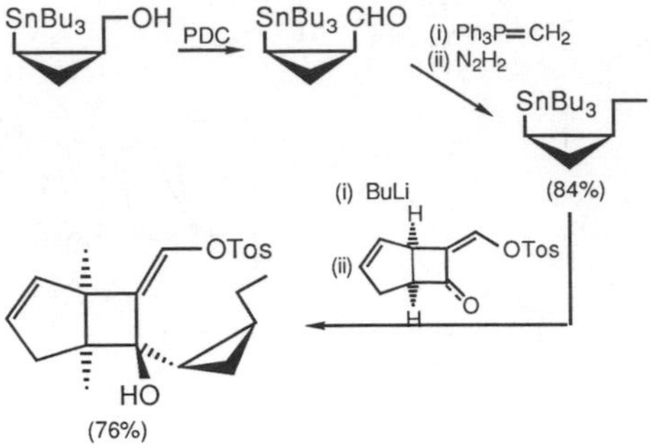

SnBu3 OH PDC SnBu3 CHO (i) Ph3P=CH2 (ii) N2H2 SnBu3 (84%) (i) BuLi (ii) H OTos O H OTos HO (76%)

Scheme 23

10.3.6 *Oxidation of organotin compounds: formation of carbon–oxygen bonds*

Oxidation of alkyltin compounds occurs readily using excess CrO_3 in pyridine (e.g. eqn 10.85).[135]

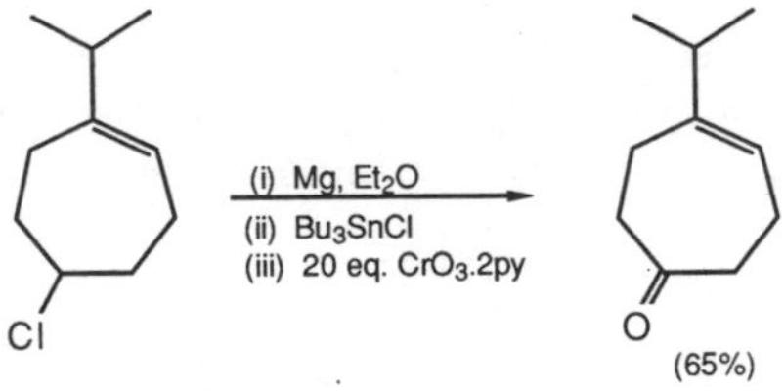

Cl (i) Mg, Et2O (ii) Bu3SnCl (iii) 20 eq. CrO3.2py O (65%) (10.85)

As shown in Scheme 24, an oxidation step is used in the synthesis of dihydrojasmone (overall yield 71%).[136]

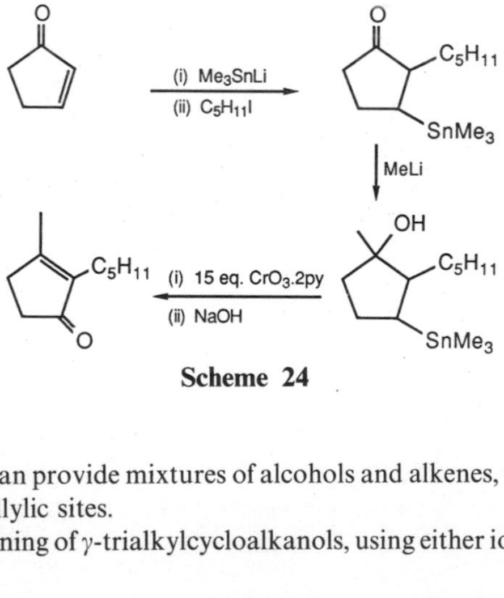

O (i) Me3SnLi (ii) C5H11I O C5H11 SnMe3 MeLi OH C5H11 SnMe3 (i) 15 eq. CrO3.2py (ii) NaOH C5H11 O

Scheme 24

Tertiary alkyltins can provide mixtures of alcohols and alkenes, with the latter being further oxidized at allylic sites.

Oxidation ring opening of γ-trialkylcycloalkanols, using either iodosylbenzene[137] or lead tetra-acetate,[138] leads to linear enones. As shown in eqn (10.85), the reaction is stereospecific.

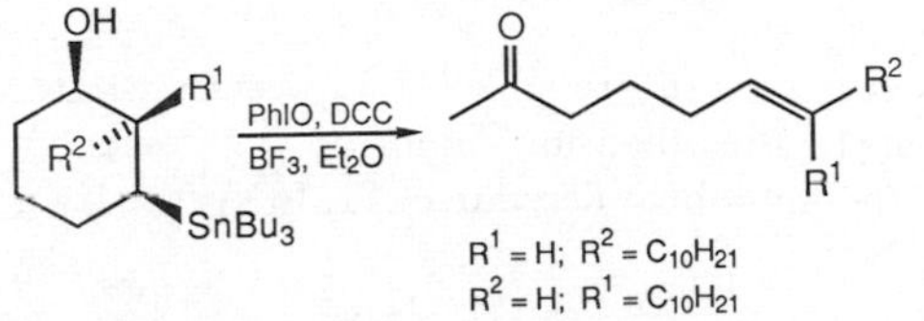

PhIO, DCC
BF_3, Et_2O

R^1 = H; R^2 = $C_{10}H_{21}$
R^2 = H; R^1 = $C_{10}H_{21}$

(10.86)

Mild oxidation of allyltins, e.g. by *m*-chloroperbenzoic acid, provides allylic rearranged alcohols.[139] A preference for antara-stereochemistry is indicated (eqn 10.87).

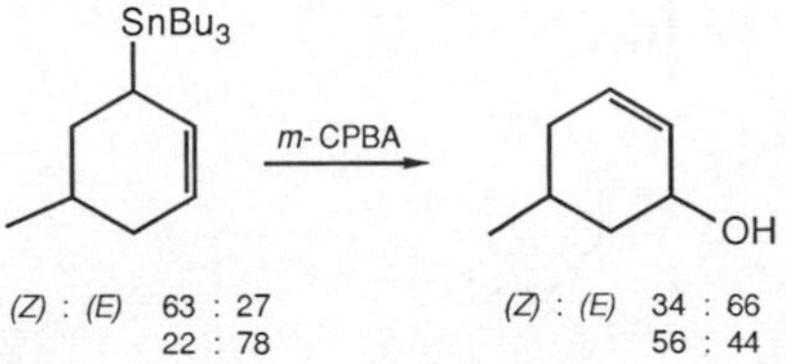

m-CPBA

(*Z*) : (*E*) 63 : 27 → (*Z*) : (*E*) 34 : 66
22 : 78 → 56 : 44

(10.87)

This oxidation reaction has been incorporated into a synthesis of β-bromo-α-enones from α,β-unsaturated aldehydes[140] with 42% yield overall (Scheme 25).

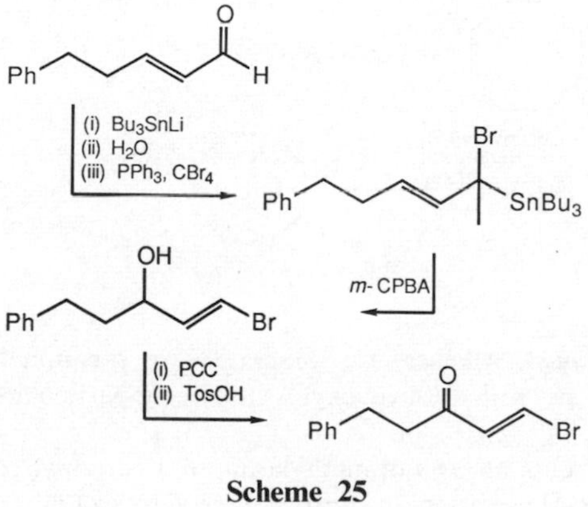

(i) Bu_3SnLi
(ii) H_2O
(iii) PPh_3, CBr_4

m-CPBA

(i) PCC
(ii) TosOH

Scheme 25

For aryltin reagents, the best route to phenols involves[141] the intermediate formation of arylboranes, which are then oxidized by H_2O_2.

10.3.7 *Eliminations*

10.3.7.1 *α-Eliminations.* Halomethyltriorganotin species, on heating in the presence of alkenes, provide cyclopropanes (e.g. eqn 10.88).[142] The formation of free carbenes in these reactions appears unlikely. Rather, an initial dissociation into

$$CF_2{=}CF_2 + Me_3SnCF_3 \xrightarrow[80°C,\ NaI,\ DME]{150°C} \begin{matrix} CF_2{-}CF_2 \\ \diagdown\ \diagup \\ CF_2 \end{matrix} \tag{10.88}$$

a cationic species seems more probable.[143] Lower temperatures are required for reaction of trihalomethyltins than for mono- or di-halomethyltin reagents.[144] Use has been made of the monohalo derivative (**12**) to synthesize a tetracyclic product (eqn 10.89).[145]

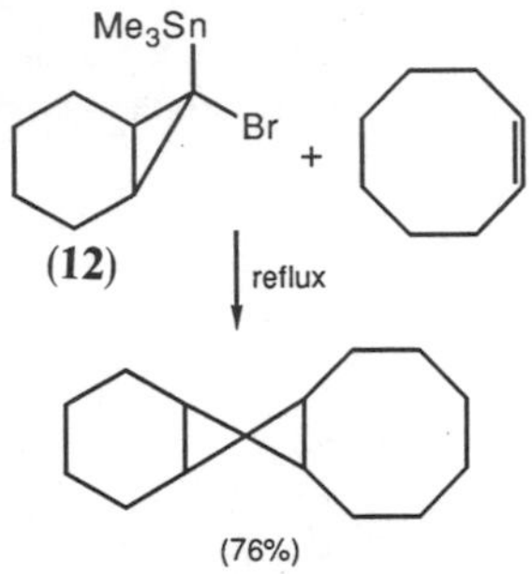

Me$_3$Sn, Br, +, (**12**), reflux, (76%)

(10.89)

Trihaloacetates, $R_3SnO_2CCX_3$, may be used as *in-situ* R_3SnCX_3 sources (e.g. eqn 10.90).[146]

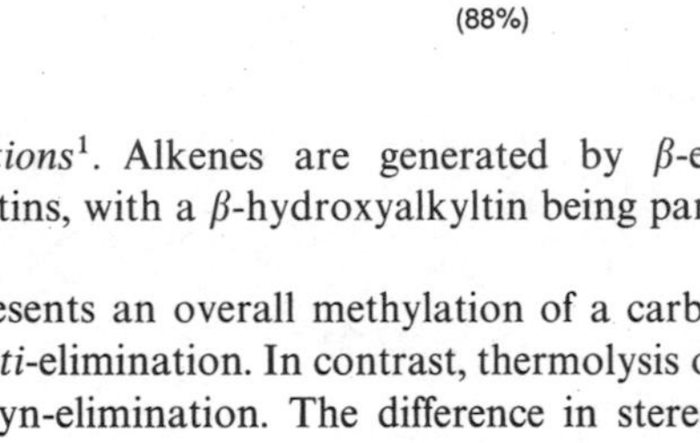

+ $Cl_3CCO_2SnMe_3$, diglyme, 140°C, - Me_3SnCl, - CO_2, Cl, Cl, (88%)

(10.90)

10.3.7.2 *β-Eliminations*[1]. Alkenes are generated by β-elimination from suitable β-substituted alkyltins, with a β-hydroxyalkyltin being particularly useful[147,148] (e.g. Scheme 26).

Scheme 26 represents an overall methylation of a carbonyl compound involving an acid-induced *anti*-elimination. In contrast, thermolysis of β-hydroxyalkyltins leads to alkenes via a syn-elimination. The difference in stereochemistry resulting from thermal and proton-generated eliminations is illustrated in Scheme 27.[149] The *erythro*-(**13**) also reacts stereospecifically. *Erythro*- and *threo*-(**13**) are obtained in equal amounts from Ph_3SnCH_2SPh on successive treatment with LDA and PhCHO. An alternative route to β-hydroxyalkyltins involves the stereospecific ring opening of epoxides (Scheme 28).[150]

By analogous metallation-addition-elimination reactions to those in Scheme 27, using appropriate aldehydes and R_3SnCH_2X compounds, it is possible to obtain other vinyl derivatives with X = selenide, sulphone, phosphine oxide.[147,151] Where X =

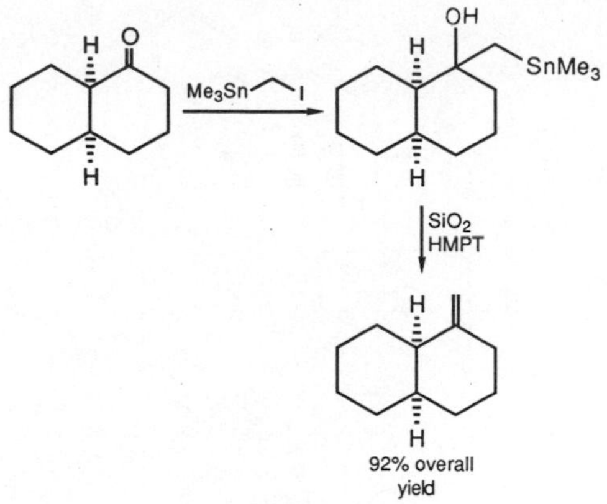

Scheme 26

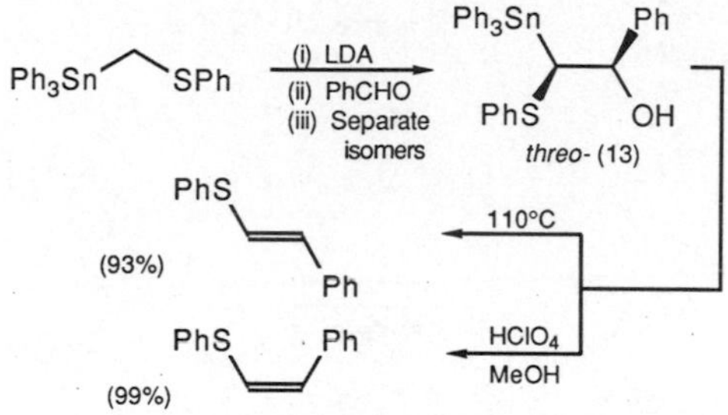

Scheme 27

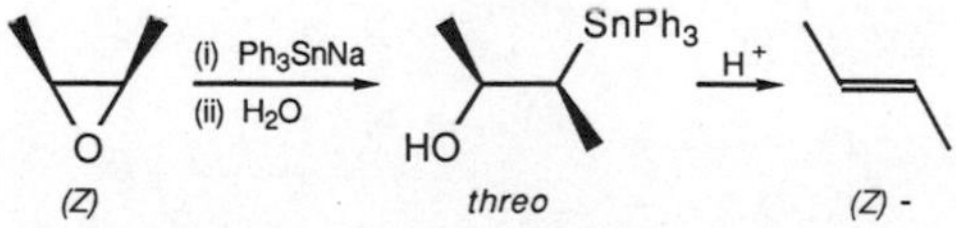

Scheme 28

Ph_2PO or $PhSO_2$, the thermal decomposition of the addition products occurs even below $-70°C$. As well as the hydroxy group, other β-substituted groups which are able to produce eliminations are sulphides, sulphoxides and sulphones (Schemes 29 and 30).[152–154]

10.3.7.3 *γ-Eliminations.* 1,3-Deoxystannylations of γ-hydroxyalkyltins provide cyclopropane derivatives,[148] (Scheme 31 and 32).

The ring closure step involves a double inversion, as indicated in Scheme 33.[155] The BF_3 catalyst system has been proved to be very successful for tertiary alcohols

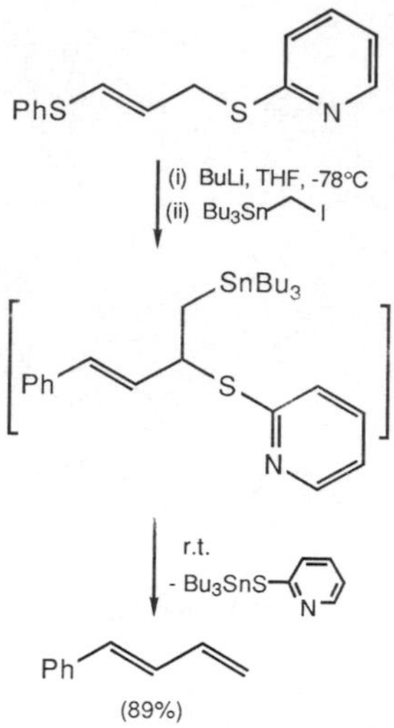

PhS / S–(2-pyridyl); (i) BuLi, THF, -78°C; (ii) Bu_3Sn / I; $SnBu_3$; Ph; S; N; r.t.; - Bu_3SnS–(2-pyridyl); Ph; (89%)

Scheme 29

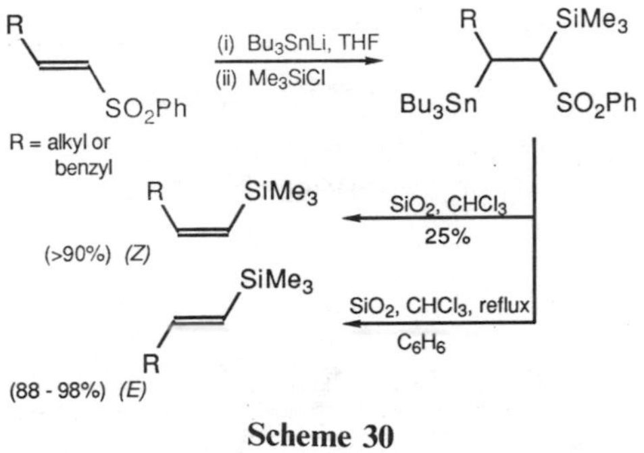

R; SO_2Ph; R = alkyl or benzyl; (i) Bu_3SnLi, THF; (ii) Me_3SiCl; R; $SiMe_3$; Bu_3Sn; SO_2Ph; SiO_2, $CHCl_3$ 25%; R; $SiMe_3$; (>90%) *(Z)*; SiO_2, $CHCl_3$, reflux C_6H_6; $SiMe_3$; R; (88 - 98%) *(E)*

Scheme 30

Ph; Ph; O; (i) Me_3Sn / Li; OH; Ph; Ph; $SnMe_3$; (75%); $BF_3.Et_2O$; Ph; Ph; (97%)

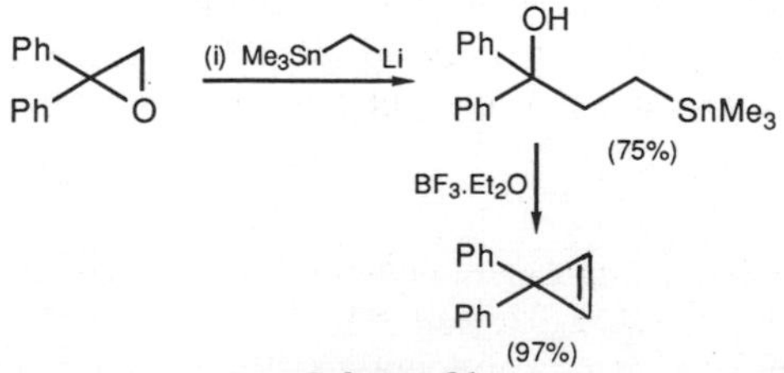

Scheme 31

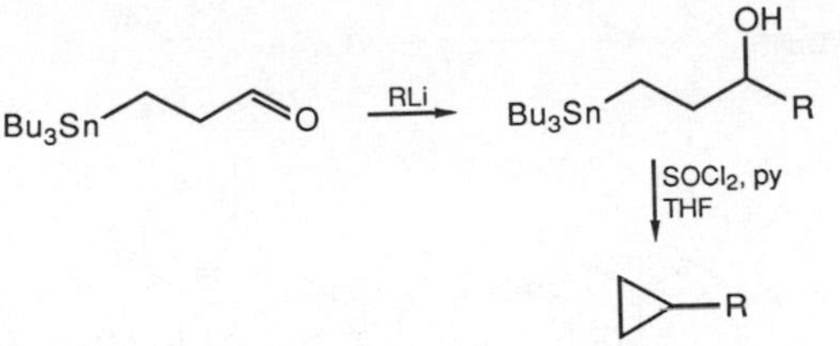

Scheme 32

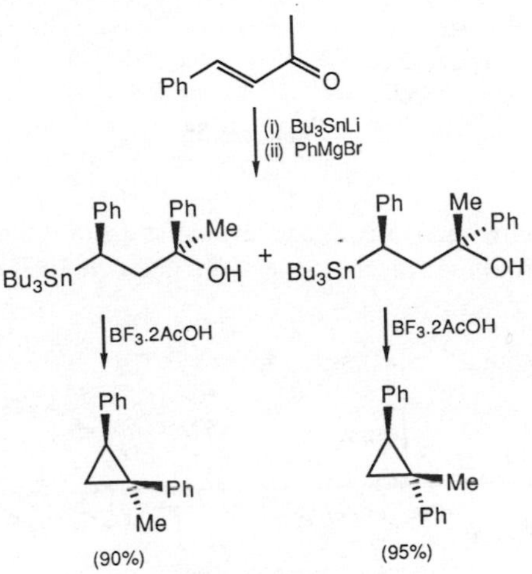

Scheme 33

and secondary benzylic alcohols.[155] For other secondary alcohols and for the synthesis of fused-ring compounds (e.g. Scheme 34),[156] $SOCl_2$ is recommended.

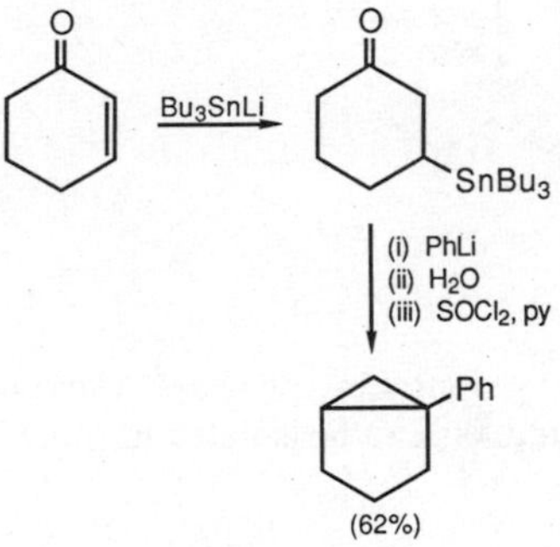

Scheme 34

Potentially valuable cyclopropanation sequences involve the use of β-cyanoethyltin derivatives[157] (Scheme 35).

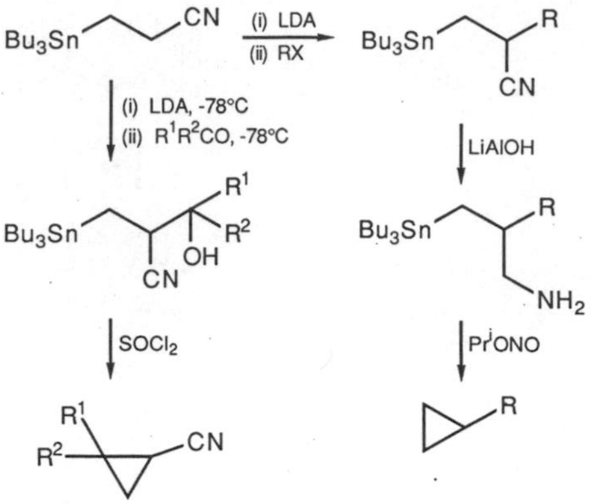

Bu_3Sn CN (i) LDA (ii) RX Bu_3Sn R CN

(i) LDA, -78°C (ii) R^1R^2CO, -78°C

LiAlOH

Bu_3Sn R^1 R^2 OH CN

Bu_3Sn R NH_2

$SOCl_2$

Pr^iONO

R^1 R^2 CN

R

Scheme 35

Cyclopropylcarbinyl derivatives are available from butenyltin derivatives on reaction with electrophilic reagent (Scheme 36 and eqn 10.91).[158–160]

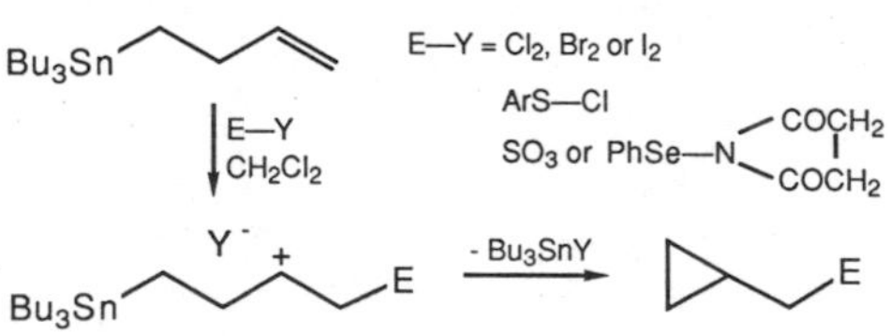

Bu_3Sn

E—Y = Cl_2, Br_2 or I_2

ArS—Cl

SO_3 or PhSe—N($COCH_2$)($COCH_2$)

E—Y CH_2Cl_2

Y^- + E Bu_3Sn

- Bu_3SnY

E

Scheme 36

Bu_3Sn O Me

CF_3CO_2H 25°C

O

(90%)

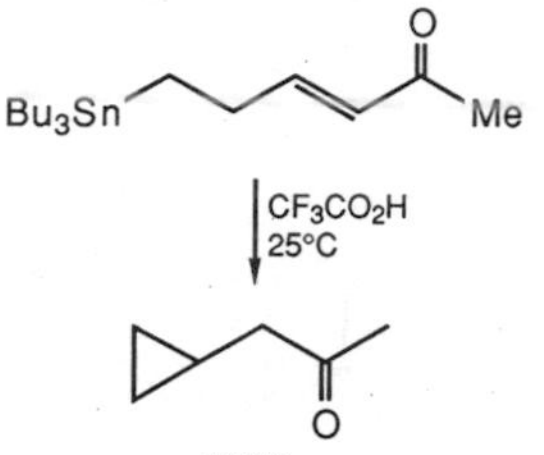

(10.91)

Thermolysis of 3,4-epoxytributyltin provides cyclopropylmethanol in good yield.[158] The intermediate adducts can be isolated in some cases,[158,161] (e.g. Scheme 37).

10.4 Organotin alkoxides

Replacing an hydroxy group by an organotin alkoxy group strongly increases the nucleophilic character of the oxygen (much more than the increase in the basicity),

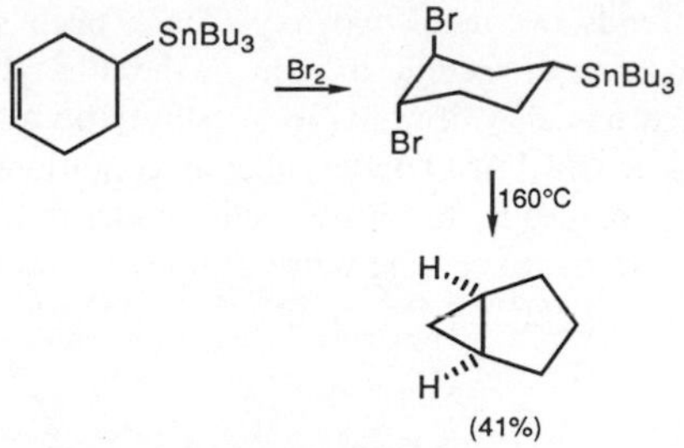

SnBu3 Br2 Br SnBu3 Br 160°C H H (41%)

Scheme 37

thereby allowing more and easier reactions with acyl and alkyl halides.[1] Use has been made of $(Bu_3Sn)_2O$ and Bu_2SnO to form organotin alkoxides and bisalkoxides respectively. The transformation of 1,2-diols to cyclic tin-alkoxides has proved valuable in synthesis, e.g. in forming macrocyclic lactones (Scheme 38).[162]

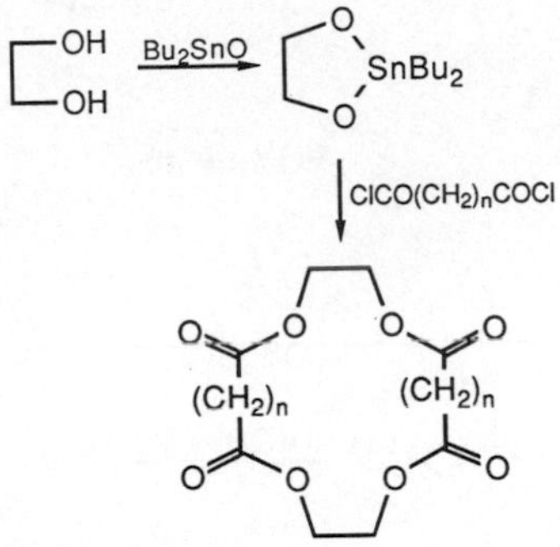

OH OH Bu2SnO O SnBu2 O ClCO(CH2)nCOCl O O O O (CH2)n (CH2)n O O O O

Scheme 38

Particularly good use has been made in the glycol[163a] and carbohydrate fields[163b] due to the high regioselectivity obtained[164] (e.g. Scheme 39), as well as the increased reactivity of the Sn–O bonds.

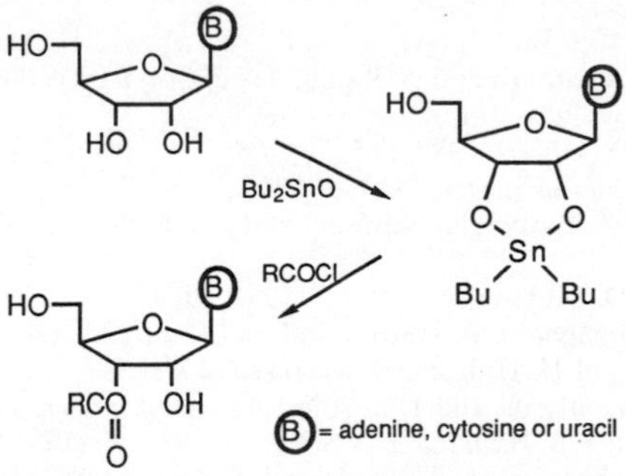

HO O B HO OH Bu2SnO HO O B O O Sn Bu Bu RCOCl HO O B RCO OH O B = adenine, cytosine or uracil

Scheme 39

Regioselectivity trends of sugar hydroxyls have been studied,[1,165] as have the reasons for the activation of specific oxygen in the dibutyltin-5-membered cycle.[166] Use in the sugar area has also been made of tributyltin alkoxides. The reactivity of the various hydroxyls to $(Bu_3Sn)_2O$ under different conditions has been investigated.[167]

Oxidation of tin alkoxides by bromine occurs under mild conditions and provides a useful route to keto derivatives. The stereo- and regio-selectivities are excellent (e.g. Scheme 40[168] and eqn 10.92[169]).

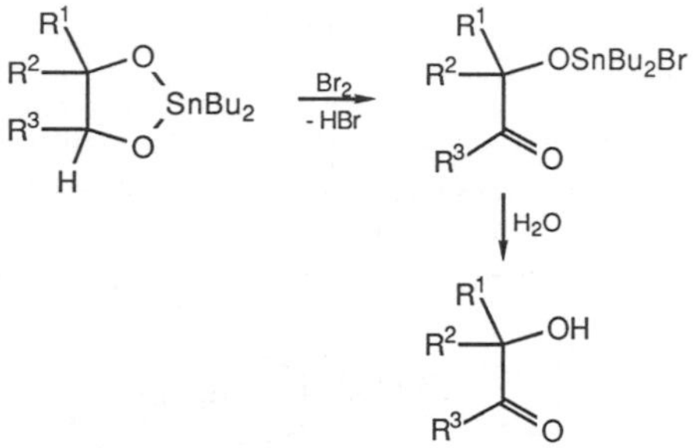

Scheme 40

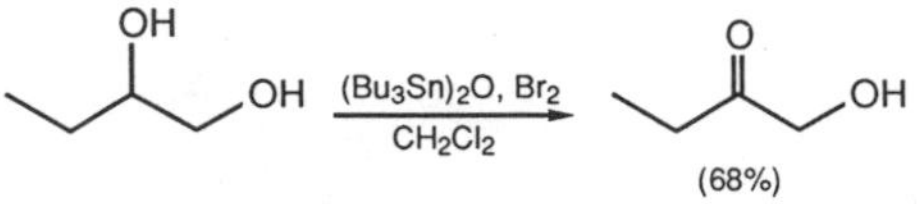

(10.92)

Alkoxides of the type $BuCl_2Sn\text{–}O\text{–}(CH_2)_n\text{–}OH$ ($n = 4, 5$) have been proposed as intermediates in the $BuSnCl_3$-catalysed cyclodehydration of 1,4-butane- and 1,5-pentanediol to tetrahydrofuran and tetrahydropyran, respectively.[170]

References

1. M. Pereyre, J.-P. Quintard and A. Rahm, *Tin in Organic Synthesis*, Butterworth, London, 1987.
2. W.P. Neumann, *Synthesis*, 1987, 665.
3. N.O. Brace, *J. Fluorine Chem.*, 1982, **20**, 313.
4. D.R. Williams, B.A. Barner, K. Nishitani and J.G. Phillips, *J. Amer. Chem. Soc.*, 1982, **104**, 4708.
5. C.R. Engel and D. Mukherjee, *Steroids*, 1981, **37**, 73.
6. A. Medici, M. Fogagnolo, P. Pedrini and A. Dondoni, *J. Org. Chem.*, 1982, **47**, 3844.
7. W.P. Neumann and H. Hillgartner, *Synthesis*, 1971, 537.
8. D. Seyferth, H. Yamazaki and D.L. Alleston, *J. Org. Chem.*, 1963, **28**, 703.
9. M.S. Kellogg and E.S. Hamanake, US Pat., 4, 397, 783 (1983).
10. S. Takano, S. Nishizawa, M. Akiyama and K. Ogasawara, *Synthesis*, 1984, 949.
11. H. Yamamaka, J. Kikui, K. Teramura and T. Ando, *J. Org. Chem.*, 1976, **41**, 3794.

12. H.G. Kuivila, L.W. Menapace and C.R. Warner, *J. Amer. Chem. Soc.*, 1962, **84**, 3584.
13. H. Schumann, B. Pachaly and B.C. Schuetze, *J. Organomet. Chem.*, 1984, **265**, 145.
14. T. Ando, H. Yamanaka, F. Namigata and W. Funasaka, *J. Org. Chem.*, 1970, **35**, 33; T. Ishihara, E. Ohtani and T. Ando, *J. Chem. Soc., Chem. Commun.*, 1975, 367.
15. T. Ando, K. Wakabayashi, H. Yamanaka and W. Funasaka, *Bull. Chem. Soc. Jpn*, 1972, **45**, 1576.
16. M.S. Manhas, M.S. Khajavi, S.S Bari and A.K. Bose, *Tetrahedron Lett.*, 1983, **24**, 2327.
17. E.C. Ashby and T.N. Pham, *Tetrahedron Lett.*, 1984, **25**, 4333.
18. A.L.J. Beckwith and W.B. Gara, *J. Chem. Soc. Perkin II*, 1975, 795.
19. A. Padwa, H. Nimmesgern and G.S.K. Wong, *Tetrahedron Lett.*, 1985, **26**, 957.
20. G. Stork, R. Mook, S.A. Biller and S.D. Rychnovsky, *J. Amer. Chem. Soc.*, 1983, **105**, 3741.
21. O. Moriya, M. Okawara and Y. Ueno, *Chem. Lett.*, 1984, 1437.
22. T.W. Smith and G.B. Butler, *J. Org. Chem.*, 1978, **43**, 6.
23. N.N. Marinovic and H. Ramanathan, *Tetrahedron Lett.*, 1983, **24**, 1871.
24. G. Stork and R. Mook, *J. Amer. Chem. Soc.*, 1983, **105**, 3720.
25. D.P. Curran and D.M. Rakiewicz, *J. Amer. Chem. Soc.*, 1985, **107**, 1448.
26. G. Stork and N.H. Bain, *J. Amer. Chem. Soc.*, 1982, **104**, 2321.
27. M.D. Bachi, F. Frolan and C. Hoornaert, *J. Org. Chem.*, 1983, **48**, 1841.
28. B. Giese, *Angew. Chem., Int. Edn. Engl.*, 1985, **24**, 553.
29. B. Giese and T. Witzel, *Angew. Chem., Ind. Edn. Engl.*, 1986, **98**, 459.
30. G. Stork and P.M. Sher, *J. Amer. Chem. Soc.*, 1986, **108**, 303.
31. N.Y.M. Fung, P. De Mayo, J.H. Schauble and A.C. Weedon, *J. Org. Chem.*, 1978, **43**, 3977.
32. T.X. Yang, P. Four, F. Guire and G. Balavoine, *Nouv. J. Chim.*, 1984, **8**, 6111.
33. E. Keinan and P.A. Gleize, *Tetrahedron Lett.*, 1982, **23**, 477.
34. H.G. Kuivila, *Adv. Organomet. Chem.*, 1964, **1**, 47; J. Luszetyk, E. Lusztyk, B. Maillard and K.U. Ingold, *J. Amer. Chem. Soc.*, 1984, **106**, 2923.
35. P. Four and F. Guibe, *J. Org. Chem.*, 1981, **46**, 4439.
36. H. Hrebabecki, J. Brokes and J. Baranek, *Coll. Czech. Chem. Commun.*, 1980, 599.
37. D.G. Norman and C.B. Reese, *Synthesis*, 1983, 304.
38. W. Hartwig, *Tetrahedron*, 1983, **39**, 2609.
39. J.R. Rasmussen, *J. Org. Chem.*, 1980, **45**, 2725.
40. P. di Cesare and B. Gross, *Synthesis*, 1980, 714.
41. C.M. Tice and C.H. Heathcote, *J. Org. Chem.*, 1981, **46**, 9.
42. J.J. Patroni and R.V. Stick, *Austr. J. Chem.*, 1979, **32**, 411.
43. D.H.R. Barton and R. Subramanian, *J. Chem. Soc., Perkin 1*, 1977, 1718.
44. Y. Ueno, C. Tanaka and M. Okawara, *Chem. Lett.*, 1983, 795.
45. W.J. Scott, G.T. Crisp and J.K. Stille, *J. Amer. Chem. Soc.*, 1984, **106**, 4630.
46. C.G. Gutierrez and L.R. Summerhays, *J. Org. Chem.*, 1984, **49**, 5206; Y. Ueno, T. Miyano and M. Okawara, *Tetrahedron Lett.*, 1982, **23**, 443; E. Vedejs and D.W. Powell, *J. Amer. Chem. Soc.*, 1982, **104**, 2046.
47. J.D. Buynak, M.N. Rao, H. Pajouhesh, R.Y. Chandraskaran, K. Finn, P. De Meester and S.C. Chu, *J. Org. Chem.*, 1985, **50**, 4245.
48. T. Kametani and T. Honda, *Heterocycles*, 1982, **19**, 1861.
49. T.H. Haskell, P.W.K. Woo and D.R. Watson, *J. Org. Chem.*, 1977, **42**, 1302.
50. C.G. Gutierrez and L.R. Summerhays, *J. Org. Chem.*, 1984, **49**, 5206.
51. C. Paulnier, *Selenium Reagents and Intermediates in Organic Synthesis*, Pergamon, Oxford, 1986.
52. N. Petragnani and J.V. Comasseto, *Synthesis*, 1986, 1.
53. D.L.J. Clive, G. Chittattu, V. Farina, W.A. Kiel, S.M. Menchen, C.G. Russell, A. Singh, C.K. Wong and N.J. Curtis, *J. Amer. Chem. Soc.*, 1980, **102**, 4438.
54. W.R. Leonard and T. Livingstone, *Tetrahedron Lett.*, 1985, **26**, 6431.
55. S.D. Burke, W.F. Fobare and D.M. Armistead, *J. Org. Chem.*, 1982, **47**, 3348.
56. D.H.R. Barton and W.B. Motherwell, *Pure & Appl. Chem.*, 1981, **53**, 15.
57. Z.J. Witczak, *Tetrahedron Lett.*, 1986, **27**, 155.
58. D.H.R. Barton, G. Bringmann and W.B. Motherwell, *J. Chem. Soc., Perkin Trans. 1*, 1980, 2665.
59. D.I. John, E.J. Thomas and N.D. Tyrrell, *J. Chem. Soc., Chem. Commun.*, 1979, 345.
60. N. Ono and A. Kaji, *Synthesis*, 1986, 693.

61. N. Ono, H. Miyaka, A. Kanimura, N. Tsukui and A. Kaji, *Tetrahedron Lett.*, 1982, **23**, 2957.
62. N. Ono, I. Hamamoto and A. Kaji, *J. Org. Chem.*, 1986, **51**, 2832.
63. J.M. Aizpurua, M. Oiarbide and C. Palomo, *Tetrahedron Lett.*, 1987, **28**, 5365.
64. N. Ono, H. Miyake and A. Kaji, *Chem. Lett.*, 1985, 635.
65. D.H.R. Barton, D. Crich and W.B. Motherwell, *J. Chem. Soc. Chem. Commun.*, 1983, 939; *Tetrahedron*, 1985, **41**, 3901.
66. W.A. Asomaning, C. Eaborn and D.R.M. Walton, *J. Chem. Soc., Perkin Trans. 2*, 1979, 203.
67. C. Eaborn, I.D. Jenkins and D.R.M. Walton, *J. Chem. Soc. C*, 1974, 870.
68. M.E. Jung and L.A. Light, *Tetrahedron Lett.*, 1982, **23**, 3851.
69. H.E. Ensley, R.R. Buescher and K. Lee, *J. Org. Chem.*, 1982, **47**, 404.
70. Y. Ueno, H. Sano and M. Okawara, *Tetrahedron Lett.*, 1980, **21**, 1767.
71. D.E. Sietz, R.A. Milius and H. El-Wakil, *Synth. Commun.*, 1981, **11**, 281.
72. M.J. Adams, T.J. Ruth, B.D. Pate and L.D. Hall, *J. Chem. Soc., Chem. Commun.*, 1982, 625.
73. G.L. Tonnesen, R.N. Hanson and D.E. Seitz, *Int. J. Appl. Radiat. Isot.*, 1981, **32**, 171.
74. G. Wickham and W. Kitching, *J. Org. Chem.*, 1983, **48**, 612.
75. A. Gambaro, D. Marton, V. Peruzzo and G. Tagliavini, *J. Organomet. Chem.*, 1981, **204**, 191.
76. Y. Yamamoto, K. Maruyama and K. Matsumoto, *J. Chem. Soc., Chem. Commun.*, 1983, 489.
77. G. Tagliavini, *Revs. Si, Ge, Sn and Pb Compounds*, 1985, **8**, 237.
78. A. Gambaro, V. Peruzzo, G. Plazzogna and G. Tagliavini, *J. Organomet. Chem.*, 1980, **197**, 45.
79. G. Tagliavini, V. Peruzzo and D. Marton, *Inorg. Chim. Acta*, 1978, **26**, L41.
80. A. Gambaro, P. Ganis, D. Marton, V. Peruzzo and G. Tagliavini, *J. Organomet. Chem.*, 1982, **231**, 307.
81. A. Boretto, D. Marton, G. Tagliavini and A. Gambaro, *J. Organomet. Chem.*, 1985, **286**, 9.
82. e.g. D. Matsubara, K. Wakamatsu, Y. Morizawa, N. Tsuboniwa, K. Oshima and H. Nozaki, *Bull. Chem. Soc. Jpn.*, 1985, **58**, 1196; C. Petrier, J. Einhorn and J.L. Luche, *Tetrahedron Lett.*, 1985, **26**, 1449; G.P. Boldrini, D. Savoia, E. Tagliavini, C. Trombini and A. Umani-Ronchi, *J. Organomet. Chem.*, 1985, **280**, 307.
83. T. Mukaiyama, T. Harada and S. Shoda, *Chem. Lett.*, 1980, 1507.
84. Y. Naruta, S. Ushida and K. Maruyama, *Chem. Lett.*, 1979, 919.
85. N. Ueno, S. Aoki and M. Okawara, *J. Chem. Soc., Chem. Commun.*, 1980, 683.
86. C. Servens and M. Pereyre, *J. Organomet. Chem.*, 1972, **35**, C20.
87. A. Gambaro, A. Boaretto, D. Marton and G. Tagliavini, *J. Organomet. Chem.*, 1984, **260**, 255.
88. S. Auge, *Tetrahedron Lett.*, 1985, **26**, 753.
89. J. Nokami, J. Otera, T. Sudo and R. Okawara, *Organometallics*, 1983, **2**, 191.
90. Y. Yamamoto, H. Yatagai, Y. Ishihara, N. Maeda and K. Maruyama, *Tetrahedron*, 1984, **40**, 2239.
91. G.E. Keck, D.E. Abbott, E.P. Boden and E.J. Enholm, *Tetrahedron Lett.*, 1984, **25**, 3927.
92. G.A. Takuwa, Y. Naruta, O. Soga and K. Maruyama, *J. Org. Chem.*, 1984, **49**, 1857; K. Mori, M. Maku and M. Sakakibura, *Tetrahedron*, 1985, **41**, 2825.
93. J.K. Stille, *Angew. Chem., Int. Edn. Engl.*, 1986, **25**, 508; *Pure Appl. Chem.*, 1985, **57**, 1771.
94. I.P. Beletskaya, *J. Organomet. Chem.*, 1983, **250**, 551.
95. R. Sustmann, J. Law and M. Zipp, *Tetrahedron Lett.*, 1986, **27**, 5207.
96. S. Kartsumura, *Tetrahedron Lett.*, 1987, **28**, 1191.
97. F.K. Sheffy, J.P. Godshalx and J.K. Stille, *J. Amer. Chem. Soc.*, 1984, **106**, 4833.
98. W.J. Scott and W.J. Stille, *J. Amer. Chem. Soc.*, 1986, **108**, 3033.
99. D.R. McKean, G. Parrinello, A.F. Renaldo and J.K. Stille, *J. Org. Chem.*, 1987, **52**, 422.
100. J.P. Quintard, B. Elissondo and M. Pereyre, *J. Org. Chem.*, 1983, **48**, 1559.
101. N.A. Bumagin, I.G. Bumagina and I.P. Beletskaya, *Dokl. Akad. Nauk, SSSR*, 1983, **272**, 1384.
102. T.R. Bailey, *Tetrahedron Lett.*, 1986, **27**, 4407.
103. A. Dondoni, M. Fogagnolo, A. Medici and E. Negrini, *Synthesis*, 1987, 185; A. Dondoni, G. Fantin, M. Fogagnolo, A. Medici and P. Pedrini, *Synthesis*, 1987, 693.
104. J. Solberg and K. Undheim, *Acta Chem. Scand. B*, 1987, **71**, 712.
105. A.M. Echavarren and J.K. Stille, *J. Amer. Chem. Soc.*, 1987, **109**, 5478.
106. M. Matsubara, M. Mitani and K. Utimoto, *Tetrahedron Lett.*, 1987, **28**, 5857.

107. J.W. Labadie and J.K. Stille, *J. Amer. Chem. Soc.*, 1983, **48**, 4634.
108. J.A. Soderquist and W.W.H. Leong, *Tetrahedron Lett.*, 1983, **24**, 2361.
109. W.F. Goure, M.E. Wright, P.D. Davis, S.S. Labadie and J.K. Stille, *J. Amer. Chem. Soc.*, 1984, **106**, 6417.
110. T. Kobayashi and M. Tanaka, *J. Organomet. Chem.*, 1981, **205**, C27.
111. M. Kosugi, T. Sumiya, Y. Obara, M. Suzuki, H. Sano and T. Migita, *Bull. Chem. Soc. Jpn*, 1987, **60**, 767.
112. J.B. Verilhac, E. Chanson, B. Jousseaume and J.-P. Quintard, *Tetrahedron Lett.*, 1985, **26**, 6075.
113. N.A. Bumagain, Y.V. Gulevich and I.P. Beletskaya, *J. Organomet. Chem.*, 1985, **282**, 421.
114. S. Kanemoto, S. Matsubara, K. Oshima, K. Utimoto and H. Nozaki, *Chem. Lett.*, 1987, 5.
115. e.g. J.H. Simpson and J.K. Stille, *J. Org. Chem.*, 1985, **50**, 1759; E. Block and M. Aslam, *J. Amer. Chem. Soc.*, 1983, **105**, 6165; N. Ono, K. Zinsmeister and A. Kaji, *Bull. Chem. Soc. Jpn*, 1985, **58**, 1069.
116. S. Chandrasekhar, S. Latour, J.D. Wuest and B. Zacharie, *J. Org. Chem.*, 1983, **48**, 3810.
117. G.E. Keck and J.B. Yates, *J. Amer. Chem. Soc.*, 1982, **104**, 5829.
118. J.E. Baldwin, R.M. Adlington and A. Basak, *J. Chem. Soc., Chem. Commun.*, 1984, 1284.
119. G.A. Russell and L.L. Herold, *J. Org. Chem.*, 1985, **50**, 1037.
120. G.A. Russell, H. Tashtoush and P. Ngoviwatchai, *J. Amer. Chem. Soc.*, 1984, **106**, 4622.
121. J.E. Baldwin and D.R. Kelley, *J. Chem. Soc., Chem. Commun.*, 1985, 682; J.E. Baldwin, D.R. Kelly and C.B. Ziegler, *Ibid.*, 1984, 133.
122. J.P. Godschalx and J.K. Stille, *Tetrahedron Lett.*, 1983, **24**, 1905.
123. G.E. Keck and E.J. Enholm, *Tetrahedron Lett.*, 1985, **26**, 3311.
124. J.L. Wardell, in *The Chemistry of the Metal–Carbon Bond*, ed. F.R. Hartley, Vol. 4, Chap. 1, Wiley, New York, 1987.
125. B.J. Wakefield, *The Chemistry of Organolithium Compounds*, Pergamon, Oxford, 1974; Compounds of the Alkali and Alkaline Earth Metals in Organic Synthesis, in *Comprehensive Organometallic Chemistry*, eds. G. Wilkinson, F.G.A. Stone and E.W. Abel, Vol. 7, Chap. 44, Pergamon, Oxford, 1982.
126. D.E. Seitz and A. Zapata, *Tetrahedron*, 1980, **21**, 3451.
127. D.Y. Curtin and W.J. Koehl. *J. Amer. Chem. Soc.*, 1964, **84**, 1967.
128. W.C. Still and C. Sreekumar, *J. Amer. Chem. Soc.*, 1980, **102**, 1201.
129. E.J. Corey and B. De, *J. Amer. Chem. Soc.*, 1984, **106**, 2735.
130. D. Seyferth and L.G. Vaughan, *J. Amer. Chem. Soc.*, 1964, **86**, 883.
131. H. Westmijer, K. Ruitenberg, J. Meijer and P. Vermeer, *Tetrahedron Lett.*, 1982, **23**, 2797.
132. H. Nemoto, X.M. Wu, H. Kurobe, M. Ihara and K. Fukumoto, *Tetrahedron Lett.*, 1983, **24**, 4257.
133. W.C. Steel and A. Mitra, *J. Amer. Chem. Soc.*, 1978, **100**, 1927.
134. E.J. Corey and T.M. Eckrich, *Tetrahedron Lett.*, 1984, **25**, 2415.
135. A. Itoh, T. Saito, K. Oshima and H. Nozaki, *Bull. Chem. Soc. Jpn*, 1981, **54**, 1456.
136. W.C. Still, *J. Amer. Chem. Soc.*, 1977, **99**, 4836.
137. M. Ochiai, T. Ukita, Y. Nagao and E. Fugita, *J. Chem. Soc. Chem. Commun.*, 1985, 637; 1984, 1007.
138. K. Nakatani and S. Isoe, *Tetrahedron Lett.*, 1985, **26**, 2209; 1984, **25**, 5335.
139. M. Pereyre and J.P. Quintard, *Pure & Appl. Chem.*, 1981, **53**, 2401.
140. M. Shibasaki, H. Suzuki, Y. Torisawa and S. Ikegami, *Chem. Lett.*, 1983, 1303.
141. G.M. Pickles, T. Spencer, F.G. Thorpe, A.B. Chopa and J.C. Podesta, *J. Organomet. Chem.*, 1984, **260**, 7.
142. H.C. Clark and C.J. Willis, *J. Amer. Chem. Soc.*, 1960, **82**, 1888; D. Seyferth, H. Dertouzos, R. Suzuki and J.Y.P. Mui, *J. Org. Chem.*, 1967, **32**, 2980.
143. P.M. Warner and R.D. Herold, *Tetrahedron Lett.*, 1984, **25**, 4897.
144. D. Seyferth and F.M. Armbrecht, *J. Amer. Chem. Soc.*, 1969, **91**, 2616.
145. D. Seyferth and R.L. Lambert, *J. Organomet. Chem.*, 1975, **91**, 31.
146. F.M. Armbrecht, W. Tronich and D. Seyferth, *J. Amer. Chem. Soc.*, 1969, **91**, 3218.
147. T. Kauffmann, *Angew. Chem., Int. Edn. Engl.*, 1982, **21**, 410.
148. E. Murayama, T. Kikuchi, K. Sasaki, N. Sootome and T. Sato, *Chem. Lett.*, 1984, 1897.

149. T. Kauffmann, R. Kriegesmann and A. Hamsen, *Chem. Ber.*, 1982, **115**, 1818.
150. D.D. Davis and C.E. Cray, *J. Org. Chem.*, 1970, **35**, 1303; W. Kitching, H. Olszowy, J. Waugh and D. Doddrell, *J. Org. Chem.*, 1978, **43**, 898.
151. H.J. Tilhard, H. Ahlers and T. Kauffmann, *Tetrahedron Lett.*, 1980, **21**, 2803.
152. R.D. Taylor and J.L. Wardell, *J. Organomet. Chem.*, 1975, **94**, 15.
153. M. Ochiai, S. Tada, K. Sumi and E. Fujita, *Tetrahedron Lett.*, 1982, **23**, 2205; M. Ochiai, T. Ukita, E. Fujita and S. Tada, *Chem. Pharm. Bull.*, 1984, **32**, 1829; M. Ochiai, K. Sumi, E. Fujita and S. Tada, *Chem. Pharm. Bull.*, 1983, **31**, 3346; B.A. Peariman, S.R. Putt and J.A. Fleming, *J. Org. Chem.*, 1985, **50**, 3622.
154. M. Ochiai, T. Ukita and E. Fujita, *Chem. Lett.*, 1983, 1457; *J. Chem. Soc., Chem. Commun.*, 1983, 619.
155. I. Fleming and C.J. Urch, *J. Organomet. Chem.*, 1985, **285**, 173.
156. J.F. Kadow and C.R. Johnson, *Tetrahedron Lett.*, 1984, **25**, 5255.
157. S. Teratake and S. Morikawa, *Chem. Lett.*, 1975, 1333; S. Teratake, *Noguchi Kenkyusho Jiho*, 1977, **20**, 33.
158. D.J. Peterson, M.D. Robbins and J.R. Hansen, *J. Organomet. Chem.*, 1974, **73**, 237.
159. K.C. Nicolaou, D.A. Claremon, W.E. Barnette and S.P. Seitz, *J. Amer. Chem. Soc.*, 1979, **101**, 3704.
160. Y. Ueno, M. Ohta and M. Okawara, *Tetrahedron Lett.*, 1982, **23**, 2577.
161. C.A. Grob and A. Waldner, *Helv. Chim. Acta*, 1979, **62**, 1736.
162. A. Shanzer, M. Mayer-Schochet, F. Frolow and D. Rabinovich, *J. Org. Chem.*, 1981, **46**, 4662.
163*a*. A. Ricci, S. Roelans and A. Vannucchi, *J. Chem. Soc., Chem. Commun.*, 1985, 1457; N. Nagashima and M. Ohno, *Chem. Lett.*, 1987, 141.
163*b*. S. David and S. Hanessian, *Tetrahedron*, 1985, **41**, 643.
164. D. Wagner, J.P.H. Verheyden and J.G. Moffatt, *J. Org. Chem.*, 1974, **39**, 24.
165. Y. Tsuda, M.E. Haque and E. Yoshimoto, *Chem. Pharm. Bull. Jpn.*, 1983, **31**, 1612.
166. B. Delmond, J.C. Pommier and J. Valade, *Tetrahedron Lett.*, 1969, 2089.
167. T. Ogawa and M. Matsu, *Tetrahedron*, 1981, **37**, 2363.
168. D.H.G. Crout and S.M. Morrey, *J. Chem. Soc., Perkin Trans. 1*, 1983, 2435.
169. Y. Oueno and M. Okawara, *Tetrahedron Lett.*, 1976, 4597.
170. G. Tagliavini and D. Marton, *Gazetta Chim. Ital.*, 1988, **118**, 483.

11 Biological chemistry of tin

M.J. SELWYN

11.1 Introduction

Tin and tin compounds are widely used and have the advantage that metallic tin, inorganic compounds and long-chain alkyl compounds, such as the organotin stabilizers in PVC, are effectively non-toxic, while the organotin biocides are selective and potentially degradable to non-toxic compounds. Knowledge of the biological chemistry of tin is needed for the rational development of more effective compounds and newer uses, for example as antitumour agents.

Three events have shown the necessity for care in the use of tin compounds and provided additional impetus to investigation of their biological chemistry. Firstly, the proprietary medicine Stalinon, which contained triethyltin iodide, caused the deaths of about a hundred people in 1954. Secondly, the discovery that inorganic mercury could be biologically methylated suggested that similar processing could occur with tin. Thirdly, tributyltin compounds from antifouling paints on boats have accumulated in estuaries and poisoned shellfish, particularly the Pacific oyster.

In consequence, there has been an explosive growth in the number of publications on the biological effects of tin compounds including reviews on general aspects[1–5], environmental problems[6–9a] and biochemistry and toxicology.[10–12]

This chapter is particularly concerned with the molecular interactions of tin compounds with enzymes and biological membranes and with the metabolic processing of tin compounds. The biochemical effects of inorganic tin compounds are dealt with separately in section 11.2. Section 11.3 provides a brief background on the overall toxicity of organotins, their applications and modes of entry into the environment to make contact with living organisms. Section 11.4 deals with the biochemical activities of organotin compounds leading from complex formation, through effects on individual enzymes and biological membranes to their inhibition of ATP synthesis and finally to their actions on whole cell systems in the nervous system, in immunosuppression and as antitumour agents. Section 11.5 describes the metabolic breakdown of organotin compounds and the possibilities for biological methylation.

11.2 Inorganic tin

Some observations suggest that tin is a trace nutrient; for example, rats fed on a tin-supplemented diet grew 60% faster than animals fed on a tin-free diet.[13] The level of tin at 5–10 μg per rat per day was comparable to human dietary intake of around 10 mg per day. However, in the absence of any known biochemical function of tin, its status as an essential micronutrient[12] remains doubtful.

At high dietary levels, 0.5 g tin ($SnCl_2$) per 100 g dry weight of food over a period of months, rats developed irritation of the intestinal tract and anaemia.[14,15] Oral administration of $SnCl_2$ at 60 mg kg^{-1} body weight for three days led to decreased calcium binding by the mucosa, and bone appeared to be the critical tissue.[16,17]

The properties of Sn^{4+} and Sn^{2+} ions provide some explanation for the low toxicity of inorganic tin.[18] Firstly, as tin ions do not have a strong preference for sulphur or nitrogen ligands over oxygen, they do not block catalytically essential sulphydryl, amino or imidazole groups in enzymes, unlike the highly toxic Hg^{2+}, Pb^{2+} and Cu^{2+} ions. Secondly, only very low concentrations of simple Sn^{2+}_{aq} and Sn^{4+}_{aq} ions exist at the neutral or slightly alkaline pH of most body fluids and cytoplasm, since both Sn (II) and Sn (IV) form complex ions with hydroxide and chloride, some of which are polymeric and some have low solubility. As a result, tin ions are poor competitors for oxygen ligands in binding sites for Mg^{2+} ions, which are essential activators for many enzymes, or Ca^{2+} ions, which control several cell processes. They are also poor competitors with essential metal ions for membrane transport systems and are thus poorly accumulated. Although Sn (IV) inhibited lead uptake by kidney slices,[19] this may have been the result of a reaction between Sn (IV) and Pb (II), rendering the lead unavailable, rather than a direct effect of tin on the uptake system. Intravenous injection of tin salts, radioactively labelled with ^{113}Sn, into rats showed that both Sn (II) and Sn (IV) were excreted via the urine, and that rather less Sn (II) but no Sn (IV) was eliminated via the bile.[20] In mice the half-life of injected inorganic tin was 29 days.[21] In short, the low toxicity of inorganic tin ions is a result of their being poorly absorbed, weakly bound and steadily excreted.

However, inorganic tin has been found to affect some enzymes involved in haem synthesis and breakdown. Haem oxygenase catalyses the first step in the breakdown of haem to bilirubin in which the porphyrin ring is broken and the iron liberated (eqn 11.1).

$$\text{Haem} + 3AH_2 + 3O_2 \rightarrow \text{Biliverdin} + Fe^{2+} + CO + 3A + 3H_2O \qquad (11.1)$$

Kappas and Maines[22] found that injection of rats with $SnCl_2$ at up to 250 μmol kg^{-1} body weight produced an increase in haem oxygenase activity in the liver and kidney. Other metal ions also induced increases, the magnitudes depending on both tissue and metal, but the largest was the twenty-five-fold increase in the kidney enzyme produced by tin.[23] Mn^{2+} or Zn^{2+} blocked the increase induced by $SnCl_2$ if administered simultaneously or shortly before, but not after, the tin.[24]

Injection of rabbits with tin (II) chloride or citrate at 5 μmol kg^{-1} body weight caused a decrease of δ-aminolaevulinate dehydratase activity in red blood cells.[25] This enzyme, also called porphobilinogen synthase, catalyses the condensation of two molecules of δ-aminolaevulinic acid to yield porphobilinogen (eqn 11.2), and is a control step in the biosynthesis of porphyrins from glycine and succinylCoA. δ-Aminolaevulinate dehydratase activity declined almost to zero within 24 hours of injection of tin salts, remained low for two days then slowly increased towards the normal level. The decline could have been due to direct inhibition or to repression of enzyme synthesis, but analogy with the action of other metal ions[26] favours the latter explanation.

Inducers or repressors are ligand molecules which bind to the gene activator or repressor proteins, and can either increase or decrease binding of these proteins to DNA, thereby controlling protein synthesis at the level of transcription. These ligand interactions are highly specific, even though the inducer or repressor may be significantly different from the substrate or immediate product of the enzyme. The

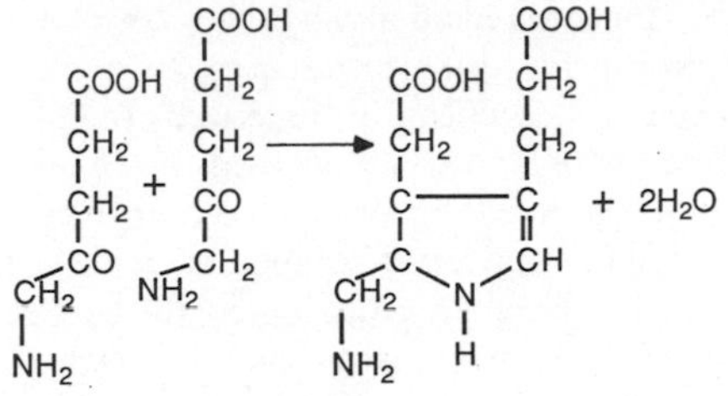

$$HOOC{-}CH_2{-}CH_2{-}CO{-}CH_2{-}NH_2 + HOOC{-}CH_2{-}CH_2{-}CO{-}CH_2{-}NH_2 \longrightarrow \text{porphobilinogen} + 2H_2O \qquad (11.2)$$

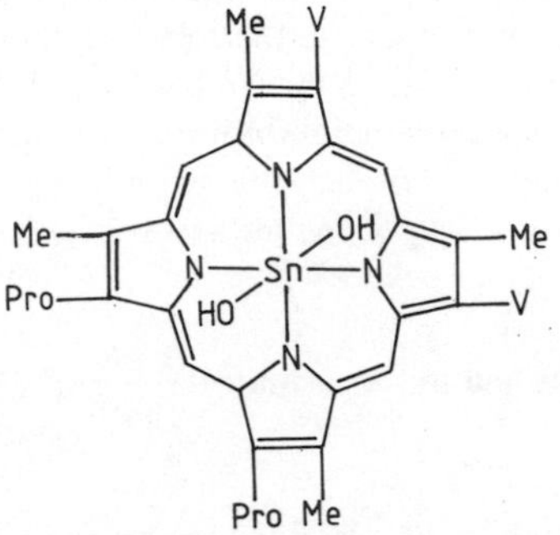

Figure 11.1 Sn-protoporphyrin IX shown with axial bonded hydroxyl groups. V represents a vinyl group, Pro a propionate group.

patterns of tissue and metal ion specificity indicate highly selective interactions of Sn^{2+} ions with the control proteins of haem oxygenase and δ-aminolaevulinate dehydratase, but there is no indication whether these are purely fortuitous or natural biochemical functions of tin.

The tin (IV) analogue of haem, Sn-protoporphyrin IX (Sn-haem) (Figure 11.1) has been found to have a highly specific interaction with haem oxygenase (eqn 11.1), which may have therapeutic applications. In newborn mammals, including human infants, large amounts of the bile pigment bilirubin are produced as a result of lysis of red blood cells and subsequent degradation of the foetal haemoglobin. The biliverdin produced from the haem is reduced to bilirubin which, if produced in the spleen or bone marrow, is transported bound to blood plasma albumen to the liver where it is normally detoxified by conjugation with glucuronic acid prior to excretion in the bile. Neonatal jaundice can develop because the liver has insufficient glucouronyl transferase activity to detoxify bilirubin as the glucuronide and, in severe cases, can cause brain damage as bilirubin can cross the incompletely developed blood–brain barrier. Sn-haem injected at 10 μmol kg^{-1} body weight produced a decline of serum bilirubin to near adult levels in 24 hours,[27–29] by preventing the post-natal increase in the activity of haem oxygenase of which it is a potent competitive inhibitor, with a K_i of 0.011 μM, K_m for the substrate Fe-haem being 12.5 μM. Other metal-haems were less effective and some, including Fe-haem, were inducers of haem oxygenase (induction by free Sn^{2+} and other metal ions has been described above). In contrast to the high affinity for haem oxygenase, Sn-haem has low affinity for both apomyoglobin, which has high affinity for the natural prosthetic group Fe-haem, and plasma albumen which binds bilirubin.[30] In the case of apomyoglobin the lowered affinity has been correlated with the existence of free Sn-haem as a monomer in aqueous solution in contrast to Fe-haem and metal-free protoporphyrin IX which are significantly dimerized. Initially the dimer binds to apomyoglobin and formation of myoglobin proceeds by an induced-fit mechanism in which a conformation change in the protein is accompanied by release of a haem monomer. No such induced fit occurs with serum albumen, but the weak binding and also the lack of dimerization have been attributed to the strong axial bonding of hydroxide ions or water[31] in Sn-haem (Figure 11.1), which makes substitution by protein ligands, such as histidine imidazole in apomyoglobin, energetically less favourable than with Fe-haem. Increased withdrawal of electrons from the porphyrin ring system in Sn-haem compared with Fe-haem also weakens ring–ring or ring–protein interactions. In haem oxygenase, the relatively polar binding site does not favour the electron-rich ring system of the Fe-haem, and it seems that axial ligand bonding favours binding. The potent inhibition of the haem oxygenase and non-interference with haem-protein interactions or bilirubin transport indicate the potential value of Sn-haem as a drug to reduce neonatal jaundice without harmful side effects.

11.3 Organotins, organisms and the environment

11.3.1 *Principles*

Albert[32] has listed the principles of selective toxicity as accumulation, cytology and biochemistry, and these can be further resolved into:

(i) Uptake by the organism
(ii) Transport and distribution within the organism
(iii) Excretion
(iv) Presence of sensitive organs, tissues, cells and sub-cellular organelles
(v) Affinity for and quantity of receptor molecules
(vi) Sequestration in harmless complexes or deposits
(vii) Metabolic conversion to less or more toxic compounds.

All of these apply to individual organisms, but the toxicology and biological chemistry must also take into account the global or environmental distribution of tin compounds. This can also be divided into several processes:

(viii) Manufacture and release into the environment
(ix) Mobility, e.g. volatility in air, solubility in water
(x) Trapping in and leaching from non-living materials, such as clay or humic acids
(xi) Distribution and concentration via food chains
(xii) Environmental, including geochemical and biological, processing.

This section give brief outlines of, firstly, the quantitative relationships between chemical structure and toxicity to individual organisms, and, secondly, the sources and processing of organotins in the environment.

11.3.2 *Structure–toxicity relationships*

There are several detailed reviews and bibliographies of the extensive literature on the toxicity of organotin compounds.[3,11,33–38] In the formula R_nSnX_{4-n} R is generally a hydrocarbon and X a halide or monovalent anionic ligand, although the oxides of triorganotin compounds $(R_3Sn)_2O$, are also important. As a rule X has relatively little effect on toxicity, unless it also has some toxic action, because the organotin compounds are hydrolysed in aqueous media (see section 11.4.1). Exceptions showing reduced toxicity occur when the X radical forms a very stable link or links with the tin atom, particularly bifunctional radicals which form bidentate complexes with the tin atom, for example dithiols with dialkyltins and radicals containing a nitrogenous ligand which can form five-coordinate complexes with triorganotins.[39–41] Radicals forming five-coordinate polymers have less effect, due to dissociation on dilution. The reduction in toxicity may reflect not only perturbation of equilibria between free organotin cation or hydroxide and binding sites but also inhibition of mobility, uptake and the halide-hydroxide exchange catalysed by triorganotins (section 11.4.3). The effects of the X groups on toxicity are greater in dialkyltins than in trialkyltins, and are particularly important in the balance between toxicity and antitumour activity (section 11.4.6). The difference may result from the dependence of the toxic action of dialkyltins on the formation of bifunctional links with dithiol compounds (section 11.4.3) comparable to the links with bifunctional X radicals, in contrast to the trialkyltins which form weaker bifunctional or single links with hydrophobic bonding to the carbon skeleton (which seldom occurs with the X radicals) contributing to the high affinity of some binding sites (section 11.4.2).

The number of R groups has a crucial effect on toxicity (Table 11.1). The general rule is that for all organisms the triorganotin compounds are the most toxic, followed by the diorganotins, while the mono-organotins are of very low toxicity. The toxic effects of tetraorganotins are delayed, and appear to be due to their metabolic conversion to the triorganotin compounds (section 11.5.1).

The effect of the R group depends on the class of organism. Triethyltin is the most toxic to mammals: both trimethyltin and tripropyltin are also highly toxic, but tributyltin is markedly less toxic; and tri-*n*-octyltin has very low toxicity. Tributyltin and tripropyltin were the most toxic to Gram positive bacteria and fungi, but Gram-negative bacteria were much less sensitive, triethyltin and tripropyltin being the most toxic.[34]

Aquatic organisms, whether fish, molluscs (such as oysters), crustaceans (Table 11.1) or algae, are particularly sensitive to tributyltin, triphenyltin and tricyclohexyltin.[38,42–44]

The terrestial higher plants are essentially insensitive to the larger triorganotins, but tributyltin is toxic, especially to seedlings.[45]

Trimethyltin is highly toxic to insects, and the higher triorganotins have an 'anti-feedant' effect.[46] The spider mites, as well as terrestial arthropods, are highly sensitive to triorganotins with large R groups, such as tricyclohexyltin.[47] The structure activity relationships of these acaricides or 'miticides' is interesting. Firstly, they all have extremely low mammalian toxicity but some have fungicidal activity. Secondly, there appears to be a requirement for steric crowding about the tin atom. In the cyclohexyl compounds a methylene group between the ring and the tin decreases, and a propylene group abolishes, both acaricidal and fungicidal activity. In the triphenyl-based series, tribenzyltin is inactive, but compounds with highly branched chains as in the neophyl group (**1**) are active. The diorganotins show a similar progressive decrease in oral toxicity to mammals when the size of the R group is larger than propyl (Table 11.1).

As well as the specific chemical interactions described in section 11.4, some physical and physiological factors have a major influence on toxicity.

Table 11.1 Toxicity of organotins

R_nSn fragment	Rat (mmol kg^{-1})	Crab larva (μM)	*Botrytis* (mM)	*B. subtilis* (mM)	*E. coli* (mM)
Me_4Sn	1.5				
Et_4Sn	0.06				
Bu_4Sn	>12				
$^{n}Oct_4Sn$	>7				
Me_3Sn	0.07	0.56	0.9	>2.2	>2.2
Et_3Sn	0.04	0.39	0.004	0.2	0.08
$^{n}Pr_3Sn$	0.3	0.19	0.002	0.007	0.16
$^{n}Bu_3Sn$	0.7	0.055	0.001	0.006	>1.4
$^{n}Pen_3Sn$			0.01	0.013	>1.3
$^{n}Hex_3Sn$	2		>1.2	0.12	>1.2
$^{c}Hex_3Sn$	1	0.020			
$^{n}Hep_3Sn$				>1.1	>1.1
$^{n}Oct_3Sn$	>8				
Ph_3Sn	0.3	0.097	0.02	0.001	>1.2
Me_2Sn	0.7	92	i	0.9	2.3
Et_2Sn	0.3	15	i	0.2	0.4
$^{n}Pr_2Sn$	0.05		i	0.07	0.2
$^{n}Bu_2Sn$	0.5	2.8	i	0.07	0.07
$^{n}Pen_2Sn$			i	0.06	1.5
$^{n}Hex_2Sn$			i	0.14	>1.4
$^{c}Hex_2Sn$		0.37			
$^{n}Hep_2Sn$				>1.3	>1.3
$^{n}Oct_2Sn$	>10				
Ph_2Sn		2.6	0.06	0.06	>1.5
Bz_2Sn		27			
MeSn	4		i		
EtSn	<0.8		i		
^{n}BuSn	8		i		
^{n}OctSn	9		i		
PhSn			0.3		

Data for: rat, Smith[35]; crab larvae, Laughlin *et al.*[42]; fungi and bacteria, Sijperstein *et al.*[34]. Where necessary, concentrations have been converted to molarities or (for the rat) to mmol kg^{-1}. This assists comparison, but it should be noted that conditions are not comparable, LD_{50} values for the rat are maximum body load while LC_{50} values for crab larvae are continuous exposure and the body load may be much higher. i, inactive.

$$(C_6H_5{-}C(CH_3)_2{-}CH_2)_3Sn{-}X$$

(1)

Terrestrial organisms generally have impermeable surfaces, skins, exoskeletons or plant cuticles, which minimize water loss and special organs for controlled gas exchange. The main route of organotin entry is ingestion with food and absorption from the alimentary canal, but volatile compounds, such as trimethyltin chloride, may enter via the lungs of vertebrates and tracheae of insects. Surface absorption may be important for small terrestrial organisms.

Many aquatic organisms circulate large volumes of water over the gills to obtain oxygen, and the contact-water volume is further increased in filter-feeders which extract particulate food from a dilute suspension. Tributyltin, triphenyltin and tricyclohexyltin compounds have low solubility in water but, being hydrophobic molecules, readily cross biological membranes and accumulate in hydrophobic binding sites and lipids. When the R groups are very large the effective toxicity may be lowered, not only by their extremely low water solubility but also by retention in, rather than passage through, lipid membranes.[42] The marine mussel, *Mytilus edulis*, accumulates tributyltin both from solution in water and, more rapidly and to a higher level, by ingestion of phytoplankton containing tributyltin.[48] The highest concentrations were in the gill or viscera rather than muscle and were correlated with the lipid content of the tissue. As neither kaolin nor humic acids which bind tributyltin had much effect on accumulation, and since the bio-concentration factor (BCF) was ten-fold greater than predicted from the octanol/water partition coefficient, accumulation involves more than simple partition into lipid. Salmon can accumulate tributyltin by a factor of around 1000-fold from water containing tributyltin at 0.1–1.0 $\mu g\,l^{-1}$, and farmed salmon may contain up to 1.0 $mg\,kg^{-1}$ in muscle tissue, probably leached from tributyltin antifouling on netting cages.[49]

The problems in comparing the bioconcentration factors (BCF) for different types of organism have been discussed by Kenaga[50], and relate not only to the different sources, soil, food or water but also to the use of different denominators of molecular properties such as water solubility, 1-octanol/water partition coefficient and organic carbon soil absorption coefficient. However, reasonable regressions were obtained not only between the BCF for beef fat and these three molecular properties, but also with BCF values for fish in flowing and static water. The toxicity of organotins to freshwater diatoms has been correlated with water solubility,[51] and uptake of trialkyltins by *E. coli* spheroplasts with the calculated total surface area of the molecules.[52]

A detailed investigation has been made by Laughlin *et al.*[42] of the quantitative structure–activity relationships of the toxicity of sets of dialkyltins and trialkyltins to crab larvae. The LC_{50} values ranged from 92 μM for dimethyltin to 20 nM for tricyclohexyltin, with some overlap between the two sets. There were linear free-energy correlations between the LC_{50} and the Hansch π parameter[53] for the diorganotin and triorganotin series, although the slopes and levels of the lines were different. Correlation with the calculated total molecular surface area also gave good correlation with separate lines for the two series of compounds. By using Hansch fragment constants to allow for the substitution of one hydrocarbon R group by a hydroxyl group in the diorganotins, the authors produced a correlation in which both series could be fitted to one line. This suggests the paramount importance of hydrophobic interactions in the toxicity to crab larvae, although the two series are not precisely co-linear. The value of this type of correlation for predictive as well as analytical purposes has been discussed by Brinckman[54] with particular reference to molecular total surface area calculations for triorganotins.

11.3.3 *Organotins in the environment*

Only a brief survey as a background to the biochemistry of organotins is given here, but several comprehensive accounts are available.[3,4,6–9a,46,55–57]

The major source of organotins in the environment is their use by man.[4,5,8,55] Table 11.2 summarizes the applications with approximate percentages of the total

Table 11.2 Applications of organotins

Application	Formula	R group	% Total use
(i) Minimal or no release in use			
PVC stabilizers	R_2SnX_2 $RSnX_3$	Me, ^{n}Bu, ^{n}Oct ^{n}BuOCOCH$_2$CH$_3$	70
Catalysts	R_2SnX_2	^{n}Bu	8
(ii) Slow or steady release			
Antifouling paint	R_3SnX	^{n}Bu, Ph	10
Wood preservation (fungicides)	R_3SnX	^{n}Bu	1
(iii) Immediate release			
Agriculture (fungicides, acaricides, anti-feedants)	R_3SnX	Ph, ^{c}Hex, Neophyl	9
Miscellaneous (disinfectants, stone preservation)	R_3SnX	^{n}Bu	1
Antihelminthics (poultry)	R_2SnX_2	^{n}Bu	1

Based on data in Blunden *et al.*[4,8] and Zuckerman *et al.*[55]

annual organotin production, which was estimated to be around 35 000 tons in 1985. Although the biocides account for only about one-fifth of organotin production, their environmental effect is proportionately much greater, since contact with living organisms is the essential feature of this use and inevitably contact is made with organisms other than the intended targets. The table also divides the organotins according to the rate at which they are released into the environment. Those directly released are a small proportion of the total, and although some rapidly enter food chains, a proportion will be trapped in soil or sediments. Release into the environment plays no part in the use of by far the largest proportion, the PVC stabilizers and polymerization catalysts, and leaching from these products is slow.[58] Disposal by incineration could lead to release of organotins to the atmosphere if incineration is incomplete. Release of organotins from plastics buried in refuse tips is slow, and some at least will be degraded to inorganic tin.

The use of tributyltin compounds in marine antifouling paints gives highly effective protection against fouling, but requires a slow steady release which leads to serious environmental pollution by organotins. The leaching of the tributyltin leads to a build-up in estuaries where large numbers of boats are moored, and the first indications of a problem were the failure of Pacific oyster to grow properly in beds in the vicinity of marinas in France and UK.[59,60] As a consequence, the use of tributyltin antifouling has been banned or severely restricted. The seasonal variation of tributyltin concentration in nine UK estuaries has been recorded and compared with the toxicity to a wide range of organisms.[38] The harmful effects are not confined to oysters and mussels, and other shellfish such as the dog-whelk are also seriously affected, the females becoming sterile with a condition called imposex.[61] This condition is sufficiently well correlated with tributyltin pollution that it can be used as an indicator of such pollution.[62,63] The sterility means that even if the tributyltin concentration is reduced to zero, the damage to the population will persist for many years.

The use of tributyltin to control the aquatic snail hosts of the parasitic flatworms which cause bilharzia[64] is associated with similar problems, since concentrations which will kill the snail will also harm other aquatic life.

Improvements in analytical methods[65,66] have enabled identification and measurement of very low environmental levels, not only of man-made organotin compounds but also of breakdown products and methylated derivatives.

At near-neutral pH, in the dark and at normal environmental temperatures alkyl- and aryltin compounds are stable, but breakdown occurs when they are exposed to uv light. This has been observed with trimethyltin[67], tributyltin[68] and triphenyltin[69,70], and proceeds by sequential removal of alkyl groups, eventually yielding inorganic Sn^{4+}. On sugar beet leaves, triphenyltin yielded about 50% Sn^{4+}, with only small amounts of diphenyltin (5%) and phenyltin(1%) (Freitag and Bock[71]).

Comparison of sterilized and unsterilized soils by Barug and Vonk[72] showed biological breakdown of tributyltin with production of dibutyltin and CO_2. Similar studies revealed microbial degradation of triphenyltin in soil and by cultures of soil micro-organisms.[69] Barug[73] has shown degradation of bis(tributyltin) oxide by pure cultures of bacteria and fungi including wood-rotting organisms. The main product was monobutyltin which was not further degraded. In coastal sea-water, tributyltin was degraded with a half-life of less than ten days, giving dibutyltin as the main product. The degradation was mostly biological, since breakdown of tributyltin by UV light measured separately was much slower.[74]

Much of the organotin released into the environment becomes trapped. Tributyltin preservatives become trapped in wood as insoluble polymerized carbonates, not, as was originally thought, bound to polysaccharide hydroxyl groups.[75] Triorganotins bind to humic acids and minerals in the soil or marine sediments[69,76] but are extracted from suspended particles by filter feeders.[48]

The only significant formation of tin–carbon bonds in the environment appears to be methylation but, as noted by Craig[7], environmental processing combining loss of alkyl (including methyl) and aryl groups with methylation could produce a variety of compounds including triorganotins (eqn 11.3). Some of these, such as methyldibutyltin, dimethylbutyltin and dimethyl-*n*-octyltin, would be highly toxic.

$$
\begin{array}{lllll}
R_4Sn \rightarrow & R_3Sn^+ \rightarrow & R_2Sn^{2+} \rightarrow & RSn^{3+} \rightarrow & Sn^{4+} \\
\swarrow\!\!\nearrow & \swarrow\!\!\nearrow & \swarrow\!\!\nearrow & \swarrow\!\!\nearrow & \updownarrow \\
R_3MeSn \rightarrow & R_2MeSn^+ \rightarrow & RMeSn^{2+} \rightarrow & MeSn^{3+} \leftarrow & Sn^{2+} \\
\swarrow\!\!\nearrow & \swarrow\!\!\nearrow & \swarrow\!\!\nearrow & & \\
R_2Me_2Sn \rightarrow & RMe_2Sn^+ \rightarrow & Me_2Sn^{2+} & & \\
\swarrow\!\!\nearrow & \swarrow\!\!\nearrow & & & \\
RMe_3Sn \rightarrow & Me_3Sn^+ & & & \\
\swarrow\!\!\nearrow & & & & \\
Me_4Sn & & & &
\end{array}
\qquad (11.3)
$$

The biochemistry of these conversions is described in section 11.5.

11.4 Biochemical effects of organotin compounds

11.4.1 *Ligands and complexes*

Studies on crystalline complexes of di- and tri-organotins with model compounds have shown that 4-coordinate monomers are tetrahedral, but in polymers or 5-

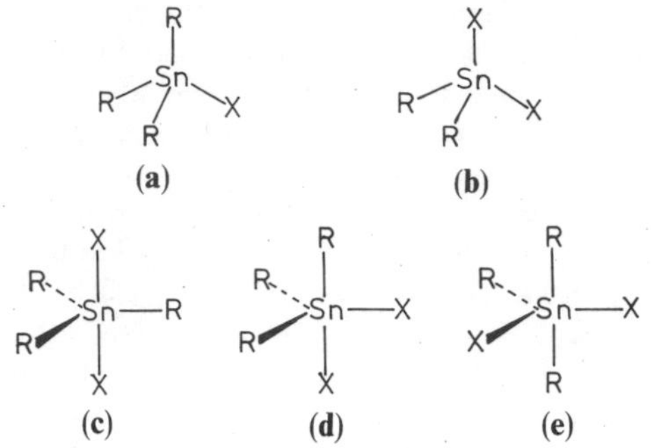

Figure 11.2 Configuration of tetra- and penta-coordinated organotins. In tetracoordinate trialkyltin (*a*) and dialkyltin (*b*), only one arrangement is possible in each case. In the case of pentacoordinate trialkyltins, three different arrangements are possible, described with reference to the X ligands as (*c*) *trans* or axial, (*d*) *cis*, and (*e*) meridional.

coordinate complexes the configuration is trigonal bipyramidal, in which several non-equivalent arrangements are possible (Figure 11.2). X-ray analysis showed that in a crystalline salt the tributyl-diaquo-tin cation is trigonal bipyramidal with apical water molecules.[77]

Information about complex formation in aqueous solution reveals the ligand preferences of organotins and allows prediction of the speciation of organotins in cytoplasm, blood plasma, experimental media or sea water.

The aqueous chemistry of organometallic cations has been described in detail by Tobias.[78] At low concentrations, 10^{-5} M, over the pH range 7–9 virtually all dimethyltin will be the dihydroxide, $Me_2Sn(OH)_2$. Outside this pH range, species such as $(Me_2SnOH)^+$ and $(Me_2Sn(OH)_3)^-$ are present, and at higher concentrations, 10^{-3} M, polynuclear complexes are formed.

Trimethyltin has been extensively investigated, but the low solubility of many organotin compounds prevents the use of potentiometric titration methods. Measurement of the pH dependence of partition into hydrophobic phases can give useful information[10,79] and nmr spectrometry has been used to measure speciation of tributyltin compounds.[80] In aqueous solution $(CH_3)_3SnCl$ is largely dissociated to give the aquo-cation $(CH_3)_3Sn(H_2O)^+$ which undergoes further hydrolysis to yield the hydroxide $(CH_3)_3SnOH{\cdot}aq$, with pK_a 6.3–6.8. At neutral or alkaline pH, most of the un-complexed trimethyltin will be the hydroxide and, as the formation constant of the chloride complex is 0.68 M^{-1}, even in blood plasma (100 mM Cl^-) or sea water (0.5 M Cl^-) only a small proportion of the organotin will be present as the chloride. The situation is similar for triethyltin.

Hynes and O'Dowd[81] have measured complex formation between trimethyltin and several small ligands of biochemical interest, and some of their data are given in Table 11.3. The low values for the formation constants can be contrasted with high values exhibited by cations or organocations of metals such as mercury or copper, especially with nitrogen or sulphur groups. Of natural ligands, glutathione, γ-glutamyl-cysteinyl-glycine, is present in some cells at 2–10 mM and could form significant amounts of trimethyltin complex. Inorganic orthophosphate and the nucleotides are

Table 11.3 Formation constants for trimethyltin complexes in aqueous solution

Ligand	Log constant B_{110}	B_{111}	B_{120}	B_{121}	K_{MHL}
Cl^-	−0.17				
F^-	2.3				
Acetate	1.25				
Succinate		6.69	3.93		1.44
Aspartate		11.58			1.55
Histidinate		11.10			1.76
2-NH_2EtOH		15.52			4.60
Cysteinate		15.21			4.67
$HSCH_2CH_2OH$	5.94				
Penicillamine		14.50			3.64
2, 3-Dimercaptopropanol	8.50	16.22			5.60
Glutathione		14.17			4.39
$(H_2PO_4)^-$	3.53		5.32		
AMP	3.31	7.92	4.73		1.59
IMP		11.41		14.26	2.52
$(P_2O_7)^{4-}$		10.80			2.71

Data for Cl^- and F^- from Tobias[78], otherwise from Hynes and O'Dowd,[81] where additional values can be found. B_{pqr} and K_{MHL} defined as in Hynes and O'Dowd.[81] M = $[Sn(CH_3)_3(H_2O)_2]^+$; L is ligand (number of ionizable protons varies), $B_{pqr} = [L_pM_qH_r]/[L]^p[M]^q[H]^r$; $K_{MHL} = [MHL]/[M][HL] = B_{111}/B_{101}$.

Log B values at 25°C and ionic strength 0.3 M ($NaClO_4$).

also present in cells in the millimolar range, and complex formation may occur. Although ADP and ATP were found to form complexes, the number of different complexes prevented computation of accurate formation constants. The extent to which organotins are complexed by ATP or ADP depends on the concentrations of Mg^{2+} and Ca^{2+} which form high-affinity complexes with nucleoside polyphosphates, and *in vivo* much of the ATP is complexed with Mg^{2+}. The moderate affinity for phosphate is in accord with the inhibitory effects of phosphate buffers on organotin action or binding to macromolecules.[82,83]

Only a few proteins have a high affinity for triorganotin compounds.[10] A soluble cytoplasmic protein in guinea-pig liver bound triethyltin with a dissociation constant of 2.5×10^{-7} M, and photo-oxidation indicated histidine residues in the binding site.[84,85] A binding site in myelin is discussed in section 12.4.5. A triorganotin-binding protein has been isolated from snails.[86] As well as the inhibition of enzymes (section 11.4.2), major interest centres on the ligands involved in binding to rat and cat haemoglobins and the F_0 component of mitochondrial ATPase.[10,87,88]

Haemoglobin from the rat or cat, but not other animals, was found to bind triethyltin with high affinity[84,89] and, although not of general toxicological importance, the ready availability of haemoglobin and wealth of knowledge of its properties and structure resulted in this binding being thoroughly investigated as a model for less accessible sites. Rat haemoglobin has two triethyltin binding sites per tetramer, which depend on the three-dimensional structure of the protein being lost on denaturation. The effect of pH and the loss of binding on photoxidation suggested that each triethyltin was bound to two histidine residues,[82] but diethylpyrocarbonate, which is specific for histidine, abolished only one site,[89] and reagents for sulphydryl groups

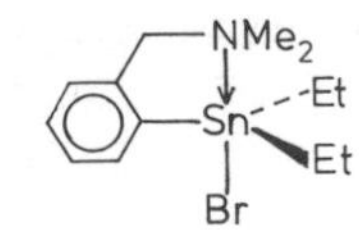

(2)

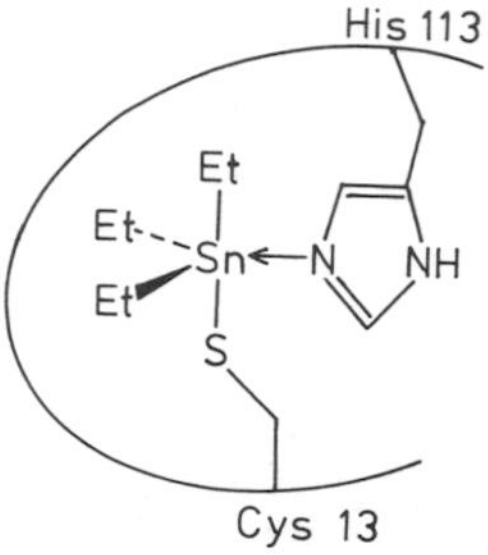

Figure 11.3 Configuration and ligands of trialkyltin bound to cat or rat haemoglobin.

indicated bonds to cysteine thiols.[90] Mössbauer spectroscopy showed only one type of site, probably either 4-coordinate or *cis*-pentacoordinate, and the failure of a compound with one intramolecular coordination **(2)**[91] to bind at these sites suggested pentacoordinate binding in the haemoglobin sites (Figure 11.3).[10,87] Taketa and Siebenlist[92,93] have compared sequence and structural data for haemoglobins with and without binding sites and found that Cys 13 and His 113 of the alpha monomers provide the ligands, and triethyltin does not bind to haemoglobins in which either is replaced by another amino acid.

Although the natural allosteric effectors bind to a different site on haemoglobin, there is allosteric interaction between triethyltin and oxygen binding. $Hb(O_2)_4$ has high affinity for triethyltin, while Hb has much lowered affinity, and bound triethyltin increases the affinity for oxygen. The natural function of the trialkyltin binding side is not known, but both cat and rat haemoglobin have low affinity for oxygen.[93]

Neither the protein nor the ligand in the mitochondrial ATPase site have yet been conclusively identified. Aldridge and Street[94] measured triethyltin binding to mitochondria and fitted their data with two types of binding site; one of low affinity and the other, which appeared to be involved in the action of triethyltin on oxidative phosphorylation, of high affinity. By using a sucrose medium to eliminate the complications produced by the triorganotin-catalysed Cl^-/OH^- exchange, it was found[95] that the effects of triorganotins on oxidative phosphorylation and ATPase activity were very similar to those of oligomycin, a diagnostic inhibitor of the F_0 component of the mitochondrial proton translocating ATPase. Rose and Aldridge[96] replaced chloride with nitrate or isethionate and showed quantitative correlation between binding to the high-affinity site and inhibition of oxidative phosphorylation and ATPase activity.

Using Mössbauer spectrometry and measurements of binding, Farrow and Dawson[97] identified not only the high- and low-affinity sites but also partition (i.e. a non-saturating uptake) into the membrane. The low-affinity site could be blocked with *N*-ethylmaleimide, which reacts with sulphydryl groups, and after washing to remove partitioned triethyltin the much simplified Mössbauer spectrum showed only one type of site. The Mössbauer quadrupole splitting and isomer shift for this, the high-affinity, site, indicated either tetrahedral or possibly *cis*-pentacoordinate bonding of the tin atom. Investigation of intramolecularly coordinated tin compounds, showed that (**2**) which has one internal coordination is a highly potent inhibitor of mitochondrial ATPase, but (**3**) with two internal coordinate bonds is a feeble inhibitor.[10,97] These data eliminate an earlier hypothesis,[98] based on the ideas then current about the haemoglobin site, that the triethyltin in the mitochondrial high-affinity site was pentacoordinated with two *trans*-histidine imidazole groups. The effectiveness of (**2**) is strong evidence that binding to the high-affinity site requires only one bond between the tin atom and the mitochondrial component, with hydrophobic bonding to the organotin carbon skeleton contributing to produce high affinity.[99]

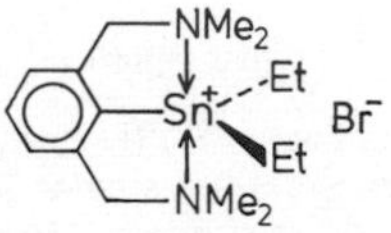

(**3**)

Using the two-site model, the amount of high-affinity site was estimated to be either[94] 0.8 or[97] 0.53 nmol mg^{-1} protein. Since the ATPase is estimated to be present at 0.12 nmol mg^{-1} protein, these figures indicate a stoichiometry of around 6:1, in agreement with that estimated by Cain and Griffiths[100] for yeast mitochondrial ATPase preparations, and suggesting a 1:1 stoichiometry with the *c*-subunit of the F_0 component (see Figure 11.4). However, after allowing for partition, the amount of high-affinity site falls to 0.19 nmol mg^{-1} protein, giving approximately 1:1 stoichiometry with the whole F_0 component and suggesting that, as with DCCD, binding to only one *c*-subunit blocks both proton conduction and further binding. Further support for this interpretation is provided by compound (**2**), which gave complete inhibition of the ATPase at 0.12 nmol mg^{-1} protein.[10,97]

Neither oligomycin nor dicyclohexylcarbodi-imide block triorganotin binding, showing that although all bind to the F_0 component of mitochondria and block proton transfer, the actual sites are different. The finding of trialkyltin-resistant mutants of *Saccharomyces cerevisiae* which retained sensitivity to oligomycin is additional evidence that these two inhibitors bind at different sites.[105]

A proteolipid which binds trialkyltin compounds has been extracted from mitochondria,[106] by a procedure similar to that used to extract the dicyclohexylcarbodi-imide (DCCD) labelled proteolipid component of the F_0 moiety of the mitochondrial ATPase.[101] The Mössbauer spectrum of triethyltin-treated proteolipid corresponded closely to the spectrum of triethyltin bound to the high-affinity site. The M_r of this proteolipid was estimated to be about 5000–6000, which is lower than the accepted value of 8000 for M_r of the DCCD-proteolipid.

Dibutylchloromethyltin chloride was designed as an analogue of simple trialkyltins which might form a covalent link with, and thus (when radioactive) label, the inhibitory

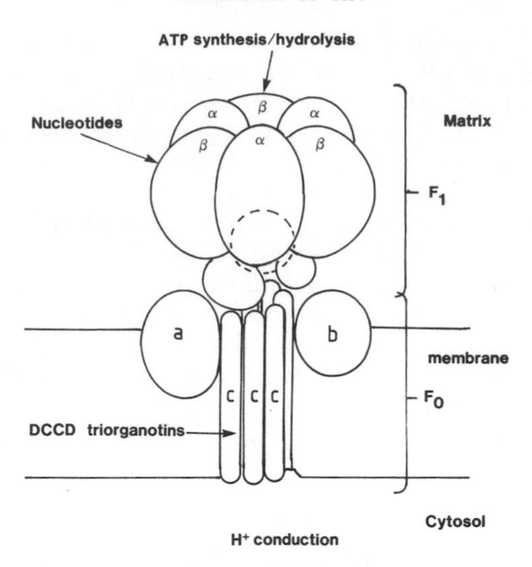

Figure 11.4 Structure of an $F_1 - F_0$ ATPase or ATP synthase based largely on the mitochondrial enzyme. Chloroplast and bacterial enzymes are similar, the α and β subunits being constant features of the F_1 components and the c subunits, probably bearing both the DCCD and trialkyltin binding sites, being consistently present in the F_0 components, though the number may vary.[101–104]

site on the F_0 component,[107] but the major labelled component obtained from ATPase preparations was lipoic acid (**4**).[108] Since lipoic-acid-deficient mutants of *E. coli* carry out oxidative phosphorylation,[109–110] lipoic acid cannot have an essential role in this process. Inhibition of the ATPase and oxidative phosphorylation can be accounted for by dibutylchloromethyltin chloride acting on the F_0 component in the same way as simple trialkyltins, since a small percentage of radioactivity was associated with a 6000–8000 M_r proteolipid.[108] Dibutylchloromethyltin chloride affects the binding and esr spectrum of a nitroxide-labelled analogue of DCCD, which indicates that, although the binding sites are different, they are sufficiently close to interact.[111] Thus dibutylchloromethyltin chloride acts not only in the same way as triethyltin or higher homologues, but also, by virtue of the reactive chloromethyl group as a bifunctional reagent, with specificity for dithiol compounds similar to that of diorganotin compounds (section 11.4.2).

H_2C—CH_2—CH—$(CH_2)_4COOH$; ring: H_2C–S–CH_2–Sn(Bu)(Bu)–S–CH

(**4**)

Although diethylpyrocarbonate inhibits binding of triethyltin to the high-affinity site,[10] there are no histidine residues in the c component of F_0.[101] A pK of 6.3 has been reported[83] for triethyltin binding to the high-affinity site, which is probably the pK of the triethyltin aquocation rather than the pK of a ligand in the binding site, and suggests that the species which binds is triethyltin hydroxide.

Thiol groups are implicated in the proton conductivity of the mitochondrial ATPase F_0 component[112] but the failure of *N*-ethylmaleimide, which inhibits proton conduction, to block triorganotin binding[97] argues against a thiol group being involved in the high-affinity site. Another protein known as Factor B or F_B enhances coupling in some submitochondrial preparations, although its relation to the F_1–F_0 complex has not been defined.[113] However, it contains thiol groups and could be the site of the low-affinity binding by trialkyltins, dibutylchloromethyltin chloride or dialkyltins.[114]

Triorganotin binding to the CF_0 component of chloroplast ATPase can be inferred from the inhibition by triphenyltin of H^+ conduction in chloroplast particles which had been stripped of CF_1 by treatment with EDTA.[115,116] The ligand in chloroplast CF_0 could be a thiol, but reversal of triphenyltin inhibition by dithiol compounds[117] is not good evidence for the chemical nature of the ligand as it reflects only affinity of triphenyltin for the dithiol and reversibility of binding to CF_0.

11.4.2 *Enzyme inhibition*

11.4.2.1 *Dialkyltins.* Although both dialkyltins and triorganotins inhibit oxidative metabolism, they act in very different ways.[118] The triorganotins act on the coupling ATPase, but the diorganotins act on substrate oxidation, and inhibition by diorganotins is relieved by dithiol compounds such as 2, 3-dimercaptopropan-1-ol (BAL, British Anti-Lewisite). In this they resemble phenylarsenious acid which is known to react with dithiols such as lipoic acid, and it is generally accepted that dialkyltins form a stable bidentate, six-membered ring, complex with lipoic acid (**5**). The tissue concentration of lipoic acid is much greater than the concentration of the F_1–F_0 ATPase, which may account for the lower toxicity of the diorganotins. Lipoic acid, bound to the enzyme by an amide link, functions in reactions such as pyruvate dehydrogenase (eqn 11.4) and 1-oxoglutarate dehydrogenase in which the substrate is oxidized, decarboxylated and the residue attached to coenzyme A.[119]

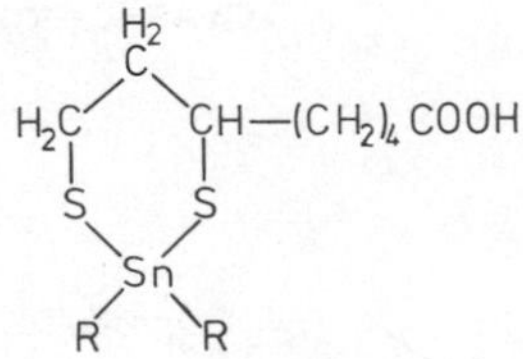

H_2
C
H_2C CH—$(CH_2)_4$COOH
S S
Sn
R R

(**5**)

Re-oxidation of dihydrolipoamide by NAD^+ is catalysed by the enzyme dihydrolipoyl dehydrogenase (E_3 in eqns 11.4*d*, 11.4*e*), which is an integral part of the multienzyme complexes forming pyruvate dehydrogenase and 1-oxoglutarate dehydrogenase. The dihydrolipoamide dehydrogenase molecule contains a dithiol/disulphide bridge formed from cysteine residues in the sequence shown in (**6**) in the enzymes from pig or *E. coli*. This undergoes oxidation and reduction in conjunction with an FAD prosthetic group as in eqns 11.4*d*, 11.4*e*.

$$E_1\text{-TPP} + CH_3\text{-}\underset{\underset{O}{\|}}{C}\text{-COOH} \longrightarrow E_1\text{-TPP-}\underset{\underset{OH}{|}}{CH}\text{-}CH_3 + CO_2 \qquad (11.4a)$$

$$E_1\text{-TPP-}\underset{\underset{OH}{|}}{CH}\text{-}CH_3 + E_2L\text{-(S–S ring)} \longrightarrow E_1\text{-TPP} + E_2L\text{-}(SH)(S\text{-}\underset{\underset{O}{\|}}{C}\text{-}CH_3) \qquad (11.4b)$$

$$E_2L\text{-}(SH)(S\text{-}\underset{\underset{O}{\|}}{C}\text{-}CH_3) + \text{CoASH} \longrightarrow E_2L\text{-}(HS)(SH) + \text{CoAS-}\underset{\underset{O}{\|}}{C}\text{-}CH_3 \qquad (11.4c)$$

$$E_2L\text{-}(HS)(SH) + E_3\text{-FAD(S–S)} \longrightarrow E_3\text{-FAD(SH)(S)} + E_2L\text{-(S–S ring)} \qquad (11.4d)$$

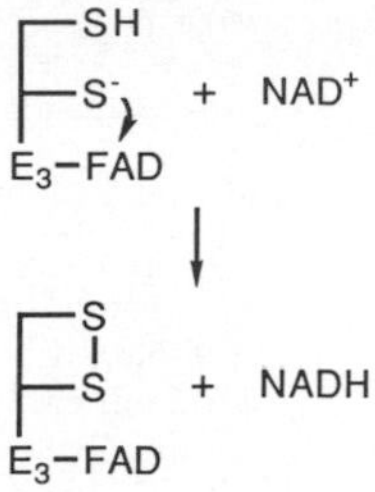

SH
S^- + NAD^+
E_3–FAD

↓

S–S + NADH
E_3–FAD (11.4*e*)

where TPP is thiamine pyrophosphate, FAD is flavin adenine dinucleotide

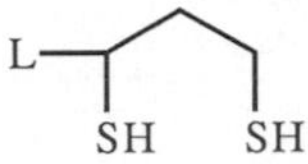

L– SH SH and L– S—S

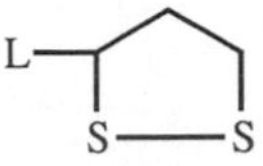

represent reduced and oxidized lipoate.

Two other enzymes contain similar systems of two thiols with FAD[119]; yeast glutathione disulphide reductase in which Leu in the sequence in (**6**) is replaced by Val, and thioredoxin dehydrogenase which has the sequence shown in (**7**) and may involve fully reduced $FADH_2$ and SH groups in the catalytic cycle. In all these reaction sequences, there are alternations between disulphide bridges and vicinal SH groups in the enzymes and lipoic acid, and diorganotins may react directly with enzyme dithiols as well as with lipoic acid.

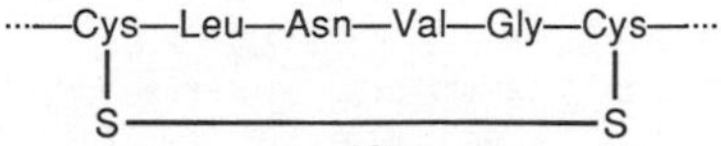

····—Cys—Leu—Asn—Val—Gly—Cys—····
S————————S

(**6**)

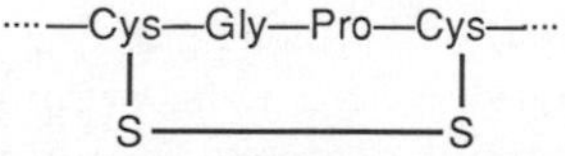

····—Cys—Gly—Pro—Cys—····
S————————S

(**7**)

Dialkyltins, but not phenylarsenious acid, have some other effect on the mitochondrial oxidation systems since, under some circumstances, the increase in 1-oxoglutatarate concentration parallels inhibition of oxidation, but under other conditions it may be little altered or even decreased.[10] These effects may be related to the inhibition of phosphorylation and mitochondrial ATPase activity by dibutyltin dichloride.[114]

11.4.2.2 *Inhibition by triorganotins.* Impairment of a metabolic function by a triorganotin can be produced not only by direct inhibition of a particular enzyme but also indirectly, for example by lowered ATP concentration, alteration of pH or an effect on the synthesis, breakdown or activation of enzymes. Inhibition by triphenyltin of superoxide production after stimulation of human neutrophil leucocytes by surface

Table 11.4 Enzymes inhibited by triorganotin compounds

Enzyme	Source
Proton-translocating ATPases	
F-type	Mitochondria, chloroplasts, bacteria
P-type	Plant and bacteria plasmalemma
V-type	Plant and fungal vacuole membranes
	Animal chromaffin granule membranes
Na^+, K^+-translocating ATPase	Animal cell membranes
Ca^{2+}-translocating ATPase	Sarcoplasmic reticulum
Cytochrome oxidase	Mitochondria
NADP transhydrogenase	*E. coli* membranes
Hexokinase	Erythrocytes, yeast
Pyruvate kinase	Rabbit muscle
Glutathione-S-transferase (ligandin)	Mammalian liver
Adenylate cyclase	Brain
Cytochrome P-450	Liver

For further description and references see text in this section, except for adenylate cyclase, section 11.4.5 and cytochrome P-450, section 11.5.1.

active agents is an example of indirect inhibition. Triphenyltin blocked the activation process but did not inhibit NADPH oxidase, the enzyme which catalyses the formation of O_2^-, isolated from activated leucocytes.[120] Enzymes for which there is good evidence, from investigation on at least partially purified preparations, for direct interaction with triorganotin compounds are listed in Table 11.4.

The action of triorganotins on cation-translocating ATPases has been reviewed previously[88], but since then it has been recognized that, besides the proton-translocating ATPases of mitochondria, chloroplasts and bacterial membranes, there are others which can be distinguished on the basis of function, structure and sensitivity to different types of inhibitor.[121–124] As shown in Table 11.4, three classes are recognized, the function of the *F*-type being primarily the synthesis of ATP, utilizing proton gradients produced by electron transfer (hence the alternative name ATP synthases), while the *P*- and *V*- types utilize ATP to generate proton fluxes which adjust intracellular or intra-organelle pH and can be coupled to other transport processes. The *F*-type ATPases are particularly important in triorganotin toxicity, not only because of their role in providing ATP for energy-requiring reactions including the *P*- and *V*-type ATPases, but also because they have a higher affinity for triorganotins.[10,88,125]

The *F*-type ATPases of mitochondria, bacteria and chloroplasts have a common molecular architecture (Figure 11.4) and catalyse the reversible hydrolysis of ATP coupled to the passage of protons through the membrane. Two major components are readily distinguished, the F_1 component, which can be detached from the membrane, contains the sites where ADP, ATP and inorganic orthophosphate bind and catalyses the hydrolysis or condensation reactions to break or form ATP; the F_0 component, which is largely embedded in the membrane, serves to conduct protons across the membrane. The mechanism by which these two processes are linked so that the energy from one is used to drive the other is not yet known, but probably each process involves changes in protein conformation. When the F_1 and F_0 components are combined, the conformation change is transmitted from one component to the other, thereby driving the process catalysed by the second component in a particular direction.[126–128] Thus, passage of two or possibly three protons through the F_0 component towards the F_1 component forces the ATPase to operate in the direction of ATP synthesis. When the F_1 and F_0 are combined, the functional coupling is very tight, so that neither synthesis nor hydrolysis of ATP occurs if protons cannot move through the F_0 component. However, when the two are separated, the F_1 component becomes a thermodynamically irreversible ATPase, which is not inhibited by triorganotin compounds, while the F_0 component becomes a proton-conducting channel. Triorganotin compounds bind to the F_0 channel with high affinity, and when bound prevent the passage of protons. This has been demonstrated by the decrease in proton permeability produced by adding tributyltin to sub-mitochondrial particles from which much of the F_1 component had been stripped[29], and similar effects have been shown in chloroplast preparations.[115,116] Because of the tight coupling between proton flow and the ATPase, this blockage of proton flow by triorganotins prevents both ATP synthesis and hydrolysis in mitochondria.[96] When yeast mitochondrial ATPase and bacterial rhodopsin (which functions as a light-driven H^+ pump) were incorporated together into phospholipid vesicles, it was found that ATP hydrolysis was sensitive to inhibition by triphenyltin, but light-driven ATP synthesis was scarcely affected.[130] The authors suggested that this indicated different modes of coupling between the ATP reactions and H^+ translocation in the forward and reverse directions. A similar effect has been observed with sub-mitochondrial particles[131], and in chloroplasts phosphorylation is more sensitive than the hydrolysis of ATP.[116] There is no need to postulate distinct mechanisms for ATP synthesis and hydrolysis, as in these systems ATP synthesis requires the intravesicular space to be acid, in contrast to the internal alkalinization of intact mitochondria. Coleman and Palmer[132] observed that at low pH triethyltin did not inhibit ATP synthesis by whole mitochondria, and triethyltin binding at the high-affinity site decreased at low pH.[83,106]

Little is known about the inhibition of *P*- or *V*-type H^+-ATPases, but the low concentration at which tributyltin acts on the *Neurospora* enzymes suggests specific binding sites.[121–122] The proton translocating ATPase of bovine adrenal medulla chromaffin granules was 50% inhibited by 2×10^{-7} M tributyltin chloride, but was much less sensitive to further inhibition by higher concentrations.[133] A preparation from the rubber tree, *Hevea brasiliensis*, equivalent to vacuolar ATPase, was 90–95% inhibited by 0.1 mM triphenyltin when bound to the membrane, but was little inhibited after solubilization.[134] The structures of the *P*- and *V*-type ATPases are different from that of the F_1–F_0 ATPases, so that comparisons of changes in sensitivity to organotins on solubilization are not meaningful.

There has been no further evidence, since the previous discussion[88] of inhibition of the Na^+, K^+-ATPases and the Ca^{2+}-ATPase of intracellular membranes, to alter the view that the poor inhibition by triethyltin and trimethyltin indicates the overriding importance of hydrophobic interactions, and that the organotins are inhibiting by interfering with the interactions between the enzymes and the immediately adjacent membrane lipids. It is also unlikely that effects on these enzymes make a major contribution to the toxic action of triorganotin compounds, at least in mammals. In other organisms, inhibition of these enzymes by the more hydrophobic triorganotins may contribute to toxicity since tricyclohexyltin hydroxide at 10^{-5} M inhibited the Mn^{2+}, Mg^{2+} and Ca^{2+} activated ATPases of lobster axon plasma membranes.[135]

The inhibition of cytochrome oxidase by triphenyltin is noteworthy, since this protein couples electron transport to proton translocation.[136] Although with triphenyltin, inhibition only of electron transport has been reported, another inhibitor of proton translocation in the F_0 components of F-type ATPases, dicyclohexylcarbodi-imide, has been found to inhibit proton translocation by cytochrome oxidase.[137]

NADPH transhydrogenase (eqn 11.5), another enzyme which pumps protons, has

$$NADH + NADP^+ \rightarrow NAD^+ + NADPH \tag{11.5}$$

been shown to be inhibite by tributyltin, although it is difficult to eliminate indirect effects via the chloride/hydroxide exchange or on the energizing process.[138]

Sequence studies give grounds for believing that the triorganotin-sensitive components of the F-type ATPases are evolutionarily related, however, this seems less likely for the P- and V-type ATPase, and the other proton pumps. It is possible that a proton-conducting triorganotin-binding domain has been conserved and distributed during evolution, but the functional requirement for both hydrophobic regions and proton bonding groups is likely to generate sites for triorganotin binding in proteins which conduct protons across phospholipid membranes.

Glutathione-S-transferase exists as several isozymes and, in addition to normal metabolic reactions, functions in detoxification not only catalytically but also by reacting with toxic agents. In this latter capacity it is sometimes called ligandin.[139] Henry and Byington[140] found that triorganotins and triorgano- germanium and lead compounds, but not diorganotins, were highly effective inhibitors. For the triorganotin series the order of effectiveness was $Et_3Sn > {}^nBu_3Sn > Ph_3Sn > Me_3Sn \gg {}^cHeX_3Sn$. The enzyme is a dimer, and the isozymes are produced by combination of pairs of four different subunits, identified by numbers 1–4. The isozymes differ in their substrate specificity and in sensitivity to inhibitors, including triethyltin.[141–143]. Subunits 2 and 3 confer sensitivity to triethyltin, and for the isozyme 3-3 with 1-chloro-2,4-dinitrobenzene as substrate the K_i for triethyltin was 3.2×10^{-8} M, while for other isozymes K_i was up to 1000-fold greater.[141] Thus measurements using selected substrates and inhibitors provide some discrimination in assays for the different isozymes.[141–143]

The very low K_i is dependent on the presence of glutathione; equilibrium dialysis gave dissociation constants of 7.1×10^{-4} M in absence of glutathione but 2.0×10^{-7} M in the presence of 5×10^{-3} M glutathione.[141] As glutathione and triethyltin form a complex in aqueous solution, there are three possible routes to formation of a complex in which both glutathione and triethyltin are bound to the enzyme. The extremely low K_i shows that there must be a very precise fit between the enzyme and triethyltin, and the concept of transition state stabilization[144] suggests that triethyltin may be an analogue for the transition state rather than for a substrate or product.

The interaction of triethyltin with pyruvate kinase is of low affinity, 50% inhibition at 1.4 mM triethyltin, and therefore unlikely to be significant in toxic effects.[145] Estimates of binding reveal two types of site, one with 14 sites and $K = 4.7 \times 10^{-4}$ M, the other with 322 sites and $K = 0.70$ M. Pyruvate kinase is a tetramer of identical subunits but has more than three triethyltin molecules bound per subunit at the higher-affinity sites, which do not, therefore, appear to be specific interactions with either catalytic or control sites. Nevertheless, triethyltin affects the catalytic activity of the enzyme, stimulating activity at low concentrations but inhibiting at higher concentrations, and also exhibits competitive interactions with activating Mn^{2+} ions.

Triethyltin bromide reacts slowly with yeast or human red blood cell hexokinase, 100 μm producing 50% inhibition after 20 minutes' incubation,[146] but at this concentration is unlikely to be toxicologically important. Substrates such as D-glucose protected against inhibition by triethyltin, but non-substrate sugars such as L-glucose and the other substrate ATP, MgATP or product ADP gave no protection.

11.4.3 *Effects on biological membranes*

A tetraorganotin analogue of palmitic acid, 12,12-dimethyl-12-stannahexadecanoic acid (**8**) has been synthesized and incorporated into phosphatidylcholine as a probe for lipid organization. Artificial membranes containing the probe were compared with those containing natural phosphatidylcholine. *Acholeplasma laidlawii* incorporated up to 39% of the fatty acid into membrane lipids without impairment of growth.[147]

$$CH_3(CH_2)_3\overset{\overset{\displaystyle CH_3}{|}}{\underset{\underset{\displaystyle CH_3}{|}}{Sn}}(CH_2)_{10}COOH$$

(8)

In addition to effects on enzymes and other proteins, triorganotin compounds have two actions in biological membranes. At high concentrations, the more hydrophobic compounds, such as triphenyltin, have a detergent-like effect disrupting the membrane structure and increasing the permeability to hydrophilic solutes and ions.[10,148]

At low concentrations, triorganotin compounds catalyse an exchange of halide or pseudohalide ions for hydroxide ions across biological and also artificial phospholipid membranes.[10,148–155] The latter showing that this is an intrinsic property of the triorganotin compounds, not modification of an existing transporter. Triorganolead and monorganomercury compounds also catalyse anion–hydroxide exchange, while triethylgermanium[156] and possibly diorganotin[10] compounds have only feeble activity.

The chemical properties of the triorganotins indicate the formation of a lipid-soluble covalent complex $R_3Sn{-}X$, but the details of the mechanism of anion transport are not known. A simple exchange of anions on the tin atom at the membrane–water interface is not plausible, but the exchange may be facilitated by phospholipid head groups (Figure 11.5*a*), since nucleophiles are known to catalyse ligand exchange in triorganotins[157], and the compound (**2**) with one internally coordinated nucleophile has very weak ionophoretic activity.[10] Alternatively, the anion exchange at the tin atom may proceed via the hydrated cation, either by release of the organotin into the aqueous phase or at the membrane–water interface (Figure 11.5*b*).

The order of effectiveness of triorganotins in Cl^-/OH^- exchange across lecithin liposome membranes is $^nBu_3Sn > {}^nPr_3Sn > Ph_3Sn \approx Et_3Sn > {}^cHex_3Sn > Me_3Sn$, the ratios of concentrations being 1:1.2:10:10:20:51. The exchanges across several natural membranes exhibit similar but not identical series.[88] Comparison of these series with the partition coefficients[151] suggests an optimum partition coefficient which could be taken to support ligand exchange in the aqueous phase (Figure 11.5*b*), since re-entry into the aqueous phases would decrease as the hydrophobicity increases. However, the decreased activity shown by compounds with larger chains could result from shielding of the tin atom or reduced mobility in the lipid. Using ^{36}Cl, Wieth and Tosteson[158,159] found that tributyltin-catalysed Cl^-/Cl^- exchange across erythrocyte membranes was

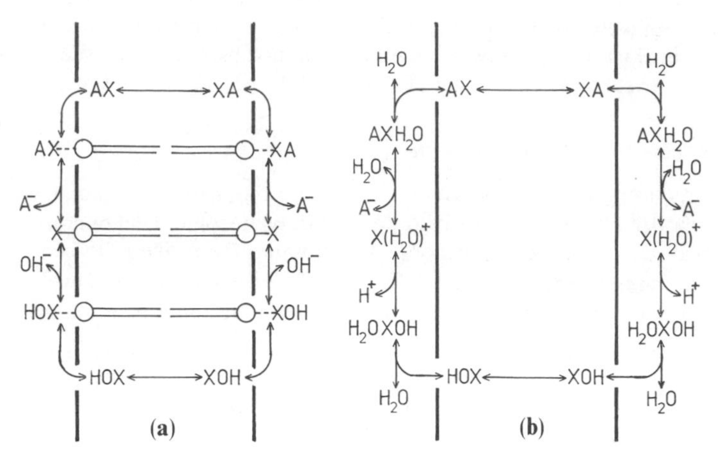

Figure 11.5 Hypothetical mechanisms for the halide–hydroxide exchange across phospholipid membranes catalysed by triorganotin compounds. These mechanisms take into account the need for a fifth coordination with either (*a*) nucleophiles in the phospholipid head groups or (*b*) water. The latter mechanism does not necessarily imply equilibration with the bulk aqueous phases. X represents the triorganotin moiety, R_3Sn.

much faster than Cl^-/OH^- exchange. They calculated that the rate of tributyltin diffusion in water was too low to allow equilibration with the bulk aqueous phase, but this observation does not preclude exchange in a layer adjacent to the membrane surface. All mechanisms predict an optimum pH for anion–hydroxide exchange which will depend on the anion concentration as well as the rate and equilibrium constants for formation of the triorganotin–anion complex and partition into the membrane. Experimentally, the pH optima are markedly dependent on the anion but vary to a lesser extent with the nature of the organic groups. The pH curves are relatively flat, indicating that steps other than those dependent on proton or hydroxyl ion concentrations are important in rate limitation. Increasing flatness with increase in size and hydrophobicity suggests rate limitation by partition into or mobility within the lipid.[156]

At the pH optima the order of effectiveness is $Br^- > Cl^- > SCN^- > I^- > F^-$, while ions such as nitrate, phosphate and isethionate are inactive. Ions which are not exchanged may fail to form complexes with the organotin, form complexes which are not lipid-soluble, or form complexes which do not readily dissociate. Nitrate appears to be in the first category and phosphate in the second. Fluoride, which is transported slowly, inhibits chloride transport, suggesting that the rate of transport is limited by dissociation of fluoride from the triorganotin complex.[151,158,159]

The ubiquity of chloride ions means that the presence of triorganotins has important consequences for the ionic equilibria of cells and organelles. At the equilibrium produced by this antiport, the ratios of the chloride and the hydroxide concentrations on the two sides, R and L, of a membrane will be the same (eqn 11.6):

$$[Cl^-]_R/[Cl^-]_L = [OH^-]_R/[OH^-]_L \qquad (11.6)$$

Rewriting this equation in the pH notation (eqn 11.7):

$$pH_R - pH_L = \log\{[Cl^-]_R/[Cl^-]_L\} \qquad (11.7)$$

shows that the chloride/hydroxide antiport will clamp the pH difference across the membrane at a value determined by the ratio of the chloride concentrations. These equations resemble those for Donnan equilibria, but in the Donnan mechanism the ions pass across the membrane by uniport, which is an important difference. The molecularly-coupled antiport in the Cl^-/OH^- exchange is an electrically neutral process, and the equilibrium is independent of the electrical potential difference across the membrane (the membrane potential). When ion movements are mediated by uniporters, although there is a requirement for macroscopic net electrical neutrality of ion transfers, each ion is in equilibrium with the membrane potential, as described by the Nernst equation, and this imposes an extra constraint on the equilibrium. In a simple system, the antiporter and uniporter systems may yield the same equilibrium, but this is not the case if some other process determines the membrane potential.

This anion hydroxide antiport has useful applications. It enables titration of the contents of vesicles such as mitochondria which are normally impermeable to protons.[88,148] It has been used as a probe to study mitochondrial Ca^{2+} and monovalent cation transport[160,161] since, depending on the mode of transport, the need for charge and pH balance or both may require concomitant anion or hydroxide transport. In the presence of thiocyanate ions, which cross membranes by a uniport process as well as by triorganotin catalysed SCN^-/OH^- exchange, uncoupling of oxidative phosphorylation is observed. This is not only an illustration of the difference between uniport and the antiport of the anion/OH^- exchange, as the two processes together (triorganotin with SCN^-) but neither by itself (triorganotin plus Cl^- or SCN^- in the absence of triorganotin) is equivalent to OH^- or H^+ uniport and produces uncoupling, but also good evidence that uncoupling of oxidative phosphorylation can be brought about by transport of protons across the coupling membrane.[88,95]

Equation (11.7) shows that when the chloride concentration is higher outside than inside a cell or organelle, addition of a triorganotin will make the internal pH lower than the external. The effects of this on mitochondrial metabolism are described in the next section.

11.4.4 *Inhibition of energy metabolism*

The sensitivity of mitochondrial oxidative phosphorylation to inhibition by triorganotin compounds has been known for many years,[10,162,163] but has proved to be a complex process in which several actions of the triorganotins are involved. Resolution required the development of fundamental understanding of coupling between biochemical reactions by proton fluxes (Mitchell's chemiosmotic hypothesis). Experimental investigation was assisted by the discovery of the triorganotin-catalysed Cl^-/OH^- exchange and elimination of complications which it produces by using media free from chloride or other exchangeable anions.

The current view of chemiosmotic coupling processes in mitochondria is outlined in Figure 11.6.[164–167] Fundamentally, the respiratory chain uses the energy available from electron transfer to oxygen to pump protons across the mitochondrial inner membrane from the matrix to the cytoplasmic side, the energy being conserved in what is called the proton motive force, PMF or $\Delta\tilde{\mu}_{H^+}$, which can be split into the pH difference, ΔpH, and the electrostatic potential difference across the membrane, $\Delta\psi$ (eqn 11.8).

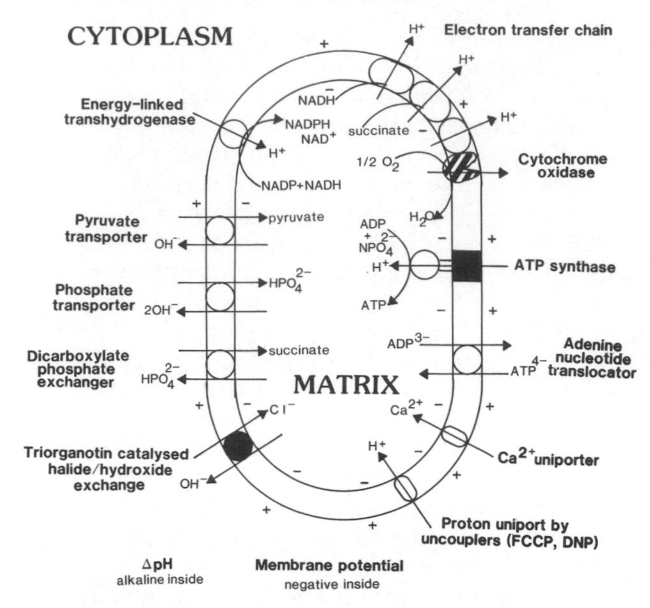

Figure 11.6 Diagram of chemiosmotic processes in mitochondria. The fundamental processes are the pumping of protons from the matrix to the cytoplasm by the respiratory chain and the synthesis of ATP driven by the return of protons through the ATP synthase ($F_1 - F_0$ ATPase). The energy-linked transhydrogenase is also coupled to proton fluxes. The membrane potential drives uptake of Ca^{2+} ions and uptake of ADP^{3-} in exchange for expulsion of ATP^{4-}. The alkaline internal pH causes uptake of pyruvate and phosphate in exchange for OH^- ions and hence succinate in exchange for phosphate. At low concentrations, shown in black, triorganotins (i) inhibit the passage of protons through the F_0 component of the ATP synthase, and (ii) if Cl^- ions are present the triorganotin-catalysed Cl^-/OH^- exchange tends to acidify the matrix, thereby decreasing substrate uptake via the hydroxide- and phosphate-linked transporters. At higher concentrations (striped), cytochrome oxidase may be inhibited directly.

$$\begin{aligned} \mathrm{PMF} &= \Delta\psi - (2.303RT/F)\Delta\mathrm{pH} \\ &= \Delta\psi - 60\Delta\mathrm{pH} \quad \mathrm{mV\ at\ 30°C} \end{aligned} \tag{11.8}$$

The protons on the cytoplasmic side are thus at higher potential energy and can flow back to the matrix through the $F_1 - F_0$ ATPase, driving it in reverse so as to synthesize ATP. However, there are doubts as to whether the protons equilibrate freely with the aqueous phases or are constrained in their passage between pumps, about the stoichiometry of the H^+-pumps, and whether this stoichiometry is fixed or varies according to the proton motive force (PMF).[166–168]

The blocking of H^+ transport through the F_0 component of the mitochondrial ATPase directly inhibits both ATP synthesis and utilization of ATP for the proton-driven energy-linked functions as well as ATP hydrolysis by uncoupled mitochondria.

In chloride media, acidification of the matrix by the Cl^-/OH^- exchange lowers the proton motive force and, although an increase in membrane potential may compensate to some extent, this may lead to uncoupling by increasing ion uniport or lowered H^+ stoichiometry in the pumps and thus lowered efficiency of energy conservation. If a freely permeant ion which is also active in the anion/OH^- exchange, for example thiocyanate, is present, the uncoupling by triorganotins can be very effective.[95]

Acidification of the matrix also decreases substrate accumulation, since the decreased internal hydroxide concentration is reflected in a decreased internal pyruvate and phosphate via their natural hydroxide antiporters. The latter leads to lowered internal malate and succinate via malate/phosphate and succinate/phosphate antiport, and in turn to lowered citrate via the citrate/malate antiport.[169,170]

Acidification of the matrix also shifts the pH away from the optimum for some internal enzymes such as those in respiration.[83] Inhibition of ATPase activity or ATP-driven processes is particularly severe, since not only is the ATPase shifted away from its pH optimum, but the increased $\Delta\psi$ increases the tendency to accumulate ADP, an inhibitory product of the ATPase, and expel the substrate ATP, via the electrically driven ATP/ADP translocator. At high concentrations of triorganotins, respiration may also be inhibited by the direct effect on cytochrome oxidase.[136]

Chloride ions are present in most living cells, and *in vivo* the triorganotins have a potent action on mitochondrial oxidative phosphorylation, the direct inhibition of the ATP synthase being enhanced by acidification of the matrix which decreases enzyme activity and substrate concentrations.

The effects of triorganotins on the energy metabolism of bacteria and photosynthetic organisms has been less well studied. Tributyltin was found to decrease photophosphorylation,[171] and this has been shown to be due both a direct effect on the coupling ATPase and uncoupling via the Cl^-/OH^- exchange.[172] The work of Singh and Bragg[138] has shown that tributyltin affects energized processes in *E. coli.*

In the integrated metabolic and bioenergetic process of a cell, the decreased ATP concentration is the most important primary effect and has secondary consequences in nearly all energy-requiring processes. Biosynthesis, including DNA and RNA replication, protein synthesis, cell signalling and control via cyclic AMP and protein phosphorylation, muscle contraction, and other cytoplasmic motility including mitosis and ion pumping (not only across the plasmalemma but also vacuole or other organelle membranes) are all dependent on ATP. The lack of ATP is exacerbated by the interference with ion balance and pH produced by the second primary effect, the Cl^-/OH^- exchange. Not only does this adversely affect enzyme activity, but compensation by regulatory processes makes extra demands on the ATP resources. At high triorganotin concentrations, particularly with the more hydrophobic compounds, the effects on ion balance will be further inhibited, since not only are the vacuolar (*V*-type) and plasmalemma (*P*-type) H^+-translocating enzymes directly inhibited by triorganotins, but so are the Na^+, K^+-ATPase (which is essential for processes such as glucose accumulation by linked transport with Na^+ and nerve impulse conduction) and the Ca^{2+}-ATPase (which sequesters Ca^{2+} ions in the endoplasmic reticulum). Some of these effects are described in more detail for nerve cells and thymocytes in the following sections.

11.4.5 *Neurotoxicity*

Although the central nervous system is absolutely dependent on oxidative phosphorylation, the relation between inhibition of this process by triorganotins,[173] which pass readily across the blood–brain barrier, and the neuropathology of triorganotin poisoning is not clear. The behavioural and physiological effects of organotin poisoning are complicated, as are the patterns of anatomical and cytological lesions which reflect biochemical differences between regions and cell-types in the brain. The situation is further complicated by the multiplicity of effects of organotins on biochemical processes in nerve cells. Furthermore, the overall effects produced by trimethyltin differ qualitatively from those of triethyltin and higher homologues, and the fine balance in chemical interactions with living cells is shown by the finding that dimethylethyltin produces changes similar to trimethyltin while methyldiethyltin resembles triethyltin in its action.[174] There is considerable interest in the use of trimethyltin as a probe for brain function and triethyltin as an agent to induce oedema of the white matter. The literature is vast, but there is a recent review of the behavioural and neurological aspects[175] and a bibliography.[176] In developing animals, organotins cause more severe and permanent neurological damage.[177,178]

Trimethyltin intoxication in the rat initially produces depression, and subsequently tremors, hyperexcitability, hyperactivity, aggression, loss of balance, convulsions, coma and eventually death. The major lesions in the brain are the death of neurones in the hippocampus, pyriform cortex and amygdaloid body.[179–181] There is limited swelling of the brain but little or no oedema of the myelin sheaths of nerve fibres. After a non-lethal dose, recovery can occur with complete disappearance of behavioural and physiological symptoms, even though there is permanent damage to parts of the brain.

Triethyltin compounds (also tripropyltin and tributyltin) produce weakness, paresis (paralysis without loss of sensation), particularly as paraplegia (paralysis of the hind legs and hind body), together with hypothermia. Anatomical damage is again evident in the central nervous system, but it is oedema of the white matter caused by the formation of fluid-filled spaces in the myelin sheaths of the nerve axons which split the myelin at the interperiod line, i.e. the fluid is in extracellular space.[182] This oedema may be caused by an increase in ion transport with consequent water movement[183] and has been monitored using proton nmr.[184] Decreased motor nerve conduction velocity[185] may be related to the myelin oedema, since the function of myelin sheaths is to increase the conduction velocity of nerve fibres. Splitting the myelin layers by insertion of aqueous spaces of high dielectric constant will increase the capacitance and thereby decrease the velocity of or even block nerve impulse conduction, but this cannot account for behavioural changes which precede oedema.[186] The compensatory capability of the brain is again apparent, because behavioural recovery can occur even though visible lesions persist.

Lock and Aldridge[187] fractionated rat brain and found that purified myelin bound triethyltin with high affinity. As hexachlorophene [2,2′ methylene-(3,4,6-trichlorophenol)] and 3,5-di-iodo-4′-chlorosalicylanilide did not compete with triethyltin for this site, but produced similar oedema, while triethyllead and trimethyltin competed but produced a different lesion, this myelin site is unlikely to be of major importance in the production of oedema. Triethyltin sulphate did not induce oedema of the myelin in mice with the quaking mutation.[188]

Organotins affect many biochemical processes in nerve tissue. They not only pass readily through the blood–brain barrier but increase its permeability to macromolecules.[189] Organotins inhibit the basal level of adenylate cyclase in rat brain homogenates, but not all produce a decrease in brain cyclic-AMP concentrations *in vivo*.[190] Triethyltin produced a decrease in the activity of cyclic 3′,5′-monophosphate phosphodiesterase, the enzyme which catalyses hydrolysis of cyclic-AMP,[191] and inhibited the incorporation of radioactively-labelled inorganic orthophosphate into phospholipids[192] and the phosphorylation of brain proteins.[193]

Trimethyltin inhibited synthesis of proteins *in vivo*,[194] including synapsin I which may be correlated with degeneration of nerve terminal regions.[195] In contrast, trimethyltin has been found to increase synthesis of membrane components in retinal neurones.[196] Organotins lowered the brain concentrations of the neurotransmitters γ-aminobutyric acid and dopamine[197], and inhibited γ-aminobutyric acid uptake[198] and efflux[199] *in vitro*. Trimethyltin inhibited acetylcholine release from synaptosomes[200], and triethyltin decreased the response at the neuromuscular junction, possibly by interfering with intracellular Ca^{2+} movement.[201] The overall pattern of effects on neurotransmitters is complicated.[202] Furthermore, the mechanisms by which triorganotins produce these effects are often uncertain, since the inhibition of oxidative phosphorylation will affect all ATP-dependent processes. Furthermore, the chloride–hydroxide exchange and consequent change in intracellular pH will affect many processes and a secondary effect, alteration in intracellular Ca^{2+} distribution, will affect the responses of excitable cells.[203]

Trialkyltin compounds have been found to bind to the protein tubulin and inhibit its polymerization to form microtubules.[204] In nerve cells the microtubules, called neurotubules, are involved in the fast transport along the nerve fibres or axons which can both away from the cell body (anterograde) and towards the cell body (retrograde). There are several microtubular-associated proteins (MAPs) the abundance of which varies in different regions of the brain. Two of these are ATPases; one, kinesin, is an ATPase producing force along the microtubule in the anterograde direction while another MAP ATPase produces force in the retrograde direction.[205] Some of the neurotoxic effects of organotins may be related to impairment of neurotubule function[206], and hence the essential transport of metabolites along the nerve axons, since this will be adversely affected not only indirectly through decreases in ATP, required for the ATPase-dependent motility and GTP, required for polymerization of tubulin, but also by the direct effect of triorganotins on tubulin polymerization.

11.4.6 *Antitumour and immunosuppressive activity*

The discovery of the antitumour activity of platinum and the clinical use of cisplatin[207,208] (*cis*-diamminedichloroplatinum). (**9**), led to the synthesis and screening of organotin compounds for antitumour activity. In the square planar platinum complexes, the general requirements are a pair of *cis*-N-ligands and two moderately good leaving groups, e.g. halides or carboxylates. Some Pt (IV) compounds with axial organo groups also show antitumour activity.

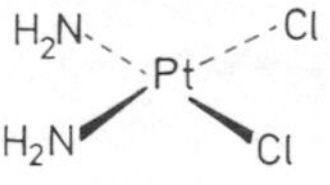

H_2N Pt Cl
H_2N Cl

(**9**)

In general triorganotin compounds are not active, but several diorganotin complexes have antitumour activity.[209–214] Cardarelli *et al.*[215] administered a variety of dibutyltin complexes to mice in drinking water, and observing significant reductions in tumour growth rates, have suggested that this is a natural function of tin.

Crowe *et al.*[209] have investigated the structure–activity relationships of series of compounds of general formula $R_2SnX_2 \cdot L_n$, in which the organic group R, the halide or pseudohalide X and the donor ligand ($n = 1$ if bidentate) or ligands ($n = 2$ if monodentate) as well as the configuration affect the antitumour activity. Bidentate ligands form *cis* complexes, and ^{119m}Sn Mössbauer spectrometry showed the X groups to be in a *cis* configuration with *trans* R groups (**10**). Compounds with ethyl or phenyl R groups were consistently more often active agents than those with methyl, propyl or butyl groups, and similarly bromide compounds were more often effective than chloride, which in turn was better than iodide or thiocyanate. Of the donor ligands, 1, 10-phenanthroline, 3, 4, 7, 8-phenanthroline and 2-(2-pyidyl)-benzimidazole were the most effective in yielding active complexes. However, it should be noted that, although these complexes were active against P388 mouse leukaemia, they were ineffective against a range of other tumours, unlike the platinum agents which are active against several types of tumour.

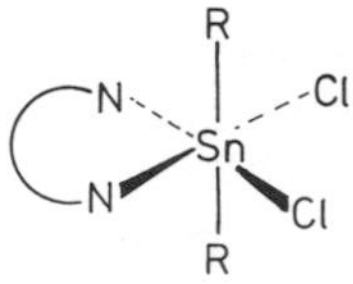

(**10**)

Barbieri, Huber and their collaborators[210,214] have investigated the activity of organotin compound complexes with purines, amino acids and related compounds. Diphenyltin and dibutyltin adenine compounds (**11**) in which the Sn–N(9) bond is covalent and the dimethyl-, dibutyl-, dioctyl- and diphenyl-tin complexes with glycylglycine (**12**) were all active, as were the diphenyltin complexes with cysteine and mercaptoethanesulphonate. None of the triorganotin complexes tested were active. In

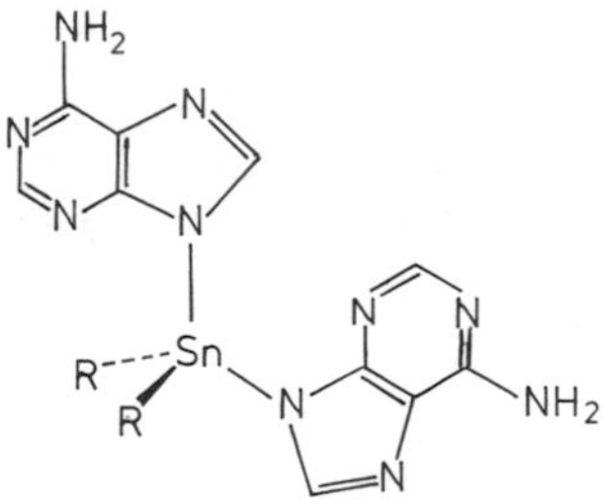

(**11**)

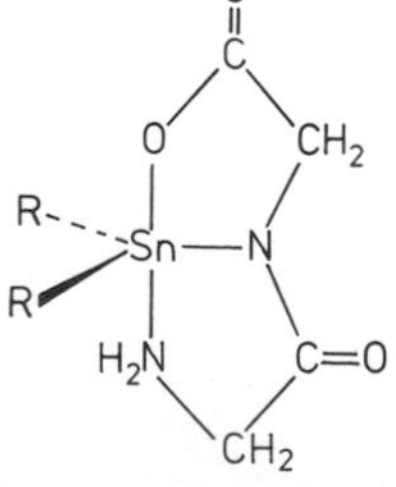

(**12**)

aqueous solution the glycylglycine and cysteine complexes are partially hydrolysed to tetrahedral four-coordinate species, with a single tin bond to the peptide nitrogen or the sulphur atom respectively. The common features of active complexes are the availability of a coordination site at Sn in any physical state and moderately stable ligand–Sn bonds which undergo slow hydrolytic decomposition. One hypothesis is that the active compounds are sufficiently stable to penetrate cells, but sufficiently unstable to react inside the cell.[209,214] It is thought that both platinum and non-platinum metal complexes react with DNA since there is a greater effect on DNA synthesis than on RNA or protein synthesis, but it is unlikely that all complexes act in the same manner, and the precise mode of action or binding site on DNA is not known definitively for any complex.[216]

Although organotins with large R groups are less toxic to mammals than triethyltin, both triorganotins and diorganotins have been found to cause atrophy of the thymus gland and impair the immune response.[217–219]

Studies on isolated thymocytes showed that as in other tissues the diorganotins blocked oxidative metabolism (but not glycolysis) by inhibiting pyruvate dehydrogenase.[220] Diethyltin dichloride inhibited uptake of the aminoacid analogue γ-aminoisobutyrate by isolated thymocytes.[221] Dioctyltin and dibutyltin as dichlorides inhibited the migration of macrophages from tissue fragments of spleen and thymus.[222]

Tributyltin compounds given orally to rats produced immunosuppression and atrophy of the thymus, and *in vitro* were highly toxic to isolated rat thymocytes.[223,224] Rather high concentrations, 1–5 μM, of tributyltin caused general membrane damage in thymocytes, allowing penetration of the dye Trypan blue.[225,226] At concentrations greater than 0.1 μM, tributyltin interfered with energy metabolism; glucose uptake was increased and ATP concentration fell, particularly if glucose was omitted from the medium.

Incorporation of [^{3}H]thymidine into DNA, [^{3}H]uridine into RNA and [^{3}H]proline into protein were all inhibited by tributyltin. Production of c-AMP was decreased, but uptake of 3-O-methyl glucose was enhanced. Uptake of the amino-acid analogues γ-aminoisobutyrate and aminocyclopentane carboxylate was not affected.[226] Similar effects were observed with other inhibitors and uncouplers of oxidative phosphorylation. Tributyltin did not inhibit key enzymes such as thymidine kinase or adenylate cyclase.[227] All the observations are consistent with the primary inhibition being the block of oxidative phosphorylation. Although glycolysis is stimulated it is unable to compensate in full, and in consequence energy-dependent processes are inhibited. The moderately lipophilic tripropyltin, tributyltin and tri-*n*-hexyltin were more toxic and effective in reducing ATP concentrations in thymocytes than trimethyltin, triethyltin or tri-*n*-octyltin.[227] The authors suggested that, as triethyltin is much more effective than tributyltin in inhibiting isolated liver mitochondria, movement across the cell membrane may be an important factor which favours for the higher homologues. However, triethyltin is also more toxic than tributyltin to whole animals, which suggests that some special factor may be involved in the thymus and thymocytes.

11.5 Metabolism of organotins

11.5.1 *Metabolic degradation*

Cremer[228] showed that tetraethyltin was metabolized by the rat, *in vivo* and by a liver microsomal oxygenase enzyme system, to the triethyl derivative. Bridges *et al.*[229]

found that rats excreted ethyltin unchanged, in the faeces if it was given orally, but in the urine if its was injected intraperitoneally. Injected diethyltin was excreted in the faeces, indicating biliary excretion, and in the urine. About 50% of injected ^{14}C-diethyltin was dealkylated to ethyltin,but the ethyl group was not oxidized to CO_2. Iwai and Wada[230] found that tetraethyltin was more rapidly absorbed than tetrabutyltin, and that dealkylated derivatives were at higher concentrations in portal vein blood than in systemic venous blood, suggesting that some dealkylation occurs in the intestinal mucosa.

Thin-layer chromatography of the products showed that triethyltin was converted to diethyltin and that the tri- propyl, butyl, pentyl, *n*-hexyl and cyclohexyl derivatives, but not triphenyltin, were dealkylated by a rabbit microsomal NADPH oxidase system[231].

In this mono-oxygenase system of mammalian liver microsomes, which catalyses the oxidation of many foreign organic compounds, one atom of an oxygen molecule is incorporated into the substrate as a hydroxyl group while the other is reduced to water by a second reducing agent. In the present case, the enzyme is a cytochrome P-450 with a flavoprotein which catalyses the transfer of electrons from NADPH to reduce the second oxygen atom.[232] The reaction scheme (eqn 11.9), involves a ternary complex with the substrate and oxygen and is competitively inhibited by carbon monoxide

RH → $RH.Fe^{3+}$ → $RH.Fe^{2+}$ ($e^{\ominus}$; NADPH → FAD → $NADP^+$) → (O_2) $RH.Fe^{2+}—O_2$ → ($e^{\ominus}$ from FAD) $RH.Fe^{3+}—O_2^{2-}$ → ($2H^+$, H_2O) $RH.Fe^{3+}—O$ → $ROH.Fe^{3+}$ → ROH + Fe^{3+}

(11.9)

binding to the haem iron in place of oxygen. The transfer of electrons to the haem groups is generally slow and is exacerbated in the case of the organotin compounds by the formation of unreactive complexes.[233] Direct transfer of the oxenoid between a carbon and hydrogen of the substrate has been proposed.[230] Fish *et al.*[234] favour a free-radical mechanism for the hydroxylation of the organotins, partly because non-enzymatic models of the monooxygenase reaction break the C–Sn bond directly without hydroxylation, but also because the site of hydroxylation is not specific. Using [1-^{14}C]tributyltin acetate, Kimmel *et al.*[235] found monohydroxybutyl dibutyltin derivatives with the hydroxyl group distributed in the proportions 12% 4-OH, 16% 3-OH, 44% 2-OH and 24% 1-OH, plus a small percentage of the 3-oxo derivative. Oxidation of the first butyl chain of tetrabutyltin is slow, but yields 77% of 1-OH and 23% 2-OH monohydroxybutyltributyltin, both of which are unstable; the former breaks down to 1-butanol at neutral pH and the latter yields 1-butene under acid conditions. The instability of these hydroxylated intermediates leads to breakage of the C–Sn bond as observed in the overall reaction. Tricyclohexyltin is dealkylated to yield inorganic tin via hydroxylated derivatives.

The extent of metabolism of different triorganotins by rabbit liver oxygenase decreased in the series:

$$Et_3Sn = {}^nPr_3Sn = {}^nBu_3Sn > {}^nPen_3Sn > {}^cHex_3Sn$$
$$> {}^nHex_3Sn \gg {}^nOct_3Sn > Ph_3Sn = 0$$

and produced similar metabolites to tributyltin. Tricyclohexyltin yielded (2-OH–^{c}Hex) $^cHex_2Sn$, which destannylated to yield cyclohexanol, cyclohexanone and cyclohexene.[236] The rabbit liver system did not oxidize triphenyltin, but rats given $Ph_3{}^{114}Sn$ excreted diphenyltin and phenyltin as well as unmetabolized triphenyltin.[71]

Mixed cyclohexyl- and phenyltins gave *trans*-4-OH-cyclohexyltriphenyltin as the major product from cyclohexyltriphenyltin, while cyclohexyldiphenyltin gave the 2-OH cyclohexyl derivative as the major product plus some 3-OH and 4-OH cyclohexyl derivatives.[237]

The di-, tri-, and tetra-ethyl derivatives of tin, lead and germanium have been found to yield ethylene as the major and ethane as a minor product with rat liver cytochrome P-450 mono-oxygenase. Triethyltin, either directly or produced from tetraethyltin, caused conversion of cytochrome P-450 into cytochrome P-420, with consequent inhibition of the enzyme.[238]

A comparative investigation,[239] showed that radioactive tributyltin was metabolized by fish and crabs. After 48 h about 50% of the radioactivity in the liver or hepatopancreas was in dibutyltin, monobutyltin and other degradation products. Oysters metabolized the tributyltin much more slowly, about 10% in 72 h.

These investigations of organotin metabolism in animals confirm the data on environmental degradation by micro-organisms (section 11.3.3), the overall process being oxidation of one carbon group at a time followed by rupture of the C–Sn bond, and thus progressive dealkylation of the tin. Removal of the last alkyl group is slow. Alkyl group shortening has not been reported.

11.5.2 *Biomethylation of tin compounds*

Methyltin, and possibly mixed methylbutyltin compounds, have been found in the environment[65,240], and are almost certainly produced by biological or geochemical methylation since very little methyltin is released into the environment by man (section 11.3.3). Biological methylation of heavy elements has become well recognised[241–43] since the discovery by Challenger[244] of the methylation of arsenic, and of mercury[243] following the episode of Minamata disease in Japan.

Tetramethyltin was produced by incubating trimethyltin with either untreated or autoclaved marine sediments, but more rapidly with the untreated samples which indicated methylation by living cells.[245] Monomethyltin was produced from inorganic tin ($SnCl_4 \cdot 5H_2O$) by an inoculum of mixed micro-organisms from Chesapeake Bay sediments[246], and an aerobic strain of *Pseudomonas* (Ps 244) produced a variety of methyltins from inorganic tin (IV), but only traces from tin (II).[247] Ashby and Craig[248] have reported methylation of tin (II) complexes with oxalate or amino acids containing sulphur, but little from SnS and none from $SnCl_4$, $SnCl_2$ or metallic tin by a pure culture of *Saccharomyces cerevisiae*. Monomethyltin products predominated with traces of dimethyltin.

Enzyme-catalysed methylation of tin has not been demonstrated in cell-free systems, but other metabolic methylations and model reactions are guides to possible mechanisms of metabolic methylation of tin compounds and to reactions in which tin is methylated outside living cells by biological products.[243,249,250]

The common metabolic methyl donors are: S-adenosylmethionine, which is a methyl sulphonium compound, N5-methyltetrahydrofolic acid, methylcobalamin (vitamin B_{12}), in which the methyl is attached to the cobalt atom, and methylated quaternary nitrogen compounds such as betaine.

Methylcobalamin is the most powerful of these methyl carriers, since the methyl group can be transferred as a carbanion, free radical or a carbonium ion with corresponding changes in the oxidation state of the cobalt. When the recipient molecule is a metal ion in its highest oxidation state, the methyl must be transferred as the carbanion, and the Co corrinoids are the only known biological carbanion donors, although carbanion transfer between metals and disproportionation of methyltin compounds can also occur.[251,252] At a low state of oxidation (Sn (II)), methylation can take place by carbonium ion transfer with concomitant oxidation of the metal ion (eqn 11.10):

$$Sn^{2+} + CH_3^+ \rightarrow CH_3Sn^{3+} \qquad (11.10)$$

Non-enzymatic transfer of methyl from cobalamin to Sn^{2+}, but not Sn (IV) salts, in aqueous HCl has been demonstrated in a model reaction which required the presence of a mild oxidizing agent such as oxygen, suggesting free-radical transfer (eqn 11.11):[252,253]

$$CH_3[Co] + Sn^{2+} \rightarrow [Co] + CH_3Sn^{\cdot III} \qquad (11.11a)$$

$$CH_3Sn^{\cdot III} \xrightarrow{-e} CH_3Sn^{3+} \qquad (11.11b)$$

A very slow methylation of SnO_2 by methylcobalamin has also been reported.[254] In other model reactions, a methyl group was transferred from a dimethyl-Co analogue of cobalamin to aqueous Sn (II) in the presence of manganese dioxide; methylcobalamin was inactive in this system.[255,256]

Methylation is an essential metabolic function, and S-adenosyl methionine is the most common methyl donor in biosynthesis of structural components and cofactors (such as phosphatidyl choline, creatine and carnitine), in the control of gene expression and the recognition of foreign DNA (by methylation of histones and nucleic acids), and in the control of bacterial motility and chemotaxis.[257] Regeneration of the sulphonium S-adenosyl methionine is metabolically expensive, since it requires three ATP pyrophosphate bonds for each methyl group transferred. In animals, the source of methyl groups is largely methionine in the diet, or choline which after oxidation to betaine can donate one methyl group to homocysteine to yield methionine. Methyl groups can be derived from a variety of one-carbon units because the unit can be oxidized, deaminated or reduced to the methyl level while attached to tetrahydrofolic acid.[258] Synthesis of methionine by transfer of a methyl group from N_5-methyltetrahydrofolic acid to homocysteine is one of the few reactions in mammals in which cobalamin is a coenzyme.

The Co-corrinoids are particularly important in bacteria called methanogens which obtain energy from the reduction of CO_2 and other carbon compounds to methane using molecular hydrogen. The methanogens are members of the Archaebacteria, a major kingdom which has several biochemical differences from other organisms, including different cofactors in one-carbon unit metabolism.[259] These unusual cofactors are coenzyme M, or CoM, which is 2-mercaptoethanesulphonic acid, 2-

(methylthio)ethanesulphonate when methylated, and tetrahydromethanopterin which accepts methyl at the N5 position and strongly resembles tetrahydrofolic acid in both structure and properties. Co-corrinoid compounds have been shown to be active in methyl transfers catalysed by cell-free extracts of these organisms. Furthermore, methylated Co-corrinoids are present at high concentrations in some methanogens,[260,261] which makes them likely candidates, alive or dead, as tin methylating agents.

Environmental methylation of tin may result from a combination of non-biological reactions and biological production of methylating agents, such as methylcorrinoids or betaines released from dead organisms and methyl iodide or dimethylsulphide excreted by living organisms. In laboratory experiments methyl iodide, which is produced by seaweed, reacted with tin metal or Sn(II) cations to yield dimethyltin and trimethyltin as the major products.[262,263] Betaine and trimethylsulphonium iodide, which could be produced from dimethylsuphide in the environment and is a model for S-adenosyl methionine, were much less effective. All these reactions are slow by laboratory standards, but on the volume and time scales of environmental processes and combined with bio-accumulation they could produce significant local concentrations of methyltin compounds.

References

1. J.J. Zuckermann, ed., *Organotin Compounds: New Chemistry and Applications*, Adv. Chem. Ser., **157**, American Chemical Society, Washington DC, 1976.
2. A.G. Davies and P.J. Smith, *Adv. Inorg. Chem. Radiochem.*, 1980, **23**, 1.
3. J.S. Thayer, *Organometallic Compounds and Living Organisms*, Academic Press, Orlando and London, 1984.
4. S.J. Blunden, P.A. Cusack and R. Hill, *The Industrial Uses of Tin Chemicals*, Royal Society of Chemistry, London, 1985.
5. C.J. Evans and S. Karpel, *Organotin Compounds in Modern Technology*, J. Organomet. Chem. Library, 1985, **16**.
6. F.E. Brinckman and J.M. Bellama, eds., *Organometals and Organometalloids, Occurrence and Fate in the Environment*, ACS Symp. Ser. **82**, American Chemical Society Washington DC, 1978.
7. P.J. Craig, in *Comprehensive Organometallic Chemistry*, ed. G. Wilkinson, Pergamon, Oxford, Vol. 2, 1981, 979.
8. S.J. Blunden, L.A. Hobbs and P.J. Smith, in *Environmental Chemistry*, RSC Spec. Periodical Rept, Vol. 3, 1984, 49.
9. (*a*) P.J. Craig, ed., *Organometallic Compounds in the Environment, Principles and Reactions*, Longman., Harlow, 1986; (*b*) R.J. Maguire, *Appl. Organomet. Chem.*, 1987, **1**, 475.
10. W.N. Aldridge, in *Proc. 2nd Int. Conf. on Si, Ge, Sn and Pb Compounds*, eds. M. Gielen and P.G. Harrison, *Rev. Si, Ge, Sn and Pb Cmpds, Special Issue*, Freund Publishing House, Tel Aviv, 1978, 9.
11. A.K. Saxena, *Appl. Organomet. Chem.*, 1987, **1**, 39.
12. N.F. Cardarelli, ed., *Tin as a Vital Nutrient: Implications in Cancer Prophylaxis and Other Physiological Processes*, CRC Press, Boca Raton, Florida, 1986.
13. K. Schwarz, D.B. Milne, E. Vinyard, *Biochem. Biophys. Res. Commun.*, 1970, **40**, 22.
14. P. Fritsch, G. De Saint Blanquat and R. Derache, *Toxicology*, 1977, **8**, 165.
15. P. Fritsch, G. De Saint Blanquat and R. Derache, *Toxicol. Eur. Res.*, 1978, **1**, 253.
16. M. Yamaguchi, Y. Kubo and T. Yamamoto, *Toxicol. Appl. Pharm.*, 1979, **47**, 441.
17. M. Yamaguchi, M. Kitade and S. Okada, *Toxicology Lett.*, 1980, **5**, 275.
18. N.V. Sidgwick, *The Chemical Elements and their Compounds*, Vol. 1, Oxford University Press, Oxford, 1950, 602, 620.
19. A.J. Vander, D.R. Mouw, J. Cox and B. Johnson, *Amer. J. Physiol.*, 1979, **236**, F373.

20. R.A. Hiles, *Toxicol. Appl. Pharmacol.*, 1974, **27**, 366.
21. R.A. Brown, C.M. Nazario, R.S. De Tirado, J. Castrillon and E.T. Agard, *Environ. Res.*, **13**, 56.
22. A. Kappas and M.D. Maines, *Science*, 1976, **192**, 60.
23. M.D. Maines and A. Kappas, *Science*, 1977, **198**, 1215.
24. G.S. Drummond and A.S. Kappas, *Biochem. J.*, 1980, **192**, 637.
25. M. Chiba and M. Kibuchi, *Biochem. Biophys. Res. Commun.*, 1978, **82**, 1057.
26. M.D. Maines and P. Sinclair, *J. Biol. Chem.*, 1977, **252**, 211.
27. G.S. Drummond and A.S. Kappas, *Proc. Natl. Acad. Sci. USA*, 1981, **78**, 6466.
28. G.S. Drummond and A.S. Kappas, *Science*, 1982, **217**, 1250.
29. T. Yoshinaga, S. Sassa and A. Kappas, *J. Biol. Chem.*, 1982, **257**, 7778.
30. E. Breslow, R. Chandra and A. Kappas, *J. Biol. Chem.*, 1986, **261**, 3135.
31. J.A. Shelnutt, *J. Amer. Chem. Soc.*, 1983, **105**, 7179.
32. A. Albert, *Selective Toxicity*, 7th edn, Chapman and Hall, London and New York, 1985.
33. Ref. 4, 23.
34. A.K. Sijperstein, J.G.A. Luijten and G.J.M. van der Kerk, *Fungicides: An Advanced Treatise*, Vol. 2, ed., D.C. Torgeson, Academic Press, New York, 1969, 331.
35. P.J. Smith, *Toxicological Data on Organotin Compounds*, Publ., **538**, International Tin Research Institute, Greenford, UK, 1981.
36. O. Wada, S. Manabe, H. Iwai and Y. Arakawa, *Jpn. J. Ind. Health*, 1982, **24**, 24.
37. A. Sylph, *Bibliography on the Effect of Organotin Compounds on Aquatic Organisms*, Bibliogr. **11**, International Tin Research Council, Greenford, UK, 1987.
38. M.J. Waldock, J.E. Thain and M.E. Waite, *Appl. Organomet. Chem.*, 1987, **1**, 287.
39. S.J. Blunden, P.J. Smith and B. Sugavanam, *Pestic. Sci.*, 1984, **15**, 253.
40. K.C. Molloy, T.G. Purcell, D. Cunningham, P. McCardle and T. Higgins, *Appl. Organomet. Chem.*, 1987, **1**, 119.
41. S.J. Blunden, B.N. Patel, P.J. Smith and B. Sugavanam, *Appl. Organomet. Chem.*, 1987, **1**, 241.
42. R.B. Laughlin, R.B. Johannesen, W. French, H. Guard and F.E. Brinckman, *Environ. Toxicol. Chem.*, 1985, **4**, 343.
43. L.W. Pinkney, *CRC Critical Reviews in Toxicology*, 1984, **14**, 159.
44. A.R. Beaumont and P.B. Newman, *Mar. Pollut. Bull.*, 1986, **17**, 457.
45. S. Balabaskaran, K. Tilakavati and V.G. Kumar Das, *Appl. Organomet. Chem.*, 1987, **1**, 347.
46. A.J. Crowe, *Appl. Organomet. Chem.*, 1987, **1**, 143 and 331.
47. M.H. Gitlitz, in ref. 1, 167.
48. R.B. Laughlin, W. French and H.E. Guard, *Environ. Sci. Technol.*, 1986, **20**, 884.
49. I.M. Davies and J.C. McKie, *Mar. Pollut. Bull.*, 1987, **18**, 405.
50. E.E. Kenaga, *Environ. Sci. Technol.*, 1980, **14**, 553.
51. P.T.S. Wong, Y.K. Chau, O. Kramar and G.A. Bengert, *Can. J. Fish Aquat. Sci.*, 1982, **39**, 483.
52. J. Yamada, *Agric. Biol. Chem.*, 1981, **45**, 997.
53. C. Hansch and A. Lee, *Substituent Constants for Correlation Analysis in Chemistry and Biology*, Wiley, New York, 1979.
54. F.E. Brinckman, in ref. 9, 49.
55. J.J. Zuckermann, R.P. Reisdorf, H.V. Ellis and R.R. Wilkinson, in ref. 6, 388.
56. S.J. Blunden and A. Chapman, in ref. 9, 111.
57. *Proc. Oceans 86 Conf. and Exposition on Science Engineering and Adventure, Washington DC*, 23–25 Sept. 1986, Organotin Symposium, IEEE, Piscataway, NJ, and Marine Technology Society, Washington DC, Vol. 4, 1986, 1101–1329.
58. G.A. Senich, *Polymer*, 1982, **23**, 1385.
59. C. Alzieu, M. Heral, Y. Thibaud, M.J. Dardignac and M. Feuillet, *Rev. Trav. Inst. Pech. Marit.*, 1982, **45**, 101.
60. M.J. Waldock and J.E. Thain, *Mar. Pollut. Bull.*, 1983, **14**, 411.
61. G.W. Bryan, P.E. Gibbs, G.R. Burt and L.G. Hummerstone, *J. Mar. Biol. Ass. U.K.*, 1987, **67**, 525.
62. P.E. Gibbs, G.W. Bryan, P.L. Pascoe and G.R. Burt, *J. Mar. Biol. Ass. U.K.*, 1987, **67**, 507.
63. I.M. Davies, S.K. Bailey and D.C. Moore, *Mar. Pollut. Bull.*, 1987, **18**, 400.
64. T.C. Cheng, ed., *Molluscicides in Schistomiasis Control*, Academic Press, New York, 1974.

65. C.H. Matthias, J.M. Bellama, G.J. Olson and F.E. Brinckman, *Environ. Sci. Technol.*, 1986, **20**, 609.
66. J.W. Short, *Bull. Environ. Contam. Toxicol.*, 1987, **39**, 412.
67. S.J. Blunden, *J. Organomet. Chem.*, 1983, **248**, 149.
68. R.J. Maguire, J.H. Carey and E.J. Hale, *J. Agric. Food Chem.*, 1983, **31**, 1060.
69. R.D. Barnes, A.T. Bull and R.C. Poller, *Pestic. Sci.*, 1973, **4**, 305.
70. C.J. Soderquist and D.G. Crosby, *J. Agric. Food Chem.*, 1980, **28**, 111.
71. K.D. Freitag and R. Bock, *Pestic. Sci.*, 1974, **5**, 731.
72. D. Barug and J.W. Vonk, *Pestic. Sci.*, 1980, **11**, 77.
73. D. Barug, *Chemosphere*, 1981, **10**, 1145.
74. P.F. Seligman, A.O. Valkirs and R.F. Lee, *Environ. Sci. Technol.*, 1986, **20**, 1229.
75. P.J. Smith, A.J. Crowe, D.W. Allen, J. Brooks and R. Formstone, *Chem. Ind. (London)*, 1977, 874.
76. O.F.X. Donard and J.H. Weber, *Environ. Sci. Technol.*, 1985, **19**, 1104.
77. A.G. Davies, J.P. Goddard, M.B. Hursthouse and N.P.C. Walker, *J. Chem. Soc., Chem. Commun.*, 1983, 597.
78. R.S. Tobias, in ref. 6, p. 130.
79. R.B. Laughlin, H.E. Guard and W.M. Coleman, *Environ. Sci. Technol.*, 1986, **20**, 201.
80. S.J. Blunden and R. Hill, *Inorg. Chim. Acta*, 1984, **87**, 83.
81. M.J. Hynes and M. O'Dowd, *J. Chem. Soc., Dalton Trans.*, 1987, 563.
82. M.S. Rose, *Biochem. J.*, 1969, **111**, 129.
83. A.P. Dawson and M.J. Selwyn, *Biochem. J.*, 1974, **138**, 349.
84. M.S. Rose and W.N. Aldridge, *Biochem. J.*, 1968, **106**, 821.
85. M.S. Rose and E.A. Lock, *Biochem. J.*, 1970, **120**, 151.
86. N.F. Cardarelli, *Controlled Release Molluscicides*, Environ. Manag. Lab. Monogr., Univ. Akron, Ohio, 1977, 34.
87. W.N. Aldridge, in ref. 1, 186.
88. M.J. Selwyn, in ref. 1, 204.
89. B.M. Elliott and W.N. Aldridge, *Biochem. J.*, 1977, **163**, 583.
90. B.M. Elliott, W.N. Aldridge and J.W. Bridges, *Biochem. J.*, 1979, **177**, 461.
91. G. van Koten and J.G. Noltes, in ref. 1, 275.
92. F. Taketa, K. Siebenlist, J. Kasten-Jolly and N. Palosaari, *Arch. Biochem. Biophys.*, 1980, **203**, 466.
93. K.R. Siebenlist and F. Taketa, *Biochem. J.*, 1986, **233**, 471.
94. W.N. Aldridge and B.W. Street, *Biochem. J.*, 1970, **118**, 171.
95. M. Stockdale, A.P. Dawson and M.J. Selwyn, *Eur. J. Biochem.*, 1970, **15**, 342.
96. M.S. Rose and W.N. Aldridge, *Biochem. J.*, 1972, **127**, 51.
97. B.G. Farrow and A.P. Dawson, *Eur. J. Biochem.*, 1978, **86**, 85.
98. W.N. Aldridge and M.S. Rose, *FEBS Lett.*, 1969, **4**, 61.
99. W.N. Aldridge, B.W. Street and J.G. Noltes, *Chem.-Biol. Interact.*, 1981, **34**, 223.
100. K. Cain and D.E. Griffiths, *Biochem. J.*, 1977, **162**, 575.
101. W. Sebald and J. Hoppe, *Curr. Top. Bioenerg.*, 1981, **12**, 1.
102. Yu. A. Ovchinnikov, N.G. Abdulaev and N.N. Modyanov, *Ann. Rev. Biophys. Bioeng.*, 1982, **11**, 445.
103. A.E. Senior and J.G. Wise, *J. Membr. Biol.*, 1983, **73**, 105.
104. D.D. Tyler in *Membrane Structure and Function*, ed. E.E Bittar, Vol. 5, Wiley, New York, 1984, 117.
105. W.E. Lancashire and D.E. Griffiths, *Eur. J. Biochem.*, 1975, **51**, 377.
106. A.P. Dawson, B.G. Farrow and M.J. Selwyn, *Biochem. J.*, 1982, **202**, 163.
107. K. Cain, M.D. Partis and D.E. Griffiths, *Biochem. J.*, 1977, **166**, 593.
108. D.E. Griffiths, K. Cain and R.L. Hyams, *Biochem. Soc. Trans.*, 1977, **5**, 205.
109. A.P. Singh and P.D. Bragg, *Biochem. Biophys. Res. Commun.*, 1978, **81**, 161.
110. S. De Chadarevjan, A. De Santis, B.A. Melandri and A. Baccarini Melandri, *FEBS Lett.*, 1979, **97**, 293.
111. M.D. Partis, E. Bertoli, D.E. Griffiths and A. Azzi, *Biochem. Biophys. Res. Commun.*, 1980, **96**, 1103.
112. F. Zanotti, F. Guerrieri, Y.W. Che, R. Scarfo and S. Papa, *Eur. J. Biochem.*, 1987, **164**, 517.
113. D.R. Sanadi, *Biochim. Biophys. Acta*, 1982, **683**, 39.

114. K. Cain, R.L. Hyams and D.E. Griffiths, *FEBS Lett.*, 1977, **82**, 23.
115. A.S. Watling-Payne and M.J. Selwyn, *FEBS Lett.*, 1975, **58**, 57.
116. J.M. Gould, *Eur. J. Biochem.*, 1976, **62**, 567.
117. J.M. Gould, *FEBS Lett.*, 1978, **94**, 90.
118. W.N. Aldridge and J.E. Cremer, *Biochem. J.*, 1955, **61**, 406.
119. C.H. Williams in *The Enzymes*, ed. P.D. Boyer, Vol. XIII, Academic Press, New York, 1976, 90.
120. H. Matsui, O. Wada, Y. Ushijima and T. Akuzawa, *FEBS Lett.*, 1983, **164**, 251.
121. E.J. Bowman, *J. Biol. Chem.*, 1983, **258**, 15238.
122. B.J. Bowman and E.J. Bowman, *J. Membrane Biol.*, 1986, **94**, 83.
123. J. Delhez, J.-P. Dufour, D. Thines and A. Goffeau, *Eur. J. Biochem.*, 1977, **79**, 319.
124. P.L. Pedersen and E. Carafoli, *Trends Biochem. Sci.*, 1987, **12**, 146 and 186.
125. K.H. Byington, *Biochem. Biophys. Res. Commun.*, 1971, **42**, 16.
126. P.C. Maloney, *J. Membr. Biol.*, 1982, **67**, 1.
127. P. Mitchell, *FEBS Lett.*, 1985, **182**, 1.
128. H.S. Penefsky, *Proc. Natl. Acad. Sci. USA*, 1985, **82**, 1589.
129. A.P. Dawson and M.J. Selwyn, *Biochem. J.*, 1975, **152**, 333.
130. R.L. van der Bend, W. Duetz, A.-M.A.F. Colen, K. van Dam and J.A. Berden, *Arch. Biochem. Biophys.*, 1985, **241**, 461.
131. E.L. Emanuel, M.A. Carver, G.C. Solani and D.E. Griffiths, *Biochem. Biophys. Acta*, 1984, **766**, 209.
132. J.O.D. Coleman and J.M. Palmer, *Biochim. Biophys. Acta*, 1971, **245**, 313.
133. D.K. Apps, J.G. Pryde, R. Sutton and J.H. Phillips, *Biochem. J.*, 1980, **190**, 273.
134. B. Marin, J. Preisser and E. Komor, *Eur. J. Biochem.*, 1985, **151**, 131.
135. J.D. Doherty, N. Salem, C.J. Lauter and E.J. Trams, *Comp. Biochem. Physiol.*, 1981, **69C**, 185.
136. J. Barranco, A. Darszon and A. Gomez-Puyou, *Biochem. Biophys. Res. Commun.*, 1981, **100**, 1402.
137. R.P. Casey, M. Thelen and A. Azzi, *J. Biol. Chem.*, 1980, **255**, 3994.
138. A.P. Singh and P.D. Bragg, *Can. J. Biochem.*, 1979, **57**, 1384.
139. B. Mannervik, *Adv. Enzymol.*, 1985, **57**, 357.
140. R.A. Henry and K.H. Byington, *Biochem. Pharmacol.*, 1976, **25**, 2291.
141. E. Tipping, B. Ketterer, L. Christodoulides, B.M. Elliott, W.N. Aldridge and J.W. Bridges, *Chem.-Biol. Interactions*, 1979, **24**, 317.
142. S. Yalcin, H. Jensson and B. Mannervik, *Biochem. Biophys. Res. Commun.*, 1983, **114**, 829.
143. W.N. Aldridge, H. Grasdalen, K. Aarstad, B.W. Street and T. Norkov, *Chem.-Biol. Interactions*, 1985, **54**, 243.
144. A. Fersht, *Enzyme Structure and Mechanisms*, 2nd edn, W.H. Freeman, New York, 1985, 314.
145. F. Davidoff and S. Carr, *Biochemistry*, 1973, **12**, 1415.
146. K.R. Siebenlist and F. Taketa, *Biochem. Biophys. Res. Commun.*, 1980, **95**, 758.
147. S.B. Andrews, J.W. Faller, R.J. Barrnett and V. Mizuhira, *Biochem. Biophys. Acta*, 1978, **506**, 1.
148. M.J. Selwyn, A.P. Dawson, M. Stockdale and N. Gains, *Eur. J. Biochem.*, 1970, **14**, 120.
149. A.S. Watling and M.J. Selwyn, *FEBS Lett.*, 1970, **10**, 139.
150. E.J. Harris, J.A. Bangham and B. Zukovic, *FEBS Lett.*, 1973, **29**, 339.
151. R.G. Wulf and K.H. Byington, *Arch. Biochem. Biophys.*, 1975, **167**, 176.
152. R. Motais, J.L. Cousin and F. Sola, *Biochim. Biophys. Acta*, 1977, **467**, 357.
153. A.P. Singh and P.D. Bragg, *Can. J. Biochem.*, 1979, **57**, 1376.
154. C.J. Chastain and J.B. Hanson, *Plant Physiol.*, 1981, **68**, 981.
155. Y. Mukohata and Y. Kaji, *Arch. Biochem. Biophys.*, 1981, **206**, 72.
156. M.J. Selwyn and A.S. Watling, in *The Biological Alkylation of Heavy Elements*, eds. P.D. Craig and F. Glockling, Royal Society of Chemistry, London, 1988, 291.
157. M. Gielen and H. Moktar-Jamai, *J. Organomet. Chem.*, 1977, **129**, 325.
158. J.O. Wieth and M.T. Tosteson, *J. Gen. Physiol.*, 1979, **73**, 765.
159. M.T. Tosteson and J.O. Wieth, *J. Gen. Physiol.*, 1979, **73**, 789.
160. F.L. Bygrave, C. Ramachandran and R.N. Robertson, *Arch. Biochem. Biophys.*, 1978, **188**, 301.
161. G.P. Brierley, M. Jurkowitz and D.W. Jung, *Arch. Biochem. Biophys.*, 1978, **190**, 181.
162. W.N. Aldridge, *Biochem. J.*, 1958, **69**, 367.
163. N. Sone and B. Hagihara, *J. Biochem.*, 1964, **56**, 151.

164. D.G. Nicholls, *Bioenergetics*, Academic Press, London and New York, 1982.
165. F.M. Harold, *The Vital Force: A Study of Bioenergetics*, W.H. Freeman, New York, 1986.
166. S.J. Ferguson, *Biochim. Biophys. Acta*, 1985, **811**, 47.
167. Y. Hatefi, *Ann. Rev. Biochem.*, 1985, **54**, 1015.
168. E.C. Slater, *Eur. J. Biochem.*, 1987, **166**, 489.
169. D.N. Skilleter, *Biochem. J.*, 1975, **146**, 465.
170. W.N. Aldridge, B.W. Street and D.N. Skilleter, *Biochem. J.*, 1977, **168**, 353.
171. J.S. Kahn, *Biochim. Biophys. Acta*, 1968, **152**, 203.
172. A.S. Watling-Payne and M.J. Selwyn, *Biochem. J.*, 1974, **142**, 65.
173. E.A. Lock, *J. Neurochem.*, 1976, **26**, 887.
174. W.N. Aldridge, R.D. Verschoyle, C.A. Thompson and A.W. Brown, *Neuropathol. Appl. Neurobiol.*, 1987, **13**, 55.
175. D.E. McMillan and G.R. Wenger, *Pharmacol. Rev.*, 1985, **37**, 365.
176. A. Sylph, *Bibliography on the Neurotoxicology of Tin and Organotin Compounds*, Library Bibliogr. **12**, International Tin Research Institute, Greenford, UK, 1986.
177. K.R. Reuhl and J.M. Cranmer, *Neurotoxicology*, 1984, **5**, 187.
178. A.D. Toews, W.D. Blaker, D.J. Thomas, J.J. Gaynor, M.R. Krigman, P. Mushak and P. Morell, *J. Neurochem.*, 1983, **41**, 816.
179. A.W. Brown, W.N. Aldridge and B.W. Street, *Neuropath. Appl. Neurobiol.*, 1979, **5**, 83.
180. L.W. Chang, T.M. Tiemeyer, G.R. Wenger and D.E. McMillan, *Environ. Res.*, 1983, **30**, 399.
181. L.W. Chang, *Fundam. Appl. Toxicol.*, 1986, **6**, 217.
182. P.N. Magee, H.B. Stoner and J.M. Barnes, *J. Pathol. Bacteriol.*, 1957, **73**, 107.
183. D.A. Kirschner and V.S. Sapirstein, *J. Neurocytol.*, 1982, **11**, 559.
184. L. Benoist, H. Allain, P. Linee, M.C. Hennon, A.M. Bernard, J. Van den Driessche, J.B. Le Polles and J. de Certaines, *J. Pharmacol.*, 1984, **15**, 143.
185. A.C. Leow, R.M. Anderson, R.A. Little and D.D. Leaver, *Acta Neuropathol. (Berl.)*, 1979, **47**, 117.
186. D.I. Graham, P.V. De-Jesus, D.E. Pleasure and N.K. Gonatas, *Arch. Neurol.*, 1976, **33**, 40.
187. E.A. Lock and W.N. Aldridge, *J. Neurochem.*, 1975, **25**, 871.
188. H. Nagara, K. Suzuki, C.W. Tiffany and K. Suzuki, *Brain Res.*, 1981, **225**, 413.
189. F. Joo, O.T. Zoltan, B. Csillik and M. Foeldi, *Arzneim. Forsch.*, 1969, **19**, 296.
190. A.C.T. Leow, K.M. Towns and D.D. Leaver, *Chem.-Biol. Interactions*, 1979, **27**, 125.
191. O. Macovachi, A.F. Prigent, G. Nemoz, J.F. Pageaux and H. Pacheco, *Biochem. Pharmacol.*, 1984, **33**, 3603.
192. M.S. Rose and W.N. Aldridge, *J. Neurochem.*, 1966, **13**, 103.
193. P.E. Neumann and F. Taketa, *Mol. Brain Res.*, 1987, **2**, 83.
194. L.G. Costa and R. Sulaiman, *Toxicol. Appl. Pharmacol.*, 1986, **86**, 189.
195. G.J. Harry, J.F. Goodrum, M.R. Krigman and P. Morell, *Brain Res.*, 1985, **326**, 9.
196. A.D. Toews, R. Ritab, N.D. Goines and T.W. Bouldin, *Brain Res.*, 1986, **398**, 298.
197. R.S. Divedi, G. Kaur, R.C. Srivastava and T.N. Srivastava, *Ind. Health*, 1985, **23**, 9.
198. L.G. Costa, *Toxicol. Appl. Pharmacol.*, 1985, **79**, 471.
199. B.I. Kanner and L. Kifer, *Biochemistry*, 1981, **20**, 3354.
200. Y. Morot-Gaudry, *Neurochem. Int.*, 1984, **6**, 531.
201. J.E. Allen, P.W. Gage, D.D. Leaver and A.C.T. Leow, *Chem.-Biol. Interactions*, 1980, **31**, 227.
202. I. Hanin, M.R. Krigman and R.B. Mailman, *Neurotoxicology*, 1985, **5**, 567.
203. H. Komulainen and S.C. Bondy, *Toxicol. Appl. Pharmacol.*, 1987, **88**, 77.
204. L.P. Tan, M.L. Ng and V.G. Kumar Das, *J. Neurochem.*, 1978, **31**, 1035.
205. B.M. Paschal and R.B. Vallee, *Nature*, 1987, **330**, 181.
206. S. Bondy and D.L. Hall, *Neurotoxicology*, 1986, **7**, 51.
207. P.J. Sadler, *Chem. in Britain*, 1982, **18**, 182.
208. S.E. Sherman and S.J. Lippard, *Chem. Rev.*, 1987, **87**, 1153.
209. A.J. Crowe, P.J. Smith and G. Atassi, *Chem.-Biol. Interactions*, 1980, **32**, 171.
210. R. Barbieri, L. Pellerito, G. Ruisi, M.T. Lo Guidice, F. Huber and G. Atassi, *Inorg. Chim. Acta*, 1982, **66**, L39.
211. A.K. Saxena and J.P. Tandon, *Cancer Letts.*, 1983, **19**, 73.
212. M. Takashi, F. Furakawa, T. Kokubo, Y. Kurato and Y. Hayashi, *Cancer Letts.*, 1983, **20**, 271.
213. A.J. Crowe, P.J. Smith and G. Atassi, *Inorg. Chim. Acta*, 1984, **93**, 179.
214. F. Huber, G. Roge, L. Carl, G. Atassi, F. Spreafico, S. Filippeschi, R. Barbieri, A. Silvestri,

E. Rivarola, G. Ruisi, F. Di Bianca and G. Alonzo, *J. Chem. Soc., Dalton Trans.*, 1985, 523.
215. N.F. Cardarelli, B.M. Cardarelli, E.P. Libby and E. Dobbins, *Austr. J. Exp. Biol. Med. Sci.*, 1984, **62**, 209.
216. P. Kopf-Maier and H. Kopf, *Chem. Rev.*, 1987, **87**, 1137.
217. W. Seinen and M.I. Willems, *Toxicol. Appl. Pharmacol.*, 1976, **35**, 63.
218. W. Seinen, J.G. Vos, R. van Kreiken, A. Penninks and H. Hooykaas, *Toxicol. Appl. Pharmacol.*, 1977, **42**, 213.
219. G. Hennighausen and P. Lange, *Arch. Toxicol. Suppl.*, 1979, **2**, 315.
220. A.H. Penninks and W. Seinen, *Toxicol. Appl. Pharmacol.*, 1980, **56**, 221.
221. R.R. Miller, R. Hartung and H.H. Cornish, *Toxicol. Appl. Pharmacol.*, 1980, **55**, 564.
222. P. Lange, G. Hennighausen and U. Karnstedt, *Arch. Toxicol. Suppl.*, 1980, **4**, 132.
223. N.J. Snoeij, A.A.J. Van Irsel, A.H. Penninks and W. Seinen, *Toxicol. Appl. Pharmacol.*, 1985, **81**, 274.
224. J.G. Vos, A. De Klerk, E.I. Krajnc, W. Kruizinga, B. Van Ommen and J. Rozing, *Toxicol. Appl. Pharmacol.*, 1984, **75**, 387.
225. N.J. Snoeij, A.A.J. Van Irsel, A.H. Penninks and W. Seinen, *Toxicology*, 1986, **39**, 71.
226. N.J. Snoeij, P.M. Punt, A.H. Penninks and W. Seinen, *Biochim. Biophys. Acta*, 1986, **852**, 234.
227. N.J. Snoeij, H.J.M. van Rooijen, A.H. Penninks and W. Seinen, *Biochim. Biophys. Acta*, 1986, **852**, 244.
228. J.E. Cremer, *Biochem. J.*, 1958, **68**, 685.
229. J.W. Bridges, D.S. Davies and R.T. Williams, *Biochem. J.*, 1967, **105**, 1261.
230. H. Iwai and O. Wada, *Industr. Health*, 1981, **19**, 247.
231. J.E. Casida, E.C. Kimmel, B. Holm and G. Widmark, *Acta Chem. Scand.*, 1971, **25**, 1497.
232. S.D. Black and M.J. Coon, *Adv. Enzymol.*, 1987, **60**, 35.
233. R.H. Fish, E.C. Kimmel and J.E. Casida, *J. Organomet. Chem.*, 1976, **118**, 41.
234. R.H. Fish, E.C. Kimmel and J.E. Casida, in ref. 1, 197.
235. E.C. Kimmel, R.H. Fish and J.E. Casida, *J. Agric. Food. Chem.*, 1977, **25**, 1.
236. E.C. Kimmel, R.H. Fish and J.E. Casida, *J. Agric. Food Chem.*, 1977, **28**, 117.
237. R.H. Fish, J.E. Casida and E.C. Kimmel, in ref. 6, 82.
238. R.A. Prough, M.A. Stalmach, P. Wiebkin and J.W. Bridges, *Biochem. J.*, 1981, **196**, 763.
239. R.F. Lee, *Marine Environ. Res.*, 1985, **17**, 145.
240. S. Tuğrul, T.I. Balkas and E.D. Goldberg, *Mar. Pollut. Bull.*, 1983, **14**, 297.
241. P.J. Craig and F. Glockling, eds., *The Biological Alkylation of Heavy Elements*, Royal Society of Chemistry, London, 1988.
242. F.E. Brinckman and G.J. Olson, in ref. 241, 168.
243. J.M. Wood, A. Cheh, L.J. Dizikes, W.P. Ridley, S. Rakow and J.R. Lakowicz, *Fed. Proc.*, 1978, **37**, 16.
244. F. Challenger, *Q. Rev. Chem. Soc.*, 1955, **9**, 255.
245. H.E. Guard, A.B. Cobet and W.M. Coleman, *Science*, 1981, **213**, 770.
246. L.E. Hallas, J.C. Means and J.J. Cooney, *Science*, 1982, **215**, 1505.
247. J.-A.A. Jackson, W.R. Blair, F.E. Brinckman and W.P. Iversen, *Env. Sci. Technol.*, 1982, **16**, 110.
248. J. Ashby and P.J. Craig, *Appl. Organomet. Chem.*, 1987, **1**, 275.
249. R.J.P. Williams, in ref. 241, 5.
250. W.P. Ridley, L.J. Dizikes and J.M. Wood, *Science*, 1977, **197**, 329.
251. Y.K. Chau, P.T.S. Wong, C.A. Mojesky and A.J. Carty, *Appl. Organomet. Chem.*, 1987, **1**, 235.
252. P.J. Craig and S. Rapsomanikis, *Inorg. Chim. Acta*, 1985, **107**, 39.
253. Y.T. Fanchiang and J.M. Wood, *J. Amer. Chem. Soc.*, 1981, **103**, 5100.
254. J.S. Thayer, in ref. 6, 188.
255. S. Rapsomanikis and J.H. Weber, *Environ. Sci. Technol.*, 1985, **19**, 352.
256. S. Rapsomanikis, O.F.X. Donard and J.H. Weber, *Appl. Organomet. Chem.*, 1987, **1**, 115.
257. D.E. Koshland, *Ann. Rev. Biochem.*, 1981, **50**, 765.
258. M. Dixon and E.C. Webb, *Enzymes*, 3rd edn., Longman, London, 1979, 762.
259. R.S. Wolfe, *Trends Biochem. Sci.*, 1985, **10**, 396.
260. V. Höllreigl, P. Scherer and P. Renz, *FEBS Lett.*, 1983, **151**, 156.
261. L. Daniels, R. Sparling and G.D. Sprott, *Biochim. Biophys. Acta*, 1984, **768**, 113.
262. J.S. Thayer, G.J. Olson and F.E. Brinckman, *Environ. Sci. Technol.*, 1984, **18**, 726.
263. P.J. Craig and S. Rapsomanikis, *Environ. Sci. Technol.*, 1985, **19**, 726.

12 Tin (IV) oxide: surface chemistry, catalysis and gas sensing

P.G. HARRISON

12.1 Tin (IV) oxide and the nature of the tin (IV) oxide surface

Tin (IV) oxide is a very stable material ($\Delta H_f^\circ - 581\,kJ\,mol^{-1}$), and exists in three different modifications (rhombic, hexagonal and tetragonal). That with the tetragonal rutile structure (see Chapter 2, Figure 2.4)[1-3] is the most common, and occurs naturally as the mineral cassiterite. The preparation and properties of the oxide have been reviewed.[2,4] Commercial preparation of bulk oxide is by one of two methods, both of which yield hydrated forms of the oxide. Hydrolysis of aqueous solutions of tin(IV) salts yields the α-form (or α-stannic acid), whereas oxidation of tin metal by hot concentrated nitric acid gives the β-form (β-stannic acid). Some differences exist between the two forms (for example, the α-form is amphoteric, dissolving in both acids and alkalis, whereas the β-form is relatively inert). Despite these differences in chemical behaviour, freshly prepared samples appear to be very similar in physical properties.[5] X-ray powder photographs of these hydrated forms are identical to these of SnO_2,[6,7] and also show that the degree of crystallinity increases with increasing calcination temperature.[8-11] On ageing, the α-form transforms into the β-form,[12] whereas on heating, both oxides lose water by a continuous process until dehydration is complete. These materials are, however, not true hydrates, and differences between the two forms are due to differences in the degree of agglomeration and the amount of water within the sample.

The porosity and specific surface area of the oxide are of great importance in applications such catalysis or ion exchange. For such uses, polycrystalline or gel forms of the oxide are employed[13] which are usually obtained by the hydrolysis route. The method of preparation, together with thermal prehistory, is critical in determining the consequent chemical behaviour. Nearly all the studies of the pore structure of tin(IV) oxide are on samples which are the α-form.[9-11] One[5] compares the behaviour of both the α- and β-forms and shows them to be essentially similar, although the α-form has larger crystallite size (*c.* 300 μm) compared to the β-modification (*c.* 1 μm). The earliest study, on an ion-free sample prepared by the hydrolysis of $Sn(OEt)_4$, showed that the surface area decreases from $163\,m^2\,g^{-1}$ to $0.8\,m^2\,g^{-1}$ as the calcination temperature of the sample is increased from 373 K to 1723 K. However, this decrease is accompanied by an almost negligible change in pore volume and lump volume. This phenomenon was attributed to growth in the crystallite size on increase in temperature (66 Å at 473 K, 82 Å at 573 K, 337 Å at 773 K, 880 Å at 1073 K, 1400 Å at 1373 K, 2610 Å at 1473 K, and 10 900 Å at 1673 K) in such a way as to alter the pore size distribution in favour of larger pores without significantly affecting the total pore volume. The increase in pore size was corroborated by CCl_4 adsorption isotherms. Low-temperature samples exhibited type I isotherms typical of pores with molecular dimensions, whereas at increased calcination temperatures, type IV isotherms, indicative of larger pores, were obtained, the hysteresis loop moving towards higher pressures with increasing temperature, indicating increasing average pore diameter.

Later studies generally confirmed these initial findings, but emphasized the changes which can occur in pore structure after different pretreatments.[8,14-22] The most extensive study of the porous structure of tin(IV) oxide has been carried out by Sharygin[15-18] on samples prepared by precipitation from HCl solutions of tin(IV) chloride using ammonium hydroxide. Three types of sample were examined: (i) the initially prepared oxide; (ii) after hydrothermal treatment at 373 K for 6 h followed by drying at 373–383 K; and (iii) after calcination at 473 K for 6 h. All three samples have a similar microporous structure in which only micropores, albeit with different micropore diameters, are present. Effective pore neck radii were calculated as 11.5 Å for the initially-prepared oxide rising marginally to 11.6 Å for the other two samples after correction for the thickness of the adsorbed layer. As observed in other investigations,[8,9] the surface area falls dramatically with increasing calcination temperature, with no change in the overall pore volume. Up to 473 K the oxide is entirely microporous (micropore volume = total pore volume), but at higher temperatures the micropore volume decreases relative to the total pore volume as intermediate pores are formed, and at temperatures above 673 K the micropore volume is a negligible fraction of the total pore volume. Above about 933 K, the oxide is non-porous or macroporous.[5] Concomitantly, both the effective pore radius and the mean pore neck sizes increase with increasing calcination temperature (Table 12.1). The widening of pores accompanies the process of crystallization which occurs at *c.* 623–673 K. The pore structure on samples heated above this temperature appears to be of the parallel plate type, whereas at lower temperatures both parallel plate and cylindrical pores are present.[21,22]

Ageing, either under water or as dry oxide, has little effect on the pore structure. Small increases in specific surface area and pore volume occur in both cases.

Table 12.1 Pore sizes, pore volumes and surface areas of tin(IV) oxide samples after various calcination periods.[a]

Calcination temperature (K)	S_{N_2} $m^2 g^{-1}$	S_i $m^2 g^{-1}$	V_s $cm^3 g^{-1}$	W_0 $cm^3 g^{-1}$	R_{eff} Å	R_K Å
373	178		0.096	0.098	9.4	7.6
473	168		0.096	0.097	9.5	8.0
573	100	72	0.092	0.050	15	12.9
673	53	46	0.101	0.014	36	26.7
773	21	24	0.102	–	76	68
873	13	14	0.099	–	119	107
973	7.7	7.5	0.106	–	277	193
1073	7.3	6.9	0.104	–	277	199
1173	4.4	4.2	0.105	–	–	334

[a] S_{N_2}, specific surface area from N_2 adsorption (B.E.T.); S_i, intermediate pore surface area; V_s, specific pore volume; W_0, micropore volume; R_{eff}, effective pore radius (calculated from the Kelvin–Thompson equation and uncorrected for the adsorbate layer); R_K, mean pore neck size (calculated from the globular model).

Table 12.2 Surface area and pore size data for coprecipitated $SnO_2 \cdot SiO_2$ gels.

Sample[a]	Temperature K	Surface area $m^2 g^{-1}$	Cumulative pore volume $cm^3 g^{-1}$	Pore radius (Å) at given pore filling 10%	50%	90%
Sn100	483	176	0.103	8	11	19
Sn100	723	53	0.084	21	38	56
Sn80-Si20	483	250	0.139	8	10	18
Sn80-Si20	723	189	0.117	9	12	21
Sn63-Si37	483	333	0.195	8	10	22
Sn63-Si37	723	256	0.159	8	12	24
Sn40-Si60	483	423	0.280	8	13	28
Sn40-Si60	723	357	0.224	7	13	28
Sn20-Si80	483	420	0.283	8	12	57
Sn20-Si80	723	384	0.256	8	13	53
Si100	483	258	–	>20	–	–
Si100	723	251	–	>20	–	–

[a] Figures refer to atom%.

Hydrothermal treatment has a greater effect, and, compared to purely thermally treated samples, hydrothermal treatment produces materials with higher surface areas. For example, a sample treated thermally at 293 K followed by hydrothermal treatment at 673 K had a specific surface area of $86 m^2 g^{-1}$, whereas a sample which had only been thermally treated at 673 K had a surface area of $61 m^2 g^{-1}$.[19,20] However, as with calcination, hydrothermal treatment at increasing temperatures also results in decreased surface area. Calcination of the hydrated oxide *in vacuo* rather than in air also has a slight effect.[21,22]

Information on mixed tin (IV) oxide materials is sparse. Although some surface area data have been reported for the systems SnO_2–PdO,[23] the SnO_2–V_2O_5, SnO_2–P_2O_5, and SnO_2–MoO_3,[24] and the SnO_2–MO (M = Co, Ni, Cu),[25] only the latter and the SnO_2–SiO_2 system have been studied in any depth.[26] The SnO_2–MO gels exhibit markedly increased resistance to thermal sintering for temperatures upto at least 873 K. Nitrogen adsorption–desorption isotherms show that a microporous structure predominates for mixed SnO_2–SiO_2 oxides prepared by coprecipitation of the two chlorides. Increasing amounts of silica produce an increasing contribution from mesopores. Specific surface areas of the mixed oxides are considerably higher than for tin (IV) oxide, and, unlike tin (IV) oxide, which undergoes a considerable reduction in surface area on calcination, the decrease in surface area of the mixed oxides is much smaller and becomes less with increasing silica content. In addition, negligible change in pore size occurs (Table 12.2). XRD data show that only the rutile phase of tin (IV) oxide is present in the mixed oxides, even when calcined at temperatures up to 1273 K, the peaks becoming broader with increasing silica content and decreasing calcination temperature. Tin (IV) oxide particle sizes estimated from the XRD data and primary particle data for the mixed oxides calcined at 723 K are shown in Table 12.3.

The changes in surface area and porous structure of hydrated tin (IV) oxide are due to a crystallization process in which primary crystallites crosslink via surface hydroxyl group condensation to produce porous secondary particles. In contrast, it would appear that the SnO_2–SiO_2 mixed oxides do not comprise random arrays of

Table 12.3 Primary particle and tin(IV) oxide crystallite sizes in coprecipitated $SnO_2 \cdot SiO_2$ gels.

Sample[a]	Temperature K	Primary particle size Å	SnO_2 crystallite size Å
Sn100	723	159	142
Sn100	483	49	33
Sn80-Si20	723	53	31
Sn63-Si37	723	45	28
Sn40-Si60	723	39	26
Sn20-Si80	723	47	21

[a] Figures refer to atom%.

interlinked $\{SnO_6\}$ and $\{SiO_4\}$ units, but rather are composed of separate primary particles of the individual components. Once formed, these primary particles can undergo heterolytic condensation with the formation of interparticle bridging Sn–O–Si bonds. Structural incompatibility between SnO_2 and SiO_2 prevents their heterolytic sintering at 723 K, and thus accounts for the observed resistance of the mixed oxides to thermal sintering.

Knowledge of the chemical nature of the surface of tin(IV) oxide and mixed oxide materials is crucial to the understanding of their behaviour in applications as heterogeneous catalysts and in gas sensing devices. These materials are invariably prepared as gels by cohydrolysis of the corresponding chlorides in aqueous solution, and hence contain a high initial water content. 'Activation' is usually by thermal pretreatment *in vacuo* and/or in the presence of oxygen, at a few Torr pressure, but the behaviour of the oxide surface is strongly dependent upon the conditions of the thermal pretreatment.

The results of various studies of the dehydration of tin(IV) oxide are summarized in Tables 12.4 and 12.5. Desorption studies show the presence of surface hydroxyl groups,]–OH, together with physisorbed and hydrogen-bonded molecular water, the relative amounts depending largely on the method of preparation of the oxide sample. Freshly prepared tin(IV) oxide gel contains a variable amount of loosely held molecular water amounting to about a two monolayer surface coverage, and is readily removed at low temperatures. A second, more strongly held type of molecular water, amounting to somewhat less than a monolayer coverage, desorbs by 423 K. Further desorption of water is due to the condensation of surface hydroxyl groups which form surface oxide and bare surface tin sites. The condensation is a gradual process and increases with temperature as the surface area decreases (Figure 12.1). Complete removal of surface hydroxyl content is quite difficult, and some residual hydroxyl content is still present even after calcination at 1000 K.

Although X-ray studies show that the crystallite size of tin(IV) oxide gel is small (~ 3 nm), the surfaces of the oxide crystallites are expected to resemble the exposed crystal faces of macroscopic crystals of the rutile modification, for which the principal ones are the [110], [101] and [100] planes.[27] Accurate schematic representations of the [110] and [100] surface planes in various stages of dehydration are illustrated in Figures 12.2 and 12.3, respectively. Assuming that all exposed tin valencies are occupied by hydroxyl groups, the totally hydroxylated [110] surface plane will have

Table 12.4 Studies involving the interaction of the tin (IV) oxide surface and water.

Sample type	Technique(s)	Aspects studied	Reference(s)
Polycrystalline film	Secondary ion mass spectroscopy; depth profiling	Nature of atmospheric hydration layer; depth resolved by sputtering	*a*
Polycrystalline film	Electron energy loss spectroscopy; X-ray photoelectron spectroscopy	Effect of atmospheric hydration; thermal decomposition of hydroxyl layer	*b*
Calcined α- and β-stannic acids (5–$10\,m^2\,g^{-1}$)	Temperature programmed desorption	Sites for H_2O bonding on various crystal faces	*c*
Calcined β-stannic acid	Temperature programmed desorption	Condensation temperatures for hydroxyls on assorted crystal faces	*d*
Calcined β-stannic acid	Temperature programmed desorption; electrical conductivity	O_2, H_2 and H_2O interactions; effect on electrical behaviour	*e*
Calcined β-stannic acid (7–$10\,m^2\,g^{-1}$)	H_2O adsorption isotherm by volumetric measurements	Anomalous discontinuity in isotherm due to features of (100) adsorption	*f*
Calcined α- and β-stannic acids (5–$10\,m^2\,g^{-1}$)	Heat of immersion technique	Differential heat of chemisorption in region of adsorption anomaly	*g*
Calcined β-stannic acid	Kr adsorption isotherm by volumetric measurements	Surface homogeneity effects; influence of hydrolysis	*h*
Calcined polycrystalline powder (7–$9\,m^2\,g^{-1}$)	H_2O adsorption isotherm by volumetric measurements	Observation of adsorption anomaly, correlation with plateau region of isosteric heat of physisorption of water, effects of dehydroxylation	*i*

Table 12.4 (*Contd.*)

Sample type	Technique(s)	Aspects studied	Reference(s)
Polycrystalline powder	Kr adsorption isotherm by volumetric measurements	Desorption mode of surface hydroxyls	*j*
Polycrystalline powder	Infrared reflectance spectroscopy	Thermal stability of hydroxyl groups	*k*
Calcined α- and β-stannic acids ($180–1\ m^2\ g^{-1}$)	H_2O adsorption isotherm by volumetric measurements	Effect of crystallinity and porosity changes on anomalous H_2O adsorption isotherms	*l*
High surface area powders ($\sim 170\ m^2\ g^{-1}$)	H_2O adsorption isotherm	Calculation of pore parameters and discussion of adsorption mechanisms	*m*
High surface area gel ($\sim 180\ m^2\ g^{-1}$)	Infrared absorption spectroscopy thermogravimetric analysis	Thermal stability of water and hydroxyl groups. Bonding sites on various crystal faces	*n*

(*a*) D.F. Cox, G.B. Hoflund and W.B. Hocking, *Appl. Surf. Sci.*, 1986, **26**, 239; (*b*) D.F. Cox, G.B. Hoflund and H.A. Laitinen, *Appl. Surf. Sci.*, 1984, **20**, 30; (*c*) M. Egashira, M. Nakashima, S. Kawasumi and T. Seiyama, *J. Phys. Chem.*, 1981, **85**, 4125; (*d*) K. Morishige, S. Kittaka and T. Morimoto, *Bull. Chem. Soc. Jpn*, 1981, **53**, 2128; (*e*) N. Yamazoe, T. Fuchigami, M. Kishikawa and T. Seiyama, *Surf. Sci.*, 1979, **86**, 335; (*f*) T. Morimoto, Y. Yokota and S. Kittaka, *J. Phys. Chem.*, 1978, **82**, 1996; (*g*) T. Morimoto, M. Kiriki, S. Kittaka, T. Kadota and M. Nagao, *J. Phys. Chem.*, 1979, **83**, 2768; (*h*) K. Morishige, S. Kittaka and T. Morimoto, *Surf. Sci.*, 1981, **109**, 291; (*i*) S. Kittaka, S. Kanemoto and T. Morimoto, *J. Chem. Soc., Farady Trans. I*, 1978, **74**, 676; (*j*) K. Morishige, S. Kittaka and T. Morimoto, *J. Colloid Int. Sci.*, 1982, **89**, 86; (*k*) N. Takezawa, *Bull. Chem. Soc. Jpn*, 1971, **44**, 3177; (*l*) S. Kittaka, K. Norishige and T. Fujimoto, *J. Colloid Int. Sci.*, 1979, **72**, 191; (*m*) V.F. Gonchar and L.M. Sharygin, *Kinet. Catal.*, 1974, **15**, 404; (*n*) P.G. Harrison and A. Guest, *J. Chem. Soc., Farady Trans. I*, 1987, **83**, 3383.

Table 12.5 Desorption studies of H_2O-derived species on tin (IV) oxide.

Surface species	Method of investigation	Approximate desorption temp. K	Desorption energy eV	Ref.
(i) Physisorbed water	TPD[a]	330	0.46	*b,c*
(ii) Hydrogen-bonded water		420	*d*	
(iii) Various surface hydroxyl groups		520	1.04(?)	
(iv) Various surface hydroxyl groups		690–715	0.94	
(v) Various surface hydroxyl groups		760–770	1.35	
(vi) Simultaneous hydroxyl groups/surface O^{2-}		880–890	1.72	
Assorted water loss regions	pmr	270–280	*d*	*e*
		520	*d*	
		600–635	*d*	
		770	*d*	
		>870	*d*	
(i) Physisorbed water	TPD[a]	380	*d*	*f*
(ii) Surface hydroxyl groups		670	*d*	
(i) Hydroxyl groups adsorbed on [100] face	TPD[a]	513	1.47	*g*
(ii) Hydroxyl groups adsorbed on [101] face		573	1.68 or 2.29	
(i) Physisorbed water	Microbalance	290–330	*d*	*h*
(ii) Hydrogen-bonded water		350–380	*d*	
(iii) Surface hydroxyl groups		400–410	*d*	
(iv) Surface hydroxyl groups		460–480	*d*	
(v) Surface hydroxyl groups		540	*d*	
(vi) Surface hydroxyl groups		590	*d*	
(vii) Surface hydroxyl groups		630	*d*	

(*a*) Thermally programmed desorption; (*b*) M. Egashira, M. Nakashima, S. Kawasumi and T. Seiyama, *J. Phys. Chem.*, 1981, **85**, 4125; (*c*) M. Egashira, M. Nakashima and S. Kawasumi, *Nippon Kagaku Kaishi*, 1982, 556; (*d*) Not determined; (*e*) E.W. Giesekke, H.S. Gutowski, P. Kirkov and H.A. Laitinen, *Inorg. Chem.*, 1967, **6**, 1294; (*f*) N. Yamazoe, T. Fuchigami, M. Kishikawa and T. Seiyama, *Surf. Sci.*, 1979, **86**, 335; (*g*) K. Morishige, S. Kittaka and T. Morimoto, *Bull. Chem. Soc. Jpn*, 1981, **53**, 2128; (*h*) P.G. Harrison and A. Guest, *J. Chem. Soc., Faraday Trans. I*, 1987, **83**, 3383.

two types of surface hydroxyl group, a unidentate hydroxyl group whose Sn–O bond axis is orthogonal to the surface, and geminal pairs of hydroxyl groups. The [101] and [100] surface planes are similar, and these totally hydroxylated surface planes will comprise triple clusters of surface hydroxyl groups attached to each tin atom in the surface.[28] Morimoto[29] has concluded that dehydration from the surface occurs principally via condensation on the [100] plane, with secondary condensation on the [101] plane. It is interesting to note that the experimentally derived surface hydroxyl group concentration (198 nm^{-2}) for freshly prepared tin (IV) oxide gel, is identical to the calculated value for the fully-hydroxylated [100] plane. Infrared data show that the different types of surface hydroxyl groups condense at different temperatures. Above *c.* 523 K, however, only isolated surface hydroxyl groups remain.[28]

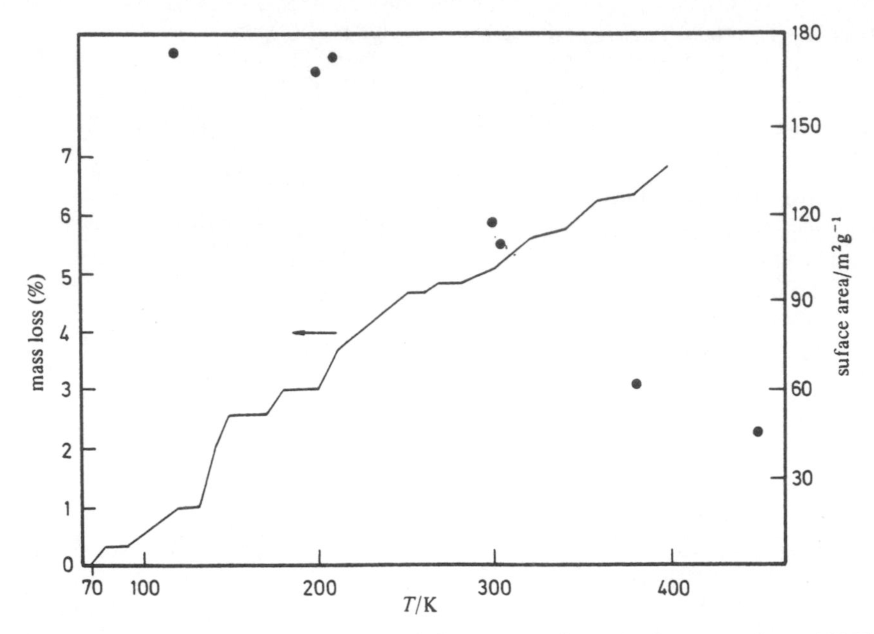

Figure 12.1 Plots of percentage mass loss (relative to mass of sample after evacuation at 333 K) and specific surface area (●) for tin (IV) oxide gel versus temperature (adapted from *J. Chem. Soc., Faraday Trans. I*, 1987, **83**, 3383).

12.2 Adsorption and chemisorption phenomena

As Figures 12.2 and 12.3 illustrate, the tin (IV) oxide surface may be considered to be a rather complex reaction medium which contains as possible sites for reaction surface hydroxyl groups,]–OH, surface oxide species which may be either terminal,]–O, or bridging,]–O–[, and Lewis acidic surface exposed tin sites. The relative ratio of these principal reaction sites is strongly dependent upon the history of the oxide sample, particularly on the thermal pretreatment. The surface hydroxyl groups and oxide species are basic in character. Although it is weakly Brønsted acidic, the surface may be made quite strongly Brønsted acidic by the inclusion in the oxide of silica.[13,30,31] The surface is quite strongly oxidizing in nature, and the oxidizing power has been attributed to the formation on the surface of highly reactive surface oxygen species which can be readily abstracted by appropriate adsorbates. The surface oxygen deficiency thus produced can be replenished subsequently by the adsorption of vapour-phase molecular oxygen (or oxide of nitrogen).

The species formed by the interaction of oxygen with the tin (IV) oxide surface has been studied intensively (Table 12.6). Several types of surface oxygen species may be present on the oxide surface depending on the history of the sample (Table 12.7). The highly reactive single-electron species,]–O–O$^{\bullet}$ and]–O$^{\bullet}$, exist only at relatively low temperatures, with]–O–O$^{\bullet}$ undergoing transformation into]–OS$^{\bullet}$ at *c.* 420 K, although some region of coexistence is highly probable. Only surface oxide (O^{2-}) is present at high temperatures.

The surface species formed by the adsorption of various molecules on tin (IV) oxide

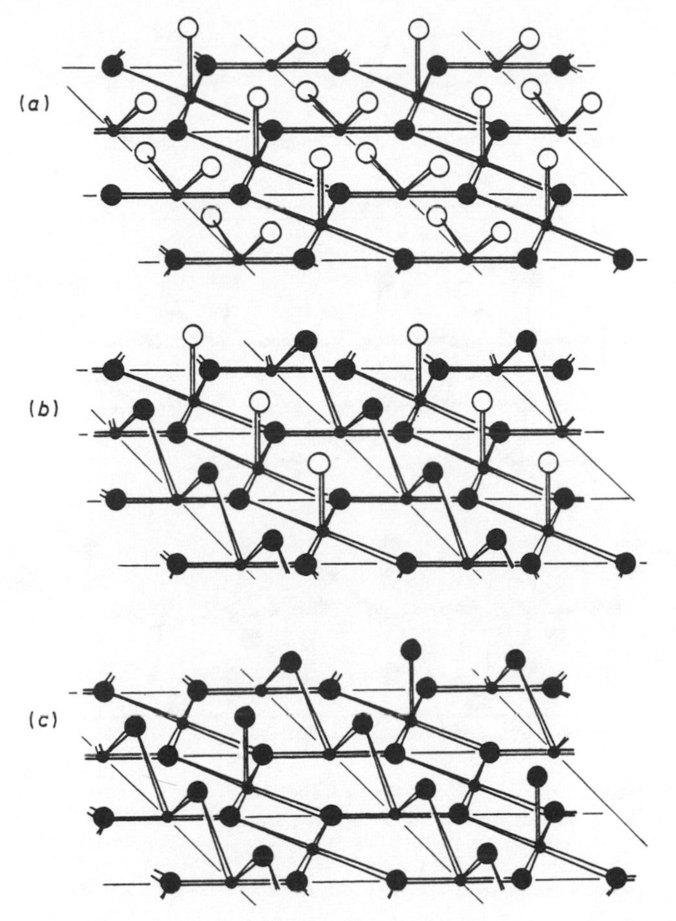

Figure 12.2 Schematic representation of the idealized [110] surface plane of the rutile structure of tin (IV) oxide. (*a*) Totally hydroxylated surface; (*b*) partially dehydrated surface; (*c*) totally dehydrated surface. Only oxygen atoms above the plane defined by the tin atoms are shown. Surface hydroxyl groups are shown as open circles, surface oxide atoms by shaded circles (reproduced with permission from *J. Chem. Soc., Faraday Trans. I*, 1987, **83**, 3383).

and related materials are shown in Table 12.8. Initial reaction at the surface can occur at surface oxide, surface hydroxyl groups, or exposed metal sites, depending upon the reactant molecule and the composition of the oxide material. In general, the tin (IV) oxide surface is quite strongly oxidizing towards oxidizable molecules. Carbon monoxide adsorbs very rapidly at surface oxide sites (reaction products have been discerned after only a few seconds' exposure of tin (IV) oxide) to give surface unidentate and bidentate carbonato- as well as surface carboxylato-species, which all appear to be formed simultaneously[13,34–36] (see also section 12.4). Carbon dioxide gives the same mixture of surface carbonato- and carboxylato-species, albeit in a somewhat different ratio. Surface bidentate carbonate and carboxylate are unstable,

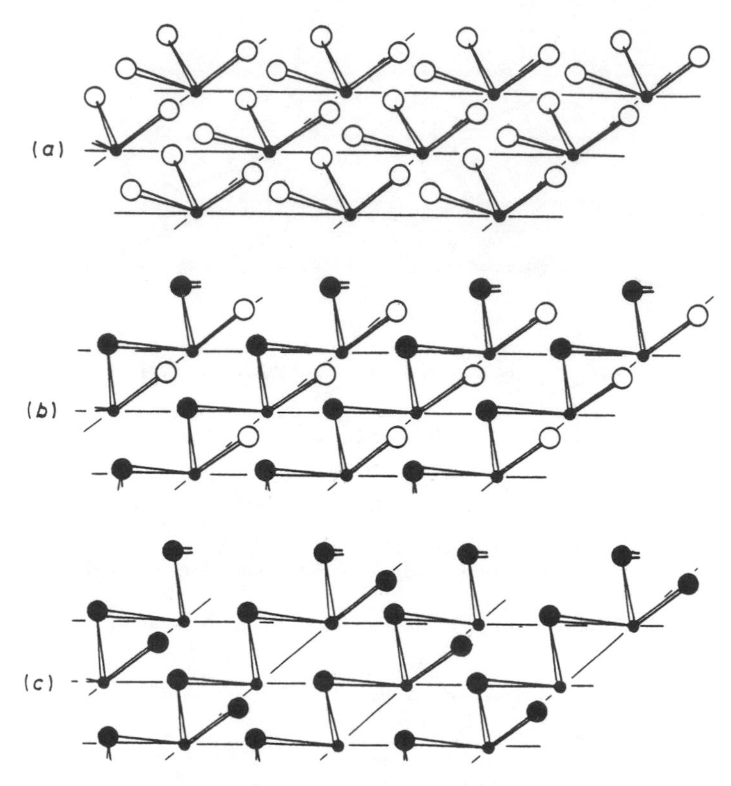

Figure 12.3 Schematic representation of the idealized [100] surface plane of the rutile structure of tin (IV) oxide. (*a*) Totally hydroxylated surface; (*b*) partially dehydrated surface; (*c*) totally dehydrated surface. Only oxygen atoms above the plane defined by the tin atoms are shown. Surface hydroxyl groups are shown as open circles, surface oxide atoms by shaded circles (reproduced with permission from *J. Chem. Soc., Faraday Trans. I*, 1987, **83**, 3383).

and both transform slowly on the surface to surface unidentate carbonate.[34] Adsorption at the transition metal to afford surface carbonyl,]–M–CO, species is observed, in addition to surface carbonate formation for transition metal-exchanged tin (IV) oxides.[37,38] Nitric oxide adsorbs reversibly on oxidized tin (IV) oxide, forming surface NO radical and $Sn(NO)_2$ species. A chelating NO_2 species formed from a NO molecule, and a surface oxide forms on transiently reduced oxide. Decomposition of NO into N_2 and N_2O occurs on deeply reduced tin (IV) oxide, restoring the surface oxygen deficiency even at room temperature.[39] Adsorption at the transition metal occurs on transition metal-exchanged tin (IV) oxides, giving fairly stable M–NO species, and the same oxides catalyse the CO/NO reaction.[37] A surface isocyanate]–NCO species is formed on $SnO_2{\cdot}O{\cdot}55$ CuO from CO/NO mixtures.[40] Highly reactive multiply-bonded molecules such as organic nitriles, R–C≡N,[41] and isocyanates, R–N=C=O,[42] react by the addition of the O–H bond of the surface hydroxyl groups to either the C≡N triple bond or C=N double bond to afford surface acetimidato,]–NH(CO)R, and surface carbamate (urethane),]–O(CO)NRH, species,

Table 12.6 Studies of surface oxygen species on tin (IV) oxide.

Sample type	Technique(s)	Aspects studied	Reference(s)
[110] single crystal + Zn-doped	Low energy electron diffraction Auger electron spectroscopy Work function measurements	O_2 adsorption	*a, b*
Rf sputtered thin film	Electrical conductivity	Oxygen vacancy diffusion	*c*
Rf sputtered thin film	Electron paramagnetic resonance	O_2 chemisorption	*d*
Polycrystalline powder	Electron spin resonance	Thermal stability of O_2^- stabilization centres	*e*
Various powders, assorted degrees of sintering (150–10 $m^2 g^{-1}$)	Temperature programmed desorption	Desorption of labile oxygen	*f*
Calcined hydrated oxide (16 $m^2 g^{-1}$)	Temperature programmed desorption Electron spin resonance	Thermal stability of O_2^-	*g*
Fine powder	Electron spin resonance Electrical conductivity	O_2, H_2 adsorption: effect on electron spin resonance signal and electrical conductivity	*h, i*
High surface area gel (200 $m^2 g^{-1}$)	Electron paramagnetic resonance	Effect of O_2 adsorption	*j*
Commercial Taguchi gas sensor (sintered material)	Electrical conductivity	Effect of O_2 adsorption	*k*
Zn-doped polycrystalline powder, sintered	Electrical conductivity	Electrical/photoconductivity as function of pretreatment temperature, O_2 adsorption	*l*
Calcined Ag^+-doped gel	Temperature programmed desorption	O_2 desorption Effect of H_2O coadsorption	*m*
Calcined oxide (20, 47–50 $m^2 g^{-1}$)	Infrared spectroscopy	O_2 adsorption	*n*

(*a*) E. de Fresart, J. Darville and J.M. Gillies, *Le Vide les Couches Mines*, 1980, **201**, 431; (*b*) E. de Fresart, J. Darville and J.M. Gillies, *Eur. Conf. Sensors and Applications*, UMIST, Manchester, Sept. 1983, paper 04.1; (*c*) G.N. Advani *et al.*, *Int. J. Electronics*, 1980, **48**, 403; (*d*) S.C. Chang, *J. Vac. Sci. Technol.*, 1980, **17**, 336; (*e*) A.M. Volodin and A.E. Cherkashin, *React. Kin. Catal. Lett.*, 1981, **17**, 329; (*f*) J.P. Joly, L. Gonzalez-Cruz and Y. Arnaud, *Bull. Soc. Chim. France*, 1986, 11; (*g*) M. Iwamoto *et al.*, *J. Phys. Chem.*, 1978, **24**, 2564; (*h*) Y. Mizokawa and S. Nakamura, *Shinku*, 1972, **15**, 292; (*i*) Y. Mizokawa and S. Nakamura, *Jap. J. Appl. Phys.*, 1975, **14**, 779; (*j*) P. Meriaudeau, C. Naccache and A.J. Tench, *J. Catal.*, 1971, **21**, 208; (*k*) L. Nanis and G. Advani, *Int. J. Electronics*, 1982, **52**, 345; (*l*) H.E. Matthews and E.E. Kohnke, *J. Phys. Chem. Solids*, 1981, 1047; (*m*) M. Egashira, M. Nakashima and S. Kawasumi, *J. Chem. Soc., Chem. Commun.*, 1981, 1047; (*n*) T.A. Gundrizer and A.A. Davydov, *React. Kin. Catal. Lett.*, 1975, **3**, 63.

Table 12.7 Desorption studies of O_2-derived species on tin(IV) oxide.

Surface species	Method of investigation	Approximate desorption temp. K	Desorption energy eV	Ref.
$O_2^-\cdot$ or $O^-\cdot$	TPD[a]	570	*b*	*c*
(i) O_2	TPD[a]	350	*b*	*d*
(ii) $O_2^-\cdot$		420		
(iii) $O^-\cdot$ or O^{2-}		790		
(iv) lattice oxygen		870		
(i) $O_2^-\cdot$	TPD[a]	320	0.28	*e*
(ii) $O^-\cdot$ or O^{2-}		670	0.48 (but possibly higher)	
(i) O_2	TPD[a]	360–435	1.12	*f*
(ii) $O_2^-\cdot$			1.29	
(iii) $O^-\cdot$		670–760	1.72	
(iv) O^{2-}		820–940	2.45	
(v) lattice O^{2-}		>1070	3.23	
$O_2^-\cdot$	epr	>430	*b*	*g*
(i) O_2^-	esr	<420	0.15–0.28[h]	*i*
(ii) O^- } (iii) O^- }		>420	*b*	

(*a*) Thermally programmed desorption; (*b*) Not determined; (*c*) M. Egashira, M. Nakashima and S. Kawasumi, *J. Chem. Soc., Chem. Commun.*, 1981, 1047; (*d*) N. Yamazoe, T. Fuchigami, M. Kishikawa and T. Seiyama, *Surf. Sci.*, **86**, 1979, 335; (*e*) B. Gillot, C. Fey and D. Delafosse, *J. Chem. Phys., Phys. Chim. Biol.*, **73**, 1976, 19; (*f*) J.P. Joly, L. Gonzalez-Cruz and Y. Arnaud, *Bull. Soc. Chim. France*, 1986, 11; (*g*) A.M. Volodin and A.E. Cherkashin, *React. Kin. Catal. Lett.*, **17**, 1981, 329; (*h*) Dependent on pressure; (*i*) S.C. Chang, *J. Vac. Sci. Technol.*, **17**, 1980, 366.

respectively. Functionally-substituted organic molecules (alcohols, aldehydes, ketones, carboxylic acids) ultimately afford surface carboxylato species,]–O_2CR, i.e. facile oxidation of the organic molecule occurs on chemisorption.[41,43,44] Surface carboxylates are also formed by chemisorption of both saturated and unsaturated hydrocarbons. Chemisorption of ethane only occurs on palladium- or silica-modified tin(IV) oxide, but not tin(IV) oxide itself. Ethene and propene are more reactive, and will also chemisorb on to all three types of oxide. The initial chemisorption step with ethane is via C–H bond fission, whereas with ethene surface hydroxyl O–H bond addition to the C=C double bond occurs; in both cases surface alkoxide,]–OR, is formed. Both pathways are observed for the adsorption of propene on to Pd–SnO_2, where surface acetate and surface acrylate are formed simultaneously.[25,46] Similar observations have been made from the adsorption of acetaldehyde, acetone and propene on a tin–molybdenum oxide.[47,48]

12.3 Heterogeneous catalysis over tin(IV) oxide-based materials

Tin(IV) oxide and materials based on tin(IV) oxide are able to function as catalysts in quite a wide range of different reactions including the oxidation of carbon monoxide, sulphur dioxide, saturated and unsaturated hydrocarbons as well as other organic

Table 12.8 Formation of surface species on tin (IV) oxide and related materials.

Oxide	Adsorbing gas	Nature of adsorbate	Ref.
SnO_2	CO, CO/O_2, or CO_2	Unidentate carbonate Bidentate carbonate Carboxylate	*a, b, c, d*
SnO_2	HCl	$[H_3O]^+$ ions	*e*
SnO_2	NH_3	NH_3 coordinated at surface Lewis acidic sites	*e*
SnO_2	(i) HCl, (ii) NH_3	$[NH_4]^+$ ions	*e*
SnO_2	Pyridine	Pyridine coordinated at surface Lewis acidic sites	*e*
SnO_2	HCO_2H	Formate	*f*
SnO_2	$(CH_3O)_2CO$	Methoxide (only at low temperature), formate	*f*
SnO_2	CH_3CHO, $(CH_3)_2CO$, CH_3CO_2H	Acetate	*f*
SnO_2	$(CCl_3)_2CO$	Trichloroacetate	*g*
SnO_2	RMeCO ($R = C_2H_5$, C_3H_7, $(CH_3)_2CH$, $(CH_3)_3C$, Ph)	Coordinated ketone, carboxylate	*h*
SnO_2	Me_3SiCl	Surface $[OSiMe_3]$ groups	*i*
SnO_2	RN=C=O (R = Et, Ph)	Carbamate	*j*
SnO_2	RC≡N ($R = CH_3$, CCl_3, $C(CH_3)_3$)	Acetimidate	*g*
SnO_2	CH_2:CHR (R = H, CH_3)	Acetate	*k, l*
SnO_2–M ($M = Cr^{III}$, Mn^{II}, Fe^{III}, Co^{II}, Ni^{II}, Cu^{II})	CO	Various types of carbonate at both tin and transition metal sites, transition metal carbonyl species	*m, n*
SnO_2–M ($M = Cr^{III}$, Fe^{III}, Co^{II}, Ni^{II}, Cu^{II})	NO	NO coordinated at transition metal sites	*m*
SnO_2–SiO_2	C_2H_6 CH_2:CHR (R = H, CH_3)	Acetate	*k, l*
SnO_2–PdO	C_2H_6, CH_2:CH_2	Acetate	*k*
SnO_2–PdO	CH_2:$CHCH_3$	Acetate, acrylate	*k, l*

(*a*) E. W. Thornton and P. G. Harrison, *J. Chem. Soc., Farad. Trans. I*, 1975, **71**, 461; (*b*) P.G. Harrison and E.W. Thornton, *J. Chem. Soc., Farad. Trans. I*, 1978, **74**, 2597; (*c*) P.G. Harrison and A. Guest, *J. Chem. Soc., Faraday Trans. I*, 1989 (in press); (*d*) P.G. Harrison and M.J. Willett, *Nature*, 1988, **332**, 337; (*e*) P.G. Harrison and E.W. Thornton, *J. Chem. Soc., Faraday Trans. I*, 1975, **71**, 1013; (*f*) P.G. Harrison and E.W. Thornton, *J. Chem. Soc., Faraday Trans. I*, 1975, **71**, 2468; (*g*) P.G. Harrison and E.W. Thornton, *J. Chem. Soc., Faraday Trans. I*, 1976, **72**, 2484; (*h*) P.G. Harrison and B. Maunders, *J. Chem. Soc., Faraday Trans. I*, 1984, **80**, 1329; (*i*) P.G. Harrison and E.W. Thornton, *J. Chem. Soc., Faraday Trans. I*, 1976, **72**, 1310; (*j*) P.G. Harrison and E.W. Thornton, *J. Chem. Soc., Faraday Trans. I*, 1976, **72**, 1317; (*k*) P.G. Harrison and B. Maunders, *J. Chem. Soc., Faraday Trans. I*, 1985, **81**, 1311; (*l*) P.G. Harrison and B. Maunders, *J. Chem. Soc., Faraday Trans. I*, 1985, **81**, 1329; (*m*) P.G. Harrison and E.W. Thornton, *J. Chem. Soc., Faraday Trans. I*, 1978, **74**, 2703; (*n*) P.G. Harrison and E.W. Thornton, *J. Chem. Soc., Faraday Trans. I*, 1979, **75**, 1487.

compounds, ammoxidation and isomerization of alkenes, and dehydration of alcohols. Examples illustrating the scope of both the range of processes and the catalyst materials are listed in Table 12.9. Tin(IV) oxide itself is mildly oxidizing, but its catalytic properties can be significantly modified by the incorporation of other elements, by cohydrolysis, ion exchange, or impregnation. The resulting materials usually need a period of calcination in order to activate the catalysts.

Unlike other common metal oxide catalyst supports such as alumina, titania and silica, tin(IV) oxide is by no means an inert support material, and the catalytic activity is derived from the subtle relationships between the chemical nature of the catalyst particle surface and the physical, chemical and electronic properties of the bulk material. Although a relatively large bank of information is available, the fundamental nature of most of the processes remains speculative. Many of the oxidation processes

Table 12.9 Examples of processes catalysed by tin(IV) oxide-based materials.

Process	Catalyst material	Ref.
Total oxidation:		
$CO + \frac{1}{2}O_2 \rightarrow CO_2$	SnO_2	*a*
	SnO_2–CuO	*b*
	SnO_2–Pd	*c, d*
	SnO_2–CrO_3	*e*
$CO + N_2O \rightarrow CO_2 + N_2$	SnO_2	*f*
$CO + NO \rightarrow CO_2 + \frac{1}{2}N_2$	SnO_2–CuO	*g*
	SnO_2–Pd	*h*
	SnO_2–Cr_2O_3	*i*
$SO_2 + \frac{1}{2}O_2 \rightarrow SO_3$	SnO_2–Cr_2O_3	*j*
$CH_4 + O_2 \rightarrow CO_2 + H_2O$	Sn–Sb–O	*k*
Partial oxidation and ammoxidation:		
$CH_4 + N_2O \rightarrow H_2CO$	SnO_2–Bi_2O_3	*l*
p-xylene → *p*-tolualdehyde, maleic anhydride *p*-toluic acid	SnO_2–V_2O_5	*m*
o-xylene → phthalic anhydride	SnO_2–V_2O_5	*n*
$CH_3{\cdot}CH{=}CH_2 \rightarrow CH_2{=}CH{\cdot}CHO$	Sn–Sb–O	*o, p*
	$SnO_2{\cdot}MoO_3$	*q*
	Sn–V–O	*r*
$CH_3{\cdot}CH{=}CH_2 \rightarrow CH_3{\cdot}CO{\cdot}CH_3$	$SnO_2{\cdot}MoO_3$	*s, t, u*
n-Butenes → $CH_3{\cdot}CO{\cdot}CH_2CH_3$	$SnO_2{\cdot}MoO_3$	*u*
$CH_2{=}C(CH_3)_2 \rightarrow CH_2{=}C(CH_3){\cdot}CHO$	Sn–Sb–O	*v*
1-Pentene → $CH_3{\cdot}CO{\cdot}CH_2CH_2CH_3$	$SnO_2{\cdot}MoO_3$	*u*
$CH_2{=}C(CH_3)_2 \rightarrow CH_2{=}C(CH_3){\cdot}CN$	Sn–Sb–Fe–O	*w*
$CH_3OH \rightarrow H_2CO$	SnO_2–MoO_3	*x, y*
	Sn–V–O	*z*
	SnO_2–Pd (electrooxidation)	*aa*
	SnO_2–Pt (electrooxidation)	*bb, cc*
Oxidative coupling and oxidative dehydrogenation:		
$CH_4 \rightarrow C_2H_4, C_2H_6$	SnO_2	*dd*
$CH_3{\cdot}CH{=}CH_2 \rightarrow$ 1,5-hexadiene	Sn–Bi–O	*ee, ff*
$C_2H_6 \rightarrow C_2H_4$	SnO_2–P_2O_5	*gg*
$C_6H_5C_2H_5 \rightarrow C_6H_5CH{=}CH_2$	SnO_2–P_2O_5	*hh, ii, jj, kk*
$RC_6H_4C_2H_5 \rightarrow RC_6H_4CH{=}CH_2$ (R = H, 2-CH_3, 4-$(CH_3)_3C$, 4-Cl)	SnO_2–P_2O_5–Al_2O_3	*ll*
n-butenes → butadiene	SnO_2–P_2O_5	*mm, nn, oo, pp*

Table 12.9 (*Contd.*)

Process	Catalyst material	Ref.
Hydrocarbon isomerization:		
n-Butenes	SnO_2	*qq*
	Sn–Sb–O	*rr, ss*
1-Butene	SnO_2–TiO_2	*tt*
Cyclopropane	Sn–Sb–O	*rr*
	SnO_2–ZrO_2	*uu*
Dehydration:		
i-$C_3H_7OH \rightarrow CH_3CH{=}CH_2$	SnO_2–P_2O_5	*vv*
	SnO_2–MoO_3	*vv*
	SnO_2–V_2O_5	*ww*
s-$C_4H_9OH \rightarrow$ butenes	SnO_2–MoO_3	*xx*

(*a*) M.J. Fuller and M.E. Warwick, *J. Catalysis*, 1973, **29**, 441; (*b*) M.J. Fuller and M.E. Warwick, *J. Catalysis*, 1974, **34**, 445; (*c*) G.C. Bond, L.R. Molloy and M.J. Fuller, *J. Chem. Soc., Chem. Commun.*, 1975, 796; (*d*) G. Croft and M.J. Fuller, *Nature*, 1977, **269**, 585; (*e*) P.G. Harrison and P. Harris, Eur. Pat. Appl., 256 822; (*f*) M.J. Fuller and M.E. Warwick, *J. Catalysis*, 1975, **39**, 412; (*g*) M.J. Fuller and M.E. Warwick, *J. Catalysis*, 1976, **42**, 418; (*h*) M.J. Fuller and M.E. Warwick, *Chem. Ind. (London)*, 1976, 787; (*i*) F. Solymosi and J. Kiss; *J. Catalysis*, 1978, **54**, 42; (*j*) G. Rienäcker and J. Schenke, *Z. anorg. allg. Chem.*, 1964, **328**, 201; (*k*) I. Brown and W.R. Patterson, *J. Chem. Soc., Faraday Trans. I*, 1983, **79**, 1431; (*l*) F. Solymosi, I. Tombácz and G. Kutsán, *J. Chem. Soc., Chem. Commun.*, 1985, 1455; (*m*) B.C. Mathur and D.S. Viswanath, *J. Catalysis*, 1974, **32**, 1; (*n*) K. Hauffe and H. Raveling, *Ber. Bunsenges. Phys. Chem.*, 1980, **84**, 912; (*o*) J.-C. Volta, G. Coudurier, I. Mutin and J.C. Védrine, *J. Chem. Soc., Chem. Commun.*, 1982, 1044; (*p*) T. Ono, T. Yamanaka, Y. Kubokawa and M. Komiyama, *J. Catalysis*, 1988, **109**, 423; (*q*) T. Ono, Y. Ikehata and Y. Kubokawa, *Bull. Chem. Soc. Jpn.*, 1983, **56**, 1284; (*r*) T. Ono, Y. Nakagawa and Y. Kubokawa, *Bull. Chem. Soc. Jpn.*, 1981, **54**, 343; (*s*) Y. Takita, Y. Moro-oka and A. Ozaki, *J. Catalysis*, 1978, **52**, 95; (*t*) J. Buiten, *J. Catalysis*, 1968, **10**, 188; (*u*) S. Tan, Y. Moro-oka and A. Ozaki, *J. Catalysis*, 1970, **17**, 132; (*v*) A. Perrard and J.-E. Germain, *Bull. Soc. Chim. France*, 1978, II-55; (*w*) A. Perrard and J.-E. Germain, *Bull. Soc. Chim. France*, 1978, II-59; (*x*) M. Niwa, M. Mizutani, M. Takahashi and Y. Murakami, *J. Catalysis*, 1981, **70**, 14; (*y*) F.J. Berry and C. Hallett, *Inorg. Chim. Acta*, 1985, **98**, 135; (*z*) P.J. Pomonis and J.C. Vickerman, *Disc. Faraday Soc.*, 1982, **72**, 247; (*aa*) A. Katayama-Aramata and I. Toyoshima, *J. Electroanal. Chem.*, 1982, **135**, 111; (*bb*) D.F. Cox, G.B. Hoflund and H.A. Laitinen, *Langmuir*, 1985, **1**, 269; (*cc*) A. Katayama, *J. Phys. Chem.*, 1980, **84**, 376; (*dd*) C.A. Andrews, J.J. Leonard and J.A. Sofranko, U.S. Pat. 4,444,984; (*ee*) E.A. Mamedov, V.P. Vislovskii and V.S. Aliev, *Kinet. Catal.*, 1978, **19**, 629; (*ff*) E.A. Mamedov, G.W. Keulks and F.A. Ruszala, *J. Catalysis*, 1981, **70**, 241; (*gg*) P.G. Harrison and A. Argent, *J. Chem. Soc., Chem. Commun.*, 1986, 1058; Eur. Pat. Appl., 85 01401; (*hh*) Y. Murakami, K. Iwayama, H. Uchida, T. Hattori and T. Tagawa, *J. Catalysis*, 1981, **71**, 257; (*ii*) Y. Murakami, K. Iwayama and H. Uchida, *Asahi Garasu Kogyo Gijutsu, Shoreikai Kenkyu Hokoku*, 1970, **16**, 285; *Chem. Abstr.*, 1971, **74**, 116405t; (*jj*) Y. Murakami, K. Iwayama, H. Uchida, T. Hattori and T. Tagawara, *Appl. Catal.*, 1982, **2**, 67; (*kk*) K. Honna, H. Sato, N. Morimoto and N. Shimizu, *Japan. Kokai*, **73 62**, 727; (*ll*) H. Sato, N. Shimizu, K. Honna, K. Kurisaki, Ger. Offen. 2,401,718; (*mm*) E.W. Pitzer, Ger. Offen 2,124,454; (*nn*) J.W. Begley, U.S. Pat. 3,502,739; (*oo*) E.W. Pitzer, U.S. Pat. 3,775,508; (*pp*) D.W. Walker and F. Farha, U.S. Pat. 4,252,680; (*qq*) S. Kodama, M. Yabuta, M. Anpo and Y. Kubokawa, *Bull. Chem. Soc. Jpn.*, 1985, **58**, 2307; (*rr*) E.A. Irvine and D. Taylor, *J. Chem. Soc., Faraday Trans. I*, 1978, **74**, 206; (*ss*) E.A. Irvine and D. Taylor, *J. Chem. Soc., Faraday Trans. I*, 1978, **74**, 1590; (*tt*) M. Itoh, H. Hattori and K. Tanabe, *J. Catalysis*, 1976, **43**, 192; (*uu*) G.-W. Wang, H. Hattori and K. Tanabe, *Bull. Chem. Soc. Jpn.*, 1983, **56**, 2407; (*vv*) M. Ai, *J. Catalysis*, 1975, **40**, 327; (*ww*) M. Ai, *J. Catalysis*, 1975, **40**, 318; (*xx*) Y. Okamoto, T. Hashimoto, T. Imanaka and S. Teranishi, *Chem. Lett.*, 1978, 1035.

are readily understood in terms of surface redox reactions exemplified in the previous section in which the molecule undergoes adsorption at a suitable surface site, abstracts a reactive surface oxygen species, and subsequently desorbs. The surface oxygen deficiency is rapidly replenished by the chemisorption of molecular oxygen or an oxide of nitrogen. Similarly, the isomerization of alkenes is facilitated by the formation of surface allyl carbonium ion due to the increased Brønsted acidity of the mixed Sn–Sb–O oxides.[49] However, a radical-type intermediate has been indicated for essentially the same isomerization on tin(IV) oxide alone,[50] illustrating the complexity of surface reactions and mechanisms which can occur on change of catalyst material and the reaction conditions. Catalytic activity towards particular reactions has been established essentially by empirical means, although the incorporation of certain elements, notably phosphorus, antimony, bismuth, molybdenum, copper, chromium (as oxides), and palladium and platinum (as elemental metals), has been found to be effective for particular types of conversion. Of the many unknowns in these catalyses, not the least is the nature of the catalyst material itself. The initial result of the manufacturing process by the cohydrolysis method is a gel in which the other element(s) are distributed evenly (or randomly). Subsequent calcination at elevated temperatures can cause sintering, the migration of the secondary elements to grain boundaries, and the separation of other oxide phases, with concomitant effects on the physical pore structure and electronic nature of the material. Thus, in most cases, an active catalyst will not be a true chemical compound, but a rather complex template on which reaction can occur. Hence the problems of reproducibility and selectivity which are frequently encountered. It also must be borne in mind that although many catalytically-active materials are represented in the literature as mixed oxides of the two metals (e.g. $SnO_2 \cdot MO_x$), such representation is at best misleading and often erroneous as to the composition of the material. A better nomenclature would be Sn–M–O followed by the Sn:M atomic ratio.

The most studied materials are the Sn–Sb–O[51] and Sn–Mo–O[52–54] catalysts, which illustrate the kind of changes which can occur in the bulk material. Calcination of the amorphous gels produced initially in the Sn–Mo–O system involves dehydration, giving highly disordered rutile-type solids which are capable of accommodating high concentrations of molybdenum. Heating to higher temperatures results in the crystallization of macroscopically distinguishable molybdenum(IV) oxide particles. Prolonged calcination at high temperatures affords materials which are best described as solid solutions of molybdenum(V) in the rutile-type lattice of tin(IV) oxide. Similarly, the nature of a Sn–Sb–O material depends critically upon the metal atom ratio and the thermal prehistory of the sample, leading to a wide range of catalytic activity. Other surface techniques such as Auger electron spectrometry can throw light on the composition of the surface of mixed catalysts.[55] In both the Sn–Sb–O and Sn–Mo–O systems, the second metal migrates and enriches the crystallite surface. Indeed, there is substantial evidence that the active phase for the oxidation of propene to acrylaldehyde over Sn–Sb–O catalysts is an oriented film of Sb_2O_4 on the Sb–SnO_2 solid solution.[55] However, other studies have indicated that pure Sb_2O_4 or Sb_6O_{13} phases are not very active for the formation of acrylaldehyde.[56] The addition of small levels of iron to these catalysts gives higher activity but somewhat lower selectivity, and has significant effects on the physicochemical nature of the catalyst particles. At low antimony contents, the presence of iron decreases the size of the Sb–SnO_2 solid solution particles, and the extraction of antimony as Sb_2O_4 is hindered.[57]

Although catalysts based on tin (IV) oxide show excellent potential particularly as oxidation catalysts, very few have as yet reached commercial operation. Nevertheless, this situation may change as understanding of the complex relationship between the nature of the bulk catalysts and the surface reactions increases. Very encouragingly, a $SnO_2 \cdot CrO_3$ material has recently been patented as a highly effective exhaust emission control catalyst.[58]

12.4 Electrical properties and gas sensing

Practical efforts to measure the intrinsic electrical properties of tin (IV) oxide have tended to concentrate on single crystals. Theoretical calculations of the electronic band structure and the influence of defect and substitutional centres upon the relevant levels have been described in several studies.[59–62] The oxide is usually described as a wide-bandgap ($E_g \approx 3.6$ eV) *n*-type semiconductor (which would therefore be expected to act as an insulator at ambient temperature in the pure form), but relatively few details of the bulk electronic structure are understood, although a simple view of the band structure[63] consists of a broad 5*s* conduction band and a 2*p* valence band, separated by a forbidden gap. It is generally accepted that the intrinsic conductivity derives from non-stoichiometry of the form $SnO_{(2-x)}$. Less attention has been paid to the theoretical study of the corresponding surfaces, although Munnix and Schmiets[64] have interpreted the results of calculated surface electronic structures for the [100], [110] and [001] faces in terms of the coordination of the surface cations. The surface effects induced by annealing single crystals have been interpreted in terms of recrystallization at particular temperatures, which explains the correlations observed between the surface chemical composition, the work function and the surface conductance.[65,66]

The general nature of semiconductivity in polycrystalline tin (IV) oxide has been considered by Vincent,[67] who has outlined the role of two mechanisms in the conduction process: non-stoichiometric (i.e. containing lattice defects) and controlled valence (i.e. incorporating additional ions). Sample morphology also has a significant effect on the electrical properties, and in particular on the relationship between mobility, charge carrier concentration and the magnitude of intercrystallite potential barriers.

The conductance behaviour of porous pellets of high purity tin (IV) oxide has been studied in detail by both McAleer and coworkers[68] (high-purity commercial samples), and Harrison and Willett[69] (tin (IV) oxide gel from hydrolysis of tin (IV) chloride), with similar results in spite of the differing origins of the samples. In both studies a sigmoidal conductance–temperature relationship was observed under conditions of high moisture levels, although the temperatures of the conductance maxima and minima vary somewhat. For the unsintered gel samples, the conductance typically reached a maximum at 420 ± 20 K followed by a minimum at 520 ± 50 K, with the conductance rising between about 290 and 420 K and between 670 and 900 K. The majority of samples studied showed Arrhenius-type behaviour in both regions for both ambient air and dry O_2/N_2 as the test atmospheres, and mean values of the activation energy for unsintered materials were determined to be 0.41 ± 0.14 eV (<450 K) and 0.78 ± 0.11 eV (>650 K) under ambient air, and 0.14 ± 0.08 eV (<450 K) and 0.89 ± 0.15 eV (>650 K) under dry O_2/N_2. However, the conductance behaviour of the unsintered oxide samples in air as a function of temperature is usually extremely complex and exhibits considerable hysteresis depending upon the history of the sample. Experimental curves for samples sintered at 1273 K become more linear as the length of presintering is increased, although in most cases there is still a region of significantly lower activation energy at a temperature (~ 500–630 K) which correlates well with the minimum of the sigmoid curves found for unsintered material. The linear regions above and below this temperature have similar gradients, and there is a trend towards higher activation energies (~ 1.7 eV) with increased length of sample presintering. Both studies indicate that the conductance mechanism is probably of the Schottky barrier type, where the magnitude of the barrier is modulated by the nature of the chemisorbed surface species, especially in samples with little or no presintering. With increasing degrees of sintering, however, the intergrain-neck model may become more important. In unsintered material in the temperature range ambient ~420 K, surface hydroxyl groups are the major conductance-modulating species. Above 670 K in the unsintered material, and in the sintered material, the conductance is primarily modulated by surface O^{2-} anions.

The use of tin (IV) oxide in gas sensors represents a specific example of the catalytic oxidation properties of the oxide in which the concomitant bulk electronic changes are of paramount importance. Although it is now only one of many materials which have been utilized in the fabrication of gas sensors, historically it was one of the first employed[70] and has remained at the forefront of the technology for over two decades. The primary factors which are responsible for this are the relatively high sensitivity coupled with low operating temperatures, versatility, and the significant changes in behaviour which may be induced after appropriate doping.

The physical forms of the oxide chosen for gas sensing applications are determined by the essentially conflicting requirements of high catalytic activity and good stability with time. This latter criterion is a more challenging one than might be imagined, because of the need to operate sensors at elevated temperatures (typically 500–600 K). Commercia! Taguchi sensors, although of various designs, all share the common feature of a sintered tin (IV) oxide body as the active element[71] (see also Chapter 13), and some 12 million such devices have been sold worldwide,[72] emphasizing the commercial importance of the field. Other designs utilizing the same general principle have been proposed,[73] some of which have the advantage of lower power consumption. In all cases, the manufacturing process consists basically of heating tin (IV) oxide, initially dispersed in a slurry with water and/or organic dispersants. Sintering and/or catalytically-active additives may also be incorporated at this stage. However, it is invariably necessary to raise the material to temperatures well above the envisaged operating range to impart the thermal stability required during use. The other well-studied form of the oxide in this context is the sputtered or printed film.[74–76] However, although offering potential advantages such as improved reproducibility in the manufacturing process, difficulties encountered in the post-deposition annealing of such films have precluded their commercial success. The use of alternative sensor architecture, for example field-effect transistors with chemically-active materials deposited on the gate, is rapidly expanding.[77] The manufacture of these devices relies heavily on intergrated circuit manufacturing techniques, and high-temperature sintering processes are unlikely to be employed. To date, tin (IV) oxide has not been widely used in such applications, although it should be emphasized that similar chemical reactions are involved in the sensing process, regardless of the method chosen to interrogate the sensitive solid. Indeed, there are strong similarities between the

Table 12.10 Applications of tin (IV) oxide-based gas sensors.

Additives	Physical form	Gas sensitivities reported (as a minor component in air)	Ref.
(?)	(Commercial) sintered body (Taguchi Gas Sensor)	CO; H_2O	*a*
	Rf sputtered thin films	CO	*b*
ThO_2; Pd	Sintered bodies, printed films	CO; H_2; C_3H_8; C_4H_{10}	*c, d, e*
	Thin polycrystalline film	AsH_3	*f*
	Sintered polycrystalline film	CH_4; SO_2; NO; CO	*g*
	Polycrystalline thin film/ single crystal	H_2	*h*
(?)	Commercial sintered body (Taguchi Gas Sensor)	CO; CH_4; H_2O	*i*
Pd; $Si(OC_2H_5)_4$	Sintered body	H_2; CH_4	*j*
Ag	Sintered body	H_2	*k*
Ag; Pd; Pt	Sintered body	CO; H_2; C_3H_8; CH_4	*l*
	Single crystal whiskers	H_2; CH_4	*m*

(*a*) J.F. Boyle and K.A. Jones, *J. Electron. Mats.*, 1977, **6**, 717; (*b*) H. Windischmann and P. Mark, *J. Electrochem. Soc.*, 1979, **126**, 627; (*c*) M. Nitta and M. Haradome, *IEEE Trans. on Electron Devices*, 1979, **26**, 219; (*d*) M. Nitta and M. Haradome, *J. Electron. Mats.*, 1979, **8**, 571; (*e*) M. Nitta, S. Ohtani and M. Haradome, *J. Electron. Mats.*, 1980, **9**, 727; (*f*) W. Mokwa, D. Kohl and G. Heiland, *Sensors and Actuators*, 1985, **8**, 101; (*g*) P. Romppainen *et al.*, *Sensors and Actuators*, 1985, **8**, 271; (*h*) T. Yamazaki, U. Mizutani and Y. Iwama, *Jap. J. Appl. Phys.*, 1983, **22**, 454; (*i*) J. Watson, *Sensors and Actuators*, 1984, **5**; (*j*) S. Yasunaga, S. Sunahara and K. Ihokura, *Sensors and Actuators*, 1986, **9**, 133; (*k*) N. Yamazoe, Y. Kurokawa and T. Seiyama, *Chem. Letts.*, 1982, **12**; (*l*) N. Yamazoe, Y. Kurokawa and T. Seiyama, *Sensors and Actuators*, 1983, **4**, 283; (*m*) M. Egashira, Y. Yoshida and S. Kawasumi, *Sensors and Actuators*, 1986, **9**, 147.

fundamental chemistry of tin (IV) oxide sensors utilizing such diverse properties as the Seebeck[78] and photoimpedance[79] effects, as well as the more common conductance modulation types.

Table 12.10 shows the considerable range of gases, both flammable (e.g. CH_4,[80] C_3H_8,[81] H_2[81]) and toxic (e.g. CO[80], SO_2,[82] NO_x,[83] NO,[82,84] H_2S[84]) which have been investigated in this context for sensors of various physical forms. In general, such studies have been performed under excess concentrations of oxygen (as usually encountered in ambient environments), and a common feature of the mechanisms proposed is the depletion of surface electron concentration by chemisorbed oxygen. Subsequent reaction with, and removal of, these charged groups is invariably assumed to be the prime source of the electrical effects on admitting the target gas. A simplified view of this process in terms of the electron energy levels near the surface of the material[85,86] is shown in Figure 12.4.

Such a model may be used as a basis of a satisfactory empirical rationalization of sensor behaviour, without more detailed understanding of the chemistry involved. However, studies in atmospheres containing water vapour in addition to the target species are more difficult to interpret, even at the empirical level. For example, adsorbed water increases the conductance of tin (IV) oxide (in contrast to the decrease observed under oxygen alone)[80] and the $CO \rightarrow CO_2$ conversion is catalysed by the presence of atmospheric moisture, whereas enhanced reaction rates have been observed on removal of surface hydroxyl groups by the incorporation of thoria in the

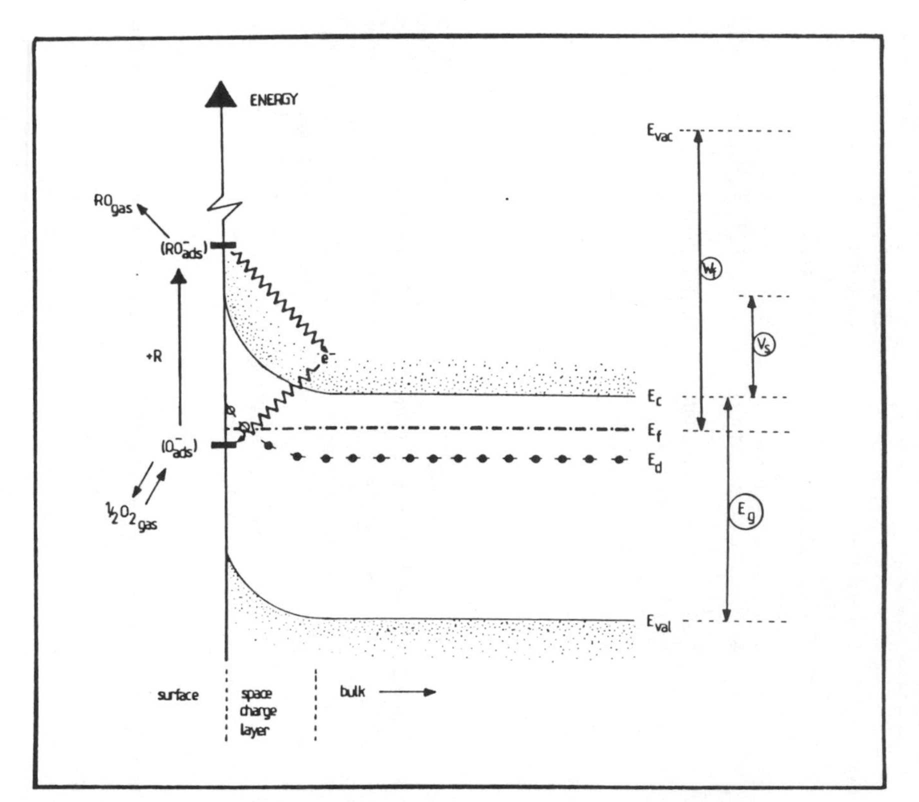

E_{vac} = Vacuum level
E_c = Conduction band
E_f = Fermi level
E_d = Donor level
E_{val} = Valence band
W_f = Work function
V_s = Schottky Barrier
E_g = Band gap
● = Occupied donor state
⊖ = Vacant donor state
▬ = Surface state
e^- = Electron
R = Reactive species (from gas phase)

Figure 12.4 Simplified representation of adsorption/desorption processes at the surface of tin(IV) oxide.

sensor.[81] Thus there is a clear requirement for further clarification of the chemical processes governing such effects, since water vapour is a prime source of interference in the signals obtained from tin (IV) oxide sensors.[87] Indeed, tin (IV) oxide has been proposed as the basis of a ceramic moisture sensor when used in conjunction with TiO_2.[88]

Only in the case of the CO–SnO_2 system has there been any attempt to correlate and understand the fundamental relationship between the electronic changes induced on exposure of tin (IV) oxide to a target gas and the nature of the adsorbed species causing such changes.[36,89] One of the earliest attempts to interpret the electrical behaviour quantitatively was by Windischmann and Mark,[89] who assumed that adsorbed CO from the atmosphere reacts with chemisorbed oxygen on the oxide surface to form chemisorbed CO_2^-, which returns an electron to the conduction band due to the exothermic nature of the reaction. The resulting neutral CO_2 is subsequently desorbed, leaving a vacant site for reoccupation by oxygen adsorbed from the atmosphere. Whilst this model offers explanations for some aspects of the observed behaviour of the sensors, such as the dependence of the sensor conductance on (CO partial pressure)$^{1/2}$ and the existence of a temperature window outside which the detection efficiency is very poor, it fails to provide information on the precise chemical nature of the intermediate surface formed between adsorption of CO and desorption of CO_2. Nor does it allow a detailed mechanism of conductance control within the gas-sensitive solid and the way in which surface chemisorption and reactions affect the measured electrical parameters. In a more comprehensive study in which infrared was employed to monitor the surface species whilst simultaneously measuring the conductance of the oxide bulk, three surface species, unidentate and bidentate carbonate and bidentate carboxylate, were observed to be formed apparently simultaneously, but their abundance on the surface is strongly affected by the temperature of the thermal pretreatment of the oxide. The abundance of the two carbonate species increases with increasing calcination temperature up to ~ 570 K, and declines rapidly thereafter. However, that of the carboxylate species increases steadily up to at least 620 K. Moreover, whereas the two carbonate species are easily desorbed, the carboxylate is held more strongly. Surprisingly, the presence of CO in the atmosphere has little effect on the absolute conductance–temperature profile, although the percentage conductance change or sensitivity (the parameter normally used to express the performance of practical sensing devices) exhibited a sharp maximum at a calcination temperature of 490 K.

The data of this study allow a much more detailed understanding of the CO sensing mechanism. The most probable adsorption sites for CO are surface $O_{ads}^{-}\cdot$ species,[32] and the processes which occur are described mechanistically by the scheme shown in Figure 12.5. The reaction of CO with adjacent pairs of $O_{ads}^{-}\cdot$ produces surface bidentate carbonate which may subsequently be transformed into surface unidentate carbonate (plus a surface oxygen vacancy). Adsorption of CO at a single $O_{ads}^{-}\cdot$ site will yield surface carboxylate. It should be noted that the formation of all three surface species occurs *without* any electron transfer to the bulk solid, and hence no conductance change would be expected if the reaction terminated at this stage. However, desorption of CO_2 as a product provides a mechanism by which electrons are returned to the conduction band of the oxide, from which an increase in conductance would be expected. The surface oxygen vacancy produced by this process is replenished by subsequent dissociative adsorption of molecular oxygen.[33]

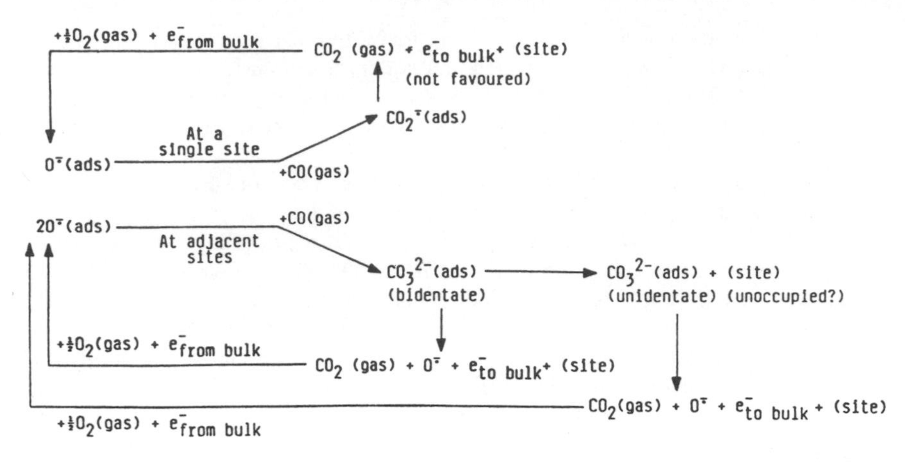

Figure 12.5 Summary of the favoured surface reactions on tin (IV) oxide at 329 K exposed to an atmosphere of dry air containing 1000 ppm CO (reproduced with permission from *Nature*, 1988, **332**, 337).

The high hydroxylation of the initial oxide surface[28] precludes a significant concentration of $O^{-}_{ads}\cdot$ adsorption sites, and so all species are sparse. As the concentration of $O^{-}_{ads}\cdot$ increases due to increasing dehydroxylation with increasing pretreatment temperature, so the concentration of all three species increases, the adsorption at adjacent $O^{-}_{ads}\cdot$ sites to give surface carbonate being favoured. At pretreatment temperatures above 570 K, however, the specific surface area of the material begins to fall very rapidly,[28] and consequently adjacent pairs of $O^{-}_{ads}\cdot$ sites become more sparse, so that surface carbonate formation is reduced and surface carboxylate formation is favoured. Thus, very little CO adsorption is expected at low pretreatment temperatures because of high surface hydroxylation; at intermediate temperatures where dehydroxylation is advanced (but prior to rapid surface area loss) and the concentration of $O^{-}_{ads}\cdot$ is high, adsorption to give surface carbonate is the predominant adsorption process; but at high pretreatment temperatures, where the abundance of $O^{-}_{ads}\cdot$ sites is low, adsorption to give surface carboxylate dominates. When the oxide is exposed to air containing CO at 329 K, the equilibrium surface coverage of carbonate species is maximized on a surface pretreated at *c.* 570 K. Even though such groups are relatively easily lost from the surface by desorption of CO_2 (thus contributing to the conductance increase observed under CO), there is no simple correlation between the carbonate coverage and the CO sensitivity as defined by $\%\Delta G$.

The use of dopants in tin (IV) oxide-based sensors, although widespread in the commercial types, has a poorly-understood effect on the properties of the devices. Dopants may act in at least two different ways: (i) either by providing 'active' sites on the oxide surface for the adsorption and spillover of reactants (e.g. Pd, Pt and other noble metals, and possibly including oxygen), and (ii) by influencing electron donors or acceptors on the host material (e.g. Ag or Sb).[90,91] Many such studies have been reported. Mizsei and Harsanyi[92] and Wada[93] have studied palladium addition in various forms, and Lalauze *et al.*[94,95] have observed significant effects on sensor properties following high-temperature SO_2 treatment. However, complex behaviour is often observed, and both increased and decreased sensor conductances on exposure to H_2 in air have been reported, depending on the precise conditions of operating temperature and gas concentration.[96] Other devices, such as tin (IV) oxide containing thoria, Pd, and hydrophobic silica, exhibit oscillatory behaviour under specific circumstances,[97,98] and, although such effects are generally attributed to rapid fluctuations in the surface coverage of oxygen species, the full implications for sensing mechanisms have not been clarified.

References

1. W.H. Baur and A.A. Khan, *Acta Crystallogr.*, 1971, **B27**, 2133.
2. Z.M. Jarzebski and J.P. Marton, *J. Electrochem. Soc.*, 1976, **123**, 199C.
3. R.W.G. Wyckoff, *Crystal Structures*, Vol. 1, Interscience, New York, 1951.
4. Z.M. Jarzebski and J.P. Marton, *J. Electrochem. Soc.*, 1976, **123**, 299C.
5. M.J. Fuller, M.E. Warwick and A. Walton, *J. Appl. Chem. Biotechnol.*, 1978, **28**, 396.
6. E. Posnjak, *J. Phys. Chem.*, 1926, **30**, 1073.
7. J.D. Donaldson and M.J. Fuller, *J. Inorg. Nucl. Chem.*, 1968, **30**, 1083.
8. M.J. Fuller and M.E. Warwick, *J. Catalysis*, 1973, **29**, 441.
9. J.F. Goodman and S.J. Gregg, *J. Chem. Soc.*, 1960, 1162.
10. C.A. Vincent and D.G.C. Weston, *J. Electrochem. Soc.*, 1972, **119**, 518.
11. M. Itch, H. Hattori and K. Tanabe, *J. Catalysis*, 1976, **43**, 192.
12. H.B. Weiser and W.O. Milligan, *J. Phys. Chem.*, 1932, **26**, 3030.
13. E.W. Thornton and P.G. Harrison, *J. Chem. Soc., Faraday Trans. I*, 1975, **71**, 461.
14. T.G. Vernardakis, Ph.D. thesis, Oklahoma State University, 1972.
15. L.M. Sharygin and V.F. Gonchar, *Kinet. Catal.*, 1974, **15**, 123.
16. V.F. Gonchar and L.M. Sharygin, *Kinet. Catal.*, 1974, **15**, 404.
17. L.M. Sharygin, V.F. Gonchar and V.M. Galkin, *Kinet. Catal.*, 1974, **15**, 1125.
18. L.M. Sharygin, V.F. Gonchar and A.P. Shtin, *Kinet. Catal.*, 1975, **16**, 178.
19. V.M. Chertov, N.T. Okopnaya and V.I. Zelentsov, *Colloid J.*, 1975, **37**, 376.
20. V.M. Chertov and N.T. Okopnaya, *Kinet. Catal.*, 1978, **19**, 1299.
21. S.A. Selim and F.I. Zeidan, *J. Appl. Chem. Biotechnol.*, 1976, **26**, 23.
22. R.Sh. Mikhail, S.A. Selim and F.I. Zeidan, *J. Appl. Chem. Biotechnol.*, 1976, **26**, 191.
23. G. Croft and M.J. Fuller, *Nature*, 1977, **269**, 585.
24. M. Ai, *J. Catalysis*, 1975, **40**, 318, 327.
25. M.J. Fuller, *J. Appl. Chem. Biotechnol.*, 1978, **28**, 539.
26. M.J. Fuller and G. Croft, private communication quoted in B.M. Maunders, Ph.D. thesis, University of Nottingham, 1982.
27. Y. Fujiki and Y. Suzuki, *J. Jpn. Assoc. Min. Econ. Geol.*, 1973, **68**, 277.
28. P.G. Harrison and A. Guest, *J. Chem. Soc., Faraday Trans. I*, 1987, **83**, 3383.
29. K. Morishige, S. Kittaka and T. Morimoto, *Bull. Chem. Soc. Jpn.*, 1980, **53**, 2128.
30. E.W. Thornton and P.G. Harrison, J. *Chem. Soc., Faraday Trans I*, 1975, **71**, 1013.
31. P.G. Harrison and B. Maunders, *J. Chem. Soc., Faraday Trans. I*, 1984, **80**, 1341.
32. J.P. Joly, L. Gonzalez-Cruz and Y. Arnaud, *Bull. Soc. Chim. France*, 1986, 11.
33. S.-C. Chang, *J. Vac. Sci. Technol.*, 1980, **17**, 366.
34. P.G. Harrison and E.W. Thornton, *J. Chem. Soc., Faraday Trans. I*, 1978, **74**, 2597.
35. P.G. Harrison and A. Guest, *J. Chem. Soc., Faraday Trans. I*, 1989 (in press).
36. P.G. Harrison and M.J. Willett, *Nature*, 1988, **332**, 337.
37. P.G. Harrison and E.W. Thornton, *J. Chem. Soc., Faraday Trans. I*, 1978, **74**, 2703.
38. P.G. Harrison and E.W. Thornton, *J. Chem. Soc., Faraday Trans. I*, 1979, **75**, 1487.
39. M. Niwa, T. Minami, H. Kodama, T. Hattori and Y. Murakami, *J. Catalysis*, 1978, **53**, 198.
40. P.G. Harrison and E.W. Thornton, *J. Chem. Soc., Faraday Trans. I*, 1978, **74**, 2604.
41. P.G. Harrison and E.W. Thornton, *J. Chem. Soc., Faraday Trans. I*, 1976, **72**, 2484.
42. P.G. Harrison and E.W. Thornton, *J. Chem. Soc., Faraday Trans. I*, 1976, **72**, 1317.
43. P.G. Harrison and E.W. Thornton, *J. Chem. Soc., Faraday Trans. I*, 1975, **71**, 2468.
44. P.G. Harrison and B. Maunders, *J. Chem. Soc., Faraday Trans. I*, 1985, **81**, 1345.
45. P.G. Harrison and B. Maunders, *J. Chem. Soc., Faraday Trans. I*, 1985, **81**, 1311.

46. P.G. Harrison and B. Maunders, *J. Chem. Soc., Faraday Trans. I*, 1985, **81**, 1329.
47. A.N. Orlov, V.I. Lygin and I.K. Kolchin, *Kinet. Catal.*, 1972, **13**, 726.
48. T. Ono and Y. Kubokawa, *J. Catalysis*, 1978, **52**, 412.
49. E.A. Irvine and K. Taylor, *J. Chem. Soc., Faraday Trans. I*, 1978, **74**, 206, 1590.
50. S. Kodama, M. Yabuta, M. Anpo and Y. Kubokawa, *Bull. Chem. Soc. Jpn.*, 1985, **58**, 2307.
51. F.J. Berry, *Adv. Catal.*, 1981, **30**, 97.
52. F.J. Berry and C. Hallett, *J. Chem. Soc., Dalton Trans.*, 1985, 451.
53. F.J. Berry and C. Hallett, *Inorg. Chim. Acta*, 1985, **98**, L69.
54. F.J. Berry and C. Hallett, *Inorg. Chim. Acta*, 1985, **98**, 135.
55. F. Figueras, M. Forissier, J.P. Lacharme and J.L. Portefaix, *Appl. Catal.*, 1985, **19**, 21.
56. J.-C. Volta, G. Coudrier, I. Martin and J.C. Védrine, *J. Chem. Soc., Chem. Commun.*, 1982, 1044.
57. J.-C. Volta, B. Benaichouba, I. Mutin and J.C. Védrine, *Appl. Catal.*, 1983, **8**, 215.
58. P.G. Harrison and P. Harris, Eur. Pat. Appl. 256 822.
59. J.L. Jacquemin, C. Alibert and G. Bordure, *Solid State Commun.*, 1972, **10**, 1295.
60. F.J. Arlinghaus, *J. Phys. C, Solid State Phys.*, 1974, **35**, 931.
61. J. Robertson, *J. Phys. C, Solid State Phys.*, 1979, **42**, 4767.
62. J. Robertson, *Phys. Rev. B*, 1984, **6**, 3520.
63. K. Hirajima, T. Sasaki and K. Hijikata, *Proc. Inst. Nat. Sci. Jpn.*, 1973, **8**, 17.
64. S. Munnix and M. Schmeits, *Phys. Rev. B*, 1983, **27**, 7624.
65. E. de Frésart, J. Darville and J.M. Gillies, *Solid State Comm.*, 1980, **37**, 13.
66. E. de Frésart, J. Darville and J.M. Gillies, *Surf. Sci.*, 1983, **126**, 518.
67. C.A. Vincent, *J. Electrochem. Soc.*, 1972, **119**, 515.
68. J.F. McAleer, P.T. Moseley, J.O.W. Norris and D.E. Williams, *J. Chem. Soc., Faraday Trans. I*, 1987, **83**, 1323.
69. P.G. Harrison and M.J. Willett, *J. Chem. Soc., Faraday Trans. I*, 1989 (in press).
70. N. Taguchi, UK Pat. 1,280,809.
71. S. Karpel, *Tin and Its Uses*, 1986, **149**, 1.
72. K. Ihokura, *New Mats. & New Procs. in Electrochem. Technol.*, 1981, **1**, 43.
73. B. Bott, UK Pat. 1,374,375.
74. L.B. Malhotra, M.Sc. thesis, University of West Virginia, 1976.
75. B. Bischof, O. Oehler and W. Baumgartner, *Vide les Couches Minces*, 1980, **201**, 754.
76. P. Tischer, H. Pink L. Treitinger, *Jap. J. Appl. Phys.*, 1980, **19**, 513.
77. B.C. Webb, 'Chemically sensitive field effect transistors', *Analysis '85*; Conf. on Chemical Sensors, October 1985.
78. J.F. McAleer, P.T. Moseley and D.E. Williams, *Sensors and Actuators*, 1985, **8**, 251.
79. H.E. Hager and J.A. Belko, *Sensors and Actuators*, 1985, **8**, 161.
80. J.F. Boyle and K.A. Jones, *J. Electron. Mats.*, 1977, **6**, 717.
81. M. Nitta and M. Haradome, *IEEE Trans. on Electron. Devices*, 1979, **26**, 219.
82. P. Romppainen, *Sensors and Actuators*, 1985, **8**, 101.
83. S.C. Chang, *IEEE Trans. on Electron. Devices*, **ED-26**, 1979, 1875.
84. T.W. Capehart and S.C. Chang, *J. Vac. Sci. Technol.*, 1981, **18**, 393.
85. S.R. Morrison, *The Chemical Physics of Surfaces*, Plenum, New York, 1977.
86. K. Ihokura, *New Mats. & New Procs. in Electrochem. Technol.*, 1981, **1**, 43.
87. J. Watson, *Sensors and Actuators*, 1984, **5**, 29.
88. T. Yamamoto and H. Shimizu, *IEEE Trans. on Components, Hybrids and Manufacturing Technol.*, **CHMT-5**, 1982, 238.
89. H. Windischmann and P. Mark, *J. Electrochem. Soc.*, 1979, **126**, 627.
90. N. Yamazoe, Y. Kurokawa and T. Seiyama, *Sensors and Actuators*, 1983, **4**, 283.
91. R.G. Egdell, W.R. Flavell and P. Tavener, *J. Solid State Chem.*, 1984, **51**, 345.
92. J. Mizsei and J. Harsanyi, *Sensors and Actuators*, 1983, **4**, 397.
93. K. Wada, N. Yamazoe and T. Seiyama, *Nippon Kagaku Kaishi*, 1980, **10**, 1597.
94. R. Lalauze, N. Bui and C. Pijolat, *Sensors and Actuators*, 1984, **6**, 119.
95. R. Lalauze, N Bui and C. Pijolat, *Solid State Ionics*, 1984, **12**, 453.
96. S. Kanefusa, M. Nitta and M. Haradome, *J. Appl. Phys.*, 1979, **50**, 1145.
97. M. Nitta and M. Haradome, *IEEE Trans. on Electron. Devices*, 1979, **26**, 219.
98. M. Nitta and M. Haradome, *J. Electron. Mats.*, 1984, **13**, 15.

13 Industrial uses of tin chemicals

C.J. EVANS

13.1 Introduction

Tin chemicals represent a significant outlet for tin, accounting for about 25 000 tons of the metal, approximately 16% of primary tin consumption. Moreover, according to some estimates, this figure is likely to increase[1]. There are various reasons for this growth. On one hand, tin chemicals exhibit a wide range of properties and these can often be tailored towards the requirements of particular end uses. Moreover, tin chemicals often exert a significant effect at very low concentrations, enhancing their cost effectiveness. In many cases there are environmental advantages over alternative materials. Inorganic tin compounds in particular are regarded as not presenting any health or environmental hazard in their use, and most of the industrially used organotin compounds have similar advantages.

In its inorganic compounds, tin can appear in two oxidation states: stannous or tin (II), and stannic or tin (IV). Industrially important compounds of each type exist. Inorganic tin compounds are generally made by reacting tin metal with oxygen or halogens or by dissolving tin or tin (IV) oxide in acids. Schematics showing routes to inorganic tin compounds have been drawn up.[2]

Many of the major uses of inorganic tin compounds are of long standing, for example in the ceramics industry, where a number of tin-based compounds find use as pigments and glazes. The largest single use is in plating baths for producing tin and tin alloy coatings. Inorganic tin chemicals are also employed as heterogeneous catalysts in industrial processes. A newer area of use and one of growing importance consists of tin (IV) oxide coatings on glass surfaces to strengthen the glass or confer other properties.

The great majority of organotin compounds have tin in the IV+ oxidation state, although a few are known in which tin is in the II+ form. The number and nature of the organic groups attached to the tin atom dramatically influence the properties of the organotin compound. Thus, whereas most mono- and diorganotin compounds are not biologically active, triorganotins are powerful biocides (Chapter 11).

Manufacture of organotin compounds involves the first significant step of creating tin–carbon bonds and this is still a relatively expensive process. The most commonly used routes are via a Grignard reagent or via an aluminium alkyl compound. The Wurtz synthesis:

$$8Na + 4RCl + SnCl_4 \rightarrow R_4Sn + 8NaCl \tag{13.1}$$

is to-date only used by one manufacturer, situated in the German Democratic Republic, and an account of this process has been given.[3] Tri-, di- and mono-organotin derivatives are obtained by coproportionation between the tetra-organotin and tin tetrachloride, whereupon redistribution occurs:

$$3R_4Sn + SnCl_4 \rightarrow 4R_3SnCl \tag{13.2}$$

$$R_4Sn + SnCl_4 \rightarrow 2R_2SnCl_2 \tag{13.3}$$

Reaction products are separated by vacuum distillation. Fuller accounts of the synthesis of organotin compounds have been published.[4,5] Mono- and diorganotin compounds are used in PVC stabilization, the first important use of organotin compounds. Diorganotin compounds are employed as homogeneous catalysts, for example in the production of polyurethane foams. Triorganotin compounds have a range of biocidal applications, for example as antifouling paint additives, pesticides and wood preservatives. A full account of all these applications has been given.[6] Growth in the use of organotin compounds has been spectacular, from virtually zero before 1950 up to 40 000 tons per year in the late 1960s.

13.2 The plastics industry

Tin chemicals, principally organotin compounds, find a number of important uses in the plastics industry; the most significant, accounting for some 60% of organotin consumption, is in the stabilization of poly(vinyl chloride), PVC.

13.2.1 *PVC stabilization*

Poly(vinyl chloride) is a very versatile polymer and is used in a diverse range of applications. However, it is subject to degradation by heat and by short-wavelength light, and to make it commercially useful, stabilizers must be incorporated in the polymer. Various stabilizing systems have been developed, and organotin compounds form one of the most important groups of these and have been used for this purpose for many years. The early development of tin-based stabilizers has been summarized.[4]

Before considering the role of the organotin stabilizers, it is helpful to look at the mechanisms which have been put forward for the degradation of PVC. The basic structure of the polymer consists of a hydrocarbon backbone with a chlorine atom attached to alternate carbon atoms along the chain. Degradation results in structures being formed which can colour the polymer and lead to embrittlement. This subject has been intensively researched and there is a large volume of literature on the subject. In spite of this, the complexities of the degradation process have not been completely resolved and a number of aspects are still disputed. Therefore, only a broad outline can be given within the scope of this chapter.

Two principal processes occur during degradation: loss of HCl and autoxidation. Some possible reaction schemes are summarized in Figure 13.1. The removal of HCl leads to the formation of a double bond which then activates adjoining atoms so that further loss occurs. As this 'unzipping' process continues, a conjugated polyene system is formed which confers colour to the molecule and deepens with increasing degradation, from pale yellow to brown and eventually to black. These polyene sequences are, however, not the only source of discoloration. The higher polyenes are basic in character and are able to form strongly-coloured carbonium ions. The presence of free HCl itself catalyses the dehydrochlorination process.

Since commercial PVC samples begin to degrade at only 90–130°C, there clearly must be active sites present in the polymer for dehydrochlorination to be initiated. These may be introduced during polymerization or in the early stages of processing. The presence of a double bond with an adjacent allylic chlorine atom has been shown to be a source of thermal instability and a certain number of such sites have been shown to

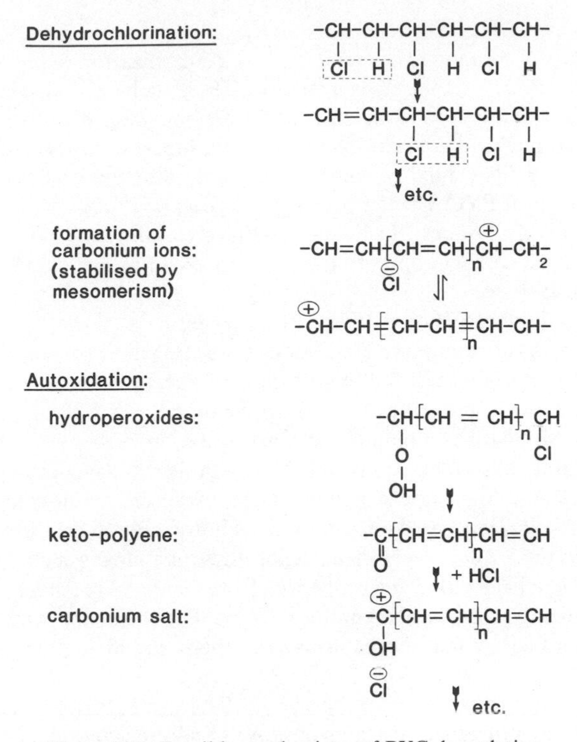

Figure 13.1 Possible mechanisms of PVC degradation.

occur in commercial PVC. Other active sites may result from oxidation processes. For example hydroperoxides may be formed which are subsequently degraded by heat to form olefinic or keto structures which can trigger off the unzipping process. The keto-allyl chloride group (**1**) has been considered to be the major source of thermal dehydrochlorination.[7]

$$-\overset{\overset{\displaystyle O}{\|}}{C}-CH{=}CH-\overset{\overset{\displaystyle Cl}{|}}{CH}$$

(**1**)

A major point of dispute has been whether a free radical or an ionic mechanism is responsible for dehydrochlorination.[8,9] It is suggested that both mechanisms may operate simultaneously, the predominant mechanism depending on the processing conditions and on the PVC formulation (e.g. types of plasticizer, lubricant, etc., present in the polymer).[9]

Much less work has been reported on the degradation of PVC by light although the subject is equally complex. Again the presence of irregularities such as polyene sequences, initiator end groups and oxidation structures can enable PVC to absorb light of longer wavelength and hence initiate degradation processes. Degradation is again accompanied by evolution of HCl, and the rate is accelerated in the presence of oxygen. In this case, it is generally agreed that the mechanism is a free-radical process.[10]

No single class of compound can fulfil the broad stabilization requirements of PVC, and a range of primary stabilizers is available. These include lead salts, mixed metal soaps (barium/cadmium, calcium/zinc, barium/zinc), and organotin compounds. These latter, although rather more expensive, give superior performance and exhibit good heat stability, light stability, compatibility with different grades of polymer and crystal clarity in clear PVC. The low toxicity and good leach resistance of certain types of stabilizers (notably di-*n*-octyltin di-*iso*-octylthioglycolate) enables them to be used in rigid PVC for food contact applications, and approval has been granted in the USA and in most of Europe.

Sulphur-containing organotins (thiotins) have the most powerful heat-stabilizing activity. A slight disadvantage is a faint odour persisting after processing, for example in the blow moulding of bottles, but this disappears with time. Octyltin compounds are used for food-contact and other critical applications and butyltin compounds in industrial uses. Methyltin compounds were introduced in 1970 and can be used in very low concentrations, allowing cost savings. These stabilizers have secured a substantial market in the USA for rigid pipe extrusions. The most recent development has been the introduction of stabilizers based on 2-butoxy-carbonylethyltin derivatives, commonly known as 'estertins'. Advantages claimed for these stabilizers include their ease of synthesis, excellent heat-stabilizing properties, lower odour, less tendency to migration and light-stabilizing properties superior to thiotins. The chemistry of these compounds has been described[11], and mechanisms for their stabilizing action have been reviewed.[12]

Principal organotin stabilizers are shown in Table 13.1. Proprietary formulations commonly contain synergistic mixtures of mono- and diorganotin stabilizers. Recent views on the function of a stabilizer in PVC have been reviewed,[13] and are summarized as follows:

(i) Scavenging of HCl: removal of HCl before it can further catalyse dehydrochlorination. With the more efficient thiotins this is probably a minor role, since these compounds delay the onset of dehydrochlorination for much longer than is suggested by a mere neutralization of HCl eliminated unimolecularly.

Table 13.1 The principal organotin stabilizers.

Compound	Formula	R
Dialkyltin di-*iso*-octylthioglycolate	$R_2Sn(SCH_2COO{\cdot}C_8H^i_{17})_2$	Methyl, *n*-butyl, *n*-octyl
Dialkyltin maleate	$[R_2SnOOC{\cdot}CH{=}CH{\cdot}COO]_n$	*n*-butyl, *n*-octyl
Dialkyltin maleate ester	$R_2Sn(OOC{\cdot}CH{=}CH{\cdot}COOR)_2$	*n*-butyl, *n*-octyl
Dialkyltin dilaurate	$R_2Sn(OOC{\cdot}C_{11}H_{23})_2$	*n*-butyl, *n*-octyl
Dibutyltin *β*-mercaptopropionate	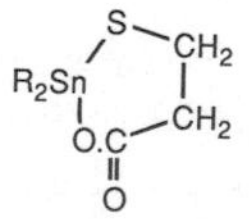	*n*-butyl
Diestertin di-*iso*-octylthioglycolate	$R_2Sn(SCH_2COO{\cdot}C_8H^i_{17})_2$	$C_4H_9OOC{\cdot}CH_2CH_2$–

(ii) Reaction with labile chlorine: exchange of the labile chlorine atoms in the PVC molecule with the stabilizer moiety which is much less easily eliminated. This is only a temporary protection, since the groups are eventually converted back to their original structures by reaction with HCl.
(iii) Destruction of polyenes: efficient stabilizers interrupt developing conjugation by adding to the unsaturated system.
(iv) Radical scavenging: compounds which can remove Cl or other radicals will contribute to overall stabilization.
(v) Destruction of hydroperoxides: these are produced by oxidation and can serve as active sites for further degradation, unless removed.

During stabilization, organotin compounds are converted to the corresponding chlorides. These are Lewis acids and, in the case of the mono-organotin chloride, can promote degradation. Processes which delay their formation or break them down can thus have a secondary stabilizing effect. This has been considered the reason for the marked synergism between mono- and dialkyltin stabilizers.[14] Dialkyltin stabilizers provide long-term stability but poor colour control in PVC, whilst mono-alkyltins give good colour control but only short-term stability. The mono-compound is the most powerful stabilizer, but once converted to the chloride promotes further degradation. By combining the two compounds, exchange takes place between *iso*-octylthioglycolate and chlorine, preventing accumulation of mono-alkyltin chloride. Dialkyltin dichloride, being a weaker Lewis acid, does not promote degradation. Abeler and Bussing[15] have also demonstrated the sensitivity of synergistic mixtures to lubricants and to the nature of the attached alkyl group.

Organotin stabilizers are the preferred system for a number of PVC products. In the packaging field, blow-moulded PVC bottles are commonly stabilized with organotin compounds, particularly for food contact applications or when extreme clarity is needed. The process involves production of a tube (a parison) by extrusion or by injection moulding and locating this in a two-part mould, air being blown into the parison until it fills the mould. A new method for producing biaxially oriented PVC bottles, by conditioning the parison to a particular temperature and then subjecting it to a controlled stretching and blowing operation, has led to better mechanical properties and reduced gas permeability.[16] The level of stabilization for blow-moulded bottles is normally 1.0–1.5 phr (parts per hundred of resin).

Thin calendered sheet is also used for packaging applications, and industrial production has been described.[17] Processing conditions are fairly rigorous, with the polymer being subjected to high temperatures and high rates of shear; thiotins are often used in PVC to be subjected to calendering operations. PVC is extruded as a homogeneous gel at about 200°C, is flattened by passage between rollers, and then led through a series of heated calendering rolls from which it emerges as a thin, uniform, continuous sheet. This sheet can be thermoformed into blister packs or other packages (Figure 13.2). Allen *et al.*[18] have examined the effect of gamma irradiation on the breakdown of organotin stabilizers in PVC, in view of the interest in irradiating packaged food as a means of preservation. With dibutyltin di-*iso*-octylthioglycolate and maleate, although some de-alkylation occurred leading to formation of mono-butyltin trichloride and tin(IV) chloride, the authors concluded that, at permitted doses, the extent of degradation was likely to be slight, with little change in physical appearance of the PVC. Migration studies of organotin stabilizers in PVC have also been conducted, and these have been reviewed.[19]

Figure 13.2 Tin-stabilized calendered rigid PVC film for packaging. Photograph courtesy Klöckner Pentaplast Ltd, Reading, UK.

Organotin-stabilized PVC also finds a number of uses outside the packaging area. Thicker PVC sheet and cladding are used in the building industry and this is normally made by extrusion, although one product is manufactured by laminating together several thin calendered sheets.[20] Transparent PVC for roofing applications is fairly widely used in the UK, and since long-term light stability is required, dibutyltin maleate is commonly used as the stabilizer because of its superior light stabilizing properties. An important market for tin-stabilized PVC in the USA, is for piping, for drain-waste and vent systems, for water and sewage mains, gas distribution, electrical and communications conduits and industrial process piping. Methyltin stabilizers are widely used in this application, at low levels down to 0.25 phr in some cases. Organotin stabilizers find limited application in PVC plastisols for vinyl floor and wall coverings. Production of vinyl wall coverings involves several heating cycles and efficient heat stabilization is essential, particularly in view of the exact colour matching required in batches.[21] Vinyl floor coverings employ a tough topcoat of clear vinyl, and it is in this layer that organotin stabilizers are chiefly used.

13.2.2 *Homogeneous catalysts*

A number of tin chemicals have important uses as homogeneous catalysts in the plastics industry. A homogeneous catalyst is one which is in the same phase as the reactants and is usually a liquid. The compounds most widely used are stannous octoate (properly termed tin(II) 2-ethylhexoate) and a number of mono- and diorganotin compounds. The most important reactions catalysed by the tin compounds are those leading to the formation of polyurethanes, silicones and polyesters, all widely used industrial plastics.

Polyurethanes are polymers in which a significant number of urethane (**2**) linkages

$$—NH—\overset{\overset{\large O}{\|}}{C}—O—$$

(**2**)

occur. These linkages serve to connect larger blocks of other polymers such as polyesters or polyethers. Urethane linkage occurs by reaction between an isocyanate group and a hydroxyl group. The isocyanate group is very reactive and can enter into reactions with other functional groups which may be present. Three principal reactions can be characterized, chain propagation, gas formation and cross-linking.

(i) *Chain propagation.* Addition of an isocyanate to an alcohol hydroxy group produces a urethane:

$$-NCO + -OH \rightarrow -NHCOO- \tag{13.4}$$

When multifunctional starting materials are used, e.g. a di-isocyanate plus a diol, a polyurethane polymer is produced with blocks connected by urethane linkages.

(ii) *Gas formation.* Isocyanates react with water to form an amine plus carbon dioxide:

$$-NCO + H_2O \rightarrow -NHCOOH \rightarrow -NH_2 + CO_2 \tag{13.5}$$

This occurs simultaneously with chain propagation and the evolved CO_2 blows the polymer into a foam.

(iii) *Cross-linking.* Additional side reactions can occur to cause branching and cross-linking between chains.

The rates of polymerization and of gas evolution must be adjusted so that a cellular foam is formed, and the growing polymer must be strong enough to maintain its structural integrity without collapse, shrinkage or splitting. The degree of branching influences the characteristics of the final polymer, and starting polyol and di-isocyanate are chosen to achieve particular end properties. An inert fluorocarbon may also be included in the formulation as a blowing agent to supplement foam expansion.

In order for these reactions to occur at commercially viable rates, the presence of a catalyst is required. Tertiary amines were the first catalysts to be used. However, these catalyse polymerization and gas formation about equally, and since the gas evolution reaction is by far the faster of the two reactions, a two-stage process had to be adopted, with addition of water for gas-blowing taking place only when a suitable degree of polymerization had been achieved. The discovery that tin compounds were powerful catalysts for the polymerization reaction meant that a 'one-shot' process could be adopted. Tin chemicals and tertiary amines are often used together in catalyst systems, and in these cases a synergistic effect is observed.[22] The mechanism of the catalytic action of the tin compound has been extensively studied. Rusch and Raden[23] consider

Table 13.2 Typical formulation for a flexible polyether urethane slab foam.

Component	Parts by weight
Polyol (trifunctional)	100.0
Toluene di-isocyanate	46.0
Tin-based catalyst	0.4
Amine catalyst	0.2
Water	3.6
Silicone surfactant	1.0
Monofluorotrichloromethane	0–15

that the tin compound activates both the polyol and the isocyanate, first by the formation of a binary complex between catalyst and polyol and then by joining of the isocyanate to form a ternary- or bridge-complex. This complex still allows the tertiary amine to approach in order to exert its catalytic influence. Amine/tin synergism is thought to be due to the extra stability of this activated intermediate complex of polyol, isocyanate, tin compound and amine.

The main application for tin chemicals in polyurethanes is in flexible foam production, although they are used to some extent for rigid foams. Stannous 2-ethylhexoate is the most widely-used tin catalyst for flexible polyether-based foams and is used commonly at the 0.1–0.4 phr level. Table 13.2 shows a typical formulation for a flexible polyether urethane slab foam. Dibutyltin dilaurate is used in high-resilience cold-cure foams. The ambient-temperature curing is achieved by the use of special, highly reactive starting materials.

Production techniques for polyurethane products have been reviewed.[24] Slab foam is produced commercially by mixing constituents in a mixing head and then dispensing the mix on to a primary conveyor belt lined on the base and sides with paper (Figure 13.3). The mixing head moves laterally across the conveyor belt, laying down a uniform layer of polymer which then begins to foam and rise as it travels along the belt, until a foam block, typically 2 m wide and about 1 m high, is achieved. The block passes to a secondary conveyor where it is cut into sections and allowed to cool. The exothermic nature of the polymerization reaction provides sufficient heat for curing. These flexible foams are used for furniture and for bedding. In reaction injection moulding, which employs a more reactive foam formulation, the polymer mix is dispensed into preheated metal moulds and then hot-cured. Moulding time may be about 10 minutes, with a total cycle time of about 20 minutes. High-resilience cold-cure formulations do not require heat for curing, and hence involve lower energy consumption. These moulding techniques are valuable when large quantities of one component have to be made to high dimensional tolerances, as in car seats and shoe soles. Rigid foams which are more highly cross-linked are widely used for insulation and for structural panels, but tin chemicals have only limited application in these systems and dibutyltin dilaurate is usually employed at low levels (up to 0.1 phr).

Another application for tin chemicals is as curing agents for room-temperature vulcanizing (RTV) silicones. Silicones are polymeric materials based on the inorganic siloxane backbone $(\text{–}\overset{|}{\underset{|}{\text{Si}}}\text{–O–})_n$, the other valencies on the silicon atom usually being taken up by organic groups or by hydrogen. Silicones may be produced as resins, oils,

Figure 13.3 Dispensing the polymer mixture into a paper-lined trough for production of polyurethane foam. Photograph courtesy ITRI and Dunlop Ltd, Dunlopillo Div., High Wycombe, UK.

gums, elastomers, pastes, etc., but the products in which tin-based catalysts find use are liquids which are fluid enough to be easily poured or spread and which then subsequently cure to a tough elastomeric solid at room temperature under the influence of a catalyst.

The chemistry of these compounds has been reviewed by Watt[25] and more recently by Karpel.[26] The linear polymer chains have to cross-link to form a network in order for elastomeric properties to develop. This process can take place by two mechanisms: addition or condensation. In the addition reaction, in which a platinum-based catalyst is used, a silicone prepolymer with vinyl end groups reacts with a silicone containing Si–H bonds, these groups adding to the vinyl double bonds forming $-CH_2-CH_2-$ cross-links between the two polymer chains. In the condensation reaction, polydialkylsiloxanes which have terminal hydroxyl groups on the chains react with a cross-linking agent, a tetra-alkoxysilane, in the presence of a catalyst. Four chains become linked in this way via new siloxane bonds, with an alcohol evolved as a by-product. The most commonly-used catalyst for this type of RTV silicone is dibutyltin dilaurate. Other organotin compounds which are used include dibutyltin diacetate, dibutyltin di(2-ethylhexoate), dioctyltin dilaurate, and tin(II) 2-ethylhexoate. The catalyst is incorporated at a level of 0.1–1%, the concentration influencing to some extent, the rate of cure. Research into the mechanism by which tin chemicals catalyse the condensation reaction suggests that the actual catalyst is a hydrolysis product (dialkyltin hydroxycarboxylate), the catalyst being regenerated as polymerization proceeds.[27]

RTV silicones are available as either two-pack or one-pack systems. In two-pack systems, one of the components contains the catalyst and the reaction begins from the time of mixing. In one-pack systems, reactive groups are blocked off until contact with moisture in the atmosphere frees them for subsequent reaction. Applications of RTV silicones have been summarized,[26] and include encapsulation and potting of electrical components for protection, sealing and caulking, moulding and casting, and for anti-stick coatings, for example, on paper to be used in food contact applications.

Carboxylate esters represent a very important class of organic compounds and find use as moulding resins, surface coatings, cosmetics, synthetic lubricants, food packaging, PVC plasticizers and precursors for polyurethanes. These esters are produced by esterification:

$$R{\cdot}COOH + R'OH \rightarrow R{\cdot}COOR' + H_2O \tag{13.6}$$

or by transesterification:

$$R{\cdot}COOR' + R''OH \rightarrow R{\cdot}COOR'' + R'OH \tag{13.7}$$

A number of tin compounds have been used commercially to catalyse these reactions, and include tin (II) oxide, chloride, oxalate, glycoxide and 2-ethylhexoate, as well as tetrabutyltin, butanestannonic acid, butylchlorotin dihydroxide, dibutyltin diacetate and dibutyltin oxide. The tin compounds have a number of advantages as esterification catalysts, particularly because of their high efficiency at low (0.05–0.3%) levels of incorporation.

Other systems in which tin compounds have been used as homogeneous catalysts include preparation of organic silicate binders in the foundry industry,[28] Friedel–Crafts alkylation and acylation,[29] and liquid-phase hydrogenation, dehydrogenation and isomerization.[2]

13.3 Glass and ceramics

Tin chemicals, principally tin (IV) oxide, find a number of uses in the glass and ceramics industries. Some of these applications are traditional and of long standing; others are at the forefront of modern technology.

13.3.1 *Glass melting*

The use of tin in the glass industry has been reviewed.[30] One of the most important applications is in the manufacture of lead-containing glass by electric melting in which tin oxide electrodes are used to conduct electricity into the melt. Molybdenum is the most widely used electrode material, but it is incompatible with glass containing easily reduced oxides such as lead, or refining agents such as arsenic. Other advantages in its favour are its good electrical conduction at high temperatures, resistance to molten glass, and lack of coloration imparted to the glass. This use of tin oxide has been reviewed recently.[31,32]

Tin oxide electrodes are fabricated either by slip casting or by isostatic pressing, followed by sintering to high density. The largest electrodes made commercially are 127 × 127 × 450 mm (rectangular) or 150 mm diameter × 500 mm long and are installed horizontally in banks as part of the furnace wall, or vertically through the floor in the case of cylindrical electrodes. A major use is for producing lead crystal glass for tableware. Other uses include tubing for electric lighting, and special optical glasses. It has been suggested recently[32] that tin oxide might serve as refractory lining material for production of optical and laser glass.

13.3.2 *Coatings on glass*

Freshly-formed glassware has considerable strength, but this strength rapidly decreases. By exposing hot, newly formed glass surfaces to a vapour or liquid spray containing certain tin chemicals, a very thin layer of tin (IV) oxide is deposited on the glass. Although only 100 nm thick and invisible to the eye, this coating has a considerable effect upon the glass surface, preventing some of the minute flaws which rapidly form as the glass cools and preserving some of its initial strength. The layer of tin (IV) oxide also promotes adhesion of a lubricant film which is applied to the glass surface when cold in order to improve its abrasion resistance and antifriction properties. The development of these strengthening treatments has been reviewed.[33,34]

Tin (IV) chloride was one of the first chemicals to be used in these treatments and is applied as a liquid spray, alone or in a solvent. A newer process uses *n*-butyltin trichloride, which is vaporized and applied to the glass surface under a specially engineered double-loop coating hood which reduces vapour losses. Less corrosion of the treatment vessel results when the organotin compound is used, and this process is being widely adopted.[35]

If the thickness of the tin (IV) oxide layer is of the same order of magnitude as visible light (100–1000 nm), thin-film interference occurs and the glass surface takes on an iridescent lustre which has been exploited for decorative effect in wall panels and glassware, particularly in Japan.[36]

Thicker coatings of tin (IV) oxide (above 1 μm) confer electrically conducting properties on the glass surface, and are still optically transparent. These coatings are normally produced by vapour phase coating, but there is growing interest in reactive sputtering as a source of tin (IV) oxide coatings.[30] The films may be doped with indium oxide for maximum conductivity. Applications include transparent heating elements for windscreens of aircraft, ships and locomotives; for example Concorde and Boeing 747 airliners use coated, specially toughened glass in the visor and main windscreen. Other uses are in transparent tube furnaces, antistatic cover glasses for sensitive scientific instruments, security alarm glazing and electroluminescent display screens. Recent papers have reported novel techniques for producing these thin films. Court *et al.*[37] have described a method based on thermal decomposition of an aerosol produced by ultrasonic wave pressure; Colombin and Sebastiano[38] described a chemical vapour deposition technique for in-line production of tin (IV) oxide coatings on glass. Frémaux[39] revealed a powder pyrolysis technique for applying tin (IV) oxide coatings directly on a float glass line. This last method is envisaged in connection with the use of low-emissivity coatings on window glazing, to prevent loss of heat through radiation. This glass is particularly valuable for greenhouses, and double glazing may incorporate a tin (IV) oxide layer as a thermal exchange barrier to improve insulation.

13.3.3 *Tin (IV) oxide in the ceramics industry*

Anhydrous tin (IV) oxide has been used as an opacifier in ceramic glazes and, to a lesser extent, in vitreous enamels since ancient times. Tin (IV) oxide is highly insoluble in glazes, giving it a high opacifying power at levels of 4–8%. Although it is challenged by cheaper alternatives, the oxide is still used for high-quality artware and for those industrial applications where the highest reflectance, purest colours, greatest strength and abrasion resistance are required. Particle size is very important, with the highest reflectance occurring with particles of 0.2–0.3 μm diameter.

However, the main application of tin (IV) oxide in the ceramics industry is as a constituent of pigments used for ceramic tiles and pottery. The characteristics of these pigments and their commercial use have been reviewed.[40,41] The oldest-known tin colours are those based on gold and tin, the so-called 'Purple of Cassius'. If a solution of tin (II) chloride is added to a very dilute solution of gold chloride, the gold is reduced and forms a colloidal dispersion of metallic gold on hydrous stannic oxide. The actual shade depends upon the composition and concentration of the solution. This pigment is still used as an on-glaze decoration for high-quality tableware, since, although costly, it provides an unsurpassed clarity and brightness.

The rutile-type structure of tin (IV) oxide is able to accommodate certain metal colorant ions in the lattice and this forms the basis of three commercial pigments used for ceramic tiles and pottery. Tin–vanadium yellows (containing 2–5% vanadium) and tin–antimony blue-greys (containing 3–8% antimony) are prepared by thermal reaction between tin (IV) oxide and either ammonium vanadate or antimony trioxide at 1200–1300°C. Chrome–tin pinks are prepared by firing a mixture of SnO_2, $K_2Cr_2O_7$, $CaCO_3$ and SiO_2 at 1150°C; the colour can range from maroon to light pink. The host lattice for this latter pigment has been shown[42] to be malayaite, $CaSnSiO_5$. Mössbauer and infrared spectroscopy and X-ray powder diffraction have been used to study the structures of these three commercial pigments.[43] Incorporation of either vanadium or antimony ions into the host lattice of SnO_2 (rutile) results in little distortion of the structure from octahedral. Incorporation of chromium ions into the lattice of synthetic malayaite makes the tin atom environment more symmetrical. A later study[44] examined the SnO_2–Cr_2O_3 binary system in an attempt to understand the mechanism of formation of chrome–tin pink. Results indicated the formation of an SnO_2 solid solution containing 17–28 mole% of Cr_2O_3 rather than a crystalline compound. Possibilities of producing new pigments by incorporating metal ions into the rutile $Sn_xTi_{1-x}O_2$ system have also been studied.[45] Incorporation of solid solutions of tin (IV) oxide into TiO_2 at up to 25 mole% SnO_2 was found to increase the opacifying power and significantly reduce the tendency of TiO_2 to yellowing, when incorporated into glazes. However, no significant colour changes were achieved by doping with metal ions.

An important use of these tin-based pigments is for underglaze decoration on heavy-duty tableware. The thermal stability of the pigments enables them to withstand the high temperatures involved in firing of glazes (above 1000°C). They also find use as glaze stains, the pigment being mixed with the glaze to produce all-over coloration, for example on ceramic tiles and on sanitary ware.[39] Certain tin pigments also have potential in vitreous enamelling.[45]

13.4 Electroplating

Plating chemicals represent the largest single application of inorganic tin chemicals, estimated annual consumption being about 6000 tonnes. The most commonly used compounds are tin (II) sulphate, tin (II) chloride, tin (II) fluoroborate, and sodium and potassium stannates. Characteristics of tin and tin alloy electrodeposits are summarized in Table 13.3. A major use of tin plating is in the production of tinplate, where a very thin tin coating is deposited continuously on rapidly-moving steel strip. Tin (II) sulphate and chloride are widely used in tinplate manufacture, whilst stannates are employed

Table 13.3 Summary of tin and tin alloy electroplating processes.

Deposit	Electrolyte	Characteristics and applications
Tin	*Alkaline bath*: sodium or potassium stannate, free hydroxide, 80°C; tin anodes.	Matt coatings which are soft, ductile, solderable and corrosion resistant. Used for electrical parts and general consumer goods.
	Acid bath: tin(II) sulphate, free sulphuric acid, cresolsulphonic acid, addition agents, room temperature.	
	Acid bath: tin(II) fluoroborate, free fluoroboric acid, addition agents, room temperature.	Coating properties as above. Used where high current densities are required, for example in continuous plating of wire and strip.
	Bright tin: acid bath plus addition agents as brighteners.	Bright deposits, somewhat harder and less ductile than matt deposits. Used for decorative applications and electronics components.
Tin–nickel	Tin(II) chloride, nickel chloride, ammonium bifluoride, 70°C; anodes can be tin–nickel alloy, separate nickel and tin, or nickel.	Brightness depends on the quality of the substrate. Hard, rather brittle, corrosion-resistant and wear-resistant deposit. Applications include instrument parts, automotive braking components, decorative uses and electronics parts.
Tin–lead	Tin(II) fluoroborate, lead fluoroborate, free fluoroboric acid, addition agent, room temperature; tin–lead anodes.	Tin and lead can be deposited in any desired proportion and deposits are fine grained. Used to protect steel, and as a bearing overlay but main use is in the electronics industry.
Tin–zinc	Sodium or potassium stannate, zinc cyanide, free hydroxide, 65°C.	The normally used deposit is tin–25% zinc and this is matt, silvery white and perfectly smooth. The deposit is fine-grained, ductile and has good anti-friction properties.
Tin–copper	Sodium or potassium stannate, copper cyanide, free hydroxide and free cyanide; sodium or potassium salts of hydroxycarboxylic acids may be added to promote anode dissolution and produce fine-grained deposits 60°C; anodes are usually cast 10% tin bronze.	Bath has very high throwing power and deposits are hard, very wear--resistant and corrosion-resistant. Used for coating moving parts in hydraulic systems and as a decorative plating.

to a lesser extent today. One line in the Federal Republic of Germany employs tin(II) fluoroborate.

13.4.1 *Tin plating*

Developments in tin and tin alloy plating have been reviewed.[46] Alkaline stannate baths are operated hot and give smooth deposits without the need for addition agents.

Potassium stannate baths give faster plating and tend to be preferred, although a limitation is the build-up of potassium ions in the electrolyte. However, a method has been developed using inert anodes and replenishing the tin content of the bath by regular additions of a hydrous tin oxide 'sol'.[47] Acid baths may be based on tin (II) sulphate or tin (II) fluoroborate. These are operated at room temperature and the baths are characterized by high current efficiency and fast deposition rates. Organic addition agents are required to ensure adequate throwing power and to obtain smooth, fine-grained deposits. Fluoroborate baths permit the fastest plating rates, and are used for high-speed continuous plating of strip and wire. Operating conditions for these baths have been described.[48] All these deposits are matt, but by the use of suitable brightening agents, bright deposits which are aesthetically pleasing and do not fingermark on handling can be produced. There are some applications, for example in the electronics industry or for food contact uses, where a matt deposit might be preferred because the use of brightening agents could lead to undesirable contaminants.

In addition to tinplate, tin coatings find a wide range of uses. They have a protective role on sieves, screws, can openers, etc. and in contact with food, as on steel baking tins for bread. The electronics industry uses tin as a protective and easily soldered finish for radio chassis, computer frames, tags, connectors and printed circuit boards. Much of the copper wire used in the electrical industry is now electrotinned for solderability. Decorative items such as jewellery are also frequently bright-tinned.

13.4.2 *Tin alloy plating*

The scope of tin-based coatings is widened by the availability of alloy electrodeposits.[49] Tin–nickel alloy can be electrodeposited with a composition corresponding to the intermetallic NiSn (65% tin–35% nickel) over a fairly wide range of plating conditions.[50] The bath is based on a mixture of tin and nickel chlorides with some free fluoride ions. Substitution of ammonium fluoride for sodium fluoride overcomes problems with formation of insoluble sodium stannifluoride. An important commercial development in electrolytes has been the work of Dillenburg,[51] who modified the standard chloride/fluoride bath by including an amine, which allows operation at a more neutral pH and at lower temperatures. Tin–nickel electrodeposits tend to be rather brittle, although proprietary baths have recently become available which are claimed to improve deposit ductility. Certainly much work is in hand to try and overcome this brittleness, for example the study of Izaki *et al.*[52] Tin–nickel electroplate finds use in the electronics industry in printed circuit boards and as a coating underneath gold for electrical connectors. Its attractive appearance and tarnish resistance have led to its use on drawing instruments and equipment controls. It is also employed on components for automotive braking systems and in beer-dispensing equipment.

Tin–lead plating serves as a solderable finish in the electronics industry. The conventional bath is based on tin (II) fluoroborate and lead fluoroborate.[53] However, in an attempt to avoid the presence of fluoride ions in effluents, other baths, based on sulphamates[54] or alkylsulphonates,[55] have been developed. Electrolytes with high throwing power for plating of printed circuit boards have low metal concentrations and high acidities.

Tin–zinc alloy coatings are produced from baths based on sodium or potassium stannate and zinc cyanide.[56] The deposits are ductile, have good wear resistance, especially when lubricated, and are hard and reasonably solderable. The coating behaves as a simple mixture of the two metals from which zinc may be selectively dissolved. Recently, work has been conducted on chemical passivation treaments for the coating to increase its corrosion resistance, and an electrolytic dichromate treatment has been found to be the most effective.[57]

Tin–copper coatings are produced from a stannate/cyanide bath, and the composition of the deposit is very dependent on the free cyanide and free alkali concentrations. The most important composition is red bronze, containing 10–15% tin. The coatings are ductile and wear-resistant and have been used in hydraulic mechanisms, such as steel pit props in mining applications and on the shafts of tip-up lorries. The most recent development in tin alloy plating has been the introduction of a tin–cobalt alloy. The bath which has received most commercial development is based on tin and cobalt sulphates and a complexing agent which solubilizes cobalt and tin salts in the operating pH range. Tin–cobalt plating is used on many types of fastener, on hinges, hand tools, bathroom and kitchen fittings and tubular furniture.

Tin plating electrolytes, including tin (II) sulphate, are used in the electrolytic colouring process for anodized aluminium. A range of colours from very pale bronze through to black, depending on deposit thickness, can be produced in an alternating-current electrolytic bath immediately following anodizing.[58]

13.5 Biocidal applications

The powerful biocidal properties of triorganotin compounds were discovered in the 1950s and commercial exploitation soon followed. Triphenyltin compounds combined fungicidal activity with very low phytotoxicity, and in the late 1950s and early 1960s a number of triphenyltin-based pesticides were introduced. In the 1960s, trialkyltin compounds began to be used in solvent-borne wood preservative formulations and in antifouling paints for ships, in view of their biological activity against a wide spectrum of fouling organisms. Since then, growth in these applications has steadily progressed and triorganotin compounds have also found use as disinfectants and protective agents for materials as diverse as stone, leather, paper and textiles.

The toxicology of organotin compounds is very complex and has been extensively studied. However, a general pattern does emerge. Progressive introduction of organic groups at the tin atom in any R_nSnX_{4-n} series produces a maximum biological activity when $n = 3$, i.e. for the triorganotin compounds R_3SnX. The nature of the group X in general has little influence on the biological activity (unless of course the group X is itself biologically active). However, the nature of the organic groups R does have an important influence. Thus, in the trialkyltin series, trimethyltins show maximum toxicity to insects and mammals, triethyltins are most toxic to mammals, tripropyltins to Gram-negative bacteria, tributyltins to Gram-positive bacteria and fungi, and triphenyltins to fish, fungi and molluscs. Further increase in the *n*-alkyl chain length leads to a sharp drop in biological activity, so that the trioctyltin compounds are virtually non-toxic to all living organisms. Because of this specificity, it has been possible to tailor an organotin biocide to meet the requirements of a specific application; for example bis(tributyltin) oxide (known widely as TBTO) has been adopted as a wood preservative because it affords the widest separation between fungicidal activity and mammalian toxicity.

Because triorganotins are so bioactive, it has been a matter of great importance to ensure that they are used safely, and a great body of information has been assembled on their toxicology. Much of this data has been reviewed.[59] Another important factor is the fate of organotins in the environment, and this too has been the subject of a number of extensive reviews.[60–62] In general, most commercially used organotins are characterized by relatively low mobilities in environmental media since they have low aqueous solubilities, low vapour pressures and high affinities for soils and organic sediments. It has been shown that organotin compounds will be degraded under environmental conditions and no serious long-term pollution hazards should occur. It is of interest to note that, since 1978, there has existed an association of world organotin manufacturers, the Organotin Environmental Programme Association (ORTEPA), which promotes and fosters the dissemination of scientific and technical information on the environmental effects of organotin compounds, thus ensuring that these compounds continue to be used safely and responsibly.

13.5.1 *Wood preservation*

Today trialkyltin compounds are used in a large number of organic solvent-based wood preservative formulations and the subject has been extensively reviewed.[2,6,63] Wood is a natural product and hence is subject to biodeterioration. Under certain conditions it can be attacked by various types of fungi as well as by wood-boring insects. The ravages of such attack can be seen in many ancient buildings. Today the building industry insists that wooden joinery and structural timbers be pretreated with solvent-borne wood preservatives, and many of these contain trialkyltin compounds, notably TBTO. Remedial treatments based on these formulations are also available for older structures. TBTO is commonly employed at concentrations up to 3% in a solvent such as kerosene, usually in conjunction with a contact insecticide. Although still often applied by spraying, brushing or immersion when hazards of decay are fairly low, for long-term protection under severe conditions more sophisticated impregnation techniques have been developed. TBTO has several advantages for treating wood. It is a colourless liquid and hence does not colour the wood, has low volatility and low water solubility so that it is not easily leached from treated wood, and wood can be painted or glued in normal fashion shortly after treatment. Once the solvent has evaporated, flammability is not increased by the organotin compound, and although TBTO has a characteristic odour, this is not noticeable in treated wood at the concentrations used. Provided that normal precautions are followed, there should be no health hazards in handling, and in fact the organotin compounds have over 25 years record of safe use in wood preservation. Schweinfurth[64] has examined the toxicology of TBTO in some detail.

Other tributyltin compounds are being employed in wood preservation to some extent and in some cases have advantages over TBTO. Tributyltin phosphate is said to offer a lower mammalian toxicity and a very low vapour pressure. Tributyltin benzoate is used in some countries in order to meet legislative requirements. Tributyltin compounds formed by reaction between TBTO and long-chain carboxylic acids have lower water solubility and volatility than TBTO, and tributyltin linoleate and abietate are being used to some extent. There is growing use, also, of tributyltin naphthenate.[63]

In order to assess the effectiveness of biocides as wood preservatives, standard wood block test methods are employed. These consist essentially of exposing treated wood blocks to pure cultures of wood-destroying fungi. The culture medium is either agar or sterile soil. In both cases, blocks can be artificially weathered (for example by leaching with water) before exposure. The degree of decay is assessed by determining weight loss in the blocks and the efficiency of the biocide expressed as toxic values or a threshold value. The toxic values represent the range between concentrations which just inhibit decay and those just permitting it. The threshold value is the lowest level of preservative in the wood to just prevent any weight loss. Small test blocks ($30 \times 10 \times 5$ mm) have been used to reduce the incubation period.[65] Typical results against two test fungi are shown in Table 13.4.

When treating timber with solvent-based preservatives, the principal requirements are that the biocide should be absorbed by the wood at a sufficient concentration to provide protection, that good penetration into the wood should be achieved, and that distribution within the wood should be as uniform as possible. Double vacuum treatments allow an optimum correlation between penetration and loading, and moreover, the wood emerges from the treatment vessel virtually dry and easy to handle. In this process, the wood to be treated is loaded into a pressure vessel and a vacuum applied. Treatment solution is then flooded into the vessel and the vacuum

Table 13.4 Toxic limits of tributyltin compounds against two wood-destroying fungi

Compound	Fungus	Toxic limits* (loading in wood blocks kg m^{-3})
TBTO	C.p.	0.19–0.70
	C.v.	0.20–0.69
Tributyltin borate	C.p.	0.38–0.96
	C.v.	0.19–0.37
Tributyltin carbonate	C.p.	0.20–0.38
	C.v.	0.20–0.38
TBTO/glucose	C.p.	0.46–1.14
	C.v.	0.47–1.17
Tributyltin nitrate	C.p.	0.48–1.13
	C.v.	0.45–1.16
Tributyltin phosphate	C.p.	0.38–0.96
	C.v.	0.39–0.99
Tributyltin ethanesulphonate	C.p.	0.49–1.21
	C.v.	0.24–0.50
	C.p.[1]	0.58–1.18
	C.v.[1]	0.58–1.17
Tributyltin sulphide	C.p.	0.40–1.03
	C.v.	0.39–1.00
Tributyltin chloride	C.p.	0.38–0.97
	C.v.	0.39–0.98
	C.p.[2]	0.16–0.31
	C.v.[2]	0.31–0.61

*Mean values from several tests
C.p., *Coniophora puteana*
C.v., *Coriolus versicolor*
(1) Applied in water
(2) Applied in acetone
In all other cases the solvent was petroleum ether.

adjusted so that a predetermined excess of solution is absorbed by the timber. The treatment solution is then drawn off from the vessel and vacuum again applied so that excess solution is removed from the wood. On removal from the treatment vessel, the wood is dry to the touch. This double vacuum treatment ensures a controlled degree of impregnation, and hence enables a specified level of protection to be achieved reproducibly. Conditions can be varied to suit the characteristics of different wood types.[66] The mechanism by which the solvent penetrates into the wood has also been the subject of study.[67] Remedial treatments are usually applied to damaged wood after physical removal of decayed portions by spraying. To counter the risk of fire when solvents are applied in confined spaces, emulsions have been produced where each sprayed droplet has an outer film of water.[68]

Common fungal causes of timber decay include dry rot (*Merulius lachrymans*) which requires moist, warm, still conditions for growth. Dry-rot spores germinate when accidental wetting of the wood occurs, followed by partial drying allowing optimum levels of humidity for growth. Hyphae (rootlings) penetrate into the wood and can even penetrate masonry. Affected wood takes on a powdery appearance and cuboidal cracking may also occur. Wet rots such as *Coniophora puteana* or *Poria vaporaria* require more moisture to thrive than does the dry-rot fungus and depend on continuous dampness or running water to keep the wood damp. White rot, *Coriolus versicolor*, occurs in hardwoods, particularly when there is ground contact. Decayed wood is lighter in colour and much weaker than unharmed wood. The processes by which wood decays and the mechanisms involved in protecting the wood have been the subject of much recent study, and this work has been summarized by Blunden *et al.*[2] There is still some difference of opinion regarding the mechanism by which organotin compounds exert their protective action, some workers favouring a protection of active sites on the wood cellulose molecule, whilst others suggest intracellular attack on the fungus.

The most obvious forms of decay result from an infiltration process by Basidiomycetes which cause brown and white rots, bacteria, moulds, staining fungi and soft rots. Different phases of this colonization present different tolerances to preservatives. The interactions between such organisms and organotin preservatives are of great interest in assessing the commercial performance of preservative-treated timber.[69] There is evidence that some of these organisms, for example *Coniophora puteana, Chaetomium globosum* and *Aureobasidium pullulans*, can cause the breakdown of tributyltin compounds in the wood. Hill *et al.*[70] made biological and chemical observations on the early fungal colonization of Swedish redwood stakes which had been treated with TBTO by double vacuum impregnation or by immersion. Loss of TBTO occurred from the outer 1 mm layer of the wood within two months of exposure, accompanied by some degradation of the tributyltin to dibutyltin, monobutyltin and inorganic tin. However, after this initial period, both these processes were very much reduced. The effect is considered to be a combination of biological, chemical and physical factors. It should be emphasized that, although these studies have demonstrated some breakdown of organotin preservative in the wood, in practice organotin compounds have been used commercially for over 25 years without any reported failures in service. There is evidence that protection continues even when considerable amounts of the tributyltin compound have been de-alkylated to dibutyltin or mono-butyltin forms.

Although TBTO is toxic to certain species of insect, the results of tests carried out against wood-boring beetles suggest that its effectiveness is limited at the concentrations normally employed. For this reason TBTO is often combined with organic contact insecticides in order to provide more complete protection. In this case there may be a synergistic effect between the two preservatives. Increasing interest is being shown in the use of synthetic pyrethroids as insecticides in wood preservative formulations, since they offer a reduced risk of environmental pollution over the chlorinated hydrocarbons used at present. However, some incompatibility has been observed between such pyrethroids and TBTO, and this reduced their effectiveness. In view of this, a study has been made of the stability of synthetic pyrethroids (cypermethrin, deltamethrin and permethrin) to TBTO and other tributyltin compounds in toluene solution.[71] It was found that, whereas TBTO did react with the synthetic pyrethroids to varying degrees, other compounds such as tributyltin naphthenate showed no such reactivity. A possible mechanism for the reaction with TBTO was proposed.

The low aqueous solubility of TBTO and other commercially used tributyltin preservatives has restricted their industrial use in waterborne systems. An approach to overcome this limitation has been to synthesize a water-soluble biocide, and to-date the most promising approach has been via the tributyltin alkanesulphonates. The trialkyltin methane and ethanesulphonates have been shown to possess excellent activity when applied in aqueous solutions to wood substrates.[72,73] Aqueous solutions of these alkanesulphonates can be prepared in distilled water with concentrations up to 3%. Although the presence of salts in the water can cause a reduction in aqueous solubility, a recent study[74] has shown that this is not likely to be a problem with the natural waters likely to be used commercially.

13.5.2 *Crop protection*

Greater efficiency in the cultivation of crops has meant, in many cases, the dedication of large areas of land solely to one crop in order to be cost effective. Paradoxically this has meant a greater vulnerability to attack by colonies of fungi and insects. A solution to the problem has been found in the careful use of pesticides, and chemical control methods have proved very effective in protecting crops, particularly when based on organometallic compounds. Organotin chemicals began to be used as pesticides in the early 1960s. Basic studies on the bioactive properties of triorganotin compounds had shown the high fungistatic activity of triphenyltins, accompanied by a low activity against plants, and a formulation based on triphenyltin acetate was introduced for crop protection.[75] Shortly afterwards, triphenyltin hydroxide, which has very similar properties, was also commercialized.[76] These compounds proved effective in protecting crops against a variety of fungal diseases, including leaf spot on celery (*Septoria apii*), rice blast (*Piricularia oryzae*), coffee leaf rust (*Hemeleia vastatrix*) and coffee berry disease (*Colletotrichum coffeanum*).

In 1968, tricyclohexyltin hydroxide, which possesses very high activity against plant-feeding mites, was introduced as an acaricide for use on apple, pear and citrus fruit trees.[77] Unfortunately, subsequent tests on animals have demonstrated possible teratogenic effects in pregnant females, and this compound has now been withdrawn despite its successful use over a period of time.

Two other miticides have been introduced more recently, bis(2-methyl-2-phenylpropyltin) oxide,[78] commonly referred to as bis(trineophyltin) oxide, and 1-tricyclohexylstannyl-1,2,4 triazole.[79] These acaricides are highly selective and give excellent

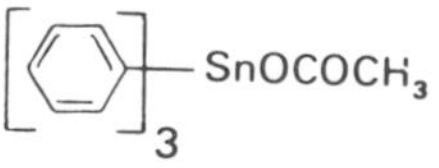

$[C_6H_5]_3SnOCOCH_3$

Fentin acetate (Brestan)
(Höechst)
mp 118–120°
LD_{50}(rats) 125 mg/kg

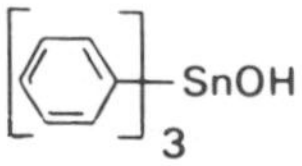

$[C_6H_5]_3SnOH$

Fentin hydroxide (Du-Ter)
(Philips Duphar)
mp 116–120°
LD_{50}(rats) 108 mg/kg

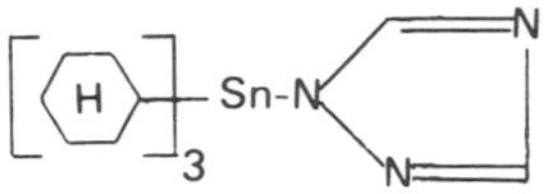

Azocyclotin (Peropal)
(Bayer)
mp 218·8°
LD_{50}(rats) 631 mg/kg

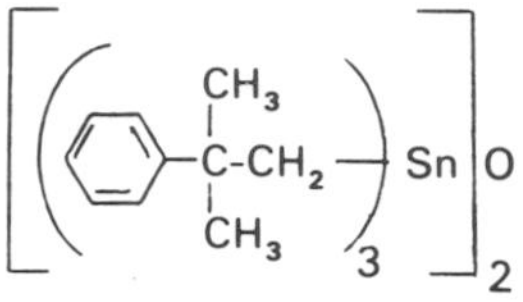

Fenbutatin Oxide (Vendex)
(Shell)
138–139°
LD_{50}(rats) 2,630 mg/kg
LD_{50}(bees) 100 μg/bee

Figure 13.4 Structures of organotin agrochemicals.

control of harmful arachnids such as the two-spotted spider mite, the European red mite and the Pacific spider mite, with little effect on predatory mites and insects. Chemical structures of the organotin pesticides are shown in Figure 13.4.

The principal advantages of organotin agrochemicals (which possess mainly prophylactic action and do not exert systemic effects) are their low toxicity to the crops to be protected, their generally low toxicity to non-target organisms, and lack of acquired resistance to these compounds by the pests. Other advantages lie in the fact that they are effective at low concentrations, and that they break down under the effects of the environment into less toxic di- and mono-derivatives and eventually into harmless inorganic forms of tin, so that their use constitutes no long-term pollution problem.

Since agricultural chemicals are involved with foodstuffs, considerable care is taken to ensure their safe application. Introduction of a new pesticide is carefully phased, allowing opportunity to seek early evidence of possible toxic effects. Safe dosage levels are recommended and the number of applications and safe waiting times between treatment and harvesting of crops are specified. In the case of the triphenyltin compounds, it is considered that the chance of residues reaching the consumer via sprayed crops is very limited, since the residues of these compounds or their metabolites occur only on the surface of the plants and are easily removed by peeling, washing, shelling, etc., and are in any case readily broken down by cooking.[80] Similar conclusions have been reached for the organotin acaricides.

A very large body of research has been conducted into organotin agrochemicals, and this is concerned both with improving the effectiveness of these compounds in known applications and with assessing their potential in new uses. Research since

1980 has been comprehensively reviewed.[81] A number of workers have looked at the possibilities of using triphenyltin compounds to control diseases in wheat, pepper plants, tobacco crops, soya beans and ground nuts. Organotin compounds can often be used effectively in binary combinations with other pesticides, often with synergistic effect. Such synergistic mixtures have been used against the cotton leafworm. Organotin acaricides have been tested against mites attacking landscape plants and cotton plants. These and other studies have been reviewed by Evans and Karpel.[6]

Modern pest management systems are concerned with keeping the level of toxic chemicals to a minimum and a number of approaches involving organotins have been reported. Hoy *et al.*[82] used tricyclohexyltin hydroxide and propargite in conjunction with natural resistant predators to control spider mites on almond orchards. Laboratory-reared pesticide-resistant strains of a predatory mite were released to control spider mites in commercial almond orchards, in conjunction with lower than normal applications of the miticides.

An interesting concept is that of using chemicals to inhibit feeding by insect pests. The possibility of using organotin chemicals as antifeedants was first demonstrated,[83] when it was shown that triphenyltin acetate present on sugar beet leaves inhibited feeding by *Spodoptera littoralis*. An antifeedant produces a cessation of feeding, preventing the insect from recognizing the normal host plant gustatory stimulus by inhibiting taste receptors. The result is that the insect pest starves to death or is eaten by predators.[81] The principal advantages of antifeedants over conventional techniques are that beneficial and non-target insects are not affected since they do not feed from the crop, and that only sublethal concentrations of the compounds are used. Moreover, control is faster than in the case of conventional insecticides, since insects may continue to feed while these are initiating their lethal action.

A further approach is the use of chemicals as chemosterilants which interfere with the reproductive cycle of the insect. The first detailed report of organotins displaying this property was by Kenager[84] who demonstrated that various types of insect showed diminished or no reproduction after feeding on triphenyltin derivatives. To date, little has been done in the way of controlled release methods for pesticides, although the technique has been used, for example, in molluscicides to control schistosomiasis. These newer techniques may well play an important part in future methods of pest control.

13.5.3 *Antifouling coatings*

An important use for triorganotin compounds is in antifouling paints. When a surface is immersed in sea-water for any length of time, it becomes part of the environment of millions of organisms which inhabit that water and as such becomes a target for their attachment. This situation represents a very real problem for boat owners. Not only does marine fouling reduce the operating efficiency of a vessel due to frictional drag and excessive weight, but in the case of large oceangoing ships such as liners and supertankers, fuel costs can be considerably increased. Removal of fouling entails periods out of service and possible mechanical damage to the hull.

The solution, which has been adopted for some years, is to coat the surfaces in contact with water with a paint containing a toxicant, which is slowly released to the surface, providing a very thin but effective barrier layer to approaching organisms. Before the 1960s, cuprous oxide was the preferred toxicant in such paints. Later, the powerful biocidal activity of triorganotin compounds became recognized and led to their being tested as antifouling toxicants. Paints containing organotin compounds first appeared in the 1960s, and today they are widely used in a variety of formulations, and sophisticated release systems have been developed to ensure that the toxicant is released at a slow and controlled rate over a long period of time. The development of these systems has been fully reviewed recently.[85]

In its simplest form, an antifouling paint contains a vehicle (for continuity and adhesion), a solvent (for ease of application), pigment (for colour, gloss and hardness), and additives to modify the paint properties, as well as the antifouling toxicant. The organotin compound, usually bis(tributyltin) oxide or tributyltin fluoride, is physically mixed with the paint. There are two types of paint system: those with a soluble matrix having a binder soluble in sea-water to release the toxicant, and those with an insoluble matrix, from which the toxicant is released by diffusion. Early developments in these systems were reviewed by Bennett and Zedler.[86]

Advantages of organotin compounds include a higher biocidal activity than copper so that lower release rates can be used, the ability to be used in white or light-coloured paints, and suitability for use on an aluminium hull where the use of copper could lead to problems with bimetallic corrosion. In these antifouling systems bis(tributyltin) oxide or tributyltin fluoride are incorporated in a vinyl/rosin or a chlorinated rubber/rosin paint. TBTO can only be used at levels up to 13 wt% in vinyl paints before it begins to affect paint performance. Tributyltin fluoride, however, can be used at levels up to 30 wt% and in general gives longer service life. The function of the rosin is to permit greater diffusion of water into the film, thereby assisting mobility of toxicant to the surface. More recently, triphenyltin compounds have been used in vinyl or rubber paints, particularly in Japan.

Because of the mechanism by which toxicant is released from these early systems, the initial release rate is high, and this reduces the effective life of the antifouling to about 1½–2 years. For small boats and for many merchant ships this is sufficient. However, in the case of naval vessels and supertankers, longer periods of protection are needed, and other paint systems have been developed which offer longer protection.[6] One approach has been to incorporate organotin compounds in rubber sheet.[87] This technique was first used to protect sonar buoys immersed in sea-water as part of tactical early-warning systems, and subsequently was adapted to seagoing vessels. The period of protection afforded by these rubbers is a function of the amount of toxicant present and the thickness of the elastomer. The leaching mechanism is a dissolution process, and protection for over 84 months in tropical waters has been claimed. A novel concept in incorporating organotin compounds into elastomers has been developed,[88] and involves addition of an unsaturated monomer such as tributyltin acrylate, followed by curing with a peroxide. The elastomer cross-links, homopolymerization of the organotin monomer occurs, and the elastomer and the organotin compound co-vulcanize. This technique makes it possible to incorporate up to 50 wt % of tributyltin acrylate into different types of elastomers.

During the late 1960s and 1970s much work took place to develop organotin polymers which could be used in long-life antifouling systems. The commercially viable systems usually incorporate a triorganotin carboxylate group chemically bound to an acrylate polymer which serves as the binder. In contact with sea-water, the organotin ester link is hydrolysed and the organotin toxicant is released. The depleted outer layer of the paint film is then eroded by motion of the sea-water past it, exposing fresh surface layers. There is an ablative effect, and turbulence occurring at peaks of roughness in the coating causes these peaks to erode faster. The result is an overall smoothing of the surface, and this 'self-polishing' facility of these paints has advantages which have been exploited commercially.[89] The significance of this smoothing effect can be appreciated when it is considered that 80% of the resistance to motion of a vessel is due to friction. Advantages of these ablative coatings include protective life proportional to coating thickness; controlled release rate; ability to overcoat without loss of activity; no depleted coating to be removed before repainting; efficient use of toxicant; continuous replenishment of active surface; and the polishing effects already described. Protection for up to 5 years can be achieved with these systems. The introduction of high-build versions with a higher solids content has reduced the number of coatings required. A third generation of ablative paints offers further savings by reducing the level of tributyltin methacrylate needed in the polymer.

Another type of polymeric coating is based on polysiloxanes in which triorganotin compounds are chemically bonded to a siloxane polymeric backbone. Hydrolysis of the tin–oxygen–silicon bond provides chemically controlled release of the toxicant. Commercial development of these coatings has been described.[90] Typically, a commercial product may consist of tributyltin groups attached to a 40% hydrolysed ethyl silicate prepolymer. However, there are many variations possible, and this allows the properties of the coatings to be modified to suit particular end-uses. Thus, varying the ratio of organotin to silicon will change the release rate, resin compatibility, longevity, and other film properties. Varying the organotin moiety, for example by introducing triphenyltins, also affords an opportunity to modify properties. Unlike organotin acrylate polymers, which are typically linear, polysiloxanes exhibit three-dimensional cross-linking, and this provides a denser, more durable film.

The release mechanism involves contact with atmospheric moisture which cures the polymer, rendering it naturally hydrophobic. As the organotin portion is removed by hydrolysis, the polymer becomes hydrophilic, allowing water to penetrate the matrix and facilitate removal of other co-toxicants which may be present. The polysiloxane coating is largely insoluble and mostly remains on the hull, so that there is no self-polishing action. Commercial coatings are available, based on this polysiloxane system.

Antifouling coatings constitute an area where organotin biocides are deliberately released into the environment. For this reason much work has been conducted into studying their fate in the environment, and this work has been reviewed.[85] The greatest risk of environmental pollution resulting from antifouling coatings exists not in the open oceans but in regions where large numbers of vessels remain stationary for relatively long periods, for example in harbours, marinas, estuaries and bays. Owing to the continuous release of organotin antifoulant, albeit in very small amounts, a significant build-up can occur, particularly if the vessels are painted with conventional systems. Thus, a short-term problem was encountered in France with oyster farms located close to estuarine harbours. The presence of very low levels of tributyltin compounds was shown to affect adversely the quality of the Pacific oyster being reared there.[91]

This was confirmed in subsequent work in the UK and led to the imposition, from January 1986, of legislation restricting sale of organotin-based antifouling paints, and from July 1986 of a total ban in the UK on retail sale of these paints.[92] The Environmental Protection Agency in the USA is requiring manufacturers to submit data on leach rates from tributyltin-containing products in view of possible risks to oysters, mussels and clams. A comprehensive survey of the effects of environmental levels of organotins on aquatic biota has been made.[93] Although some short-term effects may occur in localized areas, the bulk of evidence suggests that long-term pollution is not likely to be a problem. In areas where tributyltin species have been detected, significant levels of less toxic di- and monobutyl derivatives have also been found, presumably arising from degradation processes. The use of polymeric antifouling coatings also reduces the risk of pollution. It has been estimated that over a season, release of toxicant from copolymer paints is less than a quarter of that from conventional antifouling coatings.

13.5.4 *Materials protection*

In addition to the applications already discussed, triorganotin compounds are also used to protect other materials against biodeterioration. Biocidal protection is defined in general terms as the activation or treatment of materials in order to prevent damage by micro-organisms such as fungi, bacteria or, in some cases, algae. These micro-organisms can colonize a wide variety of materials such as wood, textiles and even plastics. A review of the use of organotins for biocidal protection has been published.[94]

Stone and masonry structures are susceptible to attachment of biological growths such as mosses and lichens which in some cases have an unwanted, disfiguring effect, for example on gravestones or pathways. The problem of controlling such growths has been reviewed by Richardson.[95] Proprietary treatments based on TBTO and a quaternary ammonium compound have proved effective against this kind of problem, providing long-term protection to treated surfaces. The quaternary ammonium compound solubilizes the organotin, and it also has a biological activity of its own, rupturing the cell walls of algae and lichens as well as exerting a toxic action. Water-soluble tributyltin compounds also have some potential in this area.[72]

Organotin biocides can be used to protect certain textile fabrics against insect attack, and also against micro-organisms causing rotting of fabric. Work has been carried out to demonstrate the possibilities of triphenyltin compounds for insect-proofing of wool.[96,97] The organotin compounds did not adversely affect textile properties and acted as a larvicide for the clothes moth and as a pronounced antifeedant for the carpet beetle during a 14-day bioassay. Textiles based on natural materials such as cotton, silk and wool are also prone to attack by micro-organisms with a resulting loss of strength and some discoloration. TBTO, tributyltin linoleate and tributyltin naphthenate are effective against a wide range of fungi and bacteria and have been tested in this application.

Tributyltin compounds have also been used to protect plastics and sealing materials such as silicones, polysulphides and polyacrylics.[94] An organotin-based formulation (TBTO plus hexachlorocyclohexane) is also available for protecting old books, manuscripts, etc., from insecticidal and fungal attack. One report[98] has assessed the possibilities of using tributyltin chloride to prevent microbial contamination of banknotes in the Federal Republic of Germany. The compound is chemically bonded to the cellulose surface so that the biologically active molecule is not released to the handler. Combinations of TBTO or tributyltin benzoate with quaternary ammonium compounds have been used to prevent slime formation during papermaking,[5] counteracting not only bacteria, but also fungi and algae. Tributyltin benzoate has also been used in disinfectant formulations applied to hospital floors and walls to prevent cross-infection, and to sanitize clothing such as footwear.

13.6 Heterogeneous catalysts

Tin (IV) oxide is used as a heterogeneous catalyst in a number of industrial processes, often in combination with a second oxide. In heterogeneous catalysis the function of the catalyst is to provide a suitable solid surface upon which the required reaction occurs, and reactants are commonly present in the gaseous state. Tin (IV) oxide with vanadium oxide can be used for the oxidation of aromatic compounds such as benzene, toluene or naphthalene to prepare acids and acid anhydrides. With antimony oxide or molybdenum oxide it can be used for the selective oxidation and ammoxidation of propylene to acrolein, acrylic acid and acrylonitrile. Tin (IV) oxide with phosphorus oxide is used in organic oxidative dehydrogenation reactions.[99] Recently a complex oxide system, $Fe_2O_3/SnO_2/Cr_2O_3/K_2O$, has been employed in the selective methylation of phenol, to form 2,6-xylenol, a precursor for the engineering plastic, poly(phenylene oxide).[100]

Another group of tin-containing catalysts which have found industrial use are the tin–platinum and tin–rhenium systems which are usually supported on alumina. The incorporation of tin with the precious metal appears to have a markedly beneficial effect on the catalytic activity for a number of reactions carried out in the petroleum industry, including dehydrogenation, dehydrocyclization, cracking, isomerization and hydrogenation of hydrocarbons.[99]

Tin (IV) oxide-based systems also act as catalysts for the low-temperature oxidation of carbon monoxide by oxygen and nitrogen oxides. Although tin (IV) oxide itself is moderately effective for the oxidation of carbon monoxide by oxygen or nitrous oxide, co-precipitated SnO_2–CuO gels (when thermally activated at temperatures up to 450°C) are considerably more active catalysts for the low-temperature oxidation of carbon monoxide by oxygen or nitric oxide.[101] The most promising approach has been found to involve precious metal/tin (IV) oxide systems in which tin (IV) oxide is used as a support for palladium or platinum. A synergistic effect occurs between tin (IV) oxide and the precious metal, and it has been suggested that the increased activity is attributable to a 'spillover' of activated carbon monoxide from the palladium on to the tin (IV) oxide surface. An additional advantage of this system is that, unlike other low-temperature-active catalysts, the catalytic activity is actually enhanced by the presence of water vapour. Possible applications for these catalyst systems include air purification systems, cigarette filters, and fuel exhaust systems.

One area where this catalyst combination has been used is in a carbon dioxide gas laser.[102,103] Laser emission is initiated by electrical discharge within an envelope containing the gas. The discharge causes dissociation of carbon dioxide into carbon monoxide and oxygen and, in a sealed system, unless these gases are recombined into carbon dioxide, some loss of output will result. It was found that palladium, finely dispersed on a tin (IV) oxide substrate, was a highly efficient recombination catalyst at ambient temperature and allowed continuous high-pulse repetition frequency to be obtained with high output power for long periods without the need to replenish the laser gas.

More fundamental aspects of the surface chemistry of these types of system are discussed in Chapter 12.

13.7 Pharmaceuticals

Tin chemicals have a limited, but important use as pharmaceutical agents in the fields of dentistry, medicine and veterinary science.

13.7.1 *Dentistry*

Tin chemicals have a place in dentistry, perhaps the most important application being the use of tin (II) fluoride as an aid in preventing tooth decay. Tin (II) fluoride is used in toothpastes, dentifrices, topical solutions, mouthwashes and, occasionally, as a constituent of dental cements of the zinc oxide/polyacrylic acid type. Tin (II) fluoride has a number of advantages over sodium fluoride for these applications: it protects dental enamel against dissolution by lactic acid to a greater extent; it produces a greater reduction in dental mottling; and it exerts a much higher inhibition of dental plaque growth. Tin (II) fluoride is compatible with acidulated phosphate fluoride and this combination has been suggested as suitable for the control of root caries and hypersensitivity.[104] A major disadvantage has been the instability of aqueous solutions of tin (II) fluoride, although this problem can now be overcome by the addition of suitable stabilizers such as glycerol, sugars and gums to the solutions.

The use of tin chemicals as curing agents in RTV silicones has already been described. Certain types of RTV silicone are used in the production of dental prosthetic devices such as crowns, bridges and inlays.[105] These require an accurate impression of the jaw, with or without teeth, as a preliminary to making the model. Special products made for dentistry transform from a pasty, spreadable consistency to a firm, elastic state in 4 or 5 minutes at mouth temperature. Shortly after making the impression a pattern or 'die' is cast in it, for subsequent work. Tin (II) 2-ethylhexoate or dibutyltin dilaurate are the preferred catalysts for these systems.

13.7.2 *Medical applications*

Tin (II) salts are used in radiopharmaceuticals. Technetium-99m is the radioisotope which has the optimum nuclear properties for clinical gamma-ray scanning, and many imaging procedures based on this nuclide are in use. Typically the technetium-99m radiopharmaceuticals (except the technetate itself) are prepared by the reduction of $^{99}TcO_4^-$ in the presence of a suitable ligand and the preferred reductant is tin (II) chloride in aqueous hydrochloric acid.

Recent work has raised the possibility that certain diorganotin compounds may have a role to play in cancer therapy, and the extensive literature on the subject has been reviewed.[106] A promising approach is to model compounds on the cisplatin drugs which are widely used in combating certain types of cancer.[107] Diorganotin dihalide and dipseudohalide octahedral complexes with monodentate or bidentate ligands have exhibited reproducible activity against P388 lymphocytic leukaemia in mice. This activity is lower than that of cisplatin and its analogues, but toxic side effects which are a major disadvantage of the platinum compounds are less severe.

Another promising field is in the prevention of neonatal jaundice (hyperbilirubinaemia) which is a severe haemolytic disorder caused by excessive plasma bilirubin levels in newborn babies. Haeme oxygenase is the agent responsible for the degradation of the haem (iron proto-porphyrin) species of the fetal haemoglobin molecule, and it was discovered that a tin (II) chelate, dichloro(protoporphyrin IX) tin (IV), also known as tin-haem, is a potent inhibitor of haem oxygenase activity in the liver, spleen, kidney and skin of rats.[108] When tin-haem was administered, an immediate (24-h) and significant lowering of serum bilirubin levels occurred and this decline continued. Studies have shown that tin-haem itself is eliminated without metabolic alteration, and no significant toxic responses to large doses in new-born animals were seen. Thus, tin-haem may well have a future in clinical treatments of humans, particularly of severely jaundiced newborn babies, and studies are in progrcss.

In the field of preventive medicine, triorganotin compounds have potential for restricting the spread of schistosomiasis (bilharzia). This is a parasitic disease encountered in tropical regions such as Central and Southern America, Asia and Africa, transmitted to humans from contact with infested water. Certain species of freshwater snail act as hosts for intermediate stages in the life cycle of these blood flukes (schistosomes), and one way of controlling spread of the disease is by killing off these snails, thus breaking the life cycle. A number of organotin compounds, notably triphenyltin chloride and acetate, TBTO and tributyltin fluoride, have been shown to be effective molluscicides at low concentrations.[76] An interesting extension of this application has been to incorporate the organotin compound in rubber pellets which can float on infested waters. The organotin is released slowly at levels toxic to the snails but not lethal to other aquatic life.[109]

13.7.3 *Veterinary applications*

Organotin compounds have been used as antheliminthic agents in poultry and as insecticides for sheep and cattle. Worm infestations in poultry, although they cause a relatively minor death toll, are responsible for retarded growth, loss of meat and egg production, and wasted feed and labour. Three types of worm are involved: large roundworms which penetrate the lining of the intestine, caecal worms which inhabit the blind ends of the caeca, and tapeworms. Dibutyltin dilaurate was shown in experimental studies to be capable of removing a wide range of species of tapeworm which affect poultry, and this compound has been used in commercial formulations in combination with piperazine (to combat roundworms) and phenothiazine (active against caecal worms).[6] Another parasitic disease, coccidiosis, can be a problem in turkeys reared in confinement, leading to poor feed utilization and stunted growth. Dibutyltin dilaurate is also employed in commercial products to prevent the development of this disease.

13.8 Fire prevention

Recent years have seen intensified research throughout the world to develop fire-retardant chemicals and treatments, particularly for fabrics made from natural and synthetic fibres and for plastics used in buildings and furniture. A number of treatments based on tin chemicals are being used commercially and others are under development.[110,111]

13.8.1 *Natural fibre treatments*

Although wool has not in the past been considered as presenting a serious flammability hazard, stricter legislation has generated interest in chemical fire retardants for this material. Flame-resistant treatments based on titanium and zirconium are widely used on woollen fabrics but have several disadvantages when applied to woolly sheepskins. Two tin-based treatments have been developed for this particular application.[112] The first, developed by the Wool Research Organisation of New Zealand (WRONZ), is a spray treatment based on $SnCl_4 \cdot 5H_2O$, ammonium bifluoride, isopropanol and a polishing agent. The treatment is said to be wash-fast and resistant to dry cleaning. About 4000 sheepskins for export are so treated each year in New Zealand.

A second process, developed by the International Wool Secretariat for use on all-wool flokati rugs, consists of K_2ZrF_6, $SnCl_2$ and HCl. The rug is treated for 30 minutes at 75°C with a liquor:wool ratio of 20:1. Flame resistance is maintained after 10 machine washings.

At the International Tin Research Institute a systematic study of several inorganic tin (II) and tin (IV) compounds as flame retardants for cotton fabrics has been carried out.[113] Two of these treatments are moderately durable. The first consists of impregnation with $Na_2Sn(OH)_6$ followed by treatment with $Zn(NO_3)_2 \cdot 4H_2O$ so that an insoluble deposit of $ZnSn(OH)_6$ forms within the fibres; the second involves a first impregnation with $Na_2WO_4 \cdot 2H_2O$ followed by treatment with $SnCl_2 \cdot 5H_2O$ to deposit insoluble tin/tungsten oxide within the fibres.

13.8.2 *Additives for synthetic polymers*

Commercially available retardants for synthetic polymers have largely been based on halogen/antimony (III) oxide combinations for hydrocarbon polymers and on halogen/phosphorus systems for oxygen-containing polymers. However, because of the increasing cost of antimony oxide and concern about the toxicity of many phosphorus additives, certain inorganic compounds are being examined as partial or total substituents. Work relating to tin compounds in this field has been summarized.[114]

An interesting finding has been the marked smoke-suppressant properties of the tin chemicals. Tin compounds may act as flame-retardant synergists with halogenated compounds in polyolefins, and a newly developed system consisting of a brominated organic compound, a phosphite and an organotin compound is claimed to be a very effective flame retardant for polyolefins. It has now reached commercialization and is used at the 4% level.[115] Researchers in the USA have found that tin (IV) oxide, used either alone or in combination with certain molybdenum compounds, is an effective flame retardant and smoke suppressant for rigid PVC.[116] Other workers[117] have reported that hydrous tin (IV) oxide incorporated into ABS plastic along with a halogen-containing organic compound such as decabromobiphenyl, is effective both as a flame retardant and a smoke suppressant.

The incorporation of either anhydrous tin (IV) oxide or its hydrous form (beta- or meta-stannic acid) at a 2% level into halogenated polyester resins results in a substantial decrease in polymer flammability and smoke generation.[118,119] There is evidence of a marked flame-retardant synergism between tin and halogen, accompanied by a significant reduction in smoke evolution from the burning polymer. The mode of action of the tin compounds appears to involve both the condensed and the vapour phases.

Another approach has been to incorporate a series of hydroxy- and anhydrous stannates, $MSn(OH)_6$ and $MSnO_3$, at a 2 phr level in glass-reinforced polyester panels.[120] Although there is no apparent increase in flame retardancy, substantial reductions in smoke evolution were observed. Their use in combination with a fire-retardant halogen would reduce the amount of halogen required as well as suppress the formation of smoke and other toxic products during burning. As such, these compounds may have a commercial future.

13.9 Miscellaneous uses

Tin chemicals have a wide range of other uses. Although some of these are quite small, others have potential for considerable growth. Again, projects now at the research and development stage may have potential for commercial development.

13.9.1 *Gas sensors*

There is an expanding need for devices that are able to detect and monitor the presence of various gases, and considerable development work is being carried out to produce sensors that are low in cost, easy to operate, sensitive and reasonably selective for particular gases. Tin(IV) oxide is an *n*-type semiconductor and has been the most widely used material for sensors of the semiconductor type, particularly in Japan (Figure 13.5). Construction of these sensors and commercial developments have been reviewed by Karpel.[121] The tin(IV) oxide in a gas detector may be in the form of a porous, sintered material or else a thin (100 nm) layer deposited on a flat ceramic substrate. Incorporation of palladium or platinum within the oxide can greatly enhance sensitivity and selectivity.[122,123] The whole technology of gas sensors is advancing rapidly and is likely to find increasing application in the future for monitoring and automatic control of a wide range of processes. This topic is discussed further in Chapter 12.

Figure 13.5 Gas sensors based on tin(IV) oxide. Photograph courtesy ITRI and Envin (Environmental) Ltd, Wantage, UK.

13.9.2 *Coal liquefaction catalysts*

High petroleum costs and the depletion of oil reserves have led to renewed interest in coal liquefaction as a source of fuel. A good catalyst is important in efficient liquefaction processes, and studies have been conducted aimed at elucidating the mechanism of attachment of tin compounds to various coals and assessing their performance as coal liquefaction catalysts.[124] Preliminary laboratory scale tests have been encouraging, particularly for tin incorporated into an alumina support. Tin appears to be converted into several different species during liquefaction, including tin (II) sulphide and tin (IV) oxide.

13.9.3 *Corrosion-resistant primer paints*

The long-term protection of steel against rusting under aggressive conditions is still a difficult problem. Prolonged tests of inorganic tin (IV) compounds as a constituent of primer paints have shown that they can perform as well as the traditionally-used zinc phosphate.[125] Chlorinated rubber paints containing the hydroxystannates of calcium and strontium show excellent protection, where a top-coat is used, after ten years' exposure of painted steel panels in an urban atmosphere. Zinc stannate and hydroxystannate have shown good performance in an alkyl resin formulation with titanium dioxide in a similar exposure test over 6 years.

13.9.4 *Water repellents*

Water-repellent properties have been exhibited by certain mono-*n*-alkyltin compounds and these have been tested on building materials (limestone, bricks and concrete)[126] and on cellulosic substrates (cotton, paper and wood).[127] Octyltin trilaurate has been shown to impart a comparable water repellency to limestone as a commercial silicone treatment. Other compounds such as octyl- and dodecyltin trichlorides show potential as damp-proofing treatments for bricks. When assessed on cotton fabrics in British Standard 3702 vertical spray test, sodium butanestannonate, butylchlorotin dihydroxide, *n*-octylchlorotin dihydroxide and butyl tris(triphenylsilanoxy)tin all imparted water repellency to the fabric.[112]

13.9.5 *Reducing agents*

The strongly reducing properties of the tin (II) cation find application in reagents for preparative organic and inorganic chemistry.[128] Reactions of industrial significance include the reduction of nitriles or acid anilides to the corresponding aldehyde, reduction of the olefinic linkage in unsaturated carbonyl compounds to form the corresponding substituted monoalkyltin trichloride, a precursor for estertin PVC stabilizers, and reduction of nitro groups to amines.

Tin (II) chloride reductions are employed in the sensitization stage of certain surface-coating techniques. Electroless plating is a technique allowing the deposition of metallic films (often copper or nickel) on plastics substrates. It is effected by first immersing the part in tin (II) chloride/hydrochloric acid solution and then in palladium chloride solution. Tin (II) ions are adsorbed in the pits of the etched plastics substrate and these reduce palladium ions to the metal, creating a catalytic layer of palladium on the surface. This subsequently causes the metallic deposition of copper or nickel on the surface when the plastic is immersed in solutions of their salts. Reductions of silver halide salts to metallic silver by tin (II) salts find limited application in the photographic field, including reduction sensitizing of silver halide emulsions, activators in reversal processing, and conventional developing of silver images.

References

1. T. Crisafulli, *American Metal Market*, Dec. 16th 1985, 22.
2. S.J. Blunden, P.A. Cusack and R. Hill, *The Industrial Uses of Tin Chemicals*, Royal Society of Chemistry, London, 1985, 10–11.
3. U. Thust, *Tin and its Uses*, 1974, **122**, 3–5.
4. R.F. Bennett, *R. Soc. Chem. Ind. Chem. Bull.*, 1983, **2** (6), 171.
5. A. Bokranz and H. Plum, *Topics Curr. Chem.*, 1971, **16**, 365.
6. C.J. Evans, *Tin and its Uses*, 1981, **130**, 5.
6. C.J. Evans and S. Karpel, *Organotin Compounds in Modern Technology*, J. Organomet. Chem. Library **16**, Elsevier, Amsterdam, 1985.
7. K.S. Minsker, M.I. Abdullin, S.V. Kolesov and G.E. Zaikov, *Developments in Polymer Stabilisation-6*, Applied Science, London, 1983, 173.
8. G. Ayrey, B.C. Head and R.C. Poller, *J. Polym. Sci., Macromol. Rev.*, 1974, **8**, 1–49.
9. J.W. Burley, *Appl. Organomet. Chem.*, 1987, **1** (2), 95.
9. M.T. Bomar and H. Müller, *Das Papier*, 1985, **39** (3), 110.
10. D. Braun, in *Degradation and Stabilisation of Polymers*, ed. G. Guesbens, Applied Science, London, 1975.
11. J.W. Burley, *Tin and its Uses*, 1977, **111**, 10–12.
12. B.B. Cooray, *Polym. Degr. and Stabil.*, 1984, **7**, 1.
13. E.D. Owen, *Degradation and Stabilisation of PVC*, ed. E.D. Owen, Chap. 5, Elsevier-Applied Science, London, 1984.
15. G. Abeler and J. Büssing, *Kunststoffe*, 1981, **71** (5), 315.
16. L.R. Brecker, *Pure Appl. Chem.*, 1981, **53** (2), 577.
17. S. Karpel, *Tin and its Uses*, 1984, **139**, 10.
18. D.W. Allen, J.S. Brooks, J. Unwin and J.D. McGuinness, *Appl. Organomet. Chem.*, 1987, **1** (4), 311.
19. A.D. Schwope, D.E. Till, D.J. Ehntholt, K.R. Sidman, R.H. Whelan, P.S. Schwartz and R.C. Reid, *Deutsche Lebensmitt-Rundsch.* 1986, **82** (9), 277.
20. S. Karpel, *Tin and its Uses*, 1981, **129**, 1.
21. S. Karpel, *Tin and its Uses*, 1984, **140**, 14.
22. S.L. Axelrood, C.W. Hamilton and K.C. Frisch, *Ind. Eng. Chem.*, 1961, **53** (11), 889.
23. T.E. Rusch and D.S. Raden, *Plastics Compounding*, 1980, **3** (4), 61, 64, 66, 69, 71.
24. S. Karpel, *Tin and its Uses*, 1980, **125**, 1.
25. J.A.C. Watt, *Chem. in Britain*, 1970, Dec., 519.
26. S. Karpel, *Tin and its Uses*, 1984, **142**, 6.
27. F.W. van der Weij, *Makromol. Chem.*, 1980, **181** (12), 2541.
28. H.G. Emblem and K. Jones, *Trans. J. Brit. Ceram. Soc.*, 1980, **79** (4), lvi.
29. M.H. Gitlitz and M.K. Moran, in *Kirk-Othmer Encyclopaedia of Chemical Technology*, 3rd edn., Vol. 23, Wiley, New York, 1983, 42.
30. C.J. Evans, *Glass*, 1984, November, **389**, 392.
31. G.B. Shaw, *Proc., Properties and Uses of Inorganic Tin Chemicals*, Brussels, Oct. 16th 1986.
32. W. Rieger, *Verre Bull. d'Information, Spec. Issue*, Jan. 1987, 3.
33. S.M. Budd, *Thin Solid Films*, 1981, **77**, 13.
34. N. Jackson and J. Ford, *Thin Solid Films*, 1981, **77**, 23.
35. S. Karpel, *Tin and its Uses*, 1987, **152**, 14.
36. C.J. Evans, *Tin and its Uses*, 1982, **132**, 5.
37. M. Court, G. Blandenet and Y. Lagard, Dopant and conductive effects of tin oxide prepared by an improved spray technique, *Verre Bull. d'Information, Spec. Issue*, Jan. 1987, 10.
38. L. Colombin and F. Sebastiano, *Verre Bull. d'Information, Spec. Issue*, Jan. 1987, 15.

39. J. Frémaux, *Verre Bull. d'Information, Spec. Issue*, Jan. 1987, 20.
40. R.R. Dean and C.J. Evans, *Tin and its Uses*, 1977, **113**, 12.
41. R.R. Dean and C.J. Evans, *Tin and its Uses*, 1977, **114**, 9.
42. T. Williamson, *Ceramic Ind. J.*, 1981, Feb., 29, 31.
43. D.V. Sanghani, G.R. Abrams and P.J. Smith, *Trans. J. Brit. Ceram. Soc.*, 1981, **80**, 210.
44. P. Escribano, M.C. Guillem and C. Guillem, *Trans. J. Brit. Ceram. Soc.*, 1983, **82**(6), 208.
45. C. Croft and M.J. Fuller, *Trans. J. Brit. Ceram. Soc.*, 1979, **78**(3), 52.
*46. A. Chapman, *Tin and Tin-Alloy Plating. A Review*, Publ. **606**, International Tin Research Institute, Greenford, UK, 1980.
47. J.C. Jongkind, *Plating*, 1968, **55**(7), 722.
48. Anon., *Instructions for Electrodepositing Tin*, Publ. **92**, International Tin Research Institute, Greenford, UK, 1959.
49. J.W. Price, *Tin and Tin Alloy Plating*, Electrochemical Publications Ltd., Scotland, 1983.
50. Anon., *Electroplated Tin–Nickel Alloy*, Publ. **235**, International Tin Research Institute, Greenford, UK, 1958.
51. H. Dillenburg, *Galvanotechnik*, 1972, **63**(4), 343.
52. M. Izaki, H. Enomoto and T. Omi, *Plating Surface Finish.*, 1987, **74**(6), 84.
53. Anon., *Electrodeposition of Tin–Lead Alloys*, Publ. **325**, International Tin Research Institute, Greenford, UK, 1961.
54. M.A.F. Samel, D.R. Gabe and D.R. Eastham, *Trans. Inst. Metal Finish.*, 1986, **64**, 119.
55. Lea Ronal Inc., US Patent 4 565 610, 1986.
56. Anon., *Tin–Zinc Alloy Plating*, Publ. **202**, International Tin Research Institute, Greenford, UK, 1952.
57. D.R. Cowieson and A.R. Scholefield, *Trans. Inst. Met. Finish.*, 1985, **63**(2), 56.
58. S. Karpel, *Tin and its Uses*, 1985, **146**, 10.
59. P.J. Smith, *Toxicological Data on Organotin Compounds*, Publ. **538**, International Tin Research Institute, Greenford, UK.
60. J.J. Zuckerman, R.P. Reisdorf, H.V. Ellis and R.R. Wilkinson, in *Organometals and Organometalloids: Occurrence and Fate in the Environment*, eds. F.E. Brinckman and J.M. Bellama, ACS Symp. Ser. **82**, American Chemical Society, Washington DC, 388.
61. Anon., *1980 Environmental Health Criteria 15: Tin and Organotin Compounds—A Preliminary Review*, World Health Organisation, Geneva, 1980.
62. S.J. Blunden and A. Chapman, in *Organometallic Compounds in the Environment*, ed. P.J. Craig, Chapter 3: Organotin Compounds in the Environment, Longman, London, 1986.
63. Anon., *Organotin Compounds: Wood Protection*, Schering AG, 1983.
64. H. Schweinfurth, *Tin and its Uses*, 1985, **143**, 9–12.
65. J.M. Baker, *J. Inst. Wood Sci.*, 1984, **10**(2), 82.
66. A.J. Crowe, R. Hill, P.J. Smith and T.R.G. Cox, *Int. J. Wood Preservation*, 1979, **1**(3), 119.
67. L.D.A. Saunders, *Brit. Wood Preserving Assn. Record Ann. Conv.*, 1982, June, 46.
68. C.J. Evans, *Tin and its Uses*, 1978, **115**, 11.
69. D.J. Dickinson and J.F. Levy, *British Wood Preserv. Assn.: Record Ann. Conv., June 26th–29th*, 1979, 33.
70. R. Hill, A.H. Chapman and A. Samuel, *15th Ann. Meeting Int. Res. Group on Wood Preservation*, Document IRG/WP/3311, 1984.
71. S.J. Blunden and R. Hill, *18th Ann. Meeting Internat. Res. Group on Wood Preservation 1987*, (to be published).
72. Anon., *Water-Soluble Organotin Biocides*, Publ. DS4, International Tin Research Institute, Greenford, UK, 1984.
73. R. Hill, P.J. Smith, J.N.R. Ruddick and K.W. Sweatman, *14th Ann. Meeting Int. Res. Group on Wood Preservation*, Document IRG/WP/3229, 1983.
74. S.J. Blunden, A.H. Chapman and R. Hill, *Internat. Pest Control*, 1987, **29**(4), 90.
75. K. Härtel, *Tin and its Uses*, 1958, **43**, 9.
76. P.J. Smith, *Metallurgie*, 1982, **22**(3), 161.
77. C.J. Evans, *Tin and its Uses*, 1970, **86**, 7.
78. C.A. Horne, US Patent 3 327 336, 1973.
79. I. Hammann, K.H. Büchel, K. Bungarz and L. Born, *Pflanz. Schutz-Nachrichten Bayer*, 1978, **31**(1), 61.

*International Tin Research Institute—present address: Kingston Lane, Uxbridge, Middlesex UB8 3PJ.

80. Anon., AEP: 1979/M/12/1, 327, FAO and WHO, Rome, 1971.
81. A.J. Crowe, *Appl. Organomet. Chem.*, 1987, **1**(2), 143; **1**(4), 331.
82. M.A. Hoy, W.W. Barnett, L.C. Hendricks, D. Castro, D. Cahn and W.J. Bentley, *Calif. Agric.*, 1984, **38**(7–8), 18.
83. K.R.S. Ascher and G. Rones, *Internat. Pest Control*, 1964, **6**, 6.
84. E.E. Kenager, *J. Econ. Entomol.*, 1965, **58**(1), 4.
85. S.J. Blunden and R. Hill, in *Surface Coatings*, eds. J.W. Nicholson and H.J. Prosser, Chapter 2: Organotin-Based Antifouling Coatings, Elsevier-Applied Science, London/New York, 1986.
86. R.F. Bennett and R.J. Zedler, *J. Oil Colour Chemists Assn.*, 1966, **49**(11), 928.
87. N.F. Cardarelli, *Controlled Release Pesticides Formulations*, CRC Press, Cleveland, 1976.
88. P. Dunn and D. Oldfield, *Rubber Ind.*, **9**(1), 1975.
89. A. Milne, *R. Inst. Chem. Ann. Chem. Congr.*, Durham, 9th–11th April 1980.
90. S. Karpel, *Tin and its Uses*, 1987, **154**, 6.
91. C. Alzieu, M. Héral, Y. Thibaud, M.J. Dardignac and M. Feuillet, *Rev. Trav. Inst. Pêches Marit.*, 1982, **45**(2), 101.
92. Anon., *The Control of Pollution (Antifouling Paints and Treatments) Regulations 1987*, UK Public Health Statutory Instruments, **783**, HMSO, London, 1987.
93. L.W. Hall and A.E. Pinkney, 1983, 42. *CRC Crit. Rev. Toxicol.*, 1984, **14**(2), 159.
94. Anon., *Organotin Compounds: Biocidal Protection of Various Materials*, Schering AG, 1983.
95. R.M. Hoskinson and I.M. Russell, *Austr. J. Textile Inst.*, 1973, **64**(9), 550.
96. B.A. Richardson, *Stone Ind.*, 1973, **8**, 2.
97. R.M. Hoskinson and I.M. Russell, *Austr. J. Textile Inst.*, 1974, **65**(9), 455.
99. M.J. Fuller, *Industrial Uses of Inorganic Tin Chemicals*, Publ. **499**, International Tin Research Institute, Greenford, UK, 1974.
100. B.E. Leach, US Patent 4 227 024, 1980.
101. Anon., *Tin and its Uses*, 1985, **144**, 10–14.
102. D.S. Stark, *CLEO 82 Conf., Phoenix*, 14th–16th April, Paper No. WJ4, 1982.
103. A.M. Coles, *Tin and its Uses*, 1983, **138**, 14.
104. H.J. Williams, I.L. Shannon and F.D. Stevens, *J. Amer. Soc. Prevent. Dent.*, 1974, **4**(4), 40.
105. C.J. Evans, *Tin and its Uses*, 1972, **91**, 3.
106. A.J. Crowe, *Drugs of the Future*, 1987, **12**(3), 255.
107. A.J. Crowe, P.J. Smith and G. Atassi, *Chem.-Biol. Interactions*, 1980, **32**, 171.
108. G.S. Drummond and A. Kappas, *Proc. Natl. Acad. Sci. USA*, 1981, **78**(10), 6466.
109. J. Duncan, *Pharmacol. Ther.*, 1980, **10**, 407.
110. P.A. Cusack and P.J. Smith, *Tin chemicals as fire retardants*, Publ. **635**, International Tin Research Institute, Greenford, UK, 1983.
111. Anon., *Tin Chemicals as Fire Retardants*, Publ. DS6, International Tin Research Institute, Greenford, UK, 1985.
112. P.A. Cusack, L.A. Hobbs, P.J. Smith and J.S. Brooks, *J. Textile Inst.*, 1980, **71**(3), 138.
113. P.A. Cusack, L.A. Hobbs, P.J. Smith and J.S. Brooks, *A study of flame-resist and water-repellent treatments of cotton by tin chemicals*, Publ. **641**, International Tin Research Institute, Greenford, UK, 1984.
115. H.W. Finck and G. Tscheulin, *Kunststoffe*, 1981, **71**(5), 320.
116. W.J. Kroenke, *J. Appl. Polym. Sci.*, 1981, **26**, 1167.
117. I. Touval, *J. Fire and Flammability*, 1972, **3**, 130.
118. P.A. Cusack, *Investigations into tin-based flame retardants and smoke suppressants*, Publ. **648**, International Tin Research Institute, Greenford, UK, 1984.
119. P.A. Cusack, *Fire and Materials*, 1986, **10**, 41.
120. P.A. Cusack, P.J. Smith and L.T. Arthur, *J. Fire Retardant Chem.*, 1980, **7**, 9.
121. S. Karpel, *Tin and its Uses*, 1986, **149**, 1.
122. M. Watanabe, S. Venkatusan and H.A. Laitinen, *J. Electrochem. Soc.*, 1983, **130**(1), 59.
123. S. Kanefusa, M. Nitta and M. Haradome, *J. Electrochem. Soc.*, 1985, **132**(7), 1770.
124. Anon., *International Tin Research Institute—Ann. Rep. for 1986*, Publ. **677**, International Tin Research Institute, Greenford, UK, 1986.
125. T. Williamson, *Polym. Paint and Colour J.*, 1981, 630, 635.
126. H. Plum, *Tin and its Uses*, 1981, **127**, 7.
127. L.A. Hobbs, *Tin and its Uses*, 1982, **131**, 10.
128. G.A. Krulik, *Platinum Metals Rev.*, 1982, **26**(2), 58.

Index